FUNDAMENTALS OF
ELECTRICAL ENGINEERING

FUNDAMENTALS OF ELECTRICAL ENGINEERING

Second Edition

Giorgio Rizzoni
The Ohio State University

James Kearns
York College of Pennsylvania

FUNDAMENTALS OF ELECTRICAL ENGINEERING, SECOND EDITION

Some ancillaries, including electronic and print components, may not be available to customers outside the United States.

This book is printed on acid-free paper.

1 2 3 4 5 6 7 8 9 LWI 24 23 22 21

ISBN 978-0-07-338056-8 (bound edition)
MHID 0-07-338056-3 (bound edition)
ISBN 978-1-259-91443-0 (loose-leaf edition)
MHID 1-259-91443-7 (loose-leaf edition)

Portfolio Manager: *Beth Bettcher*
Product Developers: *Heather Ervolino and Joan Weber*
Senior Marketing Manager: *Shannon O'Donnell*
Content Project Managers: *Jane Mohr and Samantha Donisi*
Buyer: *Susan K. Culbertson*
Designer: *Beth Blech*
Content Licensing Specialist: *Lorraine Buczek*
Cover Image: *Mason Hayes*
Compositor: *Aptara®, Inc.*

All credits appearing on page or at the end of the book are considered to be an extension of the copyright page.

Library of Congress Cataloging-in-Publication Data

Names: Rizzoni, Giorgio, author. | Kearns, James (Associate professor of
    electrical & computer engineering) author.
Title: Fundamentals of electrical engineering / Giorgio Rizzoni, The Ohio
    State University, James Kearns, York College of Pennsylvania.
Description: Second edition. | New York, NY : McGraw Hill Education, [2022]
    | Includes index.
Identifiers: LCCN 2020029814 (print) | LCCN 2020029815 (ebook) | ISBN
    9780073380568 (hardcover) | ISBN 9781259914430 (spiral bound) | ISBN
    9781259914461 (ebook) | ISBN 9780077510411 (ebook other)
Subjects: LCSH: Electrical engineering.
Classification: LCC TK146 .R4725 2022 (print) | LCC TK146 (ebook) | DDC
    621.3—dc23
LC record available at https://lccn.loc.gov/2020029814
LC ebook record available at https://lccn.loc.gov/2020029815

mheducation.com/highered

To our families

About the Authors

Giorgio Rizzoni, the *Ford Motor Company Chair in ElectroMechanical Systems*, is a Professor of Mechanical and Aerospace Engineering and of Electrical and Computer Engineering at The Ohio State University (OSU). He received his B.S. in 1980, his M.S. in 1982, and his Ph.D. in 1986, in Electrical and Computer Engineering, all from the University of Michigan. Since 1999 he has been the director of the Ohio State University Center for Automotive Research (CAR), an interdisciplinary research center in the OSU College of Engineering.

Dr. Rizzoni's research interests are in the dynamics and control of future ground vehicle propulsion systems, including advanced engines, alternative fuels, electric and hybrid-electric drivetrains, energy storage systems, and fuel cell systems. He has contributed to the development of a graduate curriculum in these areas and has served as the director of three U.S. Department of Energy Graduate Automotive Technology Education Centers of Excellence: *Hybrid Drivetrains and Control Systems* (1998–2004), *Advanced Propulsion Systems* (2005–2011), and *Energy Efficient Vehicles for Sustainable Mobility* (2011–2016).

In 1999 Dr. Rizzoni established an automotive industry research consortium that today sees the participation of over 20 automotive OEMs and suppliers; in 2008 he created the SMART@CAR consortium, focusing on plug-in hybrid and electric vehicles and vehicle-grid interaction, with funding from electric utilities, automotive OEMS, and electronics suppliers. Through the Ohio Third Frontier Wright Project Program he created a *Center of Excellence for Commercial Hybrid Vehicles* in 2009, and a *Center of Excellence for Energy Storage Technology* in 2010.

Dr. Rizzoni is a Fellow of IEEE (2004), a Fellow of SAE (2005), a recipient of the 1991 National Science Foundation Presidential Young Investigator Award, and of several other technical and teaching awards.

The OSU Center for Automotive Research

The OSU Center for Automotive Research, CAR, is an interdisciplinary research center in the OSU College of Engineering founded in 1991 and located in a 50,000 ft^2 building complex on the west campus of OSU. CAR conducts interdisciplinary research in collaboration with the OSU colleges of Engineering, Medicine, Business, and Arts and Sciences, and with industry and government partners. CAR research aims to: develop efficient vehicle propulsion and energy storage systems; develop new sustainable mobility concepts; reduce the impact of vehicles on the environment; improve vehicle safety and reduce occupant and pedestrian injuries; increase vehicle autonomy and intelligence; and create quieter and more comfortable automobiles. A team of 50 administrative and research staff supports some 40 faculty, 120 graduate and 300 undergraduate students and maintains and makes use of advanced experimental facilities. Dr. Rizzoni has led CAR for two decades, growing its research expenditures from $1M per year to over $15M today, and engaging CAR in a broad range of technology commercialization activities, start-up company incubation and spin-out, as well as providing a broad range of engineering services to the automotive industry.

CAR is also the home of the OSU Motorsports program, which supports the activities of five student vehicle competition programs of several student vehicle competition programs including: the Buckeye Bullet (holder of all current U.S. and FIA electric vehicle land speed records), the EcoCAR hybrid-electric vehicle team, the Formula Buckeyes and Baja Buckeyes SAE teams, and the Buckeye Current electric motorcycle racing team.

Jim Kearns is an Associate Professor of Electrical & Computer Engineering at York College of Pennsylvania. He received a B.S. in Mechanical Engineering (SEAS) and a B.S. in Economics (Wharton) from the University of Pennsylvania in 1982. Subsequently, he received his M.E. from Carnegie-Mellon University in 1984, and his Ph.D. from the Georgia Institute of Technology in 1990, both in Mechanical Engineering. While at Georgia Tech he was the recipient of a Presidential Fellowship. Subsequently, he worked as a Postdoctoral Fellow at the Applied Research Laboratory of the University of Texas—Austin.

In 1992, Dr. Kearns took his first teaching position at the Universidad del Turabo in Gurabo, Puerto Rico, where he worked with a small group of faculty and staff to build and develop a new school of mechanical engineering. In addition to other duties, he was tasked with developing a curriculum on electromechanics. During this time Dr. Kearns spent his summers at Sandia National Laboratories as a University Fellow.

In 1996, Dr. Kearns was the second full-time engineering faculty member hired by York College of Pennsylvania to (once again) develop a new mechanical engineering program with an emphasis on Mechatronics. As a result of that work, Jim was asked in 2003 to develop a new electrical and computer engineering program at YCP. Jim served as program coordinator until July 2010.

Throughout Dr. Kearns professional career he has been involved in teaching and research related to physical acoustics and electromechanical systems. His interest in electrical engineering began during his Ph.D. studies, when he built spark generators, DC power supplies, and signal amplifiers for his experiments in physical acoustics. His steady pursuit of electromechanical engineering education has been the hallmark of his professional career. Dr. Kearns has been involved in a variety of pedagogical activities, including the development and refinement of techniques in electrical engineering education, and recently, the application of Thévenin's theorem to determine characteristic parameters of electrical networks.

Dr. Kearns is a member of IEEE and ASEE. He is active in faculty governance at York College, where he is a past chair of its Tenure and Promotion committee and its Student Welfare committee. Dr. Kearns recently completed a four-year term as Vice-President and then President of the York College Academic Senate.

About the Cover

The Buckeye Current electric race motorcycle is an award-winning collegiate motorcycle racing team founded in 2010 by two students with a big dream. Over the course of the team's 10-year history, hundreds of students have had the opportunity to help design and race electric race motorcycles. Based at The Ohio State University's Center for Automotive Research, the team pushes the limits of electric vehicles by developing motorcycles that have raced, and won, across the world. The Buckeye Current team has set speed records and has raced at the Isle of Man and Pike's Peak. Go to the website http://org.osu.edu/buckeyecurrent for more information.

Brief Contents

Contents

Preface

The pervasive presence of electronic devices and instrumentation in all aspects of engineering design and analysis is one of the manifestations of the electronic revolution that has characterized the last sixty years. Every aspect of engineering practice, and of everyday life, has been affected in some way or another by electrical and electronic devices and instruments. Laptop and tablet computers along with so-called "smart" phones, and touchscreen interfaces are perhaps the most obvious manifestations. These devices, and their underlying technology, have brought about a revolution in computing, communication, and entertainment. They allow us to store, process, and share professional and personal data and to access audio (most notably, music) and video of every variety. These advances in electrical engineering technology have had enormous impacts on all other fields of engineering, including mechanical, industrial, computer, civil, aeronautical, aerospace, chemical, nuclear, materials, and biological engineering. This rapidly expanding electrical and electronic technology has been adopted, leveraged, and incorporated in engineering designs across all fields. As a result, engineers work on projects requiring effective communication across multiple disciplines, one of which is nearly always electrical engineering.

0.1 OBJECTIVES

Engineering education and professional practice continue to undergo profound changes in an attempt to best utilize relevant advances in electronic technology. The need for textbooks and other learning resources that relate these advances to engineering disciplines beyond electrical and computer engineering continues to grow. This fact is evident in the ever-expanding application and integration of electronics and computer technologies in commercial products and processes. This textbook and its associated learning resources represent one effort to make the principles of electrical and computer engineering accessible to students in various engineering disciplines.

The principal objective of the book is to present the *fundamentals* of electrical, electronic, and electromechanical engineering to an audience of engineering majors enrolled in introductory and more advanced or specialized electrical engineering courses.

A second objective is to present these *fundamentals* with a focus on important results and common yet effective *analytical and computational tools* to solve practical problems.

Finally, a third objective of the book is to illustrate, by way of concrete, fully worked examples, a number of relevant *applications* of electrical engineering. These examples are drawn from the authors' industrial research experience and from ideas contributed by practicing engineers and industrial partners.

These three objectives are met through the use of various pedagogical features and methods.

0.2 ORGANIZATION

The second edition contains several significant organizational changes. However, the substance of the book, while updated, is essentially unchanged. The most obvious organizational change is the location of example problems within each chapter. In the previous edition, examples were mixed in with the text so that students would encounter examples immediately after each key concept. While this type of organization works well for a first read, it has the disadvantage of making example problems difficult to locate for review. Since it is critical that students be able to easily and efficiently locate example problems when preparing for exams, in this edition of the book, with few exceptions, all example problems have been placed at the end of each section within a chapter.

A continued and enhanced emphasis on problem solving can be found in this edition. All the highlighted *Focus on Methodology* boxes found in the first edition were renamed *Focus on Problem Solving*, and many of them were rewritten to clarify and add additional detail to the steps needed by students to successfully complete end-of-chapter homework problems.

An effort was also made to reduce the aesthetic complexity of the book, without sacrificing technical content or overall aesthetic appeal. We believe that effective reading is promoted by less clutter and visual "noise," if you will. For example, a careful comparison of the first and second editions will reveal our effort to produce cleaner and sharper figures that retain only that information relevant to the issue or problem being discussed.

In addition, a thorough, exhaustive, page-by-page search was made to locate errors in the text, equations, figures, references to equations and figures, examples, and homework problems. Speaking of homework problems, the second edition contains 861 homework problems, of which over 300 are new to this edition, and, where necessary and appropriate, example problems were updated.

The book remains divided into three major parts:

  I. **Circuits**
 II. **Electronics**
III. **Electromechanics**

The pedagogical enhancements made within each part are discussed below.

0.3 PEDAGOGY AND CONTENT

Part I: Circuits

The first part of the book has undergone major revision from the first edition.

Chapter 1 begins with an emphasis on developing a student's ability to recognize structure within a circuit diagram. It is the authors' experience that this ability is key to student success. Yet, many books contain little content on developing this ability; the result is that many students wander into more difficult topics still viewing a circuit as simply an unruly collection of wires and elements.

The approach taken in this book is to encourage students to initially *focus on nodes*, rather than elements, in a circuit. For example, some of the earliest exercises in this book simply ask students to count the number of nodes in a circuit diagram. One immediate advantage of this patient approach is that it teaches students to disregard the particular aesthetic structure shown in a circuit diagram and instead to recognize and focus on the technical structure and content.

Methods of Problem Solving were enhanced and clarified. Throughout these chapters students are encouraged to think of problem solving in two steps: first **simplify**; then **solve**. In addition to being an effective problem-solving method, this method provides context for the power and importance of equivalent circuits, in general, and Thévenin's theorem, in particular. In the chapters on transient analysis and frequency response, foundational first- and second-order circuit *archetypes* are identified. Students are encouraged to simplify, when possible, transient circuit problems to these archetypes. In effect, they become clear targets for students when problem solving. Thévenin's and Norton's theorems and the principle of superposition are used throughout these chapters to simplify complicated circuits to the archetypes.

Finally, a greater emphasis was placed on visualizing phasors in the complex plane and understanding the key role of the unit phasor and Euler's theorem. Throughout the chapters on AC circuits and power students are encouraged to focus on the concepts of impedance and power triangles, and their similarity.

Part II: Electronics

While much of the content on electronics in Part II is unchanged from the first edition, the problem-solving strategies and techniques for transistor circuits were enhanced and clarified. The focus on simple but useful circuit examples was not changed.

Similar to the approach taken in Part I, Chapter 7 on operational amplifiers emphasizes three *amplifier archetypes* (the unity-gain buffer, the inverting amplifier, and the non-inverting amplifier) before introducing variations and applications.

The emphasis in Chapters 9 and 10 on large-signal models of BJTs and FETs and their applications was retained; however, an appropriate, but limited, presentation of small-signal models was included to support the discussion of AC amplifiers. These two chapters present an uncomplicated and practical treatment of the analysis and design of simple amplifiers and switching circuits using large-signal models.

Chapter 11 presents an overview of combinational and sequential logic modules, providing a comprehensive overview of digital logic circuits.

Part III: Electromechanics

Part III on electromechanics has been revised for accuracy and pedagogy, but its contents are largely unchanged. This part has been used by the first author for many years as a supplement in a junior-year System Dynamics course for mechanical engineers.

0.4 NOTATION

The notation used in this book for various symbols (variables, parameters, and units) has been updated but still follows generally accepted conventions. Distinctions in notation can be subtle. Luckily, very often the context in which a symbol appears makes its meaning clear. When the meaning of a symbol is not clear from its context a correct reading of the notation is important. A reasonably complete listing of the symbols used in this book and their notation is presented below.

For example, an uppercase roman font is used for units such as volts (V) and amperes (A). An uppercase italics math font is used for real parameters and variables such as resistance (R) and DC voltage (V). Notice the difference between the variable V and the unit V. Further, an uppercase bold math font is used for complex quantities such as voltage and current phasors ($\mathbf{V}$ and $\mathbf{I}$) as well as impedance ($\mathbf{Z}$), conductance ($\mathbf{Y}$), and frequency response functions ($\mathbf{H}$ and $\mathbf{G}$). Lowercase italic symbols are, in general, time dependent variables, such as voltage (v or $v(t)$) and current (i or $i(t)$), where (t) is an explicit indication of time dependence. Lowercase italic variables may represent constants in specific cases. Uppercase italic variables are reserved for constant (time-invariant) values exclusively.

Various subscripts are also used to denote particular instances or multiple occurrences of parameters and variables. Exponents are italicized superscripts.

Finally, in electrical engineering the imaginary unit $\sqrt{-1}$ is always represented by j rather than i, which is used by mathematicians. The reason for the use of j instead of i should be obvious!

Quantity	Symbol	Description
Voltage	v or $v(t)$	Time Dependent and Real
	V	Time Invariant and Real
	$\mathbf{V}$	Complex Phasor
Effective (rms) voltage	$\tilde{V}$	Time Invariant and Real
Current	i or $i(t)$	Time Dependent and Real
	I	Time Invariant and Real
	$\mathbf{I}$	Complex Phasor
Effective (rms) current	$\tilde{I}$	Time Invariant and Real
Volts	V	Unit of voltage
Amperes	A	Unit of current
Resistance	R	Real
Inductance	L	Real
Capacitance	C	Real
Reactance	X	Frequency Dependent and Real
Impedance	$\mathbf{Z}$	Frequency Dependent and Complex
Conductance	$\mathbf{Y}$	Frequency Dependent and Complex
Transfer Function	$\mathbf{G}$ or $\mathbf{H}$	Frequency Dependent and Complex
Cyclical Frequency	f	Time Invariant and Real
Angular Frequency	ω	Time Invariant and Real
Angle	θ	Time Invariant and Real
Amplitude	A	Time Invariant and Real

0.5 SYSTEM OF UNITS

This book employs the International System of Units (also called SI, from the French Système International des Unitès). SI units are adhered to by virtually all professional engineering societies and are based upon the seven fundamental quantities listed in Table 0.1. All other units are derived from these base units. An example of a derived unit is the radian, which is a measure of plane angles. In this book, angles are in units of radians unless explicitly given otherwise as degrees.

Since quantities often need to be described in large multiples or small fractions of a unit, the standard prefixes listed in Table 0.2 are used to denote SI units in powers of 10. In general, engineering units are expressed in powers of 10 that are multiples of 3. For example, 10^{-4} s would be expressed as 100×10^{-6} s, or 100 μs.

Tables 0.1 and 0.2 are useful references when reading this book.

Table 0.1 **SI units**

Quantity	Unit	Symbol
Length	Meter	m
Mass	Kilogram	kg
Time	Second	s
Electric current	Ampere	A
Temperature	Kelvin	K
Substance	Mole	mol
Luminous intensity	Candela	cd

Table 0.2 **Standard prefixes**

Prefix	Symbol	Power
atto	a	10^{-18}
femto	f	10^{-15}
pico	p	10^{-12}
nano	n	10^{-9}
micro	μ	10^{-6}
milli	m	10^{-3}
centi	c	10^{-2}
deci	d	10^{-1}
deka	da	10
kilo	k	10^{3}
mega	M	10^{6}
giga	G	10^{9}
tera	T	10^{12}

0.6   FEATURES OF THE SECOND EDITION

Pedagogy

The second edition continues to offer all the time-tested pedagogical features available in the earlier editions.

- **Learning Objectives** offer an overview of key chapter ideas. Each chapter opens with a list of major objectives, and throughout the chapter the learning objective icon indicates targeted references to each objective.
- **Focus on Problem Solving** sections summarize important methods and procedures for the solution of common problems and assist the student in developing a methodical approach to problem solving.
- **Clearly Illustrated Examples** illustrate relevant applications of electrical engineering principles. The examples are fully integrated with the Focus on Problem Solving material, and each one is organized according to a prescribed set of logical steps.
- **Check Your Understanding** exercises follow each set of examples and allow students to confirm their mastery of concepts.
- **Make the Connection** sidebars present analogies that illuminate electrical engineering concepts using other concepts from engineering disciplines.
- **Focus on Measurements** boxes emphasize the great relevance of electrical engineering to the science and practice of measurement.

Instructor Resources on Connect:

Instructors have access to these files, which are housed in Connect.

- **PowerPoint presentation slides** of important figures from the text
- **Instructor's Solutions Manual** with complete solutions

Writing Assignment

Available within McGraw Hill Connect®, the Writing Assignment tool delivers a learning experience to help students improve their written communication skills and conceptual understanding. As an instructor you can assign, monitor, grade, and provide feedback on writing more efficiently and effectively.

Remote Proctoring & Browser-Locking Capabilities

New remote proctoring and browser-locking capabilities, hosted by Proctorio within Connect, provide control of the assessment environment by enabling security options and verifying the identity of the student.

Seamlessly integrated within Connect, these services allow instructors to control students' assessment experience by restricting browser activity, recording students' activity, and verifying students are doing their own work.

Instant and detailed reporting gives instructors an at-a-glance view of potential academic integrity concerns, thereby avoiding personal bias and supporting evidence-based claims.

0.7 ACKNOWLEDGMENTS

The authors would like to recognize the help and assistance of reviewers, student's, and colleagues who have provided invaluable support. In particular, Dr. Ralph Tanner of Western Michigan University has painstakingly reviewed the book for accuracy and has provided rigorous feedback, and Ms. Jiyu Zhang, PhD student at Ohio State, has been generous in her assistance with the electromechanical systems portion of the chapter. Dr. Isabel Fernandez Puentes of The Ohio State University has reviewed the final manuscript for accuracy, with great attention to detail. The authors are especially grateful to Dr. Domenico Bianchi and Dr. Gian Luca Storti for creating many new homework problems and solutions and for their willingness to pitch in whenever needed. This second edition is much improved due to their efforts. Finally, Mr. Riccardo Palomba has painstakingly assembled the Instructor's Solutions Manual.

Throughout the preparation of this edition, Kathryn Rizzoni has provided editorial support and has served as an interface to the editorial staff at MHHE. We are grateful for her patience, her time invested in the project, her unwavering encouragement, her kind words, and her willingness to discuss gardening and honeybees.

The book has been critically reviewed by:

- Riadh Habash—The University of Ottawa
- Ahmad Nafisi—California Polytechnic State University
- Raveendra Rao—The University of Western Ontario
- Belinda Wang—The University of Toronto
- Brian Peterson—United States Air Force Academy
- John Durkin—University of Akron
- Chris Klein—Ohio State University
- Ting-Chung Poon—Virginia Tech
- James R. Rowland—University of Kansas

- N. Jill Schoof—Maine Maritime Academy
- Shiva Kumar—McMaster University
- Tom Sullivan—Carnegie Mellon University
- Dr. Kala Meah—York College of Pennsylvania

In addition, we would like to thank the many colleagues who have pointed out errors and inconsistencies and who have made other valuable suggestions.

Comments by Giorgio Rizzoni

As always, a new edition represents a new era. I am truly grateful to my friend and co-author, Jim Kearns, for taking on a new challenge and for bringing his perspective and experience to the book. Jim and I share a passion for teaching, and throughout this project we have invariably agreed on which course to take. It is not easy to find a suitable co-author in the life of a project of this magnitude, and I have been fortunate to find a friend willing to undertake a new journey with me.

The years go by, but my family continues to be an endless source of joy, pleasant surprises and, always, smiles. Many thanks to Kathryn, Alex, Cat, and Michael for always being there to support and encourage me.

Comments by James Kearns

My association with this remarkable book continues to be a great privilege and honor. Its contents continue to reflect the enormous effort and expertise of the principal author, and my dear friend, Dr. Giorgio Rizzoni. His leadership and vision were essential to the creation of this new edition. I remain awestruck by his seemingly unbounded energy and enthusiasm and humbled by his kind, considerate, and generous ways.

As with all things, the love and support of my family and friends sustained me throughout this work. My children, Kevin, Claire, and Caroline, continue to bless, inspire, and inform my daily life.

Finally, I wish to once again thank my parents for their many years of unconditional love and support. Despite having lost both of them in recent years they remain very much alive within me and present in my work. Can anyone ever begin to measure or repay the gift of loving parents?

steady state of any variable?

4. How fast or slow is that transition?

5. What is the final steady state of any variable?

Two types of circuits are examined in this chapter: first-order RC and RL circuits, which contain a single storage element, and second-order circuits, which contain two irreducible storage elements. The simplest of the second-order circuits to analyze are the series RLC and parallel RLC circuits. Other more complicated second-order circuits exist, as do higher-order circuits; however, since all the fundamental behaviors of transient circuits are revealed in the types just mentioned, they are the focus of this chapter.

A first-order circuit contains a single storage element. A second-order circuit contains two irreducible storage elements.

Throughout this chapter, practical applications of first- and second-order circuits are introduced. Numerous analogies are presented to emphasize the general nature of the solution methods and their applicability to a wide range of physical systems, including hydraulics, mechanical systems, and thermal systems.

Learning Objectives offer an overview of key chapter ideas. Each chapter opens with a list of major objectives, and throughout the chapter the learning objective icon indicates targeted references to each objective.

 Learning Objectives

Students will learn to...

1. Understand the fundamental qualities of transient responses. *Section 4.1*
2. Write differential equations in standard form for circuits containing inductors and capacitors. *Section 4.2*
3. Determine the steady state of DC circuits containing inductors and capacitors. *Section 4.2*
4. Determine the complete solution of first-order circuits excited by switched DC sources. *Section 4.3*
5. Determine the complete solution of second-order circuits excited by switched DC sources. *Section 4.4*
6. Understand analogies between electric circuits and hydraulic, thermal, and mechanical systems. *Sections 4.1–4.4*

FOCUS ON PROBLEM SOLVING

ROOTS OF SECOND-ORDER SYSTEMS

The general form of the roots s_1 and s_2 is
$$s_{1,2} = -\zeta\omega_n \pm \omega_n\sqrt{\zeta^2 - 1}.$$
The nature of these roots depends upon the argument of the square root.

Case 1: **Distinct, negative, real roots.** This case occurs when $\zeta > 1$ since the term under the square root sign is positive. The result is $s_{1,2} = -\omega_n\left[\zeta \pm \sqrt{\zeta^2 - 1}\right]$ and a second-order **overdamped response**.

Case 2: **Identical, negative, real roots.** This case occurs when $\zeta = 1$ since the term under the square root is zero. The result is a repeated root $s = -\zeta\omega_n = -\omega_n$ and a second-order **critically damped response**.

Case 3: **Complex conjugate roots.** This case holds when $\zeta < 1$ since the term under the square root is negative. The result is a pair of complex conjugate roots $s_{1,2} = -\omega_n\left[\zeta \pm j\sqrt{1 - \zeta^2}\right]$ and a second-order **underdamped response**.

Focus on Problem Solving sections summarize important methods and procedures for the solution of common problems and assist the student in developing a methodical approach to problem solving.

EXAMPLE 4.15   Complete Response of Critically Damped
Second-Order Circuit

Problem

Determine the complete response for the voltage v_C shown in Figure 4.45.

Figure 4.45 Circuit for
Example 4.15.

Solution

Known Quantities: I_S; R; R_S; C; L.

Find: The complete response of the differential equation in v_C describing the circuit in
Figure 4.45.

Schematics, Diagrams, Circuits, and Given Data: $I_S = 5$ A; $R = R_S = 500$ Ω; $C = 2$ μF;
$L = 500$ mH.

Assumptions: None.

Analysis:

Step 1: Steady-state response. With the switch open for a long time, any energy stored
in the capacitor and inductor has had time to be dissipated by the resistor; thus, the currents
and voltages in the circuit are zero: $i_L(0^-) = 0$, $v_C(0^-) = v(0^-) = 0$.

After the switch has been closed for a long time and all the transients have died, the
capacitor becomes an open-circuit, and the inductor behaves as a short-circuit. With the
inductor behaving as a short-circuit, all the source current will flow through the inductor,
and $i_L(\infty) = I_S = 5$ A. On the other hand, the current through the resistor is zero, and
therefore $v_C(\infty) = v(\infty) = 0$ V.

Step 2: Initial conditions. Two initial conditions are needed to solve a second-order
circuit. These two initial conditions always rely on two continuity conditions: the current
through an inductor and the voltage across a capacitor are continuous. That is, $i_L(0^-) =
i_L(0^+) = 0$ A and $v_C(0^-) = v_C(0^+) = 0$ V. Since the differential equation is in the variable
v_C, the two needed initial conditions are $v_C(0^+)$ and $dv_C(0^+)/dt$. These can be found by
applying KCL at $t = 0^+$:

$$I_S - \frac{v_C(0^+)}{R_S} - i_L(0^+) - \frac{v_C(0^+)}{R} - C\frac{dv_C(0^+)}{dt} = 0$$

Since $v_C(0^+) = 0$ and $i_L(0^+) = 0$, the result is: $dv_C(0^+)/dt$:

$$\frac{dv_C(0^+)}{dt} = \frac{I_S}{C} = \frac{5}{2 \times 10^{-6}} = 2.5 \times 10^6 \frac{V}{s}$$

Step 3: Differential equation. Apply KCL at the upper node to find a first-order differential equation in the two state variables v_C and i_L:

$$I_S - \frac{v_C}{R_S} - i_L - \frac{v_C}{R} - C\frac{dv_C}{dt} = 0 \qquad t \geq 0$$

Differentiate both sides to obtain a second-order differential equation. Then, note that v_C is
also the voltage across the inductor such that the constitutive i-v relation for the inductor
can be written as:

$$v_C = v_L = L\frac{di_L}{dt}$$

CHECK YOUR UNDERSTANDING

Use the differential i-v relations for capacitors and inductors along with KVL or KCL to
write the differential equation for each of the circuits shown below.

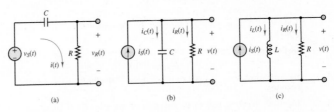

(a)   (b)   (c)

Answer: (a) $RC\dfrac{dv_c(t)}{dt} + v_c(t) = v_S(t)$; (b) $RC\dfrac{dv(t)}{dt} + v(t) = Ri_S(t)$;

(c) $\dfrac{R}{L}\dfrac{di_L(t)}{dt} + i_L(t) = i_S(t)$

Clearly illustrated examples
present relevant applications of
electrical engineering principles.
The examples are fully integrated
with the Focus on "Problem"
Solving material, and each one is
organized according to a
prescribed set of logical steps.

Check Your Understanding
exercises follow each set of
examples and allow students to
confirm their mastery of concepts.

Make the Connection sidebars present analogies that illuminate electrical engineering concepts using concepts form other engineering disciplines.

Second-order circuit response is more complex and described by one of three possible cases, each of which is determined by a parameter ζ known as the *dimensionless damping ratio*, as explained in detail in Section 4.4. When $\zeta > 1$, the response is said to be *overdamped* and is the sum of two first-order decaying exponentials, each with its own distinct time constant. When $\zeta = 1$, the response is said to be *critically damped*. When $\zeta < 1$, the response is said to be *underdamped*. As shown in Figure 4.5, the responses in these latter two cases *cannot* be simply described by two decaying exponentials.

Long-Term Steady State

The long-term steady state is that which remains after the transient response has decayed completely. For the first-order decaying exponential shown in Figure 4.7 the long-term steady state is $x(\infty)$. The long-term steady state depends upon the independent sources present in the $t > 0$ circuit and is commonly expressed in terms of a *gain K* multiplied by a forcing function $F(t)$ that represents the contributions of those sources. For simplicity, only circuits with DC independent sources are considered in this chapter, with the result that only DC long-term steady states occur.

Complete Response

The complete response is simply the sum of the transient response and the long-term steady state. In general, the transient response will contain one unknown constant for each state variable in the circuit. Thus, the complete response will also contain the same number of unknown constants. The values of these unknown constants are determined by the initial conditions on the circuit at $t = 0^+$.

A common mistake when learning to solve transient circuit problems is to apply the initial conditions to the transient response alone rather than to the complete solution. Forewarned, forearmed; don't make this error!

Natural and Forced Responses

Often, it is useful to express the complete response as the sum of *natural* and *forced* responses instead of the sum of a transient response and long-term steady state. Either way the complete response is unchanged. The natural response is that part of the complete system response due to the initial energy stored in the system at $t = 0$. The forced response is that part due to independent sources present in the $t > 0$ circuit.

As will be shown in the following section, equation 4.9 expresses the complete response $x(t)$ of an arbitrary first-order circuit variable as the sum of a transient response, with its characteristic exponential decay, and a long-term steady state $x(\infty)$.

$$x(t) = \left[x(0^+) - x(\infty)\right]e^{-t/\tau} + x(\infty) \qquad (4.9)$$

The transient response portion includes the difference between the initial condition $x(0^+)$ and the long-term steady state. This expression can be reconstructed as:

$$x(t) = x_N(t) + x_F(t) = x(0^+)e^{-t/\tau} + x(\infty)(1 - e^{-t/\tau}) \qquad (4.10)$$

The first and second terms in equation 4.10 are known as the *natural* and *forced* ~~~~~~~~~~~~~~~~~~~~~~~~ruction can be made for the

MAKE THE CONNECTION

Thermal System Dynamics

To describe the dynamics of a thermal system, we write a differential equation based on energy balance. The difference between the heat added to the mass by an external source and the heat leaving the same mass (by convection or conduction) must be equal to the heat stored in the mass:

$$q_{in} - q_{out} = q_{stored}$$

An object is internally heated at the rate q_{in} in ambient temperature $T = T_o$; the thermal capacitance and thermal resistance are C_t and R_t. From energy balance:

$$q_{in}(t) - \frac{T(t) - T_o}{R_t} = C_t \frac{dT(t)}{dt}$$

$$R_t C_t \frac{dT(t)}{dt} + T(t) = R_t q_{in}(t) + T_o$$

$$\tau_t = R_t C_t \qquad K_{St} = R_t$$

This first-order system is identical in its form to an electric RC circuit, as shown below.

Thermal system

Equivalent electric circuit

Capacitive Displacement Transducer and Microphone

As shown in Figure 3.2, the capacitance of a flat parallel-plate capacitor is:

$$C = \frac{\varepsilon A}{d} = \frac{\kappa \varepsilon_0 A}{d}$$

where ε is the **permittivity** of the dielectric material, κ is the dielectric constant, $\varepsilon_0 = 8.854 \times 10^{-12}$ F/m is the permittivity of a vacuum, A is the area of each of the plates, and d is their separation. The dielectric constant for air is $\kappa_{air} \approx 1$. Thus, the capacitance of two flat parallel plates of area 1 m^2, separated by a 1-mm air gap, is 8.854 nF, a very small value for such large plates. As a result, flat parallel-plate capacitors are impractical for use in most electronic devices. On the other hand, parallel-plate capacitors find application as motion transducers, that is, as devices that can measure the motion or displacement of an object. In a capacitive motion transducer, the plates are designed to allow relative motion when subjected to an external force. Using the capacitance value just derived for a parallel-plate capacitor, one can obtain the expression

$$C = \frac{8,854 \times 10^{-3} A}{x}$$

where C is the capacitance in picofarads, A is the area of the plates in square millimeters, and x is the separation distance in millimeters. Note that the change in C due to a change in x is nonlinear, since $C \propto 1/x$. However, for small changes in x, the change in C is

Focus on Measurements boxes emphasize the great relevance of electrical engineering to the science and practice of measurement.

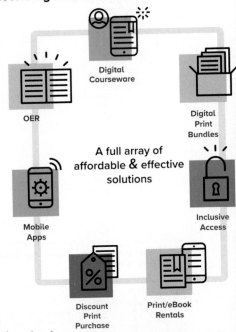

FUNDAMENTALS OF
ELECTRICAL ENGINEERING

PART I

CIRCUITS

Arthur S. Aubry/Stockbyte/Getty Images

C H A P T E R

1

FUNDAMENTALS OF ELECTRIC CIRCUITS

hapter 1 is the foundation for the entire book and presents the fundamental laws that govern the behavior of electric circuits. Basic features and terminology of electric circuits, such as nodes, branches, meshes, and loops, are defined, and the three fundamental laws of circuit analysis. Kirchhoff's current and voltage laws and Ohm's law, are introduced. The concept of electric power and the passive sign convention are introduced along with basic circuit elements—sources and resistors. Basic analytic techniques—voltage and current division—are introduced along with some engineering applications. Examples include a description of strain gauges, circuits related to the measurement of force and other mechanical variables, and a study of an automotive throttle position sensor. A brief discussion of measurement instruments is also included. Finally, the chapter closes with a discussion of the source-load perspective and an application of it to find the same voltage and current division results obtained earlier in the chapter.

Learning Objectives

Students will learn to...

1. Identify the principal *features of electric circuits or networks*: nodes, loops, meshes, and branches. *Section 1.1.*
2. Apply *Kirchhoff's laws* to simple electric circuits. *Sections 1.2–1.3.*
3. Apply the *passive sign convention* to compute the power consumed or supplied by circuit elements. *Section 1.4.*
4. Identify *sources* and *resistors* and their *i-v characteristics. Sections 1.5–1.6.*
5. Apply *Ohm's law* and *voltage and current division* to calculate unknown voltages and currents in simple series, parallel, and series-parallel circuits. *Sections 1.6–1.8.*
6. Understand the impact of internal resistance in practical models of voltage and current sources as well as of voltmeters, ammeters, and wattmeters. *Sections 1.9–1.10.*

1.1 FEATURES OF NETWORKS AND CIRCUITS

A *network* can be defined as a collection of interconnected objects. In an electric network, *elements*, such as resistors, are connected by wires. An electric *circuit* can be defined as an electric network within which at least one closed path exists and around which electric charge may flow. All electric circuits are networks, but not all electric networks contain a circuit. In this book, a circuit is any network that contains at least one complete and closed path.

There are two principal quantities within a circuit: current and voltage. *The primary objective of circuit analysis is to determine one or more unknown currents and voltages.* Once these currents and voltages are determined, any other aspect of the circuit, such as its power requirements, efficiency, and speed of response, can be computed.

Two useful concepts for circuit analysis are those of a source and of a load. In general, the load is the circuit element or segment of interest to the designer or user of the circuit. By default, the source is everything else not included in the load. Typically, the source provides energy and the load consumes it for some purpose. For example, consider the simple physical circuit of a headlight attached to a car battery as shown in Figure 1.1(a). For the driver of the car, the headlight may be the circuit element of interest since it enables the driver to see the road at night. From this perspective, the headlight is the load and the battery is the source as shown in Figure 1.1(b), which is intuitively appealing because power flows from the source (the battery) to the load (the headlight). However, in general, it is not required nor necessarily true that power flows in this manner. Electric power is discussed later in this chapter.

The use of the term *source* can be confusing at times because, as will be discussed later in this chapter, there are circuit elements known as *ideal voltage and current sources*, which have well-defined attributes and circuit symbols. These ideal sources, along with other circuit elements, are often the constituents of the source portion of a circuit, as well as the load portion. In this book, ideal sources are referred to as either voltage or current sources, explicitly, to avoid confusion.

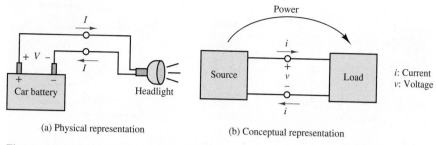

(a) Physical representation   (b) Conceptual representation

Figure 1.1 (a) Physical, and (b) conceptual representations of an electrical system

Other key conceptual features of electric circuits are the *ideal wire, node, branch, loop,* and *mesh.* The concept of a node is particularly useful for correctly interpreting circuit diagrams. Many students struggle with circuit analysis simply because they lack an organizing perspective with which to interpret circuit diagrams. One particularly helpful perspective is to see electric circuits as comprised of elements situated between nodes. Once the concept of a node is well understood this perspective simplifies and clarifies many circuits that otherwise appear complicated.

Ideal Wire

Electric circuit and network *diagrams* are used to represent (approximately) actual electric circuits and networks. These diagrams contain *elements* connected by *ideal wires.* An ideal wire is able to conduct electric charge without any loss of electric potential. In other words, no work is required to move an electric charge along an ideal wire. Luckily, in many applications, actual wires are well approximated by ideal wires. However, there are applications where wiring accounts for significant losses of potential (e.g., long-distance transmission lines and microscopic integrated circuits). In these applications, the ideal wire approximation must be augmented and/or used with care. In this book, all wires in circuit and network diagrams are ideal, unless indicated otherwise.

Node

A **node** consists of one or more ideal wires connected together such that an electric charge can travel between any two points on the node without traversing a circuit element, such as a resistor. It is important to recognize that since a node consists of ideal wires only, every point on a node has the same electric potential, which is known as the *node voltage* and its value is relative to the other nodes in the network.

The *junction* of two or more ideal wires is often used to represent an entire node; however, it is important to recognize that a wire junction is not the entire node and that a node may contain multiple wire junctions. It is crucial to correctly identify and count nodes in the analysis of electric circuits. Figure 1.2 illustrates a helpful way to mark nodes. There are three nodes in Figure 1.2(a) and two nodes in Figure 1.2(b). It is sometimes convenient to use the concept of a **supernode**, which is simply a closed boundary enclosing two or more nodes, as shown in Figure 1.2(c). In the next section, you will learn that one of the two fundamental laws of network analysis, Kirchhoff's current law (KCL), is valid for any closed boundary; that is, it is valid for any node or supernode.

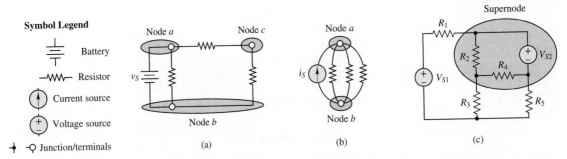

Figure 1.2 Illustrating nodes and supernodes in circuit diagrams

> Elements that sit between the same two nodes are said to be in *parallel* and have the same voltage across them.

It is also important to realize that since no work is required to move an electric charge along an ideal wire, the length and shape of an ideal wire has no impact on the behavior of a circuit. Likewise, since nodes are comprised of ideal wires, the extent and shape of a node has no impact on the behavior of a circuit. As a result, a node may be redrawn in any manner as long as the newly drawn node is attached to the same elements as the original node. Circuit diagrams are typically drawn, by convention, in a rectangular manner, with all wires drawn either side to side or up and down. However, many students find it helpful to redraw circuits so as to clarify the number and location of nodes in a circuit. Figure 1.3 shows two identical circuits drawn in two different ways. Can you tell that these circuits have the same number of nodes?

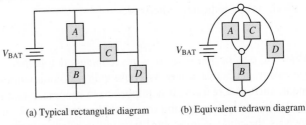

(a) Typical rectangular diagram   (b) Equivalent redrawn diagram

Figure 1.3 (a) A typical rectangular circuit diagram and (b) an equivalent redrawn diagram. A circuit can be redrawn to have almost any appearance; however, the nature of the circuit is unchanged if the number of nodes and the elements between those nodes remain unchanged.

Keep in mind that all forms of potential, including voltage, are relative quantities. For this reason, it is common to refer to the *change* in voltage *across* an element, or simply the voltage *across* an element. In circuit diagrams, a change in voltage across an element is indicated by the paired symbols + and −. Taken together as a single symbol they indicate the *assumed* direction of the change in voltage. However, as mentioned, it is also common to refer to a node voltage. To quantify a node voltage it is first necessary to select a *reference node*. Then, one can refer to the voltage of a node with the understanding that the value of that voltage is relative to the chosen reference node.

Any one node in a network can serve as the reference. The reference node and its value can be chosen freely, although a value of zero is usually chosen, for simplicity. It is often true that a smart choice of reference node will simplify the analysis that follows. A good rule of thumb is to select a node that is connected to a large number of elements.

A reference node is designated by the symbol shown in Figure 1.4(a). This symbol is also used to designate *earth ground* in applications. It is common for this symbol to appear multiple times in complicated circuits. Still, there is only one reference node per circuit. To reduce the apparent complexity of such circuits, multiple reference symbols are used to minimize the amount of displayed reference node wiring. It is simply understood that all nodes to which these symbols are attached are, in fact, connected by ideal wires and therefore part of one large reference node. Figure 1.4(b) and (c) illustrate this practice. The concepts of reference node, earth ground, and *chassis ground* are discussed later in this chapter.

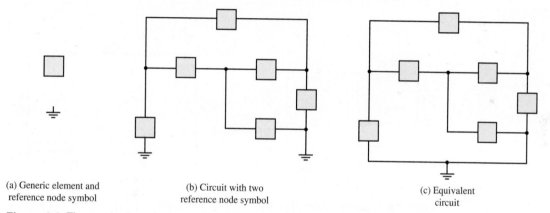

(a) Generic element and reference node symbol

(b) Circuit with two reference node symbol

(c) Equivalent circuit

Figure 1.4 There can be one and only one reference node in a network although the reference node symbol may appear more than once in order to reduce the amount of displayed reference node wiring.

Branch

A **branch** is a single electrical pathway, consisting of wires and elements. A branch may contain one or more circuit elements as shown in Figure 1.5. By definition,

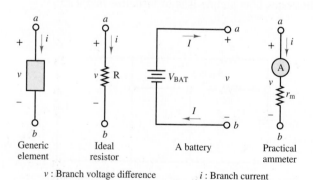

v : Branch voltage difference i : Branch current

Figure 1.5 Examples of circuit branches

the current *through* any one element in a branch is the same as the current through any other element in that branch; that is, there is one current in a branch, the *branch current*.

Elements that sit along the same branch are said to be in *series* and have the same current through them.

Loop

A **loop** is any closed pathway, physical or conceptual, as illustrated in Figure 1.6. Figure 1.6(a) shows that two different loops in the same circuit may share common elements and branches. It is interesting, and perhaps initially confusing, to note that a loop does not necessarily have to correspond to a closed electrical pathway, consisting of wires and elements. Figure 1.6(b) shows one example in which a loop passes directly from node a to node c.

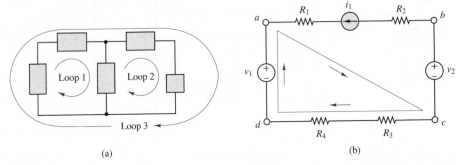

Figure 1.6 Examples of loops. How many nodes are in each of these circuits? [Answers: (a) 4; (b) 7]

Mesh

A **mesh** is a closed electrical pathway that does not contain other closed physical pathways. In Figure 1.6(a), loops 1 and 2 are meshes, but loop 3 is not a mesh because it encircles the other two loops. The circuit in Figure 1.6(b) has one mesh. Figure 1.7 illustrates how simple it is to visualize meshes.

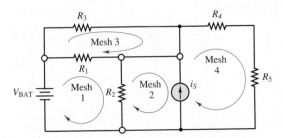

Figure 1.7 Circuit with four meshes. How many different closed electrical pathways are in this circuit? [Answer: 14]

EXAMPLE 1.1   Counting Nodes in a Network

Problem

Find the total number of nodes in each of the four networks of Figures 1.8–1.11.

Solution

Known Quantities: Wires and elements.

Find: The number of nodes in each network diagram in Figures 1.8 through 1.11.

Schematics, Diagrams, Circuits, and Given Data: Figure 1.8 contains four elements: two resistors and two ideal voltage sources, one independent and one dependent. Figure 1.9 contains five elements: four resistors and one independent ideal current source. Figure 1.10 contains five elements: four resistors and one operational amplifier. Figure 1.11 contains three elements: two headlamps and one 12-V battery.

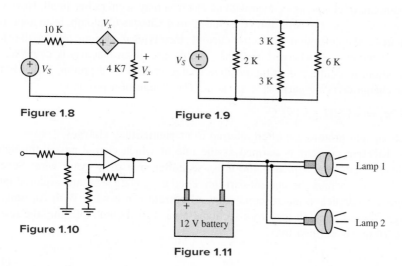

Figure 1.8

Figure 1.9

Figure 1.10

Figure 1.11

Assumptions: All wires are ideal.

Analysis: In Figure 1.8, all four elements are in a single electrical loop. There is one node between each pair of elements. Thus, there are *four nodes* in this network.

In Figure 1.9, all elements are connected between two nodes, one at the top, the other at the bottom of the circuit. In addition, there is a third node between the two 3-kΩ resistors. Thus, there are *three nodes* in this network.

In Figure 1.10, the nodes are expressly indicated by the black and white circles (note that the two circles on the far right denote the same node). In addition, the ground symbol repeated twice at the bottom of the circuit is also a node—the same node. Thus, there are *five nodes* in this network.

In Figure 1.11, there is one node between the positive + battery terminal and the two headlamps and one node between the negative − battery terminal and the two headlamps. Thus, there are *two nodes* in this network.

Comments: Notice that no knowledge of the elements is required to identify and count the nodes in a network.

1.2 CHARGE, CURRENT, AND KIRCHHOFF'S CURRENT LAW

Charles Coulomb (1736–1806)
(*INTERFOTO/Personalities/
Alamy Stock Photo*)

The earliest accounts of electricity date from about 2,500 years ago, when it was discovered that static charge on a piece of amber was capable of attracting very light objects, such as feathers. The word *electricity* originated about 600 B.C.; it comes from *elektron*, which was the ancient Greek word for amber. The true nature of electricity was not understood until much later, however. Following the work of Alessandro Volta and his invention of the copper-zinc battery, it was determined that static electricity and the current in metal wires connected to a battery are due to the same fundamental mechanism: the atomic structure of matter, consisting of a nucleus—neutrons and protons—surrounded by electrons.

The fundamental electric quantity is **charge**. The electron carries the smallest discrete unit of charge equal to:

$$q_e = -1.602 \times 10^{-19} \text{ C} \tag{1.1}$$

The amount of charge associated with an electron may seem rather small. However, the unit of charge, the **coulomb (C)**, named after Charles Coulomb, is an appropriate unit for the definition of electric current since typical currents involve the flow of large numbers of charged particles. The charge of an electron is negative, by convention, to contrast it to the positive charge carried by a proton, which is the other charge-carrying particle in an atom. The charge of a proton is:

$$q_p = +1.602 \times 10^{-19} \text{ C} \tag{1.2}$$

Electrons and protons are often referred to as **elementary charges**.

Electric current is defined as the rate at which charge passes through a predetermined area, typically the cross-sectional area of a metal wire. Several other cases in which the current-carrying conduit is not a wire are explored later. Figure 1.12 depicts a macroscopic view of current i in a wire. With Δq units of charge flowing through the cross-sectional area A in Δt units of time, the resulting current i is defined by:

Figure 1.12 Current in an electric conductor is defined as the net flow rate of charge through the cross-sectional area A.

$$i \equiv \frac{\Delta q}{\Delta t} \qquad \frac{\text{C}}{\text{s}} \tag{1.3}$$

The arrow symbol associated with the current i is its *assumed* direction through the wire segment. A negative value for i would indicate that the actual direction is opposite the assumed direction. When large numbers of discrete charges cross A in a very small period, this relationship can be written in differential form as:

$$i \equiv \frac{dq}{dt} \qquad \frac{\text{C}}{\text{s}} \tag{1.4}$$

The unit of current is the **ampere**, where 1 ampere (A) = 1 coulomb/second (C/s). The name of the unit is a tribute to the French scientist André-Marie Ampère. The electrical engineering convention is that the direction of positive current is the direction of positive charge flow. This convention is sensible; however, it can be confusing at first since the mobile charge carriers in metallic conductors are, in fact, electrons from the *conduction band* of the metal. It may help to realize that when an electron travels in one direction the effect on the distribution of *net charge* is the same as if a proton had travelled in the opposite direction. In other words, positive current is used to represent the *relative* flow of positive charges.

Current in a Closed Path

Earlier in this chapter, a circuit was defined as "a complete and closed path around which a circulating electric current can flow." In fact, conservation of electric charge requires a closed path for any nonzero current.

> To have a nonzero current, there must be a closed electrical path (i.e., a circuit).

For example, Figure 1.13 depicts a simple circuit, composed of a battery (e.g., a dry-cell or alkaline 1.5-V battery) and a lightbulb. Conservation of charge requires that the current i from the battery to the lightbulb is equal to the current from the lightbulb to the battery. No current (nor charge) is "lost" around the closed circuit. This principle was observed by the German scientist G. R. Kirchhoff[2] and is known as **Kirchhoff's current law (KCL)**. This law states that *the net sum of the currents crossing any closed boundary (a node or supernode) must equal zero.* In mathematical terms:

$$\sum_{n=1}^{N} i_n = 0 \qquad \text{Kirchhoff's current law (KCL)} \tag{1.5}$$

where the sign of currents entering the region surrounded by the closed boundary must be opposite to the sign of currents exiting the same region. In other words, the sum of currents "in" must equal the sum of currents "out." This statement leads to an alternate expression for KCL as:

$$\sum_{\text{in}} i = \sum_{\text{out}} i \qquad \text{Alternate KCL} \tag{1.6}$$

An application of Kirchhoff's current law is illustrated in Figure 1.14, where the simple circuit of Figure 1.13 has been augmented by the addition of two light-bulbs. One can find a relationship between the currents in the circuit by applying either version of KCL. To express the net sum of currents it is necessary to select a sign convention for currents entering and exiting a node. One possibility is to consider all currents entering a node as positive and all currents exiting a node as negative. (This particular sign convention is completely arbitrary.) The result of using this sign convention and applying the first version of KCL to node 1 is

$$i - i_1 - i_2 - i_3 = 0 \qquad \text{which is equivalent to} \qquad i = i_1 + i_2 + i_3$$

Note that the latter expression is exactly what would have been found if the alternate version of KCL had been applied. Also note that the result is the same if the opposite sign convention (i.e., currents entering and exiting the node are negative and positive, respectively) is used.

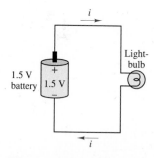

Figure 1.13 A simple electric circuit composed of a battery, a lightbulb, and two nodes

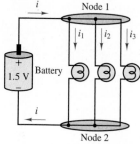

Figure 1.14 KCL applied at node 1 results in $i - i_1 - i_2 - i_3 = 0$, or equivalently $i = i_1 + i_2 + i_3$.

---

[2]Gustav Robert Kirchhoff (1824–1887), a German scientist, published the first systematic description of the laws of circuit analysis. His contribution—though not original in terms of its scientific content—forms the basis of all circuit analysis.

EXAMPLE 1.2   Charge and Current in a Conductor

Problem

Find the total charge in a cylindrical conductor (solid wire) and compute the current through the wire.

Solution

Known Quantities: Conductor geometry, charge density, charge carrier velocity.

Find: Total charge of carriers Q; current in the wire I.

Schematics, Diagrams, Circuits, and Given Data:

Conductor length: $L = 1$ m.

Conductor diameter: $2r = 2 \times 10^{-3}$ m.

Charge density: $n = 10^{29}$ carriers/m^3.

Charge of one electron: $q_e = -1.602 \times 10^{-19}$.

Charge carrier velocity: $u = 19.9 \times 10^{-6}$ m/s.

Assumptions: None.

Analysis: To compute the total charge in the conductor, we first determine the volume of the conductor:

$$\text{Volume} = \text{length} \times \text{cross-sectional area}$$

$$\text{Vol} = L \times \pi r^2 = (1 \text{ m}) \left[\pi \left(\frac{2 \times 10^{-3}}{2} \right)^2 \text{m}^2 \right] = \pi \times 10^{-6} \text{ m}^3$$

Next, we compute the number of carriers (electrons) in the conductor and the total charge:

$$\text{Number of carriers} = \text{volume} \times \text{carrier density}$$

$$N = \text{Vol} \times n = (\pi \times 10^{-6} \text{ m}^3) \left(10^{29} \frac{\text{carriers}}{\text{m}^3} \right) = \pi \times 10^{23} \text{ carriers}$$

$$\text{Charge} = \text{number of carriers} \times \text{charge/carrier}$$

$$Q = N \times q_e = (\pi \times 10^{23} \text{ carriers})$$

$$\times \left(-1.602 \times 10^{-19} \frac{\text{C}}{\text{carrier}} \right) = -50.33 \times 10^3 \text{ C}$$

To compute the current, we consider the velocity of the charge carriers and the charge density per unit length of the conductor:

$$\text{Current} = \text{carrier charge density per unit length} \times \text{carrier velocity}$$

$$I = \left(\frac{Q}{L} \ \frac{\text{C}}{\text{m}} \right) \times \left(u \ \frac{\text{m}}{\text{s}} \right) = \left(-50.33 \times 10^3 \frac{\text{C}}{\text{m}} \right) \left(19.9 \times 10^{-6} \frac{\text{m}}{\text{s}} \right) = -1 \text{ A}$$

Comments: Charge carrier density is a function of material properties. Carrier velocity is a function of the applied electric field.

EXAMPLE 1.3 **Kirchhoff's Current Law Applied to an Automotive Electrical Harness**

Problem

Figure 1.15 shows an automotive battery connected to a variety of elements in an automobile. The elements include headlights, taillights, starter motor, fan, power locks, and dashboard panel. The battery must supply enough current to satisfy each of the "load" elements. Apply KCL to find a relationship between the currents in the circuit.

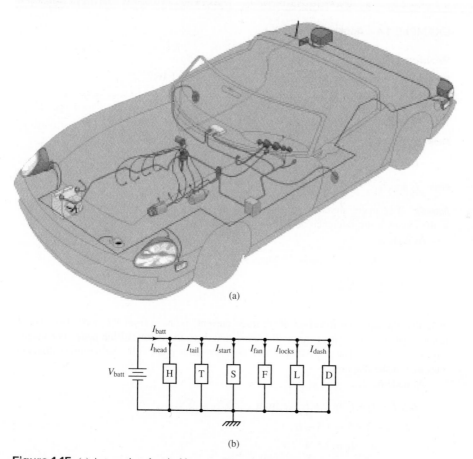

(a)

(b)

Figure 1.15 (a) Automotive electrical harness; (b) equivalent electric circuit diagram

Solution

Known Quantities: Components of electrical harness: headlights, taillights, starter motor, fan, power locks, and dashboard panel.

Find: Expression relating battery current to load currents.

Schematics, Diagrams, Circuits, and Given Data: Figure 1.15.

Assumptions: None.

Analysis: Figure 1.15(b) depicts the equivalent electric circuit, illustrating that the current supplied by the battery is divided among the various elements. The application of KCL to the upper node yields

$$I_{\text{batt}} - I_{\text{head}} - I_{\text{tail}} - I_{\text{start}} - I_{\text{fan}} - I_{\text{locks}} - I_{\text{dash}} = 0$$

or

$$I_{\text{batt}} = I_{\text{head}} + I_{\text{tail}} + I_{\text{start}} + I_{\text{fan}} + I_{\text{locks}} + I_{\text{dash}}$$

EXAMPLE 1.4   Application of KCL

Problem

Determine the unknown currents in the circuit of Figure 1.16.

Solution

Known Quantities:

$$I_S = 5 \text{ A} \qquad I_1 = 2 \text{ A} \qquad I_2 = -3 \text{ A} \qquad I_3 = 1.5 \text{ A}$$

Find: I_0 and I_4.

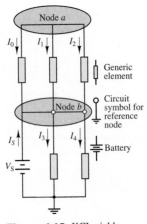

Figure 1.16 KCL yields $I_0 + I_1 + I_2 = 0$ at node a and $I_0 + I_1 + I_2 + I_S = I_3 + I_4$ at node b.

Analysis: Two nodes are clearly shown in Figure 1.16 as node a and node b; the third node in the circuit is the reference node. Apply KCL at each of the three nodes.

At node a:

$$I_0 + I_1 + I_2 = 0 \qquad \text{from} \qquad \sum i_{\text{out}} = \sum i_{\text{in}}$$

$$I_0 + 2 - 3 = 0$$

$$\therefore \qquad I_0 = 1 \text{ A}$$

Note that the assumed direction of all three currents is away from the node. However, I_2 has a negative value, which means that its actual direction is toward the node. The magnitude of I_2 is 3 A. The sign simply indicates direction relative to the assumed direction indicated in the diagram.

At node b:

$$I_0 + I_1 + I_2 + I_S = I_3 + I_4 \qquad \text{from} \qquad \sum i_{\text{in}} = \sum i_{\text{out}}$$

$$1 + 2 - 3 + 5 = 1.5 + I_4$$

$$\therefore \qquad I_4 = 3.5 \text{ A}$$

At the reference node: If we use the convention that currents entering a node are positive and currents exiting a node are negative, we obtain the following equations:

$$-I_S + I_3 + I_4 = 0$$

$$-5 + 1.5 + I_4 = 0$$

$$\therefore \qquad I_4 = 3.5 \text{ A}$$

Comments: The result obtained at the reference node is exactly the same as that calculated at node b. This fact suggests that some redundancy may result when we apply KCL at all nodes in a circuit. In Chapter 2 we develop a method called *node analysis* that ensures the derivation of the smallest possible set of independent equations.

EXAMPLE 1.5   Application of KCL

Problem

Apply KCL to the circuit of Figure 1.17, using the concept of a supernode to determine the source current i_{S1}.

Solution

Known Quantities:

$$i_3 = 2 \text{ A} \qquad i_5 = 0 \text{ A}$$

Find: i_{S1}.

Analysis: Apply KCL at the boundary of the supernode to obtain

$$i_{S1} = i_3 + i_5 \qquad \text{from} \qquad \sum i_{\text{in}} = \sum i_{\text{out}}$$
$$i_{S1} = 2 + 0 = 2 \text{ A}$$

Comments: Notice that the same result for i_{S1} is obtained by applying KCL at the bottom node. This fact is another example of a redundant result that is sometimes obtained by applying KCL at two different nodes, including supernodes. When applied correctly, the *node analysis* method discussed in Chapter 2 prevents redundant equations.

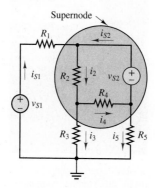

Figure 1.17 KCL applied at the boundary of the supernode yields $i_{S1} = i_3 + i_5$.

CHECK YOUR UNDERSTANDING

Repeat the exercise of Example 1.4 when $I_0 = 0.5$ A, $I_2 = 2$ A, $I_3 = 7$ A, and $I_4 = -1$ A. Find I_1 and I_S.

Answer: $I_1 = -2.5$ A and $I_S = 6$ A

CHECK YOUR UNDERSTANDING

Use the result of Example 1.5 and the following data to compute the current i_{S2} in the circuit of Figure 1.17.

$$i_2 = 3 \text{ A} \qquad i_4 = 1 \text{ A}$$

Answer: $i_{S2} = 1$ A

1.3   VOLTAGE AND KIRCHHOFF'S VOLTAGE LAW

Typically, work is required to move charge between two nodes in a circuit. The total *work per unit charge* is called **voltage**, and the unit of voltage is the **volt** in honor of Alessandro Volta.

$$1 \text{ volt (V)} = 1 \frac{\text{joule (J)}}{\text{coulomb (C)}} \qquad\qquad (1.7)$$

Gustav Robert Kirchhoff
(1824–1887) (*bilwissedition
Ltd. & Co. KG/Alamy Stock
Photo*)

The voltage, or **potential difference**, across two nodes in a circuit is the energy (in joules) per unit charge (1 coulomb) needed to move charge from one node to the other. The direction, or *polarity*, of the voltage is related to whether energy is being gained or lost by the charge in the process.

Consider again the simple circuit of a battery and a lightbulb as shown in Figure 1.18. Experimental observations led Kirchhoff to formulate the second of his laws, **Kirchhoff's voltage law** (**KVL**), which states that that *the net change in electric potential around a closed loop is zero.* In mathematical terms:

$$\sum_{n=1}^{N} v_n = 0 \qquad \text{Kirchhoff's voltage law} \tag{1.8}$$

Here, v_n are the changes in voltage from one node to another around a closed loop.

When summing these changes in voltage, it is necessary to account for the polarity of the change. Changes in voltage from the minus sign $-$ to the plus sign $+$ are considered positive (i.e., a rise in voltage), while those from plus to minus are considered negative (i.e., a drop in voltage). These two symbols act together to indicate the *assumed* direction of the change in voltage *across* an element, just as the arrow symbol is used to indicate the *assumed* direction of the current *through* an element or wire segment. A negative value indicates that the actual direction is opposite to the assumed direction.

An alternate but equivalent expression for KVL is that the sum of all voltage rises around a loop must equal the sum of all voltage drops around the same loop.

$$\sum_{\text{rises}} v = \sum_{\text{drops}} v \qquad \text{Alternate KVL} \tag{1.9}$$

In Figure 1.18, the *voltage across the lightbulb* is the change in voltage from node a to node b. This change can also be expressed as the difference between two node voltages, v_a and v_b. As stated earlier, the values of node voltages are relative to some reference node. Any single node may be chosen as the reference with its value set to zero, for simplicity, or any other convenient number. The circuit in Figure 1.18 has only two nodes, a and b, one of which can serve as the reference node. Select node b as the reference and set its value as $v_b = 0$. Then, observe that the battery's positive terminal is 1.5 V *above the reference*, so that $v_a = 1.5$ V. In general, the battery guarantees that node a will always be 1.5 V above node b. Mathematically, this fact is simply expressed as

$$v_a = v_b + 1.5 \text{ V}$$
$$v_a = 1.5 \text{ V} \qquad \text{when} \qquad v_b = 0 \text{ acts as the reference.}$$

The syntax used to express the *change* in voltage across the lightbulb, *from* node b *to* node a, is v_{ab}, where

$$v_{ab} \equiv v_a - v_b = 1.5 \text{ V}$$

This syntax is in accord with the $+$ and $-$ polarity indicator in that if v_{ab} is positive, then $v_a > v_b$ and, in fact, node a is at a higher potential than node b, as suggested by the $+$ and $-$ syntax. It may be helpful to think of v_{ab} as v_a relative to node b.

Note that the work done in moving charge from node a to node b is directly proportional to the voltage across the lightbulb. Likewise, the work done moving

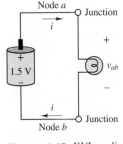

Figure 1.18 KVL applied clockwise from node b around the single loop circuit results in $1.5 \text{ V} - v_{ab} = 0$, or equivalently $v_{ab} = 1.5$ V.

charge back from b to a is directly proportional to the voltage across the battery. Let Q be the total charge that moves around the circuit per unit time, giving rise to current i. Then, the work W done *by* the battery *on* Q, from b to a (i.e., across the battery) is

$$W_{ab} = Q \times v_{ab} = Q \times 1.5 \text{ V}, \qquad \text{work done by the battery on } Q$$

which is also equal to the work done *by* Q *on* the lightbulb, from a to b (i.e., across the lightbulb). One could express this work in the negative as the work done *by* the lightbulb *on* Q, from a to b.

$$W_{ba} = Q \times v_{ba} = -Q \times v_{ab} = -Q \times 1.5 \text{ V}$$

Note that the word *potential* is quite appropriate as a synonym of voltage, in that voltage is the potential energy per unit charge between two nodes in a circuit. If the lightbulb is disconnected from the circuit, a voltage v_{ab} still exists across the battery terminals, as illustrated in Figure 1.19. This voltage represents the ability of the battery to *supply* energy to the circuit. Likewise, the voltage across the lightbulb is associated with the work done by the lightbulb to *consume* or *dissipate* energy from the circuit. The rate at which charge is moved once a closed circuit is established depends upon the circuit element connected to the battery.

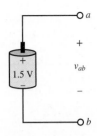

Figure 1.19 The voltage v_{ab} across the open terminals of the battery represents the potential energy available to move charge from a to b once a closed circuit is established.

The Reference Node and Ground

The concept of a reference node finds a practical use in the *ground* node of a circuit. Ground represents a specific, and usually clearly marked, reference node and voltage in a circuit. For example, the ground reference voltage can be identified with the enclosure or case of an instrument, or with the earth itself. In residential electric circuits, the ground reference is a large conductor, such as a copper spike or water pipe, that is buried in the earth. As mentioned, it is convenient and typical to assign the ground voltage reference a value of zero.

In practice, the term *ground* should not be applied to a node arbitrarily. However, the voltage value that is assigned to ground, while typically zero, is not consequential. A simple analogy with fluid flow illustrates this rule. Consider a tank of water, as shown in Figure 1.20, located at a certain height above the ground. The potential energy difference per unit mass due to gravity $u_{12} = g(h_1 - h_2)$ will cause water to flow out of the pipe at a certain flow rate. This quantity is completely analogous to the potential energy difference per unit charge $v_a - v_b$. Now assume that the height h_3 at ground level is chosen to be the zero potential energy

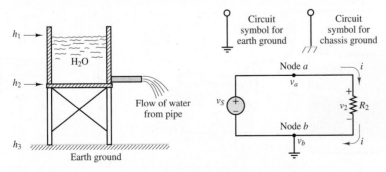

Figure 1.20 An analogy between water flow and electric current illustrates the relation between potential differences and a ground reference potential.

reference. Is the flow of water in the pipe changed due to this choice? Of course not. Is the flow of water in the pipe dependent upon the height $h_2 - h_3$ of the support structure? Again, the answer is no. The truth of these statements is demonstrated by rewriting the *head* of the water tank $h_1 - h_2$ as $(h_1 - h_3) - (h_2 - h_3)$ and by noting that the potential energy difference per unit mass can be written as the difference in potential energy per unit mass relative to the ground. That is,

$$u_{12} = g(h_1 - h_2) = [g(h_1 - h_3)] - [g(h_2 - h_3)] = u_{13} - u_{23}$$

Note that the values of u_{13} and u_{23} depend upon the value assigned to h_3; however, the value of u_{12} does *not* depend upon the value assigned to h_3. If this result were not true, our experience of the physical world would be very strange indeed, and not only because the flow of water from a tank would depend significantly upon the height of the tank itself. It is the relative difference in potential energy that matters in the water tank problem. So it is with electric circuits. The current through an element depends upon the potential difference (i.e., voltage) *across* the element and not on the selection of a reference node nor the arbitrary value of the ground reference node.

Another familiar scenario is that of a skydiver leaping from an airplane and parachuting to the surface below (see Figure 1.21). To quantify the potential energy U of the skydiver it is first necessary to choose a reference height h_0 such that $U = mg\Delta h = mg(h - h_0)$, where h represents the position of the skydiver. One possible reference is the height of the airplane such that the potential energy of the skydiver is negative ($U < 0$). However, that reference is not particularly meaningful. The surface of the earth is a more meaningful reference to the skydiver, who knows that a soft landing depends upon dissipating most of the initial potential energy through collisions with air molecules rather than through a collision with the surface. The skydiver knows that her fate is unchanged by her choice of reference; however, some choices are more meaningful than others. So it often is with electric circuits.

Figure 1.21 A skydiver understands all too well that her fate is unchanged by the choice of reference potential.

EXAMPLE 1.6   Kirchhoff's Voltage Law—Electric Vehicle Battery Pack

Problem

Figure 1.22(a) depicts the battery pack in the Smokin' Buckeye electric race car. In this example we apply KVL to the series connection of thirty-one 12-V batteries that make up the battery supply for the electric vehicle.

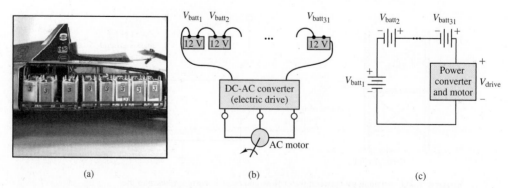

(a)   (b)   (c)

Figure 1.22 Electric vehicle battery pack illustrates KVL. (*Courtesy: David H. Koether Photography*)

Solution

Known Quantities: Nominal characteristics of **Optima™ lead-acid batteries.**

Find: Expression relating battery and electric motor drive voltages.

Schematics, Diagrams, Circuits, and Given Data: $V_{\text{batt}} = 12$ V; Figure 1.22(a), (b), and (c).

Assumptions: None.

Analysis: Figure 1.22(b) depicts the equivalent electric circuit, illustrating how the voltages supplied by the battery are applied across the electric drive that powers the vehicle's 150-kW three-phase induction motor. The application of KVL around the closed circuit of Figure 1.22(c) requires that:

$$\sum_{n=1}^{31} V_{\text{batt}_n} - V_{\text{drive}} = 0$$

Thus, the electric drive is nominally supplied by a $31 \times 12 = 372$-V battery pack. In practice, the voltage supplied by lead-acid batteries varies depending on the state of charge of the battery. When fully charged, the battery pack of Figure 1.22(a) supplies closer to 400 V (i.e., roughly 13 V per battery).

EXAMPLE 1.7   Application of KVL

Problem

Determine the unknown voltage v_2 by applying KVL to the circuit of Figure 1.23.

Solution

Known Quantities:

$$v_S = 12 \text{ V} \qquad v_1 = 6 \text{ V} \qquad v_3 = 1 \text{ V}$$

Find: v_2.

Analysis: Apply KVL starting at the reference node and proceeding clockwise around the large outer loop (the outer perimeter) of the circuit to find

$$v_S - v_1 - v_2 - v_3 = 0$$
$$v_S - v_1 - v_3 = v_2$$
$$12 - 6 - 1 = v_2 = 5 \text{ V}$$

Comments: Note that v_2 is the voltage across elements 2 and 4. These two elements are in *parallel* because they are located between the same two nodes. One can also say that the two branches that contain these elements are in parallel.

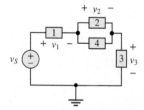

Figure 1.23 A circuit with four generic elements and one ideal voltage source

---

EXAMPLE 1.8   Application of KVL

Problem

Use KVL to determine the unknown voltages v_1 and v_4 in the circuit of Figure 1.24.

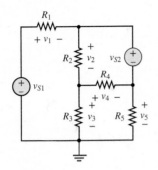

Figure 1.24 Circuit for Example 1.7

Solution

Known Quantities:

$$v_{S1} = 12 \text{ V} \qquad v_{S2} = -4 \text{ V} \qquad v_2 = 2 \text{ V} \qquad v_3 = 6 \text{ V} \qquad v_5 = 12 \text{ V}$$

Find: v_1, v_4.

Analysis: To determine the unknown voltages, apply KVL clockwise around the left and upper-right meshes:

$$v_{S1} - v_1 - v_2 - v_3 = 0$$
$$v_2 - v_{S2} + v_4 = 0$$

After substituting numerical values, the equations become:

$$12 - v_1 - 2 - 6 = 0$$
$$v_1 = 4 \text{ V}$$
$$2 - (-4) + v_4 = 0$$
$$v_4 = -6 \text{ V}$$

It is possible to solve for v_1 and v_4 using other loops in the circuit. For instance, apply KVL around the lower-right mesh to find v_4:

$$v_3 - v_4 - v_5 = 0$$
$$6 - v_4 - 12 = 0$$
$$v_4 = -6 \text{ V}$$

Or apply KVL around the outer most loop to find v_1:

$$v_{S1} - v_1 - v_{S2} - v_5 = 0$$
$$12 - v_1 - (-4) - 12 = 0$$
$$v_1 = 4 \text{ V}$$

Comments: Notice that there are seven closed wire loops in the circuit. KVL could be applied around any of these loops to find an equation. The key is to find two linearly independent equations that involve the two unknowns. In Chapter 2 a systematic procedure called *mesh analysis* is developed that yields the minimum number of linearly independent equations required to solve *all* unknown voltages and currents in a well-defined circuit.

CHECK YOUR UNDERSTANDING

Apply KVL to each of the other three closed wire loops in Figure 1.24 that were not explored in Example 1.8. Compare the results to those found in the example. Are the results consistent?

1.4 POWER AND THE PASSIVE SIGN CONVENTION

The definition of voltage as work per unit charge lends itself very conveniently to the introduction of power. Recall that power is defined as the work done per unit time. Thus, the power P either supplied or dissipated by a circuit element can be represented by the following relationship:

$$\text{Power} = \frac{\text{work}}{\text{time}} = \frac{\text{work}}{\text{charge}}\frac{\text{charge}}{\text{time}} = \text{voltage} \times \text{current} \qquad (1.10)$$

Thus,

> Electric power, P, is the product of voltage, v, *across* an element and current, i, *through* it.

$$\boxed{P = vi} \qquad (1.11)$$

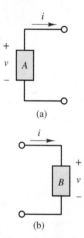

The units of voltage (joules per coulomb) multiplied by those of current (coulombs per second) equal the units of power (joules per second, or watts).

The power associated with a circuit element can be positive or negative, depending (by convention) upon whether the element consumes or supplies energy, respectively. Consider Figure 1.25(a), in which electric charge flows from low to high potential. Clearly, work has been done *by* element A *on* the flowing charge as its potential is raised. The rate at which this work is done *by* element A is its power. In this case, the power is negative because energy is either *supplied* or *released* by the element to the charge in the circuit. The other possibility is shown in Figure 1.25(b), in which electric charge flows from high to low potential. Here, work has been done *on* element B *by* the flowing charge as its potential is lowered. The rate at which this work is done *on* element B is its power. In this case, the power is positive because energy is either *dissipated* or *released* by the element from the charge in the circuit.

Figure 1.25 Assuming positive values for i and v, (a) energy is supplied by element A, while (b) energy is consumed by element B, which is labeled with the passive sign convention.

> In the *passive sign convention*, charge is assumed to flow from high to low potential such that energy is consumed or stored by the element and power is positive.

Passive elements are defined as those that do not require an external source of energy to *enable* them. Common passive elements are resistors, capacitors, inductors, diodes, and electric motors. Passive elements can dissipate energy (e.g., resistors) and/or store and release energy (e.g., capacitors and inductors). *Active* elements, on the other hand, are defined as those that do require an external source of energy to be enabled or function. Common active elements are transistors, amplifiers, and voltage and current sources.

The electrical engineering community has uniformly adopted the passive sign convention. All the constitutive laws (e.g., Ohm's law) introduced in this book assume and are based upon the convention. It is important to keep this fact in mind when using these laws to solve problems. Specifically, it is often necessary to assume directions for unknown currents and/or assume polarities for unknown voltages when solving circuit problems. It is important that these assumptions be made in accord with the passive sign convention. Violating the convention will often lead to incorrect results. On the other hand, as long as the passive sign convention is observed it is not necessary to foresee actual current directions nor actual voltage polarities. Instead, when a current direction or voltage polarity is assumed incorrectly, the solution will yield the negative of the true value for that current or voltage to indicate that the assumed direction or polarity is opposite the actual.

FOCUS ON PROBLEM SOLVING

THE PASSIVE SIGN CONVENTION

1. Assign a current through each passive element. The direction of each current can be assumed arbitrarily.

2. For each *passive* element, assign a voltage across the element such that the assigned current through the element is directed from high to low potential. Other valid descriptions are that current enters the + terminal or exits the − terminal of the element.

3. For each *active* element, assign a current through and/or a voltage across the element according to the following guidelines:

    (a) If the current through the element is given, assign a voltage across the element such that the current is directed from low to high potential.

    (b) If the voltage across the element is given, assign a current through the element such that the current is directed from low to high potential.

4. The power associated with each element is computed according to the following rules:

    (a) For passive elements, the power is positive and equal to vi. Positive power indicates that the element is either dissipating or storing energy.

    (b) For active elements, the power is usually negative and equal to $-vi$. Negative power indicates that the element is either supplying or releasing energy. Occasionally, an active element may consume energy, as when a battery is being charged, and its power would then be positive since current would be entering the positive terminal.

EXAMPLE 1.9   Use of the Passive Sign Convention

Problem

Apply the passive sign convention to solve for the voltages and *mesh current* in the circuit of Figure 1.26.

Solution

Known Quantities: Voltage of the battery and the power dissipated by load 1 and load 2.

Find: Mesh current and the voltage across each load.

Schematics, Diagrams, Circuits, and Given Data: Figure 1.27. The voltage of the battery is $V_B = 12\,\text{V}$. The power dissipated by load 1 is $P_1 = 0.8\,\text{W}$ and by load 2 is $P_2 = 0.4\,\text{W}$.

Assumptions: None.

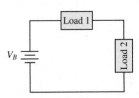

Figure 1.26 Circuit for Example 1.9

Analysis: This problem can be solved using the passive sign convention in two different approaches. The first approach assumes a clockwise mesh current, while the second approach assumes a counterclockwise current. For either approach, the passive sign convention is used to label the change in voltage across each load. Figure 1.27 shows the result of these two approaches. Notice that the change in voltage across each load was chosen so that the assumed current through each load is directed from high to low potential.

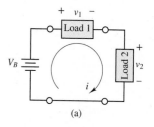

The polarity of the battery is indicated by the alternating sequence of long and short bars. The positive and negative terminals of the battery are connected to a long and short bar, respectively.

A four-step solution using the first approach, as depicted in Figure 1.27(a), is given below.

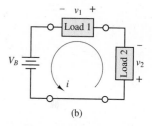

Figure 1.27 Solution steps for Example 1.9

1. Assume a clockwise direction for the current.

2. Label the change in voltage across each (passive) load so that the current through each load is directed from high to low potential.

3. Express the power dissipated by each load using the relation $P = vi$, which is valid when the passive sign convention is observed.

$$P_1 = v_1 i = 0.8 \text{ W}$$

$$P_2 = v_2 i = 0.4 \text{ W}$$

The power associated with the battery is expressed as $P_B = -V_B i$, which requires a negative sign $-vi$ because the current through the battery is directed from low to high potential, opposite of the passive sign convention.

4. Conservation of energy requires that the total power associated with the circuit be zero. Thus,

$$P_1 + P_2 + P_B = 0$$

$$P_B = -P_1 - P_2 = -0.8 \text{ W} - 0.4 \text{ W} = -1.2 \text{ W} = -V_B i$$

It is now possible to use the three vi equations to solve for the three unknown variables i, v_1, and v_2. Since $V_B = 12 \text{ V}$, the current i is:

$$i = \frac{-1.2 \text{ W}}{-12 \text{ V}} = 0.1 \text{ A}$$

As a result, the change in voltage across each load is:

$$v_1 = \frac{0.8 \text{ W}}{0.1 \text{ A}} = 8 \text{ V}$$

$$v_2 = \frac{0.4 \text{ W}}{0.1 \text{ A}} = 4 \text{ V}$$

A four-step solution using the second approach, as depicted in Figure 1.27(b), is given below.

1. Assume a counterclockwise direction for the current.

2. Label the change in voltage across each (passive) load so that the current through each load is directed from high to low potential.

3. Express the power dissipated by each load using the relation $P = vi$, which is valid when the passive sign convention is observed.

$$P_1 = v_1 i = 0.8 \text{ W}$$

$$P_2 = v_2 i = 0.4 \text{ W}$$

The power associated with the battery is expressed here as $P_B = +V_B i$, which now requires a positive sign $+vi$ because the current through the battery is directed from high to low potential, in accord with the passive sign convention.

4. Conservation of energy requires that the total power associated with the circuit be zero. Thus,

$$P_1 + P_2 + P_B = 0$$
$$P_B = -P_1 - P_2 = -0.8\,\text{W} - 0.4\,\text{W} = -1.2\,\text{W} = V_B i$$

It is now possible to use the three vi equations to solve for the three unknown variables i, v_1, and v_2. Since $V_B = 12$ V, the current i is:

$$i = \frac{-1.2\,\text{W}}{12\,\text{V}} = -0.1\,\text{A}$$

As a result, the change in voltage across each load is:

$$v_1 = \frac{0.8\,\text{W}}{-0.1\,\text{A}} = -8\,\text{V}$$
$$v_2 = \frac{0.4\,\text{W}}{-0.1\,\text{A}} = -4\,\text{V}$$

Comments: Notice that the *actual* current present in the circuit and the *actual* change in voltage across each load was found to be the same for each solution approach. For instance, using the first approach the current was found to be 0.1 A clockwise, while using the second approach the current was found to be −0.1 A counterclockwise. The negative sign found for the current in the second approach indicates that the actual current is directed clockwise, not counterclockwise. This example provides a good demonstration of the fact that it is not necessary to foresee the actual direction of unknown currents and voltages when solving a circuit problem. The important point is to observe the passive sign convention.

Also note that conservation of energy is required for electric circuits, just as it is for any other physical system. For electric circuits: *Power supplied always equals power consumed.*

EXAMPLE 1.10   Power Calculations

Problem

For the circuit shown in Figure 1.28, determine which components are consuming power and which are supplying power. Is conservation of power satisfied. Explain your answer.

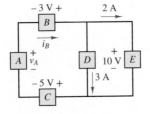

Figure 1.28 Circuit for Example 1.10

Solution

Known Quantities: Current through elements D and E; voltage across elements B, C, D, E.

Find: Which components are consuming power, and which are supplying power? Verify the conservation of power.

Analysis: Apply KCL to the node connecting elements B, D, and E to find the current through element B.

$$i_B = 2\,\text{A} + 3\,\text{A} = 5\,\text{A}$$

Apply KVL counterclockwise around the outer perimeter loop to find the voltage across element A.

$$-v_A - 3 + 10 + 5 = 0 \qquad v_A = 12\,\text{V}$$

The power associated with each element can now be computed using $P = vi$ when the passive sign convention is observed or $P = -vi$ when it is not observed.

$$P_A = -(12\,\text{V})(5\,\text{A}) = -60\,\text{W}$$
$$P_B = -(3\,\text{V})(5\,\text{A}) = -15\,\text{W}$$
$$P_C = (5\,\text{V})(5\,\text{A}) = 25\,\text{W}$$
$$P_D = (10\,\text{V})(3\,\text{A}) = 30\,\text{W}$$
$$P_E = (10\,\text{V})(2\,\text{A}) = 20\,\text{W}$$

Notice that the total power sums to zero. The same results can be expressed more literally as:

- A supplies 60 W
- B supplies 15 W
- C dissipates (consumes) 25 W
- D dissipates (consumes) 30 W
- E dissipates (consumes) 20 W
- Total power supplied equals 75 W
- Total power dissipated (consumed) equals 75 W
- Total power supplied = total power dissipated

Comments: Notice that whether power is calculated using $P = vi$ or $P = -vi$ depends entirely upon whether the passive sign convention is observed for any particular element.

CHECK YOUR UNDERSTANDING

Solve for the currents and voltages shown in Figure 1.27(a) and (b) using KVL around each mesh instead of using conservation of energy for the circuit, which gave the battery power as $P_B = -1.2$ W.

CHECK YOUR UNDERSTANDING

Compute the current through each of the headlamps shown in Figure 1.11 assuming each headlamp consumes 50 W. How much power is the battery providing?

Answers: $I_1 = I_2 = 4.17$ A; 100 W.

CHECK YOUR UNDERSTANDING

Determine which circuit element, A or B, in the top figure is supplying power and which is dissipating power. Also determine how much power is dissipated and supplied.

If the voltage source in the bottom figure supplies a total of 10 mW and $i_1 = 2$ mA and $i_2 = 1.5$ mA, what is the current i_3? If $i_1 = 1$ mA and $i_3 = 1.5$ mA, what is i_2?

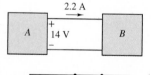

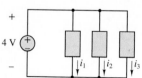

Answers: A supplies 30.8 W; B dissipates 30.8 W; $i_3 = -1$ mA; $i_2 = 0$ mA.

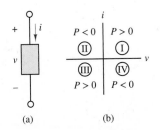

Figure 1.29 (a) The passive sign convention for a generalized circuit element, and (b) the relationship between v, i, and the power P, as depicted on a typical i-v plot

1.5 i-v CHARACTERISTICS AND SOURCES

As noted in the previous section, the power supplied or consumed by elements is defined by $P = vi$, where v and i observe the passive sign convention as shown in Figure 1.29(a). Using this convention, positive power implies that energy is consumed (dissipated or stored), while negative power $P < 0$ implies that energy is provided (supplied or released), by the element. The relationship between v, i, and the sign of the power P is depicted in Figure 1.29(b), which is a typical i-v plot.

It is possible to create an i-v plot for any particular circuit element and compare it to Figure 1.29(b) to determine whether the element supplies or consumes power for any particular values of i and/or v. The functional relationship between i and v for any particular circuit element may be quite complex and not easily expressed in a closed mathematical form, such as $i = f(v)$. However, the plot of the **i-v characteristic** (or **volt-ampere characteristic**) for most circuit elements is either known or can be determined experimentally.

For example, consider the circuit shown in Figure 1.30(a), where a conventional incandescent (tungsten filament) lightbulb is in a simple loop with a variable voltage source and a meter for measuring current. Notice the passive sign convention used to define the voltage across and the current through the lightbulb. The i-v characteristic of the lightbulb can be determined by varying the voltage over some predetermined range and recording the resulting current for each particular voltage in that range. The plot of the i-v data will be similar to that shown in Figure 1.30(b). Notice that the i-v characteristic runs from quadrant III through the origin and into quadrant I. A positive voltage across the bulb results in a positive current through it, and conversely, a negative voltage across the bulb results in a negative current through it. In both cases the bulb power is positive. Thus, the incandescent lamp always dissipates energy.

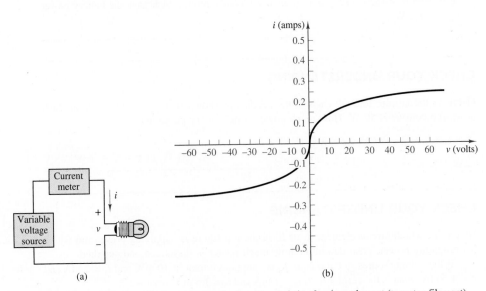

Figure 1.30 (a) Depiction of how to measure the i-v characteristic of an incandescent (tungsten filament) lightbulb; (b) typical i-v plot of such a lightbulb

There are electronic devices that can operate, for example, in three of the four quadrants of the *i-v* characteristic and can therefore act as sources of energy for specific combinations of voltages and currents. An example of this dual behavior is introduced in Chapter 8, where it is shown that the photodiode can act either in a passive mode (as a light sensor) or in an active mode (as a solar cell).

The *i-v* characteristics of ideal voltage and current sources are simple yet helpful visual aids. An ideal source is one that can provide any amount of energy without affecting the behavior of the source itself. **Ideal sources** are divided into two types: voltage sources and current sources.

Ideal Voltage Sources

An **ideal voltage source** generates a prescribed voltage across its terminals independent of the current through its terminals. The circuit symbol for an ideal voltage source is shown in Figure 1.31(a). Notice the *active* sign convention, which is opposite the passive sign convention. A vertical *i-v* characteristic in quadrant I is shown in Figure 1.31(b) and indicates that the source *supplies* energy. The amount of current and energy supplied by the source is determined by the circuit connected to it. It is important to recognize that an ideal voltage source guarantees a particular *change* in voltage from the node attached to its − terminal to the node attached to its + terminal. The + and − polarity markers do *not* indicate positive and negative voltage values relative to some zero reference. Do not make this mistake when solving problems!

> An ideal voltage source provides a prescribed voltage across its terminals independent of the current through those terminals. The amount of current through the source is determined by the circuit connected to it.

Various types of batteries, electronic power supplies, and function generators approximate ideal voltage sources when used in proper circumstances. However, all such real devices have limits on the amount of current that can be supplied without impacting the voltage across the source. This behavior can be seen in a typical 12-V car battery. A digital voltmeter can be used to observe the voltage across a car battery as various electrical devices in the car are turned on and off. Very little change in the battery voltage will be observed, even when power windows are engaged. However, when the car is started, the battery voltage will drop significantly during the short period needed for the engine to start.

Figure 1.32 depicts various symbols for voltage sources that are employed throughout this book. Note that the output voltage of an ideal source can be a function of time. In general, the following notation is employed in this book, unless otherwise noted. A generic voltage source is denoted by a lowercase *v*. If it is necessary to emphasize that the source produces a time-varying voltage, then the notation *v(t)* is employed. Finally, a constant, or *direct-current*, or *DC*, voltage

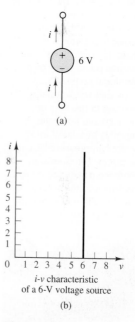

(a)

i-v characteristic
of a 6-V voltage source

(b)

Figure 1.31 (a) An ideal voltage source, shown with the *active* sign convention; and (b) a typical *i-v* characteristic for that sign convention, which indicates power is supplied by the source

MAKE THE
CONNECTION

Hydraulic Analog of a Voltage Source

The role played by a voltage source in an electric circuit is very similar to that played by a velocity pump in a hydraulic circuit. In a velocity or roto-dynamic pump, such as a centrifugal pump, impeller vanes add kinetic energy (velocity) to the fluid flow. This increase in kinetic energy is translated to an increase in pressure *across* the pump. The pressure difference *across* the pump is analogous to the voltage, or potential difference, *across* the voltage source.

A centrifugal pump
(*Giorgio Rizzoni*)

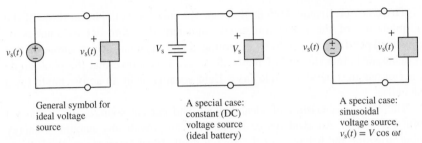

General symbol for ideal voltage source

A special case: constant (DC) voltage source (ideal battery)

A special case: sinusoidal voltage source, $v_s(t) = V \cos \omega t$

Figure 1.32 Three common ideal voltage sources

source is denoted by the uppercase character V. By convention, the direction of positive current is from low to high potential; that is, current enters the − terminal and exits the + terminal.

Ideal Current Sources

An **ideal current source** generates a prescribed current through its terminals independent of the voltage across its terminals. The circuit symbol for an ideal current source is shown in Figure 1.33(a). Notice the sign convention, which is opposite the passive sign convention. A typical straight horizontal line i-v characteristic in quadrant I is shown in Figure 1.33(b) and indicates that the source *supplies* energy. The amount of current and energy supplied by the source is determined by the circuit connected to it. It is important to recognize that an ideal current source guarantees a particular current through its terminals, such that the current entering the − terminal is the same as the current exiting the + terminal.

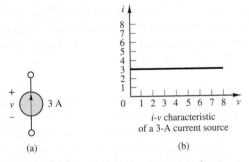

i-v characteristic of a 3-A current source

(a)   (b)

Figure 1.33 (a) An ideal current source, shown with the *active* sign convention; and (b) a typical i-v characteristic for that sign convention, which indicates power is supplied by the source.

An ideal current source provides a prescribed current through its terminals independent of the voltage across those terminals. The amount of voltage across the source is determined by the circuit connected to it.

Practical approximations to ideal current sources are not as common or numerous as those for ideal voltage sources. However, in general, an ideal voltage source in series with a large output resistance provides a nearly constant—though small—current and thus approximates an ideal current source. A battery charger is a common and approximate example of an ideal current source.

Figure 1.34 depicts a circuit that contains the generic symbol for an ideal current source. The same uppercase and lowercase convention used for voltage sources is employed in denoting constant (DC) and time-varying current sources, respectively.

Dependent (Controlled) Sources

The ideal sources described earlier are able to generate a prescribed voltage or current independent of any other element within the circuit. Thus, they are known as *independent sources*. Another category of sources, whose output (current or voltage) depends on some other voltage or current in a circuit, is known as **dependent** (or **controlled**) **sources**. As shown in Figure 1.35, the circuit symbols for these sources are diamonds, to distinguish them from independent sources. The table illustrates the relationship between the source voltage v_S or source current i_S and the circuit voltage v_x or circuit current i_x, which they depend upon and which can be any voltage or current elsewhere in the circuit.

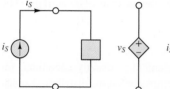

Source type	Relationship
Voltage controlled voltage source (VCVS)	$v_S = \mu v_x$
Current controlled voltage source (CCVS)	$v_S = r i_x$
Voltage controlled current source (VCCS)	$i_S = g v_x$
Current controlled current source (CCCS)	$i_S = \beta i_x$

Figure 1.34 This simple circuit contains an generic ideal time-varying current source.

Figure 1.35 Symbols for dependent sources

Dependent sources are very useful in describing certain types of electronic circuits. You will encounter dependent sources again in Chapters 7, 9, and 10, when electronic amplifiers are discussed.

1.6   RESISTANCE AND OHM'S LAW

When charge flows through a wire or circuit element, it encounters a certain amount of **resistance**, the magnitude of which depends on the *resistivity* of the material and the geometry of the wire or element. In practice, all circuit elements exhibit some resistance, which leads to energy dissipation in the form of heat. Whether this loss of electrical energy as heat is detrimental depends upon the purpose of the circuit element. For example, a typical electric toaster relies on the conversion of electrical energy to heat within its resistive coils to accomplish its purpose, the making of toast. All electric heaters rely upon this process, in one form or another. On the other hand, heat loss due to resistance in residential wiring is costly, and potentially dangerous. Resistance in microcircuitry generates heat that effectively limits the speed of microprocessors and the number and scale of transistors that can be packed into a given volume.

The resistance of a cylindrical wire segment, as shown in Figure 1.36(a), is given by

$$R = \rho \frac{l}{A} = \frac{l}{\sigma A} \tag{1.12}$$

Hydraulic Analog of a Current Source

The role played by a current source in an electric circuit is very similar to that of a positive displacement pump in a hydraulic circuit. In a positive displacement pump, such as a peristaltic or reciprocating pump, an internal mechanism, such as a roller, piston, or diaphragm, forces a particular volume of fluid to be pumped *through* a hydraulic line. The volume flow rate *through* the pump is analogous to the charge flow rate *through* the current source.

Positive displacement pump

Pump symbols

Left: Fixed capacity pump. **Right:** Fixed capacity pump with two directions of flow.

Left: Variable capacity pump. **Right:** Variable capacity pump with two directions of flow.

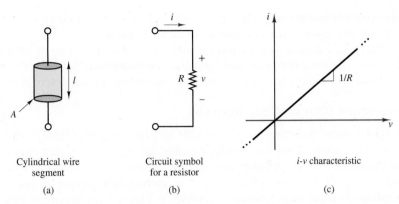

Cylindrical wire
segment

(a)

Circuit symbol
for a resistor

(b)

i-v characteristic

(c)

Figure 1.36   (a) Resistive wire segment; (b) ideal resistor circuit symbol; (c) the i-v relationship (Ohm's law) for an ideal resistor

where ρ and σ are the material properties *resistivity* and *conductivity*, respectively, and l and A are the segment length and cross-sectional area, respectively. As evident in the above equation, conductivity is simply the inverse of resistivity. The units of resistance R are **ohms (Ω)**, where

$$1\ \Omega = 1\ \text{V/A} \tag{1.13}$$

The resistance of an actual wire or circuit element is usually accounted for in a circuit diagram by an **ideal resistor**, which lumps the entire distributed resistance R of the wire or element into one single element. Ideal resistors exhibit a linear i-v relationship known as **Ohm's law**, which is

$$\boxed{v = iR \qquad \text{Ohm's law}} \tag{1.14}$$

In other words, the voltage across an ideal resistor is directly proportional to the current through it. The constant of proportionality is the resistance R. The circuit symbol and i-v characteristic for an ideal resistor are shown in Figure 1.36(b) and (c), respectively. Notice the passive sign convention used in the circuit symbol diagram, as appropriate, since a resistor is a passive element.

It is often convenient to define the *conductance*, G (in units of siemens, S), of a circuit element as the inverse of its resistance.

$$G = \frac{1}{R} \qquad \text{siemens (S)} \qquad \text{where} \qquad 1\ \text{S} = \frac{1A}{V} \tag{1.15}$$

In terms of conductance, Ohm's law is

$$i = Gv \tag{1.16}$$

Ohm's law is an *empirical* relationship that finds widespread application in electrical engineering. It is a simple yet powerful approximation of the physics of electrical conductors. However, the linear i-v relationship usually does not apply over very large ranges in voltage or current. For some conductors, Ohm's law does not approximate the i-v relationship even over modest ranges in voltage or current. Nonetheless, most conductors exhibit piecewise linear i-v characteristics for one or more ranges of voltage and current, as shown in Figure 1.37 for an incandescent lightbulb and a semiconductor diode.

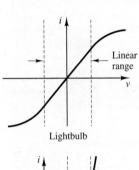

Linear range

Lightbulb

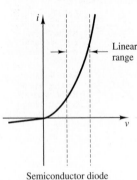

Linear range

Semiconductor diode

Figure 1.37 Piecewise linear segments within nonlinear i-v characteristics

Short- and Open-Circuits

Two convenient idealizations, the **short-circuit** and the **open-circuit**, are limiting cases of Ohm's law as the resistance approaches zero or infinity, respectively. Formally, a short-circuit is an element *across* which the voltage is zero, regardless of the current *through* it. Figure 1.38 depicts the circuit symbol for an ideal short-circuit.

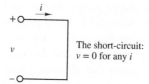

The short-circuit:
$v = 0$ for any i

Figure 1.38 The short-circuit

In practice, any conductor will exhibit some resistance. For practical purposes, however, many elements approximate a short-circuit quite accurately under certain conditions. For example, a large-diameter copper pipe is effectively a short-circuit in the context of a residential electric power supply, while in a low-power microelectronic circuit (e.g., an iPhone) a typical ground plane is 35×10^{-6} m thick, which is adequate for a short-circuit in that context. A typical solderless breadboard is designed to accept 22-gauge solid jumper wires, which act effectively as short-circuits between elements on the breadboard. Table 1.1 lists the resistance per 1,000 ft of some commonly used wire, as specified by the *American Wire Gauge Standards*.

Table 1.1 Resistance of copper wire

AWG size	Number of strands	Diameter per strand (in)	Resistance per 1,000 ft (Ω)
24	Solid	0.0201	28.4
24	7	0.0080	28.4
22	Solid	0.0254	18.0
22	7	0.0100	19.0
20	Solid	0.0320	11.3
20	7	0.0126	11.9
18	Solid	0.0403	7.2
18	7	0.0159	7.5
16	Solid	0.0508	4.5
16	19	0.0113	4.7
14	Solid	0.0641	2.52
12	Solid	0.0808	1.62
10	Solid	0.1019	1.02
8	Solid	0.1285	0.64
6	Solid	0.1620	0.4
4	Solid	0.2043	0.25
2	Solid	0.2576	0.16

The limiting case for Ohm's law when $R \to \infty$ is called an **open-circuit**. Formally, an open-circuit is an element *through* which the current is zero, regardless of the voltage *across* it. Figure 1.39 depicts the circuit symbol for an ideal short-circuit.

MAKE THE CONNECTION

Hydraulic Analog of Electrical Resistance

A useful analogy can be made between the electric current through electric components and the flow of incompressible fluids (e.g., water, oil) through hydraulic components. The fluid flow rate *through* a pipe is analogous to current *through* a conductor. Similarly, pressure drop *across* a pipe is analogous to voltage *across* a resistor. The resistance of the pipe to fluid flow is analogous to electrical resistance: The pressure difference across the pipe causes fluid flow, much as a potential difference across a resistor causes charge to flow. The figure below depicts how pipe flow is often modeled as current through a resistance.

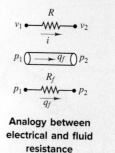

Analogy between electrical and fluid resistance

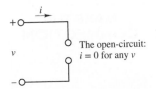

Figure 1.39 The open-circuit

In practice, it is easy to approximate an open-circuit. For moderate voltage levels, any gap or break in a conducting path amounts to an open-circuit. However, at sufficiently high voltages such a gap will become ionized and thereby enable charge to flow across the gap. Even an insulating material between two terminals will break down at a sufficiently high voltage. For an air gap between two conducting elements, ionized particles near the exposed surfaces of the elements may lead to arcing in which a pulse of charge jumps the gap and causes the ionized path to collapse. This phenomenon is employed in spark plugs to ignite the air-fuel mixture in a spark-ignition internal combustion engine. The *dielectric strength* is a measure of the maximum electric field (voltage per unit distance) that an insulating material can sustain without breaking down and allowing charge to flow. This measure is somewhat dependent upon temperature, pressure, and the material thickness; however, typical values are 3 kV/mm for air at sea level and room temperature, 10 kV/mm for window glass, 20 kV/mm for neoprene rubber, 30 kV/mm for pure water, and 60 kV/mm for PTFE, commonly known as Teflon.

Discrete Resistors

Various types of *discrete resistors* are used in laboratory experiments, tinkering projects, and commercial hardware, and are available in a wide range of nominal values, tolerances, and power ratings. Furthermore, each type of resistor has a particular temperature range within which it is designed to operate. In fact, some discrete resistors (known as thermistors) are designed to be highly sensitive to temperature and to be used as temperature transducers.

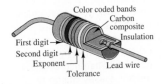

Figure 1.40 Carbon composite resistor

The majority of discrete resistors have a cylindrical shape and are color coded for their nominal value and tolerance. Several common types of resistors are *carbon composites*, in which the resistance is set by a mixture of carbon and ceramic powder (Figure 1.40); *carbon film*, in which the resistance is set by the length and width of a thin strip of carbon wrapped around an insulating core; and thin metal film, in which the resistance is set by the characteristics of a thin metal film also wrapped around an insulating core (Figure 1.41).

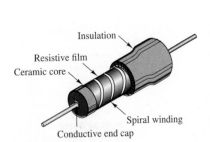

Figure 1.41 Thin-film resistor

Figure 1.42 Typical $\frac{1}{4}$-W resistors (*Jim Kearns*)

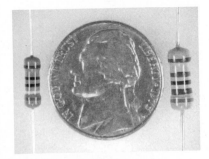

Figure 1.43 Typical $\frac{1}{2}$-W resistors (*Jim Kearns*)

Discrete resistors are available with various power ratings, where the power rating scales with the size of the resistor itself. Figures 1.42 and 1.43 show (to scale) typical $\frac{1}{4}$-W and $\frac{1}{2}$-W resistors, respectively. Notice the bands along the length of each resistor. Discrete resistors are also available with typical power ratings of 1, 2, 5, 10 W, and larger. Many industrial power resistors are manufactured by winding wire, such as Nichrome, around a nonconducting core, such as ceramic, plastic, or fiberglass. Others are made of cylindrical sections of carbon. Power resistors are available in a variety of packages, such as cement or molded plastic, aluminum encasements with fins for wicking away heat, and enamel coatings. Typical power resistors are shown in Figure 1.44.

The value of a discrete resistor is determined by the resistivity, shape, and size of the conducting element. Table 1.2 lists the resistivity of many common materials.

Table 1.2 **Resistivity of common materials at room temperature**

Material	Resistivity (Ω-m)
Aluminum	2.733×10^{-8}
Copper	1.725×10^{-8}
Gold	2.271×10^{-8}
Iron	9.98×10^{-8}
Nickel	7.20×10^{-8}
Platinum	10.8×10^{-8}
Silver	1.629×10^{-8}
Carbon	3.5×10^{-5}

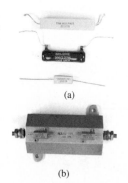

(a)

(b)

Figure 1.44 (a) 25-W, 20-W, and 5-W, and (b) two 5-W resistors sitting atop one 100-W resistor (*Jim Kearns*)

The nominal value and tolerance are often color-coded on a discrete resistor. Typically, discrete resistors have four color bands, where the first two designate a two-digit integer, the third designates a multiplier of 10, and the fourth designates the tolerance. Occasionally, discrete resistors have five bands, where the first three designate a three-digit integer, and the remaining two designate the multiplier and the tolerance. The value of each color band is decoded using the system displayed in Figure 1.45 and Table 1.3.

$$(\text{Two- or three-digit integer}) \times 10^{\text{multiplier}}, \quad \text{in ohms } (\Omega)$$

Table 1.3 b_1b_2 indicates the two-digit significand; b_3 indicates the multiplier.

b_1b_2	Code	b_3	Code	Ω	b_3	Code	kΩ	b_3	Code	kΩ	b_3	Code	kΩ
10	Brn-blk	1	Brown	100	2	Red	1.0	3	Orange	10	4	Yellow	100
12	Brn-red	1	Brown	120	2	Red	1.2	3	Orange	12	4	Yellow	120
15	Brn-grn	1	Brown	150	2	Red	1.5	3	Orange	15	4	Yellow	150
18	Brn-gry	1	Brown	180	2	Red	1.8	3	Orange	18	4	Yellow	180
22	Red-red	1	Brown	220	2	Red	2.2	3	Orange	22	4	Yellow	220
27	Red-vlt	1	Brown	270	2	Red	2.7	3	Orange	27	4	Yellow	270
33	Org-org	1	Brown	330	2	Red	3.3	3	Orange	33	4	Yellow	330
39	Org-wht	1	Brown	390	2	Red	3.9	3	Orange	39	4	Yellow	390
47	Ylw-vlt	1	Brown	470	2	Red	4.7	3	Orange	47	4	Yellow	470
56	Grn-blu	1	Brown	560	2	Red	5.6	3	Orange	56	4	Yellow	560
68	Blu-gry	1	Brown	680	2	Red	6.8	3	Orange	68	4	Yellow	680
82	Gry-red	1	Brown	820	2	Red	8.2	3	Orange	82	4	Yellow	820

b_4 b_3 b_2 b_1

Color bands

black	0	blue	6
brown	1	violet	7
red	2	gray	8
orange	3	white	9
yellow	4	silver	10%
green	5	gold	5%

Resistor value $= (b_1 \, b_2) \times 10^{b_3}$; $b_4 = \%$ tolerance in actual value

Figure 1.45 Resistor color code

For example, a resistor with four bands (yellow, violet, red, gold) has a nominal value of:

$$(\text{yellow})(\text{violet}) \times 10^{\text{red}} = 47 \times 10^2 = 4{,}700 \ \Omega = 4.7 \ \text{k}\Omega$$

and a "gold" tolerance of ± 5 percent. 4.7 kΩ is often shortened in practice to 4K7, where the letter K indicates the placement of the decimal point as well as the unit of kΩ. Likewise, 3.3 MΩ is often shortened to 3M3. Table 1.3 lists the standard nominal values established by the Electronic Industries Association (EIA) for a tolerance of 10 percent, commonly referred to as the E12 series. The number 12 indicates the number of logarithmic steps per decade of resistor values. Notice that the values in adjacent decades (columns) are different by a factor of 10.

Due to imperfect manufacturing the actual value of a discrete resistor is only approximately equal to its nominal value. The tolerance is a measure of the likely variation between the actual value and the nominal value. Other EIA series are E6, E24, E48, E96, and E192 for tolerances of 20%, 5%, 2%, 1%, and even finer tolerances, respectively.

Variable Resistors

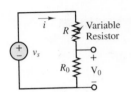

Figure 1.46 A variable resistor R in a series loop

The resistance of a variable resistor is not fixed but can vary with some other quantity. Examples of variable resistors are a photoresistor and a thermistor, in which the resistance varies with light intensity and temperature, respectively. Many useful sensors are based upon variable resistors.

Figure 1.46 shows a simple loop with a voltage source, a variable resistor R, and a fixed resistor R_0. Apply KVL around the loop to find:

$$v_S = iR + iR_0 = i(R + R_0)$$
$$= iR + v_0$$

Solve for i and substitute for it in the above equation to find:

$$i = \frac{v_S}{R + R_0} \quad \text{and} \quad v_0 = iR_0 = v_S\frac{R_0}{R + R_0}$$

Now assume that the variable resistor has a range from 0 Ω to some value R_{max} that is much larger than R_0. When $R = 0$:

$$v_0 = v_S\frac{R_0}{R + R_0} = v_S\frac{R_0}{R_0} = v_S \quad (R = 0)$$

(a)

(b)

Figure 1.47 (a) A typical cadmium sulfide (CdS) cell. (b) A nightlight relies on a CdS cell to detect dark conditions. (*Jim Kearns*)

When $R = R_{max}$:

$$v_0 = v_S\frac{R_0}{R + R_0} = v_S\frac{R_0}{R_{max} + R_0} \approx v_S\frac{R_0}{R_{max}} \approx 0 \quad (R = R_{max})$$

Thus, as R varies from 0 to R_{max}, v_0 varies from v_S to 0. The changes in R can be observed as changes in v_0. Imagine that the variable resistor in Figure 1.46 is a photoresistor, such as a cadmium sulfide (CdS) cell shown in Figure 1.47(a), that has a very small resistance when the incident light intensity is bright and has a very large resistance when the incident light intensity is dim or dark. The result is that under bright conditions, $v_0 \approx v_S$, while under dark conditions, $v_0 \approx 0$. To make a nightlight, such as that shown in Figure 1.47(b), all we need is a device that turns on the nightlight when $v_0 \ll v_{ref}$ and turns off the nightlight when $v_0 \gg v_{ref}$, where v_{ref} is some appropriate reference voltage, such as $v_S/2$.

Figure 1.48 shows a typical thermistor, which can be used in exactly the same manner as a CdS cell but which responds to changes in temperature.

Potentiometers

Figure 1.48 A typical negative temperature coefficient (NTC) thermistor (*Jim Kearns*)

A potentiometer is a three-terminal device. Figure 1.49 depicts a potentiometer and its circuit symbol. A potentiometer has a fixed resistance R_0, formed by a tightly wound coil of wire, between terminals A and C. Terminal B is connected to a *wiper* that slides along the coil as the knob is turned. The arrow in the circuit symbol represents the position of the slider along the length of the coil R_0. The resistance from terminal B to the other two terminals is determined by the wiper position. As R_{BA} increases, R_{BC} decreases, and vice versa, such that the sum $R_{BA} + R_{BC}$ always equals R_0.

Figure 1.50(a) illustrates the use of a potentiometer symbol in a simple circuit. The meter represents an ideal voltmeter that is capable of measuring the voltage across two nodes without impacting the behavior of the circuit. Figure 1.50(b) is an equivalent representation of the circuit, where the resistance between terminals A and B and that between terminals B and C are depicted as discrete resistors. Notice that there are effectively three nodes in these circuits.

The ideal voltmeter reading v_{bc} can be calculated in a manner similar to that used in the preceding section on variable resistors. Apply KVL around the loop containing the voltage source and the two discrete resistors, using Ohm's law and the current i in the loop to express the resistor voltages. The result is:

$$v_{BC} = v_S \frac{R_{BC}}{R_{BC} + R_{AB}}$$

This important result for two resistors in series is an example of *voltage division*, which is discussed fully in the next section. When the wiper is turned all the way to terminal C, $R_{BC} = 0$ and so $v_{BC} = 0$. When the wiper is turned all the way to terminal A, $R_{AB} = 0$ and so $v_{BC} = v_S$. In general, as the wiper is turned from terminal A to terminal C, the voltage across terminals B and C falls continuously from v_S to 0.

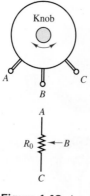

Figure 1.49 A potentiometer is a three-terminal resistive device with a fixed resistance R_0 between terminals A and C. The resistances between terminal B (the "wiper") and the other two terminals is set by the knob.

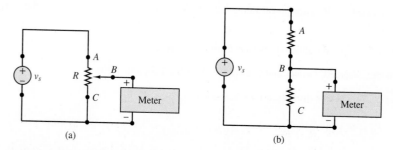

Figure 1.50 (a) A potentiometer in a simple circuit; (b) an equivalent circuit of (a), where $R = R_{AB} + R_{BC} = R_{AC}$

Power Dissipation in Resistors

All discrete resistors have a power rating, which is not designated by a color band, but which tends to scale with the size of the resistor itself. Larger resistors typically have a larger power rating. The power consumed or dissipated by a resistor R is given by:

$$P = vi = (iR)i = i^2 R > 0$$

$$= v\left(\frac{v}{R}\right) = \frac{v^2}{R} > 0 \qquad (1.17)$$

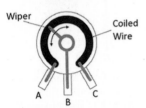

Remember that the voltage v and the current i are defined and linked by the *passive sign convention* and that power consumed by an element is positive. In the case of resistors, power is always positive and dissipated as heat to the surrounding environment. The implication is that if the current through (or the voltage across) a resistor is too large, the power will exceed the resistor's rating and result in a smoking and/or burning resistor! The smell of an overheating resistor is well known to technicians and hobbyists alike.

> Positive power is power dissipated (i.e., consumed) by an element.

Example 1.11 illustrates how to use the power rating to determine whether a given resistor is suitable for a particular application.

Figure 1.51 A typical $\frac{1}{2}$-W potentiometer and its internal construction (*Jim Kearns*)

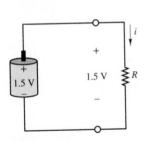

Figure 1.52 Figure for Example 1.11

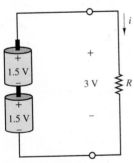

Figure 1.53 Figure for Example 1.11

EXAMPLE 1.11 Using Resistor Power Ratings

Problem

For a given voltage across a resistor, determine the minimum allowed resistance for a $\frac{1}{4}$-W power rating.

Solution

Known Quantities: Resistor power rating 0.25 W. Voltages due to a battery across the resistor: 1.5 V and 3 V.

Find: The minimum allowed resistance for a $\frac{1}{4}$-W resistor.

Schematics, Diagrams, Circuits, and Given Data: Figures 1.52 and 1.53.

Analysis: The power dissipated by a resistor is

$$P_R = vi = v \cdot \frac{v}{R} = \frac{v^2}{R}$$

Setting P_R equal to the resistor power rating yields $v^2/R \leq 0.25$, or $R \geq v^2/0.25$. For a 1.5-V battery, the minimum size resistor will be $R = 1.5^2/0.25 = 9\ \Omega$. For a 3-V battery, the minimum size resistor will be $R = 3^2/0.25 = 36\ \Omega$.

Comments: Sizing resistors on the basis of power rating is very important since, in practice, resistors eventually fail when the power rating is exceeded. Also, notice that the minimum resistor size was quadrupled when the voltage was doubled, which reflects the fact that power increases with the square of the voltage. Another implication of the relationship between power and voltage is that the power dissipated by a particular R for the 3-V battery is not twice, but four times, the power dissipated by the same R for the 1.5-V battery. In other words, the power dissipated by R in Figure 1.53 cannot be computed by assuming that each of the two 1.5-V batteries supplies the same amount of power as the single 1.5-V battery in Figure 1.52. In fact, each battery in Figure 1.53 supplies twice as much power as the single battery in Figure 1.52. In mathematical terms, power is not linear and so does not satisfy the principle of superposition, which is an important concept discussed in Chapter 2.

CHECK YOUR UNDERSTANDING

A typical three-terminal electronic power supply (see the first illustration) provides ±12 V, such that the change in voltage from terminal C to B is +12 V and that from terminals B to A is also +12 V. What is the minimum size (value) of a $\frac{1}{4}$-W resistor placed across terminals A and C? (*Hint:* The voltage from terminal C to A is +24 V.)

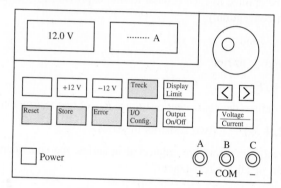

The single loop circuit in the illustration on the right contains a battery, a resistor, and an unknown circuit element.

1. If the voltage V_{battery} is 1.45 V and $i = 5$ mA, find the power supplied to or by the battery.
2. Repeat part 1 if $i = -2$ mA.

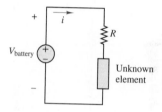

The battery in the triple mesh circuit shown below supplies power to resistors R_1, R_2, and R_3. Use KCL to determine the current i_B, and find the power supplied by the battery if $V_{\text{battery}} = 3$ V.

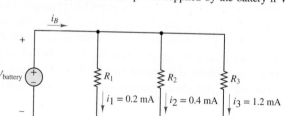

1.7 RESISTORS IN SERIES AND VOLTAGE DIVISION

It is common to find two or more circuit elements situated along a single current path or branch; that is, the elements are in *series*. When elements are in series, the voltage across the entire branch is divided among the elements in the branch. This important observation is known as **voltage division**.

> The ratio of the voltages across any two resistances in series equals the ratio of those resistances.

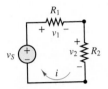

Figure 1.54 The current i flows through each of the three elements in the series loop. KVL requires $v_S = v_1 + v_2$.

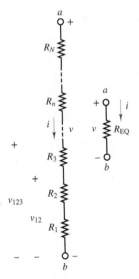

Figure 1.55 The equivalent resistance of three or more resistances in series equals the sum of those resistances.

The most fundamental instance of voltage division occurs when two resistors are in series, as shown in Figure 1.54. KVL applied around the series loop requires the voltage drop v_S across the source to be equal to the sum of the voltage drops v_1 and v_2 across the two resistors.

$$v_S = v_1 + v_2 \qquad \text{KVL}$$

Ohm's law can be applied to each resistor to find expressions for v_1 and v_2. (Notice the passive sign convention.)

$$v_1 = iR_1 \qquad \text{and} \qquad v_2 = iR_2$$

Plug in for v_1 and v_2 to find:

$$v_S = iR_1 + iR_2 = i(R_1 + R_2) \equiv iR_{EQ}$$

This expression defines the *equivalent resistance* R_{EQ} of two resistors in series, where:

$$R_{EQ} = (R_1 + R_2) \qquad \text{(two resistors in series)}$$

When three or more resistors are connected in series, the equivalent resistance is equal to the sum of all the resistances.

$$R_{EQ} = \sum_{n=1}^{N} R_n \qquad \text{resistors in series} \tag{1.18}$$

Clearly, R_{EQ} is greater than any of the individual resistances in the series. It is often useful to replace a series of two or more resistances with a single equivalent resistance, as indicated in Figure 1.55. To do so correctly, remove all the resistances in series along the branch and replace them with a single equivalent resistance along the same branch. This simple procedure illustrates a very important principle: From the perspective of whatever else is eventually attached to that branch (e.g., the voltage source in Figure 1.54), the resistances in series are *seen* as a single resistance R_{EQ}.

An expression for how the voltage across the entire branch is divided among the individual resistances along that branch can be found by using Ohm's law and noting that the current is the same through each resistance. Consider the series loop in Figure 1.54:

$$i = \frac{v_1}{R_1} = \frac{v_2}{R_2} = \frac{v_S}{R_{EQ}}$$

which yields the following relationships:

$$\frac{v_1}{v_S} = \frac{R_1}{R_{EQ}} \qquad \text{and} \qquad \frac{v_2}{v_S} = \frac{R_2}{R_{EQ}} \qquad \text{and} \qquad \frac{v_1}{v_2} = \frac{R_1}{R_2}$$

These results, known as *voltage division*, indicate that *for resistors in series the ratio of voltages equals the ratio of the corresponding resistances.* The voltage drops v_1 and v_2 are fractions of the total voltage v_S because R_1 and R_2 are both less than R_{EQ}.

When series connections are encountered in circuit diagrams, one should immediately think of voltage division.

$$\text{Series Connection} \Rightarrow \text{Voltage Division} \tag{1.19}$$

It is important to realize that the voltage division rule applies to any two resistances in series, not just any two discrete resistors. For example, consider the series of resistors shown in Figure 1.55. The ratio of the voltage across

$R_1 + R_2$ to the voltage across $R_1 + R_2 + R_3$ equals the ratio of $R_1 + R_2$ to $R_1 + R_2 + R_3$. That is:

$$\frac{v_{12}}{v_{123}} = \frac{R_1 + R_2}{R_1 + R_2 + R_3} \qquad \text{Voltage division}$$

Example 1.12 illustrates the voltage division rule.

EXAMPLE 1.12   Voltage Division

Problem

Determine the voltage v_3 in the circuit of Figure 1.56.

Solution

Known Quantities: Source voltage, resistance values.

Find: Unknown voltage v_3.

Schematics, Diagrams, Circuits, and Given Data: $R_1 = 10\ \Omega$; $R_2 = 6\ \Omega$; $R_3 = 8\ \Omega$; $v_S = 3$ V. Figure 1.56.

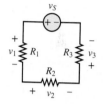

Figure 1.56 Figure for Example 1.12

Analysis: The circuit is a simple series loop; that is, all the elements are along the same (the only) current pathway. Apply voltage division directly to solve for v_3:

$$\frac{v_3}{v_S} = \frac{R_3}{R_1 + R_2 + R_3} = \frac{8}{10 + 6 + 8} = \frac{1}{3}$$

Thus: $v_3 = v_S/3 = 1$ V.

Comments: The application of voltage division to a series of elements along a branch is fairly straightforward. Occasionally, it may be difficult to determine which elements in a circuit are, in fact, in series. This issue is explored in Example 1.14.

CHECK YOUR UNDERSTANDING

Repeat Example 1.12 by reversing the polarity of each resistor voltage, and show that the same result is obtained when the meaning of a negative sign is taken into account.

1.8   RESISTORS IN PARALLEL AND CURRENT DIVISION

It is common to find two or more circuit elements situated between the same two nodes; that is, the elements are in *parallel*. When elements are in parallel, the current entering either of the two nodes is divided among the parallel elements. This important observation is known as **current division**.

> The ratio of the currents through any two resistances in parallel equals the inverse ratio of those resistances.

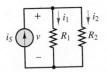

Figure 1.57 The voltage v is across each of the three elements in parallel. KCL requires $I_S = i_1 + i_2$.

The most fundamental instance of current division occurs when two resistors are in parallel, as shown in Figure 1.57. KCL applied at the upper node requires the current i_S through the source to be equal to the sum of the currents i_1 and i_2 through the two resistors.

$$i_S = i_1 + i_2 \qquad \text{KCL}$$

Ohm's law can be applied to each resistor to find expressions for i_1 and i_2. (Notice the passive sign convention.)

$$i_1 = \frac{v}{R_1} \qquad \text{and} \qquad i_2 = \frac{v}{R_2}$$

Plug in for i_1 and i_2 to find:

$$i_S = \frac{v}{R_1} + \frac{v}{R_2} = v\left(\frac{1}{R_1} + \frac{1}{R_2}\right) \equiv v\frac{1}{R_{EQ}}$$

This expression defines the *equivalent resistance* R_{EQ} of two resistors in parallel, where:

$$\frac{1}{R_{EQ}} = \frac{1}{R_1} + \frac{1}{R_2} \qquad \text{(two resistors in parallel)}$$

However, this inverted expression for the equivalent resistance is awkward and nonintuitive. Often, a more useful form is:

$$R_{EQ} = R_1 \| R_2 = \frac{R_1 R_2}{R_1 + R_2} \qquad \text{(two resistors in parallel)}$$

The notation $R_1 \| R_2$ indicates a parallel combination of R_1 and R_2. The same notation can be used to indicate a parallel combination of three or more resistors by writing:

$$R_1 \| R_2 \| R_3 \cdots$$

It is easy to show that R_{EQ} is smaller than either R_1 or R_2. To do so, simply write R_{EQ} as:

$$R_{EQ} = R_1 \| R_2 = R_1 \frac{R_2}{R_1 + R_2} = R_2 \frac{R_1}{R_1 + R_2}$$

Both fractions in the above equation are less than 1; thus, $R_{EQ} < R_1$ and $R_{EQ} < R_2$.

When three or more resistors are connected in parallel, as shown in Figure 1.58, the inverse of the equivalent resistance is equal to the sum of the inverses of all the resistances.

$$\frac{1}{R_{EQ}} = \frac{1}{R_1} + \frac{1}{R_2} + \cdots + \frac{1}{R_N} \qquad (1.20)$$

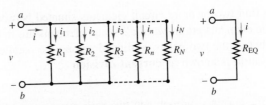

Figure 1.58 The inverse of the equivalent resistance of three or more resistances in parallel equals the sum of the inverses of those resistances.

or

$$R_{EQ} = \frac{1}{1/R_1 + 1/R_2 + \cdots + 1/R_N} \qquad \text{Equivalent parallel resistance} \qquad (1.21)$$

Notice that R_{EQ} is smaller than any of the individual resistances in parallel. It is often useful to replace two or more resistances in parallel with a single equivalent resistance, as indicated in Figure 1.58. To do so correctly, remove all the resistances between nodes a and b and replace them with a single equivalent resistance attached between these same two nodes. This simple procedure illustrates a very important principle: From the perspective of whatever else is eventually attached to nodes a and b (e.g., the current source in Figure 1.57), the parallel resistances are *seen* as a single resistance of value R_{EQ}.

An expression for how the current entering either of the two nodes is divided among the individual resistances in parallel can be found by using Ohm's law and noting that the voltage is the same across each resistance. Consider the parallel circuit in Figure 1.57:

$$v = i_1 R_1 = i_2 R_2 = i_S R_{EQ}$$

which yields the following relationships:

$$\frac{i_1}{i_S} = \frac{R_{EQ}}{R_1} \qquad \text{and} \qquad \frac{i_2}{i_S} = \frac{R_{EQ}}{R_2} \qquad \text{and} \qquad \frac{i_1}{i_2} = \frac{R_2}{R_1}$$

These results, known as *current division*, indicate that *for resistors in parallel the ratio of currents equals the inverse ratio of the corresponding resistances*. The currents i_1 and i_2 are fractions of the total current i_S because R_1 and R_2 are both greater than R_{EQ}.

The current division results shown in the previous equation can be rewritten by plugging in for R_{EQ} to find:

$$\frac{i_1}{i_S} = \frac{R_2}{R_1 + R_2} \qquad \text{and} \qquad \frac{i_2}{i_S} = \frac{R_1}{R_1 + R_2} \qquad \text{and} \qquad \frac{i_1}{i_2} = \frac{R_2}{R_1}$$

In these forms, an equivalent description of current division for two resistances in parallel is that the ratio of i_1 to i_S equals the ratio of the "other" resistance R_2 to the sum of the two resistances $R_1 + R_2$. Likewise, the ratio of i_2 to i_S equals the ratio of the "other" resistance R_1 to the sum of the two resistances $R_1 + R_2$. These forms may be appealing since they resemble the expressions used to compute voltage division for two resistances in series.

When parallel connections are encountered in circuit diagrams, one should immediately think of current division.

$$\text{Parallel Connection} \Rightarrow \text{Current Division} \qquad (1.22)$$

It is important to realize that the current division rule applies to any two resistances in parallel, not just any two discrete resistors. For example, consider the parallel resistors shown in Figure 1.58. The ratio of the combined current through R_1 and R_2 to the current through R_3 equals the ratio of R_3 to $(R_{12})_{EQ}$. That is:

$$\frac{i_1 + i_2}{i_3} = \frac{R_3}{(R_{12})_{EQ}} \qquad \text{Current division}$$

where

$$(R_{12})_{EQ} = \frac{R_1 R_2}{R_1 + R_2}$$

Likewise:

$$\frac{i_n}{i} = \frac{(R_{1\cdots N})_{EQ}}{R_n} \qquad \text{Current division}$$

where

$$\frac{1}{(R_{1\cdots N})_{EQ}} = \frac{1}{R_1} + \frac{1}{R_2} + \cdots + \frac{1}{R_N}$$

These last two expressions can be combined to yield:

$$\boxed{\frac{i_n}{i} = \frac{1/R_n}{1/R_1 + 1/R_2 + \cdots + 1/R_n + \cdots + 1/R_N} \qquad \begin{array}{l}\text{Current}\\ \text{divider}\end{array}} \qquad (1.23)$$

Example 1.13 illustrates the current division rule. Many practical circuits contain resistors in parallel and in series. Examples 1.14 and 1.15 illustrate how such circuits can be analyzed using voltage and current division. These principles are useful even in very complicated circuits, such as those presented in Chapter 2, which introduces a variety of techniques and methods for analyzing resistive networks.

EXAMPLE 1.13   Current Division

Problem

Determine the current i_1 in the circuit of Figure 1.59.

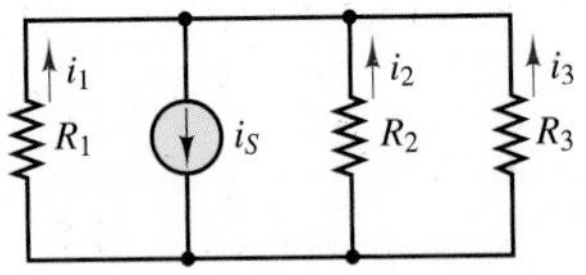

Figure 1.59 Figure for Example 1.13

Solution

Known Quantities: Source current, resistance values.

Find: Unknown current i_1.

Schematics, Diagrams, Circuits, and Given Data: $R_1 = 10\ \Omega$; $R_2 = 2\ \Omega$; $R_3 = 20\ \Omega$; $i_S = 4$ A. Figure 1.59.

Analysis: Apply current division directly to find:

$$\frac{i_1}{i_S} = \frac{1/R_1}{1/R_1 + 1/R_2 + 1/R_3} = \frac{\frac{1}{10}}{\frac{1}{10} + \frac{1}{2} + \frac{1}{20}} = \frac{2}{13}$$

Thus:

$$i_1 = 4\ \text{A} \times \frac{2}{13} \approx 0.62\ \text{A}$$

An alternative approach is to find the equivalent resistance of $R_2\|R_3$ and then apply one of the simpler expressions for current division between two resistances in parallel.

$$R_2\|R_3 = \frac{R_2 R_3}{R_2 + R_3} = \frac{(2)(20)}{2 + 20} \approx 1.82\ \Omega$$

(Notice that $R_2 \| R_3$ is less than both R_2 and R_3.)

$$\frac{i_1}{i_S} = \frac{R_2 \| R_3}{R_2 \| R_3 + R_1} \approx \frac{1.82}{1.82 + 10} = \frac{2}{13}$$

The result is the same as that found by applying current division directly:

$$i_1 = 4 \text{ A} \times \frac{2}{13} \approx 0.62 \text{ A}$$

Comments: The application of current division to elements in parallel between two nodes is fairly straightforward. Occasionally, it may be difficult to determine which elements are, in fact, in parallel. This issue is explored in Example 1.14.

EXAMPLE 1.14   Resistors in Series and Parallel

Problem

Determine the voltage v in the circuit of Figure 1.60.

Solution

Known Quantities: Source voltage, resistance values.

Find: Unknown voltage v.

Schematics, Diagrams, Circuits, and Given Data: See Figures 1.60, 1.61.

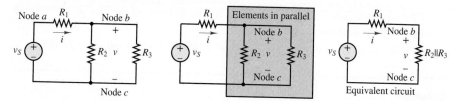

Figure 1.60 Three-node circuit   **Figure 1.61** Simplified three-node circuit

Analysis: The circuit of Figure 1.60 contains three resistors that are not completely in series nor in parallel with each other. This fact may not be apparent at first glance, but consider whether the conditions for series and parallel are met for all three resistors.

1. Are all three resistors situated along the same single current path? Is there one common current through all three resistors? Clearly, the current i entering node b will be divided on its way to node c. Some of it will pass through R_2 while the rest will pass through R_3. Thus, there is *not* one common current through all three resistors; that is, they are *not* in series.

2. Are all three resistors situated between the same two nodes? R_1 sits between nodes a and b, while R_2 and R_3 sit between nodes b and c. Thus, the three resistors do *not* sit between the same two nodes; that is, they are *not* in parallel.

However, it is possible to simplify the circuit by noting, as mentioned above and depicted in Figure 1.61, that R_2 and R_3 sit between the same two nodes and are, therefore, in parallel. These two resistors can be removed from the circuit and replaced by the equivalent resistance between nodes b and c, which is:

$$R_{EQ} = R_2 \| R_3 = \frac{R_2 R_3}{R_2 + R_3}$$

An equivalent circuit can now be drawn as shown in Figure 1.61. The result is a simple series loop. Voltage division can be applied directly to solve for v:

$$v = \frac{R_2 \| R_3}{R_1 + R_2 \| R_3} v_S$$

The current can also be found:

$$i = \frac{v}{R_2 \| R_3} = \frac{v_S}{R_1 + R_2 \| R_3}$$

Comments: Notice that the expression for i is exactly what would be found by applying Ohm's law to the equivalent resistance of R_1 in series with $R_2 \| R_3$. See Figure 1.61.

EXAMPLE 1.15   The Wheatstone Bridge

Problem

The **Wheatstone bridge** is a resistive circuit that is frequently encountered in a variety of measurement circuits. The general form of the bridge circuit is shown in Figure 1.62(a), where R_1, R_2, and R_3 are known and R_x is to be determined. The circuit can be redrawn, as shown in Figure 1.62(b), to clarify that R_1 and R_2 are in series, as are R_3 and R_x. The two branches from node c to the reference node are in parallel.

In the figures, v_a and v_b are *node voltages* relative to the common reference node. The value of the reference node can be chosen arbitrarily; however, it may be helpful to consider its value to be zero.

1. Find an expression for the voltage $v_{ab} = v_a - v_b$ in terms of the four resistances and the source voltage v_S.

2. Find the value of R_x when $R_1 = R_2 = R_3 = 1 \text{ k}\Omega$, $v_S = 12 \text{ V}$, and $v_{ab} = 12 \text{ mV}$.

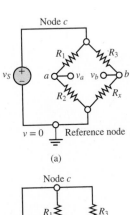

(a)

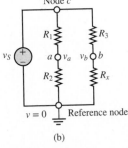

(b)

Figure 1.62 A Wheatstone bridge is a mixed series-parallel circuit.

Solution

Known Quantities: Source voltage, resistance values, bridge voltage.

Find: Unknown resistance R_x.

Schematics, Diagrams, Circuits, and Given Data: See Figure 1.62.

$$R_1 = R_2 = R_3 = 1 \text{ k}\Omega; \ v_S = 12 \text{ V}; \ v_{ab} = 12 \text{ mV}.$$

Analysis:

1. The circuit consists of three parallel branches: the voltage source v_S branch, the $R_1 + R_2$ branch, and the $R_3 + R_x$ branch. All three branches sit between node c and the reference node, with the same voltage v_S across each branch.

    In the analysis that follows it is important to keep in mind that all node voltages are understood to be relative to the reference node. That is, v_a is the voltage across R_2, v_b is the voltage across R_x, and $v_c = v_S$.

    Since R_1 and R_2 are in series, voltage division can be applied to find v_a in terms of v_c. Likewise, since R_3 and R_x are in series, voltage division can also be applied to find v_b in terms of v_c.

$$\frac{v_a}{v_c} = \frac{R_2}{R_1 + R_2} \qquad \text{and} \qquad \frac{v_b}{v_c} = \frac{R_x}{R_3 + R_x}$$

Plug in $v_c = v_S$ to find that $v_{ab} = v_a - v_b$ is given by:

$$v_{ab} = v_S \left(\frac{R_2}{R_1 + R_2} - \frac{R_x}{R_3 + R_x} \right)$$

This result is very useful and quite general.

2. Plug in numerical values for v_{ab}, v_S, R_1, R_2, and R_3 in the preceding equation to find:

$$0.012 = 12 \left(\frac{1 \text{ k}\Omega}{2 \text{ k}\Omega} - \frac{R_x}{1 \text{ k}\Omega + R_x} \right)$$

Divide both sides by -12 and add 0.5 to both sides to find:

$$0.499 = \frac{R_x}{1 \text{ k}\Omega + R_x}$$

Multiply both sides by $1 \text{ k}\Omega + R_x$ to find:

$$0.499(1 \text{ k}\Omega + R_x) = R_x \quad \text{or} \quad 499.0 = 0.501 R_x \quad \text{or} \quad R_x = 996 \ \Omega$$

Comments: The Wheatstone bridge finds application in many measuring instruments.

CHECK YOUR UNDERSTANDING

For the circuit in Figure 1.59, apply current division to find i_2 and i_3 and verify that KCL at either node is satisfied by the results. Also, verify that the ratio of any two branch currents equals the inverse ratio of their associated resistances. Finally, verify that $i_2 = 5 \times i_1$ because $R_1 = 5 \times R_2$ and that $i_1 = 2 \times i_3$ because $R_3 = 2 \times R_1$. (These results should not be a surprise since larger currents are expected through the smaller resistances.)

CHECK YOUR UNDERSTANDING

Consider the circuit of Figure 1.60, with resistor R_3 replaced by an open-circuit. Calculate the voltage v when the source voltage is $v_S = 5$ V and $R_1 = R_2 = 1$ kΩ.

Repeat when resistor R_3 is in the circuit and its value is $R_3 = 1$ kΩ.

Repeat when resistor R_3 is in the circuit and its value is $R_3 = 0.1$ kΩ.

Answers: $v = 2.50$ V; $v = 1.67$ V; $v = 0.4167$ V

CHECK YOUR UNDERSTANDING

Use the result of part 1 of Example 1.15 to find the relationship between R_x and the other three resistors such that $v_{ab} = 0$. Using the data in Example 1.15, what is the value of R_x that satisfies $v_{ab} = 0$, the so-called balanced condition for the bridge? Does the balanced bridge condition require that all four resistors be identical?

Answers: $R_1 R_x = R_2 R_3$; 1 kΩ; No

Resistive Throttle Position Sensor

Problem:

A typical **automotive resistive throttle position sensor** is shown in Figure 1.63(a). Figure 1.63(b) and (c) depict the geometry of the throttle plate and the equivalent circuit of the throttle sensor. A typical throttle plate has a useful measurement range of just under 90°, from closed throttle to wide-open throttle. The possible mechanical range of rotation of the sensor is usually somewhat greater. It is always necessary to *calibrate* any sensor to determine the actual relationship between the input variable (e.g., the throttle position) and the output variable (e.g., the sensor voltage). The following example illustrates such a procedure.

Figure 1.63(a) 500 series resistive throttle position sensors (*Courtesy: CTS Corporation*)

Figure 1.63(b) Throttle blade geometry

Solution:

Known Quantities: Functional specifications of throttle position sensor.

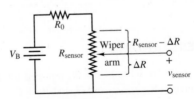

Figure 1.63(c) Throttle position sensor equivalent circuit

Find: Calibration of sensor in volts per degree of throttle plate opening.

Schematics, Diagrams, Circuits, and Given Data:

Functional specifications of throttle position sensor

Total resistance $= R_{sensor} + R_0$	12 kΩ
R_0	3 kΩ
Input V_B	5 V $\pm$ 4% regulated
Output V_{sensor}	5% to 95% V_B
Current draw I_S	$\leq$ 20 mA
Recommended load R_L	$\leq$ 220 kΩ
Electrical travel,[1] maximum	112°

[1]Note that in actual operation the sensor will only be actuated between 2° and 90°.

(Continued)

(*Concluded*)

Assumptions: Assume a nominal supply voltage of 5 V and total throttle plate travel of 88°, with a closed-throttle angle of 2° and a wide-open throttle angle of 90°.

Analysis: The equivalent circuit of the sensor is a series loop with a battery, a fixed resistor, and a potentiometer, as shown in Figure 1.43(c). The sensor output voltage is determined by the position of the *wiper arm*, whose actual displacement is angular; however, it is convention to depict all potentiometers in circuit diagrams as having straight line displacement, as shown in the figure. The range of the potentiometer (see specifications above) is 2° to 112° for a resistance of 3 to 12 kΩ; thus, assuming a linear sensor response, the *calibration constant* of the potentiometer is:

$$k_{\text{pot}} = \frac{112 - 2}{12 - 3} = 12.22°/\text{k}\Omega, \text{ such that } \theta = k_{\text{pot}} \Delta R$$

Voltage division requires that the sensor voltage be proportional to the ratio of the series resistances.

$$v_{\text{sensor}} = V_B \left(\frac{\Delta R}{R_0 + R_{\text{sensor}}} \right) = (5 \text{ V}) \left(\frac{\Delta R}{12} \right)$$

$$= 0.417 \, \Delta R \quad \text{V} \quad (\Delta R \text{ in k}\Omega)$$

For a 5-V battery, the linear calibration of the throttle position sensor is

$$v_{\text{sensor}} = (5 \text{ V}) \frac{\Delta R}{12 \text{ k}\Omega} = 0.417 \frac{\theta}{k_{\text{pot}}}$$

The *calibration curve* for the sensor is shown in Figure 1.63(d).

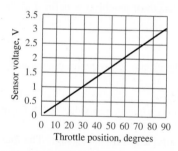

Figure 1.63(d) Calibration curve for throttle position sensor

When the throttle is closed,

$$v_{\text{sensor}} = 0$$

and when the throttle is wide open,

$$v_{\text{sensor}} = 0.417 \, \Delta R = 0.417 \frac{\theta}{k_{\text{pot}}} = 0.417 \frac{\text{V}}{\text{k}\Omega} \frac{90°}{12.22°/\text{k}\Omega} = 3.07 \text{ V}$$

Comments: The fixed resistor R_0 prevents the wiper arm from inadvertently connecting the + terminal of the battery directly to its − terminal, which would occur if the wiper were shorted to the lower node and $\theta = 112°$. Note that the intended operational range of the sensor is from 2° to 90°, specifically to avoid a harmful short-circuit scenario.

Resistance Strain Gauges

A **strain gauge** is a resistive element that has many applications in engineering measurements. A strain gauge contains one or more thin conductive strips, usually encased in an epoxy matrix. These strips shrink or stretch with the surface to which the strain gauge is bonded. Since the resistance of a thin conducting strip is dependent upon its geometry, it is possible to calibrate a strain gauge to relate changes in resistance to material *strain* along the surface. Surface strain can then be related to stress, force, torque, and pressure through various constitutive relations, such as Hooke's law. A variety of strain gauges are available to *transduce* the principal strains (extensional and shear) along a surface. The most versatile and popular strain gauge is a planar rosette, with which all three planar strains can be deduced simultaneously.

Recall that the resistance of a cylindrical conductor of cross-sectional area A, length L, and conductivity σ is given by the expression

$$R = \frac{L}{\sigma A}$$

When the conductor is compressed or elongated, both the length L and (due to the Poisson effect) cross-sectional area A will change, and with them the resistance of the conductor. In particular, when the length of the conductor is increased, its cross-sectional area will decrease, with both changes causing its resistance to increase.

Likewise, when the length of the conductor is decreased, its cross-sectional area will increase, with both changes causing its resistance to decrease. The empirical relationship between change in resistance and change in length is defined as the gauge factor GF:

$$\text{GF} = \frac{\Delta R / R}{\Delta L / L}$$

The fractional change in length of an object is defined as the strain ε:

$$\varepsilon = \frac{\Delta L}{L}$$

Using these definitions, the change in resistance due to an applied strain ε is given by

$$\Delta R = R_0 \text{GF} \varepsilon$$

where R_0 is the *zero strain resistance*. The value of GF for metal foil resistance strain gauges is usually about 2.

Figure 1.64 depicts a typical foil strain gauge. The maximum strain that can be measured by a foil gauge is $\Delta L/L_{\max} \approx 0.005$, which corresponds to a maximum change in resistance of approximately 1.2 Ω for a 120-Ω gauge. Because of the small scale of the change in resistance, strain gauges are usually incorporated in a Wheatstone bridge, which increases the sensitivity of the resistance measurement.

R_G — Circuit symbol for the strain gauge

The foil is formed by a photo-etching process and is less than 0.00002 in thick. Typical resistance values are 120, 350, and 1,000 Ω. The wide areas are bonding pads for electrical connections.

Figure 1.64 Metal-foil resistance strain gauge

Comments: Resistance strain gauges are used in many measurement applications. One such application is the measurement of a force on a cantilever beam, which is discussed in the next example.

The Wheatstone Bridge and Force Measurements

One of the simplest applications of a strain gauge is the measurement of a force applied to a cantilever beam, as illustrated in Figure 1.65.

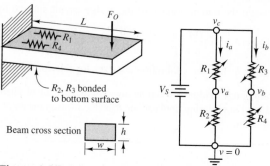

Figure 1.65 A force-measuring instrument

Four strain gauges are employed in this case, of which two are bonded to the upper surface of the beam at a distance L from the point where the external force F_0 is applied, and two are bonded on the lower surface, also at a distance L. Under the influence of the external force, the beam deforms and causes the upper gauges to extend and the lower gauges to compress. Thus, the resistance of the upper gauges will increase by an amount ΔR, and that of the lower gauges will decrease by an equal amount, assuming that the gauges are symmetrically placed. Let R_1 and R_4 be the upper gauges and R_2 and R_3 the lower gauges. Thus, under the influence of the external force, we have

$$R_1 = R_4 = R_0 + \Delta R$$
$$R_2 = R_3 = R_0 - \Delta R$$

where R_0 is the zero strain resistance of the gauges. It can be shown from elementary strength of materials and statics that the relationship between the strain ε and a force F_0 applied at a distance L for a cantilever beam is

$$\varepsilon = \frac{6LF_0}{wh^2Y}$$

where h and w are as defined in Figure 1.65 and Y is Young's modulus for the beam.

In the circuit of Figure 1.65, the currents i_a and i_b are given by

$$i_a = \frac{V_S}{R_1 + R_2} \qquad \text{and} \qquad i_b = \frac{V_S}{R_3 + R_4}$$

The bridge output voltage is defined by $v_o = v_b - v_a$ and may be found from the following expression:

$$v_0 = i_b R_4 - i_a R_2 = \frac{V_S R_4}{R_3 + R_4} - \frac{V_S R_2}{R_1 + R_2}$$

Plug in $R_1 = R_4 = R_0 + \Delta R$ and $R_2 = R_3 = R_0 - \Delta R$,

$$v_0 = V_S \frac{R_0 + \Delta R}{R_0 + \Delta R + R_0 - \Delta R} - V_S \frac{R_0 - \Delta R}{R_0 + \Delta R + R_0 - \Delta R}$$

$$= \frac{V_S}{2R_0}[R_0 + \Delta R - (R_0 - \Delta R)]$$

$$= V_S \frac{\Delta R}{R_0} = V_s(\text{GF})\varepsilon$$

(*Continued*)

(*Concluded*)

where GF is the gauge factor and $\Delta R/R_0 = \text{GF}\varepsilon$ was obtained in the previous Focus on Measurements box, "Resistance Strain Gauges." Thus, it is possible to obtain a relationship between the output voltage of the bridge circuit and the force F_0 as:

$$v_o = v_S(\text{GF})\varepsilon = V_S(\text{GF})\frac{6LF_0}{wh^2 Y} = \frac{6V_S(\text{GF})L}{wh^2 Y}F_0$$

where $6V_S(\text{GF})L/wh^2 Y$ is the calibration constant for this force transducer.

Comments: **Strain gauge bridges** are commonly used in mechanical, chemical, aerospace, biomedical, and civil engineering applications to make measurements of force, pressure, torque, stress, or strain.

CHECK YOUR UNDERSTANDING

Compute the full-scale (i.e., largest) output voltage for the force-measuring apparatus of the Focus on Measurements box, "The Wheatstone Bridge and Force Measurements." Assume that the strain gauge bridge is to measure forces ranging from 0 to 500 newtons (N), $L = 0.3$ m, $w = 0.05$ m, $h = 0.01$ m, GF = 2, and Young's modulus for the beam is 69×10^9 N/m^2 (aluminum). The source voltage is 12 V. What is the calibration constant of this force transducer?

Answer: v_o (full scale) = 62.6 mV; k = 0.125 mV/N

1.9   PRACTICAL VOLTAGE AND CURRENT SOURCES

Ideal sources were defined such that their prescribed output, a voltage or current, is completely independent of other factors. An ideal voltage source maintains a prescribed voltage across its terminals independent of the current through those terminals; likewise, an ideal current source maintains a prescribed current through its terminals independent of the voltage across those terminals. Neither of these ideal sources account for the effective *internal resistance*, which is characteristic of practical voltage and current sources, and which makes the output of a practical source dependent on the load that is *seen* by the source.

Consider, for example, a conventional car battery rated at 12 V, 450 ampere-hours (A-h). The latter rating implies that there is a limit (albeit a large one) to the amount of current the battery can deliver to a load and that, to some extent, the voltage output of the battery is dependent on the current drawn from it. This dependency can be observed as a drop in battery voltage when starting an automobile. Fortunately, a detailed understanding and analysis of the battery's physics are not necessary to model its behavior. Instead, the concept of internal resistance allows practical sources to be approximated by either of two different yet simple and effective models.

A practical voltage source can be approximated by a *Thévenin* model, which is composed of an ideal voltage source v_S in series with an internal resistance r_S. In practice, r_S is designed to be small compared to a typical equivalent resistance seen by the source.

A practical current source can be approximated by a *Norton* model, which is composed of an ideal current source i_S in parallel with an internal resistance r_S. In practice, r_S is designed to be large compared to a typical equivalent resistance seen by the source.

The shaded portion of Figure 1.66 depicts the so-called *Thévenin* model, which is composed of an ideal voltage source v_S in series with an internal resistance r_S. With this model, the source output current i_S depends upon the ideal voltage source v_S, the internal resistance r_S, and the load R_o. The maximum current is found in the limit that the load $R_o \to 0$ (i.e., a short-circuit load). The equivalent resistance "seen" by the ideal voltage source is $r_S + R_o$. Therefore, the source current i_S is given by Ohm's law as simply:

$$i_S = \frac{v_S}{r_S + R_o} \qquad \text{such that} \qquad i_{S\,\text{max}} = \frac{v_S}{r_S}$$

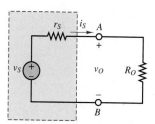

Figure 1.66 The Thévenin model of a real voltage source

The load voltage v_L can be found by direct application of voltage division.

$$\frac{v_o}{v_S} = \frac{R_o}{r_S + R_o}$$

In practice, the internal resistance r_S of a real voltage source is designed to be small compared to a typical load resistance R_o. In such cases, the load voltage v_o is approximately equal to the ideal source voltage v_S and the current requirements of a broad range of loads may be satisfied. Often, the effective internal resistance of a real voltage source is listed in its technical specifications. In cases where R_o is comparable to or smaller than r_S, the load voltage v_o will be significantly less than v_S. This result is known as a *loading effect*, such as when an automotive battery is required to start its engine.

The shaded portion of Figure 1.67 depicts the so-called *Norton* model, which is composed of an ideal current source i_S in parallel with an internal resistance r_S. With this model, the source output voltage v_S depends upon the ideal current source i_S, the internal resistance r_S, and the load R_o. The maximum current is found in the limit that the load $R_o \to \infty$ (i.e., an open-circuit load). The equivalent resistance "seen" by the ideal current source is $r_S \| R_o$. Therefore, the source voltage v_S is given by Ohm's law as simply:

$$v_S = i_S \frac{r_S R_o}{r_S + R_o} \qquad \text{such that} \qquad v_{S\,\text{max}} = i_S r_S$$

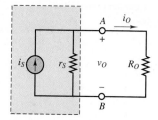

Figure 1.67 The Norton model of a real current source

The load current can be found by direct application of current division.

$$\frac{i_o}{i_S} = \frac{r_S}{r_S + R_o}$$

In practice, the internal resistance r_S of a real current source is designed to be large compared to a typical load resistance R_o. In such cases, the load current i_o is approximately equal to the ideal source current i_S and the voltage requirements of a broad range of loads may be satisfied. Often, the effective internal resistance of a real current source is listed in its technical specifications. In cases where R_o is comparable to or larger than r_S, the load current i_o will be significantly less than i_S. This result is also known as a *loading effect*.

1.10   MEASUREMENT DEVICES

In practice, the most commonly required measurements are of resistance, current, voltage, and power. An *ideal* measurement device would have no effect upon the quantity being measured. Of course, when a real measurement device is attached to a network, the network itself is changed (it now includes the measurement device) and it is quite possible that the quantity being measured is changed from what it was before the device was attached. At first glance, this problem may seem like a classic catch-22 scenario. That is, a quantity needs to be measured, so a measurement device must be used; but when the measurement device is used, the quantity is no longer what is was. To restore the quantity to its original state, the measurement device must be removed, but then . . . and so on and so on, around and around.

Luckily, if the characteristics of the measurement device are known, it is often possible to estimate the qualitative and quantitative impacts of a device on the measured quantity. In this section, simple models of real measurement devices are introduced that allow reasonable estimates of both.

The Ohmmeter

An **ohmmeter** measures the equivalent resistance *across* two nodes. In particular, an ohmmeter measures the resistance across an element when connected in *parallel* with it. Figure 1.68 depicts an ohmmeter connected across a resistor. One important rule needs to be remembered when using an ohmmeter:

> When using an ohmmeter to measure resistance, at least one terminal of the element must be disconnected from its network.

Figure 1.68 An ideal ohmmeter connected across a resistor

If the element is not disconnected from its network, the ohmmeter will measure the effective resistance of the element in parallel with the rest of its network. A common mistake made by inexperienced users of an ohmmeter is to attempt to measure the value of a discrete resistor by using one's fingers to clamp each end of the resistor to the ohmmeter probes. At best, this approach results in head scratching when the measured value is far off the nominal value of the resistor. At worst, the measured value is simply accepted as accurate with the user completely unaware that the ohmmeter measurement represents the equivalent resistance of the discrete resistor in parallel with the resistance of the user's own body.

The Ammeter

An **ammeter** measures the current *through* an element when connected in *series* with it. Figure 1.69(a) shows an *ideal* ammeter inserted into a simple series loop to measure its current. An ideal ammeter has zero resistance and therefore is able to measure the current without alteration due to the presence of the ammeter. A more realistic model of an actual ammeter has an internal resistance in series with an ideal ammeter, as shown in Figure 1.69(b). To obtain an accurate measurement the internal resistance of the ammeter must be significantly *smaller* than the total equivalent resistance of the branch to which it is attached in series. For example, the internal resistance r_m of the ammeter must be significantly smaller than $R_1 + R_2$ in the series loop of Figure 1.69(a). In practice, it is necessary to observe two rules when using an ammeter:

1. When using an ammeter to measure the current *through* an element, the ammeter must be in series with the element.

2. When using an ammeter, its internal resistance should be much *smaller* than the total equivalent resistance in series with the ammeter.

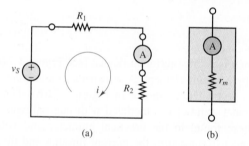

(a)   (b)

Figure 1.69 (a) An ideal ammeter in series with R_1 and R_2; (b) a practical model for an actual ammeter. r_m is the meter's internal resistance.

The Voltmeter

A **voltmeter** measures the voltage *across* an element when connected in *parallel* with it. Figure 1.70(a) shows an *ideal* voltmeter attached across resistor R_2, which is otherwise in a simple series loop. An ideal voltmeter has infinite resistance and therefore is able to measure the voltage without alteration due to the presence of the voltmeter. A more realistic model of an actual voltmeter has an internal resistance in parallel with an ideal voltmeter, as shown in Figure 1.70(b). To obtain an accurate measurement the internal resistance of the voltmeter must be significantly *larger* than the total equivalent resistance between the two nodes to which it is attached in parallel. For example, the internal resistance r_m of the voltmeter must be significantly larger than R_2 in

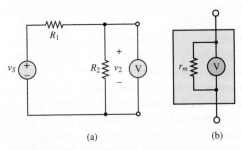

Figure 1.70 (a) An ideal voltmeter in parallel with R_2; (b) a practical model for an actual voltmeter. r_m is the meter's internal resistance.

the series loop of Figure 1.70(a). In practice, it is necessary to observe two rules when using a voltmeter:

1. When using a voltmeter to measure the voltage *across* an element, the voltmeter must be in parallel with the element.
2. When using a voltmeter, its internal resistance should be much *larger* than the total equivalent resistance in parallel with the voltmeter.

The Wattmeter

A **wattmeter**, which is a three-terminal device [see Figure 1.71(a)], measures the power dissipated by a circuit element. A wattmeter is essentially a combination of an ammeter and a voltmeter, as shown in Figure 1.71(b). Thus, it should be no surprise that an actual wattmeter is modeled with internal resistances at its terminals similar to those found in the practical ammeter and voltmeter models. A wattmeter simultaneously measures the current through and the voltage across an element and computes the product of these two quantities to determine the power dissipated.

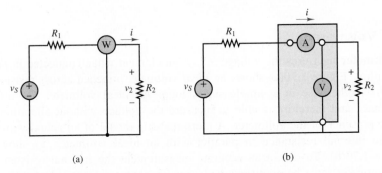

Figure 1.71 (a) An ideal wattmeter in series and parallel with R_2. (b) A model of an ideal wattmeter as a combination of an ideal ammeter and an ideal voltmeter. A practical model would replace the ideal meters with their own practical models.

EXAMPLE 1.16   Impact of a Practical Voltmeter

Problem

Use the tabulated data below to determine the effective internal resistance of the voltmeter shown in Figure 1.70(a), where the voltmeter is modeled as shown in Figure 1.70(b).

Solution

Known Quantities: $v_S = 5.0$ V; various values of $R_1 = R_2$; voltmeter data.

Find: The effective internal resistance r_m of the voltmeter.

Schematics, Diagrams, Circuits, and Given Data: Figure 1.70 and Table 1.4.

Analysis: Substitute the practical model of a voltmeter shown in Figure 1.70(b) for the ideal voltmeter shown in Figure 1.70(a). Notice that the internal resistance r_m of the voltmeter is in parallel with R_2. Their parallel equivalent resistance is:

Table 1.4 **Voltmeter data for determining internal resistance**

$R_1 = R_2$	v_2 (V)
10 kΩ	2.49
470 kΩ	2.44
1 MΩ	2.38
4.7 MΩ	2.02
10 MΩ	1.67

$$R_{EQ} = r_m \| R_2 = \frac{r_m R_2}{r_m + R_2}$$

The voltage across R_2 and the voltmeter can be found directly by voltage division:

$$\frac{v_2}{v_S} = \frac{R_{EQ}}{R_1 + R_{EQ}} \tag{1.24}$$

$$= \frac{r_m R_2}{R_1(r_m + R_2) + r_m R_2} \tag{1.25}$$

Divide the numerator and denominator by R_1 and gather coefficients of r_m.

$$= \frac{r_m(R_2/R_1)}{r_m(1 + R_2/R_1) + R_2} \tag{1.26}$$

Multiply both sides by the denominator on the right and gather coefficients of r_m to find:

$$r_m = \frac{(v_2/v_S)R_2}{R_2/R_1 - (v_2/v_S)(1 + R_2/R_1)} \tag{1.27}$$

When $R_2 = R_1$:

$$= \frac{v_2/v_S}{1 - 2v_2/v_S} R_2 \tag{1.28}$$

Notice that $r_m = R_2$ when:

$$\frac{v_2/v_S}{1 - 2v_2/v_S} = 1$$

Solve for v_2/v_S to find:

$$\frac{v_2}{v_S} = \frac{1}{3}$$

Since $v_S = 5.0$ V, the previous condition is satisfied when $v_2 = 5.0/3 = 1.67$ V. This value for v_2 is found in Table 1.4 for $R_1 = R_2 = 10$ MΩ. Thus, the internal resistance of the voltmeter is:

$$r_m = 10 \text{ M}\Omega$$

This value is typical of many handheld digital multimeters in voltmeter mode.

Comments: It is possible to acquire a separate estimate of r_m for each pair of values $R_1 = R_2$ and v_2 found in Table 1.4 by simply plugging in for v_2, v_S, R_1, and R_2, and solving for r_m. However, in practice, the calculated estimates for r_m will not be the same. The reason for the different estimates is that the measurement of v_2 is much more *sensitive* to experimental error when $R_2 \ll r_m$, as is the case for the first few pairs of data in the table. The least sensitivity to experimental error occurs when $R_2 = r_m$.

CHECK YOUR UNDERSTANDING

Find a separate estimate of r_m for each pair of values $R_1 = R_2$ and v_2 found in Table 1.4. Make a plot of r_m versus R_2.

Answer: Estimates for r_m are: 1.25 M; 9.56 M; 9.92 M; 9.89 M; 10.06 M

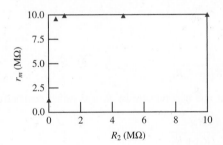

Conclusion

This chapter introduced the fundamentals student need in later chapters in the book to successfully analyze electric circuits. Upon successful completion of this chapter, a student will have learned to:

1. Identify the principal *features of electric circuits or networks*: nodes, loops, meshes, and branches. *Section 1.1.*
2. Apply *Kirchhoff's laws* to simple electric circuits. *Sections 1.2–1.3.*
3. Apply the *passive sign convention* to compute the power consumed or supplied by circuit elements. *Section 1.4.*
4. Identify *sources* and *resistors* and their *i-v characteristics*. *Sections 1.5–1.6.*
5. Apply *Ohm's law* and *voltage and current division* to calculate unknown voltages and currents in simple series, parallel, and series-parallel circuits. *Sections 1.6–1.8.*
6. Understand the impact of internal resistance in practical models of voltage and current sources as well as of voltmeters, ammeters, and wattmeters. *Sections 1.9–1.10.*

HOMEWORK PROBLEMS

Section 1.1: Charge, Current, and Voltage

1.1 A free electron has an initial potential energy per unit charge (*voltage*) of 17 kJ/C and a velocity of 93 Mm/s. Later, its potential energy per unit charge is 6 kJ/C. Determine the change in velocity of the electron.

1.2 The units for voltage, current, and resistance are the volt (V), the ampere (A), and the ohm (Ω), respectively. Express each unit in fundamental MKS units.

1.3 A particular fully charged battery can deliver $2.7 \cdot 10^6$ coulombs of charge.

a. What is the capacity of the battery in ampere-hours?

b. How many electrons can be delivered?

1.4 The charge cycle shown in Figure P1.4 is an example of a three-rate charge. The current is held constant at 30 mA for 6 h. Then it is switched to 20 mA for the next 3 h. Find:

a. The total charge transferred to the battery.

b. The energy transferred to the battery.

Hint: Recall that energy w is the integral of power, or $P = dw/dt$.

1.5 Batteries (e.g., lead-acid batteries) store chemical energy and convert it to electric energy on demand. Batteries do not store electric charge or charge carriers. Charge carriers (electrons) enter one terminal of the battery, acquire electrical potential energy, and exit from the other terminal at a lower voltage. Remember the electron has a negative charge! It is convenient to think of positive carriers flowing in the opposite direction, that is, conventional current, and exiting at a higher voltage. (Benjamin Franklin caused this mess!) For a battery rated at 12 V and 350 A-h, determine:

a. The rated chemical energy stored in the battery.

b. The total charge that can be supplied at the rated voltage.

1.6 What determines:

a. The current *through* an ideal voltage source?

b. The voltage *across* an ideal current source?

1.7 An automotive battery is rated at 120 A-h. This means that under certain test conditions it can output 1 A at 12 V for 120 h (under other test conditions, the battery may have other ratings).

a. How much total energy is stored in the battery?

b. If the headlights are left on overnight (8 h), how much energy will still be stored in the battery in the morning? (Assume a 150-W total power rating for both headlights together.)

1.8 A car battery kept in storage in the basement needs recharging. If the voltage and the current provided by the charger during a charge cycle are shown in Figure P1.8,

a. Find the total charge transferred to the battery.

b. Find the total energy transferred to the battery.

1.9 Suppose the current through a wire is given by the curve shown in Figure P1.9.

a. Find the amount of charge q that flows through the wire between $t_1 = 0$ and $t_2 = 1$ s.

b. Repeat part a for $t_2 = 2, 3, 4, 5, 6, 7, 8, 9,$ and 10 s.

c. Sketch $q(t)$ for $0 \leq t \leq 10$ s.

1.10 The charge cycle shown in Figure P1.10 is an example of a two-rate charge. The current is held constant at 70 mA for 1 h. Then it is switched to 60 mA for the next 1 h. Find:

a. The total charge transferred to the battery.

b. The total energy transferred to the battery.

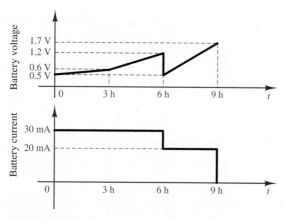

Figure P1.4

Hint: Recall that energy w is the integral of power, or $P = dw/dt$. Let

$$v_1(t) = 5 + e^{t/5194.8} \text{ V}$$

$$v_2(t) = \left(6 - \frac{4}{e^{1h} - 1}\right) + \frac{4}{e^{2h} - e^{1h}} \cdot e^t \text{ V}$$

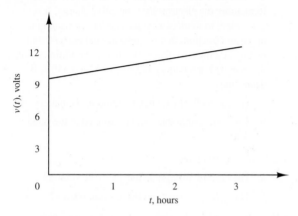

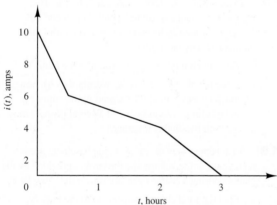

Figure P1.8

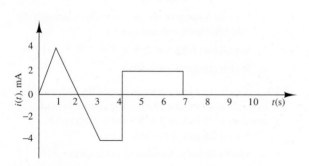

Figure P1.9

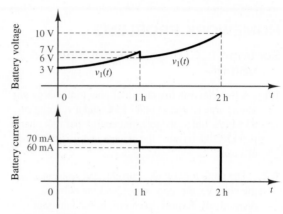

Figure P1.10

1.11 The charging scheme used in Figure P1.11 is an example of a constant-current charge cycle. The charger voltage is controlled such that the current into the battery is held constant at 40 mA, as shown in Figure P1.11. The battery is charged for 6 h. Find:

a. The total charge delivered to the battery.

b. The energy transferred to the battery during the charging cycle.

Hint: Recall that the energy w is the integral of power, or $P = dw/dt$.

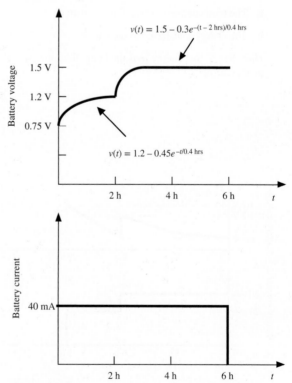

Figure P1.11

1.12 The charging scheme used in Figure P1.12 is called a *tapered-current charge cycle*. The current starts at the highest level and then decreases with time for the entire charge cycle, as shown. The battery is charged for 12 h. Find:

a. The total charge delivered to the battery.

b. The energy transferred to the battery during the charging cycle.

Hint: Recall that the energy w is the integral of power, or $P = dw/dt$.

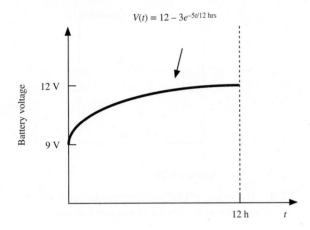

$V(t) = 12 - 3e^{-5t/12 \text{ hrs}}$

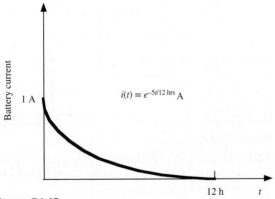

$i(t) = e^{-5t/12 \text{ hrs}} \text{ A}$

Figure P1.12

Sections 1.2, 1.3: KCL, KVL

1.13 Use KCL to determine the unknown currents in the circuit of Figure P1.13. Assume $i_0 = 2$ A and $i_2 = -7$ A.

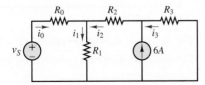

Figure P1.13

1.14 Use KCL to find the current i_1 and i_2 in Figure P1.14. Assume that $i_a = 3$ A, $i_b = -2$ A, $i_c = 1$ A, $i_d = 6$ A and $i_e = -4$ A.

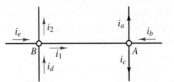

Figure P1.14

1.15 Use KCL to find the current i_1, i_2, and i_3 in the circuit of Figure P1.15. Assume that $i_a = 2$ mA, $i_b = 7$ mA and $i_c = 4$ mA.

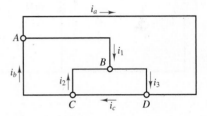

Figure P1.15

1.16 Use KVL to find the voltages v_1, v_2, and v_3 in Figure P1.16. Assume that $v_a = 2$ V, $v_b = 4$ V, and $v_c = 5$ V.

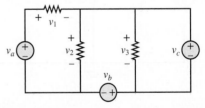

Figure P1.16

1.17 Use KCL to determine the current i_1, i_2, i_3, and i_4 in the circuit of Figure P1.17. Assume that $i_a = -2$ A, $i_b = 6$ A, $i_c = 1$ A and $i_d = -4$ A.

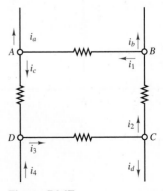

Figure P1.17

Section 1.4: Power and the Passive Sign Convention

1.18 In the circuits of Figure P1.18, the directions of current and polarities of voltage have already been defined. Find the actual values of the indicated currents and voltages.

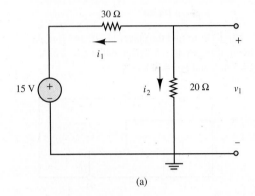

(a)

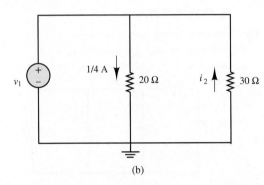

(b)

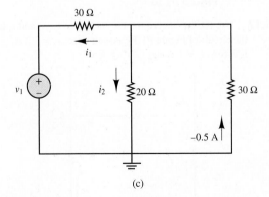

(c)

Figure P1.18

1.19 Find the power delivered by each source in Figure P1.19.

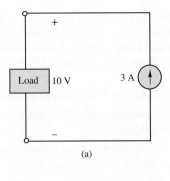

(a)

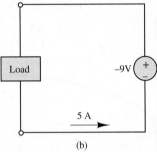

(b)

Figure P1.19

1.20 Determine whether each element in Figure P1.20 is supplying or dissipating power, and how much.

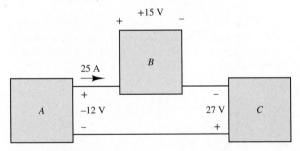

Figure P1.20

1.21 In the circuit of Figure P1.21, determine the power absorbed by the resistor R_4 and the power delivered by the current source.

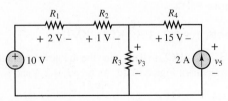

Figure P1.21

1.22 For the circuit shown in Figure P1.22:

a. Determine whether each component is absorbing or delivering power.

b. Is conservation of power satisfied? Explain your answer.

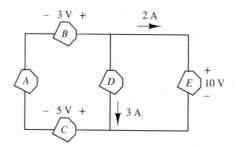

Figure P1.22

1.23 For the circuit shown in Figure P1.23, determine the power absorbed by the 5-Ω resistor.

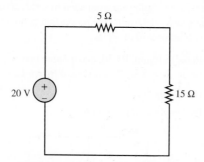

Figure P1.23

1.24 For the circuit shown in Figure P1.24, determine which components are supplying power and which are dissipating power. Also determine the amount of power dissipated and supplied.

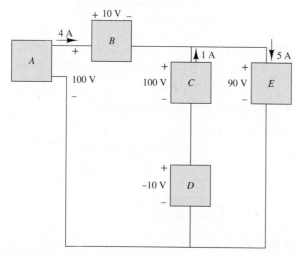

Figure P1.24

1.25 For the circuit shown in Figure P1.25, determine which components are supplying power and which are

dissipating power. Also determine the amount of power dissipated and supplied.

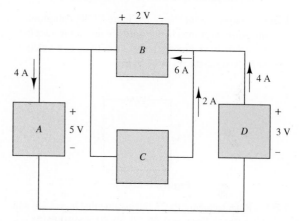

Figure P1.25

1.26 If an electric heater requires 23 A at 110 V, determine

a. The power it dissipates as heat or other losses.

b. The energy dissipated by the heater in a 24-h period.

c. The cost of the energy if the power company charges at the rate 6 cents/kWh.

Sections 1.5–1.6: Sources, Resistance, and Ohm's Law

1.27 In the circuit shown in Figure P1.27, determine the terminal voltage v_T of the source, the power absorbed by R_o, and the efficiency of the circuit. Efficiency is defined as the ratio of load power to source power.

$$v_S = 12 \text{ V} \qquad R_S = 5 \text{ k}\Omega \qquad R_o = 7 \text{ k}\Omega$$

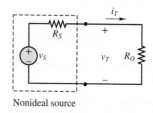

Nonideal source

Figure P1.27

1.28 A 24-V automotive battery is connected to two headlights that are in parallel, similar to that shown in Figure 1.11. Each headlight is intended to be a 75-W load; however, one 100-W headlight is mistakenly installed. What is the resistance of each headlight? What is the total resistance seen by the battery?

1.29   What is the equivalent resistance seen by the battery of Problem 1.28 if two 15-W tail lights are added (in parallel) to two 75-W headlights?

1.30   For the circuit shown in Figure P1.30, determine the power absorbed by the variable resistor R, ranging from 0 to 30 Ω. Plot the power absorption as a function of R. Assume that $v_S = 15$ V, $R_S = 10$ Ω.

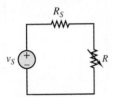

Figure P1.30

1.31   Refer to Figure P1.27 and assume that $v_S = 15$ V and $R_S = 100$ Ω. For $i_T = 0$, 10, 20, 30, 80, and 100 mA:

   a. Find the total power supplied by the ideal source.

   b. Find the power dissipated within the nonideal source.

   c. How much power is supplied to the load resistor?

   d. Plot the terminal voltage v_T and power supplied to the load resistor as a function of terminal current i_T.

1.32   In the circuit of Figure P1.32, assume $v_2 = v_S/6$ and the power delivered by the source is 150 mW. Also assume that $R_1 = 8$ kΩ, $R_2 = 10$ kΩ, $R_3 = 12$ kΩ. Find R, v_S, v_2, and i.

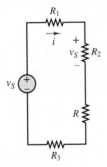

Figure P1.32

1.33   A GE SoftWhite Longlife lightbulb is rated as follows:

P_R = rated power = 60 W

P_{OR} = rated optical power = 820 lumens (lm) (average)

1 lumen = $\dfrac{1}{680}$ W

Operating life = 1,500 h (average)

V_R = rated operating voltage = 115 V

   The resistance of the filament of the bulb, measured with a standard multimeter, is 16.7 Ω.

When the bulb is connected into a circuit and is operating at the rated values given above, determine:

   a. The resistance of the filament.

   b. The efficiency of the bulb.

1.34   An incandescent lightbulb rated at 100 W will dissipate 100 W as heat and light when connected across a 110-V ideal voltage source. If three of these bulbs are connected in series across the same source, determine the power each bulb will dissipate.

1.35   An incandescent lightbulb rated at 60 W will dissipate 60 W as heat and light when connected across a 100-V ideal voltage source. A 100-W bulb will dissipate 100 W when connected across the same source. If the bulbs are connected in series across the same source, determine the power that either one of the two bulbs will dissipate.

1.36   Refer to Figure P1.36, and assume that $v_S = 12$ V, $R_1 = 5$ Ω, $R_2 = 3$ Ω, $R_3 = 4$ Ω, and $R_4 = 5$ Ω. Find:

   a. The voltage v_{ab}.

   b. The power dissipated in R_2.

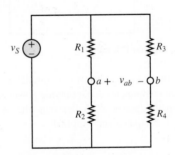

Figure P1.36

1.37   Refer to Figure P1.37, and assume that $V_S = 7$ V, $I_S = 3$ A, $R_1 = 20$ Ω, $R_2 = 12$ Ω, and $R_3 = 10$ Ω. Find:

   a. The currents i_1 and i_2.

   b. The power supplied by the source V_s.

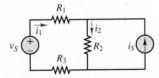

Figure P1.37

1.38   Refer to Figure P1.38, and assume $v_1 = 15$ V, $v_2 = 6$ V, $R_1 = 18$ Ω, $R_2 = 10$ Ω. Find:

   a. The currents i_1, i_2.

   b. The power delivered by the sources v_1 and v_2.

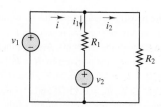

Figure P1.38

1.39 Consider NiMH hobbyist batteries depicted in Figure P1.39.

    a. If $V_1 = 12.0$ V, $R_1 = 0.15\,\Omega$, and $R_o = 2.55\,\Omega$, find the load current I_o and the power dissipated by the load.

    b. If battery 2 with $V_2 = 12$ V and $R_2 = 0.28\,\Omega$ is placed in parallel with battery 1, will the load current I_o increase or decrease? Will the power dissipated by the load increase or decrease? By how much?

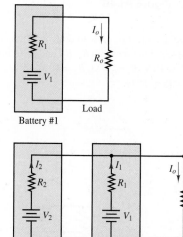

Figure P1.39

1.40 With no load attached, the voltage at the terminals of a particular power supply is 50.8 V. When a 10-W load is attached, the voltage drops to 49 V.

    a. Determine v_S and R_S for this nonideal source.

    b. What voltage would be measured at the terminals in the presence of a 15-Ω load resistor?

    c. How much current could be drawn from this power supply under short-circuit conditions?

1.41 A 220-V electric heater has two heating coils that can be switched such that either coil can be used independently or the two can be connected in series or parallel, for a total of four possible configurations. If the warmest setting corresponds to 2,000-W power

dissipation and the coolest corresponds to 300 W, find the resistance of each coil.

Sections 1.7–1.8: Voltage and Current Division

1.42 For the circuits of Figure P1.42, determine the resistor values (including the power rating) necessary to achieve the indicated voltages. Resistors are available in ⅛-, ¼-, ½-, and 1-W ratings.

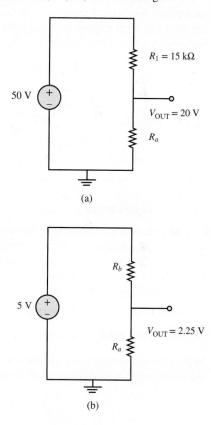

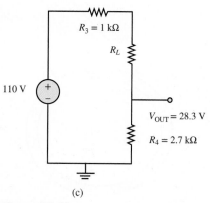

Figure P1.42

1.43 At an engineering site, a 1-hp motor is placed a distance d from a portable generator, as depicted in Figure P1.43. The generator can be modeled as an ideal DC source $V_G = 110$ V. The nameplate on the motor gives the following rated voltages and full-load currents:

$$V_{M\,min} = 105 \text{ V} \rightarrow I_{MFL} = 7.10 \text{ A}$$
$$V_{M\,max} = 117 \text{ V} \rightarrow I_{MFL} = 6.37 \text{ A}$$

If $d = 150$ m and the motor must deliver its full-rated power, determine the minimum AWG conductors that must be used in a rubber-insulated cable. Assume that losses occur only in the wires.

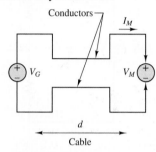

Figure P1.43

1.44 Cheap resistors are fabricated by depositing a thin layer of carbon onto a nonconducting cylindrical substrate (see Figure P1.44). If such a cylinder has radius a and length d, determine the thickness of the film required for a resistance R if:

$$a = 1 \text{ mm} \qquad\qquad R = 33 \text{ k}\Omega$$
$$\sigma = \frac{1}{\rho} = 2.9 \text{ M} \frac{\text{S}}{\text{m}} \qquad d = 9 \text{ mm}$$

Neglect the end surfaces of the cylinder and assume that the thickness is much smaller than the radius.

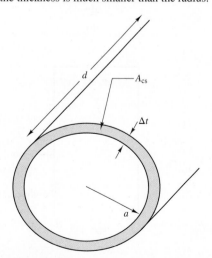

Figure P1.44

1.45 The resistive elements of fuses, lightbulbs, heaters, etc., are nonlinear (i.e., the resistance is dependent on the current through the element). Assume the resistance of a fuse (Figure P1.45) is given by $R = R_0[1 + A(T - T_0)]$ where: $T - T_0 = kP$; $T_0 = 25°\text{C}$; $A = 0.7[°\text{C}]^{-1}$; $k = 0.35°\text{C/W}$; $R_0 = 0.11\,\Omega$; and P is the power dissipated in the resistive element of the fuse. Determine the rated current at which the fuse will melt (that is, "blow") and thus act as an open-circuit. (*Hint:* The fuse blows when R becomes infinite.)

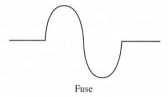

Fuse

Figure P1.45

1.46 Use KCL and Ohm's law to determine the current through each of the resistors R_4, R_5, and R_6 in Figure P1.46. $V_S = 10$ V, $R_1 = 20\,\Omega$, $R_2 = 40\,\Omega$, $R_3 = 10\,\Omega$, $R_4 = R_5 = R_6 = 15\,\Omega$.

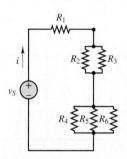

Figure P1.46

1.47 Refer to Figure P1.13. Assume $R_0 = 1\,\Omega$, $R_1 = 2\,\Omega$, $R_2 = 3\,\Omega$, $R_3 = 4\,\Omega$, and $v_S = 10$ V. Use KCL and Ohm's law to find the unknown currents.

1.48 Apply KCL and Ohm's law to find the power supplied by the voltage source in Figure P1.48. Assume $k = 0.25$ A/A^2.

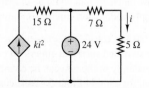

Figure P1.48

1.49   Refer to Figure P1.49 and assume $R_0 = 2\,\Omega$, $R_1 = 1\,\Omega$, $R_2 = \frac{4}{3}\,\Omega$, $R_3 = 6\,\Omega$, and $V_S = 12$ V. Use KVL and Ohm's law to find:

a. The mesh currents i_a, i_b, and i_c.

b. The current through each resistor.

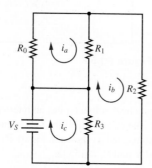

Figure P1.49

1.50   Refer to Figure P1.49 and assume $R_0 = 2\,\Omega$, $R_1 = 2\,\Omega$, $R_2 = 5\,\Omega$, $R_3 = 4$ A, and $V_S = 24$ V. Use KVL and Ohm's law to find:

a. The mesh currents i_a, i_b, and i_c.

b. The voltage across each resistor.

1.51   Assume that the voltage source in Figure P1.49 is now replaced by a DC current source I_S, and $R_0 = 1\,\Omega$, $R_1 = 3\,\Omega$, $R_2 = 2\,\Omega$, $R_3 = 4$ A, and $I_S = 12$ A, directed positively upward. Use KVL and Ohm's law to determine the voltage across each resistor.

1.52   The voltage divider network of Figure P1.52 is designed to provide $v_{out} = v_S/2$. However, in practice, the resistors may not be perfectly matched; that is, their tolerances are such that the resistances are unlikely to be identical. Assume $v_S = 10$ V and nominal resistance values of $R_1 = R_2 = 5$ kΩ.

a. If the resistors have ± 10 percent tolerance, find the expected range of possible output voltages.

b. Find the expected output voltage range for a tolerance of ± 5 percent.

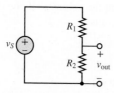

Figure P1.52

Sections 1.9–1.10: Practical Sources and Measuring Devices

1.53   A *thermistor* is a nonlinear device that changes its terminal resistance value as its surrounding temperature changes. The resistance and temperature generally have a relation in the form of:

$$R_{th}(T) = R_0 e^{-\beta(T-T_0)}$$

where R_{th} = resistance at temperature T, Ω

R_0 = resistance at temperature $T_0 = 298$ K, Ω

β = material constant, K^{-1}

T, T_0 = absolute temperature, K

a. If $R_0 = 300\,\Omega$ and $\beta = -0.01$ K^{-1}, plot $R_{th}(T)$ as a function of the surrounding temperature T for $350 \leq T \leq 750$.

b. If the thermistor is in parallel with a 250-Ω resistor, find the expression for the equivalent resistance and plot $R_{th}(T)$ on the same graph for part a.

1.54   A moving-coil meter movement has a meter resistance $r_M = 200\,\Omega$, and full-scale deflection is caused by a meter current $i_m = 10\,\mu$A. The meter is to be used to display pressure, as measured by a sensor, up to a maximum of 100 kPa. Models of the meter and pressure sensor are shown in Figure P1.54 along with the relationship between measured pressure and the sensor output v_T.

a. Devise a circuit that will produce the desired behavior of the meter, showing all appropriate connections between the terminals of the sensor and the meter.

b. Determine the value of each component in the circuit.

c. What is the linear range, that is, the minimum and maximum pressure that can accurately be measured?

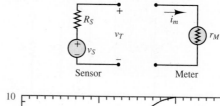

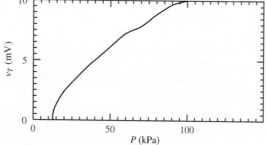

Figure P1.54

1.55   The circuit of Figure P1.55 is used to measure the internal impedance of a battery. The battery being tested is a NiMH battery cell.

a. A fresh battery is being tested, and it is found that the voltage v_{out} is 2.28 V with the switch open and 2.27 V with the switch closed. Find the internal resistance of the battery.

b. The same battery is tested one year later. v_{out} is found to be 2.2 V with the switch open but 0.31 V with the switch closed. Find the internal resistance of the battery.

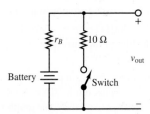

Figure P1.55

1.56   Consider the practical ammeter, depicted in Figure P1.56, consisting of an ideal ammeter in series with a 1-kΩ resistor. The meter sees a full-scale deflection when the current through it is 30 μA. If we desire to construct a multirange ammeter reading full-scale values of 10 mA, 100 mA, and 1 A, depending on the setting of a rotary switch, determine appropriate values of R_1, R_2, and R_3.

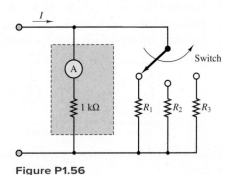

Figure P1.56

1.57   A circuit that measures the internal resistance of a practical ammeter is shown in Figure P1.57, where $R_S = 50,000$ Ω, $v_S = 12$ V, and R_p is a variable resistor that can be adjusted at will.

a. Assume that $r_a \ll 50,000$ Ω. Estimate the current i.

b. If the meter displays a current of 150 μA when $R_p = 15$ Ω, find the internal resistance of the meter r_a.

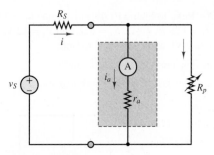

Figure P1.57

1.58   A practical voltmeter has an internal resistance r_m. What is the value of r_m if the meter reads 11.81 V when connected as shown in Figure P1.58? Assume $V_S = 12$ V and $R_S = 25$ kΩ.

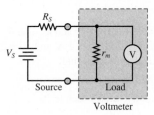

Figure P1.58

1.59   Using the circuit of Figure P1.58, find the voltage that the meter reads if $V_S = 24$ V and R_S has the following values:

$$R_S = 0.2r_m, 0.4\,r_m, 0.6r_m, 1.2r_m, 4\,r_m, 6\,r_m, \text{ and } 10\,r_m.$$

How large (or small) should the internal resistance of the meter be relative to R_S?

1.60   A voltmeter is used to determine the voltage across a resistive element in the circuit of Figure P1.60. The instrument is modeled by an ideal voltmeter in parallel with a 120-kΩ resistor, as shown. The meter is placed to measure the voltage across R_4. Assume $R_1 = 8$ kΩ, $R_2 = 22$ kΩ, $R_3 = 50$ kΩ, $R_S = 125$ kΩ, and $i_S = 120$ mA. Find the voltage across R_4 with and without the voltmeter in the circuit for the following values:

a. $R_4 = 100 \,\Omega$

b. $R_4 = 1$ kΩ

c. $R_4 = 10$ kΩ

d. $R_4 = 100$ kΩ

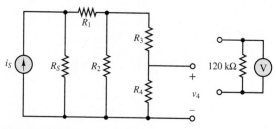

Figure P1.60

1.61 An ammeter is used as shown in Figure P1.61. The ammeter model consists of an ideal ammeter in series with a resistance. The ammeter model is placed in the branch as shown in the figure. Find the current through R_5 both with and without the ammeter in the circuit for the following values, assuming that $R_S = 20\,\Omega$, $R_1 = 800\,\Omega$, $R_2 = 600\,\Omega$, $R_3 = 1.2\,\text{k}\Omega$, $R_4 = 150\,\Omega$, and $v_S = 24$ V.

a. $R_5 = 1\,\text{k}\Omega$

b. $R_5 = 100\,\Omega$

c. $R_5 = 10\,\Omega$

d. $R_5 = 1\,\Omega$

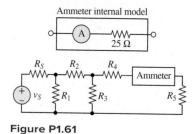

Figure P1.61

1.62 Figure P1.62 shows an *aluminum* cantilevered beam loaded by the force F. Strain gauges R_1, R_2, R_3, and R_4 are attached to the beam as shown in Figure P1.62 and connected into the circuit shown.

The force causes a tension stress on the top of the beam that causes the length (and therefore the resistance) of R_1 and R_4 to increase and a compression stress on the bottom of the beam that causes the length (and therefore the resistance) of R_2 and R_3 to decrease. The result is a voltage of 50 mV at node B with respect to node A. Determine the force if

$$R_o = 1\,\text{k}\Omega \qquad v_S = 12\,\text{V} \qquad L = 0.3\,\text{m}$$
$$w = 25\,\text{mm} \qquad h = 100\,\text{mm} \qquad Y = 69\,\text{GN/m}^2$$

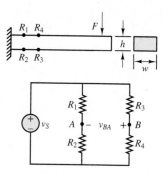

Figure P1.62

1.63 Refer to Figure P1.62 but assume that the cantilevered beam loaded by a force F is made of *steel*. Strain gauges R_1, R_2, R_3, and R_4 are attached to the beam and connected in the circuit shown. The force causes a tension stress on the top of the beam that causes the length (and therefore the resistance) of R_1 and R_4 to increase and a compression stress on the bottom of the beam that causes the length (and therefore the resistance) of R_2 and R_3 to decrease. The result is a voltage v_{BA} across nodes B and A. Determine this voltage if $F = 1.3$ MN and

$$R_o = 1\,\text{k}\Omega \qquad v_S = 24\,\text{V} \qquad L = 1.7\,\text{m}$$
$$w = 3\,\text{cm} \qquad h = 7\,\text{cm} \qquad Y = 200\,\text{GN/m}^2$$

C H A P T E R

2

RESISTIVE NETWORK ANALYSIS

C hapter 2 illustrates the fundamental techniques for the analysis of resistive circuits. The chapter begins with the definition of network variables and of network analysis problems. Next, the two most widely applied methods— *node analysis* and *mesh analysis*—are introduced. These are the most generally applicable circuit solution techniques used to derive the equations of all electric circuits; their application to resistive circuits in this chapter is intended to acquaint you with these methods, which are used throughout the book. The second solution method presented is based on the *principle of superposition,* which is applicable only to linear circuits. Next, the concept of *Thévenin and Norton equivalent circuits* is explored, which leads to a discussion of *maximum power transfer* in electric circuits and facilitates the ensuing discussion of nonlinear loads and *load-line analysis.* At the conclusion of the chapter, you should have developed confidence in your ability to compute numerical solutions for a wide range of resistive circuits. The following box outlines the principal learning objectives of the chapter.

(LO) Learning Objectives

Students will learn to...
1. Compute the solution of circuits containing linear resistors and independent and dependent sources by using *node analysis. Section 2.2.*
2. Compute the solution of circuits containing linear resistors and independent and dependent sources by using *mesh analysis. Section 2.3.*
3. Apply the *principle of superposition* to linear circuits containing independent sources. *Section 2.4.*
4. Compute *Thévenin and Norton equivalent circuits* for networks containing linear resistors and independent and dependent sources. *Section 2.5.*
5. Use equivalent-circuit ideas to compute the *maximum power transfer* between a source and a load. *Section 2.6.*

2.1 NETWORK ANALYSIS

The analysis of an electric network consists of determining one or more of the unknown branch currents and node voltages. It is important to define all the relevant variables as clearly as possible and in a systematic fashion. Once the known and unknown variables have been identified, a set of equations relating these variables is constructed, and these equations are solved by means of suitable techniques. The procedures and conventions required to write these equations are the subject of Chapter 2 and are documented and codified in the form of simple rules.

Example 2.1 defines voltages and currents associated with a typical circuit.

EXAMPLE 2.1

Problem

Identify the branch and node voltages and the loop and mesh currents in the circuit of Figure 2.1.

Solution

The following node and branch voltages may be identified:

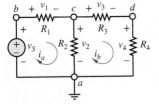

Figure 2.1 Figure for Example 2.1

Node voltages	Branch voltages	Relationship
$v_a = 0$ (reference)		
v_b	v_S	$v_S = v_b - v_a$
	v_1	$v_1 = v_b - v_c$
v_c	v_2	$v_2 = v_c - v_a$
	v_3	$v_3 = v_c - v_d$
v_d	v_4	$v_4 = v_d - v_a$

Comments: Currents i_a and i_b are known as mesh currents.

Nine variables were identified in the example! Methods are needed to organize the wealth of information that can be generated even in simple circuits. Ideally, these methods would produce only the minimum number of equations needed and result in n equations in n unknowns. The remainder of the chapter is devoted to an exploration of systematic circuit analysis and solution methods.

2.2   THE NODE VOLTAGE METHOD

Node voltage analysis is the most general method for the analysis of electric circuits. Its application to linear resistive circuits is illustrated in this section. The **node voltage method** is based on defining the voltage at each node as an independent variable. One of the nodes is freely chosen as a **reference node** (usually—but not necessarily—ground), and each of the other node voltages is relative to this node. Ohm's law is used to express resistor currents in terms of node voltages, such that *each branch current is expressed in terms of one or more node voltages.* Finally, KCL is applied to each nonreference node to generate one equation for each node voltage. The result is that only node voltages and known parameters appear explicitly in the equations. Figures 2.2 and 2.3 illustrate how to apply Ohm's law and KCL in this method.

In the node voltage method, the branch current flowing from a to b is expressed in terms of the node voltages v_a and v_b using Ohm's law.

$$i = \frac{v_a - v_b}{R}$$

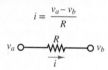

Figure 2.2 Branch current formulation in node analysis

By KCL: $i_1 - i_2 - i_3 = 0$. In the node voltage method, we express KCL by

$$\frac{v_a - v_b}{R_1} - \frac{v_b - v_c}{R_2} - \frac{v_b - v_d}{R_3} = 0$$

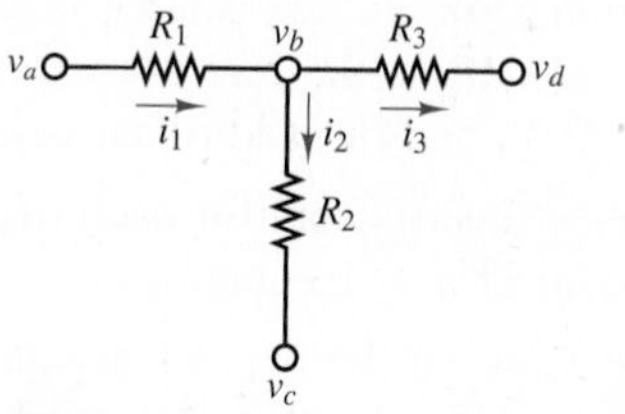

Figure 2.3 Use of KCL in node analysis

Once each branch current is defined in terms of the node voltages, Kirchhoff's current law is applied at each node:

$$\sum i = 0 \tag{2.1}$$

The systematic application of this method to a circuit with n nodes leads to $n - 1$ equations. However, one node must be chosen as a reference, which is freely and conveniently assigned a value of 0 V. (Recall the discussion of reference voltage from Chapter 1.) Thus, the result is $n - 1$ variables (the node voltages) in $n - 1$ *independent linear equations.* Node analysis provides the minimum number of equations required to solve the circuit. A systematic method for applying node analysis is outlined in the Focus on Problem Solving box, "Node Analysis."

FOCUS ON PROBLEM SOLVING

NODE ANALYSIS

1. Select a reference node. Often, the best choice for this node is the one with the most elements attached to it. The voltage associated with each nonreference node will be relative to the reference node, which is (for simplicity) typically assigned a value of 0 V.

2. Define voltage variables v_1, v_2, . . . , v_{n-1} for the remaining $n - 1$ nodes.

   • If the circuit contains no voltage sources, then all $n - 1$ node voltages are treated as independent variables.

   • If the circuit contains m voltage sources:

      ○ There are only $(n - 1) - m$ independent voltage variables.

      ○ There are m dependent voltage variables.

      ○ For each voltage source, one of the two adjacent node voltage variables (e.g., v_j or v_k) must be treated as a dependent variable.

3. Apply KCL at each node associated with an independent variable, using Ohm's law to express each resistor current in terms of the adjacent node voltages. The convention used consistently in this book is that currents entering a node are positive and currents exiting a node are negative; however, the opposite convention could be used.

   • For each voltage source v_S there will be one additional dependent equation (e.g., $v_k = v_j + v_S$).

   • When a voltage source is adjacent to the reference node, either v_j or v_k will be the reference node value, which is typically assigned as 0 V, and the additional dependent equation is particularly simple.

4. Collect coefficients for each of the $n - 1$ variables, and solve the linear system of $n - 1$ equations.

   • Some of the dependent equations may have the simple form $v_k = v_S$. In this case, the total number of equations and variables is reduced by direct substitution.

This procedure can be used to find a solution for any circuit. A good approach is to first practice solving circuits without any voltage sources and then learn to deal with the added complexity of circuits with voltage sources. The remainder of this section is organized in this fashion.

For some readers it is advantageous to redraw circuits in an equivalent but nonrectangular manner by viewing the circuit as a collection of circuit elements located between nodes. The right-hand portion of Figure 2.4 is constructed by first drawing three node circles and then adding in the elements that sit between each pair of nodes. To successfully redraw a circuit or to apply node analysis it is imperative that the correct number of nodes is known. Thus, it is worthwhile to practice recognizing and counting nodes!

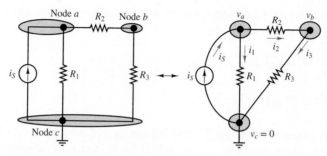

Figure 2.4 Illustration of node analysis

Details and Examples

Consider the circuit shown in Figure 2.4. The directions of currents i_1, i_2, and i_3 may be selected arbitrarily; however, it is often helpful to select directions that appear to conform with one's expectations. In this case, i_S is directed into node a and therefore one might guess that i_1 and i_2 should be directed out of that same node. Application of KCL at node a yields

$$i_S - i_1 - i_2 = 0 \tag{2.2}$$

whereas at node b

$$i_2 - i_3 = 0 \tag{2.3}$$

It is not necessary nor appropriate to apply KCL at the reference node since that equation is dependent on the other two. It may be useful to demonstrate this fact for this example. The equation obtained by applying KCL at node c is

$$i_1 + i_3 - i_S = 0 \tag{2.4}$$

The sum of the equations obtained at nodes a and b yields this same equation. This observation confirms the statement made earlier:

> KCL yields $n - 1$ independent equations for a circuit of n nodes. The nth equation is usually chosen to be $v = 0$ for the reference.

When applying the node voltage method, the currents must be expressed in terms of the node voltages. For the example above, i_1, i_2, and i_3 can be expressed in terms of v_a, v_b, and v_c by applying Ohm's law. Between nodes a and c the current is

$$i_1 = \frac{v_a - v_c}{R_1} \tag{2.5}$$

Similarly, for the other two branch currents

$$i_2 = \frac{v_a - v_b}{R_2}$$
$$i_3 = \frac{v_b - v_c}{R_3}. \tag{2.6}$$

These expressions for i_1, i_2, and i_3 can be substituted into the two nodal equations (equations 2.2 and 2.3) to obtain:

$$i_S - \frac{v_a}{R_1} - \frac{v_a - v_b}{R_2} = 0 \qquad (2.7)$$

$$\frac{v_a - v_b}{R_2} - \frac{v_b}{R_3} = 0 \qquad (2.8)$$

With a little practice, equations 2.7 and 2.8 can be obtained by direct observation of the circuit. These equations can be solved for v_a and v_b, assuming that i_S, R_1, R_2, and R_3 are known. The same equations may be reformulated as follows:

$$\left(\frac{1}{R_1} + \frac{1}{R_2}\right)v_a + \left(-\frac{1}{R_2}\right)v_b = i_S$$

$$\left(-\frac{1}{R_2}\right)v_a + \left(\frac{1}{R_2} + \frac{1}{R_3}\right)v_b = 0 \qquad (2.9)$$

Examples 2.2 to 2.6 further illustrate the details of this method.

EXAMPLE 2.2   Node Analysis: Solving for Branch Currents

Problem

Solve for all currents and voltages in the circuit of Figure 2.5.

Solution

Given: Source currents, resistor values.

Find: All node voltages and branch currents.

Schematics, Diagrams, Circuits, and Given Data: $i_1 = 10$ mA; $i_2 = 50$ mA; $R_1 = 1\,\text{k}\Omega$; $R_2 = 2\,\text{k}\Omega$; $R_3 = 10\,\text{k}\Omega$; $R_4 = 2\,\text{k}\Omega$.

Analysis: Follow the steps outlined in the Focus on Problem Solving box, "Node Analysis."

1. The node at the bottom of the circuit is chosen as the reference.
2. The circuit of Figure 2.5 is shown again in Figure 2.6, with two nonreference nodes and the associated two independent variables v_1 and v_2.
3. Apply KCL at each node and use Ohm's law to express branch currents in terms of node voltages to obtain:

$$i_1 - \frac{v_1 - 0}{R_1} - \frac{v_1 - v_2}{R_2} - \frac{v_1 - v_2}{R_3} = 0 \qquad \text{node 1}$$

$$\frac{v_1 - v_2}{R_2} + \frac{v_1 - v_2}{R_3} - \frac{v_2 - 0}{R_4} - i_2 = 0 \qquad \text{node 2}$$

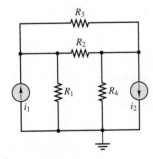

Figure 2.5 Figure for Example 2.2

The same equations can be written more systematically as a function of the node voltages, as was done in equation 2.9.

$$\left(\frac{1}{R_1} + \frac{1}{R_2} + \frac{1}{R_3}\right) v_1 + \left(-\frac{1}{R_2} - \frac{1}{R_3}\right) v_2 = i_1 \qquad \text{node 1}$$

$$\left(-\frac{1}{R_2} - \frac{1}{R_3}\right) v_1 + \left(\frac{1}{R_2} + \frac{1}{R_3} + \frac{1}{R_4}\right) v_2 = -i_2 \qquad \text{node 2}$$

4. Solve the system of equations. With some manipulation, the equations can be represented in the following form:

$$1.6v_1 - 0.6v_2 = 10$$
$$-0.6v_1 + 1.1v_2 = -50$$

These equations may be solved simultaneously to obtain

$$v_1 = -13.57 \text{ V}$$
$$v_2 = -52.86 \text{ V}$$

Knowing the node voltages, each branch current can be determined. For example, the current through R_3 (the 10-kΩ resistor) is given by

$$i_{R_3} = \frac{v_1 - v_2}{10,000} = 3.93 \text{ mA}$$

The positive value for i_{R_3} indicates that the initial (arbitrary) choice of direction for this current is the same as its actual direction. As another example, consider the current through R_1, the 1-kΩ resistor:

$$i_{R_1} = \frac{v_1}{1,000} = -13.57 \text{ mA}$$

Here, the value is negative, which indicates that the actual direction of this current is from ground to node 1, opposite of what was assumed, but as it must be, since the voltage at node 1 is negative with respect to ground. The branch-by-branch analysis may be continued to verify that $i_{R_2} = 19.65$ mA and $i_{R_4} = -26.43$ mA.

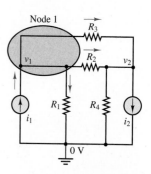

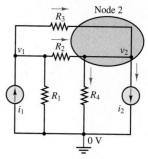

Figure 2.6 Figures for Example 2.2

EXAMPLE 2.3   Node Analysis: Solving for Node Voltages

Problem

Write the node equations and solve for the node voltages in the circuit of Figure 2.7.

Solution

Given: Source currents, resistor values.

Find: All node voltages.

Schematics, Diagrams, Circuits, and Given Data: $i_a = 1$ mA; $i_b = 2$ mA; $R_1 = 1 \text{ k}\Omega$; $R_2 = 500 \ \Omega$; $R_3 = 2.2 \text{ k}\Omega$; $R_4 = 4.7 \text{ k}\Omega$.

Analysis: Follow the steps outlined in the Focus on Problem Solving box, "Node Analysis."

1. The bottom node of the circuit is chosen as the reference (ground) node.

2. See Figure 2.8. There are two nonreference nodes, labeled v_a and v_b, in the circuit. The two nodes are associated with two independent variables, the node voltages v_a and v_b.

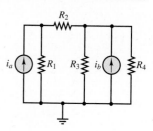

Figure 2.7 Figure for Example 2.3

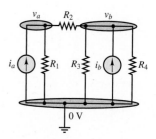

Figure 2.8 Figure for Example 2.3

3. Apply KCL at each node and use Ohm's law to express branch currents in terms of node voltages to obtain:

$$i_a - \frac{v_a}{R_1} - \frac{v_a - v_b}{R_2} = 0 \qquad \text{node } a$$

$$\frac{v_a - v_b}{R_2} + i_b - \frac{v_b}{R_3} - \frac{v_b}{R_4} = 0 \qquad \text{node } b$$

These equations may be rewritten as the following linear system:

$$\left(\frac{1}{R_1} + \frac{1}{R_2}\right) v_a + \qquad \left(-\frac{1}{R_2}\right) v_b = i_a$$

$$\left(-\frac{1}{R_2}\right) v_a + \left(\frac{1}{R_2} + \frac{1}{R_3} + \frac{1}{R_4}\right) v_b = i_b$$

4. Numerical values can be plugged into these equations to find:

$$3 \times 10^{-3} v_a \qquad - 2 \times 10^{-3} v_b = 1 \times 10^{-3}$$

$$-2 \times 10^{-3} v_a \ + 2.67 \times 10^{-3} v_b = 2 \times 10^{-3}$$

or

$$3 v_a \qquad - 2 v_b = 1$$

$$-2 v_a \ + 2.67 v_b = 2$$

Multiply the second equation by $\frac{3}{2}$ and add the result to the first equation to find $v_b = 2$ V. Plug v_b into either equation to find $v_a = 1.667$ V.

EXAMPLE 2.4   **Using Cramer's Rule to Solve a 2 × 2 System of Linear Equations**

Problem

Use Cramer's rule (see Appendix A) to solve the circuit equations obtained in Example 2.3.

Solution

Given: A linear system of equations.

Find: Node voltages.

Analysis: The 2 × 2 system of equations in Example 2.3 may be written in matrix form and solved, in terms of ratios of determinants, using Cramer's rule from linear algebra.

$$\begin{bmatrix} 3 & -2 \\ -2 & 2.67 \end{bmatrix} \begin{bmatrix} v_a \\ v_b \end{bmatrix} = \begin{bmatrix} 1 \\ 2 \end{bmatrix}$$

Cramer's rule provides the solution for v_a and v_b as follows:

$$v_a = \frac{\begin{vmatrix} 1 & -2 \\ 2 & 2.67 \end{vmatrix}}{\begin{vmatrix} 3 & -2 \\ -2 & 2.67 \end{vmatrix}} = \frac{(1)(2.67) - (-2)(2)}{(3)(2.67) - (-2)(-2)} = \frac{6.67}{4} = 1.667 \text{ V}$$

$$v_b = \frac{\begin{vmatrix} 3 & 1 \\ -2 & 2 \end{vmatrix}}{\begin{vmatrix} 3 & -2 \\ -2 & 2.67 \end{vmatrix}} = \frac{(3)(2) - (-2)(1)}{(3)(2.67) - (-2)(-2)} = \frac{8}{4} = 2 \text{ V}$$

Voila! The result is the same as in Example 2.3.

Comments: Cramer's rule is an efficient solution method for simple circuits (e.g., those with only two nonreference nodes); however, it is not recommended for larger circuits. Once the nodal equations have been set in the general form (e.g., equation 2.9), various computer applications, such as Matlab®, may be employed to compute the solution. See Example 2.5 for an example of using Matlab.

EXAMPLE 2.5   Using MatLab to Solve a 3 × 3 System of Linear Equations

Problem

Use the node voltage analysis to determine the voltage v in the circuit of Figure 2.9(a). Assume that $R_1 = 2\,\Omega$, $R_2 = 1\,\Omega$, $R_3 = 4\,\Omega$, $R_4 = 3\,\Omega$, $i_1 = 2$ A, and $i_2 = 3$ A.

Solution

Given: Values of the resistors and the current sources.

Find: Voltage across R_3.

Analysis: Refer to Figure 2.9a and the steps in the Focus on Problem Solving box, "Node Analysis."

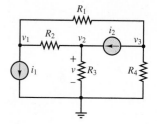

Figure 2.9(a) Circuit for Example 2.5

1. Select one node as the reference and label it.

2. Define node voltages v_1, v_2, v_3 for the three nonreference nodes.

3. Apply KCL at each of the $n-1$ nodes, using Ohm's law to express the current through a resistor as the difference between the two adjacent node voltages divided by the resistance.

$$\frac{v_3 - v_1}{R_1} + \frac{v_2 - v_1}{R_2} - i_1 = 0 \qquad \text{node 1}$$
$$\frac{v_1 - v_2}{R_2} - \frac{v_2}{R_3} + i_2 = 0 \qquad \text{node 2}$$
$$\frac{v_1 - v_3}{R_1} - \frac{v_3}{R_4} - i_2 = 0 \qquad \text{node 3}$$

4. Collect the coefficients of each independent variable (node voltage) to express the system of equations as:

$$-\left(\frac{1}{R_1} + \frac{1}{R_2}\right)v_1 \qquad +\left(\frac{1}{R_2}\right)v_2 \qquad +\left(\frac{1}{R_1}\right)v_3 = \quad i_1$$
$$\left(\frac{1}{R_2}\right)v_1 \quad -\left(\frac{1}{R_2} + \frac{1}{R_3}\right)v_2 \qquad\qquad = -i_2$$
$$\left(\frac{1}{R_1}\right)v_1 \qquad\qquad -\left(\frac{1}{R_1} + \frac{1}{R_4}\right)v_3 = \quad i_2$$

5. Multiply both sides of each equation by the common denominator on the left side. The common denominators are R_1R_2 for node 1, R_2R_3 for node 2, and R_1R_4 for node 3. Plug in values for the resistors and current sources to obtain:

$$(-1\ -2)\quad v_1 + \qquad 2v_2 + \qquad 1v_3 = \quad 4 \qquad \text{node 1}$$
$$4v_1 + \quad (-1\ -4)v_z + \qquad 0v_3 = -12 \qquad \text{node 2}$$
$$3v_1 + \qquad 0v_2 + \quad (-2\ -3)v_3 = \quad 18 \qquad \text{node 3}$$

By including zero coefficients explicitly, all three voltage variables are now present in each equation. The resulting system of three equations in three unknowns can be solved by many handheld calculators. An alternative is Matlab.

Command Window

ⓘ New to MATLAB? Watch this <u>Video</u>, see <u>Examples</u>, or read <u>Getting Started</u>.

```
            This is a Classroom License for instructional use only.
            Research and commercial use is prohibited.
>> A = [-3 2 1; 4 -5 0; 3 0 -5]

A =

      -3       2       1
       4      -5       0
       3       0      -5

>> b = [4; -12; 18]

b =

       4
     -12
      18

>> x = a\b
Undefined function or variable 'a'.

Did you mean:
>> x = A\b

x =

    -3.5000
    -0.4000
    -5.7000

fx >>
```

Figure 2.9(b) Typical Matlab command window. User-entered data follows the ≫ prompt. Note that Matlab is case sensitive, as shown at the fourth prompt. (*The MathWorks, Inc.*)

6. To solve using Matlab it is necessary to write the equations in matrix form.

$$\begin{bmatrix} -3 & 2 & 1 \\ 4 & -5 & 0 \\ 3 & 0 & -5 \end{bmatrix} \begin{bmatrix} v_1 \\ v_2 \\ v_3 \end{bmatrix} = \begin{bmatrix} 4 \\ -12 \\ 18 \end{bmatrix}$$

In general, these equations can be written using compact notation as

$$Ax = b$$

where x is a 3×1 column vector whose elements are the node voltages v_1, v_2, and v_3. In Matlab the 3×3 A matrix and the 3×1 b column vector are entered as shown in Figure 2.9(b).

$$A = [-3\ 2\ 1\ ;\ 4\ -5\ 0\ ;\ 3\ 0\ -5]$$

$$b = [4\ ;\ -12\ ;\ 18] \quad (\text{or} \quad b = [4\ -12\ 18]')$$

The apostrophe at the far right of the above equation is the Matlab transpose operator. It is used here to change a 1×3 row matrix into a 3×1 column vector. The solution for x is computed in Matlab by writing $x = A\backslash b$ to yield

$$x = [-3.5 \text{ V} \quad -0.4 \text{ V} \quad -5.7 \text{ V}]'$$

which are the three node voltages $[v_1 \ v_2 \ v_3]'$. The solution for the voltage drop v across R_3 is

$$v = v_2 = -0.4 \text{ V}$$

CHECK YOUR UNDERSTANDING

Find i_o and v_x in the circuits on the left and right, respectively, using the node voltage method.

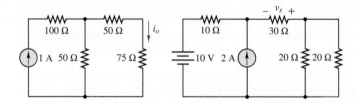

CHECK YOUR UNDERSTANDING

In Example 2.3, use the two node voltages to verify that KCL is indeed satisfied at each node.

CHECK YOUR UNDERSTANDING

Repeat Example 2.5 when the directions of the current sources are opposite those shown in Figure 2.9(a). Find v.

Node Analysis With Voltage Sources

The circuits in the preceding examples did not contain voltage sources. However, in practice, it is quite common to encounter them in circuits. To illustrate how node analysis is applied to such circuits, consider the circuit in Figure 2.10. Verify that this circuit has $n = 4$ total nodes. The relevant solution steps found in the Focus on Problem Solving box, "Node Analysis," are listed below with added comments.

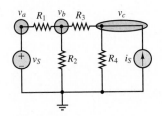

Figure 2.10 Node analysis
with voltage sources

Step 1: *Select a reference node. Often, the best choice for this node is the one with the most elements attached to it. The voltage associated with each nonreference node will be relative to the reference node, which is (for simplicity) typically assigned a value of* 0 *V.*

When voltage sources are present, it is also advantageous to pick the reference node so that at least one of those voltage sources is attached to it. In Figure 2.10, the reference node, denoted by the ground symbol, is assumed to have a value of 0 V.

Step 2: *Label the remaining* $n - 1$ *nodes with voltage variables* $v_1, v_2, \ldots, v_{n-1}$. *If the circuit contains m voltage sources:*

- *There are only* $(n - 1) - m$ *independent voltage variables.*
- *There are m dependent voltage variables.*
- *For each voltage source, one of the two adjacent node voltage variables (e.g.,* v_j *or* v_k*) must be treated as a dependent variable.*

The remaining three $(4 - 1 = 3)$ node voltages are labeled v_a, v_b, and v_c as shown Figure 2.10. Since the circuit contains one $(m = 1)$ voltage source, there are two $(4 - 1 - 1 = 2)$ independent variables and one dependent variable. Only v_a is adjacent to the voltage source; thus, it is the one dependent variable.

Step 3: *Apply KCL at each node associated with an independent variable, using Ohm's law to express each resistor current in terms of the adjacent node voltages. Currents entering a node are assumed to be positive while those exiting a node are assumed to be negative.*

- *For each voltage source* v_S *there will be one additional dependent equation (e.g.,* $v_k = v_j + v_S$*).*
- *When a voltage source is adjacent to the reference node, either* v_j *or* v_k *will be the reference node value, which is typically assigned as* 0 *V, and the additional dependent equation is particularly simple.*

Apply KCL at the two nodes associated with the independent variables v_b and v_c:

At node *b:*

$$\frac{v_a - v_b}{R_1} - \frac{v_b - 0}{R_2} - \frac{v_b - v_c}{R_3} = 0 \tag{2.10a}$$

At node *c:*

$$\frac{v_b - v_c}{R_3} - \frac{v_c}{R_4} + i_S = 0 \tag{2.10b}$$

The equation for the dependent variable v_a is simply:

$$v_a = 0 + v_S = v_S \tag{2.10c}$$

Step 4: *Collect coefficients for each of the* $n - 1$ *variables, and solve the linear system of* $n - 1$ *equations.*

- *Some of the dependent equations may have the simple form* $v_k = v_S$. *In this case, the total number of equations and variables is reduced by direct substitution.*

Substitute for v_a in equation 2.10. At node *b:*

$$\frac{v_S - v_b}{R_1} - \frac{v_b}{R_2} - \frac{v_b - v_c}{R_3} = 0 \qquad (2.11)$$

Finally, collect the coefficients of the two independent variables to express the system of two equations as:

$$\left(\frac{1}{R_1} + \frac{1}{R_2} + \frac{1}{R_3}\right)v_b \quad + \left(-\frac{1}{R_3}\right)v_c = \frac{1}{R_1}v_S$$

$$\left(-\frac{1}{R_3}\right)v_b \quad + \left(\frac{1}{R_3} + \frac{1}{R_4}\right)v_c = i_S \qquad (2.12)$$

The resulting system of two equations in two unknowns is now ready to be solved.

EXAMPLE 2.6   Solution When a Voltage Source Is Not Adjacent to the Reference Node

Problem

Use node analysis to determine the current i through the voltage source in the circuit of Figure 2.11(a). Assume that $R_1 = 2\,\Omega, R_2 = 2\,\Omega, R_3 = 4\,\Omega, R_4 = 3\,\Omega,\ i_S = 2$ A, and $v_S = 3$ V.

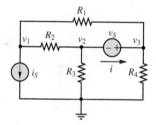

Figure 2.11(a) Circuit for Example 2.6

Solution

Given: Resistance values; current and voltage source values.

Find: The current i through the voltage source.

Analysis: Refer to Figure 2.11(a) and the steps in the Focus on Problem Solving box, "Node Analysis."

1. Select the reference node and label it.

2. Define three nonreference node voltages v_1, v_2, and v_3. Because of the voltage source, we must treat either v_2 or v_3 as a dependent variable. Observe that $v_3 = v_2 + v_S$ because the voltage source requires that the potential at node 3 be v_S higher than the potential at node 2. This expression for v_3 makes it convenient to choose v_3 as the dependent variable. This choice requires the other two node voltages (v_1 and v_2) to be treated as independent variables, with one KCL equation needed for each.

3. We apply KCL at the two nodes associated with the independent variables v_1 and v_2:

$$\frac{v_3 - v_1}{R_1} + \frac{v_2 - v_1}{R_2} - i_S = 0 \qquad \text{KCL at node 1}$$

$$\frac{v_1 - v_2}{R_2} + \frac{0 - v_2}{R_3} - i = 0 \qquad \text{KCL at node 2}$$

where $\qquad i = \dfrac{v_3 - v_1}{R_1} + \dfrac{v_3 - 0}{R_4} \qquad$ KCL at node 3

Rearranging the node 2 equation by substituting the value of i yields

$$\frac{v_1 - v_2}{R_2} + \frac{0 - v_2}{R_3} - \frac{v_3 - v_1}{R_1} - \frac{v_3 - 0}{R_4} = 0$$

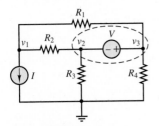

Figure 2.11(b) Circuit for Example 2.6 with "supernode"

It turns out that the somewhat cumbersome process of introducing a new variable i, applying KCL to find an expression for it in terms of node voltages, and then eliminating it immediately by substitution is not necessary. In fact, this problem illustrates that while KCL can certainly be applied at any node, it also can be applied to any closed boundary. For instance, imagine a closed boundary surrounding nodes v_2, v_3, and the voltage source, as shown in Figure 2.11(b). There are four currents that cross this boundary, namely those through the four resistors. When KCL is applied to this boundary, the result is

$$\frac{v_1 - v_2}{R_2} + \frac{0 - v_2}{R_3} + \frac{v_1 - v_3}{R_1} + \frac{0 - v_3}{R_4} = 0$$

Rearranging yields

$$\frac{v_1 - v_2}{R_2} + \frac{0 - v_2}{R_3} - \frac{v_3 - v_1}{R_1} - \frac{v_3 - 0}{R_4} = 0$$

which is identical to the node 2 equation obtained after the substitution for i. A closed boundary that surrounds two or more nodes is known as a *supernode*.

4. Finally, substitute known values for the resistors, and the current and voltage sources, collect coefficients of v_1, v_2, and v_3, and write the system of equations as:

$$-2v_1 + 1v_2 + 1v_3 = 4 \qquad \text{node 1}$$
$$12v_1 + (-9)v_2 + (-10)v_3 = 0 \qquad \text{node 2 (supernode)}$$
$$-v_2 + v_3 = 3 \qquad \text{dependent equation}$$

This system of linear equations may be solved directly on most modern calculators. However, it is also possible to simplify the equations by substituting $v_2 + 3$ V for v_3 in the node 1 and 2 equations to yield

$$-2v_1 \qquad + 2v_2 = 1$$
$$12v_1 \quad + (-19)v_2 = 30$$

This system of two equations in two unknowns can be solved analytically by multiplying the first equation by 6 and adding the result to the second equation to eliminate v_1. The result is

$$-7v_2 = 36 \qquad \text{or} \qquad v_2 = \frac{-36}{7} = -5.14 \text{ V}$$

Substitute this value into either equation to find

$$2v_1 = -72/7 - 1 \qquad \text{or} \qquad v_1 = \frac{-79}{14} = -5.64 \text{ V}$$

And

$$v_3 = v_2 + 3 = \frac{-36}{7} + 3 \qquad \text{or} \qquad v_3 = \frac{-15}{7} = -2.14 \text{ V}$$

The current through the voltage source i is

$$i = \frac{v_3 - v_1}{R_1} + \frac{v_3}{R_4} = \frac{-2.14 + 5.64}{2} + \frac{-2.14}{3} = 1.04 \text{ A}$$

Comments: Knowing all three node voltages, the current through each resistor can be computed as follows: $i_1 = |v_3 - v_1|/R_1$ (to left), $i_2 = |v_2 - v_1|/R_2$ (to left), $i_3 = |v_2|/R_3$ (upward), and $i_4 = |v_3|/R_4$ (upward).

CHECK YOUR UNDERSTANDING

Repeat the exercise of Example 2.6 when the direction of the current source is opposite that shown in Figure 2.11(a). Find the node voltages and i.

Answer: $v_1 = 5.21$ V, $v_2 = 1.71$ V, $v_3 = 4.71$ V, and $i = 1.32$ A

2.3   THE MESH CURRENT METHOD

Another method of circuit analysis employs **mesh currents**. The objective, similar to that of node analysis, is to generate one independent equation for each independent variable in a circuit. In this method, each mesh in a circuit is assigned a mesh current variable and Kirchhoff's voltage law (KVL) is applied at some or all of the meshes to generate a system of equations that relate these variables.

It is important to recall that mesh currents are not the same as branch currents. The perspective taken in the mesh current method is that there is one current circulating within each mesh and that branch currents in the circuit are comprised of these mesh currents. Specifically, when a branch is part of only one mesh, the branch current is the same as that mesh current. However, when a branch is shared by two meshes, the branch current is comprised of two mesh currents.

In the mesh current method it is necessary to assume a direction for the circulation of each mesh current. A helpful convention is to assume that *all mesh currents circulate in the clockwise (CW) direction*. With this convention, when a branch is shared by two meshes, the branch current is equal to the difference of two mesh currents. This result is illustrated in Figure 2.12 where the current through resistor R_2 is the difference of i_1 and i_2. The voltage drop across R_2 is given by Ohm's law as either:

$$(i_1 - i_2) R_2 \quad \text{or} \quad (i_2 - i_1) R_2$$

Which of these expressions is the right one to use? The answer depends upon the convention used when applying KVL. To avoid confusion when expressing Ohm's law it is helpful to always apply KVL around a mesh in the same direction (e.g., CW) used to define the mesh current. This approach is helpful because Ohm's law implies that current through a resistor is directed from high to low voltage, as shown in Figure 2.13, and that the change in voltage is proportional to the *net* current through the resistor. Thus, when KVL is applied in the same direction as the mesh current (e.g., i_1), the voltage *drop* across a resistor in that mesh will be represented in the same direction as the mesh current (see Figure 2.14) and equal to either:

$$v_1 = i_1 R_1 \tag{2.13}$$

or

$$v_2 = (i_1 - i_2) R_2 \tag{2.14}$$

Notice that the *net* current through R_2 in the direction of mesh current i_1 is $(i_1 - i_2)$. The following procedure outlines the steps taken in applying the mesh current method to a linear circuit.

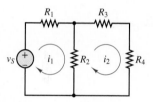

Figure 2.12 Two meshes

The current i and the passive sign convention determine the polarity of the voltage across R.

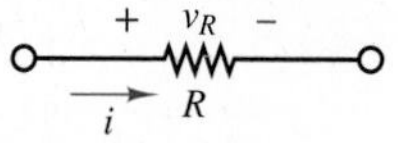

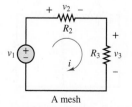

Figure 2.13 Ohm's law implies that current is directed from high (+) to low (−) potential.

KVL requires that $v_1 - v_2 - v_3 = 0$.

A mesh

Figure 2.14 Use of KVL in mesh analysis

FOCUS ON PROBLEM SOLVING

MESH ANALYSIS

1. Select a circulation convention (either CW or CCW) for the mesh currents and KVL. All the examples in this book use a CW (clockwise) convention unless there is a good reason to do otherwise.

2. Define mesh current variables $i_1, i_2, \ldots, i_n$ for each of the n meshes.
   - If the circuit contains no current sources, then all n mesh currents are treated as independent variables.
   - If the circuit contains m current sources:
     - There are $n - m$ independent variables.
     - There are m dependent variables.
     - When a current source borders only one mesh, the value of that mesh current is dictated by the current source. Treat that mesh current as a dependent variable.
     - When a current source borders two meshes, the value of the *difference* in those mesh currents is dictated by the source. Treat one of those mesh currents as a dependent variable and the other as an independent variable.

3. Apply KVL at each mesh associated with an independent variable, using Ohm's law to express each resistor voltage drop in terms of the adjacent mesh currents.
   - For each current source i_S there will be one additional dependent equation (e.g., $i_S = i_k - i_j$).

4. Collect coefficients for each of the n variables, and solve the linear system of n equations.
   - Some of the dependent equations may have the simple form $i_j = i_S$. In this case, the total number of equations and variables is reduced by direct substitution.

5. Use the known mesh currents to solve for any or all branch currents in the circuit. Any voltage drop can be found by applying Ohm's law and, when necessary, KVL.

This procedure can be used to find a solution for any planar circuit. A good approach is to first practice solving circuits without any current sources and then learn to deal with the added complexity of circuits with current sources. The remainder of this section is organized in this fashion.

Mesh 1: KVL requires
$v_S - v_1 - v_2 = 0$, where $v_1 = i_1 R_1$,
$v_2 = (i_1 - i_2)R_1$.

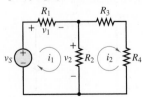

Figure 2.15 Assignment of currents and voltages around mesh 1

Details and Examples

In Figure 2.15, there are two meshes, each with a defined clockwise mesh current. There are no current sources in the circuit, so there are two independent mesh current variables i_1 and i_2. The KVL equation for mesh i_1 is:

$$v_S - i_1 R_1 - (i_1 - i_2) R_2 = 0 \qquad \text{mesh 1} \tag{2.15}$$

When KVL is applied to mesh i_2, the net current through R_2 in the direction of mesh current i_2 is $(i_2 - i_1)$. Thus, the KVL equation for mesh i_2 (see Figure 2.16) is:

$$(i_2 - i_1)R_2 + i_2R_3 + i_2R_4 = 0 \qquad \text{mesh 2} \qquad (2.16)$$

Then, collect coefficients of i_1 and i_2 in each equation to yield the following system of equations:

$$
\begin{aligned}
(R_1 + R_2)i_1 - R_2i_2 &= v_S \\
-R_2i_1 + (R_2 + R_3 + R_4)i_2 &= 0
\end{aligned} \qquad (2.17)
$$

These two equations can be solved simultaneously for the two independent mesh current variables i_1 and i_2. The branch current through R_2 can then be found as well. If the resulting numerical answer for a mesh current is negative, then the actual direction for that mesh current is opposite of the defined direction.

Note that the expressions for the voltage drop across R_2 in the two KVL equations are different because the same clockwise convention is used in both meshes for KVL. In mesh 1, the KVL loop traverses R_2 from top to bottom while in mesh 2 the KVL loop traverses R_2 from bottom to top. The result is a potential source of confusion and error when applying the mesh current method. A careful determination of the voltage drops around each mesh, one mesh at a time, and in accord with the positive sign convention for Ohm's law, is necessary for success.

Mesh 2: KVL requires

$$v_2 + v_3 + v_4 = 0$$

where

$$v_2 = (i_2 - i_1)R_2$$
$$v_3 = i_2R_3$$
$$v_4 = i_2R_4$$

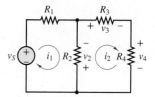

Figure 2.16 Assignment of currents and voltages around mesh 2

Examples 2.7 to 2.11 further illustrate the details of this method.

EXAMPLE 2.7   Mesh Analysis: Solving for Mesh Currents in a Circuit With Two Meshes

Problem

Find the mesh currents in the circuit of Figure 2.17.

Solution

Given: Source voltages; resistor values.

Find: Mesh currents.

Schematics, Diagrams, Circuits, and Given Data: $v_a = 10$ V; $v_b = 9$ V; $v_c = 1$ V; $R_1 = 5\,\Omega$; $R_2 = 10\,\Omega$; $R_3 = 5\,\Omega$; $R_4 = 5\,\Omega$.

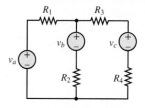

Figure 2.17 Circuit for Example 2.7

Analysis: Refer to Figures 2.17 and 2.18 and the steps in the Focus on Problem Solving box, "Mesh Analysis."

1. Select a clockwise circulation convention.

2. Note that there are two meshes in the circuit, and define clockwise mesh current variables i_1 and i_2. There are no current sources in the circuit, so both i_1 and i_2 are independent variables.

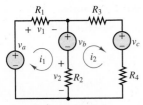

Analysis of mesh 1

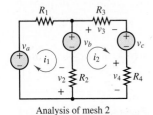

Analysis of mesh 2

Figure 2.18 Mesh analysis for Example 2.7

3. Apply KVL to each mesh associated with an independent variable, and use Ohm's law to express each resistor voltage drop in terms of the adjacent mesh currents to generate two equations.

$$v_a - R_1 i_1 - v_b - R_2(i_1 - i_2) = 0 \qquad \text{mesh 1}$$
$$-R_2(i_2 - i_1) + v_b - R_3 i_2 - v_c - R_4 i_2 = 0 \qquad \text{mesh 2}$$

4. Collect coefficients, and enter parameter values to yield the following system of linear equations:

$$15i_1 - 10i_2 = 1 \qquad \text{mesh 1}$$
$$-10i_1 + 20i_2 = 8 \qquad \text{mesh 2}$$

Multiply the mesh 1 equation by 2 and add the result to the mesh 2 equation to find i_1. Substitute for i_1 in either equation to find i_2. The results are:

$$i_1 = 0.5\,\text{A} \qquad \text{and} \qquad i_2 = 0.65\,\text{A}$$

Comments: Note that the voltage across R_2 is assigned a different polarity in the KVL expression for mesh 1 than that in the KVL expression for mesh 2. In mesh 1, R_2 is traversed top to bottom in the KVL loop; in mesh 2, R_2 is traversed bottom to top in the KVL loop.

EXAMPLE 2.8   Mesh Analysis: Finding Mesh Equations in a Circuit With Three Meshes

Problem

Find the mesh current equations for the circuit of Figure 2.19.

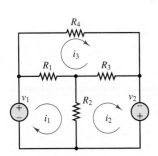

Figure 2.19 Circuit for Example 2.8

Solution

Given: Source voltages; resistor values.

Find: Mesh current equations.

Schematics, Diagrams, Circuits, and Given Data: $v_1 = 12\,\text{V}; v_2 = 6\,\text{V}; R_1 = 3\,\Omega; R_2 = 8\,\Omega; R_3 = 6\,\Omega; R_4 = 4\,\Omega.$

Analysis: Refer to Figure 2.19 and the steps in the Focus on Problem Solving box, "Mesh Analysis."

1. Select a clockwise circulation convention.

2. Note that there are three meshes in the circuit, and define clockwise mesh current variables i_1, i_2, and i_3 as shown in Figure 2.19. There are no current sources in the circuit, so i_1, i_2, and i_3 are all independent variables.

3. Apply KVL to each mesh associated with an independent variable, and use Ohm's law to express each resistor voltage drop in terms of the adjacent mesh currents. KVL applied to mesh 1 yields

$$v_1 - R_1(i_1 - i_3) - R_2(i_1 - i_2) = 0$$

KVL applied to mesh 2 yields

$$-R_2(i_2 - i_1) - R_3(i_2 - i_3) + v_2 = 0$$

while in mesh 3 KVL yields

$$-R_1(i_3 - i_1) - R_4 i_3 - R_3(i_3 - i_2) = 0$$

4. Collect coefficients and enter parameter values to obtain

$$(3 + 8)i_1 - 8i_2 - 3i_3 = 12$$
$$-8i_1 + (6 + 8)i_2 - 6i_3 = 6$$
$$-3i_1 - 6i_2 + (3 + 6 + 4)i_3 = 0$$

Check that KVL holds around any mesh to check these equations.

EXAMPLE 2.9 **Mesh Analysis: Using MatLab to Solve for Mesh Currents in a Circuit With Three Meshes**

Problem

The circuit of Figure 2.20 is a simplified DC circuit model of a three-wire electrical distribution service to residential and commercial buildings. The two ideal sources and the resistances R_4 and R_5 represent the equivalent circuit of the distribution system; R_1 and R_2 represent 110-V lighting and utility loads rated at 800 and 300 W, respectively. Resistance R_3 represents a 220-V heating load rated at 3 kW. Determine the voltages across the three loads.

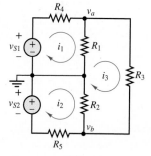

Figure 2.20 Circuit for Example 2.9

Solution

Given: The values of the voltage sources and of the resistors in the circuit of Figure 2.20 are $v_{S1} = v_{S2} = 110$ V; $R_4 = R_5 = 1.3\,\Omega$; $R_1 = 15\,\Omega$; $R_2 = 40\,\Omega$; $R_3 = 16\,\Omega$.

Find: i_1, i_2, i_3, v_a, and v_b.

Analysis: Refer to Figure 2.20 and the steps in the Focus on Problem Solving box, "Mesh Analysis."

1. Select a clockwise circulation convention.

2. Note that there are three meshes in the circuit, and define clockwise mesh current variables i_1, i_2, and i_3 as shown in Figure 2.20. There are no current sources in the circuit, so i_1, i_2, and i_3 are all independent variables.

3. Apply KVL to each mesh separately, and use Ohm's law to represent the voltage drop across each resistor in terms of the mesh currents directly.

Mesh 1: $v_{S1} - R_4 i_1 - R_1(i_1 - i_3) = 0$

Mesh 2: $v_{S2} - R_2(i_2 - i_3) - R_5 i_2 = 0$

Mesh 3: $-R_1(i_3 - i_1) - R_3 i_3 - R_2(i_3 - i_2) = 0$

4. Collect coefficients to obtain the following system of three equations in three unknown mesh currents.

$$-(R_1 + R_4)i_1 \qquad\qquad + \qquad\qquad R_1 i_3 = -v_{S1}$$
$$- (R_2 + R_5)i_2 + \qquad\qquad R_2 i_3 = -v_{S2}$$
$$R_1 i_1 + \qquad R_2 i_2 - (R_1 + R_2 + R_3)i_3 = 0$$

Enter numerical values for the parameters, and express the equations in matrix form as shown.

$$\begin{bmatrix} -16.3 & 0 & 15 \\ 0 & -41.3 & 40 \\ 15 & 40 & -71 \end{bmatrix} \begin{bmatrix} i_1 \\ i_2 \\ i_3 \end{bmatrix} = \begin{bmatrix} -110 \\ -110 \\ 0 \end{bmatrix}$$

This form can be more simply represented as the product of a resistance matrix $[R]$ and a mesh current vector $[I]$ set equal to a voltage source vector $[V]$.

$$[R][I] = [V]$$

with a solution of

$$[I] = [R]^{-1}[V]$$

The solution for the mesh current vector can be found using an analytic or numerical technique. In this problem, Matlab was used to compute the inverse $[R]^{-1}$ of the 3×3 $[R]$ matrix.

$$[R]^{-1} = \begin{bmatrix} -0.1072 & -0.0483 & -0.0499 \\ -0.0483 & -0.0750 & -0.0525 \\ -0.0499 & -0.0525 & -0.0542 \end{bmatrix}$$

The value of each mesh current is now determined.

$$[I] = [R]^{-1}[V] = \begin{bmatrix} -0.1072 & -0.0483 & -0.0499 \\ -0.0483 & -0.0750 & -0.0525 \\ -0.0499 & -0.0525 & -0.0542 \end{bmatrix} \begin{bmatrix} -110 \\ -110 \\ 0 \end{bmatrix} = \begin{bmatrix} 17.11 \\ 13.57 \\ 11.26 \end{bmatrix}$$

Therefore, we find

$$i_1 = 17.11 \text{ A} \qquad i_2 = 13.57 \text{ A} \qquad i_3 = 11.26 \text{ A}$$

The two unknown node voltages v_a and v_b are easily calculated using Ohm's law and the mesh currents. Notice the positive sign convention used in the following calculations!

$$v_a - 0 = R_1(i_1 - i_3)$$
$$v_a = 87.75 \text{ V}$$

$$v_b - 0 = R_2(i_3 - i_2)$$
$$v_b = -92.40 \text{ V}$$

The values of the node voltages v_a and v_b are relative to the reference node. Verify that KVL holds for each mesh to check your understanding.

Comments: The inverse matrix computation is numerically inefficient compared to the Matlab left division computation used in Example 2.5. However, for most problems solved by a typical personal computer, the difference in computation time will not be noticed by a user.

CHECK YOUR UNDERSTANDING

Use mesh analysis to find the unknown voltage v_x in the circuit on the left.

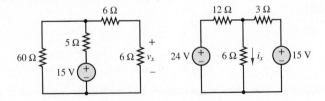

Use mesh analysis to find the unknown current i_x in the circuit on the right.

CHECK YOUR UNDERSTANDING

Repeat the exercise of Example 2.9, using node voltage analysis instead of mesh analysis.

Mesh Analysis With Current Sources

The circuits in the preceding examples contained no current sources. However, it is common, in practice, to encounter current sources in circuits. The relevant steps found in the Highlights section are listed below with added comments.

Step 1: *Select a circulation convention (either CW or CCW) for the mesh currents and KVL.*

Step 2: *Define mesh current variables $i_1, i_2, \ldots i_n$ for each of the n meshes. If the circuit contains m current sources:*

- *There are $n - m$ independent variables.*
- *There are m dependent variables.*
- *When a current source borders only one mesh, the value of that mesh current is dictated by the current source. Treat that mesh current as a dependent variable.*
- *When a current source borders two meshes, the value of the* difference *in those mesh currents is dictated by the source. Treat one of those mesh currents as a dependent variable and the other as an independent variable.*

The circuit in Figure 2.21 has two meshes and one current source. Thus, there is one independent mesh current variable i_1 and one dependent mesh current variable i_2. Note that the circulation of i_1 indicates a clockwise convention. With that convention, i_2 has the opposite direction of i_S in the rightmost branch. Thus, $i_2 = -i_S$ and Figure 2.21 shows the second mesh current as i_S circulating counterclockwise.

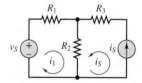

Figure 2.21 Mesh analysis with current sources

Step 3: *Apply KVL at each mesh associated with an independent variable, using Ohm's law to express each resistor voltage drop in terms of the adjacent mesh currents.*

- *For each current source i_S there will be one additional dependent equation (e.g., $i_S = i_k - i_j$).*

The one dependent equation is, of course:

$$i_2 = -i_S$$

KVL around the i_1 mesh yields:

$$v_S - R_1 i_1 - R_2(i_1 + i_S) = 0$$

or $\quad (R_1 + R_2)i_1 = v_S - R_2 i_S$ (2.18)

Step 4: *Collect coefficients for each of the n variables, and solve the linear system of n equations.*

- *Some of the dependent equations may have the simple form $i_j = i_S$. In this case, the total number of equations and variables is reduced by direct substitution.*

The presence of the current source has simplified the problem. There is only one unknown mesh current, i_1, and one equation.

$$i_1 = \frac{v_S - R_2 i_S}{R_1 + R_2} \tag{2.19}$$

Step 5: *Use the known mesh currents to solve for any branch current in the circuit. Any voltage drop can be found by applying Ohm's law and, when necessary, KVL.*

By inspection, the current through R_1 is i_1 and the current through R_3 is i_S. The current through R_2 is $i_1 + i_S$. The change in voltage across the current source, with respect to the reference node, is given by KVL as:

$$i_S R_3 + (i_1 + i_S) R_2 \tag{2.20}$$

EXAMPLE 2.10   Mesh Analysis: Three Meshes and One Current Source

Problem

Find the mesh currents in the circuit of Figure 2.22.

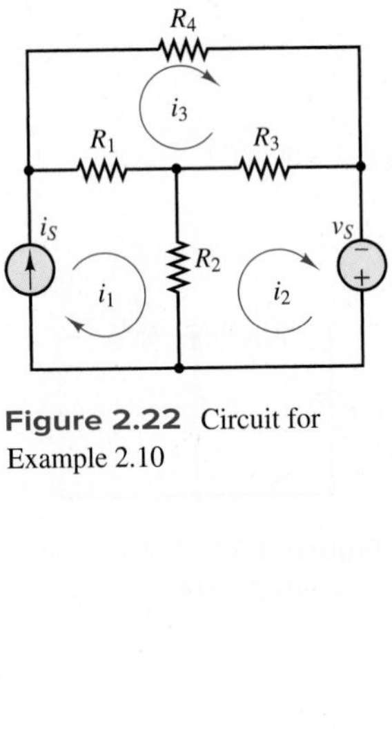

Solution

Given: Source and resistor values.

Find: Mesh currents.

Schematics, Diagrams, Circuits, and Given Data: $i_S = 0.5$ A; $v_S = 6$ V; $R_1 = 3\ \Omega$; $R_2 = 8\ \Omega$; $R_3 = 6\ \Omega$; $R_4 = 4\ \Omega$.

Analysis: Refer to Figure 2.22 and the steps in the Focus on Problem Solving box, "Mesh Analysis."

1. Select a clockwise circulation convention.
2. Define three mesh current variables i_1, i_2, and i_3. The current source is not shared by two meshes; thus, mesh current i_1 is a dependent variable and mesh currents i_2 and i_3 are the independent variables.
3. Apply KVL at each mesh associated with an independent variable, using Ohm's law to express each resistor voltage drop in terms of the adjacent mesh currents.
   - *For each current source there will be one additional dependent equation.*

   By inspection, the one dependent equation is:

   $$i_1 = i_S = 0.5\,\text{A}$$

   KVL around meshes 2 and 3 yield:

   $$-R_2(i_2 - i_1) - R_3(i_2 - i_3) + v_S = 0 \qquad \text{mesh 2}$$
   $$-R_1(i_3 - i_1) - R_4 i_3 - R_3(i_3 - i_2) = 0 \qquad \text{mesh 3}$$

4. Collect coefficients for each of the n variables, and solve the linear system of n equations.
   - *Some of the dependent equations may have the simple form $i_j = i_S$. In this case, the total number of equations and variables is reduced by direct substitution.*

   $$14 i_2 - 6 i_3 = 10$$
   $$-6 i_2 + 13 i_3 = 1.5$$

Figure 2.22 Circuit for Example 2.10

The presence of the current source has simplified the problem. There are only two unknown mesh currents, i_2 and i_3, and two equations. These equations can be solved to obtain

$$i_2 = 0.95 \, \text{A} \qquad i_3 = 0.55 \, \text{A}$$

Check these answers by verifying that KVL holds.

EXAMPLE 2.11   Mesh Analysis: Two Meshes and One Current Source

Problem

Find the unknown voltage v_x in the circuit of Figure 2.23.

Solution

Given: The values of the voltage sources and of the resistors in the circuit of Figure 2.23: $v_S = 10 \, \text{V}$; $i_S = 2 \, \text{A}$; $R_1 = 5 \, \Omega$; $R_2 = 2 \, \Omega$; and $R_3 = 4 \, \Omega$.

Find: v_x.

Analysis: Refer to Figure 2.23 and the steps in the Focus on Problem Solving box, "Mesh Analysis."

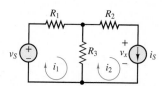

Figure 2.23 Illustration of mesh analysis in the presence of current sources

1. Select a clockwise circulation convention.
2. Define two mesh current variables i_1 and i_2. The current source borders mesh 2 only; thus, i_2 is a dependent variable and i_1 is an independent variable.
3. Apply KVL at each mesh associated with an independent variable, using Ohm's law to express each resistor voltage drop in terms of the adjacent mesh currents.
   - For each current source there will be one additional dependent equation.

   By inspection, the one dependent equation is:

   $$i_2 = i_S$$

   KVL around mesh 1 yields:

   $$v_S - i_1 R_1 - (i_1 - i_2) R_3 = 0$$

4. Collect coefficients for each of the n variables, and solve the linear system of n equations.
   - *Some of the dependent equations may have the simple form $i_j = i_S$. In this case, the total number of equations and variables is reduced by direct substitution.*

   The presence of the current source has simplified the problem. There is only one unknown mesh current, i_1, and one equation. Substitute $i_2 = i_S$ into the mesh 1 equation to find:

   $$i_1 = \frac{v_S + i_S R_3}{R_1 + R_3} = \frac{10 + 2 \times 4}{5 + 4} = 2 \, \text{A}$$

5. Use the known mesh currents to solve for any or all branch currents in the circuit. Any voltage drop can be found by applying Ohm's law and, when necessary, KVL. Apply KVL around mesh 2 to find:

   $$-(i_2 - i_1) R_3 - i_2 R_2 - v_x = 0$$

   or

   $$v_x = i_1 R_3 - i_2 (R_2 + R_3) = (2 \times 4) - [2 \times (2 + 4)] = -4 \, \text{V}$$

Comments: The presence of the current source reduced the number of unknown mesh currents by one. Thus, we were able to find v_x without solving simultaneous equations.

CHECK YOUR UNDERSTANDING

Use the mesh currents to find the branch currents in Example 2.10. Apply KCL at each node to validate.

CHECK YOUR UNDERSTANDING

Find the value of the current i_1 in Example 2.11 when the value of the current source is 1 A.

Answer: 1.56 A

2.4   THE PRINCIPLE OF SUPERPOSITION

The *principle of superposition* is a valid, and frequently used, analytic tool for any linear circuit. It is also a powerful conceptual aid for understanding the behavior of circuits with multiple sources.

For any linear circuit, the principle of superposition states that each *independent* source contributes to each voltage and current present in the circuit. Moreover, the contributions of one source are independent of those from the other sources. In this way, each voltage and each current in a circuit of N independent sources is the sum of N component voltages and N component currents, respectively.

As a problem-solving tool, the principle of superposition permits a problem to be decomposed into two or more simpler problems. The efficiency of this "divide and conquer" tactic depends upon the particular problem being solved. However, it may enable a simple closed-form solution of an otherwise complicated symbolic circuit problem, where node and mesh analyses may offer little help.

The method is to turn off (set to zero) all independent sources but one, and then solve for voltages and currents due to the lone remaining independent source. This procedure may be repeated successively for each source until the contributions due to all the sources have been computed. The components for a particular voltage or current can be summed to find its value in the original complete circuit.

1. When a voltage source equals to zero, replace it with a short-circuit.

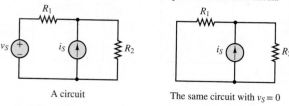

A circuit   The same circuit with $v_S = 0$

2. When a current source equals to zero, replace it with an open-circuit.

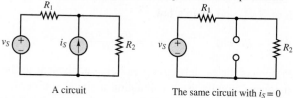

A circuit   The same circuit with $i_S = 0$

Figure 2.24 Zeroing voltage and current sources

A zero voltage source is equivalent to a short-circuit and a zero current source is equivalent to an open-circuit. When using the principle of superposition, it is necessary, and helpful, to replace each zero source with its equivalent short- or open-circuit and thus simplify the circuit. These substitutions are summarized in Figure 2.24.

Superposition may be applied to circuits containing *dependent* sources; however, the dependent sources must not be set to zero. They are not independent sources and must not be treated as such. To do so, would lead to an incorrect result.

FOCUS ON PROBLEM SOLVING

SUPERPOSITION

1. Define the voltage V or current I to be solved in the circuit.

2. For each of the N sources, define a component voltage v_k or current i_k such that

$$V = v_1 + v_2 + \cdots + v_N \qquad \text{or} \qquad I = i_1 + i_2 + \cdots + i_N$$

3. Turn off all sources except source S_k and solve for the component voltage v_k or current i_k. Find components for all k where $k = 1, 2, \ldots, N$.

4. Find the complete solution for the voltage V or current I by summing all of the components as defined in step 2.

Details and Examples

An elementary application of the principle is to find the current in a single loop with two sources connected in series, as shown in Figure 2.25.

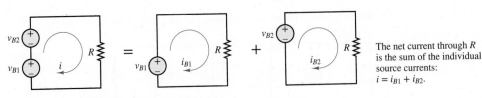

Figure 2.25 The principle of superposition

The current in the far left circuit of Figure 2.25 is easily found by a direct application of KVL and Ohm's law.

$$v_{B1} + v_{B2} - iR = 0 \quad \text{or} \quad i = \frac{v_{B1} + v_{B2}}{R} \tag{2.21}$$

Figure 2.25 depicts the far left circuit as being equivalent to the combined effects of two component circuits, each containing a single source. In each of these two circuits, one battery (which is a DC voltage source) has been set to zero and replaced with a short-circuit.

KVL and Ohm's law can be applied directly to each of these component circuits.

$$i_{B1} = \frac{v_{B1}}{R} \quad \text{and} \quad i_{B2} = \frac{v_{B2}}{R} \tag{2.22}$$

According to the principle of superposition

$$i = i_{B1} + i_{B2} = \frac{v_{B1}}{R} + \frac{v_{B2}}{R} = \frac{v_{B1} + v_{B2}}{R} \tag{2.23}$$

Voila! The complete solution is found, as expected. This simple example illustrates the essential method; however, more challenging examples are needed to reinforce it.

Examples 2.14 and 2.15 further illustrate the details of this method.

EXAMPLE 2.12   Principle of Superposition

Problem

Determine the current i_2 in the circuit of Figure 2.26(a), using the principle of superposition.

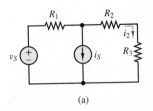

(a)

Figure 2.26(a) Circuit for the illustration of the principle of superposition

Solution

Known Quantities: Voltage and current values of each source; resistor values.

Find: Unknown current i_2.

Given Data: $v_S = 10$ V; $i_S = 2$ A; $R_1 = 5$ Ω; $R_2 = 2$ Ω; $R_3 = 4$ Ω.

Analysis: Refer to Figure 2.26(a) and the steps in the Focus on Problem Solving box, "Superposition."

1. The objective is to find current i_2.

2. There are two independent sources in the circuit, so there will be two components of i_2.

$$i_2 = i_2' + i_2''$$

3. *Part 1:* Turn off the current source and replace it with an open-circuit. The resulting circuit is a simple series loop shown in Figure 2.26(b). Here, i_2' is the same as the loop current because of the open-circuit. The total series resistance is $5 + 2 + 4 = 11\ \Omega$, such that $i_2' = 10\ \text{V}/11\ \Omega = 0.909\ \text{A}$.

   Part 2: Turn off the voltage source and replace it with a short-circuit. The resulting circuit consists of three parallel branches, as shown in Figure 2.26(c): i_S, R_1, and $R_2 + R_3$. By current division, we find

$$i_2'' = (-i_S)\frac{R_1}{R_1 + R_2 + R_3} = (-2\,\text{A})\frac{5}{5 + 2 + 4} = -0.909\ \text{A}$$

4. The complete i_2 is found to be

$$i_2 = i_2' + i_2'' = 0.909\ \text{A} - 0.909\ \text{A} = 0\ \text{A}$$

Comments: Superposition is not always a very efficient tool. Beginners may find it preferable to rely on more systematic methods, such as node analysis, to solve circuits. Eventually, experience will suggest the preferred method for any given circuit.

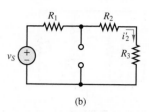

Figure 2.26(b) Circuit with current source set to zero

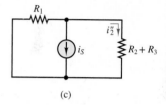

Figure 2.26(c) Circuit with voltage source set to zero

EXAMPLE 2.13   Principle of Superposition

Problem

Determine the voltage v_R across resistor R in the circuit of Figure 2.27(a).

Solution

Known Quantities: The values of the sources and resistors in the circuit of Figure 2.27(a) are $i_B = 12\ \text{A}$; $v_G = 12\ \text{V}$; $R_B = 1\ \Omega$; $R_G = 0.3\ \Omega$; $R = 0.23\ \Omega$.

Find: v_R.

Analysis: Refer to Figure 2.27(a) and the steps in the Focus on Problem Solving box, "Superposition."

1. The objective is to find voltage v_R.

2. There are two independent sources in the circuit, so there will be two components of v_R.

$$v_R = v_R' + v_R''$$

3. *Part 1:* Turn off the voltage source and replace it with a short-circuit. Redraw the circuit as shown in Figure 2.27(b), find the equivalent resistance of all three resistors in parallel, and apply Ohm's law to find v_R'.

$$R_{\text{eq}} = (R_B \| R_G \| R) = \frac{1}{1/R_B + 1/R_G + 1/R} = \frac{1}{1/1 + 1/0.3 + 1/0.23}$$

$$= \frac{(0.3)(0.23)}{(0.3)(0.23) + 0.23 + 0.3} = \frac{0.069}{0.599}\ \Omega$$

$$v_R' = i_B R_{\text{eq}} = (12\ \text{A})\frac{0.069}{0.599} = 1.38\ \text{V}$$

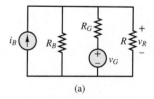

Figure 2.27(a) Circuit used to demonstrate the principle of superposition

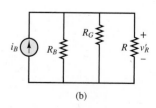

Figure 2.27(b) Circuit obtained by suppressing the voltage source

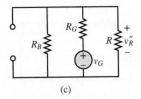

(c)

Figure 2.27(c) Circuit obtained by suppressing the current source

Part 2: Turn off the current source and replace it with an open-circuit. Redraw the circuit, as shown in Figure 2.27(c), and apply KCL at the upper node:

$$-\frac{v_R''}{R_B} - \frac{v_R'' - v_G}{R_G} - \frac{v_R''}{R} = -v_R''\left[\frac{1}{R_B} + \frac{1}{R_G} + \frac{1}{R}\right] + \frac{v_G}{R_G} = 0$$

$$v_R'' = \frac{v_G}{R_G}\frac{1}{1/R_B + 1/R_G + 1/R} = \frac{12}{0.3}\frac{1}{1/1 + 1/0.3 + 1/0.23} = 4.61 \text{ V}$$

This same result can be found by finding the equivalent resistance of R_B in parallel with R and applying voltage division.

$$R_{eq} = R_B \| R = \frac{R_B R}{R_B + R} = \frac{0.23}{1.23} \approx 0.187 \ \Omega$$

$$v_R'' = v_G\frac{R_{eq}}{R_{eq} + R_G} = (12 \text{ V})\frac{0.23}{0.23 + (1.23)(0.3)} = 4.61 \text{ V}$$

4. Compute the voltage across R as the sum of the two component voltages:

$$v_R = v_R' + v_R'' = 5.99 \text{ V}$$

Comments: The only advantage offered by the principle of superposition in this problem is that it clearly reveals the contributions to v_R from each source. However, the work required to solve this problem is nearly doubled. The voltage across R can easily be determined by applying KCL at the upper node.

CHECK YOUR UNDERSTANDING

In Example 2.12, verify that the same answer is obtained by mesh or node analysis.

CHECK YOUR UNDERSTANDING

In Example 2.13, verify that the same answer can be obtained by a single application of KCL.

CHECK YOUR UNDERSTANDING

Find the voltages v_a and v_b for the circuits of Example 2.3 by superposition.

CHECK YOUR UNDERSTANDING

Solve Example 2.7, using superposition.

CHECK YOUR UNDERSTANDING

Solve Example 2.10, using superposition.

2.5   EQUIVALENT NETWORKS

Recall from the discussion of ideal sources in Chapter 1 that a circuit can be thought of as having two parts, a *source* and a *load*, each attached to the other at two terminals such that power flows from the source to the load. It is important to note that the term *source* used in this context (and throughout this chapter) is a generalization of the source terms (i.e., *ideal, practical, dependent,* and *independent*) defined in Chapter 1. Figure 1.1 shows two circuit diagrams in which each circuit is divided into two parts, a *source* and a *load*. Essentially, the load is that circuit portion of interest to the analyst. Everything else is, by default, the source for that load. A simple example and an abstract representation of this perspective are shown in Figure 1.1(a) and (b), respectively. The small circles along the upper and lower wires in these figures represent connection points between the two *one-port* networks, the source and load. Figure 1.1(a) shows a simple practical example of an automotive battery as the source attached to a headlight as the load. Such figures vary in the detail with which the source and load are represented. Nonetheless, the impact of the source on the load is completely determined by the *i-v* characteristic of the source. If the *i-v* characteristic is the same for two sources, then the sources are said to be *electrically equivalent*. The details inside each source do not matter so far as the load is concerned.

A network (a circuit or part of a circuit) that has two, and only two, terminals at which it can attach to other networks, as in Figure 2.28, is known as a **one-port network**. Such a network is characterized by the relationship between the current *i* through and voltage *v* across its terminals for various loads (e.g., open-circuit, short-circuit). The key concepts are:

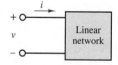

Figure 2.28 One-port network

- The impact of a one-port source on a one-port load is completely represented by the *i-v characteristic* of the source.

- Two one-port networks are *electrically equivalent* if their *i-v* characteristics are equivalent.

- Equivalent networks are those for which the voltage across and current through their terminals are the same *for any load.*

This concept of *equivalence* was introduced for networks of resistors in Chapter 1. The central idea was that an entire network of resistors can be replaced by a single equivalent resistor. Here, the concept of equivalence is generalized to networks that include resistors, ideal sources, and other linear circuit elements.

Thévenin and Norton Equivalent Circuits

Any one-port linear network, no matter how complicated, can always be represented by either of two simple equivalent networks, and the transformations leading to these equivalent representations are easily managed, with a little practice. In this section, techniques are presented for computing these equivalent networks, which reveal some simple—yet general—results for linear networks, and are useful for analyzing basic nonlinear circuits.

H I G H L I G H T S

Any one-port linear network can be represented by either of two simple equivalent network forms. They are:

- A *Thévenin source*, comprised of an independent voltage source v_T in series with a resistor R_T, as shown in Figure 2.29.
- A *Norton source*, comprised of an independent current source i_N in parallel with a resistor R_N, as shown in Figure 2.30.

Moreover, since each of these equivalent network forms is equivalent to the original linear network, the forms themselves must be equivalent to each other. As a result, a Thévenin source is interchangeable with its equivalent Norton source, which leads to a solution technique known as *source transformation*.

The equivalent network of any specific one-port linear network is comprised of specific values for v_T and R_T, or i_N and R_N, which are known as:

- Thévenin voltage v_T.
- Thévenin equivalent resistance R_T.
- Norton current i_N.
- Norton equivalent resistance R_N.

In addition, for any specific linear source network $R_T = R_N$ and $v_T = i_N R_T$. Furthermore, the specific values of v_T and i_N are the open-circuit voltage V_{OC} across and the short-circuit current I_{SC} through, respectively, the source network terminals. Read on to learn how to compute their values for various cases.

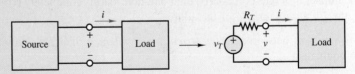

Figure 2.29 Illustration of Thévenin theorem

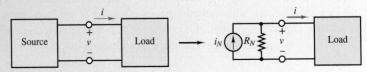

Figure 2.30 Illustration of Norton theorem

Linearity, Thévenin's Theorem, and Norton's Theorem

In general, the criteria for a linear function are:

Superposition: If $y_1 = f(x_1)$ and $y_2 = f(x_2)$, then $y_1 + y_2 = f(x_1 + x_2)$.

Homogeneity: If $y = f(x)$, then $\alpha y = f(\alpha x)$.

where x is the function input and y is the function output.

Linear networks obey the same rules. Superposition implies that each source (e.g., x_1 and x_2) makes its own independent contribution (e.g., y_1 and y_2) to each current and voltage in a network, and that the total value of each current and voltage is the sum (e.g., $y_1 + y_2$) of these contributions.

Homogeneity implies that the contribution due to any one source scales linearly with the value of the source. For example, if the contribution due to source x_1 is y_1, then the contribution due to the same source doubled $2x_1$ is also doubled $2y_1$, where $\alpha = 2$ is an example scaling factor.

In general, to determine if a network is linear, it is necessary to verify that these two criteria are satisfied for all possible inputs, or at least to verify a range of inputs within which the network is linear. Luckily, it is not always necessary to verify superposition and homogeneity directly. A sufficient, but not necessary, alternative condition is:

Any network composed of linear elements only is itself linear. Common linear elements are ideal sources, resistors, capacitors, and inductors.

The essence of Section 2.5 is captured in the statement of two very important theorems about linear networks.

Thévenin's Theorem

When viewed from its terminals, any linear one-port network may be represented by an equivalent circuit consisting of an ideal voltage source v_T in series with an equivalent resistance R_T.

Norton's Theorem

When viewed from its terminals, any linear one-port network may be represented by an equivalent circuit consisting of an ideal current source i_N in parallel with an equivalent resistance R_N.

The next few sections illustrate how to compute R_T (and its equivalent R_N), v_T, and i_N. The only way to master the computation of Thévenin and Norton equivalent circuits is by patient repetition.

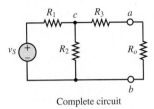

Complete circuit

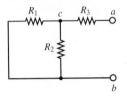

Circuit with load removed
for computation of R_T. The voltage
source is replaced by a short-circuit.

Figure 2.31 Computation of
Thévenin resistance

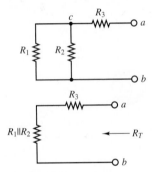

Figure 2.32 Equivalent
resistance seen by the load

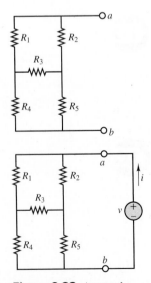

Figure 2.33 A general
method of determining the
Thévenin resistance

Computation of R_T or R_N: Networks Without Dependent Sources

The first step to calculate the Thévenin (or Norton) equivalent resistance of a one-port linear source network with no dependent sources is to identify the two terminals (e.g., a and b) of the source network. Sometimes just the one-port source network is given in a problem, in which case the network terminals should be readily apparent. Other times a complete circuit is given such that it is necessary to define and/or identify the load and, by default, the source network. In Figure 2.31, the resistor R_o is chosen as the load such that terminals a and b define the one-port (two-terminal) connection between the load and the source network.

The second step is to remove the load and set all independent sources in the source network to zero; that is, replace all independent voltage sources with short-circuits and all independent current sources with open-circuits. The source network of Figure 2.31 is shown with the voltage source replaced by a short-circuit.

Finally, apply series and parallel equivalent resistance substitutions to find the effective equivalent resistance "seen" by the load R_o across terminals a and b. For example, in the circuit of Figure 2.32, R_1 and R_2 are in parallel since they are connected between the same two nodes, b and c. The total resistance between terminals a and b is simply:

$$R_T = R_3 + R_1 \| R_2 \tag{2.24}$$

When series and parallel equivalent resistance substitutions are not sufficient, find R_T by attaching an independent voltage source v to the source network terminals, finding the current i through the voltage source, and computing v/i, which is, by definition, the equivalent resistance. For example, consider the resistor network between terminals a and b shown in Figure 2.33. Notice that there are no series nor parallel connections in this network. To determine R_T a voltage source v attached across terminals a and b results in a current i through the voltage source. Then, R_T is simply:

$$R_T = \frac{v}{i} \tag{2.25}$$

Figure 2.33 shows a particularly well known network in which the resistors are neither in series nor in parallel.

Computation of R_T or R_N: Networks With Dependent Sources

When a dependent source is present in the source network, it is not possible to calculate the Thévenin equivalent resistance R_T directly after setting *independent* sources to zero. The result of that method would be a network of resistors with one or more *dependent* sources still present. Dependent sources must not be set to zero. Instead, as above, one must rely on the definition of equivalent resistance:

$$R_{eq} \equiv \frac{v}{i} \tag{2,26}$$

As depicted in Figure 2.33, the equivalent resistance between two terminals of an arbitrary resistive network can be found by attaching a voltage source v across those terminals, solving for the current i, and computing the ratio v/i. This approach also applies for source networks containing resistors and dependent sources. Thus, *when the source network contains a dependent source*, follow the steps below to find R_T.

Step 1: *Set independent sources to zero. Replace them with short- and open-circuits.*

Step 2: *Attach an independent voltage source v_S across the source network terminals.*

Step 3: *Compute the current i_S through the voltage source, as in Figure 2.33.*

Step 4: *Compute $R_T = v_S/i_S$.*

If the original source network has no independent sources, this method will still work. Finally, to apply Thévenin's or Norton's theorem to circuits containing dependent sources, the following rule must be obeyed.

> Each dependent source and its associated dependent variable must be collocated in either the source network or the load when applying Thévenin's or Norton's theorem.

FOCUS ON PROBLEM SOLVING

THÉVENIN RESISTANCE

Use the following steps to compute the Thévenin equivalent resistance of source networks.

1. Identify the source network and label its terminals *a* and *b*.

2. Turn off all independent voltage and current sources in the source network, and replace them with short- and open-circuits, respectively.

3. For source networks without dependent sources:

   (a) Use series and parallel equivalent resistance substitutions to find R_T.

   (b) When series and parallel equivalent resistance substitutions are not sufficient, find R_T by attaching a voltage source v_S to the terminals to produce a current i_S through those terminals. Then, $R_T = v_S/i_S$.

   For source networks with dependent sources:

   (a) Attach an arbitrary independent voltage source v_S across the source network terminals.

   (b) Compute the resulting current i_S through the voltage source, as in Figure 2.33.

   (c) Compute $R_T = v_S/i_S$.

When a dependent source is present in the source network, its associated dependent variable must also be part of the source network.

The source network and its equivalents are independent of the load and, thus, are valid for any load. Also, the Thévenin and Norton equivalent resistances are always equivalent to each other:

$$\boxed{R_T = R_N}$$

(2.27)

As a result, often only the R_T notation is used.

Examples 2.14 to 2.16 further illustrate the procedure.

EXAMPLE 2.14   Computing R_T for a Network Without a Dependent Source

Problem

Find the Thévenin equivalent resistance "seen" by the load R_o in the circuit of Figure 2.34.

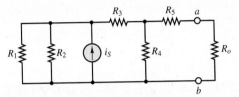

Figure 2.34 Circuit for Example 2.14

Solution

Known Quantities: Resistor and current source values.

Find: The Thévenin equivalent resistance R_T.

Schematics, Diagrams, Circuits, and Given Data: $R_1 = 20\ \Omega$; $R_2 = 20\ \Omega$; $i_S = 5$ A; $R_3 = 10\ \Omega$; $R_4 = 20\ \Omega$; $R_5 = 10\ \Omega$.

Analysis: Refer to Figure 2.34 and the steps in the Focus on Problem Solving box, "Thévenin Resistance."

1. The source network is everything except the load R_o; remove it. The source network terminals are marked a and b.

2. Turn off the current source and replace it with an open-circuit. The result is shown in Figure 2.35.

3. There are four nodes remaining in the source network and no dependent source. Clearly, R_1 and R_2 are in parallel since they sit between nodes d and b. Their parallel equivalent resistance is in series with R_3. Thus, there are two parallel resistances from $c \rightarrow b$: $R_3 + (R_1 \parallel R_2)$ and R_4. Finally, the equivalent resistance from $a \rightarrow c \rightarrow b$ is:

$$R_T = R_5 + \left\{ \left[(R_1 \parallel R_2) + R_3 \right] \parallel R_4 \right\}$$

$$= 10 + \left\{ \left[(20 \parallel 20) + 10 \right] \parallel 20 \right\} = 20\ \Omega$$

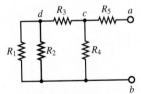

Figure 2.35 Equivalent circuit with current source set to zero

Comments: The network in this example is drawn in an uncomplicated rectangular manner. However, it is always possible to draw the same network in a more confusing manner. In any case, it is easy to correctly calculate the equivalent resistance of the network by focusing on the network as a collection of nodes, between which sit various resistances, and applying the rules for equivalent parallel and series resistances.

EXAMPLE 2.15   Computing R_T for a Network Without a Dependent Source

Problem

Compute the Thévenin equivalent resistance seen by the load R_o in Figure 2.36.

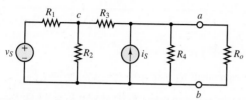

Figure 2.36 Circuit for Example 2.15

Solution

Known Quantities: Resistor values.

Find: The Thévenin equivalent resistance R_T.

Schematics, Diagrams, Circuits, and Given Data: $v_S = 5$ V; $R_1 = 2$ Ω; $R_2 = 2$ Ω; $R_3 = 1$ Ω; $i_S = 1$ A; $R_4 = 2$ Ω.

Analysis: Refer to Figure 2.36 and the steps in the Focus on Problem Solving box, "Thévenin Resistance."

1. The source network is everything except the load R_o; remove it. The source network terminals are marked a and b.

2. Turn off the voltage and current sources and replace them with a short- and open-circuit, respectively. The result is shown in Figure 2.37.

3. There are four nodes remaining in the source network and no dependent source. Clearly, R_1 and R_2 are in parallel since they sit between nodes c and b. Their parallel equivalent resistance is in series with R_3. Thus, there are two parallel resistances from $a \rightarrow b$: $R_3 + (R_1 \| R_2)$ and R_4. Finally, the equivalent resistance from $a \rightarrow b$ is:

$$R_T = \{ [(R_1 \| R_2) + R_3] \| R_4 \}$$

$$= \{ [(2 \| 2) + 1] \| 2 \} = 1 \ \Omega$$

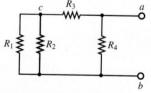

Figure 2.37 Circuit modified to compute equivalent resistance

Comments: Note the similarity of Figures 2.35 and 2.37. Could you have anticipated these similar resistive networks simply by observing Figures 2.34 and 2.36? Try it.

EXAMPLE 2.16   Computing R_T for a Network With a Dependent Source

Problem

Compute the Thévenin equivalent resistance seen by the load R_o in Figure 2.38.

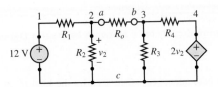

Figure 2.38 Circuit for Example 2.16

Solution

Known Quantities: Source and resistor values.

Find: The Thévenin equivalent resistance R_T seen by the load R_o.

Schematics, Diagrams, Circuits, and Given Data: $R_1 = 24 \text{ k}\Omega$; $R_2 = 8 \text{ k}\Omega$; $R_3 = 9 \text{ k}\Omega$; $R_4 = 18 \text{ k}\Omega$.

Analysis: Refer to Figure 2.38 and the steps in the Focus on Problem Solving box, "Thévenin Resistance."

1. The source network is everything except the load R_o; remove it. The source network terminals are marked a and b.
2. Turn off the independent voltage source in Figure 2.38 and replace it with a short-circuit. As a result, R_1 and R_2 are in parallel and can be replaced by a single equivalent resistance.
3. The source network contains a dependent source. Attach an arbitrary independent voltage source v_S across terminals a and b, and label its current i_S as shown in Figure 2.39. Included in this figure are two mesh currents i_1 and i_2 that can be used to solve the circuit using mesh analysis.

The counterclockwise circulation was chosen so that $i_1 = i_S$. Apply KVL around each mesh:

$$v_S - (R_1 \| R_2)i_1 - R_3(i_1 - i_2) = 0 \qquad \text{mesh 1}$$
$$2v_2 - R_4 i_2 - R_3(i_2 - i_1) = 0 \qquad \text{mesh 2}$$

Note that $v_2 = i_1(R_1 \| R_2)$ such that the equations can be rewritten as:

$$v_S - (R_1 \| R_2)i_1 - R_3(i_1 - i_2) = 0 \qquad \text{mesh 1}$$
$$2(R_1 \| R_2)i_1 - R_4 i_2 - R_3(i_2 - i_1) = 0 \qquad \text{mesh 2}$$

Collect coefficients of i_1 and i_2 and substitute values for the resistors.

$$15i_1 - 9i_2 = v_S \qquad \text{mesh 1}$$
$$21i_1 - 27i_2 = 0 \qquad \text{mesh 2}$$

Divide both sides of the mesh 2 equation by 3 and subtract the result from the mesh 1 equation.

$$8i_1 = v_S \qquad \text{or} \qquad \frac{v_S}{i_1} = R_T = 8 \text{ k}\Omega$$

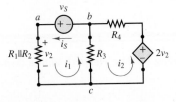

Figure 2.39 Reduced circuit for Example 2.16

Comments: This result can be computed by an alternate method. First, remove the load and compute the open-circuit voltage V_{OC} across terminals a and b. Second, connect terminals a and b with a wire and compute the short-circuit current I_{SC} through that wire. Last, compute R_T from its definition (see the section highlights):

$$R_T \equiv \frac{v_{OC}}{i_{SC}}$$

Try it. Does it work?

Computing the Thévenin Voltage

This section describes the computation of the Thévenin voltage v_T for an arbitrary linear network containing sources, both independent and dependent, and linear resistors. The *Thévenin voltage* is defined as follows:

> The Thévenin voltage v_T is equal to the **open-circuit voltage** v_{OC} across the source network terminals.

To compute v_T, it is sufficient to remove the load from the source network and compute the open-circuit voltage across the source network terminals. Figure 2.40 illustrates that the open-circuit voltage v_{OC} and the Thévenin voltage v_T are the same. That is, when terminals a-b are open, the current i through R_T is zero, thus, the voltage across R_T is also zero. KVL around the network gives

$$v_T = iR_T + v_{OC} = v_{OC} \tag{2.28}$$

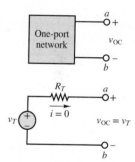

Figure 2.40 Equivalence of open-circuit and Thévenin voltage

FOCUS ON PROBLEM SOLVING

THÉVENIN VOLTAGE

Follow the steps below to compute the Thévenin voltage for a source network.

1. Identify the source network and label its terminals (e.g., a and b).
2. Define the open-circuit voltage v_{OC} across those terminals.
3. (a) For source networks with at least one independent source: Apply any preferred method (e.g., node analysis) to solve for v_{OC}.

   (b) For source networks without an independent source: The open-circuit voltage v_{OC} is simply zero even when a dependent source is present.
4. The Thévenin equivalent voltage v_T of the source network is, by definition, v_{OC}.

The actual computation of the open-circuit voltage is best illustrated by examples. As shown again in Figure 2.41, the equivalent resistance from $a \rightarrow c \rightarrow b$ is given by $R_T = R_3 + R_1 \| R_2$.

To compute v_{OC}, remove the load R_o, as shown in Figure 2.42, and observe that the current through R_3 must be zero. Thus, R_1 and R_2 are in series and, as

Figure 2.41 Circuit for illustration of open circuit voltage calculation

illustrated in Figure 2.43, v_{OC} is equal to the voltage across R_2, which can be found by voltage division in the series loop $v_S \rightarrow R_1 \rightarrow R_2 \rightarrow v_S$.

$$v_{OC} = v_{R2} = v_S \frac{R_2}{R_1 + R_2}$$

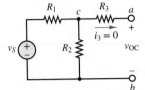
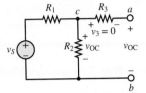

Figure 2.42 Open circuit voltage calculation - details

Figure 2.43 Open circuit voltage calculation - details

Now, consider, side by side, the original circuit and the circuit with the source network replaced by its Thévenin equivalent, as shown in Figure 2.44. The current i_o through the load R_o must be the same in both circuits.

$$i_o = v_T \cdot \frac{1}{R_T + R_o} = v_S \frac{R_2}{R_1 + R_2} \cdot \frac{1}{(R_3 + R_1 \| R_2) + R_o} \tag{2.29}$$

Notice that the latter portion of this expression is rather complicated. However, if you focus on the source network alone, it is often possible, with some practice, to readily determine R_T and v_T, by observation, and then apply Ohm's law to the simplified version of the original circuit to find the same complicated expression for i_o. Practice! Practice!! Practice!!!

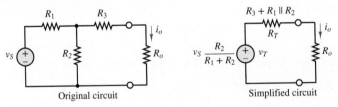

Figure 2.44 Two circuits with equivalent source networks for the load R_o

It is possible for v_T to be zero. In such a case, R_T may still be nonzero even though R_T is defined by $v_T = i_N R_T$. The implication is that when v_T is zero, i_N may also be zero, and vice versa, allowing finite, nonzero values for R_T. In this case the Thévenin equivalent of the source network is a simple resistor R_T. There are two other exceptional cases:

1. When v_T and R_T are both zero, i_N can be any value. Such a source network is trivial, being equivalent to a short-circuit. Do you see why?

2. When i_N is zero and R_T is infinitely large, v_T can be any value. Such a source network is also trivial, being equivalent to an open-circuit. Do you see why? If not, see the next section "Computing the Norton Current."

Examples 2.17 to 2.19 further illustrate the process of finding v_T.

EXAMPLE 2.17   Computing v_T for a Network With One Independent Source

Problem

Compute the open-circuit voltage v_{OC} in the circuit of Figure 2.45.

Solution

Known Quantities: Source voltage, resistor values.

Find: Open-circuit voltage v_{OC}.

Schematics, Diagrams, Circuits, and Given Data: $v_S = 12$ V; $R_1 = 1\ \Omega$; $R_2 = 10\ \Omega$; $R_3 = 10\ \Omega$; $R_4 = 20\ \Omega$.

Analysis: Refer to Figure 2.45 and the steps in the Focus on Problem Solving box, "Thévenin Voltage."

1. In this problem, the source network is given as everything to the left of terminals a and b.
2. The open-circuit voltage v_{OC} is across terminals a and b, as shown in the figure.
3. There are four nodes in the network. Node b is selected as the reference with a voltage $v_b = 0$. Another node is fixed at v_S by the voltage source. For the other two nodes, nodal analysis will yield two KCL equations in the two unknown node voltages, v and v_a. Apply KCL to obtain the following two equations:

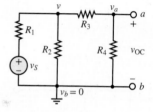

Figure 2.45 Circuit for Example 2.17

$$\frac{v_S - v}{R_1} - \frac{v - 0}{R_2} - \frac{v - v_a}{R_3} = 0 \qquad \text{node } v$$

$$\frac{v - v_a}{R_3} - \frac{v_a - 0}{R_4} = 0 \qquad \text{node } v_a$$

Collect terms to find:

$$\left(\frac{1}{R_1} + \frac{1}{R_2} + \frac{1}{R_3}\right)v \qquad\quad - \frac{1}{R_3}v_a = \frac{v_S}{R_1} \qquad \text{node } v$$

$$\frac{1}{R_3}v - \left(\frac{1}{R_3} + \frac{1}{R_4}\right)v_a = 0 \qquad \text{node } v_a$$

Substitute numerical values and write the equations in matrix form as:

$$\begin{bmatrix} 1.2 & -0.1 \\ 0.1 & -0.15 \end{bmatrix} \begin{bmatrix} v \\ v_a \end{bmatrix} = \begin{bmatrix} 12 \\ 0 \end{bmatrix}$$

Solving the above matrix equations yields $v = 10.6$ V and $v_a = 7.1$ V. Thus, $v_{OC} = v_a - 0 = 7.1$ V.

Comments: In problems involving Thévenin's theorem it can be unclear whether the circuit is to be viewed as a source network plus load or as just a source network. A common mistake would be to assume that R_4 is the load in this example even though there is no mention of a load. The fact that the voltage drop across R_4 is given as the open-circuit voltage v_{OC} indicates that the entire circuit to the left of terminals a and b is to be treated as the source network.

EXAMPLE 2.18 Computing v_T and R_T for a Network With Two Independent Sources

Problem

Find the Thévenin equivalent of the source network and use it to compute the load current i in the circuit of Figure 2.46.

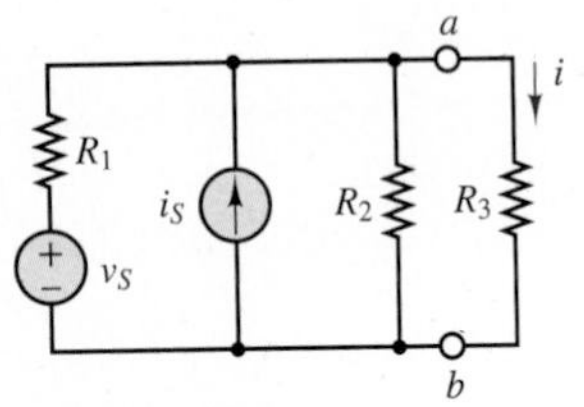

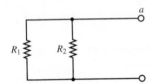

Figure 2.46 Circuit for Example 2.18

Solution

Known Quantities: Source and resistor values.

Find: v_T and R_T for the source network and the load current i.

Schematics, Diagrams, Circuits, and Given Data: $v_S = 24$ V; $i_S = 3$ A; $R_1 = 4\ \Omega$; $R_2 = 12\ \Omega$; $R_3 = 6\ \Omega$.

Analysis: Refer to Figure 2.46 and the steps in the Focus on Problem Solving box, "Thévenin Voltage."

1. R_3 is the load. Everything else is the source network.
2. The open-circuit voltage v_{OC} is defined in Figure 2.48.
3. Solve for R_T and v_T of the source network, and use them to find the load current i.

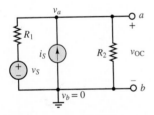

Figure 2.47 Circuit for Example 2.18 with sources set to zero

   • Find R_T: Set both voltage and current sources to zero and replace them with short- and open-circuits, respectively, as shown in Figure 2.47. The resulting equivalent resistance between terminals a and b is simply $R_T = R_1 \| R_2 = 4 \| 12 = 3\ \Omega$.

   • Find v_T: The circuit shown in Figure 2.48 has only three nodes. Node b is selected as the reference with a voltage $v_b = 0$. Of the remaining two nodes, one is fixed at a voltage v_S by the voltage source. Thus, only a single node equation is needed for a solution:

$$\frac{v_S - v_a}{R_1} + i_S - \frac{v_a}{R_2} = 0 \qquad \text{or} \qquad v_a = (v_S + i_S R_1)\frac{R_2}{R_1 + R_2}$$

Substitute numerical values to find: $v_a = v_{OC} = 27$ V. Of course, the Thévenin voltage v_T is the open-circuit voltage v_{OC} across terminals a and b.

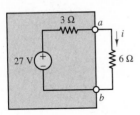

Figure 2.48 Equivalent circuit for calculation of open circuit voltage for Example 2.18

   • Find i: Construct the Thévenin equivalent of the source network and reattach the load R_3, as shown in Figure 2.49. The load current is easily computed using voltage division.

$$i = \frac{27}{3 + 6} = 3 \text{ A}$$

Comments: Equivalent circuit analysis has several key advantages. By reducing any complicated linear source network to a simple structure, one can quickly determine:

   • The voltage across and current through *any* load.
   • The maximum possible load current v_T/R_T (for loads approaching short-circuits).
   • The maximum possible load voltage v_T (for loads approaching open-circuits).
   • The value of the load that gives maximum power transfer to the load (see Section 2.6).

Figure 2.49 Equivalent circuit

EXAMPLE 2.19   Computing v_T for a Network With a Dependent Source

Problem

Find the Thévenin voltage v_T of the source network seen by the load R_o in Figure 2.50.

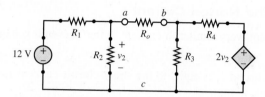

Figure 2.50 Circuit for example 2.19

Solution

Known Quantities: Source and resistor values.

Find: v_T for the source network.

Schematics, Diagrams, Circuits, and Given Data: $R_1 = 24$ kΩ; $R_2 = 8$ kΩ; $R_3 = 9$ kΩ; $R_4 = 18$ kΩ.

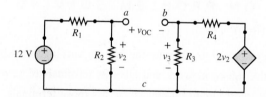

Figure 2.51 Circuit of Example 2.19 with load resistance removed

Analysis: Refer to Figure 2.50 and the steps in the Focus on Problem Solving box, "Thévenin Voltage." This circuit is identical to the one from Example 2.16, where the Thévenin equivalent resistance R_T seen by R_o was found to be 8 kΩ. In this example, the Thévenin voltage v_T seen by R_o is found.

1. The source network is everything except the load R_o; remove it. The source network terminals are marked a and b.

2. Define the open-circuit voltage v_{OC} as in Figure 2.51.

3. The resulting circuit has two series loops sharing one common node c. Define the voltage v_3 across R_3. Then KVL around the middle portion of the circuit yields:

$$v_2 = v_{OC} + v_3$$

Voltage division can be applied to the series loop on the left to solve for v_2.

$$v_2 = 12 \text{ V} \frac{R_2}{R_1 + R_2} = 12 \text{ V} \frac{8}{24 + 8} = 3 \text{ V}$$

Voltage division can also be applied to the series loop on the right to find v_3 in terms of v_2.

$$v_3 = 2v_2 \frac{R_3}{R_3 + R_4} = 6 \text{ V} \frac{9}{9 + 18} = 2 \text{ V}$$

Plug these values into the KVL equation to find:

$$v_{OC} = v_2 - v_3 = 1\,\text{V}$$

4. The Thévenin voltage is $v_T = v_{OC} = 1\,\text{V}$.

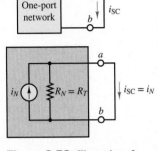

Figure 2.52 Illustration of Norton equivalent circuit

Computing the Norton Current

The Norton current, very similar in concept to the Thévenin voltage, is defined as:

> The Norton current i_N is equal to the **short-circuit current** i_{SC} through the source network terminals.

Consider an arbitrary linear one-port network and its Norton equivalent, each attached to a short-circuit, as shown in Figure 2.52. The current i_{SC} through the short-circuit is exactly the Norton current i_N because all the source current must pass through the short-circuit. This simple observation implies the basic method for finding the Norton current for any arbitrary linear source network. Attach a short-circuit wire to its terminals to determine the Norton current through the wire.

FOCUS ON PROBLEM SOLVING

NORTON CURRENT

Follow the steps below to compute the Norton current for a source network.

1. Identify the source network and label its terminals (e.g., a and b).
2. Define the short-circuit current i_{SC} across those terminals.
3. For source networks with at least one independent source:
   - Apply any preferred method (e.g., node analysis) to solve for i_{SC}.

   For source networks without an independent source:
   - The short-circuit current i_{SC} is simply zero even when a dependent source is present.
4. The Norton current i_N of the source network is, by definition, i_{SC}.

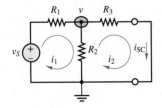

Figure 2.53 Computation of Norton current

Consider the circuit of Figure 2.53, shown with a short-circuit attached to the source network (i.e., in place of the load). The short-circuit current i_{SC} can be found easily using any of the solution techniques presented in this chapter. Both node and mesh analysis work very well here.

In terms of the mesh currents i_1 and i_2, the KVL mesh equations are:

$$v_S - R_1 i_1 - R_2(i_1 - i_2) = 0 \qquad \text{mesh 1}$$
$$-R_2(i_2 - i_1) - R_3 i_2 = 0 \qquad \text{mesh 2}$$

Collect terms to find:

$$(R_1 + R_2)i_1 - R_2 i_2 = v_S \qquad \text{mesh 1}$$
$$-R_2 i_1 + (R_2 + R_3)i_2 = 0 \qquad \text{mesh 2}$$

Multiply the mesh 2 equation by $(R_1 + R_2)/R_2$ and add the result to the mesh 1 equation to find:

$$\left[\frac{(R_1 + R_2)(R_2 + R_3)}{R_2} - R_2\right] i_2 = v_S$$

Finally, multiply both sides of the equation by R_2 to obtain:

$$i_{SC} = i_2 = \frac{v_S R_2}{(R_1 + R_2)(R_2 + R_3) - R_2^2} = \frac{v_S R_2}{R_1 R_2 + R_1 R_3 + R_2 R_3}$$

Alternatively, in terms of the node voltage v, the one KCL node equation is:

$$\frac{v_S - v}{R_1} = \frac{v}{R_2} + \frac{v}{R_3}$$

Multiply both sides of the equation by $R_1 R_2 R_3$ and collect terms to find:

$$v_S R_2 R_3 = v(R_1 R_2 + R_1 R_3 + R_2 R_3)$$

or

$$v = v_S \frac{R_2 R_3}{R_1 R_2 + R_1 R_3 + R_2 R_3}$$

Finally, the short-circuit current is:

$$i_{SC} = \frac{v - 0}{R_3} = \frac{v_S R_2}{R_1 R_2 + R_1 R_3 + R_2 R_3}$$

Of course, the results are the same for both methods. Great! Thus, the Norton current i_N is:

$$i_N = i_{SC} = \frac{v_S R_2}{R_1 R_2 + R_1 R_3 + R_2 R_3}$$

Why solve for i_{SC} twice, using two separate methods? When time allows, it is always a good idea to validate your results!

Examples 2.20 and 2.21 further illustrate the process of finding i_N.

EXAMPLE 2.20   Computing i_N for a Network With Two Independent Sources

Problem

Determine the Norton current i_N and the Norton equivalent for the network in Figure 2.54.

Solution

Known Quantities: Source voltage and current; resistor values.

Find: Norton current $i_N = i_{SC}$; Equivalent resistance R_T.

Schematics, Diagrams, Circuits, and Given Data: $v_S = 6$ V; $i_S = 2$ A; $R_1 = 6\ \Omega$; $R_2 = 3\ \Omega$; $R_3 = 2\ \Omega$.

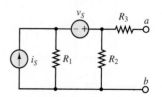

Figure 2.54 Circuit for Example 2.20

Assumptions: Assume the reference node is at the bottom of the circuit.

Analysis: Refer to Figure 2.54 and the steps in the Focus on Problem Solving box, "Norton Current."

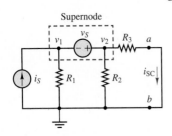

Figure 2.55 Supernode in Circuit for Example 2.20

- Find i_N: In Figure 2.55, the source network terminals a and b are defined and a short-circuit wire is attached to them. Mesh analysis would work very well in this problem (do you see why?), but node analysis will also work, and this circuit provides a good opportunity to practice the use of a "supernode," as discussed earlier in this chapter.

  (a) There are three nodes in this circuit. The reference node is labeled in the figure.

  (b) There are two nonreference nodes labeled v_1 and v_2.

  (c) There is one voltage source in the circuit; thus, only one of the node voltage variables is independent. The other variable is dependent.

  (d) Apply KCL at the boundaries of the supernode shown in the figure to find:

$$i_S - \frac{v_1 - 0}{R_1} - \frac{v_2 - 0}{R_2} - \frac{v_2 - 0}{R_3} = 0 \qquad \text{supernode}$$

$$v_2 - v_1 = v_S \qquad \text{constraint equation}$$

  (e) v_2 is the primary objective since $v_2 = i_{SC} R_3$. Use the constraint equation to substitute for v_1 in the supernode equation.

$$i_S = \frac{v_2 - v_S}{R_1} + v_2 \frac{R_2 + R_3}{R_2 R_3}$$

$$i_S + \frac{v_S}{R_1} = v_2 \left[\frac{1}{R_1} + \frac{R_2 + R_3}{R_2 R_3} \right]$$

Form the common denominator $R_1 R_2 R_3$ for the bracketed term and find:

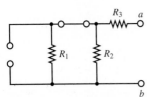

Figure 2.56 Circuit of Example 2.20 with sources set to zero

$$v_2 = \left(i_S + \frac{v_S}{R_1} \right) \left[\frac{R_1 R_2 R_3}{R_1 R_2 + R_1 R_3 + R_2 R_3} \right]$$

$$= \left(2 + \frac{6}{6} \right) \left[\frac{6 \cdot 3 \cdot 2}{6 \cdot 3 + 6 \cdot 2 + 3 \cdot 2} \right]$$

$$= (2 + 1) \left[\frac{18}{18} \right] = 3 \, \text{V}$$

  (f) Finally, the short-circuit current is given by:

$$i_{SC} = \frac{v_2}{R_3} = \frac{3}{2} = 1.5 \, \text{A} = i_N$$

- Find R_T: To compute the Thévenin equivalent resistance, set the independent voltage and current sources to zero and replace them with short- and open-circuits, respectively. The resulting resistor network is shown in Figure 2.56. It is easy to see that $R_T = R_1 \| R_2 + R_3 = 6 \| 3 + 2 = 4 \, \Omega$.

The Norton equivalent of the original source network is shown in Figure 2.57.

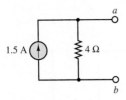

Figure 2.57 Norton equivalent network

Comments: Superposition is a reasonable alternative method for solving for i_{SC}. Take another look at Figure 2.55 and note that current division will quickly yield the component of i_{SC} due to the current source i_S. Also note that voltage division will quickly yield the component of v_2, and then by Ohm's law the component of i_{SC}, due to the voltage source v_S.

EXAMPLE 2.21   Computing i_N for a Network With a Dependent Source

Problem

Find the Norton current i_N of the source network seen by the load R_o in Figure 2.58.

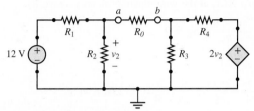

Figure 2.58 Circuit for Example 2.21

Solution

Known Quantities: Source and resistor values.

Find: i_N for the source network.

Schematics, Diagrams, Circuits, and Given Data: $R_1 = 24$ kΩ; $R_2 = 8$ kΩ; $R_3 = 9$ kΩ;
$R_4 = 18$ kΩ.

Assumptions: Assume the reference node is at the bottom of the circuit.

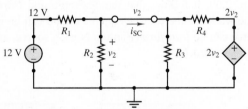

Figure 2.59 Defining the reference node

Analysis: Refer to Figure 2.58 and the steps in the Focus on Problem Solving box, "Norton
Current." This circuit is identical to the one from Example 2.19, where the Thévenin voltage seen by R_o was found to be 1 V. In this example, the Norton current i_N seen by R_o is
found.

1. The source network is everything except the load R_o; remove it. The source network
    terminals are marked a and b.

2. Define the short-circuit current i_{SC} as in Figure 2.59.

3. The resulting circuit has three nonreference nodes; however, the voltage of one
    is known while the voltages of the other two are both determined by v_2. Thus,
    there is only one independent variable v_2, which can be found by applying KCL at
    node v_2.

$$\frac{12 - v_2}{R_1} - \frac{v_2 - 0}{R_2} - \frac{v_2 - 0}{R_3} - \frac{v_2 - 2v_2}{R_4} = 0$$

Plug in values for the resistors and multiply both sides of the equation by the common denominator to get:

$$3(12 - v_2) - 9v_2 - 8v_2 - 4(-v_2) = 0$$

or

$$16v_2 = 36 \quad \text{which yields} \quad v_2 = \frac{9}{4} = 2.25\,\text{V}$$

4. To find i_{SC} apply KCL at the wire junction directly above R_2.

$$\frac{12 - v_2}{R_1} - \frac{v_2 - 0}{R_2} - i_{SC} = 0$$

Plug in for v_2 to find

$$i_{SC} = \frac{9.75}{24} - \frac{2.25}{8} = \frac{3}{24} = \frac{1}{8} = 0.125\,\text{mA} = i_N$$

Comments: Note that the circuit in this example is identical to the one used in Examples 2.16 and 2.19. In these three example problems, the Thévenin equivalent resistance R_T, the Thévenin voltage v_T, and the Norton current i_N were found to be $8\,\text{k}\Omega$, 1 V, and 0.125 mA, respectively, for the same source network. Although all three values were found by independent means, the result is that $v_T = i_N \cdot R_T$. Check it out! Amazing!! Read on to learn about a powerful and popular (among students) solution technique based upon this important relationship.

Source Transformations

The Norton and Thévenin theorems state that any linear one-port network has an equivalent representation as a voltage source in series with a resistance (a Thévenin source), or as a current source in parallel with a resistance (a Norton source), as illustrated in Figure 2.60. It follows that the Thévenin and Norton equivalents of a specific source network are themselves equivalent and can be interchanged. That is, a Thévenin source can be *transformed* into a Norton source, and vice versa. The parameter values are related by:

$$v_T = i_N R_T \tag{2.30}$$

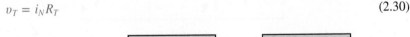
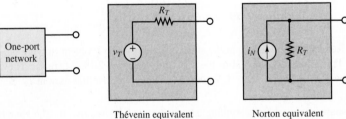

Thévenin equivalent   Norton equivalent

Figure 2.60   Simplified equivalent representations of a linear one-port network

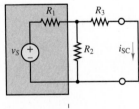

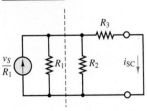

Figure 2.61   Result of source transformation

The Thévenin source in the shaded box in Figure 2.61 may be replaced by its Norton equivalent. The computation of i_{SC} is straightforward since a simple current divider may be used to compute the short-circuit current. Observe that the short-circuit current is the current through R_3; therefore,

$$i_{SC} = i_N = \frac{1/R_3}{1/R_1 + 1/R_2 + 1/R_3}\frac{v_S}{R_1} = \frac{v_S R_2}{R_1 R_3 + R_2 R_3 + R_1 R_2} \tag{2.31}$$

which is the identical result obtained for the same circuit in the preceding section. Source transformations can be very useful, if employed correctly. Figure 2.62 shows how to recognize subnetworks that are amenable to these transformations. Example 2.22 illustrates the procedure.

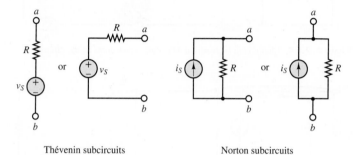

Thévenin subcircuits   Norton subcircuits

Figure 2.62 Networks amenable to source transformation

EXAMPLE 2.22   Source Transformations

Problem

Use source transformations to find the Norton equivalent of the network seen by the load R_o, as shown in Figure 2.63.

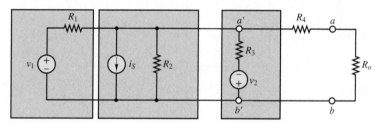

Figure 2.63 Circuit for Example 2.22

Solution

Known Quantities: Source voltages and current; resistor values.

Find: Thévenin equivalent resistance R_T; Norton current $i_N = i_{SC}$.

Schematics, Diagrams, Circuits, and Given Data: $v_1 = 50$ V; $i_S = 0.5$ A; $v_2 = 5$ V; $R_1 = 100$ Ω; $R_2 = 100$ Ω; $R_3 = 200$ Ω; $R_4 = 160$ Ω.

Assumptions: Assume the reference node is at the bottom of the circuit.

Analysis: Highlight key terminals in the circuit to emphasize the Thévenin and Norton sources present in the circuit, as shown in Figure 2.64. The Thévenin source consisting of v_1 and R_1, which appears between terminals a'' and b'', can be replaced with a Norton source consisting of a current source v_1/R_1 in parallel with R_1. Similarly, the Thévenin source between terminals a' and b' can be replaced with a Norton source consisting of a current source v_2/R_3 in parallel with R_3. Both of these transformations are shown in Figure 2.65. The order of elements in parallel can be interchanged without changing the behavior of the overall circuit, as is shown in Figure 2.66(a) with numerical values of the elements included.

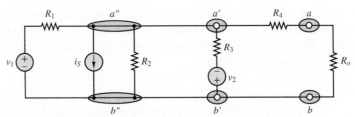

Figure 2.64 Identifying Norton and Thevenin equivalents in the circuit

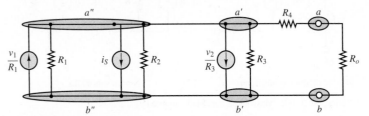

Figure 2.65 Converting all equivalent pairs to Norton form

The three current sources in parallel can be replaced by a single equivalent 0.025-A current source $(0.5 - 0.025 - 0.5 = 0.025$ A) directed downward, and the three parallel resistors $200 \parallel 100 \parallel 100$ can be replaced by a single equivalent 40-Ω resistor, as shown in Figure 2.66(b).

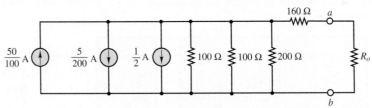

Figure 2.66(a) Transformed, but not yet simplified, circuit

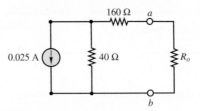

Figure 2.66(b) Simplified circuit

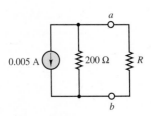

Figure 2.67 Equivalent circuit

The Norton source on the left can be transformed into a Thévenin source such that the 40-Ω resistor is in series with the 160-Ω resistor and the current source is transformed into a 0.025 A $\times$ 40 Ω = 1-V voltage source. Finally, combine the two series resistors to form a single equivalent 200-Ω resistor and transform the resulting Thévenin source into an equivalent Norton source with a 0.005-A current source directed downward, as shown in Figure 2.67.

Comments: The Thévenin equivalent resistance seen by the load R_o in Figure 2.63 is easily computed once each current and voltage source is set to zero such that:

$$R_T = R_1 \parallel R_2 \parallel R_3 + R_4 = 200 \parallel 100 \parallel 100 + 160 = 200 \ \Omega$$

which is the same (as it must be) as the Thévenin equivalent resistance seen by R_o in Figure 2.67.

It is not always possible to use source transformations to simplify an entire network to a Norton or Thévenin source as was done in this example. However, source transformations can often be used to partially simplify a network so that another solution method is readily applied.

Experimental Determination of Thévenin and Norton Equivalents

Thévenin and Norton equivalent networks are often used as linear models of practical devices, such as batteries, power supplies, voltmeters, and ammeters, over a limited range of operation. While it is usually not possible nor feasible, because of the internal complexity of the devices, to determine those models analytically, simple experimental methods can be used instead. In practice, it is very useful to measure, for example, the equivalent internal (Thévenin) resistance of an instrument, so as to understand its operating limits and power requirements. Essentially, the linear model of a device is completely determined by its Thévenin (open-circuit) voltage v_T and its Norton (short-circuit) current i_N. The equivalent internal (Thévenin) resistance R_T is

$$R_T = \frac{v_T}{i_N} \tag{2.32}$$

In cases where the short-circuit current can be measured directly by an ammeter, Figure 2.68 illustrates the measurement of the short-circuit current and the open-circuit voltage. The figure clearly illustrates that the finite meter resistances r_A and r_V must be accounted for in the computation of the short-circuit current i_{SC} and the open-circuit voltage v_{OC}, respectively.

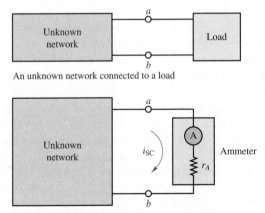

An unknown network connected to a load

Network connected for measurement of short-circuit current

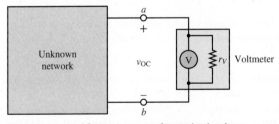

Network connected for measurement of open-circuit voltage

Figure 2.68 Measurement of open-circuit voltage and short-circuit current

See the section in chapter 1 on measurement devices to confirm that the "true" short-circuit current i_N and the "true" open-circuit voltage v_T are related to the measured values i_{SC} and v_{OC} by:

$$i_N = i_{SC}\left(1 + \frac{r_A}{R_T}\right)$$

$$v_T = v_{OC}\left(1 + \frac{R_T}{r_V}\right)$$

(2.33)

where R_T is the Thévenin equivalent resistance across terminals a and b of the unknown network. For an ideal ammeter, the internal resistance r_A is zero (a short-circuit). For an ideal voltmeter, the internal resistance r_V is infinite (an open-circuit). The two expressions in Equation 2.22 determine the "true" Thévenin and Norton equivalent networks using imperfect measurements of the short-circuit current and the open-circuit voltage, provided that the internal meter resistances are known. In practice, when the equivalent resistance seen by a voltmeter is much smaller than r_V, the measured v_{OC} will closely approximate the "true" v_{OC}. Likewise, when the equivalent resistance seen by an ammeter is much larger than r_A, the measured i_{SC} will closely approximate the "true" i_{SC}.

It is often not advisable to measure i_{SC} directly with an ammeter since its magnitude is not known. An ammeter is designed to approximate a short-circuit when inserted in a network, such that a large current may result and destroy an overcurrent protection fuse and perhaps damage the ammeter itself.

An alternative to measuring i_{SC} directly is to collect data along the device *load line* and extrapolate i_{SC} from that data. Experimental load line data can be acquired by inserting resistive loads between the device terminals. The first load should be an open-circuit to determine directly the open-circuit voltage. The second load should be very large and followed by successively smaller loads. The load voltage can be measured by a voltmeter and the load current deduced by applying Ohm's law to the resistive load. For an ideal linear device, these data points will trace a straight line from the intersection with the voltage axis (v_{OC}) to the intersection with the current axis, which is the short-circuit current i_{SC}. In practice, experimental errors should be accounted for by using the load line data to compute a "best fit" trendline.

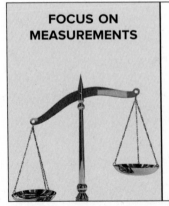

FOCUS ON MEASUREMENTS

Experimental Determination of Thévenin Equivalent Networks

Problem:
Determine the Thévenin equivalent of an unknown source network from measurements of open-circuit voltage and short-circuit current.

Solution:
Known Quantities—Short-circuit current i_{SC}, open-circuit voltage v_{OC}, ammeter internal resistance r_A, and voltmeter internal resistance r_V.

Find—Equivalent resistance R_T; Thévenin voltage $v_T = v_{OC}$.

Schematics, Diagrams, Circuits, and Given Data—Measured $v_{OC} = 6.5$ V; measured $i_{SC} = 3.25$ mA; $r_A = 25\ \Omega$; $r_V = 10$ MΩ.

(*Continued*)

(*Continued*)

Assumptions—The unknown network is linear containing ideal sources and resistors only. The short-circuit current was able to be measured directly using an ammeter without damaging the instrument or its fuse.

Analysis—The unknown circuit shown in Figure 2.69 is replaced by its Thévenin equivalent and is connected to an ammeter to measure the short-circuit current and to a voltmeter to measure the open-circuit voltage. Ohm's law can be applied to the current measurement to find:

$$i_{SC} = \frac{v_T}{R_T + r_A}$$

Voltage division can be applied to the voltage measurement to find:

$$v_{OC} = \frac{r_V}{R_T + r_V} v_T$$

These expressions can be solved for v_T to yield:

$$v_T = i_{SC}(R_T + r_A)$$

$$= v_{OC}\left(1 + \frac{R_T}{r_V}\right)$$

Or

$$i_{SC}R_T\left(1 + \frac{r_A}{R_T}\right) = v_{OC}\left(1 + \frac{R_T}{r_V}\right)$$

Since r_V is typically on the order of 10^6 times larger than r_A, one or both of the fractions in the previous expression will be negligible for a given R_T. Under the assumption that $R_T \ll r_V$ the above expression is approximated by:

$$i_{SC}R_T\left(1 + \frac{r_A}{R_T}\right) = v_{OC}$$

Under the assumption that $R_T \gg r_A$ the above expression is instead approximated by:

$$i_{SC}R_T = v_{OC}\left(1 + \frac{R_T}{r_V}\right)$$

If both assumptions are true, the Thévenin equivalent resistance is approximated by:

$$i_{SC}R_T = v_{OC}$$

which is the calculation that many inexperienced users make for every measurement, regardless of the relative values of R_T, r_A, and r_V. Of course, R_T is not known a priori so it is important to consider whether either or both of the limiting assumptions is reasonable.

Consider the example measurement data listed above. The measured values of the short-circuit current and open-circuit voltage are:

$$i_{SC} = 3.25\,\text{mA} \qquad \text{and} \qquad v_{OC} = v_T = 6.5\,\text{V}$$

If both limiting assumptions are made, then the Thévenin equivalent resistance R_T between terminals a and b of the unknown network is approximately:

$$R_T \approx \frac{v_{OC}}{i_{SC}} = 2.0\,\text{k}\Omega$$

This value is 80 times larger than r_A but 5,000 times smaller than r_V. Thus, one might expect that the impact of r_A is more significant than the impact of r_V for this particular network.

If only $R_T \ll r_V$ is assumed, then using the appropriate expression above yields:

$$R_T \approx \frac{v_{OC}}{i_{SC}} - r_A = 2.0\,\text{k}\Omega - 25\,\Omega = 1975\,\Omega$$

(*Continued*)

(*Concluded*)

which is a 1.25% change from 2.0 kΩ. If only $R_T \gg r_A$ is assumed, then using the appropriate expression above yields:

$$R_T \approx \frac{v_{OC}}{i_{SC}} \frac{r_V}{r_V - \frac{v_{OC}}{i_{SC}}} = (2.0 \text{ k}\Omega) \frac{10^7}{10^7 - 2.0 \text{ k}\Omega} = 2000.4 \ \Omega$$

which is a negligibly small 0.02% change from 2.0 kΩ. If neither limiting assumption is made, then R_T is:

$$R_T = \frac{v_{OC} - i_{SC} r_A}{i_{SC} - \frac{i_{SC}}{r_V}} = 1975.4 \ \Omega$$

As expected it is important in this example to include the impact of r_A when calculating the "true" value of R_T. The impact of r_V on the calculation is negligible.

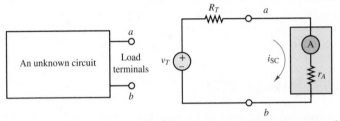

Network connected for measurement of
short-circuit current (practical ammeter)

Network connected for measurement of
open-circuit voltage (ideal voltmeter)

Figure 2.69 Experimental set-up for measurement of short-circuit current and open-circuit voltage

CHECK YOUR UNDERSTANDING

Find the Thévenin equivalent resistance of the circuit below, as seen by the load resistor R_o.

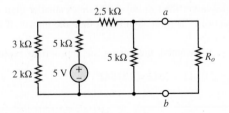

Find the Thévenin equivalent resistance seen by the load resistor R_o in the following circuit.

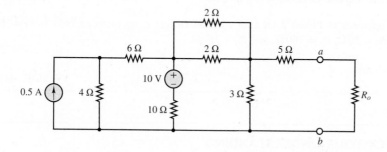

CHECK YOUR UNDERSTANDING

For the circuit below, find the Thévenin equivalent resistance seen by the load resistor R_o.

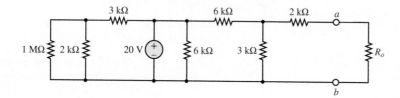

For the circuit below, find the Thévenin equivalent resistance seen by the load resistor R_o.

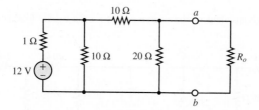

CHECK YOUR UNDERSTANDING

Find the open-circuit voltage v_{OC} for the circuit of Figure 2.45 if $R_1 = 5\,\Omega$.

CHECK YOUR UNDERSTANDING

With reference to Figure 2.41, find the load current i_o by mesh analysis if $v_S = 10$ V, $R_1 = R_3 = 50\,\Omega$, $R_2 = 100\,\Omega$, and $R_0 = 150\,\Omega$.

Answer: 28.57 mA

CHECK YOUR UNDERSTANDING

Find the Thévenin equivalent of the source network seen by the load resistor R_o.

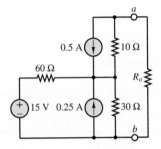

Answers: $R_T = 30\,\Omega$; $v_{OC} = v_T = 5$ V

CHECK YOUR UNDERSTANDING

Find the Thévenin equivalent of the source network seen by the load resistor R_o.

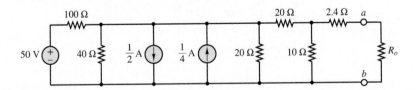

Answers: $R_T = 10\,\Omega$; $v_{OC} = v_T = 0.704$ V

CHECK YOUR UNDERSTANDING

Repeat Example 2.20, using mesh analysis. Note that in this case one of the three mesh currents is known, and therefore the complexity of the solution will be unchanged.

2.6   MAXIMUM POWER TRANSFER

The reduction of any linear resistive circuit to its Thévenin or Norton equivalent form is a very convenient conceptualization, as far as the computation of load-related quantities is concerned. One such computation is that of the power absorbed by the load. The Thévenin and Norton models imply that some of the power generated by the source will necessarily be dissipated by the internal circuits within the source. Given this unavoidable power loss, a logical question to ask is, How much power can be transferred to the load from the source under the most ideal conditions? Or, alternatively, what is the value of the load resistance that will absorb maximum power from the source? The answer to these questions is contained in the **maximum power transfer theorem**, which is the subject of this section.

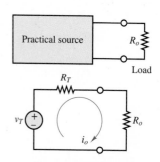

Given v_T and R_T, what value of R_o will allow for maximum power transfer?

Figure 2.70 Power transfer between source and load

The model employed in the discussion of power transfer is illustrated in Figure 2.70, where a practical source is represented by means of its Thévenin equivalent circuit. The power absorbed by the load P_o is:

$$P_o = i_o^2 R_o \tag{2.34}$$

and the load current is:

$$i_o = \frac{v_T}{R_o + R_T} \tag{2.35}$$

Combining the two expressions, the load power can be computed as

$$P_o = \frac{v_T^2}{(R_o + R_T)^2} R_o \tag{2.36}$$

The expression for P_o can be differentiated with respect to R_o and set to zero to find the value of R_o that gives the maximum power absorbed by the load. (Here, v_T and R_T are assumed constant.)

$$\frac{dP_o}{dR_o} = 0 \tag{2.37}$$

Plug in for P_o and solve to obtain:

$$\frac{dP_o}{dR_o} = \frac{v_T^2 (R_o + R_T)^2 - 2 v_T^2 R_o (R_o + R_T)}{(R_o + R_T)^4} \tag{2.38}$$

Thus, at the maximum value of P_o the following expression must be satisfied.

$$(R_o + R_T)^2 - 2 R_o (R_o + R_T) = 0 \tag{2.39}$$

The solution of this equation is:

$$\boxed{R_o = R_T} \tag{2.40}$$

Thus, to transfer maximum power to a load, the equivalent source and load resistances must be **matched**, that is, equal to each other. Figure 2.71 depicts a plot of the load power divided by v_T^2 versus the ratio of R_o to R_T. Note that this value is maximum when $R_o = R_T$.

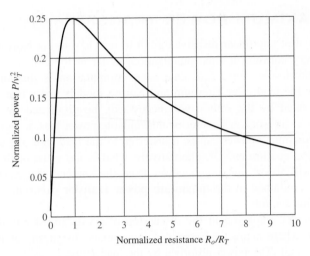

Figure 2.71 Graphical representation of maximum power transfer

This analysis shows that to transfer maximum power to a load, given a fixed equivalent source resistance, the load resistance must match the equivalent source resistance. What if the problem statement were reversed such that the maximum power transfer to the source resistance is sought for a fixed load resistance? What would be the value of the source resistance that maximizes the power transfer in this case? The answer can be found by solving the Check Your Understanding exercises at the end of the section.

A problem related to power transfer is that of **source loading**. This phenomenon, which is illustrated in Figure 2.72, may be explained as follows: When a practical voltage source is connected to a load, the source current through the load will cause a voltage drop across the internal source resistance v_{int}; as a consequence, the voltage actually seen by the load will be somewhat lower than the *open-circuit voltage* of the source. As stated earlier, the open-circuit voltage is the Thévenin voltage. The extent of the internal voltage drop within the source depends on the amount of current drawn by the load. With reference to Figure 2.72, this internal drop is equal to iR_T, and therefore the load voltage will be

$$v_o = v_T - iR_T \qquad (2.41)$$

It should be apparent that it is desirable to have as small an internal resistance as possible in a practical voltage source.

In the case of a current source, the internal resistance will draw some current away from the load because of the presence of the internal source resistance; this current is denoted by i_{int} in Figure 2.72. Thus the load will receive only part of the *short-circuit current* (the Norton current) available from the source:

$$i_o = i_N - \frac{v}{R_T} \qquad (2.42)$$

It is therefore desirable to have a very large internal resistance in a practical current source. Refer to the discussion of practical sources in Chapter 1 to see that these sources are themselves Thévenin and Norton equivalent sources.

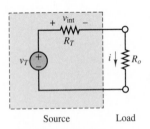

Source   Load

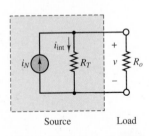

Source   Load

Figure 2.72 Source loading effects

EXAMPLE 2.23   Maximum Power Transfer

Problem

Use the maximum power transfer theorem to determine the increase in power delivered to a loudspeaker resulting from matching the speaker resistance to the amplifier source resistance, as depicted in the simplified model of Figure 2.73.

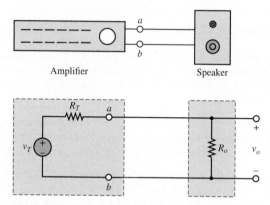

Figure 2.73 A simplified model of an audio system

Solution

Known Quantities: Source equivalent resistance R_T; unmatched speaker load resistance R_U; matched loudspeaker load resistance R_M.

Find: Difference between power delivered to loudspeaker with unmatched and matched loads, and corresponding percentage increase.

Schematics, Diagrams, Circuits, and Given Data: $R_T = 8\ \Omega$; $R_U = 16\ \Omega$; $R_M = 8\ \Omega$.

Assumptions: The amplifier can be modeled as a linear resistive device.

Analysis: Consider connecting (unwittingly) an 8-Ω amplifier to a 16-Ω speaker. The power delivered to the speaker can be computed using voltage division as follows:

$$v_U = \frac{R_U}{R_U + R_T} v_T = \frac{2}{3} v_T$$

and the load power is then computed to be

$$P_U = \frac{v_U^2}{R_U} = \frac{4}{9} \frac{v_T^2}{R_U} = 0.0278\, v_T^2$$

Repeat the calculation for the case of a matched 8-Ω speaker resistance R_M. The new load voltage v_M and the corresponding load power P_M are calculated as follows:

$$v_M = \frac{1}{2} v_T$$

and

$$P_M = \frac{v_M^2}{R_M} = \frac{1}{4} \frac{v_T^2}{R_M} = 0.03125\, v_T^2$$

The increase in load power is therefore

$$\Delta P = \frac{0.03125 - 0.0278}{0.0278} \times 100 = 12.5\%$$

Comments: In practice, an audio amplifier and a speaker are not well represented by the simple resistive models used in this example. Circuits that are appropriate to model amplifiers and loudspeakers are presented in later chapters. The audiophile can find further information concerning hi-fi circuits in Chapter 6.

CHECK YOUR UNDERSTANDING

A practical voltage source has an internal resistance of 1.2 Ω and generates a 30-V output under open-circuit conditions. What is the smallest load resistance we can connect to the source if we do not wish the load voltage to drop by more than 2 percent with respect to the source open-circuit voltage?

A practical current source has an internal resistance of 12 kΩ and generates a 200-mA output under short-circuit conditions. What percentage drop in load current will be experienced (with respect to the short-circuit condition) if a 200-Ω load is connected to the current source?

Repeat the derivation leading to equation 2.30 for the case where the load resistance is fixed and the source resistance is variable. That is, differentiate the expression for the load power P_o with respect to R_S instead of R_o. What is the value of R_S that results in maximum power transfer to the load?

Answers: 58.8 Ω; 1.64%; $R_s = 0$

CONCLUSION

This chapter provides a practical introduction to the analysis of linear resistive circuits and the important two-step problem-solving method of simplifying, and then solving. The emphasis on examples is important at this stage since familiarity with basic circuit analysis techniques greatly eases the task of learning more advanced concepts. In this chapter, a student should have mastered five analysis methods, summarized as follows:

1. Recognizing *nodes* and learning the *source-load perspective* are important first steps toward acquiring the ability to see meaningful structure in circuit diagrams.
2. *Node voltage and mesh analysis* are analogous in concept. They are generally applicable to the circuits analyzed in this book and are amenable to solution by matrix methods.
3. *The principle of superposition* is an important concept as well as a useful simplification method in problem solving.
4. *Thévenin and Norton equivalent networks* are also important concepts as well as invaluable simplification methods for problem solving. Mastery of these two simplification methods is essential for successful further study.
5. *Maximum power transfer.* Equivalent circuits provide a very clear explanation of how power is transferred from a source to a load.

The material covered in this chapter is essential to the development of more advanced techniques throughout the remainder of the book.

HOMEWORK PROBLEMS

Sections 2.2–2.3: Nodal and Mesh Analyses

2.1 Use node voltage analysis to find the voltages V_1 and V_2 for the circuit of Figure P2.1.

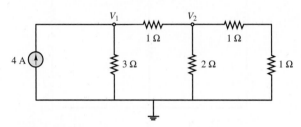

Figure P2.1

2.2 Using node voltage analysis, find the voltages V_1 and V_2 for the circuit of Figure P2.2.

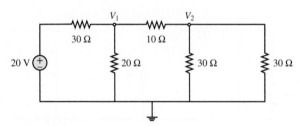

Figure P2.2

2.3 Using node voltage analysis in the circuit of Figure P2.3, find the voltage v across the 0.25-Ω resistance.

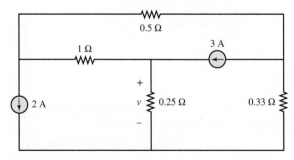

Figure P2.3

2.4 Using node voltage analysis in the circuit of Figure P2.4, find the current i through the voltage source.

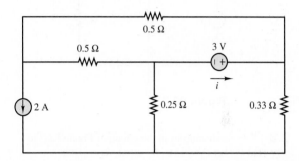

Figure P2.4

2.5 In the circuit shown in Figure P2.5, the mesh currents are

$$I_1 = 5\,\text{A} \qquad I_2 = 3\,\text{A} \qquad I_3 = 7\,\text{A}$$

Determine the branch currents through:

  a. R_1.   b. R_2.   c. R_3.

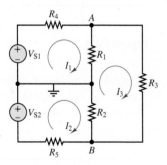

Figure P2.5

2.6 In the circuit shown in Figure P2.5, the source and node voltages are

$$V_{S1} = V_{S2} = 110\,\text{V}$$

$$V_A = 103\,\text{V} \qquad V_B = -107\,\text{V}$$

Determine the voltage across each of the five resistors.

2.7 Use nodal analysis in the circuit of Figure P2.7 to find V_a. Let $R_1 = 12\,\Omega$, $R_2 = 6\,\Omega$, $R_3 = 10\,\Omega$, $V_1 = 4$ V, $V_2 = 1$ V.

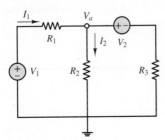

Figure P2.7

2.8 Use mesh analysis in the circuit of Figure P2.7 to find V_a. Let $R_1 = 12\,\Omega$, $R_2 = 6\,\Omega$, $R_3 = 10\,\Omega$, $V_1 = 4$ V, $V_2 = 1$ V.

2.9 Use nodal analysis in the circuit of Figure P2.9 to find v_1, v_2, and v_3. Let $R_1 = 10\,\Omega$, $R_2 = 8\,\Omega$, $R_3 = 10\,\Omega$, $R_4 = 5\,\Omega$, $i_S = 2$ A, $v_S = 1$ V.

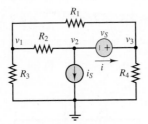

Figure P2.9

2.10 Use nodal analysis in the circuit of Figure P2.10 to find the voltages at nodes A, B, and C. Let $V_1 = 12$ V, $V_2 = 10$ V, $R_1 = 2\,\Omega$, $R_2 = 8\,\Omega$, $R_3 = 12\,\Omega$, $R_4 = 8\,\Omega$.

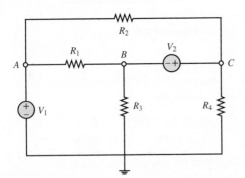

Figure P2.10

2.11 Use nodal analysis in the circuit of Figure P2.11 to find V_a and V_b. Let $R_1 = 10\,\Omega$, $R_2 = 4\,\Omega$, $R_3 = 6\,\Omega$, $R_4 = 6\,\Omega$, $V_1 = 2$ V, $V_2 = 4$ V, $I_1 = 2$ A.

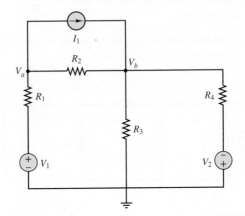

Figure P2.11

2.12 Find the power delivered to the load resistor R_0 for the circuit of Figure P2.12, using node voltage analysis, given that $R_1 = 2\,\Omega$, $R_V = R_2 = R_0 = 4\,\Omega$, $V_S = 4$ V, and $I_S = 0.5$ A.

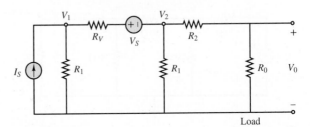

Figure P2.12

2.13 For the circuit of Figure P2.13, write the node equations necessary to find voltages V_1, V_2, and V_3. Note that $G = 1/R =$ conductance. From the results, note the interesting form that the matrices $[G]$ and $[I]$ have taken in the equation $[G][V] = [I]$ where

$$[G] = \begin{bmatrix} g_{11} & g_{12} & g_{13} & \cdots & g_{1n} \\ g_{21} & g_{22} & \cdots & \cdots & g_{2n} \\ g_{31} & & \ddots & & \\ \vdots & & & \ddots & \\ g_{n1} & g_{n2} & \cdots & \cdots & g_{nn} \end{bmatrix} \quad \text{and} \quad [I] = \begin{bmatrix} I_1 \\ I_2 \\ \vdots \\ \vdots \\ I_n \end{bmatrix}$$

Write the matrix form of the node voltage equations again, using the following formulas:

$$g_{ii} = \sum \text{conductances connected to node } i$$
$$g_{ij} = -\sum \text{conductances } shared \text{ by nodes } i \text{ and } j$$
$$I_i = \sum \text{all } source \text{ currents into node } i$$

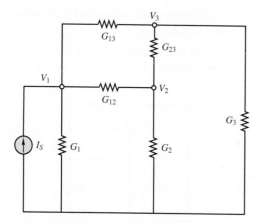

Figure P2.13

2.14   Using mesh analysis, find the currents i_1 and i_2 for the circuit of Figure P2.14.

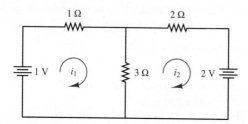

Figure P2.14

2.15   Using mesh analysis, find the currents i_1 and i_2 and the voltage across the upper 10-Ω resistor in the circuit of Figure P2.15.

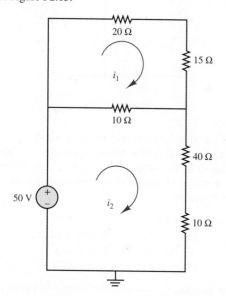

Figure P2.15

2.16   Using mesh analysis, find the voltage v across the 3-Ω resistor in the circuit of Figure P2.16.

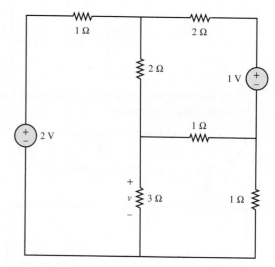

Figure P2.16

2.17   Using mesh analysis, find the currents I_1, I_2, and I_3 in the circuit of Figure P2.17 (assume polarity according to I_2).

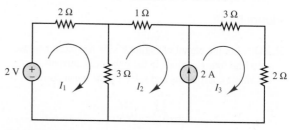

Figure P2.17

2.18   Using mesh analysis, find the voltage V across the current source in Figure P2.18.

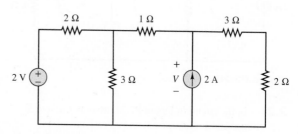

Figure P2.18

2.19   For the circuit of Figure P2.19, write the mesh
equations in matrix form. Notice the form of the [R]
and [V] matrices in [R][I] = [V], where

$$[R] = \begin{bmatrix} r_{11} & r_{12} & r_{13} & \cdots & r_{1n} \\ r_{21} & r_{22} & \cdots & \cdots & r_{2n} \\ r_{31} & & \ddots & & \\ \vdots & & & \ddots & \\ r_{n1} & r_{n2} & \cdots & \cdots & r_{nn} \end{bmatrix} \quad \text{and} \quad [V] = \begin{bmatrix} V_1 \\ V_2 \\ \vdots \\ \vdots \\ V_n \end{bmatrix}$$

Write the matrix form of the mesh equations again by
using the following formulas:

$$r_{ii} = \sum \text{resistances around loop } i$$
$$r_{ij} = -\sum \text{resistances } shared \text{ by loops } i \text{ and } j$$
$$V_i = \sum \text{source voltages around loop } i$$

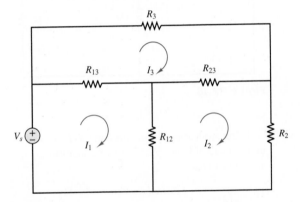

Figure P2.19

2.20   For the circuit of Figure P2.20, use mesh analysis
to find four equations in the four mesh currents.
Collect coefficients and solve for the mesh currents.

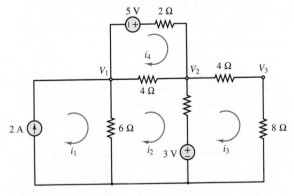

Figure P2.20

2.21   In the circuit in Figure P2.21, assume the source
voltage and source current and all resistances are
known.

a. Write the node equations required to determine the
node voltages.

b. Write the matrix solution for each node voltage in
terms of the known parameters.

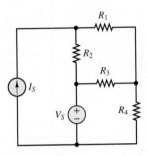

Figure P2.21

2.22   For the circuit of Figure P2.22 determine:

a. The most efficient way to solve for the voltage
across R_3. Prove your case.

b. The voltage across R_3.

$$V_{S1} = V_{S2} = 110 \text{ V}$$
$$R_1 = 500 \text{ m}\Omega \qquad R_2 = 167 \text{ m}\Omega$$
$$R_3 = 700 \text{ m}\Omega$$
$$R_4 = 200 \text{ m}\Omega \qquad R_5 = 333 \text{ m}\Omega$$

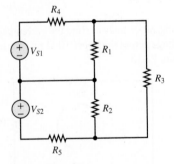

Figure P2.22

2.23   Figure P2.23 represents a temperature
measurement system, where temperature T is linearly
related to the voltage source V_{S2} by a transduction
constant k. Use nodal analysis to determine the
temperature.

$$V_{S2} = kT \qquad k = 10 \text{ V/°C}$$
$$V_{S1} = 24 \text{ V} \qquad R_S = R_1 = 12 \text{ k}\Omega$$
$$R_2 = 3 \text{ k}\Omega \qquad R_3 = 10 \text{ k}\Omega$$
$$R_4 = 24 \text{ k}\Omega \qquad V_{ab} = -2.524 \text{ V}$$

In practice, V_{ab} is used as the measure of temperature,
which is introduced to the circuit through a

temperature sensor modeled by the voltage source V_{S2} in series with R_S.

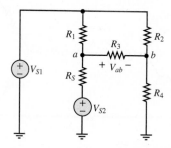

Figure P2.23

2.24   Use mesh analysis to find the mesh currents in Figure P2.24. Let $R_1 = 10\ \Omega$, $R_2 = 5\ \Omega$, $V_1 = 2$ V, $V_2 = 1$ V, $I_s = 2$ A.

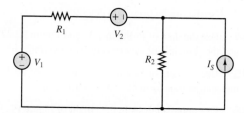

Figure P2.24

2.25   Use mesh analysis to find the mesh currents in Figure P2.25. Let $R_1 = 6\ \Omega$, $R_2 = 3\ \Omega$, $R_3 = 3\ \Omega$ $V_1 = 4$ V, $V_2 = 1$ V, $V_3 = 2$ V.

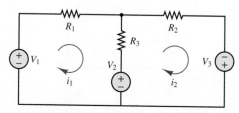

Figure P2.25

2.26   Use mesh analysis to find V_4 in Figure P2.26. Let $R_2 = 6\ \Omega$, $R_3 = 3\ \Omega$, $R_4 = 3\ \Omega$, $R_5 = 3\ \Omega$, $v_S = 4$ V, $i_S = 2$ A.

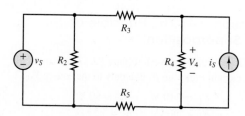

Figure P2.26

2.27   Use mesh analysis to find mesh currents in Figure P2.27. Let $R_1 = 8\ \Omega$, $R_2 = 3\ \Omega$, $R_3 = 5\ \Omega$, $R_4 = 2\ \Omega$, $R_5 = 4\ \Omega$, $R_6 = 3\ \Omega$, $V_1 = 4$ V, $V_2 = 2$ V, $V_3 = 1$ V, $V_4 = 2$ V, $V_5 = 3$ V, $V_6 = 2$ V.

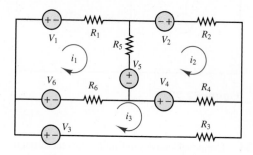

Figure P2.27

2.28   Use mesh analysis to find the current i in Figure P2.28. Assume $i_S = 2$ A.

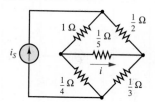

Figure P2.28

2.29   Use nodal analysis to find node voltages V_1, V_2, and V_3 in Figure P2.29. Let $R_1 = 10\ \Omega$, $R_2 = 6\ \Omega$, $R_3 = 7\ \Omega$, $R_4 = 4\ \Omega$, $I_1 = 2$ A, $I_2 = 1$ A.

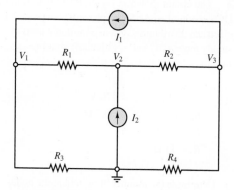

Figure P2.29

2.30   Use mesh analysis to find the currents through every branch in Figure P2.30. Let $R_1 = 10\ \Omega$, $R_2 = 5\ \Omega$, $R_3 = 4\ \Omega$, $R_4 = 1\ \Omega$, $V_1 = 5$ V, $V_2 = 2$ V.

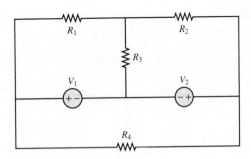

Figure P2.30

2.31 Use nodal analysis to find the current through R_4 in Figure P2.31. Let $R_1 = 10\ \Omega$, $R_2 = 6\ \Omega$, $R_3 = 4\ \Omega$, $R_4 = 3\ \Omega$, $R_5 = 2\ \Omega$, $R_6 = 2\ \Omega$, $I_1 = 2$ A, $I_2 = 3$ A, $I_3 = 5$ A.

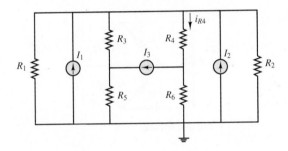

Figure P2.31

2.32 The circuit shown in Figure P2.32 is a simplified DC version of an AC three-phase wye-wye (Y-Y) electrical distribution system commonly used to supply industrial loads, particularly rotating machines.

$$V_{S1} = V_{S2} = V_{S3} = 170\text{ V}$$
$$R_{w1} = R_{w2} = R_{w3} = 0.7\ \Omega$$
$$R_1 = 1.9\ \Omega \qquad R_2 = 2.3\ \Omega$$
$$R_3 = 11\ \Omega$$

a. Determine the number of nonreference nodes.

b. Determine the number of unknown node voltages.

c. Compute v_1', v_2', v_3', and v_n'.

Notice that once v_n' is known the other unknown node voltages can be computed directly by voltage division.

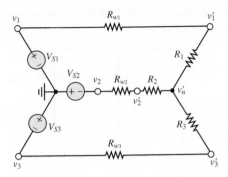

Figure P2.32

2.33 Using the data of Problem 2.32 and Figure P2.32

a. Determine the number of meshes.

b. Compute the mesh currents.

c. Use the mesh currents to determine v_n'.

2.34 Use the data of Problem 2.32 and Figure P2.32 and the principle of superposition to determine v_n'.

2.35 Use the data of Problem 2.32 and Figure P2.32 and source transformations to determine v_n'.

2.36 Use nodal analysis in the circuit of Figure P2.36 to find the three indicated node voltages and the current i. Assume: $R_1 = 10\ \Omega$, $R_2 = 20\ \Omega$, $R_3 = 20\ \Omega$, $R_4 = 10\ \Omega$, $R_5 = 10\ \Omega$, $R_6 = 10\ \Omega$, $R_7 = 5\ \Omega$, $V_1 = 20$ V, $V_2 = 20$ V.

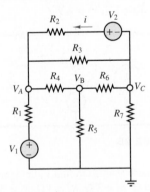

Figure P2.36

Section 2.4: The Principle of Superposition

2.37 With reference to Figure P2.37 determine the current i through R_1 due only to the source V_{S2}.

$$V_{S1} = 110\text{ V} \qquad V_{S2} = 90\text{ V}$$
$$R_1 = 560\ \Omega \qquad R_2 = 3.5\text{ k}\Omega$$
$$R_3 = 810\ \Omega$$

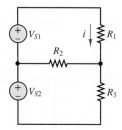

Figure P2.37

2.38 Refer to Figure P2.10 and use the principle of superposition to find the voltages at nodes A, B, and C. Assume $V_1 = 12$ V, $V_2 = 10$ V, $R_1 = 2\,\Omega$, $R_2 = 8\,\Omega$, $R_3 = 12\,\Omega$, $R_4 = 8\,\Omega$.

2.39 Use the principle of superposition to determine the voltage v across R_2 in Figure P2.39.

$$V_{S1} = V_{S2} = 12 \text{ V}$$
$$R_1 = R_2 = R_3 = 1 \text{ k}\Omega$$

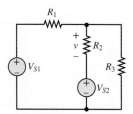

Figure P2.39

2.40 Refer to Figure P2.40 and use the principle of superposition to determine the component of the current i through R_3 that is due to V_{S2}.

$$V_{S1} = V_{S2} = 450 \text{ V}$$
$$R_1 = 7\,\Omega \qquad R_2 = 5\,\Omega$$
$$R_3 = 10\,\Omega \qquad R_4 = R_5 = 1\,\Omega$$

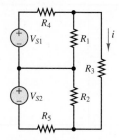

Figure P2.40

2.41 Refer to Figure P2.41 and use the principle of superposition to determine the current i through R_4 due to the current source I_S. Assume: $R_1 = 12\,\Omega$, $R_2 = 8\,\Omega$, $R_3 = 5\,\Omega$, $R_4 = 3\,\Omega$, $V_S = 3$ V, and $I_S = 2$ A.

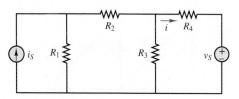

Figure P2.41

2.42 Refer to Figure P2.41 and use the principle of superposition to determine the current i through R_4 due to the voltage source V_S. Assume: $R_1 = 12\,\Omega$, $R_2 = 8\,\Omega$, $R_3 = 5\,\Omega$, $R_4 = 3\,\Omega$, $V_S = 3$ V, and $I_S = 2$ A.

2.43 Use the principle of superposition node to determine the voltages V_a and V_b in Figure P2.11. Let $R_1 = 10\,\Omega$, $R_2 = 4\,\Omega$, $R_3 = 6\,\Omega$, $R_4 = 6\,\Omega$, $V_1 = 2$ V, $V_2 = 4$ V, $I_1 = 2$ A.

2.44 Use the principle of superposition to determine the current i through R_3 in Figure P2.44. Let $R_1 = 10\,\Omega$, $R_2 = 4\,\Omega$, $R_3 = 2\,\Omega$, $R_4 = 2\,\Omega$, $R_5 = 2\,\Omega$, $V_S = 10$ V, $I_S = 2$ A.

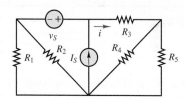

Figure P2.44

2.45 Figure P2.23 represents a temperature measurement system, where temperature T is linearly related to the voltage source V_{S2} by a transduction constant k. Use the principle of superposition to determine the temperature.

$$V_{S2} = kT \qquad\qquad k = 1 \text{ V/}°\text{C}$$
$$V_{S1} = 24 \text{ V} \qquad\quad R_s = R_1 = 12 \text{ k}\Omega$$
$$R_2 = 3 \text{ k}\Omega \qquad\quad R_3 = 10 \text{ k}\Omega$$
$$R_4 = 24 \text{ k}\Omega \qquad\quad V_{ab} = -2.524 \text{ V}$$

In practice, the voltage across R_3 is used as the measure of temperature, which is introduced to the circuit through a temperature sensor modeled by the voltage source V_{S2} in series with R_s.

2.46 Use the principle of superposition to determine the power P supplied by V_S in Figure P2.46. Let $R_1 = 12\,\Omega$, $R_2 = 10\,\Omega$, $R_3 = 5\,\Omega$, $R_4 = 5\,\Omega$, $V_S = 10$ V, $I_S = 5$ A. (Hint: Is power linear?)

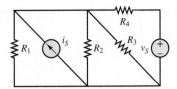

Figure P2.46

2.47   Use the principle of superposition to determine the current i_o through R_1 in Figure P2.47. Let $R_1 = 8\ \Omega$, $R_2 = 2\ \Omega$, $R_3 = 3\ \Omega$, $R_4 = 4\ \Omega$, $R_5 = 2\ \Omega$, $V_1 = 15$ V, $I_1 = 2$ A, $I_2 = 3$ A.

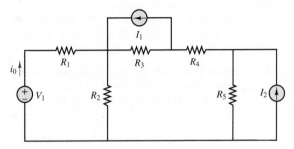

Figure P2.47

Section 2.5: Equivalent Networks

2.48   Find the Thévenin equivalent of the network seen by the 3-Ω resistor in Figure P2.48.

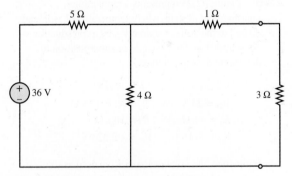

Figure P2.48

2.49   Find the Thévenin equivalent of the network seen by the 3-Ω resistor in Figure P2.49. Use it and voltage division to find the voltage v across the 3-Ω resistor.

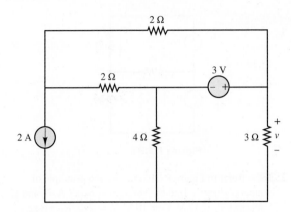

Figure P2.49

2.50   Find the Norton equivalent of the network seen by R_2 in Figure P2.50. Use it and current division to compute the current i through R_2. Assume $I_1 = 10$ A, $I_2 = 2$ A, $V_1 = 6$ V, $R_1 = 3\ \Omega$, and $R_2 = 4\ \Omega$.

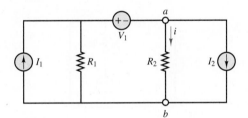

Figure P2.50

2.51   Find the Norton equivalent of the network between nodes a and b in Figure P2.51.

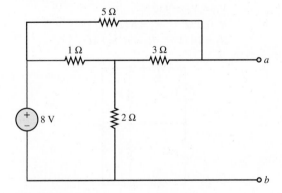

Figure P2.51

2.52   Find the Thévenin equivalent of the network seen by R in Figure P2.52 and use the result to compute the current i_R. Assume $V_o = 10$ V, $I_o = 5$ A, $R_1 = 2\ \Omega$, $R_2 = 2\ \Omega$, $R_3 = 4\ \Omega$, and $R = 3\ \Omega$.

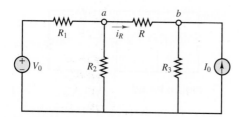

Figure P2.52

2.53 Find the Thévenin equivalent resistance seen by the load R_o in Figure P2.53.

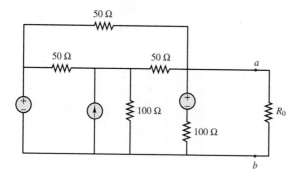

Figure P2.53

2.54 Find the Thévenin equivalent of the network seen by the load R_o in Figure P2.54.

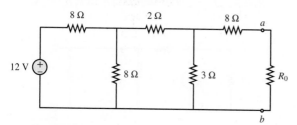

Figure P2.54

2.55 Find the Thévenin equivalent network seen by the load R_o in Figure P2.55, where $R_1 = 10\ \Omega$, $R_2 = 20\ \Omega$, $R_g = 0.1\ \Omega$, and $R_p = 1\ \Omega$.

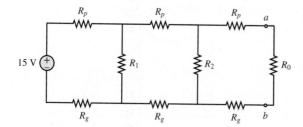

Figure P2.55

2.56 A Wheatstone bridge such as that shown in Figure P2.56 is used in numerous practical applications, such as determining the value of an unknown resistor R_X.

Determine $V_{ab} = V_a - V_b$ in terms of R, R_X, and V_S. If $R = 1\ \text{k}\Omega$, $V_S = 12$ V, and $V_{ab} = 12$ mV, what is the value of R_X?

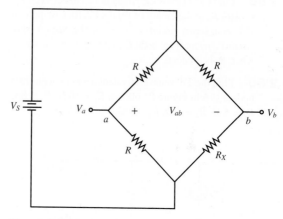

Figure P2.56

2.57 Thévenin's theorem can be useful when dealing with a Wheatstone bridge. For the circuit of Figure P2.57:

a. Express the Thévenin equivalent resistance seen by the load resistor R_o in terms of R_1, R_2, R_3, and R_X.

b. Determine the Thévenin equivalent network seen by R_o and use it to compute the power dissipated by R_o. Assume $R_o = 500\ \Omega$, $V_S = 12$ V, $R_1 = R_2 = R_3 = 1\ \text{k}\Omega$, and $R_X = 996\ \Omega$.

c. Find the power dissipated by the Thévenin equivalent resistance seen by R_o.

d. Find the power dissipated by the entire bridge when R_o is replaced by an open-circuit.

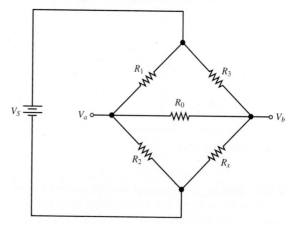

Figure P2.57

2.58 Find the Thévenin equivalent resistance seen by resistor R_3 in the circuit of Figure P2.5. Compute the Thévenin (open-circuit) voltage and the Norton (short-circuit) current from node A to node B when R_3 is the load.

2.59 Find the Thévenin equivalent resistance seen by resistor R_4 in the circuit of Figure P2.10. Compute the Thévenin (open-circuit) voltage and the Norton (short-circuit) current from node C to the reference node when R_4 is the load.

2.60 Find the Thévenin equivalent network seen from node a to b in Figure P2.60. Let $R_1 = 10\ \Omega$, $R_2 = 8\ \Omega$, $R_3 = 5\ \Omega$, $R_4 = 4\ \Omega$, $R_5 = 1\ \Omega$, $V_S = 10$ V, $I_S = 2$ A.

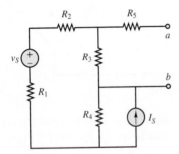

Figure P2.60

2.61 Find the Thévenin equivalent resistance seen by R_3 in Figure P2.23. Compute the Thévenin (open-circuit) voltage V_T and the Norton (short-circuit) current I_N from node a to node b when R_3 is the load.

2.62 Find the Norton equivalent of the network seen by R_5 in Figure P2.62. Use it and current division to compute the current through R_5. Assume $R_1 = 15\ \Omega$, $R_2 = 8\ \Omega$, $R_3 = 4\ \Omega$, $R_4 = 4\ \Omega$, $R_5 = 2\ \Omega$, $I_1 = 2$ A, $I_2 = 3$ A.

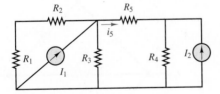

Figure P2.62

2.63 Find the Norton equivalent of the network seen by R_3 in Figure P2.63. Use it to determine the power dissipated by R_3. Assume $R_1 = 10\ \Omega$, $R_2 = 9\ \Omega$, $R_3 = 4\ \Omega$, $R_4 = 4\ \Omega$, $I_S = 2$ A.

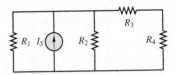

Figure P2.63

2.64 Find the Thévenin equivalent resistance seen by R in Figure P2.64. Compute the Thévenin (open-circuit) voltage V_T and the Norton (short-circuit) current I_N from node a to node b when R is the load. Assume:

$$I_B = 12\text{ A} \qquad R_B = 1\ \Omega$$
$$V_G = 12\text{ V} \qquad R_G = 0.3\ \Omega$$
$$R = 0.23\ \Omega$$

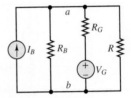

Figure P2.64

2.65 Find the Norton equivalent network between terminals a and b in Figure P2.65. Let $R_1 = 6\ \Omega$, $R_2 = 3\ \Omega$, $R_3 = 2\ \Omega$, $R_4 = 2\ \Omega$, $V_s = 10$ V, $I_S = 3$ A.

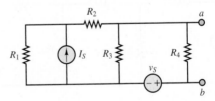

Figure P2.65

2.66 Find the Norton equivalent of the network seen by R_4 in Figure P2.66. Use it to determine the current through R_4. Assume $R_1 = 8\ \Omega$, $R_2 = 5\ \Omega$, $R_3 = 4\ \Omega$, $R_4 = 3\ \Omega$, $V_o = 10$ V, and $I_o = 2$ A.

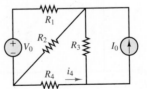

Figure P2.66

2.67 Find the Norton and Thévenin equivalent networks seen from node a to b in Figure P2.67. Assume $R_1 = 12\ \Omega$, $R_2 = 10\ \Omega$, $R_3 = 5\ \Omega$, $R_4 = 2\ \Omega$, $I_S = 3$ A.

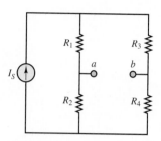

Figure P2.67

2.68 A real voltage source is modeled in Figure P2.68 as an ideal source V_S in series with a resistance R_S. This model accounts for internal power losses found in a real voltage source. The following data characterizes the real (nonideal) source:

When $R \to \infty$ $V_R = 20$ V
When $R = 2.7\ \text{k}\Omega$ $V_R = 18$ V

Determine the internal resistance R_S and the ideal voltage V_S.

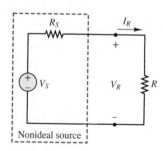

Figure P2.68

Section 2.6: Maximum Power Transfer

2.69 The Thévenin equivalent network seen by a load R_o is depicted in Figure P2.69. Assume $V_T = 10$ V, $R_T = 2\ \Omega$, and that the value of R_o is such that maximum power is transferred to it. Determine:

a. The value of R_o.

b. The power P_o dissipated by R_o.

c. The efficiency (P_o/P_{V_T}) of the circuit.

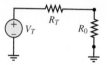

Figure P2.69

2.70 The Thévenin equivalent network seen by a load R_o is depicted in Figure P2.69. Assume $V_T = 25$ V, $R_T = 100\ \Omega$, and that the value of R_o is such that maximum power is transferred to it. Determine:

a. The value of R_o.

b. The power P_o dissipated by R_o.

c. The efficiency (P_o/P_{V_T}) of the circuit.

2.71 A real voltage source is modeled in Figure P2.68 as an ideal source V_S in series with a resistance R_S. This model accounts for internal power losses found in a real voltage source. A load R is connected across the terminals of the model. Assume:

$$V_S = 12\ \text{V} \qquad R_S = 0.3\ \Omega$$

a. Plot the power dissipated in the load as a function of the load resistance. What can you conclude?

b. Prove that your conclusion is valid in general.

C H A P T E R

3

AC NETWORK ANALYSIS

hapter 3 introduces capacitors and inductors, which are energy storage elements, and methods for solving circuits that contain them. This chapter also introduces AC circuits, which contain time-dependent sinusoidal voltage and current sources, as opposed to DC circuits, which contain constant sources only. Solutions of AC circuits containing capacitors and/or inductors result in differential equations because the i-v relationships for capacitors and inductors involve time derivatives. Luckily, the method of phasor analysis can be used to convert differential equations into algebraic equations, which are much easier to solve. However, the price ("there is no such thing as a free lunch") of using phasor analysis is that the algebraic equations contain complex quantities, which must be added, subtracted, multiplied, and divided. (Most calculators can perform these operations.) More important, it is necessary to understand the meaning of and relationships among complex quantities. With some practice and patience even those students with no prior experience using complex numbers will soon become comfortable with phasor analysis.

Sinusoids are an especially important class of signals for two reasons. First, nearly all residential and industrial electric power is generated, transmitted, and distributed as a sinusoidal waveform. All turbine-based power systems (e.g., coal-fired power stations, solar power arrays, hydroelectric dams, wind turbines) produce periodic rotating motion, which is represented mathematically by a sinusoid.

Second, all periodic waveforms (e.g., sawtooth, triangle, square) can be reconstructed as the sum of component sinusoidal waves (Fourier's theorem).

Sinusoidal signals (voltages and currents) have three basic characteristics: amplitude (or magnitude), frequency, and phase (or phase angle). The frequency of all voltages and currents in an AC circuit is constant, uniform, and determined by an independent voltage or current source. As a result, it is not necessary to compute frequency in phasor analysis. On the other hand, while the amplitude and phase of voltages and currents in an AC circuit are constant, they are not uniform and are determined not only by an independent source but also by the elements present in the circuit. Consequently, AC circuit analysis is concerned with the computation of the amplitude and phase of one or more voltages and currents. (DC circuit analysis was only concerned with amplitude.) Phasor analysis is well suited for AC circuit analysis because a phasor bundles together amplitude and phase into a single quantity known as a phasor.

In phasor analysis, resistors, capacitors, and inductors are represented as impedance elements. Impedance allows Ohm's law to be generalized as a phasor relationship applicable to resistors, capacitors, and inductors. Kirchhoff's laws can also be generalized as phasor relationships. Consequently, AC circuits can be solved using the same DC methods (e.g., voltage division, current division, nodal analysis, superposition, Thévenin's and Norton's theorem, and source transformations) discussed in Chapters 1 and 2. The only difference is that these relationships now involve phasors, that is, complex quantities.

The average and effective (root-mean-square) amplitude of a waveform are introduced in this chapter. An effective value represents the equivalent DC value required to supply or dissipate the same power as the AC waveform and thus provides a means of comparing different waveforms.

In this chapter and throughout the book, angles are given in units of radians, unless indicated otherwise.

Learning Objectives

Students will learn to...
1. Compute current, voltage, and energy of capacitors and inductors. *Section 3.1.*
2. Calculate the average and effective (root-mean-square) value of an arbitrary periodic waveform. *Section 3.2.*
3. Write the differential equation(s) for circuits containing inductors and capacitors. *Section 3.3.*
4. Convert time-domain sinusoidal voltages and currents to phasor notation, and vice versa; and represent circuits using impedances. *Section 3.4.*
5. Apply DC circuit analysis methods to AC circuits in phasor form. *Section 3.5.*

3.1 CAPACITORS AND INDUCTORS

The ideal resistor was introduced in Chapter 1 as a useful approximation of many practical electrical devices. However, in addition to resistance, which always dissipates energy, an electric circuit may also exhibit capacitance and inductance, which act to store and release energy, in the same way that an expansion tank and flywheel, respectively, act in a mechanical system. These two distinct energy storage mechanisms are represented in electric circuits by two ideal circuit elements: the ideal capacitor and the ideal inductor, which approximate the behavior of actual discrete capacitors and inductors. They also approximate the bulk properties of capacitance and inductance that are present in any physical system. In practice, any element of an electric circuit will exhibit some resistance, some inductance, and some capacitance, that is, some ability to dissipate and store energy.

The energy of a capacitor is stored within the electric field between two conducting plates, while the energy of an inductor is stored within the magnetic field of a conducting coil. Both elements can be charged (i.e., stored energy is increased) or discharged (i.e., stored energy is decreased). Ideal capacitors and inductors can store energy indefinitely; however, in practice, discrete capacitors and inductors exhibit "leakage," which typically results in a gradual reduction in the stored energy over time.

All the relationships for capacitors and inductors exhibit duality, which means that the capacitor relations are mirror images of the inductor relations. Examples of duality are apparent in Table 3.1.

Table 3.1 **Properties of capacitors and inductors**

	Capacitors	**Inductors**
Differential i-v	$i = C\dfrac{dv}{dt}$	$v = L\dfrac{di}{dt}$
Integral i-v	$v_C(t) = \dfrac{1}{C}\displaystyle\int_{-\infty}^{t} i_C(\tau)\,d\tau$	$i_C(t) = \dfrac{1}{L}\displaystyle\int_{-\infty}^{t} v_L(\tau)\,d\tau$
DC equivalent	Open-circuit	Short-circuit
Two in series	$C_{eq} = \dfrac{C_1 C_2}{C_1 + C_2}$	$L_{eq} = L_1 + L_2$
Two in parallel	$C_{eq} = C_1 + C_2$	$L_{eq} = \dfrac{L_1 L_2}{L_1 + L_2}$
Stored energy	$W_C = \dfrac{1}{2} C v_C^2$	$W_L = \dfrac{1}{2} L i_L^2$

The Ideal Capacitor

A capacitor is a device that can store energy due to a charge separation. In general, a capacitor (and thus, capacitance) is present when any two conducting surfaces are separated by a distance. A simple example is two parallel plates of shared cross-sectional area A separated by a distance d. The gap between the plates may be a vacuum or filled with some dielectric material, such as air, mica, or Teflon. The impact of the dielectric material on the capacitance is represented by the dielectric constant κ.[1] Figure 3.2 depicts a typical configuration and the circuit symbol for a capacitor.

The capacitance C of an *ideal* parallel-plate capacitor such as the one described above is:

$$C = \frac{\kappa \varepsilon_0 A}{d} \tag{3.1}$$

where $\varepsilon_0 = 8.85 \times 10^{-12}\ F/m$ is the permittivity constant of a vacuum.

---

[1]A dielectric material is a material that is not an electrical conductor but contains a large number of electric dipoles, which become polarized in the presence of an electric field.

MAKE THE CONNECTION

Hydraulic Analog of a Capacitor

If the walls of a vessel have some elasticity, energy is stored in the walls when the vessel is filled by a fluid or gas (e.g., an inflated balloon). The ratio of the mass of the fluid or gas to the *potential energy* stored in the walls per unit mass is the **fluid capacitance** of the vessel, a property similar to **electrical capacitance**. Figure 3.1 depicts a gas bag accumulator, such as an expansion tank attached to the hot water line in many residential homes. The two-chamber arrangement permits fluid to displace a membrane separating an incompressible fluid (e.g., water) from a compressible fluid (e.g., air). The analogy shown in Figure 3.1 assumes that the reference pressure p_0 and the reference voltage v_2 are both zero.

$$q_f = C_f \frac{d\Delta p}{dt} = C_f \frac{dp}{dt}$$

$$i = C \frac{d\Delta v}{dt} = C \frac{dv_1}{dt}$$

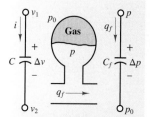

Figure 3.1 Analogy between electrical and fluid capacitance

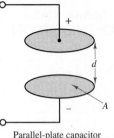

Parallel-plate capacitor

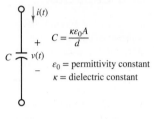

Circuit
symbol

Figure 3.2 Structure of
parallel-plate capacitor

The presence of a dielectric or vacuum between the conducting plates does not permit charge to pass directly from one plate to the other. However, if the applied voltage across a capacitor changes, so will the accumulated charge. Thus, although no charge can literally pass from one plate of an ideal capacitor directly through to the other, a change in voltage will cause the accumulated charge to change, which is the equivalent effect of a current through the capacitor.

At all times the charge separation is proportional to the applied voltage

$$Q_C = Cv_C \tag{3.2}$$

where the parameter C is the capacitance and is a measure of the ability of the device to accumulate charge. The unit of capacitance is coulomb per volt, or farad (F). The farad is an impractically large unit for many common electronic applications; units of microfarads (1 μF = 10^{-6} F) and picofarads (1 pF = 10^{-12} F) are more common in practice.

The current through a capacitor is defined as the time rate of change of its stored charge. That is,

$$i_C(t) = \frac{dQ_C(t)}{dt} \tag{3.3}$$

The *i-v* relationship for a capacitor is obtained from equation 3.3 by using equation 3.2 to plug in for $q_C(t)$. The result is:

$$\boxed{i_C(t) = C\frac{dv_C(t)}{dt} \qquad \textit{i-v relationship for capacitor}} \tag{3.4}$$

Equation 3.4 can be integrated to yield an equivalent *i-v* relationship for a capacitor:

$$v_C(t) = \frac{1}{C}\int_{-\infty}^{t} i_C(\tau)\,d\tau \tag{3.5}$$

One immediate implication of equation 3.4 is that the current through a capacitor in a DC circuit is zero. Why? Since the voltage across a capacitor in a DC circuit must, by definition, be constant, the time derivative of the voltage must be zero. Thus, equation 3.4 requires the current through the capacitor to also be zero.

A capacitor in a DC circuit is equivalent to an open-circuit.

Equation 3.5 indicates that the voltage across a capacitor depends on the history of the current through it. To calculate that voltage it is necessary to know the initial voltage V_0 (i.e., an initial condition) across the capacitor at some previous time t_0. Then:

$$v_C(t) = V_0 + \frac{1}{C}\int_{t_0}^{t} i_C(\tau)\,d\tau \qquad t \geq t_0 \tag{3.6}$$

The significance of the initial voltage $v_C(t_0) = V_0$ is simply that at time t_0 some charge was stored in the capacitor, resulting in V_0, according to the relationship $Q = Cv$.

Equivalent Capacitance

Just as resistors can be in series and parallel to yield an equivalent resistance, so capacitors can also be in series and parallel to yield an equivalent capacitance. The calculation rules are given below.

For two capacitors in series and parallel, the equivalent capacitances are, respectively:

$$C_{eq} = \frac{C_1 C_2}{C_1 + C_2} \qquad \text{and} \qquad C_{eq} = C_1 + C_2 \qquad (3.7)$$

Notice that the rule for the equivalent capacitance of two capacitors in series is the product divided by the sum, which is the same rule used for two resistors in parallel. Likewise, the equivalent capacitance of two capacitors in parallel is simply the sum of the two, which is the same rule used for two resistors in series. The more general rules are illustrated in Figure 3.3.

> **LO** When calculating equivalent capacitance, capacitors in series combine like resistors in parallel and capacitors in parallel combine like resistors in series.

Discrete Capacitors

Actual capacitors are rarely constructed of two parallel plates separated by air because this configuration either yields very low values of capacitance or requires very large plate areas. To increase the capacitance (i.e., the ability to store energy), physical capacitors are often made of tightly rolled sheets of metal film, with a dielectric (e.g., paper or Mylar) sandwiched in between. Table 3.2 illustrates typical values, materials, maximum voltage ratings, and useful frequency ranges for various types of capacitors. The voltage rating is important because any insulator will break down if a sufficiently high voltage is applied across it.

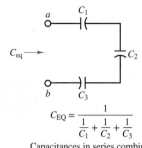

$$C_{EQ} = \frac{1}{\dfrac{1}{C_1} + \dfrac{1}{C_2} + \dfrac{1}{C_3}}$$

Capacitances in series combine like resistors in parallel

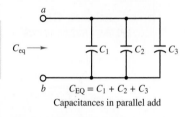

$$C_{EQ} = C_1 + C_2 + C_3$$

Capacitances in parallel add

Figure 3.3 Equivalent capacitance in a circuit

Table 3.2 **Capacitors**

Material	Capacitance range	Maximum voltage (V)	Frequency range (Hz)
Mica	1 pF to 0.1 μF	100–600	10^3–10^{10}
Ceramic	10 pF to 1 μF	50–1,000	10^3–10^{10}
Mylar	0.001 μF to 10 μF	50–500	10^2–10^8
Paper	1,000 pF to 50 μF	100–10,000	10^2–10^8
Electrolytic	0.1 μF to 0.2 F	3–600	10–10^4

In practice, actual capacitors exhibit some leakage between the plates. Imperfect construction techniques invariably provide some capability for charge to pass from one plate to the other. This imperfection is often represented by an equivalent resistance in parallel with an ideal capacitor.

Energy Storage in Capacitors

The energy stored in a capacitor $W_C(t)$ may be derived easily from its definition as the time integral of power, which is the product of voltage and current:

$$P_C(t) = i_C(t)v_C(t) = C\frac{dv_C(t)}{dt}v_C(t) = \frac{d}{dt}\left[\frac{1}{2}Cv_C^2(t)\right] \qquad (3.8)$$

The total energy stored in the inductor is found by integrating the power, as shown below:

$$W_C(t) = \int P_C(\tau)d\tau = \int \frac{d}{d\tau}\left[\frac{1}{2}Cv_C^2(\tau)\right]d\tau \tag{3.9}$$

$$W_C(t) = \frac{1}{2}Cv_C^2(t) \qquad \text{Energy, in joules, stored in a capacitor}$$

Example 3.4 illustrates the calculation of the energy stored in a capacitor.

FOCUS ON MEASUREMENTS

Capacitive Displacement Transducer and Microphone

As shown in Figure 3.2, the capacitance of a flat parallel-plate capacitor is:

$$C = \frac{\varepsilon A}{d} = \frac{\kappa\varepsilon_0 A}{d}$$

where ε is the **permittivity** of the dielectric material, κ is the dielectric constant, $\varepsilon_0 = 8.854 \times 10^{-12}$ F/m is the permittivity of a vacuum, A is the area of each of the plates, and d is their separation. The dielectric constant for air is $\kappa_{air} \approx 1$. Thus, the capacitance of two flat parallel plates of area 1 m^2, separated by a 1-mm air gap, is 8.854 nF, a very small value for such large plates. As a result, flat parallel-plate capacitors are impractical for use in most electronic devices. On the other hand, parallel-plate capacitors find application as motion transducers, that is, as devices that can measure the motion or displacement of an object. In a capacitive motion transducer, the plates are designed to allow relative motion when subjected to an external force. Using the capacitance value just derived for a parallel-plate capacitor, one can obtain the expression

$$C = \frac{8,854 \times 10^{-3}A}{x}$$

where C is the capacitance in picofarads, A is the area of the plates in square millimeters, and x is the separation distance in millimeters. Note that the change in C due to a change in x is nonlinear, since $C \propto 1/x$. However, for small changes in x, the change in C is approximately linear.

The sensitivity S of the transducer is defined as the rate of change in capacitance C with respect to a change in separation distance x.

$$S = \frac{dC}{dx} = -\frac{8.854 \times 10^{-3}A}{x^2} \quad \frac{\text{pF}}{\text{mm}}$$

Thus, the sensitivity is itself a function of the separation distance, as shown in Figure 3.4. Note that as $x \to 0$, the slope of $C(x)$ increases and so the sensitivity S increases as well. Figure 3.4 depicts this behavior for a transducer with area equal to 10 mm^2. This type of capacitive displacement transducer is used in the popular condenser microphone, in which sound pressure waves act to displace a thin metallic foil. The change in capacitance can then be converted to a change in voltage or current by means of a suitable circuit. An extension of this concept that permits measurement of differential pressures is shown in Figure 3.5. A three-terminal variable capacitor is made of two fixed surfaces with a single deflecting plate, often made of steel, between them. Typically, the fixed surfaces are spherical depressions ground into glass disks and coated with a conducting material.

(*Continued*)

(Concluded)

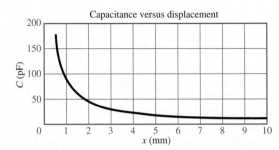

Figure 3.4 Response of a capacitive displacement transducer

Inlet orifices expose the deflecting plate to the outside fluid or gas. When the pressure on both sides of the deflecting plate is the same, the capacitance between terminals b and d, denoted by C_{db}, will be equal to that between terminals b and c, denoted by C_{bc}. If any pressure differential exists, the two capacitances will change, with an increase on the side where the deflecting plate has come closer to the fixed surface and a corresponding decrease on the other side.

A Wheatstone bridge circuit, such as that shown in Figure 3.5, is ideally suited to precisely balance the output voltage v_{out} when the differential pressure across the transducer is zero. The output voltage will deviate from zero whenever the two capacitances are not equal because of a pressure differential across the transducer. The bridge circuit will be analyzed later.

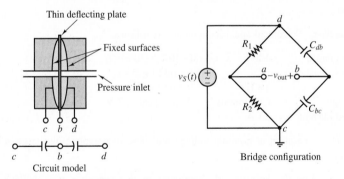

Figure 3.5 Capacitive pressure transducer and related bridge circuit

---

EXAMPLE 3.1   Charge Separation in Ultracapacitors

Problem

Ultracapacitors are finding application in a variety of fields, including as a replacement or supplement for batteries in hybrid-electric vehicles. In this example you will make your first acquaintance with these devices.

An ultracapacitor, or "supercapacitor," stores energy electrostatically by polarizing an electrolytic solution. Although it is an electrochemical device (also known as an

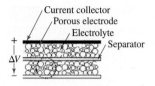

Figure 3.6 Ultracapacitor structure

electrochemical double-layer capacitor), there are no chemical reactions involved in its energy storage mechanism. This mechanism is highly reversible, allowing the ultracapacitor to be charged and discharged hundreds of thousands of times. An ultracapacitor can be viewed as two nonreactive porous plates suspended within an electrolyte, with a voltage applied across the plates. The applied potential on the positive plate attracts the negative ions in the electrolyte while the potential on the negative plate attracts the positive ions. This effectively creates two layers of capacitive storage, one where the charges are separated at the positive plate and another at the negative plate.

Recall that capacitors store energy in the form of separated electric charge. The greater the area for storing charge and the closer the separated charges, the greater the capacitance. A conventional capacitor gets its area from plates of a flat, conductive material. To achieve high capacitance, this material can be wound in great lengths, and sometimes a texture is imprinted on it to increase its surface area. A conventional capacitor separates its charged plates with a dielectric material, sometimes a plastic or paper film, or a ceramic. These dielectrics can be made only as thin as the available films or applied materials.

An ultracapacitor gets its area from a porous carbon-based electrode material, as shown in Figure 3.6. The porous structure of this material allows its surface area to approach 2,000 square meters per gram (m^2/g), much greater than can be accomplished using flat or textured films and plates. An ultracapacitor's charge separation distance is determined by the size of the ions in the electrolyte, which are attracted to the charged electrode. This charge separation [less than 10 angstroms (Å)] is much smaller than can be achieved using conventional dielectric materials. The combination of enormous surface area and extremely small charge separation gives the ultracapacitor its outstanding capacitance relative to conventional capacitors.

Use the data provided to calculate the charge stored in an ultracapacitor and calculate how long it will take to discharge the capacitor at the maximum current rate.

Solution

Known Quantities: Technical specifications are as follows:

Capacitance	100 F	$(-10\%/+30\%)$
Series resistance	DC	15 mΩ ($\pm25\%$)
	1 kHz	7 mΩ ($\pm25\%$)
Voltage	Continuous	2.5 V; peak 2.7 V
Rated current	25 A	

Find: Charge separation at nominal voltage and time to complete discharge at maximum current rate.

Analysis: Based on the definition of charge storage in a capacitor, we calculate

$$Q = Cv = 100 \text{ F} \times 2.5 \text{ V} = 250 \text{ C}$$

To calculate how long it would take to discharge the ultracapacitor, approximate the current as:

$$i = \frac{dQ}{dt} \approx \frac{\Delta Q}{\Delta t}$$

Since the available charge is 250 C, the time to completely discharge the capacitor, assuming a constant 25-A discharge, is:

$$\Delta t = \frac{\Delta Q}{i} = \frac{250 \text{ C}}{25 \text{ A}} = 10 \text{ s}$$

Comments: Ultracapacitors will be explored further in Chapter 5. The charging and discharging behavior of these devices is examined, taking into consideration their internal resistance.

EXAMPLE 3.2   Calculating Capacitor Current From Voltage

Problem

Calculate the current through a capacitor from knowledge of its terminal voltage.

Solution

Known Quantities: Capacitor terminal voltage; capacitance value.

Find: Capacitor current.

Assumptions: The initial current through the capacitor is zero.

Schematics, Diagrams, Circuits, and Given Data: $v(t) = 5(1 - e^{-t/10^{-6}})$ V; $t \geq 0$ s; $C = 0.1$ μF. The terminal voltage is plotted in Figure 3.7.

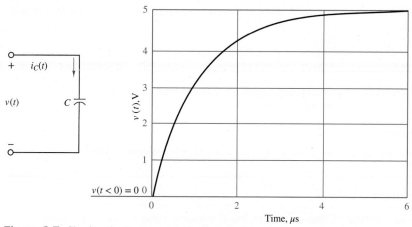

Figure 3.7 Circuit and voltage response for Example 3.2.

Assumptions: The capacitor is initially discharged: $v(t = 0) = 0$.

Analysis: Using the defining differential relationship for the capacitor, we may obtain the current by differentiating the voltage:

$$i_C(t) = C\frac{dv(t)}{dt} = 10^{-7}\frac{5}{10^{-6}}\left(e^{-t/10^{-6}}\right) = 0.5e^{-t/10^{-6}} \quad \text{A} \qquad t \geq 0$$

A plot of the capacitor current is shown in Figure 3.8. Note how the current jumps to 0.5 A instantaneously as the voltage rises exponentially: The ability of a capacitor's current to change instantaneously is an important property of capacitors.

Comments: As the voltage approaches the constant value 5 V, the capacitor reaches its maximum charge storage capability for that voltage (since $Q = Cv$) and no more current flows through the capacitor. The total charge stored is $Q = 0.5 \times 10^{-6}$ C. This is a fairly small amount of charge, but it can produce a substantial amount of current for a brief time. For example, the fully charged capacitor could provide 100 mA of current for a time equal to 5 μs:

$$I = \frac{\Delta Q}{\Delta t} = \frac{0.5 \times 10^{-6}}{5 \times 10^{-6}} = 0.1 \text{ A}$$

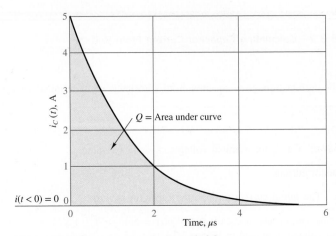

Figure 3.8 Current response for Example 3.2.

There are many useful applications of this energy storage property of capacitors in practical circuits.

EXAMPLE 3.3   Calculating Capacitor Voltage From Current and
                 an Initial Condition

Problem

Solve for the voltage across a capacitor from knowledge of its current and initial charge.

Solution

Known Quantities: Capacitor current; initial capacitor voltage; capacitance value.

Find: Capacitor voltage as a function of time.

Schematics, Diagrams, Circuits, and Given Data:

$$i_C(t) = I = \begin{cases} 0 & t < 0\,\text{s} \\ 10\,\text{mA} & 0 \le t \le 1\,\text{s} \\ 0 & t > 1\,\text{s} \end{cases}$$

$$v_C(t = 0) = 2\ \text{V} \qquad C = 1{,}000\ \mu\text{F}$$

The capacitor current is plotted in Figure 3.9(a).

Assumptions: The capacitor is initially charged such that $V_0 = v_C(t = 0) = 2$ V.

Analysis: The integral relationship between voltage and current for a capacitor can be used to find voltage when current is known.

$$v_C(t) = \frac{1}{C}\int_{t_0}^{t} i_C(t')\,dt' + v_C(t_0) \qquad t \ge t_0$$

$$= \begin{cases} \dfrac{1}{C}\displaystyle\int_{0}^{1} I\,dt' + V_0 = \dfrac{I}{C}t + V_0 = 10t + 2\ \text{V} & 0 \le t \le 1\,\text{s} \\ 12\ \text{V} & t > 1\,\text{s} \end{cases}$$

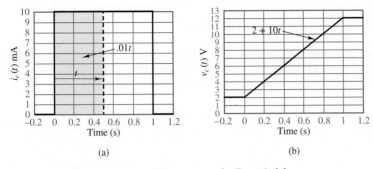

Figure 3.9 Current input and voltage response for Example 3.3.

Comments: Once the current stops, at $t = 1$ s, the capacitor voltage remains constant because the charge remains constant. That is, $v = Q/C = \text{constant} = 12$ V at $t = 1$ s. Remember, the final value of the capacitor voltage depends on two factors: (1) the initial value of the capacitor voltage and (2) the history of the capacitor current. Figure 3.9(a) and (b) depict the two waveforms.

EXAMPLE 3.4   Energy Storage in Ultracapacitors

Problem

Determine the energy stored in the ultracapacitor of Example 3.1.

Solution

Known Quantities: See Example 3.1.

Find: Energy stored in capacitor.

Analysis: To calculate the energy, use equation 3.9:

$$W_C = \frac{1}{2}Cv_C^2 = \frac{1}{2}(100 \text{ F})(2.5 \text{ V})^2 = 312.5 \text{ J}$$

CHECK YOUR UNDERSTANDING

Compare the energy stored in the ultracapacitor of Example 3.4 with a (similarly sized) electrolytic capacitor used in power electronics applications. Calculate the energy stored for a 2,000-μF electrolytic capacitor rated at 400 V.

Answer: 160 J

CHECK YOUR UNDERSTANDING

Compare the charge separation achieved in the ultracapacitor of Example 3.1 with a (similarly sized) electrolytic capacitor used in power electronics applications, by calculating the charge separation for a 2,000-μF electrolytic capacitor rated at 400 V.

Answer: 0.8 C

CHECK YOUR UNDERSTANDING

Find the maximum current through the capacitor of Example 3.3 if the capacitor voltage is described by $v_C(t) = 5t + 3$ V for $0 \leq t \leq 5$ s.

Answer: 5 mA

CHECK YOUR UNDERSTANDING

The voltage waveform shown below appears across a 1,000-μF capacitor. Plot the capacitor current $i_C(t)$.

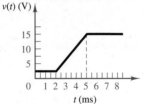

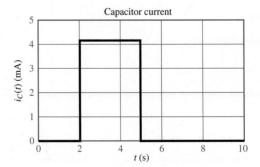

The Ideal Inductor

An inductor is an element that can store energy in a magnetic field within and around a conducting coil. In general, an inductor (and thus, inductance) is present whenever a conducting wire is turned to form a loop. A simple example is a solenoid, which is a narrow and tightly wound coil of length ℓ, cross-sectional area A, and N turns. Inductors are typically made by winding wire around a **core**, which

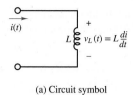

(a) Circuit symbol

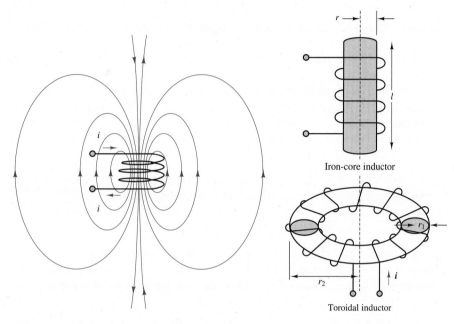

(b) Magnetic flux lines in the vicinity of a current-carrying coil   (c) Practical inductors

Figure 3.10 Inductance and practical inductors

can be an insulator or a ferromagnetic material, as shown in Figure 3.10. A current through the coil establishes a magnetic field through and around the core. In an ideal inductor, the resistance of the wire is zero.

The inductance L is defined by the following ratio:

$$L \equiv \frac{N\Phi}{i_L}$$

where Φ is the magnetic flux through the inductor core and i_L is the current through the inductor coil. The inductance of an ideal solenoid is:

$$L = \frac{\mu N^2 A}{\ell}$$

where μ is the permeability of the core. Another inductor found in many applications is the toroid, which is also depicted in Figure 3.10. Expressions for the inductance of toroids with rectangular and circular cross sections are readily found.

The inductance of a coil is measured in henrys (H) where

$$1\ \text{H} = 1\ \text{V·s/A} \tag{3.10}$$

Henrys are reasonable units for practical inductors although millihenrys (mH) are very common and microhenrys (μH) are occasionally found.

The *i-v* relationship for an inductor is derived directly from Faraday's law of induction but with the total flux $N\Phi$ replaced by Li from the definition of inductance L. The result is:

$$v_L(t) = L\frac{d\,i_L(t)}{dt} \qquad \text{\textit{i-v} relation for inductor} \tag{3.11}$$

Equation 3.11 can be integrated to yield an equivalent *i-v* relationship for an inductor:

$$i_L(t) = \frac{1}{L}\int_{-\infty}^{t} v_L(\tau)\,d\tau \tag{3.12}$$

One immediate implication of equation 3.11 is that the voltage across an inductor in a DC circuit is zero. Why? Since the current through an inductor in a DC circuit must, by definition, be constant, the time derivative of the current must be zero. Thus, equation 3.11 requires the voltage across an inductor to also be zero.

An inductor in a DC circuit is equivalent to a short-circuit.

Equation 3.12 indicates that the current through an inductor depends on the history of the voltage across it. To calculate the current it is necessary to know the initial current I_0 (i.e., an initial condition) through the inductor at some previous time t_0. Then:

$$i_L(t) = I_0 + \frac{1}{L}\int_{t_0}^{t} v_L(\tau)\,d\tau \qquad t \geq t_0 \tag{3.13}$$

Equivalent Inductance

Just as resistors can be in series and parallel to yield an equivalent resistance, so inductors can also be in series and parallel to yield an equivalent inductance. The calculation rules are given below.

For two inductors in series and parallel, the equivalent inductances are, respectively,

$$L_{\text{eq}} = L_1 + L_2 \qquad \text{and} \qquad L_{\text{eq}} = \frac{L_1 L_2}{L_1 + L_2} \tag{3.14}$$

Notice that the equivalent inductance of two inductors in series is simply the sum of the two, which is the same rule used for two resistors in series. Likewise, the rule for the equivalent inductance of two inductors in parallel is the product divided by the sum, which is the same rule used for two resistors in parallel. The more general rules are illustrated in Figure 3.11.

When calculating equivalent inductance, inductors in series combine like resistors in series and inductors in parallel combine like resistors in parallel.

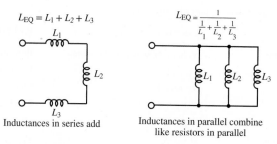

$$L_{EQ} = L_1 + L_2 + L_3$$

Inductances in series add

$$L_{EQ} = \frac{1}{\frac{1}{L_1} + \frac{1}{L_2} + \frac{1}{L_3}}$$

Inductances in parallel combine like resistors in parallel

Figure 3.11 Equivalent inductance in a circuit

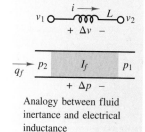
Duality

All the relationships for capacitors and inductors exhibit duality, which means that the capacitor relations are mirror images of the inductor relations. Specifically, the roles played by voltage and current in a capacitor relation are reversed in the analogous inductor relation. For example, the *i-v* relationships for capacitors and inductors, respectively, are:

$$i = C\frac{dv}{dt} \quad \text{and} \quad v = L\frac{di}{dt}$$

Notice that the inductor relation is obtained from the capacitor relation by replacing *i* with *v* and *v* with *i*. It is also necessary, of course, to replace the capacitance *C* with the inductance *L*. Another example of duality is found in the energy storage relations for capacitors and inductors (see the next section).

Duality is also at work in other relations not involving voltage and current explicitly. For example, consider the rules for calculating equivalent capacitance and equivalent inductance. Capacitors in series combine like inductors in parallel while capacitors in parallel combine like inductors in series. Another example of duality is seen in the DC behavior of capacitors and inductors. In a DC circuit, a capacitor acts like an open-circuit while an inductor acts like a short-circuit. Other examples of duality can be found in later chapters. Duality provides a powerful mnemonic for students trying to memorize equations for a test!

Energy Storage in Inductors

The energy stored in an inductor $W_L(t)$ may be derived easily from its definition as the time integral of power, which is the product of voltage and current:

$$P_L(t) = i_L(t)v_L(t) = i_L(t)L\frac{di_L(t)}{dt} = \frac{d}{dt}\left[\frac{1}{2}Li_L^2(t)\right] \tag{3.15}$$

$$W_L(t) = \int P_L(\tau)\,d\tau = \int \frac{d}{d\tau}\left[\frac{1}{2}Li_L^2(\tau)\right]d\tau \tag{3.16}$$

$$W_L(t) = \frac{1}{2}Li_L^2(t) \qquad \text{Energy, in joules, stored in an inductor}$$

Note, once again, the duality with the expression for the energy stored in a capacitor, in equation 3.9.

Table 3.3 **Analogy between electric and fluid circuits**

Property	Electric element or equation	Hydraulic analogy
Potential variable	Voltage or potential difference	Pressure difference
Flow variable	Current	Fluid volume flow rate
Resistance	Resistor R	Fluid resistor R_f
Capacitance	Capacitor C	Fluid capacitor C_f
Inductance	Inductor L	Fluid inertor I_f
Power dissipation	$P = i^2R$	$P_f = q_f^2 R_f$
Stored energy (potential)	$W_C = \frac{1}{2}Cv^2$	$W_p = \frac{1}{2}C_f p^2$
Stored energy (kinetic)	$W_L = \frac{1}{2}Li^2$	$W_k = \frac{1}{2}I_f q_f^2$

EXAMPLE 3.5   Calculating Inductor Voltage From Current

Problem

Calculate the voltage across an inductor from knowledge of its current.

Solution

Known Quantities: Inductor current; inductance value.

Find: Inductor voltage.

Schematics, Diagrams, Circuits, and Given Data:

$$
i_L(t) = \begin{cases}
0\,\text{mA} & t \le 1\,\text{ms} \\[2mm]
-\dfrac{0.1}{4} + \dfrac{0.1}{4}t \quad \text{mA} & 1 \le t \le 5\,\text{ms} \\[2mm]
0.1\,\text{mA} & 5 \le t \le 9\,\text{ms} \\[2mm]
13 \times \dfrac{0.1}{4} - \dfrac{0.1}{4}t \quad \text{mA} & 9 \le t \le 13\,\text{ms} \\[2mm]
0\,\text{mA} & t \ge 13\,\text{ms}
\end{cases}
$$

$$L = 10\,\text{H}$$

where time t is in milliseconds. The inductor current is plotted in Figure 3.12.

Assumptions: $i_L(t = 0) \le 0$.

Analysis: The voltage across the inductor is obtained by differentiating the current and multiplying by the inductance L.

$$
v_L(t) = L\frac{di_L(t)}{dt}
$$

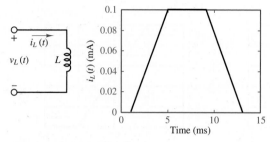

Figure 3.12 Inductor current input.

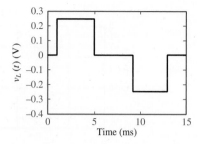

Figure 3.13 Inductor voltage response.

Piecewise differentiating the expression for the inductor current, we obtain

$$v_L(t) = \begin{cases} 0\,\text{V} & t < 1\,\text{ms} \\ 0.25\,\text{V} & 1 < t < 5\,\text{ms} \\ 0\,\text{V} & 5 < t < 9\,\text{ms} \\ -0.25\,\text{V} & 9 < t < 13\,\text{ms} \\ 0\,\text{V} & t > 13\,\text{ms} \end{cases}$$

The inductor voltage is plotted in Figure 3.13.

Comments: Note how the inductor voltage has the ability to change instantaneously!

EXAMPLE 3.6   Calculating Inductor Current From Voltage

Problem

Use a time plot of the voltage across an inductor and its initial current to calculate the current through it as a function of time.

Solution

Known Quantities: Inductor voltage; initial condition (current at $t = 0$); inductance value.

Find: Inductor current.

Schematics, Diagrams, Circuits, and Given Data:

$$v(t) = \begin{cases} 0\,\text{V} & t < 0\,\text{s} \\ -10\,\text{mV} & 0 < t < 1\,\text{s} \\ 0\,\text{V} & t > 1\,\text{s} \end{cases}$$

$$L = 10\,\text{mH}; \qquad i_L(t = 0) = I_0 = 0\,\text{A}$$

The voltage across the inductor is plotted in Figure 3.14(a).

Analysis: Use the integral i-v relationship for an inductor to obtain the current through it:

$$i_L(t) = i_L(t_0) + \frac{1}{L}\int_{t_0}^{t} v(\tau)\,d\tau \qquad t \geq t_0$$

$$= \begin{cases} I_0 + \dfrac{1}{L}\displaystyle\int_{0}^{t}(-10 \times 10^{-3})\,d\tau = 0 + \dfrac{-10^{-2}}{10^{-2}}t = -t\,\text{A} & 0 \leq t \leq 1\,\text{s} \\[4mm] -1\,\text{A} & t \geq 1\,\text{s} \end{cases}$$

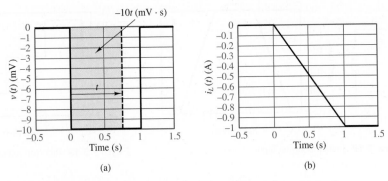

Figure 3.14 Inductor voltage input and current response.

The inductor current is plotted in Figure 3.14b.

Comments: Note that the inductor voltage can change instantaneously!

EXAMPLE 3.7   **Energy Storage in an Ignition Coil**

Problem

Determine the energy stored in an automotive ignition coil.

Solution

Known Quantities: Inductor current initial condition (current at $t = 0$); inductance value.

Find: Energy stored in inductor.

Schematics, Diagrams, Circuits, and Given Data: $L = 10$ mH; $i_L = I_0 = 8$ A.

Analysis:

$$W_L = \frac{1}{2}Li_L^2 = \frac{1}{2} \times 10^{-2} \times 64 = 32 \times 10^{-2} = 320 \text{ mJ}$$

Comments: A more detailed analysis of an automotive ignition coil is presented in Chapter 4 to accompany the discussion of transient voltages and currents.

CHECK YOUR UNDERSTANDING

The waveform below shows the current through a 50-mH inductor. Plot the inductor voltage $v_L(t)$.

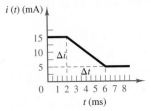

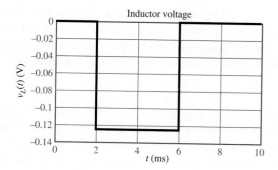

Inductor voltage

CHECK YOUR UNDERSTANDING

Find the maximum voltage across a 10-mH inductor when the inductor current is $i_L(t) = -2t(t - 2)$ A for $0 \le t \le 2$ s and zero otherwise.

Answer: 40 mV

CHECK YOUR UNDERSTANDING

Calculate and plot the inductor energy and power for a 50-mH inductor subject to the current waveform shown below. What is the energy stored at $t = 3$ ms?

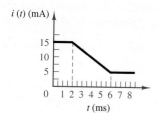

$i(t)$ (mA)

Answer:

$$w(t) = \begin{cases} 5.625 \times 10^{-6} \text{ J} & 0 \le t < 2 \text{ ms} \\ 0.156t^2 - (2.5 \times 10^{-3})t + 10^{-5} & 2 \le t < 6 \text{ ms} \\ 0.625 \times 10^{-6} & t \ge 6 \text{ ms} \end{cases}$$

$$p(t) = \begin{cases} 0 \\ (20 \times 10^{-3})(-0.125t - 2.5t) \text{ W} & 2 \le t < 6 \text{ ms} \\ 0 & \text{otherwise} \end{cases}$$

$$w(t = 3 \text{ ms}) = 3.9 \text{ } \mu\text{J}$$

3.2   TIME-DEPENDENT SOURCES

Time-dependent periodic waveforms appear frequently in practical applications and are a useful approximation of many physical phenomena. For example, electric power worldwide is generated and delivered to industrial and household users in the form

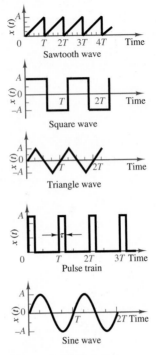

Figure 3.15 Periodic waveforms

of periodic (i.e., 50- or 60-Hz sinusoidal) voltages and currents. The methods developed in this chapter apply to many engineering systems, not just to electric circuits, and will be encountered again in the study of dynamic systems and control systems.

In general, a periodic waveform $x(t)$ satisfies the equation

$$x(t) = x(t + nT) \qquad n = 1, 2, 3, \ldots \tag{3.17}$$

where T is the period of $x(t)$. Figure 3.15 illustrates a number of periodic waveforms that are typically encountered in the study of electric circuits. Waveforms such as the sine, triangle, square, pulse, and sawtooth waves are provided in the form of voltages (or, less frequently, currents) by commercially available signal generators.

In this chapter, time-varying voltages and currents and, in particular, sinusoidal (AC) sources are introduced. Figure 3.16 illustrates the convention employed to denote time-dependent sources.

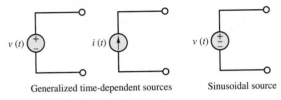

Generalized time-dependent sources   Sinusoidal source

Figure 3.16 Time-dependent sources

Sinusoids constitute the most important class of time-dependent waveforms. A generalized sinusoid is defined as

$$x(t) = A \cos(\omega t + \phi) \tag{3.18}$$

where A is the peak amplitude, ω the angular frequency, and ϕ the phase angle. Figure 3.17 summarizes the definitions of A, ω, and ϕ for the waveforms

$$x_1(t) = A \cos(\omega t) \qquad \text{and} \qquad x_2(t) = A \cos(\omega t + \phi)$$

where

$$f = \text{cyclical frequency} = \frac{1}{T} \qquad \text{cycles/s, or Hz}$$

$$\omega = \text{angular frequency} = 2\pi f \qquad \text{rad/s}$$

$$\phi = 2\pi \frac{\Delta t}{T} \qquad \text{rad} \tag{3.19}$$

$$= 360 \frac{\Delta t}{T} \qquad \text{deg}$$

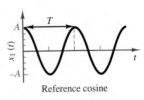

Reference cosine

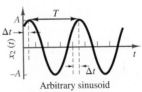

Arbitrary sinusoid

Figure 3.17 Sinusoidal waveforms

The value of the phase shift ϕ is a measure of the time shift of a sinusoid relative to a reference sinusoid, typically a cosine waveform. For example, a sine wave can be represented in terms of a cosine wave by introducing a phase shift of $\pi/2$ radians:

$$A \sin(\omega t) = A \cos\left(\omega t - \frac{\pi}{2}\right) \tag{3.20}$$

Notice that a negative phase angle represents a time shift to the right.

Although angular frequency ω, in units of radians per second, is commonly used to denote sinusoidal frequency, it is also common to employ the cyclical frequency f in units of cycles per second, or hertz (Hz). In music theory, a sinusoid is a pure tone; an A-440, for example, is a tone at a frequency of 440 Hz. The cyclical frequency is related to the angular frequency by the factor 2π.

$$\omega = 2\pi f \quad \text{Angular frequency} \tag{3.21}$$

Average (Mean) Value

Various measures exist for quantifying the amplitude of a time-varying electric signal. One of these measures is the average or mean value (also called the DC value). The average value of a waveform is computed by integrating it over a suitably chosen period, as shown below.

$$\langle x(t) \rangle = \frac{1}{T} \int_0^T x(\tau)\, d\tau \quad \text{Average or mean value} \tag{3.22}$$

where T is the period of integration. Figure 3.18 illustrates the average amplitude of $x(t)$ over a period of T seconds. It can be shown that the average or mean value of a sinusoid is zero.

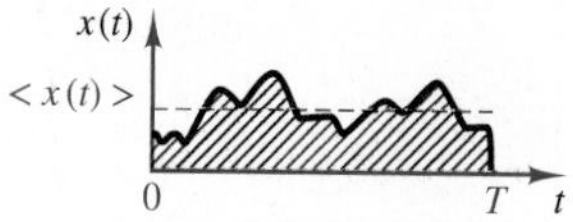

$$\langle A \cos (\omega t + \phi) \rangle = 0 \tag{3.23}$$

Figure 3.18 Average of a waveform

This result might be perplexing at first: If any sinusoidal voltage or current has zero average value, is its average power equal to zero? Clearly, the answer must be no. Otherwise, it would be impossible to illuminate households and streets and power industrial machinery with 60-Hz sinusoidal current!

Effective or RMS Value

A more useful measure of the amplitude of an AC waveform $x(t)$ is the root-mean-square (or rms) value, which takes into account fluctuations of a waveform about its mean, and which is defined as:

$$x_{\text{rms}} = \sqrt{\frac{1}{T} \int_0^T x^2(t')\, dt'} \quad \text{Root-mean-square value} \tag{3.24}$$

Notice that the argument of the square root is the mean value of $x^2(t)$. Thus, the RMS value is literally the square root of the mean of the square. Also note that the unit of the "mean of the square" is the unit of $x^2(t)$. Thus, the unit of the "root of the mean of the square" x_{rms} is the unit of $x(t)$.

Why are rms values useful? Consider two similar circuits, each with a resistor R connected to a source: one with a DC source, and one with an AC source, as shown in Figure 3.19. The effective value of the AC source is the DC source value such that the average power dissipated by the resistor R is the same in both circuits. Thus, the effective value of an AC source provides a clear

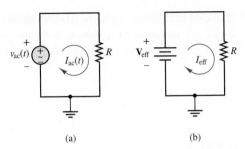

Figure 3.19 AC and DC circuits used to illustrate the concept of effective and rms values

measure of the power associated with the source, which can now be simply expressed as:

$$P_{\text{avg}} = I_{\text{eff}}^2 R = \frac{V_{\text{eff}}^2}{R} \tag{3.25}$$

But how is this discussion of effective values related to rms values? It can be shown that the effective value of an AC source is equivalent to its rms value! That is,

$$I_{\text{eff}} = I_{\text{rms}} = \sqrt{\frac{1}{T}\int_0^T i_{\text{ac}}^2(\tau)\,d\tau} \quad \text{and} \quad V_{\text{eff}} = V_{\text{rms}} = \sqrt{\frac{1}{T}\int_0^T v_{\text{ac}}^2(\tau)\,d\tau} \tag{3.26}$$

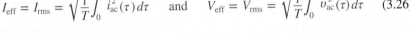

The rms, or effective, value of an AC source is the DC value that produces the same average power to be dissipated by a common resistor.

The effective (or rms) value of a voltage or current is indicated by the notation V_{rms}, or $\tilde{V}$, and I_{rms}, or $\tilde{I}$. Example 3.9 shows that the ratio of the rms value of a sinusoid to its peak value is $1/\sqrt{2} \approx 0.707$. In general, the same ratio for a different type of waveform, such as a square wave, triangle wave, or sawtooth wave, will yield a different value. Table 3.4 lists the value of this ratio for these waveforms. The table also lists a Fourier sine series for each waveform to demonstrate that each is a summation of sine waves.

Table 3.4 **Ratio of RMS value to peak value**

Waveform	$x(t)$	$x_{\text{rms}}/x_{\text{pk}}$
Sinusoid	$A\sin(\omega t)$	$\dfrac{\sqrt{2}}{2} \approx 0.707$
Square	$\dfrac{8A}{\pi}\sum_{k=1}^{\infty}\dfrac{\sin[(2k-1)\omega t]}{2k-1}$	1
Triangle	$\dfrac{8A}{\pi^2}\sum_{k=1}^{\infty}(-1)^k\dfrac{\sin[(2k-1)\omega t]}{(2k-1)^2}$	$\dfrac{\sqrt{3}}{3} \approx 0.577$
Sawtooth	$\dfrac{2A}{\pi}\sum_{k=1}^{\infty}\dfrac{\sin(k\omega t)}{k}$	$\dfrac{\sqrt{3}}{3} \approx 0.577$

MAKE THE CONNECTION

Why Do We Use Units of Radians for the Phase Angle ϕ?

The engineer finds it frequently more intuitive to refer to the phase angle in units of degrees; however, to use consistent units in the argument (the quantity in the parentheses) of the expression $x(t) = A\sin(\omega t + \phi)$, we must express ϕ in units of radians, since the units of ωt are $[\omega] \cdot [t] = (\text{rad/s}) \cdot \text{s} = \text{rad}$. Thus, we will consistently use units of radians for the phase angle ϕ in all expressions of the form $x(t) = A\sin(\omega t + \phi)$. To be consistent is especially important when one is performing numerical calculations; if one used units of degrees for ϕ in calculating the value of $x(t) = A\sin(\omega t + \phi)$ at a given t, the answer would be incorrect.

EXAMPLE 3.8   Average Value of Sinusoidal Waveform

Problem

Compute the average value of the signal $x(t) = 10 \cos(100t)$.

Solution

Known Quantities: Functional form of the periodic signal $x(t)$.

Find: Average value of $x(t)$.

Analysis: The signal is periodic with period $T = 2\pi/\omega = 2\pi/100$; thus we need to integrate over only one period to compute the average value:

$$\langle x(t) \rangle = \frac{1}{T} \int_0^T x(t') dt' = \frac{100}{2\pi} \int_0^{2\pi/100} 10 \cos(100t) \, dt$$

$$= \frac{10}{2\pi} \langle \sin(2\pi) - \sin(0) \rangle = 0$$

Comments: The mean value of a sinusoidal is zero, independent of its amplitude and frequency.

EXAMPLE 3.9   RMS Value of Sinusoidal Waveform

Problem

Compute the rms value of the sinusoidal current $i(t) = I \cos(\omega t)$.

Solution

Known Quantities: Functional form of the periodic signal $i(t)$.

Find: RMS value of $i(t)$.

Analysis: Applying the definition of rms value in equation 3.26, we compute

$$i_{\text{rms}} = \sqrt{\frac{1}{T} \int_0^T i^2(t') \, dt'} = \sqrt{\frac{\omega}{2\pi} \int_0^{2\pi/\omega} I^2 \cos^2(\omega t') \, dt'}$$

$$= \sqrt{\frac{\omega}{2\pi} \int_0^{2\pi/\omega} I^2 \left[\frac{1}{2} + \frac{1}{2} \cos(2\omega t') \right] dt'}$$

$$= \sqrt{\frac{1}{2}I^2 + \frac{\omega}{2\pi} \int_0^{2\pi/\omega} \frac{I^2}{2} \cos(2\omega t') \, dt'}$$

At this point, we recognize that the integral under the square root sign is equal to zero (see Example 3.8), because we are integrating a sinusoidal waveform over two periods. Hence,

$$i_{\text{rms}} = \frac{I}{\sqrt{2}} = 0.707I$$

where I is the peak value of the waveform $i(t)$.

Comments: The rms value of a sinusoidal signal is independent of its amplitude and frequency.

CHECK YOUR UNDERSTANDING

Express the voltage $v(t) = 155.6 \sin(377t + \pi/6)$ in cosine form. Note that the angular frequency $\omega = 377$ rad/s is equivalent to the cyclical frequency 60 Hz, which is the frequency of the electric power generated in North America.

Answer: $v(t) = 155.6 \cos(377t - \pi/3)$

CHECK YOUR UNDERSTANDING

Compute the mean (average) and rms values of the sawtooth waveform shown below.

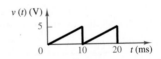

Answers: $v_{avg} = 2.5$ V; $v_{rms} = 2.89$ V

CHECK YOUR UNDERSTANDING

Compute the mean (average) and rms values of the triangle waveform shown below.

$v(t)$ (V)

3

0   5   10 t (ms)

Answers: $v_{avg} = 1.5$ V; $v_{rms} = \sqrt{3}$ V

CHECK YOUR UNDERSTANDING

Compute the mean (average) and rms values of the clipped cosine waveform shown below.

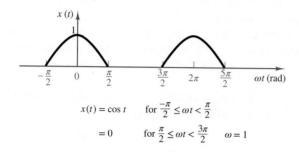

$$x(t) = \cos t \quad \text{for } \frac{-\pi}{2} \leq \omega t < \frac{\pi}{2}$$

$$= 0 \quad \text{for } \frac{\pi}{2} \leq \omega t < \frac{3\pi}{2} \quad \omega = 1$$

Answers: $x_{\text{ave}} = 1/\pi$; $x_{\text{rms}} = 0.5$

3.3 CIRCUITS CONTAINING ENERGY STORAGE ELEMENTS

The resistive circuits studied in Chapters 1 and 2 had no dependence on time. The sources had constant (DC) values and the i-v relationship for resistors (Ohm's law) had no time dependence. As a result, all the equations obtained in those chapters were algebraic and the voltages and currents were all constants. If a sinusoidal source is present in a resistive circuit, the voltages and currents in the circuit will no longer be constant but instead will vary sinusoidally in time with the same frequency and phase angle as the source. Also, just as with DC circuits, the amplitudes of the voltages and currents will depend upon the resistive network. A circuit with a sinusoidal source is known as an **AC circuit**.

Purely resistive AC circuits offer no new challenges compared to DC circuits. However, when capacitors and/or inductors are introduced into an AC circuit, the resulting behavior is significantly more interesting and challenging. The reason is that the i-v relationships for capacitors and inductors are time dependent. The result, in general, is that the amplitudes and phase angles of voltages and currents in the circuit can be different from those of the source. Consequently, in the solution of AC circuits it is necessary to keep track of two variables (amplitude and phase) for each voltage and current. By contrast, when solving DC circuits, it was necessary to keep track of only one variable (amplitude). Note that the frequency of all voltages and currents in an AC circuit equals the source frequency.

To clarify this discussion, consider the simple series loop shown in Figure 3.20, which consists of a known sinusoidal voltage source, a resistor, and a capacitor. Apply KVL around the loop to obtain the governing equation:

$$v_S - v_R - v_C = 0 \quad \text{or} \quad v_R + v_C = v_S \tag{3.27}$$

The so-called *state variable* for this circuit is the voltage v_C across the capacitor. The state variables in a circuit are the voltages across capacitors and the currents

A circuit containing energy storage elements is described by a differential equation.

$$\frac{di}{dt} + \frac{1}{RC} i = \frac{dv_S}{dt} \quad i_R = i_C = i$$

Figure 3.20 Circuit containing energy storage element

through inductors. Note that capacitors in parallel share the same state variable. The same is true for inductors in series. Capacitors in series and inductors in parallel can also be reduced to a single equivalent capacitance and inductance, respectively, with a single state variable for each equivalent capacitance and inductance. As a result, the number of state variables in a circuit is equal to the number of irreducible capacitors and inductors. In general, it is a good idea to learn to solve for state variables first because they play an important role in the complete solution of time-dependent circuits (see Chapter 4). It is also true that the state variables completely describe the behavior of a circuit. All other variables can be found readily from them.

To find v_C, it is necessary to employ the constitutive i-v relationships for the resistor and the capacitor, which are, respectively:

$$v_R = i_R R \qquad \text{and} \qquad i_C = C \frac{dv_C}{dt} \tag{3.28}$$

Notice that the resistor current and the capacitor current are the same for this simple loop. Thus:

$$v_R = i_R R = i_C R = RC \frac{dv_C}{dt} \tag{3.29}$$

Plug this result into equation 3.27 to obtain:

$$RC \frac{dv_C}{dt} + v_C = v_S \tag{3.30}$$

Divide both sides of equation 3.30 to find the standard form:

$$\frac{dv_C}{dt} + \frac{1}{RC} v_C = \frac{1}{RC} v_S \tag{3.31}$$

The result is a first-order, linear, ordinary differential equation. The solution for v_C has two parts, as usual: (1) a transient solution, and (2) a steady-state solution. The complete solution of the differential equation is the sum of these two parts. It is important to note that once the complete solution for v_C is found, it is a simple matter to find $i(t)$ and $v_R(t)$ from equations 3.28 and 3.29, respectively.

It is also possible to find similar differential equations in the variables $i(t)$ and $v_R(t)$. For $i(t)$, the result is:

$$\frac{di}{dt} + \frac{1}{RC} i = \frac{1}{R} \frac{dv_S}{dt} \tag{3.32}$$

Notice that the left-hand side of this equation is identical to that found in equation 3.31. Only the right-hand side is different. The differential equation for $v_R(t)$ follows the same rule. The constant RC has units of time and is a common example of an important class of parameters known as time constants.

For more complicated circuits, the process is largely the same except that KVL and KCL may have to be applied multiple times and the circuit may contain multiple resistors, capacitors, and inductors. The result will be multiple first- and perhaps second-order linear, ordinary differential equations. It is not difficult to imagine that for even modest circuits the procedure and results may become quite complicated and cumbersome. In fact, in some cases the above procedure would effectively require that all state variables in a circuit be solved simultaneously, when only one particular voltage or current may be sought.

To avoid these complications, an alternative approach is to dispense with time derivatives as often as possible and solve for the steady-state (particular) and transient (homogeneous) solutions separately using the following two methods:

- *Steady-state solution.* To solve for the steady-state solution, Euler's formula is employed to represent sinusoids as complex exponentials and to eliminate the time derivatives in the constitutive *i-v* relations for capacitors and inductors. The result is algebraic equations with complex constants and variables. These equations can be solved using standard algebra techniques. The only complication is that the arithmetic involves complex numbers rather than real numbers. As an added bonus, many of the intermediate details of this approach need only be understood but not actually executed during the solution of a circuit problem.

- *Transient solution.* Whenever possible, Thévenin's and Norton's theorems are used to simplify complicated circuits and to focus on solving for the state variables. The result will often be simple first- or second-order circuit archetypes with previously established solutions, such that no formal solving of differential equations is necessary. This approach is beneficial whenever all capacitors and inductors can be isolated within a relatively simple load. The remainder of the circuit will then be comprised of resistors and independent sources and can be simplified using techniques explored in Chapter 2. The transient solution is explored in greater detail in Chapter 4.

The remainder of this chapter is focused on steady-state solutions of circuits like the one in Figure 3.20 and others more complicated. Although these steady-state solutions do not contain sinusoidal functions explicitly, it is important to keep in mind that they nonetheless represent sinusoids and can be converted to explicit sinusoids using Euler's formula in the end. A typical depiction of these sinusoidal solutions is shown in Figure 3.21. The dark black curve represents the sinusoidal voltage source that is driving the circuit in Figure 3.20. The gray curve represents the sinusoidal voltage across the capacitor responding to the driving source. Notice that the capacitor voltage is a scaled version of the source and is time shifted (i.e., phase shifted) with respect to it. The driving source acts as the reference against which all other voltages and currents are compared. The results shown in Figure 3.21 are typical of a steady-state solution and can be summarized as follows:

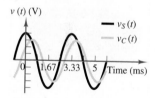

Figure 3.21 Waveforms for the AC circuit of Figure 3.20

> In a linear circuit with a sinusoidal source, all voltages and currents are sinusoids at the same frequency as the source. These voltages and currents are *scaled* versions of the source and may be shifted in time (i.e., phase shifted) with respect to it.

CHECK YOUR UNDERSTANDING

Find the differential equation for $v_R(t)$ for the circuit in Figure 3.20.

Answer: $\dfrac{dv_R}{dt} + \dfrac{1}{RC}v_R = \dfrac{dv_s}{dt}$

3.4 PHASOR SOLUTION OF CIRCUITS WITH SINUSOIDAL SOURCES

In this section, an efficient notation is introduced to represent sinusoidal signals as complex numbers and to eliminate the need for solving differential equations. (A reasonably complete treatment of complex algebra is presented in Appendix A, including examples and exercises.) The remainder of the chapter assumes that the reader is familiar with both rectangular and polar forms of complex numbers; with the conversion between these two forms; and with the basic operations of addition, subtraction, multiplication, and division of complex numbers.

Any sinusoidal signal may be represented either in the time domain:

$$v(t) = A \cos(\omega t + \theta)$$

or in the frequency domain (also known as the phasor) form:

$$V(j\omega) = A e^{j\theta} = A\angle\theta = A(\cos\theta + j\sin\theta)$$

The argument $j\omega$ indicates the $e^{j\omega t}$ time dependence of the phasor.

A phasor is a complex number consisting of a magnitude equal to the peak amplitude of a sinusoid and a phase angle equal to the phase shift of the same sinusoid with respect to a reference sinusoid, usually a source. A phase shift in the frequency domain is equivalent to a time delay in the time domain.

Since the sinusoidal source frequency ω is common to all phasors in an AC circuit, the complex exponential $e^{j\omega t}$ is not expressed explicitly. Thus, it is important to note the specific frequency ω of each sinusoidal source in a circuit.

Euler's Formula

Named after the famous Swiss mathematician Leonhard Euler, this formula is the basis of phasor notation. A phasor is similar to a vector in that it has an amplitude and direction in the complex plane. Also, just as a vector can be decomposed into x and y components, a phasor can be decomposed into real and imaginary components. Euler's formula defines a complex exponential $e^{j\theta}$ as a unit phasor in the complex plane, with real and imaginary components given by:

$$e^{j\theta} = \cos\theta + j\sin\theta \tag{3.33}$$

where $j \equiv \sqrt{-1}$ is the imaginary unit. The symbol θ is simply a place holder in Euler's formula. Any quantity or expression can be substituted for θ in the formula. In the next section on phasors, θ takes on the physical meaning of the phase shift of a sinusoid.

The **dark** black arrow in Figure 3.22 represents the complex exponential in the complex plane. The real and imaginary components of the complex exponential are shown as $\cos\theta$ and $\sin\theta$, respectively. These two components and the complex exponential itself form the three legs of a right triangle. The Pythagorean theorem requires:

$$\left|e^{j\theta}\right|^2 = \cos^2\theta + \sin^2\theta = 1 \tag{3.34}$$

Thus, the magnitude of $e^{j\theta}$ is unity, which is why it is also known as a unit phasor. The angle of inclination of the unit phasor is θ. As θ increases or decreases the unit phasor rotates counterclockwise or clockwise, respectively, about the origin of the complex plane.

Leonhard Euler (1707–1783)
(*Oxford Science Archive/ Heritage Images/The Print Collector/Alamy Stock Photo*)

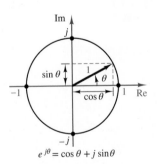

$$e^{j\theta} = \cos\theta + j\sin\theta$$

Figure 3.22 Euler's formula

It is difficult to overstate the power of the visualization presented in Figure 3.22. For example, when $\theta = \pi/2$, the unit phasor points straight up along the imaginary axis. Thus:

$$e^{j\pi/2} = 1\angle\frac{\pi}{2} = j \tag{3.35}$$

where the notation $1\angle\frac{\pi}{2}$ indicates a magnitude of 1 and a phase angle $\theta = \pi/2$. When $\theta = \pi$, the unit phasor points to the left along the negative real axis. Thus:

$$e^{j\pi} = 1\angle\pi = -1 \tag{3.36}$$

Likewise:

$$e^{j3\pi/2} = 1\angle\frac{3\pi}{2} = -j \quad \text{and} \quad e^{j2\pi} = 1\angle 2\pi = 1 \tag{3.37}$$

Each of these expressions equates the two polar forms on the left to the rectangular form on the far right side. In polar form, a phasor is represented by a magnitude (or amplitude) and a phase angle, whether as $Ae^{j\theta}$ or $A\angle\theta$. In rectangular form, a phasor is represented by real and imaginary components. Table 3.5 lists a few other commonly encountered phasors in polar and rectangular forms.

Table 3.5 **Polar and rectangular forms of common phasors**

Complex exponential	Polar	Rectangular
$Ae^{\pm j(\pi/6)}$	$A\angle \pm \pi/6$	$A(\sqrt{3}/2 \pm j/2)$
$Ae^{\pm j\pi/4}$	$A\angle \pm \pi/4$	$A(\sqrt{2}/2 \pm j\sqrt{2}/2)$
$Ae^{\pm j\pi/3}$	$A\angle \pm \pi/3$	$A(1/2 \pm j\sqrt{3}/2)$
$Ae^{\pm j \arctan(3/4)}$	$A\angle \pm \arctan(3/4)$	$A(0.8 \pm j0.6)$
$Ae^{\pm j \arctan(4/3)}$	$A\angle \pm \arctan(4/3)$	$A(0.6 \pm j0.8)$

In general, the polar and rectangular forms are related by:

$$Ae^{j\theta} = A\angle\theta = A\cos\theta + jA\sin\theta \tag{3.38}$$

In effect, Euler's identity is simply a trigonometric relationship in the complex plane.

Phasors

To see how complex numbers can be used to represent sinusoidal signals, rewrite the expression for a generalized sinusoid in light of Euler's equation:

$$A\cos(\omega t + \theta) = \text{Re}\left\{Ae^{j(\omega t+\theta)}\right\} \tag{3.39}$$

Notice that it is possible to express any sinusoid as the real part of a complex exponential with an argument of $\omega t + \theta$ and a magnitude or amplitude of A. The expression can be further simplified by remembering that the angular frequency ω is common to all voltages and currents. Thus, the $e^{j\omega t}$ portion of the complex exponential is understood to be present in every phasor, but not written explicitly. The same perspective is taken with regard to the real part operator Re so that the complex exponential is simplified as shown below.

$$\text{Re}\left\{Ae^{j(\omega t+\theta)}\right\} = \text{Re}\left\{Ae^{j\omega t}e^{j\theta}\right\} \Rightarrow Ae^{j\theta} \tag{3.40}$$

In this expression, the relational operator $\Rightarrow$ indicates equality but with the real part operator Re and the sinusoidal portion $e^{j\omega t}$ of the complex exponential hidden but understood implicitly. In general, this simplification will be used to express a phasor in polar and rectangular form as:

$$Ae^{j\theta} = A\angle\theta = A(\cos\theta + j\sin\theta) \qquad \text{Phasor notation} \qquad (3.41)$$

The reason for this simplification is convenience, as will become apparent in the examples. It is imperative to remember that the $e^{j\omega t}$ term is still present implicitly.

In Section 3.5, a new quantity known as *impedance* is introduced. Impedance is defined as the ratio of a voltage phasor to a current phasor. At this point, it is worth mentioning five key rules of complex arithmetic that will be used to resolve complex multiplication and division:

1. The magnitude of the ratio of two phasors is the ratio of the individual magnitudes. For example, $|\mathbf{V/I}| = |\mathbf{V}|/|\mathbf{I}|$.

2. The phase angle of the ratio of two phasors is the difference of the individual phase angles. For example, $\angle(\mathbf{V/I}) = \angle(\mathbf{V}) - \angle(\mathbf{I})$.

3. The complex conjugate $\bar{\mathbf{A}}$ of a phasor $\mathbf{A}$ is found by changing the sign of the imaginary unit, j, everywhere in the phasor. The magnitude of the complex conjugate of a phasor equals the magnitude of the phasor itself. The angle of the complex conjugate of a phasor equals the negative of the angle of the phasor itself.

4. The product of a phasor and its complex conjugate is a real number equal to the square of the magnitude of the phasor, which is equal to the sum of the square of the real part of the phasor and the square of the imaginary part of the phasor.

5. The angle of a phasor is the inverse tangent of the ratio of the imaginary part to the real part. That is, $\angle\mathbf{A} = \arctan([\,\text{Im}(\mathbf{A})/\text{Re}(\mathbf{A})\,])$.

A bold uppercase font indicates a phasor quantity.

Superposition of AC Signals

As explained later, Example 3.10 explores the effect of having two sinusoidal sources of different phase and amplitude, but of the same frequency, in a circuit. It is important to realize that the approach used there *does not apply* to the superposition of two (or more) sinusoidal sources that *are not at the same frequency*. The more general case of two sinusoidal sources of two different frequencies is explored here.

Consider the circuit depicted in Figure 3.23 with a load excited by two current sources in parallel.

$$i_1(t) = A_1\cos(\omega_1 t + \theta_1)$$
$$i_2(t) = A_2\cos(\omega_2 t + \theta_2) \qquad\qquad (3.42)$$

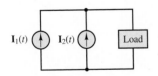

Figure 3.23 Superposition of AC

By KCL, the load current is equal to the sum of the two source currents; that is,

$$i_{\text{load}}(t) = i_1(t) + i_2(t) \qquad\qquad (3.43)$$

So far, so good. However, the expression in equation 3.43 *cannot* be expressed in phasor form without masking the fact that i_1 has a different frequency than that of i_2. For example, it may be tempting to write:

$$\mathbf{I}_{\text{load}} = \mathbf{I}_1 + \mathbf{I}_2$$
$$= A_1 e^{j\theta_1} + A_2 e^{j\theta_2} \qquad\qquad (3.44)$$

However, it is imperative to remember that the $e^{j\omega_1 t}$ and $e^{j\omega_2 t}$ terms are present implicitly in $\mathbf{I}_1$ and $\mathbf{I}_2$, respectively, as shown below.

$$
\begin{aligned}
i_1(t) &= \text{Re}\left\{\mathbf{I}_1 e^{j\omega_1 t}\right\} \\
i_2(t) &= \text{Re}\left\{\mathbf{I}_2 e^{j\omega_2 t}\right\}
\end{aligned}
\tag{3.45}
$$

The two phasors of equation 3.44 cannot be added, but must be kept separate; the only unambiguous expression for the load current is equation 3.43. To analyze a circuit with multiple sinusoidal sources at different frequencies, it is necessary to solve the circuit separately for each source and add the individual answers. Example 3.11, later in the Examples section, illustrates the use of AC superposition to determine the response of a circuit with two sources of different frequencies.

3.5   IMPEDANCE

The i-v relationships of resistors, capacitors, and inductors can be expressed in phasor notation. As phasors, each i-v relationship takes the form of a generalized Ohm's law:

$$\mathbf{V} = \mathbf{IZ}$$

where the phasor quantity $\mathbf{Z}$ is known as impedance. For a resistor, inductor, and capacitor, the impedances are, respectively:

$$\mathbf{Z}_R = R \qquad \mathbf{Z}_L = j\omega L \qquad \mathbf{Z}_C = \frac{1}{j\omega C} = \frac{-j}{\omega C}$$

Combinations of resistors, inductors, and capacitance can be represented by a single equivalent impedance of the form:

$$\mathbf{Z}(j\omega) = R(j\omega) + jX(j\omega) \qquad \text{units of } \Omega \text{ (ohms)}$$

where $R(j\omega)$ and $X(j\omega)$ are known as the "resistance" and "reactance" portions, respectively, of the equivalent impedance $\mathbf{Z}$. Both terms are, in general, functions of frequency ω.

The admittance is defined as the inverse of impedance.

$$\mathbf{Y} \equiv \frac{1}{\mathbf{Z}} \qquad \text{units of S (siemens)}$$

Consequently, all the DC circuit relations and techniques introduced in Chapter 2 can be extended to AC circuits. Thus, it is not necessary to learn new techniques and formulas to solve AC circuits; it is only necessary to learn to use the same techniques and formulas with phasors.

Generalized Ohm's Law

The impedance concept reflects the fact that capacitors and inductors act as frequency-dependent resistors. Figure 3.24 depicts a generic AC load. If this load impedance is excited by a sinusoidal voltage source $\mathbf{V}_S$ phasor and the load impedance is $\mathbf{Z}$, which represents the effect of a generic network of resistors, capacitors, and inductors. The resulting current $\mathbf{I}$ is a phasor determined by:

$$\boxed{\mathbf{V}_Z = \mathbf{IZ} \qquad \text{Generalized Ohm's law}} \tag{3.46}$$

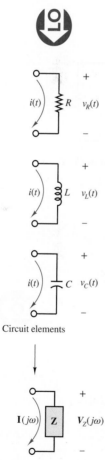

Circuit elements

Generalized impedance element

Figure 3.24 The impedance concept

A specific expression for the impedance $\mathbf{Z}$ is found for each specific network of resistors, capacitors, and inductors attached to the source. To determine $\mathbf{Z}$ it is first necessary to determine the impedance of resistors, capacitors, and inductors using the definition of impedance:

$$\mathbf{Z} \equiv \frac{\mathbf{V}}{\mathbf{I}} \qquad \text{Definition of impedance} \qquad (3.47)$$

Once the impedance of each resistor, capacitor, and inductor in a network is known, they can be combined in series and parallel (using the usual rules for resistors) to form an equivalent impedance "seen" by the source.

Impedance of a Resistor

The i-v relationship for a resistor is, of course, Ohm's law, which in the case of sinusoidal sources is written as (see Figure 3.25):

$$v_R(t) = i_R(t)R$$

or, in phasor form,

$$\mathbf{V}_R e^{j\omega t} = \mathbf{I}_R e^{j\omega t} R$$

(3.48)

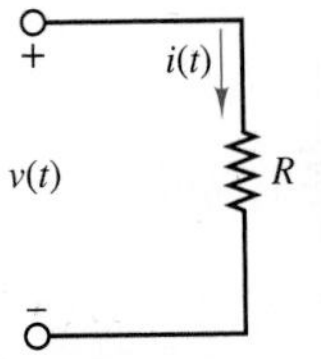

where $\mathbf{V}_R = V_R e^{j\theta_V}$ and $\mathbf{I}_R = I_R e^{j\theta_I}$ are phasors.

Both sides of equation 3.48 can be divided by $e^{j\omega t}$ to yield:

$$\mathbf{V}_R = \mathbf{I}_R R \qquad (3.49)$$

The impedance of a resistor is then determined from the definition of impedance:

$$\mathbf{Z}_R \equiv \frac{\mathbf{V}_R}{\mathbf{I}_R} = R \qquad (3.50)$$

Thus:

$$Z_R = R \qquad \text{Impedance of a resistor} \qquad (3.51)$$

The impedance of a resistor is a real number; that is, it has a magnitude R and a zero phase, as shown in Figure 3.26. The phase of the impedance is equal to the phase difference between the voltage across an element and the current through the same element. In the case of a resistor, the voltage is completely in phase with the current, which means that there is no time delay or time shift between the voltage waveform and the current waveform in the time domain.

It is important to keep in mind that the phasor voltages and currents in AC circuits are functions of frequency, $\mathbf{V} = \mathbf{V}(j\omega)$ and $\mathbf{I} = \mathbf{I}(j\omega)$. This fact is crucial for determining the impedance of capacitors and inductors, as shown below.

Impedance of an Inductor

The i-v relationship for an inductor is (see Figure 3.27):

$$v_L(t) = L\frac{di_L(t)}{dt} \qquad (3.52)$$

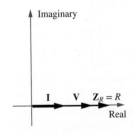

Figure 3.25 For a resistor, $v_R(t) = i_R(t)R$

Figure 3.26 Phasor diagram of the impedance of a resistor. Remember that $\mathbf{Z} = \mathbf{V}/\mathbf{I}$.

At this point, it is important to proceed carefully. The time-domain expression for the current through the inductor is:

$$i_L(t) = I_L \cos(\omega t + \theta)$$

such that

$$\frac{d}{dt} i_L(t) = -I_L \omega \sin(\omega t + \theta)$$

$$= I_L \omega \cos(\omega t + \theta + \pi/2)$$

$$= \mathrm{Re}\{I_L \omega e^{j\pi/2} e^{j\omega t+\theta}\}$$

$$= \mathrm{Re}\{I_L (j\omega) e^{j\omega t+\theta}\}$$

(3.53)

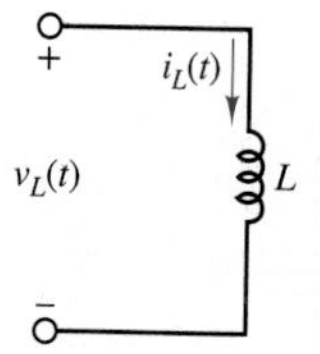

Figure 3.27 For an inductor, $v_L(t) = L\frac{di_L}{dt}(t)$

Notice that the net effect of the time derivative is to produce an extra $(j\omega)$ term along with the complex exponential expression of $i_L(t)$. That is:

Time domain	Frequency domain
$\dfrac{d}{dt}$	$j\omega$

Therefore, the phasor equivalent of the $i\text{-}v$ relationship for an inductor is:

$$\mathbf{V}_L = L(j\omega) I_L$$

(3.54)

The impedance of an inductor is then determined from the definition of impedance:

$$\mathbf{Z}_L \equiv \frac{\mathbf{V}_L}{I_L} = j\omega L$$

(3.55)

Thus:

$$\boxed{\mathbf{Z}_L = j\omega L = \omega L \angle \frac{\pi}{2} \qquad \text{Impedance of an inductor}}$$

(3.56)

The impedance of an inductor is a positive, purely imaginary number; that is, it has a magnitude of ωL and a phase of $\pi/2$ radians or $90°$, as shown in Figure 3.28. As before, the phase of the impedance is equal to the phase difference between the voltage across an element and the current through the same element. In the case of an inductor, the voltage *leads* the current by $\pi/2$ radians, which means that a feature (e.g., a zero crossing point) of the voltage waveform occurs $T/4$ seconds earlier than the same feature of the current waveform. T is the common period.

Note that the inductor behaves as a complex frequency-dependent resistor and that its magnitude ωL is proportional to the angular frequency ω. Thus, an inductor will "impede" current flow in proportion to the frequency of the source signal. At low frequencies, an inductor acts like a short-circuit; at high frequencies, it acts like an open-circuit.

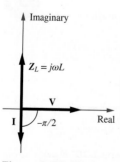

Figure 3.28 Phasor diagram of the impedance of an inductor. Remember that $\mathbf{Z} = \mathbf{V}/\mathbf{I}$.

Figure 3.29 For a capacitor, $i_C(t) = C\frac{d}{dt}v_C(t)$

Impedance of a Capacitor

The principle of duality suggests that the procedure to derive the impedance of a capacitor should be a mirror image of the procedure shown earlier for an inductor. The i-v relationship for a capacitor is (see Figure 3.29):

$$i_C(t) = C\frac{dv_C(t)}{dt} \tag{3.57}$$

The time-domain expression for the voltage across the capacitor is:

$$v_C(t) = V_C\cos(\omega t + \theta)$$

such that

$$\frac{d}{dt}v_C(t) = -V_C\omega\sin(\omega t + \theta) \tag{3.58}$$

$$= V_C\omega\cos(\omega t + \theta + \pi/2)$$

$$= \mathrm{Re}\{V_C\omega\, e^{j\pi/2}e^{j\omega t+\theta}\}$$

$$= \mathrm{Re}\{V_C(j\omega)\, e^{j\omega t+\theta}\}$$

Notice that the *net* effect of the time derivative is to produce an extra $(j\omega)$ term along with the complex exponential expression of $v_C(t)$. Therefore, the phasor equivalent of the i-v relationship for a capacitor is:

$$\mathbf{I}_C = C(j\omega)\mathbf{V}_C \tag{3.59}$$

The impedance of an inductor is then determined from the definition of impedance:

$$\mathbf{Z}_C \equiv \frac{\mathbf{V}_C}{\mathbf{I}_C} = \frac{1}{j\omega C} = \frac{-j}{\omega C} \tag{3.60}$$

Thus:

$$\boxed{\mathbf{Z}_C = \frac{1}{j\omega C} = \frac{-j}{\omega C} = \frac{1}{\omega C}\angle\frac{-\pi}{2}} \qquad \text{Impedance of a capacitor} \tag{3.61}$$

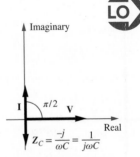

Figure 3.30 Phasor diagram of the impedance of a capacitor. Remember that $\mathbf{Z} = \mathbf{V}/\mathbf{I}$.

The impedance of a capacitor is a negative, purely imaginary number; that is, it has a magnitude of $1/\omega C$ and a phase of $-\pi/2$ radians or $-90°$, as shown in Figure 3.30. As before, the phase of the impedance is equal to the phase difference between the voltage across an element and the current through the same element. In the case of a capacitor, the voltage *lags* the current by $\pi/2$ radians, which means that a feature (e.g., a zero crossing point) of the voltage waveform occurs $T/4$ seconds *later* than the same feature of the current waveform. T is the common period of each waveform.

Note that the capacitor also behaves as a complex frequency-dependent resistor, except that its magnitude $1/\omega C$ is inversely proportional to the angular frequency ω.

Thus, a capacitor will "impede" current flow in inverse proportion to the frequency of the source. At low frequencies, a capacitor acts like an open-circuit; at high frequencies, it acts like a short-circuit.

Generalized Impedance

The impedance concept is very useful in solving AC circuit analysis problems. It allows network theorems developed for DC circuits to be applied to AC circuits. Examples 3.12 to 3.14, in the Examples section, illustrate how circuits with impedance elements in series and parallel are reduced to a single equivalent impedance, in much the same way as was done in resistive circuits. The only difference is that complex arithmetic, rather than scalar arithmetic, must be employed to find the equivalent impedance.

Figure 3.31 depicts $\mathbf{Z}_R(j\omega)$, $\mathbf{Z}_L(j\omega)$, and $\mathbf{Z}_C(j\omega)$ in the complex plane. It is important to emphasize that although the impedance of resistors is purely real and the impedance of capacitors and inductors is purely imaginary, the equivalent impedance seen by a source in an arbitrary circuit can be complex.

$$\mathbf{Z}(j\omega) = R + X(j\omega) \tag{3.62}$$

Here, R is *resistance* and X is *reactance*. The unit of R, X, and $\mathbf{Z}$ is the ohm.

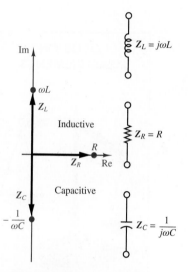

Figure 3.31 The impedances of R, L, and C are shown in the complex plane. Impedances in the upper right quadrant are inductive while those in the lower right quadrant are capacitive.

Admittance

In Chapter 2, it was suggested that the solution of certain circuit analysis problems was handled more easily in terms of conductances than resistances. This is true, for example, when one is using node analysis, or in circuits with many parallel elements, since conductances in parallel add as resistors in series do. In AC circuit analysis, an analogous quantity may be defined—the reciprocal of complex impedance. Just as conductance G was defined as the inverse of resistance, admittance $\mathbf{Y}$ is defined as the inverse of impedance.

$$\mathbf{Y} \equiv \frac{1}{\mathbf{Z}} \qquad \text{units of S (siemens)} \tag{3.63}$$

Whenever the impedance $\mathbf{Z}$ is purely real, the admittance $\mathbf{Y}$ is identical to the conductance G. In general, however, $\mathbf{Y}$ is complex.

$$\mathbf{Y} = G + jB \tag{3.64}$$

where G is the AC conductance and B is the susceptance, which is analogous to reactance. Clearly, G and B are related to R and X; however, the relationship is not a simple inverse. If $\mathbf{Z} = R + jX$, then the admittance is:

$$\mathbf{Y} = \frac{1}{\mathbf{Z}} = \frac{1}{R + jX} \tag{3.65}$$

Multiply the numerator and denominator by the complex conjugate $\overline{\mathbf{Z}} = R - jX$:

$$\mathbf{Y} = \frac{\overline{\mathbf{Z}}}{\overline{\mathbf{Z}}\mathbf{Z}} = \frac{R - jX}{R^2 + X^2} \tag{3.66}$$

and conclude that

$$G = \frac{R}{R^2 + X^2}$$

$$B = \frac{-X}{R^2 + X^2}$$

(3.67)

Notice in particular that G is not the reciprocal of R in the general case! Example 3.15, in the Examples section, illustrates the determination of Y for some common circuits.

FOCUS ON MEASUREMENTS

Capacitive Displacement Transducer

As introduced in the previous Focus on Measurements section, a capacitive displacement transducer consists of a parallel-plate capacitor with a variable separation distance x. The capacitance was shown to be:

$$C = \frac{8.854 \times 10^{-3} A}{x} \quad \text{pF}$$

where C is the capacitance in picofarads, A is the area of the plates in square millimeters, and x is the (variable) distance in millimeters. The impedance of the capacitor is:

$$\mathbf{Z}_C = \frac{1}{j\omega C} = \frac{x}{j\omega(8.854 \times 10^{-3})A} \quad \text{T}\Omega$$

Thus, at a given frequency ω, the impedance of the capacitor varies linearly with the separation distance. This result can be exploited in a bridge circuit, as shown in Figure 3.5 where half of the bridge is a differential pressure transducer in which a thin diaphragm (plate) is situated between two fixed plates and subject to variations in pressure across the diaphragm. The result is that when the capacitance of one leg of the bridge, shown here again as Figure 3.32, increases, the capacitance of the other leg decreases. Assume the bridge is excited by a sinusoidal source.

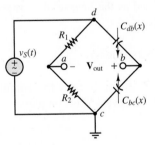

Figure 3.32 Bridge circuit for capacitive displacement transducer

Apply voltage division and KVL to express the output voltage in phasor notation as:

$$\mathbf{V}_{\text{out}}(j\omega) = \mathbf{V}_S(j\omega)\left(\frac{\mathbf{Z}_{C_{bc}}(x)}{\mathbf{Z}_{C_{db}}(x) + \mathbf{Z}_{C_{bc}}(x)} - \frac{R_2}{R_1 + R_2}\right)$$

(Continued)

(Concluded)

When the diaphragm is not displaced from its center position, the nominal capacitance of each half of the transducer is given by:

$$C = \frac{\varepsilon A}{d}$$

where d is the nominal separation distance between the diaphragm and the fixed surfaces (in millimeters). Thus, when the diaphragm is displaced an effective distance x, the capacitance of each leg of the bridge is given by:

$$C_{db} = \frac{\varepsilon A}{d-x} \quad \text{and} \quad C_{bc} = \frac{\varepsilon A}{d+x}$$

Therefore, the corresponding impedance of each leg is:

$$\mathbf{Z}_{C_{db}} = \frac{d-x}{j\omega(8.854 \times 10^{-3})A} \quad \text{and} \quad \mathbf{Z}_{C_{bc}} = \frac{d+x}{j\omega(8.854 \times 10^{-3})A}$$

such that the phasor output voltage is:

$$\mathbf{V}_{out}(j\omega) = \mathbf{V}_S(j\omega)\left(\frac{\dfrac{d+x}{j\omega(8.854 \times 10^{-3})A}}{\dfrac{d-x}{j\omega(8.854 \times 10^{-3})A} + \dfrac{d+x}{j\omega(8.854 \times 10^{-3})A}} - \frac{R_2}{R_1 + R_2} \right)$$

$$= \mathbf{V}_S(j\omega)\left(\frac{1}{2} + \frac{x}{2d} - \frac{R_2}{R_1 + R_2} \right)$$

$$= \mathbf{V}_S(j\omega)\frac{x}{2d} \quad \text{(assuming } R_1 = R_2\text{)}$$

Thus, the output voltage will vary as a scaled version of the input voltage in proportion to the displacement. A typical $v_{out}(t)$ is displayed in Figure 3.33 for a 0.05-mm "triangular" diaphragm displacement, with $d = 0.5$ mm and $\mathbf{V}_S$ a 25-Hz sinusoid with 1-V amplitude.

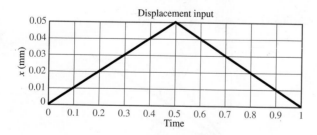

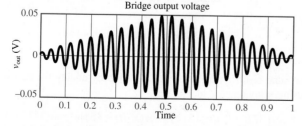

Figure 3.33 Displacement input and bridge output voltage for capacitive displacement transducer

EXAMPLE 3.10 Addition of Two Sinusoidal Sources Using Phasor Notation

Problem

Compute the phasor voltage across a series connection of two sinusoidal voltage sources (Figure 3.34).

Solution

Known Quantities:

$$v_1(t) = 15\cos\left(377t + \frac{\pi}{4}\right) \quad \text{V}$$

$$v_2(t) = 15\cos\left(377t + \frac{\pi}{12}\right) \quad \text{V}$$

Find: Equivalent phasor voltage $v_S(t)$.

Analysis: Write the two voltages in phasor form:

$$\mathbf{V}_1(j\omega) = 15\angle\frac{\pi}{4} \quad \text{V}$$

$$\mathbf{V}_2(j\omega) = 15\,e^{j\pi/12} = 15\angle\frac{\pi}{12} \quad \text{V}$$

The phasor diagram of Figure 3.35 shows $\mathbf{V}_1$ and $\mathbf{V}_2$ in the complex plane. Convert the phasor voltages from polar to rectangular form:

$$\mathbf{V}_1(j\omega) = 10.61 + j10.61 \quad \text{V}$$
$$\mathbf{V}_2(j\omega) = 14.49 + j3.88 \quad \text{V}$$

Then, by KVL:

$$\mathbf{V}_S(j\omega) = \mathbf{V}_1(j\omega) + \mathbf{V}_2(j\omega) = 25.10 + j14.49 = 28.98\,e^{j\pi/6} = 28.98\angle\frac{\pi}{6} \quad \text{V}$$

Finally, convert $\mathbf{V}_S(j\omega)$ to its time-domain form:

$$v_S(t) = 28.98\cos\left(377t + \frac{\pi}{6}\right) \quad \text{V}$$

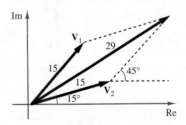

Figure 3.35 Phasor diagram showing the addition of two voltage phasors

Comments: The same result could have been obtained by adding the two sinusoids in the time domain, using trigonometric identities:

$$v_1(t) = 15\cos\left(377t + \frac{\pi}{4}\right) = 15\cos\frac{\pi}{4}\cos(377t) - 15\sin\frac{\pi}{4}\sin(377t) \quad \text{V}$$

$$v_2(t) = 15\cos\left(377t + \frac{\pi}{12}\right) = 15\cos\frac{\pi}{12}\cos(377t) - 15\sin\frac{\pi}{12}\sin(377t) \quad \text{V}$$

Figure 3.34 Circuit with two sinusoidal sources.

Combine like terms to obtain:

$$v_1(t) + v_2(t) = 15\left(\cos\frac{\pi}{4} + \cos\frac{\pi}{12}\right)\cos(377t) - 15\left(\sin\frac{\pi}{4} + \sin\frac{\pi}{12}\right)\sin(377t)$$

$$= 15[1.673\cos(377t) - 0.966\sin(377t)]$$

$$= 15\sqrt{(1.673)^2 + (0.966)^2} \times \cos\left[377t + \arctan\left(\frac{0.966}{1.673}\right)\right]$$

$$= 15\left[1.932\cos\left(377t + \frac{\pi}{6}\right)\right] = 28.98\cos\left(377t + \frac{\pi}{6}\right) \quad \text{V}$$

The above expression is, of course, identical to the one obtained using phasor notation, but it required more computation. Phasor analysis often simplifies calculations.

EXAMPLE 3.11 **AC Superposition**

Problem

Compute the voltages $v_1(t)$ and $v_2(t)$ in the circuit of Figure 3.36.

Solution

Known Quantities:

$$i_S(t) = 0.5\,\cos\left[2\pi(100t)\right] \quad \text{A}$$

$$v_S(t) = 20\,\cos\left[2\pi(1{,}000t)\right] \quad \text{V}$$

Find: $v_1(t)$ and $v_2(t)$.

Analysis: Since the two sources are at different frequencies, we must compute a separate solution for each. Consider the current source first, with the voltage source set to zero (short-circuit) as shown in Figure 3.37. The circuit thus obtained is a simple current divider. Write the source current in phasor notation:

$$\mathbf{I}_S(j\omega) = 0.5e^{j0} = 0.5\angle 0 \text{ A} \qquad \omega = 2\pi 100 \text{ rad/s}$$

Then

$$\mathbf{V}_{R1}(\mathbf{I}_S) = \mathbf{I}_S\frac{R_2}{R_1 + R_2}R_1 = 0.5\angle 0\left(\frac{50}{150 + 50}\right)150 = 18.75\angle 0 \text{ V}$$

$$\omega = 2\pi(100)\,\text{rad/s}$$

$$\mathbf{V}_{R2}(\mathbf{I}_S) = \mathbf{I}_S\frac{R_1}{R_1 + R_2}R_2 = 0.5\angle 0\left(\frac{150}{150 + 50}\right)50 = 18.75\angle 0 \text{ V}$$

$$\omega = 2\pi(100)\,\text{rad/s}$$

Next, consider the voltage source, with the current source set to zero (equivalent to an open-circuit), as shown in Figure 3.38. First, write the source voltage in phasor notation:

$$\mathbf{V}_S(j\omega) = 20e^{j0} = 20\angle 0 \text{ V} \qquad \omega = 2\pi(1{,}000)\,\text{rad/s}$$

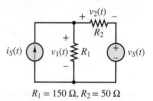

$$R_1 = 150\ \Omega,\ R_2 = 50\ \Omega$$

Figure 3.36 Circuit for Example 3.11.

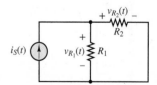

Figure 3.37 Circuit for Example 3.11 with voltage source set to zero.

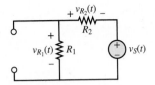

Figure 3.38 Circuit for Example 3.11 with current source set to zero.

Then, apply the voltage divider law, to obtain:

$$\mathbf{V}_{R1}(\mathbf{V}_S) = \mathbf{V}_S \frac{R_1}{R_1 + R_2} = 20\angle 0\left(\frac{150}{150 + 50}\right) = 15\angle 0 \text{ V}$$

$$\omega = 2\pi(1{,}000) \text{ rad/s}$$

$$\mathbf{V}_{R2}(\mathbf{V}_S) = -\mathbf{V}_S \frac{R_2}{R_1 + R_2} = -20\angle 0\left(\frac{50}{150 + 50}\right) = -5\angle 0 = 5\angle \pi \text{ V}$$

$$\omega = 2\pi(1{,}000) \text{ rad/s}$$

The voltage across each resistor is obtained by adding the contributions from each source and converting the equivalent phasor to the time domain:

$$\mathbf{V}_{R1} = \mathbf{V}_{R1}(\mathbf{I}_S) + \mathbf{V}_{R1}(\mathbf{V}_S)$$
$$v_1(t) = 18.75 \cos[2\pi(100t)] + 15 \cos[2\pi(1{,}000t)] \quad \text{V}$$

and

$$\mathbf{V}_{R2} = \mathbf{V}_{R2}(\mathbf{I}_S) + \mathbf{V}_{R2}(\mathbf{V}_S)$$
$$v_2(t) = 18.75 \cos[2\pi(100t)] + 5 \cos[2\pi(1{,}000t) + \pi] \quad \text{V}$$

Comment: It is impossible to further simplify the final expression because the two components are at different frequencies.

EXAMPLE 3.12   Impedance of a Practical Capacitor

Problem

A practical capacitor can be modeled by an ideal capacitor in parallel with a resistor. The parallel resistance represents leakage losses in the capacitor that can be quite significant. Find the impedance of a practical capacitor at the radian frequency $\omega = 377$ rad/s (60 Hz). How will the impedance change if the capacitor is used at a much higher frequency, say, 800 kHz?

Solution

Known Quantities: Figure 3.39; $C_1 = 0.001 \ \mu\text{F} = 1 \times 10^{-9}$ F; $R_1 = 1$ MΩ.

Find: The equivalent impedance $\mathbf{Z}_1$ across the parallel elements.

Analysis: Combine the two impedances in parallel to determine the equivalent impedance.

$$\mathbf{Z}_1 = R_1 \left\| \frac{1}{j\omega C_1} = \frac{R_1(1/j\omega C_1)}{R_1 + 1/j\omega C_1} = \frac{R_1}{1 + j\omega C_1 R_1} \right.$$

Substitute numerical values to find:

$$\mathbf{Z}_1(\omega = 377) = \frac{10^6}{1 + j377 \times 10^6 \times 10^{-9}} = \frac{10^6}{1 + j0.377}$$

$$= 9.3571 \times 10^5 \angle(-0.3605) \ \Omega$$

Figure 3.39 Impedance of a practical capacitor.

The impedance of the capacitor alone at this frequency is:

$$\mathbf{Z}_{C1}(\omega = 377) = \frac{1}{j377 \times 10^{-9}} = 2.6525 \times 10^6 \angle(-1.5708)\,\Omega$$

When the frequency is increased to 800 kHz, or $1600\pi \times 10^3$ rad/s—a radio frequency in the AM range—the impedance changes to:

$$\mathbf{Z}_1(\omega = 1600\pi \times 10^3) = \frac{10^6}{1 + j1600\pi \times 10^3 \times 10^{-9} \times 10^6}$$

$$= \frac{10^6}{1 + j1600\pi} = 198.9\angle(-1.5706)\,\Omega$$

The impedance of the capacitor alone at this frequency would be

$$\mathbf{Z}_{C1}(\omega = 1600\pi \times 10^3) = \frac{1}{j1600\pi \times 10^3 \times 10^{-9}} = 198.9\angle(-1.5708)\,\Omega$$

Now, the impedances $\mathbf{Z}_1$ and $\mathbf{Z}_{C1}$ are virtually identical. Thus, the effect of the parallel resistance is negligible at high frequencies.

Comments: For elements in parallel, the element with the smallest impedance tends to dominate the equivalent impedance across two nodes. At the lower frequency (corresponding to the well-known 60-Hz AC power frequency) the impedance of the resistor is roughly 38 percent smaller than that of the ideal capacitor. Thus, the resistor tends to dominate the equivalent impedance; in fact, the equivalent impedance is only 6.5 percent smaller than the resistance and so the practical and ideal capacitors are substantially different. At the higher frequency, the impedance of the ideal capacitor is much smaller than the resistance, which effectively acts as an open-circuit. The equivalent impedance is dominated by the ideal capacitor. At frequencies above and below $\omega = 1/RC$, the network is capacitive and resistive, respectively. This example suggests that the behavior of a network may depend heavily on frequency.

EXAMPLE 3.13   Impedance of a Practical Inductor

Problem

Figure 3.40 shows a toroidal (doughnut-shaped) inductor. A practical inductor can be modeled as an ideal inductor in series with a resistor, as shown in Figure 3.41. The series resistance represents the resistance of the wire. Find the range of frequencies over which the impedance of the practical inductor is largely inductive (i.e., due to the inductance in the circuit). Consider the impedance to be inductive if it is at least 10 times larger than the resistance.

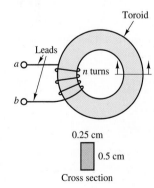

Solution

Known Quantities: $L = 0.098$ H; lead length $= 2 \times 10$ cm; $n = 250$ turns; wire is 30 gauge. Resistance of 30-gauge wire $= 0.344\ \Omega/\text{m}$.

Find: The range of frequencies over which the practical inductor acts nearly as an ideal inductor.

Analysis: To determine the equivalent resistance of the practical inductor, use the cross section of the toroid to estimate the length l_w of the wire coil:

$$l_w = 250(2 \times 0.25 + 2 \times 0.5) = 375 \text{ cm}$$

$$\text{Total length} = 375 + 20 = 395 \text{ cm}$$

Figure 3.40 A practical inductor

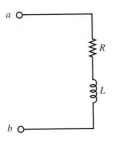

Figure 3.41 Equivalent circuit of a practical inductor.

Thus, the total resistance is:

$$R = 0.344 \ \Omega/\text{m} \times 3.95 \ \text{m} = 1.36 \ \Omega$$

To determine the range of frequencies, ω, over which the impedance $j\omega L$ of the ideal inductor is 10× greater than 1.36 Ω:

$$\omega L > 13.6 \quad \text{or} \quad \omega > \frac{13.6}{L} = \frac{13.6}{0.098} = 139 \ \text{rad/s}$$

In terms of cyclical frequency, the range is $f = \omega/2\pi > 22$ Hz.

Comments: For elements in series, the element with the largest impedance tends to dominate the equivalent impedance across two nodes. At frequencies above 139 rad/s the impedance of the inductor is at least 10× greater than the resistance and the resistance is insignificant. (Remember the 10:1 rule.) At lower frequencies, the resistance is significant; at very low frequencies ($\omega L \ll R$), the impedance of the inductor effectively acts as a short-circuit and is negligible. At high frequencies, the separation between the insulated coil wires begins to exhibit significant capacitance and so the model should be modified accordingly.

EXAMPLE 3.14   Impedance of a Series-Parallel Network

Problem

Find the equivalent impedance of the circuit shown in Figure 3.42.

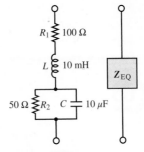

Figure 3.42 Circuit for Example 3.14.

Solution

Known Quantities: $\omega = 10^4$ rad/s; $R_1 = 100 \ \Omega$; $L = 10$ mH; $R_2 = 50 \ \Omega$; $C = 10 \ \mu$F.

Find: The equivalent impedance of the series-parallel circuit.

Analysis: The equivalent impedance $\mathbf{Z}_\parallel$ of R_2 in parallel with C is:

$$\mathbf{Z}_\parallel = R_2 \left\| \frac{1}{j\omega C} = \frac{R_2(1/j\omega C)}{R_2 + 1/j\omega C} = \frac{R_2}{1 + j\omega C R_2} \right.$$

$$= \frac{50}{1 + j\,10^4 \times 10 \times 10^{-6} \times 50} = \frac{50}{1 + j5} = 1.92 - j9.62$$

$$= 9.81\angle(-1.3734) \ \Omega$$

To determine the equivalent impedance $\mathbf{Z}_{\text{eq}}$ across the entire network:

$$\mathbf{Z}_{\text{eq}} = R_1 + j\omega L + \mathbf{Z}_\parallel = 100 + j\,10^4 \times 10^{-2} + 1.92 - j9.62$$

$$= 101.92 + j90.38 = 136.2\angle 0.725 \ \Omega$$

Comment: At $\omega = 10^4$ rad/s, the impedance across the network is inductive since the reactance is positive (or, equivalently, the phase angle is positive). (See Figure 3.31.)

EXAMPLE 3.15   Admittance

Problem

Find the equivalent admittance across each of the two networks shown in Figure 3.43.

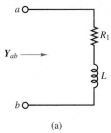

Solution

Known Quantities: $\omega = 2\pi \times 10^3$ rad/s; $R_1 = 50\ \Omega$; $L = 16$ mH; $R_2 = 100\ \Omega$; $C = 3\ \mu$F.

Find: The equivalent admittance across each of the two networks.

Analysis: Network (a): First, determine the equivalent impedance across the network *ab:*

$$\mathbf{Z}_{ab} = R_1 + j\omega L$$

To obtain the admittance, compute the inverse of $\mathbf{Z}_{ab}$ by multiplying the numerator and denominator by the complex conjugate of the denominator:

$$Y_{ab} = \frac{1}{\mathbf{Z}_{ab}} = \frac{1}{R_1 + j\omega L} = \frac{R_1 - j\omega L}{R_1^2 + (\omega L)^2}$$

Substitute numerical values to find:

$$Y_{ab} = \frac{1}{50 + j2\pi \times 10^3 \times 0.016} = \frac{50 - j(2\pi \times 10^3)(0.016)}{50^2 + (2\pi \times 10^3)^2(0.016)^2}$$

$$= 3.966 \times 10^{-3} - j7.975 \times 10^{-3}\quad \text{S}$$

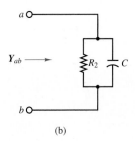

Figure 3.43 Circuits for Example 3.15.

Network (b): First, determine the equivalent impedance across the network *ab*:

$$\mathbf{Z}_{ab} = R_2 \left\| \frac{1}{j\omega C} = \frac{R_2(1/j\omega C)}{R_2 + (1/j\omega C)} \right.$$

Multiply the numerator and denominator by $j\omega C$ to find:

$$\mathbf{Z}_{ab} = \frac{R_2}{1 + j\omega R_2 C}$$

The inverse of $\mathbf{Z}_{ab}$ is the admittance:

$$Y_{ab} = \frac{1}{\mathbf{Z}_{ab}} = \frac{1 + j\omega R_2 C}{R_2} = \frac{1}{R_2} + j\omega C = 0.01 + j0.019\quad \text{S}$$

Comment: The units of admittance, siemens (S), are the same as the units of conductance.

CHECK YOUR UNDERSTANDING

Add the sinusoidal voltages $v_1(t) = A \cos(\omega t + \phi)$ and $v_2(t) = B \cos(\omega t + \theta)$ using phasor notation, and then convert back to time-domain form.

   a. $A = 1.5$ V, $\phi = 10°$; $B = 3.2$ V, $\theta = 25°$.

   b. $A = 50$ V, $\phi = -60°$; $B = 24$ V, $\theta = 15°$.

Answers: (a) $v_1 + v_2 = 4.67 \cos(\omega t + 0.353 \text{ rad})$;
(b) $v_1 + v_2 = 60.8 \cos(\omega t - 0.656 \text{ rad})$

CHECK YOUR UNDERSTANDING

Add the sinusoidal currents $i_1(t) = A \cos(\omega t + \phi)$ and $i_2(t) = B \cos(\omega t + \theta)$ for

a. $A = 0.09$ A, $\phi = 72°$; $B = 0.12$ A, $\theta = 20°$.
b. $A = 0.82$ A, $\phi = -30°$; $B = 0.5$ A, $\theta = -36°$.

Answers: (a) $i_1 + i_2 = 0.19 \cos(\omega t + 0.733)$; (b) $i_1 + i_2 = 1.32 \cos(\omega t - 0.5633)$

CHECK YOUR UNDERSTANDING

Compute the equivalent impedance across the network of Example 3.14 for $\omega = 1,000$ and 100,000 rad/s.

Find the reactance across the parallel R_2C network of Example 3.14 at the frequency $\omega = 10$ rad/s and calculate its equivalent capacitance.

Answers: $\mathbf{Z}(1,000) = 140 - j10$; $\mathbf{Z}(100,000) = 100 + j999$; $X_{\parallel} = 0.25$; $C = 0.4$ F

CHECK YOUR UNDERSTANDING

Compute the equivalent admittance across the network of Example 3.14.

Answer: $Y_{eq} = 5.492 \times 10^{-3} - j4.871 \times 10^{-3}$

Conclusion

This chapter introduced concepts and tools useful in the analysis of AC circuits. The importance of AC circuit analysis cannot be overemphasized, for a number of reasons. First, circuits made up of resistors, inductors, and capacitors constitute reasonable models for more complex devices, such as transformers, electric motors, and electronic amplifiers. Second, sinusoidal signals are ever-present in the analysis of many physical systems, not just circuits. Upon completion of this chapter a student will have learned to:

1. *Compute currents, voltages, and energy stored in capacitors and inductors.*
2. *Calculate the average and root-mean-square value of an arbitrary (periodic) signal.*
3. *Write differential equations for circuits containing inductors and capacitors.*
4. *Convert time-domain sinusoidal voltages and currents to phasor notation, and vice versa, and represent circuits using impedances.*

HOMEWORK PROBLEMS

Section 3.1: Capacitors and Inductors

3.1 The current through a 0.8-H inductor is given by $i_L = \sin(100t + \frac{\pi}{4})$. Write the expression for the voltage across the inductor.

3.2 For each case shown below, derive the expression for the current through a 200-μF capacitor. $v_C(t)$ is the voltage across the capacitor.

    a. $v_C(t) = 22 \cos(20t - \frac{\pi}{3})$ V

    b. $v_C(t) = -40 \cos(90t + \frac{\pi}{2})$ V

    c. $v_C(t) = 28 \cos\left(15t + \frac{\pi}{8}\right)$ V

    d. $v_C(t) = 45 \cos(120t + \frac{\pi}{4})$ V

3.3 Derive the expression for the voltage across a 200-mH inductor when its current is:

    a. $i_L = -2 \sin 10t$ A

    b. $i_L = 2 \cos 3t$ A

    c. $i_L = -10 \sin(50t - \frac{\pi}{4})$ A

    d. $i_L = 7 \cos(10t + \frac{\pi}{4})$ A

3.4 In the circuit shown in Figure P3.4, assume $R = 1\ \Omega$ and $L = 2$ H. Also, let:

$$i(t) = \begin{cases} 0 & -\infty < t < 0 \\ t & 0 \leq t < 10\ \text{s} \\ 10 & 10\ \text{s} \leq t < \infty \end{cases}$$

Find the energy stored in the inductor for all time.

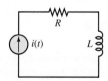

Figure P3.4

3.5 Refer to Problem 3.4 and find the energy delivered by the source for all time.

3.6 In the circuit shown in Figure P3.4, assume $R = 2\ \Omega$ and $L = 4$ H. Also, let:

$$i(t) = \begin{cases} 0 & -\infty < t < 0 \\ 2t & 0 \leq t < 5\ \text{s} \\ 10 - 4t & 5 \leq t < 12\ \text{s} \\ 2 & 12\ \text{s} \leq t < \infty \end{cases}$$

Find:

a. The energy stored in the inductor for all time.

b. The energy delivered by the source for all time.

3.7 In the circuit shown in Figure P3.7, assume $R = 2\ \Omega$ and $C = 0.1$ F. Also, let:

$$v(t) = \begin{cases} 0 & \text{for } -\infty < t < 0 \\ t & \text{for } 0 \leq t < 10\ \text{s} \\ 10 & \text{for } 10\ \text{s} \leq t < \infty \end{cases}$$

Find the energy stored in the capacitor for all time.

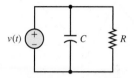

Figure P3.7

3.8 Refer to Problem 3.7 and find the energy delivered by the source for all time.

3.9 In the circuit shown in Figure P3.7, assume $R = 4\ \Omega$ and $C = 0.2$ F. Also, let:

$$v(t) = \begin{cases} 0 & -\infty < t < 0 \\ 4t & 0 \leq t < 4\ \text{s} \\ 2 - 0.5t & 4 \leq t < 10\ \text{s} \\ 0 & t > 10\ \text{s} \end{cases}$$

Find:

a. The energy stored in the inductor for all time.

b. The energy delivered by the source for all time.

3.10 The voltage waveform shown in Figure P3.10 is piecewise linear and continuous. Assume the voltage is across a 20-mH inductor. Determine the current $i_L(t)$ through the inductor, assuming $i_L(0) = 50$ mA.

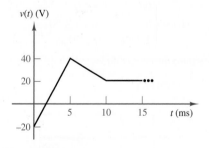

Figure P3.10

3.11   The voltage waveform shown in Figure P3.10 is piecewise linear and continuous. Assume the voltage is across a 100-μF capacitor. Determine the current $i_C(t)$ through the capacitor. Explain the concept of current through a capacitor even when the space between the capacitor plates is a perfect insulator. How is current through a capacitor different from leakage current?

3.12   The voltage across a 0.5-mH inductor, plotted as a function of time, is shown in Figure P3.12. Determine the current through the inductor at $t = 6$ ms.

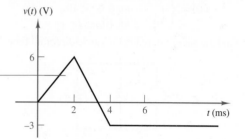

Figure P3.12

3.13   Figure P3.13 shows the voltage across a capacitor plotted as a function of time where:

$$v_{PK} = 20 \text{ V} \qquad T = 40 \ \mu s \qquad C = 680 \text{ nF}$$

Determine and plot the waveform for the current through the capacitor as a function of time. How is the current affected by the discontinuities in slope in the voltage waveform?

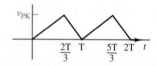

Figure P3.13

3.14   The current through a 16-μH inductor is zero at $t = 0$, and the voltage across the inductor (shown in Figure P3.14) is:

$$v(t) = \begin{cases} 0 & t \leq 0 \\ 3t^2 & 0 \leq t \leq 20 \ \mu s \\ 1.2 \text{ nV} & t \geq 20 \ \mu s \end{cases}$$

Determine the current through the inductor at $t = 30 \ \mu s$.

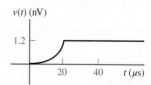

Figure P3.14

3.15   The voltage across a generic element X has the waveform shown in Figure P3.15. For $0 < t < 10$ ms, determine and plot the current through X when it is a:

a.   7-Ω resistor.

b.   0.5-μF capacitor.

c.   7-mH inductor.

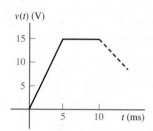

Figure P3.15

3.16   The plots shown in Figure P3.16 are the voltage across and the current through an ideal capacitor. Determine its capacitance.

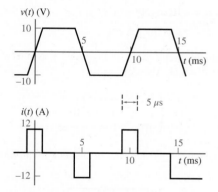

Figure P3.16

3.17   The plots shown in Figure P3.17 are the voltage across and the current through an ideal inductor. Determine its inductance.

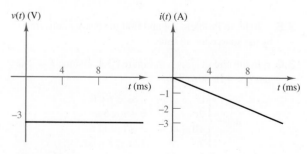

Figure P3.17

3.18   The plots shown in Figure P3.18 are the voltage across and the current through an ideal capacitor. Determine its capacitance.

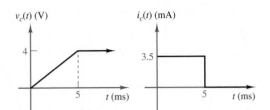

Figure P3.18

3.19   The plots shown in Figure P3.19 are the voltage across and the current through an ideal capacitor. Determine its capacitance.

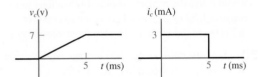

Figure P3.19

3.20   The voltage $v_L(t)$ across a 10-mH inductor is shown in Figure P3.20. Find the current $i_L(t)$ through the inductor. Assume $i_L(0) = 0$ A.

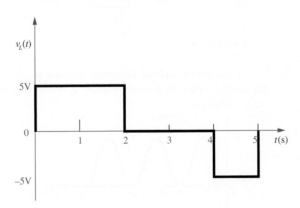

Figure P3.20

3.21   The current through a 2-H inductor is plotted in Figure P3.21. Plot the inductor voltage $v_L(t)$.

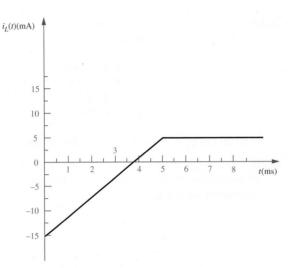

Figure P3.21

3.22   The voltage across a 100-mH inductor and a 500-μF capacitor is shown in Figure P3.22. Plot the inductor and capacitor currents, $i_L(t)$ and $i_C(t)$, for $0 < t < 6$ s, assuming $i_L(0) = 0$ A.

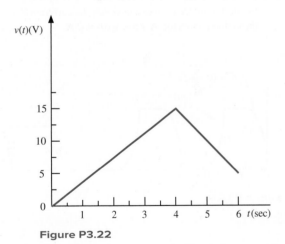

Figure P3.22

3.23   In the circuit shown in Figure P3.4, assume $R = 1$ Ω and $L = 2$ H. Also, let:

$$i(t) = \begin{cases} 0 & -\infty < t < 0 \\ t & 0 \leq t < 1 \text{ s} \\ -(t-2) & 1 \leq t < 2 \text{ s} \\ 0 & 2 \text{ s} \leq t < \infty \end{cases}$$

Find the energy stored in the inductor for all time.

3.24   In the circuit shown in Figure P3.7, assume $R = 2\ \Omega$ and $C = 0.1$ F. Also, let:

$$v(t) = \begin{cases} 0 & -\infty < t \le 0 \\ 2t & 0 \le t \le 1\ \text{s} \\ -(2t-4) & 1 \le t \le 2\ \text{s} \\ 0 & 2\ \text{s} \le t < \infty \end{cases}$$

Find the energy stored in the capacitor for all time.

3.25   The voltage $v_C(t)$ across a capacitor is shown in Figure P3.25. Determine and sketch the current $i_C(t)$ through the capacitor.

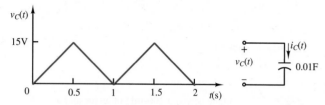

Figure P3.25

3.26   The voltage $v_L(t)$ across an inductor is shown in Figure P3.26. Determine and sketch the current $i_L(t)$ through the inductor. Assume $i_L(0) = 0$ A.

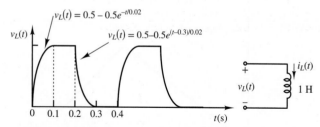

Figure P3.26

Section 3.2: Time-Dependent Sources

3.27   Find the average and rms values of $x(t)$ when:

$$x(t) = 3\cos(7\omega t) + 4$$

3.28   The output voltage waveform of a controlled rectifier is shown in Figure P3.28. The input voltage waveform was a sinusoid of amplitude 110 V rms. Find the average and rms voltages of the output waveform in terms of the firing angle θ.

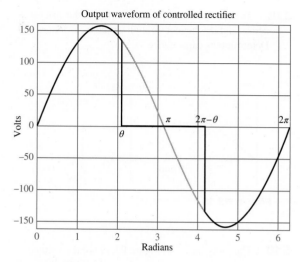

Figure P3.28

3.29   Refer to Problem 3.28 and find the angle θ that would cause the rectified waveform to deliver to a resistive load exactly one-half of the total power delivered to the same load by the input waveform.

3.30   Find the ratio between the average and rms value of the waveform shown in Figure P3.30.

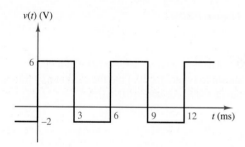

Figure P3.30

3.31   The current through a 1-Ω resistor is shown in Figure P3.31. Find the power dissipated by the resistor.

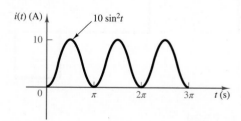

Figure P3.31

3.32   Derive the ratio between the average and rms value of the voltage waveform of Figure P3.32.

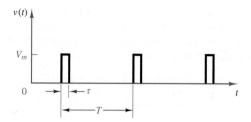

Figure P3.32

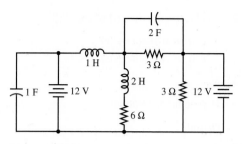

Figure P3.36

3.33 Find the rms value of the current waveform shown in Figure P3.33.

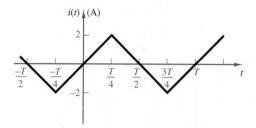

Figure P3.33

3.34 Determine the rms (or effective) value of
$v(t) = V_{DC} + v_{ac} = 35 + 63 \sin(215t^2)$ V.

Section 3.3: Circuits Containing Energy Storage Elements

3.35 Assume steady-state conditions and find the energy stored in each capacitor and inductor shown in Figure P3.35.

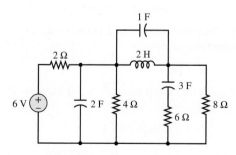

Figure P3.35

3.36 Assume steady-state conditions and find the energy stored in each capacitor and inductor shown in Figure P3.36.

Section 3.4: Phasor Solution of Circuits with Sinusoidal Sources

3.37 Find the phasor form of the following functions:

  a. $v(t) = 155 \cos (377t - 25°)$ V

  b. $v(t) = 5 \sin (1{,}000t - 40°)$ V

  c. $i(t) = 10 \cos (10t + 63°) + 15 \cos (10t - 42°)$ A

  d. $i(t) = 460 \cos (500\pi t - 25°)$
        $- 220 \sin (500\pi t + 15°)$ A

3.38 Convert the following complex numbers to polar form:

  a. $7 + j9$

  b. $-2 + j7$

  c. $j\frac{2}{3} + 4 - j\frac{1}{3} + 3$

3.39 Convert the rectangular factors to polar form and compute the product. Also compute the product directly using the rectangular factors. Compare the results.

  a. $(50 + j10)\,(4 + j8)$

  b. $(j2 - 2)\,(4 + j5)\,(2 + j7)$

3.40 Complete the following exercises in complex arithmetic.

  a. Find the complex conjugate of $(4 + j4)$, $(2 - j8)$, $(-5 + j2)$.

  b. Multiply the numerator and denominator of each ratio by the complex conjugate of the denominator. Use the result to express each ratio in polar form.

$$\frac{1 + j7}{4 + j4} \qquad \frac{j4}{2 - j8} \qquad \frac{1}{-5 + j2}$$

  c. Convert the numerator and denominator of each ratio in part b to polar form. Use the result to express each ratio in polar form.

3.41 Convert the following expressions to rectangular form:

$$j^{+j} \qquad e^{-j\pi} \qquad e^{+j2\pi}$$

3.42   Find $v(t) = v_1(t) + v_2(t)$ where

$$v_1(t) = 10 \cos(\omega t + 30°)$$
$$v_2(t) = 20 \cos(\omega t + 60°)$$

using:

a. Trigonometric identities.

b. Phasors.

3.43   The current through and the voltage across a circuit element are, respectively,

$$i(t) = 8 \cos\left(\omega t + \frac{\pi}{4}\right) \quad \text{A}$$
$$v(t) = 2 \cos\left(\omega t - \frac{\pi}{4}\right) \quad \text{V}$$

where $\omega = 600$ rad/s. Determine:

a. Whether the element is a resistor, capacitor, or inductor.

b. The value of the element in ohms, farads, or henrys.

3.44   Express the sinusoidal waveform shown in Figure P3.44 using time-dependent and phasor notation.

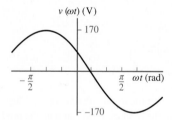

Figure P3.44

3.45   Express the sinusoidal waveform shown in Figure P3.45 using time-dependent and phasor notation.

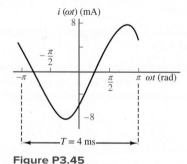

Figure P3.45

Section 3.5: Impedance

3.46   Convert the following pairs of voltage and current waveforms to phasor form. Each pair of waveforms

corresponds to an unknown element. Determine whether each element is a resistor, a capacitor, or an inductor, and compute the value of the corresponding parameter **R**, **C**, or **L**.

a. $v(t) = 20 \cos(400t + 1.2)$, $i(t) = 4 \sin(400t + 1.2)$

b. $v(t) = 9 \cos\left(900t - \frac{\pi}{3}\right)$, $i(t) = 4 \sin\left(900t + \frac{2}{3}\pi\right)$

c. $v(t) = 13 \cos\left(250t + \frac{\pi}{3}\right)$, $i(t) = 7 \sin\left(250t + \frac{5}{6}\pi\right)$

3.47   Determine the equivalent impedance seen by the source v_S in Figure P3.47 when:

$$v_S(t) = 10 \cos(4,000t + 60°) \text{ V}$$

$$R_1 = 800 \text{ mH}, \Omega \qquad R_2 = 500 \text{ nF}\Omega$$

$$L = 200 \text{ mH} \qquad C = 70 \text{ nF}$$

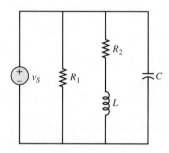

Figure P3.47

3.48   Determine the equivalent impedance seen by the source v_S in Figure P3.47 when:

$$v_S(t) = 5 \cos(1,000t + 30°) \text{ V}$$

$$R_1 = 300 \text{ mH}, \Omega \qquad R_2 = 300 \text{ nF}\Omega$$

$$L = 100 \text{ mH} \qquad C = 50 \text{ nF}.$$

3.49   The generalized version of Ohm's law for impedance elements is:

$$\mathbf{V} = \mathbf{IZ}$$

Assume the current through a 0.5-μF capacitor is given by:

$$i_s(t) = I_o \cos\left(\omega t + \frac{\pi}{6}\right)$$
$$I_o = 13 \text{ mA} \qquad \omega = 1,000 \text{ rad/s}$$

a. Express the source current in phasor notation.

b. Determine the impedance of the capacitor.

c. Determine the voltage across the capacitor, in phasor notation.

3.50   Determine $i_2(t)$ in the circuit shown in Figure P3.50. Assume:

$$i_1(t) = 100\cos(\omega t + 4)\text{ mA}$$
$$i_3(t) = 80\sin(\omega t - 1.2)\text{ mA}$$
$$i_4(t) = 150\sin(\omega t + 2)\text{ mA}$$
$$\omega = 377\text{ rad/s}$$

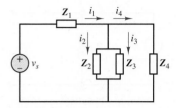

Figure P3.50

3.51   Determine the voltage $v_2(t)$ across R_2 in the circuit of Figure P3.51.

$$i(t) = 20\cos(533.33t)\text{ A}$$
$$R_1 = 8\ \Omega \qquad R_2 = 16\ \Omega$$
$$L = 15\text{ mH} \qquad C = 117\ \mu\text{F}$$

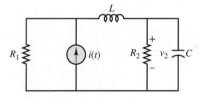

Figure P3.51

3.52   Determine the frequency so that the current $\mathbf{I}_i$ and the voltage $\mathbf{V}_o$ in Figure P3.52 are in phase.

$$\mathbf{Z}_s = 13{,}000 + j\omega 3\ \ \Omega$$
$$R = 120\ \Omega$$
$$L = 19\text{ mH} \qquad C = 220\text{ pF}$$

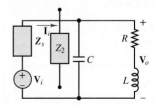

Figure P3.52

3.53   A common model for a practical inductor is a coil resistance in series with an inductance L. The coil resistance accounts for the internal losses of an

inductor. Figure P3.53 shows an ideal capacitor in parallel with a practical inductor. Determine the current supplied by the source v_S. Assume:

$$v_S(t) = V_o\cos(\omega t + 0)$$
$$V_o = 10\text{ V} \qquad \omega = 6\text{ Mrad/s} \qquad R_S = 50\ \Omega$$
$$R_C = 40\ \Omega \qquad L = 20\ \mu\text{H} \qquad C = 1.25\text{ nF}$$

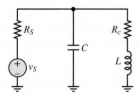

Figure P3.53

3.54   Use phasor techniques to solve for the current $i(t)$ shown in Figure P3.54.

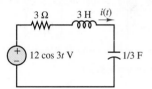

Figure P3.54

3.55   Use phasor techniques to solve for the voltage $v(t)$ shown in Figure P3.55.

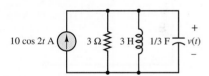

Figure P3.55

3.56   Solve for $\mathbf{I}_1$ in the circuit shown in Figure P3.56.

$$\mathbf{I} = 20\angle -\frac{\pi}{4}\text{ A} \qquad R = 3\ \Omega$$
$$\mathbf{Z}_1 = -j3\ \Omega \qquad \mathbf{Z}_2 = -j7\ \Omega$$

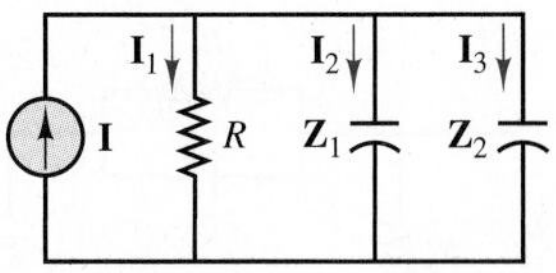

Figure P3.56

3.57   Solve for $\mathbf{V}_R$ shown in Figure P3.57. Assume:

$$\omega = 3\text{ rad/s} \qquad \mathbf{V}_s = 13\angle 0\text{ V}$$
$$R = 15\ \Omega \qquad L_1 = 7\text{ H} \qquad L_2 = 2\text{ H}$$

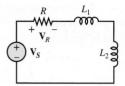

Figure P3.57

3.58 With reference to Problem 3.55, find the value of ω for which the current through the resistor is maximum.

3.59 Find the current $i_R(t)$ through the resistor shown in Figure P3.59.

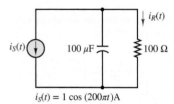

$i_S(t) = 1 \cos(200\pi t)$A

Figure P3.59

3.60 Find $v_{out}(t)$ shown in Figure P3.60.

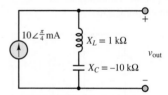

Figure P3.60

3.61 Find the impedance **Z** shown in Figure P3.61, assuming $\omega = 2$ rad/s, $R_1 = R_2 = 2\ \Omega$, $C = 0.25$ F, and $L = 1$ H.

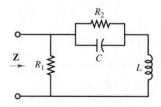

Figure P3.61

3.62 Find the sinusoidal steady-state output $v_{out}(t)$ for each circuit shown in Figure P3.62.

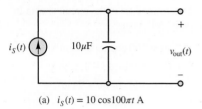

(a) $i_S(t) = 10\ \cos 100\pi t$ A

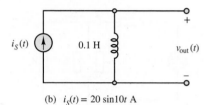

(b) $i_S(t) = 20\ \sin 10t$ A

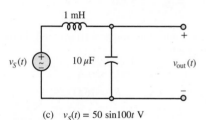

(c) $v_S(t) = 50\ \sin 100t$ V

Figure P3.62

3.63 Determine the voltage $v_L(t)$ across the inductor shown in Figure P3.63.

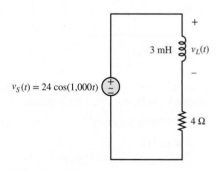

Figure P3.63

3.64 Determine the current $i_R(t)$ through the resistor shown in Figure P3.64.

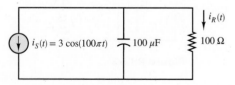

Figure P3.64

3.65 Find the frequency that causes the equivalent impedance $\mathbf{Z}_{eq}$ in Figure P3.65 to be purely resistive.

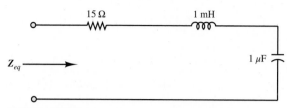

Figure P3.65

3.66

a. Find the equivalent impedance $\mathbf{Z}_o$ seen by the source in Figure P3.66(a). Assume the frequency is 377 rad/s.

b. What capacitance should be placed between terminals a and b, as shown in Figure P3.66(b), to make the equivalent impedance $\mathbf{Z}_o$ purely resistive? [*Hint:* Find C so that the phase angle of $\mathbf{Z}_o$ is zero.]

c. What is the amplitude of $\mathbf{Z}_o$ when the capacitor is included?

3.67   A common model for a practical capacitor has a "leakage" resistance, R_C, in parallel with an ideal capacitor, as shown in Figure P3.67. The effects of lead wires are also represented by resistances R_1 and R_2 and inductances L_1 and L_2.

a. Assume $C = 1\ \mu\mathrm{F}$, $R_C = 100\ \mathrm{M\Omega}$, $R_1 = R_2 = 1\ \mu\Omega$, and $L_1 = L_2 = 0.1\ \mu\mathrm{H}$, and find the equivalent impedance $\mathbf{Z}_{ab}$ seen across terminals a and b as a function of frequency ω.

b. Find the range of frequencies for which $\mathbf{Z}_{ab}$ is capacitive.

[*Hint:* Assume that R_C is much greater than $1/\omega C$ such that R_C can be ignored in part b.]

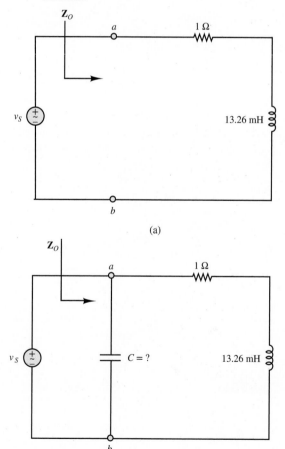

Figure P3.66

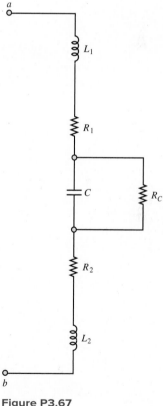

Figure P3.67

Design Credits: Mini DVI cable adapter isolated on white background: Robert Lehmann/Alamy Stock Photo; Balance scale: Alex Slobodkin/E+/Getty Images; Icon for "Focus on measurements" weighing scales: Media Bakery.

C H A P T E R

4

TRANSIENT ANALYSIS

hapter 4 focuses on the *transient* portion of the complete response of a time-dependent circuit. Recall from Chapter 3 that the complete response is composed of two parts: (1) the transient response and (2) the *steady-state* response. These parts can also be rearranged as natural and forced responses, respectively. Chapter 3 explored the latter part for circuits with AC sources; Chapter 4 explores the former for circuits that experience a *transient event*, such as the throwing of a switch. The general qualities of a transient response are independent of the type of event. For simplicity, this chapter explores transient responses due to switching.

The fundamental quality of any transient response is that it eventually vanishes to zero. Once this occurs, only a steady-state solution remains. The role of the transient solution is to provide a *transition over time* from one state (i.e., an "old" or "initial" steady state) to another (i.e., a "new" or "final" steady state). The Latin root of the adjective *transient* is *trans*, meaning "across." Literally, the transient solution is a bridge across time from one steady state to another. In most of the examples presented in this chapter, both the "old" and the "new" are DC steady states. However, transient analysis is applicable to a transition between two AC steady states or any other pair of states, which need not be steady.

When a switch opens or closes in an electric circuit, the voltages and currents in that circuit will, in general, transition to a new state. The throwing of a switch is a transient event because it causes a short-circuit (a closed switch) to be replaced by an open-circuit (an open switch), or vice versa. These two switch positions produce two distinctly different circuits. The abrupt change from one to the other provokes a transient response.

The transition from the "old" state to the "new" state does not happen instantaneously because capacitors and inductors store energy. Sometimes a finite time is required to charge and discharge the energy storage elements to reach the "new" steady state. The transition may take place quickly, but it cannot take place instantaneously.

The objectives of transient analysis can be expressed by the following questions:

1. What are the *initial conditions* on the *state variables* at the moment of the transient event?
2. How are the initial conditions on the state variables related to the initial conditions on other variables?
3. What is the manner of the transition from the initial conditions to the final steady state of any variable?
4. How fast or slow is that transition?
5. What is the final steady state of any variable?

Two types of circuits are examined in this chapter: first-order *RC* and *RL* circuits, which contain a single storage element, and second-order circuits, which contain two irreducible storage elements. The simplest of the second-order circuits to analyze are the series *RLC* and parallel *RLC* circuits. Other more complicated second-order circuits exist, as do higher-order circuits; however, since all the fundamental behaviors of transient circuits are revealed in the types just mentioned, they are the focus of this chapter.

A first-order circuit contains a single storage element. A second-order circuit contains two irreducible storage elements.

Throughout this chapter, practical applications of first- and second-order circuits are introduced. Numerous analogies are presented to emphasize the general nature of the solution methods and their applicability to a wide range of physical systems, including hydraulics, mechanical systems, and thermal systems.

 Learning Objectives

Students will learn to...

1. Understand the fundamental qualities of transient responses. *Section 4.1*
2. Write differential equations in standard form for circuits containing inductors and capacitors. *Section 4.2*
3. Determine the steady state of DC circuits containing inductors and capacitors. *Section 4.2*
4. Determine the complete solution of first-order circuits excited by switched DC sources. *Section 4.3*
5. Determine the complete solution of second-order circuits excited by switched DC sources. *Section 4.4*
6. Understand analogies between electric circuits and hydraulic, thermal, and mechanical systems. *Sections 4.1–4.4*

4.1   TRANSIENT ANALYSIS

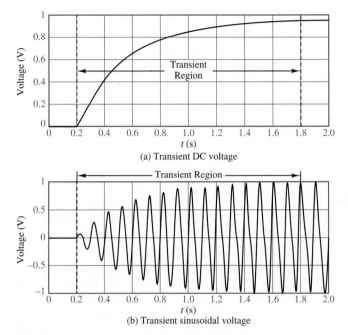

Figure 4.1 First- and second-order transient responses

Figure 4.1 shows two typical results due to a transient event at $t = 0.2$ s in a DC circuit [Figure 4.1(a)] and an AC circuit [Figure 4.1(b)], respectively. Each waveform has three parts:

- The *initial steady state* for $0 \le t \le 0.2$ s.

- The *transient response* for $0.2 \le t \le 1.8$ s (approximately).

- The *final steady state* for $t > 1.8$ s.

The objective of **transient analysis** is to determine the manner and speed with which voltages and currents transition from one steady state to another.

Figure 4.2 shows a typical circuit used to explore transient responses. The single-pole, single-throw (SPST) switch connects the battery to the *RLC* network suddenly at $t = 0$ initiating a transient response. The complexity of transient analysis increases with the number of irreducible energy storage elements in the circuit; the analysis can be quite involved for higher-order circuits. In this chapter, only first- and second-order circuits are analyzed. Luckily, these two cases exhibit all the fundamental aspects of transient analysis.

The discussion and analysis in this chapter is focused on circuits that conform to the general circuit models shown in Figure 4.3, where the network in the box acts as the load and consists of either one or two *storage elements* and possibly various resistors. In Figure 4.3(a), R_T is the Thévenin equivalent resistance seen by the load and V_T is the open-circuit voltage across terminals a and b. In Figure 4.3(b), R_N is the Norton (i.e., Thévenin) equivalent resistance R_N seen by the load and I_N is the short-circuit current from terminal a to terminal b.

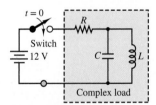

Figure 4.2 Circuit with switched DC excitation

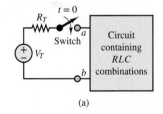

(a)

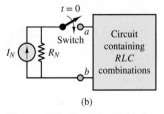

(b)

Figure 4.3 General models of the transient analysis problem. The load may contain *RLC* combinations while the source is either a (a) Thévenin or (b) Norton equivalent network.

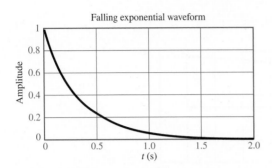

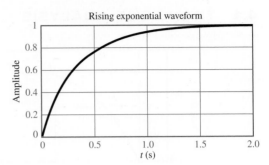

Figure 4.4 Falling and rising exponential responses

When the load is first order, containing either an inductor or capacitor, the transient response will be either a **rising** or **falling exponential** waveform, such as those shown in Figure 4.4. Both of these waveforms *decay* over time; that is, the transient response goes to zero leaving only the new steady-state response.

In the case of two storage elements, series and parallel *RLC* networks are considered in detail although a method for solving more complicated arrangements is also presented. The analysis of second-order circuits is complicated because there are three distinctly different transient responses possible, depending upon the magnitude of a **dimensionless damping ratio** ζ. When $\zeta > 1$, the transient response is *overdamped* and is represented by the sum of two exponentially decaying waveforms, either rising or falling. When $\zeta < 1$, the transient response is *underdamped* and is represented by a *decaying sinusoid*. When $\zeta = 1$, the transient response is *critically damped* and is represented by a waveform that has aspects of both the overdamped and underdamped waveforms. The impact of ζ is exemplified in the transient response to the sudden switching of a DC source, as shown in Figure 4.5.

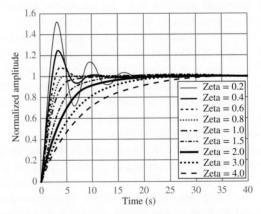

Figure 4.5 Typical second-order transient responses for various values of the dimensionless damping ratio ζ (zeta)

4.2   ELEMENTS OF TRANSIENT PROBLEM SOLVING

The key elements involved in the solution of a first- or second-order transient problem are outlined below. Keep in mind that circuits containing DC sources are considered in this chapter only. The mathematics for circuits containing AC sources is somewhat more complicated; however, the fundamental ideas are the same. Examples of transient circuit problems are found in Sections 4.3 and 4.4.

Time Intervals

The moment of a transient event is defined as $t = 0$. The moments immediately before and after the event are denoted as $t = 0^-$ and $t = 0^+$, respectively. The initial steady state is the behavior of the circuit for the time interval $t < 0$. The final steady state is the behavior of the circuit as $t \to \infty$, which should be understood to mean "t gets very large." In between the initial and final steady states is the transient response.

In practice, the final steady state is reached when $t \geq t_\infty$, where t_∞ marks the effective end of the transient response. The most common choice for t_∞ is 5τ, where τ is a *time constant* associated with the circuit. Further discussion of time constants is found below and in later sections.

Initial Steady State ($t < 0$)

During the initial steady state the voltages and currents in the circuit may be constant (DC), sinusoidal (AC), or some other waveform. In the case of an initial DC steady state, capacitors and inductors are equivalent to open- and short-circuits, respectively, and the circuit can be solved using the methods of Chapter 2. In the case of an initial AC steady state, the circuit can be solved using the impedance methods of Chapter 3.

In a DC steady state, a capacitor acts as an open-circuit and an inductor acts as a short-circuit.

State Variables

The state variables in electric circuits are the currents through inductors and the voltages across capacitors. The number of state variables equals the number of irreducible storage elements. Thus, first- and second-order circuits have one and two state variables, respectively. It is often convenient to first solve for the transient response of the state variables and then solve for other variables through their relationships to the state variables. Regardless of the solution method employed, it is always necessary to know the values of the state variables at $t = 0^-$, as explained next.

Initial Conditions

The initial conditions on the transient response of a circuit are determined by its stored energy at the instant of the transient event. Recall that energy is stored in capacitors, as expressed by their voltages, and in inductors, as expressed by their currents. Consequently, since it is not physically possible for the energy stored in a capacitor or inductor to change instantaneously, the voltage across a capacitor and the current through an inductor also cannot change instantaneously. In other words, the state variables are continuous functions of time.

The continuity requirement on the state variables is evident in the v-i relationships for capacitors and inductors.

$$i_C = C\frac{dv_C}{dt} \quad \text{and} \quad v_L = L\frac{di_L}{dt} \tag{4.1}$$

A discontinuity in v_C or i_L would require i_C or v_L, respectively, to be infinite. Since it is not physically possible to achieve an infinite current or voltage, v_C and i_L must always be continuous.

The same is not guaranteed for other nonstate variables in a circuit. The current through a resistor or capacitor, and the voltage across a resistor or inductor, may be discontinuous. An important implication of these results is that only the state variables are guaranteed to be continuous across a transient event.

Only the current through an inductor and the voltage across a capacitor are always continuous. Consequently, these two state variables are also continuous across a transient event. In mathematical terms:

$$v_C(0^+) = v_C(0^-) \tag{4.2}$$
$$i_L(0^+) = i_L(0^-) \tag{4.3}$$

Nonstate variables may or may not be continuous across a transient event and are therefore unreliable as initial conditions. Only state variables should be used to develop initial conditions on a transient event.

Energy and the Transient Response

During a transient response, energy is, in general, continually supplied, exchanged, and dissipated within a circuit until a new steady state is reached. Independent voltage and current sources, if present, will supply energy; storage elements, if more than one are present, may exchange energy; and resistors will dissipate energy. These processes will continue until a new steady state is reached, in which the time-averaged energy supplied continually equals the time-averaged energy dissipated. It is instructive to consider the interaction between these processes during a transient response.

Consider the circuit shown in Figure 4.6. For $t < 0$, assume that the capacitor has been connected to the battery for a long time so that the capacitor voltage v_C equals the battery voltage V_B and the energy stored in the capacitor is $W_C = Cv_C^2/2$ (see Chapter 3). Also notice that the current through each resistor is zero.

At $t = 0$ the two switches are thrown such that the capacitor is disconnected from the battery loop but simultaneously connected to R_2 in a simple series loop. Since the energy stored by the capacitor must be continuous with time, $W_C = Cv_C^2/2$ at $t = 0^+$. At the same moment, the voltage across the resistor has changed from zero to v_C and, therefore, the current through R_2 has also changed from zero to some finite nonzero value. Since KCL requires $i_C + i_{R_2} = 0$ for the series loop, the capacitor current is:

$$i_C = -i_{R_2} = -\frac{v_{R_2}}{R_2} = -\frac{v_C}{R_2} \tag{4.4}$$

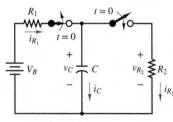

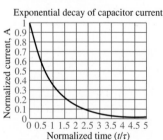

Figure 4.6 Energy stored in a capacitor is dissipated by a resistor.

where the expression for i_{R_2} is simply Ohm's law. Recall from Chapter 3 that the capacitor current is defined as the time rate of change of the stored charge, which, in turn, is proportional to the voltage across the capacitor. That is:

$$i_C(t) = \frac{dQ_C}{dt} \quad \text{and} \quad Q_C = Cv_C \tag{4.5}$$

Substitution of these expressions into equation 4.4 results in:

$$C\frac{dv_C}{dt} = -\frac{v_C}{R_2} \quad \text{or} \quad \frac{dv_C}{dt} = -\frac{1}{R_2 C}v_C \tag{4.6}$$

Equation 4.6 indicates that the rate of change of the voltage across the capacitor is proportional to the voltage across the capacitor itself. That is, at $t = 0^+$ the capacitor is *discharging* at its maximum rate because v_C itself and, thus, i_{R_2} are both maximums at that moment. As the capacitor continues to discharge, v_C and i_{R_2} continue to decrease such that the rate of decrease in v_C decreases as well. In theory, the capacitor never quite discharges fully because the rate of discharge becomes smaller and smaller as time goes by. Note that the energy in the $R_2 C$ series loop can only decrease because there is no independent source present to offset the energy dissipated by the resistor.

The graph of Figure 4.6 shows the normalized transient response of i_{R_2}. One can easily check that the slope at any point on the curve is proportional to the value at the same point. This type of relationship wherein the rate of change of a variable is proportional to the value of the variable itself is the fundamental feature of the exponential function. Thus, the transient response of the $R_2 C$ series loop shown in Figure 4.6 is characterized by:

$$\frac{dv_C}{dt} \propto v_C(t) \propto e^{-st} \quad \text{where } s = \frac{1}{RC} \tag{4.7}$$

The constant s is commonly expressed as $1/\tau$, where τ is known as a *time constant*. Such decaying exponentials, whether rising or falling, are ubiquitous in the mathematical representations of transient responses of physical systems. The exponential rate of decay is determined by the time constant.

Notice that the normalized transient response of i_{R_2} is shown in Figure 4.6 up to $t = 5\tau$, at which time v_C and i_{R_2} are less than 1 percent of their original values. For most practical purposes, the capacitor can be considered fully discharged for $t \geq 5\tau$.

Now consider what happens if the switches in the circuit of Figure 4.6 are returned to their original positions at some moment after $t = 5\tau$. The capacitor will then be disconnected from R_2 and reconnected in a series loop with the battery V_B and the resistor R_1. At that moment, the capacitor is now *charging* at its maximum rate because the voltage $(V_B - v_C)$ across R_1 and, thus, i_{R_1} are maximums. As the capacitor continues to charge, $(V_B - v_C)$ and i_{R_1} continue to decrease such that the rate of increase in v_C decreases as well. The result is another decaying, but rising, exponential, such as that shown on the right in Figure 4.4. The time constant τ for the $V_B R_1 C$ series loop is $R_1 C$.

These fundamental behaviors also occur for circuits containing an inductor and one or more resistors and independent sources.

It is also worth noting that the currents i_{R_1} and i_{R_2} in this illustrative example were discontinuous across the transient events. As emphasized earlier in this

MAKE THE CONNECTION

Thermal Capacitance

Just as an electric capacitor can store energy and a hydraulic capacitor can store fluid (see the Make the Connection sidebar, "Fluid Capacitance" in Chapter 3), the thermal capacitance C_t of an object is related to two physical properties: mass and specific heat:

$C_t = mc$; m = mass [kg]

c = specific heat [J/°C-kg]

Physically, thermal capacitance is related to the ability of a mass to store heat and describes how much the temperature of the mass will rise for a given addition of heat. If we add heat at the rate q (in Watts) for time Δt and the resulting temperature rise is ΔT, then we can define the thermal capacitance to be

$$C_t = \frac{\text{heat added}}{\text{temperature rise}}$$

$$= \frac{q\,\Delta t}{\Delta T}$$

If the temperature rises from value T_0 at time t_0 to T_1 at time t_1, then we can write

$$T_1 - T_0 = \frac{1}{C_t}\int_{t_0}^{t_1} q(t)\,dt$$

or, in differential form,

$$C_t \frac{dT(t)}{dt} = q(t)$$

section, only state variables (e.g., v_C) are guaranteed to be continuous across a transient event.

Finally, for circuits with two or more storage elements it is possible that those elements will exchange energy back and forth with each other during the transient response. When this phenomenon occurs, the result is oscillating voltages and currents in the circuit even as the average values of the oscillations decay exponentially over time.

Keep in mind that the purpose of this discussion is to reveal the physical basis for the fundamental behaviors observed in transient responses. Sections 4.3 and 4.4 contain numerous detailed examples of transient responses with mathematical derivations of the governing differential equations and associated problem-solving methods. Two examples of writing differential equations of first-order circuits are included at the end of this section for those interested in having a look ahead.

Time Constants

First-order circuits have one time constant τ, which is a measure of the speed of response of the circuit to a transient event. A small or large time constant indicates a fast or slow response, respectively. The time constant τ of a first-order circuit is either:

$$R_T C \qquad \text{or} \qquad \frac{L}{R_N} \tag{4.8}$$

depending upon whether the storage element is a capacitor or an inductor. Here, R_T and R_N are the Thévenin and Norton equivalent resistances seen by the capacitor and inductor, respectively.

Figure 4.7 shows a typical first-order decaying exponential. The time constant τ can be found graphically by two methods. The simplest and most common method is to determine τ as the time required for the exponential curve to fall $(e-1)/e$ (or approximately 63 percent) of the difference between its initial value $x(0)$ and its long-term steady state $x(\infty)$. An alternate method is to determine τ as the time marked by the intersection of the tangent to the exponential curve at $t = 0$ and the horizontal asymptote $x(\infty)$.

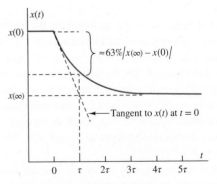

Figure 4.7 Generic first-order response $x(t)$ suggesting two graphical methods for finding a time constant

Second-order circuit response is more complex and described by one of three possible cases, each of which is determined by a parameter ζ known as the *dimensionless damping ratio*, as explained in detail in Section 4.4. When $\zeta > 1$, the response is said to be *overdamped* and is the sum of two first-order decaying exponentials, each with its own distinct time constant. When $\zeta = 1$, the response is said to be *critically damped*. When $\zeta < 1$, the response is said to be *underdamped*. As shown in Figure 4.5, the responses in these latter two cases *cannot* be simply described by two decaying exponentials.

Long-Term Steady State

The long-term steady state is that which remains after the transient response has decayed completely. For the first-order decaying exponential shown in Figure 4.7 the long-term steady state is $x(\infty)$. The long-term steady state depends upon the independent sources present in the $t > 0$ circuit and is commonly expressed in terms of a *gain K* multiplied by a forcing function $F(t)$ that represents the contributions of those sources. For simplicity, only circuits with DC independent sources are considered in this chapter, with the result that only DC long-term steady states occur.

Complete Response

The complete response is simply the sum of the transient response and the long-term steady state. In general, the transient response will contain one unknown constant for each state variable in the circuit. Thus, the complete response will also contain the same number of unknown constants. The values of these unknown constants are determined by the initial conditions on the circuit at $t = 0^+$.

A common mistake when learning to solve transient circuit problems is to apply the initial conditions to the transient response alone rather than to the complete solution. Forewarned, forearmed; don't make this error!

Natural and Forced Responses

Often, it is useful to express the complete response as the sum of *natural* and *forced* responses instead of the sum of a transient response and long-term steady state. Either way the complete response is unchanged. The natural response is that part of the complete system response due to the initial energy stored in the system at $t = 0$. The forced response is that part due to independent sources present in the $t > 0$ circuit.

As will be shown in the following section, equation 4.9 expresses the complete response $x(t)$ of an arbitrary first-order circuit variable as the sum of a transient response, with its characteristic exponential decay, and a long-term steady state $x(\infty)$.

$$x(t) = \left[x(0^+) - x(\infty)\right]e^{-t/\tau} + x(\infty) \tag{4.9}$$

The transient response portion includes the difference between the initial condition $x(0^+)$ and the long-term steady state. This expression can be reconstructed as:

$$x(t) = x_N(t) + x_F(t) = x(0^+)e^{-t/\tau} + x(\infty)(1 - e^{-t/\tau}) \tag{4.10}$$

The first and second terms in equation 4.10 are known as the *natural* and *forced* responses, $x_N(t)$ and $x_F(t)$, respectively. A similar construction can be made for the complete response of a second-order circuit.

MAKE THE CONNECTION

Thermal System Dynamics

To describe the dynamics of a thermal system, we write a differential equation based on energy balance. The difference between the heat added to the mass by an external source and the heat leaving the same mass (by convection or conduction) must be equal to the heat stored in the mass:

$$q_{in} - q_{out} = q_{stored}$$

An object is internally heated at the rate q_{in} in ambient temperature $T = T_a$; the thermal capacitance and thermal resistance are C_t and R_t. From energy balance:

$$q_{in}(t) - \frac{T(t) - T_a}{R_t} = C_t\frac{dT(t)}{dt}$$

$$R_t C_t\frac{dT(t)}{dt} + T(t) = R_t q_{in}(t) + T_a$$

$$\tau_t = R_t C_t \quad K_{St} = R_t$$

This first-order system is identical in its form to an electric *RC* circuit, as shown below.

Thermal system

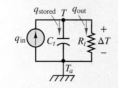

Equivalent electric circuit

EXAMPLE 4.1 **Initial Conditions**

Problem

For the circuit shown in Figure 4.8(a), determine the current through the inductor just before the switch is opened.

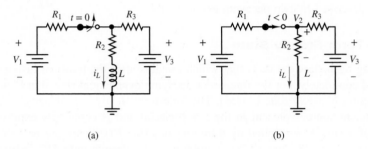

(a)   (b)

Figure 4.8 (a) Circuit for Example 4.1; (b) the same circuit just before the switch is opened

---

Solution

Known Quantities: $R_1 = 1 \text{ k}\Omega$; $R_2 = 5 \text{ k}\Omega$; $R_3 = 3.33 \text{ k}\Omega$; $L = 0.1 \text{ H}$; $V_1 = 12 \text{ V}$; $V_3 = 4 \text{ V}$.

Find: The current i_L through the inductor.

Assumptions: Assume the switch has been closed for a long time prior to $t = 0$.

Analysis: Because the switch has been closed for a long time prior to $t = 0$, the circuit is in a DC steady-state condition, and the inductor acts as a short-circuit, as shown in Figure 4.8(b). The current i_L through the inductor can be found quickly by applying KCL at node 2:

$$\frac{V_2 - V_1}{R_1} + \frac{V_2 - 0}{R_2} + \frac{V_2 - V_3}{R_3} = 0$$

Collect the coefficients of V_1, V_2, and V_3 to find:

$$\left(\frac{1}{R_1} + \frac{1}{R_2} + \frac{1}{R_3}\right) V_2 - \frac{V_1}{R_1} - \frac{V_3}{R_3} = 0$$

Finally, rearrange the terms to find:

$$V_2 = \left(\frac{1}{R_1} + \frac{1}{R_2} + \frac{1}{R_3}\right)^{-1} \left(\frac{V_1}{R_1} + \frac{V_3}{R_3}\right) = 8.80 \text{ V}$$

To determine the current through the inductor, observe that

$$i_L(0) = \frac{V_2}{R_2} = \frac{8.80}{5,000} = 1.76 \text{ mA}$$

Comments: The current $i_L(0)$ is *the* initial condition for the circuit in Figure 4.8(a). Only the state variables (i.e., the current through an inductor and the voltage across a capacitor) are guaranteed to be continuous across a transient event, such as the opening or closing of a switch.

EXAMPLE 4.2 Continuity of Inductor Current and Capacitor Voltage

Problem

Find the initial conditions at $t = 0$ on the current through the inductor and the voltage across the capacitor in the circuit in Figure 4.9.

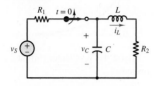

Figure 4.9 Circuit for Example 4.2.

Solution

Known Quantities: v_S; R_1; R_2; L; C

Find: The current through the inductor and the voltage across the capacitor at $t = 0^+$.

Assumptions: The switch has been closed for a very long time prior to $t = 0$.

Analysis: In a DC steady state, the inductor acts as a short-circuit while the capacitor acts as an open-circuit. Then, the circuit is effectively a single loop with a current i equal to the inductor short-circuit current and given by:

$$i = i_L = \frac{v_S}{R_1 + R_2} \qquad t < 0$$

The voltage across the capacitor open-circuit is given by voltage division.

$$v_C = v_S \frac{R_2}{R_1 + R_2} \qquad t < 0$$

Since neither the current through an inductor nor the voltage across a capacitor can change instantaneously, the initial conditions on the inductor current and capacitor voltage are:

$$i_L(t = 0^+) = i_L(t = 0^-) = \frac{v_S}{R_1 + R_2}$$

$$v_C(t = 0^+) = v_C(t = 0^-) = v_S \frac{R_2}{R_1 + R_2}$$

EXAMPLE 4.3 Continuity of Inductor Current

Problem

Find the initial condition and final value of the inductor current in the circuit in Figure 4.10.

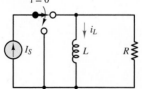

Figure 4.10 Circuit for Example 4.3.

Solution

Known Quantities: Source current I_S; inductor and resistor values.

Find: Inductor current at $t = 0^+$ and as $t \to \infty$.

Schematics, Diagrams, Circuits, and Given Data: $I_S = 10$ mA.

Assumptions: The current source has been connected to the circuit for a very long time.

Analysis: For $t < 0$, the inductor acts as a short-circuit, and $i_L = I_S$. Since all the current flows through the inductor short-circuit, the voltage across the resistor R must be zero. At $t = 0^+$, the switch opens and since the inductor current must be continuous

$$i_L(0^+) = i_L(0^-) = I_S$$

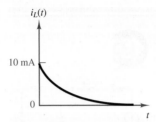

Figure 4.11 Inductor current response.

For $t > 0$, the current source is in its own isolated loop, cut off from the inductor and resistor. The inductor and resistor are in series in a separate isolated loop. Since this loop has no source, the loop current will eventually decay to zero (the long-term steady state) due to the energy dissipation of the resistor. A qualitative sketch of the current as a function of time is shown in Figure 4.11.

Comments: Note that the direction of the current in the circuit in Figure 4.11 is dictated by the initial condition since the inductor current cannot change instantaneously.

LO

EXAMPLE 4.4   Long-Term DC Steady State

Problem

Determine the capacitor voltage in the circuit in Figure 4.12(a) a long time after the switch has been closed.

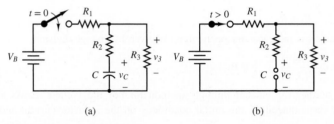

(a)   (b)

Figure 4.12 (a) Circuit for Example 4.4; (b) same circuit a long time after the switch is closed

---

Solution

Known Quantities: The values of the circuit elements are $R_1 = 100\ \Omega$; $R_2 = 75\ \Omega$; $R_3 = 250\ \Omega$; $C = 1\ \mu\text{F}$; $V_B = 12$ V.

Analysis: After the switch has been closed for a long time ($t \to \infty$), the transient response has decayed away and the circuit has reached a new DC steady state. In a DC state the capacitor acts as an open-circuit, as shown in Figure 4.12(b). As a result, no current flows through resistor R_2, and so resistors R_1 and R_3 are in series. Apply voltage division to find:

$$v_3(\infty) = \frac{R_3}{R_1 + R_3}\, V_B = \frac{250}{350}(12) = 8.57 \text{ V}$$

Since the current through R_2 is zero, the voltage across R_2 is also zero. Then, v_C equals the voltage drop from the upper right node to the bottom node, which is, of course, also equal to v_3. Thus:

$$v_C(\infty) = V_3(\infty) = 8.57 \text{ V}$$

Comments: The voltage $v_C(\infty)$ is the long-term steady-state value of the voltage across the capacitor.

EXAMPLE 4.5   Writing the Differential Equation of an *RC* Circuit

Problem

Derive a differential equation for the voltage across the capacitor shown in Figure 4.13.

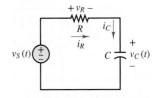

Figure 4.13 Circuit for Example 4.5.

Solution

Known Quantities: R; C; $v_S(t)$.

Find: The differential equations in $v_C(t)$ and $i(t)$.

Assumptions: None.

Analysis: Apply KVL around the loop to obtain:

$$v_S - iR - v_C = 0$$

Use the *i-v* relationship for a capacitor

$$i_C = C\frac{dv_C}{dt}$$

to substitute for i, where $i = i_C$ for the series loop:

$$v_S - RC\frac{dv_C}{dt} - v_C = 0$$

After dividing both sides of this equation by RC and rearranging terms, the result is:

$$\frac{dv_C}{dt} + \frac{1}{RC}v_C = \frac{1}{RC}v_S$$

Notice that the first term in the differential equation has dimensions of voltage per time. Since the other terms in the sum must also have the same dimensions, we can infer that the dimension of RC must be time!

A differential equation in the loop current i can also be found by differentiating both sides of the KVL equation above to obtain:

$$\frac{dv_S}{dt} - R\frac{di}{dt} - \frac{dv_C}{dt} = 0$$

Again, use the *i-v* relationship for a capacitor to substitute for the derivative of v_C to obtain:

$$\frac{dv_S}{dt} - R\frac{di}{dt} - \frac{i}{C} = 0$$

Divide both sides of the equation by R and rearrange to obtain:

$$\frac{di}{dt} + \frac{1}{RC}i = \frac{1}{R}\frac{dv_S}{dt}$$

Keep in mind that both the current i and the voltage v_C are functions of time as they transition from an old to a new steady state.

Note the similarity between the differential equations for $v_C(t)$ and $i(t)$. The left-hand sides have the same form and same coefficients while the right-hand side depends upon the voltage source v_S. The solution of either equation is sufficient to determine any other variable in the circuit.

Comments: First-order *RC* circuits have one state variable, v_C, the voltage across the capacitor.

EXAMPLE 4.6   Writing the Differential Equation of an *RL* Circuit

Problem

Derive differential equations from the circuit shown in Figure 4.14.

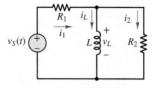

Figure 4.14 Circuit for
Example 4.6.

Solution

Known Quantities: $R_1 = 10\ \Omega$; $R_2 = 5\ \Omega$; $L = 0.4$ H.

Find: The differential equations for i_L and v_L.

Assumptions: None.

Analysis: Apply KCL at the top right node to obtain:

$$i_1 - i_L - i_2 = 0$$

Apply KVL around the left mesh to obtain:

$$v_S - i_1 R_1 - v_L = 0$$

Use the KCL to substitute for i_1 in the KVL equation and rearrange terms to obtain:

$$(i_L + i_2)R_1 + v_L = v_S$$

Use the Ohm's law expression $v_L = i_2 R_2$ to substitute for i_2 to find:

$$\left[i_L + \frac{v_L}{R_2}\right]R_1 + v_L = v_S$$

To obtain a differential equation for i_L simply use the differential *i-v* relationship for an inductor

$$v_L = L\frac{di_L}{dt}$$

to substitute for v_L. The result is:

$$\left[i_L + \frac{L}{R_2}\frac{di_L}{dt}\right]R_1 + L\frac{di_L}{dt} = v_S$$

Collect terms and divide both sides of the equation by R_1 to find:

$$L\frac{R_1 + R_2}{R_1 R_2}\frac{di_L}{dt} + i_L = \frac{v_S}{R_1}$$

Divide both sides of the equation by the coefficient of the first derivative term to yield:

$$\frac{di_L}{dt} + \frac{R_1 R_2}{R_1 + R_2}\frac{i_L}{L} = \frac{R_2}{R_1 + R_2}\frac{v_S}{L}$$

or

$$\frac{di_L}{dt} + \frac{R_T}{L}i_L = \frac{R_T}{LR_1}v_S$$

where R_T is the Thévenin equivalent resistance seen by the inductor. Notice that the first term in the differential equation has dimensions of current per time. Since the other terms in the sum must also have the same dimensions, we can infer that the dimension of L/R must be time!

Substitute numerical values to obtain:

$$\frac{di_L}{dt} + 8.33\,i_L = 0.833\,v_S$$

To obtain a differential equation for v_L take the derivative of both sides of

$$\left[i_L + \frac{v_L}{R_2}\right]R_1 + v_L = v_S$$

and use the differential *i-v* relation for an inductor to substitute for the derivative of i_L. The result is:

$$\left[\frac{v_L}{L} + \frac{1}{R_2}\frac{dv_L}{dt}\right]R_1 + \frac{dv_L}{dt} = \frac{dv_S}{dt}$$

Collect terms and divide both sides of the equation by R_1 to find:

$$\frac{R_1 + R_2}{R_1 R_2}\frac{dv_L}{dt} + \frac{v_L}{L} = \frac{1}{R_1}\frac{dv_S}{dt}$$

Notice that the coefficient of the first-derivative term is the inverse of the Thévenin equivalent resistance R_T seen by the inductor. Multiply both sides of the equation by R_T to obtain:

$$\frac{dv_L}{dt} + \frac{R_T}{L}v_L = \frac{R_T}{R_1}\frac{dv_S}{dt}$$

Comments: First-order *RL* circuits have one state variable, i_L, the current through the inductor.

EXAMPLE 4.7   Charging a Camera Flash—Capacitor Energy and Time Constants

Problem

A capacitor is used to store energy in a camera flash light. The camera operates on a 6-V battery. Determine the time required for the energy stored to reach 90 percent of its maximum. Compute the time in seconds and as a multiple of the time constant.

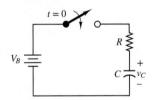

Figure 4.15 Equivalent circuit of camera flash charging circuit

Solution

Known Quantities: V_B; R; C.

Find: Time required to reach 90 percent of the total energy storage.

Schematics, Diagrams, Circuits, and Given Data: $V_B = 6$ V; $C = 1{,}000\ \mu$F; $R = 1$ kΩ.

Assumptions: The capacitor is completely discharged prior to $t = 0$.

Analysis: In the long-term steady state ($t \to \infty$) the total energy stored in the capacitor would be:

$$E_{\text{total}} = \tfrac{1}{2}Cv_C^2 = \tfrac{1}{2}CV_B^2 = 18 \times 10^{-3}\ \text{J}$$

Thus, 90 percent of the total energy will be reached when

$$E_{90\%} = 0.9 \times 18 \times 10^{-3} = 16.2 \times 10^{-3}\ \text{J}.$$

The corresponding capacitor voltage is calculated as follows:

$$\tfrac{1}{2}Cv_C^2 = 16.2 \times 10^{-3}$$

$$v_C = \sqrt{\frac{2 \times 16.2 \times 10^{-3}}{C}} = 5.692 \text{ V}$$

The Thévenin equivalent resistance seen by the capacitor for $t > 0$ is simply R, and, thus, the time constant of the circuit is $\tau = R_T C = 10^3 \times 10^{-3} = 1$ s. The time constant is a standard measure of the speed of a transient response. In this example, the transient response of the charging circuit is:

$$v_C = 6(1 - e^{-t/\tau}) = 6(1 - e^{-t})$$

Notice that this expression satisfies the initial condition $v_C(0) = 0$ and also satisfies the long-term DC steady-state value $v_C(t \to \infty) = V_B$. The time required to reach 90 percent of the energy is found by solving the previous expression when $v_C = 5.692$ V. Thus:

$$5.692 = 6(1 - e^{-t})$$
$$0.949 = 1 - e^{-t}$$
$$0.051 = e^{-t}$$
$$t = -\ln 0.051 = 2.97 \text{ s}$$

This period is approximately 3τ.

Comments: The fact that the capacitor charges to 90 percent of its total energy in a period of roughly 3τ is not limited to this example. All first-order systems have the same functional form and, therefore, have the same result. What percentage of the voltage change has occurred in this same 3τ period? How many time constants are required for the voltage to reach 99 percent of its ultimate value? Answers: 95 percent and 4.6τ

CHECK YOUR UNDERSTANDING

The single-pole, single-throw (SPST) switch in part (a) of Example 4.1 is opened at $t = 0$. What is the inductor current after a long time has passed?

$$\text{Answer: } i_L(\infty) = \frac{V_3}{R_2 + R_3} = 0.48 \text{ mA}$$

CHECK YOUR UNDERSTANDING

Use the principle of superposition to find the initial condition $i_L(t_0^+)$ in Example 4.1.

$$\text{Answer: } i_L(t_0^+) = i_L(t_0^-) = \frac{V_1}{R_2\|R_3} \frac{R_2\|R_3}{R_2\|R_3 + R_1} + \frac{V_3}{R_3} \frac{R_2\|R_3 + R_1}{R_2\,R_1\|R_2 + R_3} + \frac{R_1\|R_2}{R_2\,R_1\|R_2 + R_3} = 1.76 \text{ mA}$$

CHECK YOUR UNDERSTANDING

The single-pole, double-throw (SPDT) switch in the circuit of Example 4.3 is thrown at $t = 0$. Suppose that after a long time $t = t_\infty$ the switch is thrown again, back to its original position. What is the initial current through the inductor at $t = t_\infty$? What is the eventual long-term steady state current through the inductor for $t > t_\infty$?

Answer: $i_L(t_\infty) = 0$; $i_L(t \gg t_\infty) = I_s = 10$ mA.

CHECK YOUR UNDERSTANDING

Suppose that the single-pole, single-throw (SPST) switch in part (b) of Example 4.4 is eventually opened again. What is the capacitor voltage after an additional long time has passed?

Answer: $v_C(t \rightarrow \infty) = 0$ V. The capacitor will discharge through R_2 and R_3.

CHECK YOUR UNDERSTANDING

Use the differential i-v relations for capacitors and inductors along with KVL or KCL to write the differential equation for each of the circuits shown below.

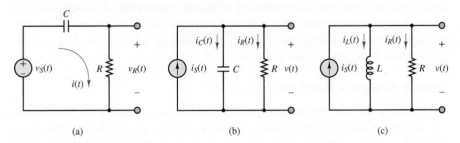

(a)   (b)   (c)

Answer: (a) $RC\dfrac{dv_C(t)}{dt} + v_C(t) = v_S(t)$; (b) $RC\dfrac{dv(t)}{dt} + v(t) = Ri_S(t)$;

(c) $\dfrac{R}{L}\dfrac{di_L(t)}{dt} + i_L(t) = i_S(t)$

CHECK YOUR UNDERSTANDING

Apply KVL twice to derive a differential equation for v_C for $t > 0$ in the circuit of Example 4.5.

Answer: $\dfrac{d^2 v_C}{dt^2} + \dfrac{L}{R_2}\dfrac{dv_C}{dt} + \dfrac{1}{LC}v_C = 0$

CHECK YOUR UNDERSTANDING

If another identical capacitor is placed in parallel with the capacitor in Example 4.7, how would the charging time change? How would the total stored energy change?

Answer: Both would double, as C_{eq} would be twice as large, thus doubling τ and E_{total}.

4.3   FIRST-ORDER TRANSIENT ANALYSIS

First-order systems are important in all engineering disciplines and occur frequently in nature. Such systems are characterized by a single state variable, where the system energy is proportional to the square of the state variable. That energy is dissipated by the system such that the rate of change of the state variable is proportional to the state variable itself. The fundamental result is that the transient response of a first-order system is a *decaying exponential* function of time. Ideal first-order electrical systems possess either capacitance or inductance (but not both) along with resistance and (perhaps) energy sources. Ideal first-order mechanical systems possess mass and damping (e.g., sliding or viscous friction) but no elasticity or compliance. An ideal first-order fluid system possesses fluid capacitance and viscous dissipation, such as a hydraulic system with a liquid-filled tank and a variable orifice. Many conductive and convective thermal systems also exhibit first-order behavior.

In general, when solving transient circuit problems it is necessary to determine three elements: (1) the steady-state response prior to a transient event, (2) the transient response immediately following the transient event, and (3) the long-term steady-state response remaining after the transient response has decayed away. The steps involved in computing the complete response of a first-order circuit with *constant sources* are outlined below. The methodology is straightforward and can be mastered with only modest practice.

Circuit Simplification for $t > 0$

The first step to solve for the response after the transient event ($t > 0$) is to partition the circuit into a source network and load, with the energy storage element as the load, as shown in Figure 4.16. If the source network is linear, it can be replaced by its Thévenin or Norton equivalent network.

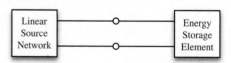

Figure 4.16 Generalized first-order circuit seen as a source network attached to an energy storage element as the load

Consider the case when the load is a capacitor and the source network is replaced by its Thévenin equivalent network, as shown in Figure 4.17. KVL can be applied around the loop to yield:

$$V_T - i R_T - v_C = 0$$

MAKE THE CONNECTION

First-Order Thermal System

An automotive transmission generates heat, when engaged, at the rate $q_{in} = 2{,}125$ W. The thermal capacitance of the transmission is $C_t = mc = 12$ kJ/°C. The effective convection resistance through which heat is dissipated is $R_t = 0.04$°C/W.

1. What is the steady-state temperature the transmission will reach when the initial (ambient) temperature is 5°C?

With reference to the Make the Connection sidebar "Thermal Capacitance," we write the differential equation based on energy balance:

$$R_t C_t \frac{dT}{dt} + T = R_t q_{in}$$

At steady state, the rate of change of temperature is zero; hence, $T(\infty) = R_t q_{in}$. Using the numbers given, $T(\infty) = 0.04 \times 2{,}125 = 85$°C

(Continued)

Of course, $i = i_C$ and for a capacitor $i_C = C\,dv_C/dt$. After substituting and rearranging the terms, the result is:

$$R_T C \frac{dv_C}{dt} + v_C = V_T \qquad \text{Capacitor load with Thévenin source} \qquad (4.11)$$

For a DC source network, the long-term steady-state solution is simply $v_C = V_T$.

Likewise, consider the case when the load is an inductor and the source network is replaced by its Norton equivalent network, as shown in Figure 4.18. KCL can be applied at either node to yield:

$$I_N - \frac{v}{R_N} - i = 0$$

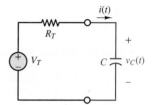

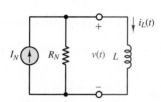

Figure 4.17 Generalized first-order circuit with a capacitor load and a Thévenin source

Figure 4.18 Generalized first-order circuit with an inductor load and a Norton source

Of course, $v = v_L$ and for an inductor $v_L = L\,di_L/dt$. After substituting and rearranging the terms, the result is:

$$\frac{L}{R_N} \frac{di_L}{dt} + i_L = I_N \qquad \text{Inductor load with Norton source} \qquad (4.12)$$

For a DC source network, the long-term steady-state solution is simply $i_L = I_N$. It is also possible to use a Norton or Thévenin source when the load is a capacitor or inductor, respectively. The results can be shown to be identical to those found above by substituting the source transformation expressions $V_T = I_N R_T$ and $R_T = R_N$.

It is important to keep in mind that these solutions are for $t > 0$, that is, the transient response. It is possible that the equivalent source network seen by the load after the transient event is different from that seen by the load before the event. Equivalent network methods can be used for both domains but *do not assume* that the equivalent network seen by the load is unchanged by the event.

First-Order Differential Equation

Both equations 4.11 and 4.12 have the same general form:

$$\tau \frac{dx(t)}{dt} + x(t) = K_S f(t) \qquad \text{First-order system equation} \qquad (4.13)$$

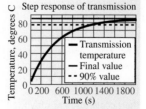

Hydraulic Tank

The analogy between electric and hydraulic circuits illustrated in earlier chapters can be applied to the hydraulic tank shown in Figure 4.21. The tank is cylindrical with cross-sectional area A, and the liquid contained in the tank exits the tank through a valve, which is modeled by a fluid resistance R. Initially, the level, or head, of the liquid is h_0. The principle of conservation of mass can be applied to the liquid in the tank in Figure 4.21 to determine the rate at which the tank will empty. For mass to be conserved, the following equation must apply:

$$q_{in} - q_{out} = q_{stored}$$

In this equation, the variable q represents a volumetric flow rate in cubic meters per second. The flow rate into the tank is zero in this particular case, and the flow rate out is given by the pressure difference across the valve, divided by the resistance:

$$q_{out} = \frac{\Delta p}{R} = \frac{\rho g h}{R}$$

The expression $\Delta p = \rho g h$ is obtained from basic fluid mechanics: $\rho g h$ is the static pressure at the bottom of the tank, where ρ is the density of the liquid, g is the acceleration of gravity, and h is the (changing) liquid level.

(Continued)

where the constants τ and K_S are the **time constant** and the **DC gain**, respectively. In this chapter, $f(t)$ is assumed equal to a constant F, which represents the contribution of one or more DC sources. With that assumption in mind, the general first-order differential equation is:

$$\tau \frac{dx(t)}{dt} + x(t) = K_S F \qquad t \geq 0 \tag{4.14}$$

The solution for $x(t)$ has two parts: the *transient* response and the *long-term steady-state* response. These two parts can also be rearranged in terms of **natural** and **forced** responses, as shown in Section 4.2. The sum of both parts is known as the **complete response**. One initial condition $x(0^+)$ is needed to specify the complete response.

First-Order Transient Response

The transient response x_{TR} is found by setting $F = 0$ in equation 4.14 such that:

$$\tau \frac{dx_{TR}(t)}{dt} + x_{TR}(t) = 0 \tag{4.15}$$

The solution for x is found by assuming a solution of the form:

$$x_{TR}(t) = \alpha e^{st} \tag{4.16}$$

Substitution of this assumed solution into equation 4.15 results in a characteristic equation.

$$\tau s + 1 = 0 \qquad \text{Characteristic equation} \tag{4.17}$$

The solution for s is simply:

$$s = \frac{-1}{\tau} \tag{4.18}$$

which is known as the root of the characteristic equation. Plugging in for s in equation 4.16 yields a decaying exponential.

$$\boxed{x_{TR}(t) = \alpha e^{-t/\tau} \qquad \text{Transient response}} \tag{4.19}$$

The constant α in equation 4.19 cannot be evaluated until the complete response has been found. If the system does not have an external forcing function, the transient response is also the complete response, and the constant α is equal to the initial condition $x(0^+)$. The Make the Connection sidebar "Hydraulic Tank" illustrates this case by considering a fluid tank emptying through an orifice. Once the valve is open, the volume of liquid in the tank decreases exponentially with a time constant τ determined by the physical properties of the system.

The amplitude of $x_{TR}(t)$ at $t = n\tau$ for $n = 0, 1, \ldots, 5$ is shown in Figure 4.20. The data show that x_{TR} has decayed by roughly 95 percent at three time constants and by over 99 percent at five time constants.

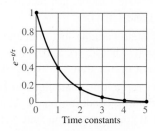

Figure 4.20 Normalized first-order exponential decay

Long-Term Steady-State Response

Still assuming that the first-order circuit contains only DC sources, such that $f(t)$ is a constant F, the long-term steady-state response of a first-order system is the solution to:

$$\tau \frac{dx_{SS}(t)}{dt} + x_{SS}(t) = K_S F \qquad t \geq 0 \tag{4.20}$$

For constant F, $x_{SS} = K_S F$ is the solution. It is a worthwhile exercise to show that this solution satisfies equation 4.20. Thus:

$$\boxed{x_{SS}(t) \equiv x(\infty) = K_S F \qquad F = \text{constant}} \tag{4.21}$$

Complete First-Order Response

The complete response is the sum of the transient and long-term steady-state responses:

$$x(t) = x_{TR}(t) + x_{SS}(t) = \alpha e^{-t/\tau} + x(\infty) = \alpha e^{-t/\tau} + K_S F \qquad t \geq 0 \tag{4.22}$$

Apply the one initial condition $x(0^+)$ to solve for the unknown constant α:

$$\begin{aligned} x(0^+) &= \alpha + x(\infty) \\ \alpha &= x(0^+) - x(\infty) \end{aligned} \tag{4.23}$$

Substitute for α in equation 4.22 to find the complete response:

$$\boxed{x(t) = [x(0^+) - x(\infty)]e^{-t/\tau} + x(\infty) \qquad t \geq 0} \tag{4.24}$$

MAKE THE CONNECTION

(Concluded)

The flow rate stored is related to the rate of change of the fluid volume contained in the tank (the tank stores energy in the mass of the fluid):

$$q_{\text{stored}} = A \frac{dh}{dt}$$

Thus, we can describe the emptying of the tank by means of the first-order linear ordinary differential equation

$$0 - q_{\text{out}} = q_{\text{stored}}$$
$$\Rightarrow \quad -\frac{\rho g h}{R} = A \frac{dh}{dt}$$
$$\frac{RA}{\rho g} \frac{dh}{dt} + h = 0$$
$$\Rightarrow \quad \tau \frac{dh}{dt} + h = 0$$
$$\tau = \frac{RA}{\rho g}$$

We know from the content of the present section that the solution of the first-order equation with zero input and initial condition h_0 is

$$h(t) = h_0 e^{-t/\tau}$$

Thus, the tank will empty exponentially, with the time constant determined by the fluid properties, that is, by the resistance of the valve and by the area of the tank.

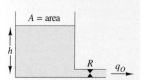

Figure 4.21 Analogy between electrical and fluid capacitance

FOCUS ON PROBLEM SOLVING

FIRST-ORDER TRANSIENT CIRCUIT ANALYSIS

1. Find the value of the state variable just before the transient event at $t = 0^-$. That is, find $v_C(0^-)$ and $i_L(0^-)$.

2. Set the value of the state variable just after the transient event equal to the value just before it. That is, set $v_C(0^+) = v_C(0^-)$ or $i_L(0^+) = i_L(0^-)$ as the initial condition on the transient response.

   Note: Only the state variable is guaranteed to be continuous across the transient event. The initial condition on an arbitrary variable $x(t)$ *must* be found from the initial condition on the state variable.

3. For $t > 0$, treat the storage element as the load and simplify the remaining source network. Assuming the source network is linear, when the storage element is:

   - a capacitor, replace the source network with its Thévenin equivalent (V_T and R_T), as shown in Figure 4.17.

   - an inductor, replace the source network with its Norton equivalent (I_N and R_N), as shown in Figure 4.18.

4. For $t > 0$, the governing differential equation for the state variable is found by applying either KVL or KCL.

   - When the load is a capacitor, apply KVL to find:

     $$\tau \frac{dv_C}{dt} + v_C = V_T$$

   - When the load is an inductor, apply KCL to find:

     $$\tau \frac{di_L}{dt} + i_L = I_N$$

5. For $t > 0$, the complete solution for the state variable is found by solving the governing differential equation and applying its initial condition.

   - When the load is a capacitor, the complete solution for the state variable is:

     $$v_C(t) = \left[v_C(0^+) - V_T\right] e^{-t/\tau} + V_T \qquad \text{with } \tau = R_T C$$

     For an arbitrary variable the complete solution is:

     $$x(t) = \left[x(0^+) - x(\infty)\right] e^{-t/\tau} + x(\infty) \qquad \text{with } \tau = R_T C$$

   - When the load is an inductor, the complete solution for the state variable is:

     $$i_L(t) = \left[i_L(0^+) - I_N\right] e^{-t/\tau} + I_N \qquad \text{with } \tau = L/R_N$$

     For an arbitrary variable the complete solution is:

     $$x(t) = \left[x(0^+) - x(\infty)\right] e^{-t/\tau} + x(\infty) \qquad \text{with } \tau = L/R_N$$

   Note: The left side of the governing differential equation for an arbitrary variable $x(t)$ is the same as that for the state variable. The right side of the governing differential equation for an arbitrary variable $x(t)$ is simply the long-term steady-state value for $x(t)$, which can be found by applying methods from Chapters 1 to 3. One particularly important observation is the time constant is the same for all variables; that is, the time constant is a characteristic of the entire original circuit.

EXAMPLE 4.8 Simplifying a First-Order Transient Circuit

Problem

Determine a symbolic solution for the first-order circuit shown in Figure 4.22.

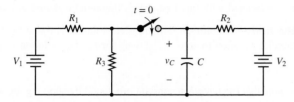

Figure 4.22 Circuit for Example 4.8.

Solution

Known Quantities: V_1; V_2; R_1; R_2; R_3; C.

Find: Capacitor voltage as a function of time $v_C(t)$ for all t.

Schematics, Diagrams, Circuits, and Given Data: Figure 4.22.

Assumptions: Assume the switch was open for a very long time prior to closing, such that the circuit is in a DC steady state prior to the transient event at $t = 0$.

Analysis:

Step 1: Find v_C for $t < 0$. For $t < 0$, the circuit is in a DC steady state such that the capacitor acts as an open-circuit. Thus, there is no current through R_2 and its voltage drop is zero. Consequently, the voltage across the capacitor is V_2, as required by KVL.

$$v_C(t) = V_2 \qquad t < 0$$

Remember that it is always necessary to solve for the value of the state variable prior to the transient event even if the state variable is not the variable of ultimate interest.

Step 2: Find the initial condition on v_C. Since the variable of interest is also the state variable v_C, its initial condition is already known from step 1; that is, the initial condition on v_C at $t = 0$ is V_2.

$$v_C(0^-) = v_C(0^+) = V_2 \qquad \text{Continuity of capacitor voltage}$$

Step 3: Simplify the circuit for $t > 0$. After the switch is closed, the resulting circuit is as shown in Figure 4.23, which was drawn to emphasize the two Thévenin sources (V_1, R_1) and (V_2, R_2) present. The approach is to select the capacitor as the load and then simplify the rest of the network to its Thévenin equivalent network.

Each Thévenin source in Figure 4.23 can be transformed to its equivalent Norton source as shown in Figure 4.24. The result is a network of resistors and independent current sources all in parallel. The current sources are combined (summed) to a single equivalent current source, and the resistors are combined to form a single equivalent resistance R_T. The resulting Norton source is then transformed to a Thévenin source, to which the load is reattached as shown in Figure 4.25.

Step 4: Find the differential equation. Apply KVL around the loop in Figure 4.25 to yield the differential equation for $t > 0$:

$$V_T - iR_T - v_C = V_T - R_T C \frac{dv_C}{dt} - v_C = 0 \qquad t > 0$$

$$R_T C \frac{dv_C}{dt} + v_C = V_T \qquad t > 0$$

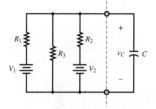

Figure 4.23 The circuit in Figure 4.22 for $t > 0$

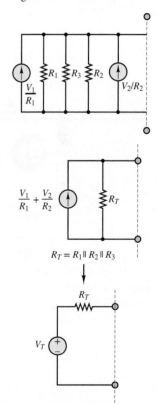

Figure 4.24 Simplification of the source network in Figure 4.23 to its Thévenin equivalent

where

$$R_T = R_1 \| R_2 \| R_3$$

$$V_T = \left(\frac{V_1}{R_1} + \frac{V_2}{R_2}\right) R_T$$

The time constant associated with this first-order differential equation is $\tau = R_T C$.

Step 5: Find the transient solution. The transient solution is found by setting the right side of the differential equation to zero and solving for v_C. The solution is always:

$$(v_C)_{TR} = \alpha e^{-t/\tau}$$

It is important to note that the unknown constant α is found by applying the initial condition to the *complete* solution, not to the transient solution alone.

Step 6: Find the long-term steady-state solution. The long-term DC steady-state solution for v_C is found after the switch has been closed for a very long time (practically $t \geq 5\tau$). The capacitor acts like an open-circuit such that $(v_C)_{SS} \equiv v_C(\infty) = V_T$.

Step 7: Complete solution. The complete solution is the sum of the transient and long-term steady-state solutions.

$$v_C(t) = (v_C)_{TR} + (v_C)_{SS} = \alpha e^{-t/\tau} + V_T$$

The unknown constant α is found by applying the initial condition $v_C(0^+) = V_2$. The result is:

$$V_2 = v_C(0^+) = \alpha + V_T \qquad \text{or} \qquad \alpha = V_2 - V_T$$

Finally, the complete solution is:

$$v_C(t) = (V_2 - V_T)e^{-t/R_T C} + V_T$$

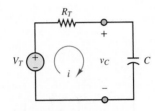

Figure 4.25 The circuit in Figure 4.23 simplified using Thévenin's theorem for $t > 0$

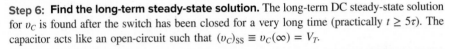

EXAMPLE 4.9   Starting Transient of DC Motor

Problem

A DC motor can be modeled approximately as an equivalent first-order series RL circuit, as shown in Figure 4.26. Find the complete solution for i_L.

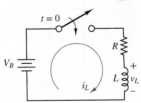

Figure 4.26 Circuit for Example 4.9

Solution

Known Quantities: Battery voltage V_B; resistance R; and inductance L.

Find: The inductor current as a function of time $i_L(t)$ for all t.

Schematics, Diagrams, Circuits, and Given Data: $R = 4\,\Omega$; $L = 0.1$ H; $V_B = 50$ V. Figure 4.26.

Assumptions: None.

Analysis:

Step 1: Find v_C for $t < 0$. The current through the inductor prior to the closing of the switch must be zero; thus,

$$i_L(t) = 0 \text{ A} \qquad t < 0$$

When the switch has been closed for a long time, the current through the inductor is constant and can be calculated by replacing the inductor with a short-circuit.

$$i_L(\infty) = \frac{V_B}{R} = \frac{50}{4} = 12.5 \text{ A}$$

Step 2: Find the initial condition on i_L. Since the variable of interest is also the state variable i_L, its initial condition is already known from step 1; that is, the initial condition on i_L at $t = 0$ is 0.

$$i_L(0^+) = i_L(0^-) = 0 \qquad \text{Continuity of inductor current}$$

Step 3: Simplify the circuit for $t > 0$. For $t > 0$, the network attached to the inductor is already in the form of a Thévenin source, so no further simplification is possible.

Step 4: Find the differential equation. Apply KVL around the loop in Figure 4.26 to find the differential equation for $t > 0$:

$$V_B - i_L R - v_L = 0 \qquad t > 0$$

$$V_B - i_L R - L\frac{di_L}{dt} = 0$$

After moving the forcing function V_B to the right side:

$$\frac{L}{R}\frac{di_L(t)}{dt} + i_L(t) = \frac{1}{R}V_B \qquad t > 0$$

The time constant τ is the coefficient of the first-derivative term:

$$\tau = \frac{L}{R} = 0.025 \text{ s}$$

Step 5: Find the transient solution. The transient solution is found by setting the right side of the differential equation to zero and solving for i_L. The solution is always of the form:

$$(i_L)_{TR} = \alpha e^{-t/\tau} = \alpha e^{-t/0.025} = \alpha e^{-40t}$$

It is important to note that the unknown constant α is found by applying the initial condition to the *complete* solution, not to the transient solution alone.

Step 6: Find the long-term steady-state solution. The long-term DC steady-state solution for i_L is found after the switch has been closed for a very long time (practically $t \geq 5\tau$). In this state, the inductor acts like a short-circuit such that $(i_L)_{SS} \equiv i_L(\infty) = V_B/R = 12.5$ A.

Step 7: Complete solution. The complete solution is the sum of the *transient* and *long-term steady-state* solutions.

$$i_L(t) = (i_L)_{TR} + (i_L)_{SS} = \alpha e^{-40t} + 12.5 \text{ A}$$

The unknown constant α is found by applying the initial condition $i_L(0^+) = 0$. The result is:

$$0 = i_L(0^+) = \alpha + 12.5 \text{ A} \qquad \text{or} \qquad \alpha = 0 - 12.5 \text{ A} = -12.5 \text{ A}$$

Finally, the complete solution is:

$$i_L(t) = -12.5 e^{-40t} + 12.5 \text{ A}$$

The complete solution can also be expressed in terms of *natural* and *forced* responses:

$$i_L(t) = i_{LN}(t) + i_{LF}(t) = 0 + 12.5(1 - e^{-40t})$$

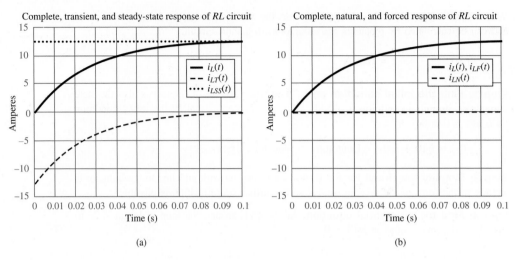

Figure 4.27 Complete response $i_L(t)$ of electric motor: (a) steady-state $i_{LSS}(t)$ plus transient $i_{LT}(t)$ responses; (b) forced $i_{LF}(t)$ plus natural $i_{LN}(t)$ responses

The complete response and its decomposition into (a) transient plus steady-state responses, and (b) natural plus forced responses are shown in Figure 4.27.

Comments: Note that in practice it is not a good idea to place a switch in series with an inductor. As the switch opens, the inductor current is forced to change suddenly, with the result that di_L/dt, and therefore $v_L(t)$, gets very large. The large voltage transient resulting from this *inductive kick* can damage circuit components. A practical solution to this problem, the freewheeling diode, is presented in Chapter 12.

EXAMPLE 4.10   Turnoff Transient of DC Motor

Problem

Determine the motor voltage for all time in the simplified electric motor circuit model shown in Figure 4.28. The motor is represented by the series *RL* circuit in the shaded box.

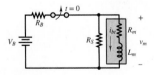

Figure 4.28 Circuit for Example 4.10.

Solution

Known Quantities: V_B; R_B; R_S; R_m; L_m.

Find: The voltage across the motor as a function of time.

Schematics, Diagrams, Circuits, and Given Data: $R_B = 2 \ \Omega$; $R_S = 20 \ \Omega$; $R_m = 0.8 \ \Omega$; $L_m = 3$ H; $V_B = 100$ V.

Assumptions: The switch has been closed for a long time.

Analysis: With the switch closed for a long time, the inductor in the circuit in Figure 4.28 behaves as a short-circuit. The current through the motor can then be calculated by the current divider rule in the modified circuit of Figure 4.29, where the inductor has been

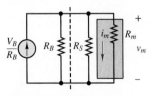

Figure 4.29 Equivalent circuit at steady state.

replaced with a short-circuit and the Thévenin circuit on the left has been replaced by its Norton equivalent:

$$i_m = \frac{1/R_m}{1/R_B + 1/R_s + 1/R_m}\frac{V_B}{R_B}$$

$$= \frac{1/0.8}{1/2 + 1/20 + 1/0.8}\frac{100}{2} = 34.72 \text{ A}$$

This current is the initial condition for the inductor current: $i_L(0) = 34.72$ A. Since the motor inductance is effectively a short-circuit, the motor voltage for $t < 0$ is equal to

$$v_m(t) = i_m R_m = 27.8 \text{ V} \qquad t < 0$$

When the switch opens and the motor voltage supply is turned off, the motor sees only the *shunt* (parallel) resistance R_S, as depicted in Figure 4.30. Remember now that the inductor current cannot change instantaneously; thus, the motor (inductor) current i_m must continue to flow in the same direction. Since all that is left is a series *RL* circuit, with resistance $R = R_S + R_m = 20.8 \ \Omega$, the inductor current will decay exponentially with time constant $\tau = L_m/R = 0.1442$ s:

$$i_L(t) = i_m(t) = i_L(0)e^{-t/\tau} = 34.7e^{-t/0.1442} \qquad t > 0$$

Figure 4.30 Circuit with switch open.

The motor voltage is then computed by adding the voltage drop across the motor resistance and inductance:

$$v_m(t) = R_m i_L(t) + L_m \frac{di_L(t)}{dt}$$

$$= 0.8 \times 34.7e^{-t/0.1442} + 3\left(-\frac{34.7}{0.1442}\right)e^{-t/0.1442} \qquad t > 0$$

$$= -694.1e^{-t/0.1442} \qquad t > 0$$

The motor voltage is plotted in Figure 4.31.

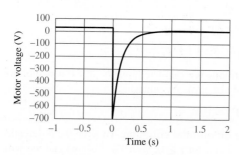

Figure 4.31 Motor voltage transient response

Comments: Notice how the motor voltage rapidly changes from the steady-state value of 27.8 V for $t < 0$ to a large negative value due to the turnoff transient. This *inductive kick* is typical of *RL* circuits and results from the fact that although the inductor current cannot change instantaneously, the inductor voltage can and does, as it is proportional to the derivative of i_L. This example is based on a simplified representation of an electric motor but illustrates effectively the need for special starting and stopping circuits in electric motors. Some of these ideas are explored in Chapter 12.

EXAMPLE 4.11   **First-Order Response due to a Pulsed Source**

Problem

The circuit in Figure 4.32 includes a switch that can be used to connect and disconnect a battery. The switch has been open for a very long time. At $t = 0$ the switch closes, and then at $t = 50$ ms the switch opens again. Determine the capacitor voltage as a function of time.

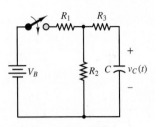

Figure 4.32 Circuit for Example 4.11.

Solution

Known Quantities: V_B; R_1; R_2; R_3; C.

Find: Capacitor voltage as a function of time $v_C(t)$ for all t.

Schematics, Diagrams, Circuits, and Given Data: $V_B = 15$ V, $R_1 = R_2 = 1{,}000$ Ω, $R_3 = 500$ Ω, and $C = 25$ μF. Figure 4.32.

Assumptions: None.

Analysis:

Part 1 $(0 \leq t < 50$ ms)

Step 1: Steady-state response. We first observe that any charge stored in the capacitor has had a discharge path through resistors R_3 and R_2. Thus, the capacitor must be completely discharged. Hence,

$$v_C(t) = 0 \text{ V} \qquad t < 0 \qquad \text{and} \qquad v_C(0^-) = 0 \text{ V}$$

To determine the steady-state response, we look at the circuit a long time after the switch has been closed. At steady state, the capacitor behaves as an open-circuit, and we can calculate the equivalent open-circuit (Thévenin) voltage and equivalent resistance to be

$$v_C(\infty) = V_B \frac{R_2}{R_1 + R_2} = 7.5 \text{ V}$$

$$R_T = R_3 + R_1 \| R_2 = 1 \text{ k}\Omega$$

Step 2: Initial condition. We can determine the initial condition for the variable $v_C(t)$ by virtue of the continuity of capacitor voltage:

$$v_C(0^+) = v_C(0^-) = 0 \text{ V}$$

Step 3: Writing the differential equation. To write the differential equation, we use the Thévenin equivalent circuit for $t \geq 0$, with $V_T = v_C(\infty)$ and we write the resulting differential equation

$$V_T - R_T i_C - v_C = V_T - R_T C \frac{dv_C(t)}{dt} - v_C(t) = 0 \qquad 0 \leq t < 50 \text{ ms}$$

$$R_T C \frac{dv_C}{dt} + v_C = V_T \qquad 0 \leq t < 50 \text{ ms}$$

Step 4: Time constant. In the above equation we recognize the following variables and parameters:

$$x = v_C \qquad \tau = R_T C = 0.025 \text{ s} \qquad K_S = 1$$

$$f(t) = V_T = 7.5 \text{ V} \qquad 0 \leq t < 50 \text{ ms}$$

Step 5: Complete solution. The complete solution is

$$v_C(t) = v_C(\infty) + [v_C(0) - v_C(\infty)]e^{-t/\tau} \qquad 0 \leq t < 50 \text{ ms}$$

$$v_C(t) = V_T + (0 - V_T)e^{-t/R_TC} = 7.5(1 - e^{-t/0.025}) \quad \text{V} \qquad 0 \leq t < 50 \text{ ms}$$

Part 2 $(t \geq 50 \text{ ms})$

As mentioned in the problem statement, at $t = 50$ ms the switch opens again, and the capacitor now discharges through the series combination of resistors R_3 and R_2. Since there is no forcing function after the switch is opened, the long-term steady-state solution $v_C(\infty)$ is zero, the transient and natural responses are identical, and the complete response has the form $v_C(t - t_1) = \alpha e^{-(t-t_1)/\tau}$, where $t_1 = 50$ ms.

1. The voltage v_C across the capacitor (a state variable) is continuous at $t = 50$ ms when the switch is opened.
2. The constant α is the initial condition on v_C at $t = 50$ ms.
3. The time constant for $t \geq 50$ ms is $\tau = (R_2 + R_3)C = 0.0375$ s.

Use the solution for $0 \leq t \leq 50$ ms to calculate $v_C(t = t_1 = 50 \text{ ms})$ and determine α.

$$\alpha = 7.5(1 - e^{-0.05/0.025}) = 6.485 \text{ V}$$

Thus, the capacitor voltage for $t \geq 50$ ms is:

$$v_C(t) = 6.485\,e^{-(t-0.05)/0.0375}$$

The composite response is plotted below.

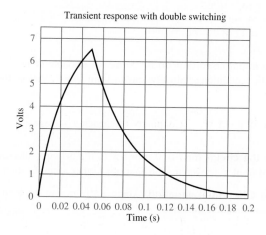

Transient response with double switching

Comments: Note that the two parts of the response are based on two different time constants and that the rising portion of the response changes faster (shorter time constant) than the falling part. Also notice that the transient solution of part 2 was expressed in terms of a *time shift* $(t - 0.05)$ ms, which accounts for the fact that the switch opened at $t = 50$ ms.

EXAMPLE 4.12   First-Order Natural and Forced Responses

Problem

Determine an expression for the capacitor voltage in the circuit of Figure 4.33.

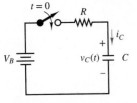

Figure 4.33

Solution

Known Quantities: $v_C(t = 0^-)$; V_B; R; C.

Find: Capacitor voltage as a function of time $v_C(t)$ for all t.

Schematics, Diagrams, Circuits, and Given Data: $v_C(t = 0^-) = 5$ V; $R = 1$ kΩ; $C = 470$ μF; $V_B = 12$ V. Figure 4.33.

Assumptions: None.

Analysis:

Step 1: Find v_C for $t < 0$. For $t < 0$, the capacitor is not part of a closed loop; therefore, the current through the capacitor must be zero for $t < 0$. In other words, its charge (and consequently, its energy) was constant. In this example, it is assumed that the capacitor has an initial charge $q = Cv_C(0^-) = C(5$ V$)$. Thus:

$$v_C(t) = 5 \text{ V} \qquad t < 0 \qquad \text{and} \qquad v_C(0^-) = 5 \text{ V}$$

When the switch has been closed for a long time, the circuit reaches its new DC steady state and the capacitor can be replaced by an open-circuit. Thus, the current in the loop eventually reaches zero.

$$v_C(\infty) = V_B = 12 \text{ V}$$

Step 2: Find the initial condition on v_C. Since the variable of interest is also the state variable v_C, its initial condition is already known from step 1; that is, the initial condition on v_C at $t = 0$ is 5 V.

$$v_C(0^-) = v_C(0^+) = 5 \text{ V} \qquad \text{Continuity of capacitor voltage}$$

Step 3: Simplify the circuit for $t > 0$. For $t > 0$, the network attached to the capacitor is already in the form of a Thévenin source, so no further simplification is possible.

Step 4: Find the differential equation. Apply KVL around the loop in Figure 4.33 to yield the differential equation for $t > 0$:

$$12 \text{ V} - Ri_C - v_C = 12 \text{ V} - RC\frac{dv_C}{dt} - v_C = 0 \qquad t > 0$$

$$RC\frac{dv_C}{dt} + v_C = 12 \text{ V} \qquad t > 0$$

The time constant τ is the coefficient of the first-derivative term:

$$\tau = RC = 0.47 \text{ s}$$

Step 5: Find the transient solution. The transient solution is found by setting the right side of the differential equation to zero and solving for v_C. The solution is always of the form:

$$(v_C)_{TR} = \alpha e^{-t/\tau} = \alpha e^{-t/0.47}$$

It is important to note that the unknown constant α is found by applying the initial condition to the *complete* solution, not to the transient solution alone.

Step 6: Find the long-term steady-state solution. The long-term DC steady-state solution for v_C is found after the switch has been closed for a very long time (practically $t \geq 5\tau$). In this state, the capacitor acts like an open-circuit such that $(v_C)_{SS} \equiv v_C(\infty) = 12$ V.

Step 7: Complete solution. The complete solution is the sum of the *transient* and *long-term steady-state* solutions.

$$v_C(t) = (v_C)_{TR} + (v_C)_{SS} = \alpha e^{-t/0.47} + 12 \text{ V}$$

The unknown constant α is found by applying the initial condition $v_C(0^+) = 5$ V. The result is:

$$5 \text{ V} = v_C(0^+) = \alpha + 12 \text{ V} \quad \text{or} \quad \alpha = 5 \text{ V} - 12 \text{ V} = -7 \text{ V}$$

Finally, the complete solution is:

$$v_C(t) = -7 e^{-t/0.47} + 12$$

The complete solution can also be expressed in terms of *natural* and *forced* responses:

$$v_C(t) = v_{CN}(t) + v_{CF}(t) = 5 e^{-t/0.47} + 12(1 - e^{-t/0.47})$$

The complete response and its decomposition into (a) transient plus steady-state responses, and (b) natural plus forced responses are shown in Figure 4.34.

Comments: Note how in Figure 4.34(a) the long-term steady-state response v_{CSS} equals the battery voltage while the transient response $v_{CTR}(t)$ rises from -7 to 0 V exponentially. In Figure 4.34(b), on the other hand, the energy initially stored in the capacitor decays to zero via its natural response v_{CN} while the external forcing function causes the capacitor voltage to rise exponentially to 12 V, as shown in the forced response v_{CF}.

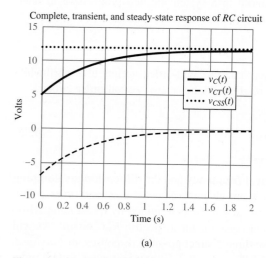

(a)

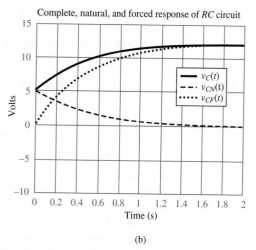

(b)

Figure 4.34 (a) Complete, transient, and steady-state responses of the circuit in Figure 4.33; (b) complete, natural, and forced responses of the same circuit

---

CHECK YOUR UNDERSTANDING

What happens if the initial condition (capacitor voltage for $t < 0$) is zero in Example 4.12?

Answer: The complete solution is equal to the forced solution.
$v_C(t) = v_{Cf}(t) = 12(1 - e^{-t/0.47})$ V

4.4 SECOND-ORDER TRANSIENT ANALYSIS

In general, a second-order circuit has two irreducible storage elements: two capacitors, two inductors, or one capacitor and one inductor. The latter case is the most important in terms of new fundamentals; however, the important aspects of all second-order system responses are discussed in this section. Since second-order circuits have two irreducible storage elements, such circuits have two state variables and their behavior is described by a second-order differential equation.

The simplest, yet arguably the most crucial, second-order circuits are those in which the capacitor and inductor are either in parallel or in series, as shown later in Figures 4.48 and 4.49. The circuits in these figures are drawn to suggest that the capacitor and inductor should be treated as a unified load. The rest of each circuit is either the Thévenin or Norton equivalent of the source network. The analysis of these circuits is somewhat less complicated than for other second-order circuits, which is appealing for anyone learning to analyze such circuits for the first time. The analysis of more complicated second-order circuits is treated in an example later in this section.

It is important to adopt a patient but determined attitude toward the material in this section, as it is notoriously challenging to students. Every effort has been made to walk you through the material in a systematic and progressive manner. Hold on to your hat! And don't panic.

General Characteristics

Before diving into the analysis of particular second-order circuits it is worthwhile to introduce the generalized differential equation for any second-order circuit.

$$\frac{1}{\omega_n^2}\frac{d^2x}{dt^2} + \frac{2\zeta}{\omega_n}\frac{dx}{dt} + x = K_S f(t) \tag{4.25}$$

The constants ω_n and ζ are the **natural frequency** and the **dimensionless damping ratio**, respectively. These parameters are characteristics of a second-order circuit and determine its response. Their values will be determined by direct comparison of equation 4.25 with the differential equation for a specific *RLC* circuit. As will be shown, second-order circuits have three distinct possible responses: *overdamped*, *critically damped*, and *underdamped*. The response for any particular second-order circuit is determined entirely by ζ.

In equation 4.25, $f(t)$ is a forcing function. K_S is the **DC gain** of a particular variable $x(t)$. Different variables in the same circuit may have different DC gains. However, all variables share the same natural frequency ω_n, the same dimensionless damping ratio ζ, and therefore also the same type of response. This fact can be an important time saver when problem solving.

Parallel *LC* Circuits

Consider the circuit shown in Figure 4.35. The two state variables are i_L and v_C, where v_C is the primary state variable because it is shared by all four circuit elements. In general, at the moment of a transient event the energy of the storage elements may be nonzero; that is, the voltage $v_C(0)$ across the capacitor and the current $i_L(0)$ through the inductor may be nonzero. As always, the two state variables are continuous such that:

$$v_C(0^+) = v_C(0^-) \quad \text{and} \quad i_L(0^+) = i_L(0^-)$$

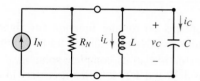

Figure 4.35 Second-order circuit with the inductor and capacitor in parallel acting as a unified load attached to a Norton equivalent network

Apply KCL to either node to find a first-order differential equation in terms of both state variables.

$$I_N - \frac{v_C}{R_N} - i_L - i_C = 0 \qquad \text{KCL}$$

The KCL equation can be transformed into a second-order differential equation in i_L by recognizing that:

$$v_C = v_L = L\frac{di_L}{dt} \quad \text{and} \quad i_C = C\frac{dv_C}{dt} = LC\frac{d^2 i_L}{dt^2}$$

Substitute for v_C and i_C in the KCL equation to find:

$$I_N - \frac{L}{R_N}\frac{di_L}{dt} - i_L - LC\frac{d^2 i_L}{dt^2} = 0$$

Rearrange the order of terms to yield:

$$LC\frac{d^2 i_L}{dt^2} + \frac{L}{R_N}\frac{di_L}{dt} + i_L = I_N \tag{4.26}$$

Alternatively, one can differentiate both sides of the KCL equation and substitute:

$$\frac{di_L}{dt} = \frac{v_C}{L} \quad \text{and} \quad \frac{di_C}{dt} = C\frac{d^2 v_C}{dt^2}$$

The result is:

$$\frac{dI_N}{dt} - \frac{1}{R_N}\frac{dv_C}{dt} - \frac{v_C}{L} - C\frac{d^2 v_C}{dt^2} = 0$$

Multiply both sides of the equation by L, and if the source I_N is a constant such that its time derivative is zero, the resulting second-order differential equation is:

$$LC\frac{d^2 v_C}{dt^2} + \frac{L}{R_N}\frac{dv_C}{dt} + v_C = 0 \qquad (4.27)$$

Equation 4.27 contains no forcing function (its right side is zero), which indicates that the long-term steady-state solution for v_C will be zero. In other words, the transient solution for v_C is also its complete solution.

To solve equations 4.26 and 4.27 it is first necessary to identify the *dimensionless damping ratio* ζ and the *natural frequency* ω_n. Notice that the left sides of both equations are identical, as they are for any variable in the circuit. Thus, ω_n and ζ can be found from either differential equation by comparing it to equation 4.25. The result is:

$$\frac{1}{\omega_n^2} = LC \quad \text{and} \quad \frac{2\zeta}{\omega_n} = \frac{L}{R_N}$$

These two equations can be solved to yield:

$$\omega_n = \frac{1}{\sqrt{LC}} \quad \text{and} \quad \zeta = \frac{1}{2R_N}\sqrt{\frac{L}{C}} \qquad (4.28)$$

The type of transient response for i_L and v_C depends upon ζ only. When ζ is greater than, equal to, or less than 1, the transient responses $(i_L)_{TR}$ and $(v_C)_{TR}$ are *overdamped*, *critically damped*, or *underdamped*, respectively. These three types of responses are described in detail later in this section. The complete solutions are:

$$i_L(t) = (i_L)_{TR} + (i_L)_{SS} = (i_L)_{TR} + I_N$$

and

$$v_C(t) = (v_C)_{TR} + (v_C)_{SS} = (v_C)_{TR} + L\frac{dI_N}{dt}$$

Note that when I_N is a constant, $(v_C)_{SS} = 0$ and $v_C(t) = (v_C)_{TR}(t)$.

Series *LC* Circuits

The development of the general solution for series *LC* circuits follows the same basic steps used above for parallel *LC* circuits. Consider the circuit in Figure 4.36 and note the duality between what follows and what was done above for the parallel *LC* circuit. In fact, the equations that follow can be found simply by starting with the equations developed above and swapping L with C, i_L with v_C, R_N with $1/R_T$, and I_N with V_T.

MAKE THE CONNECTION

Automotive Suspension

The mechanical model shown below can be analyzed using Newton's second law, $ma = \sum F$, to obtain the equation

$$m\frac{d^2 x(t)}{dt^2} = F(t) - b\frac{dx(t)}{dt} - kx(t)$$

This equation can be written in the *standard* form:

$$\frac{m}{k}\frac{d^2 x(t)}{dt^2} + \frac{b}{k}\frac{dx(t)}{dt}$$
$$+ x(t) = \frac{1}{k}f(t)$$

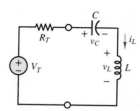

Figure 4.36 Second-order circuit with the inductor and capacitor in series acting as a unified load attached to a Thévenin equivalent network

Again, the two state variables are i_L and v_C, where i_L is the primary state variable because it is shared by all four circuit elements. Again, at the moment of a transient event the energy of the storage elements may be nonzero; that is, the voltage $v_C(0)$ across the capacitor and the current $i_L(0)$ through the inductor may be nonzero. As always, the two state variables are continuous such that:

$$v_C(0^+) = v_C(0^-) \qquad \text{and} \qquad i_L(0^+) = i_L(0^-)$$

Apply KVL around the series loop to find a first-order differential equation in terms of both state variables.

$$V_T - i_L R_T - v_C - v_L = 0 \qquad \text{KVL}$$

The KVL equation can be transformed into a second-order differential equation in v_C by recognizing that:

$$i_L = i_C = C\frac{dv_C}{dt} \qquad \text{and} \qquad v_L = L\frac{di_L}{dt} = LC\frac{d^2 v_C}{dt^2}$$

Substitute for v_L and i_L in the KVL equation to find:

$$V_T - R_T C\frac{dv_C}{dt} - v_C - LC\frac{d^2 v_C}{dt^2} = 0$$

The analogous *LC* series circuit shown below can be analyzed using KVL:

$$v_s - Ri_L - v_C - L\frac{di_L}{dt} = 0$$

$$i_L = i_C = C\frac{dv_C}{dt}$$

$$LC\frac{d^2 v_C}{dt^2} + RC\frac{dv_C}{dt} + v_C = v_s$$

Notice the similar structure of these two second-order differential equations.

(a)

(b)

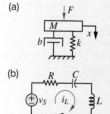

Analogy between electrical and mechanical systems

(Continued)

Rearrange the order of terms to yield:

$$LC\frac{d^2 v_C}{dt^2} + R_T C\frac{dv_C}{dt} + v_C = V_T \qquad (4.29)$$

Alternatively, one can differentiate both sides of the KVL equation and substitute:

$$\frac{dv_C}{dt} = \frac{i_L}{C} \qquad \text{and} \qquad \frac{dv_L}{dt} = L\frac{d^2 i_L}{dt^2}$$

The result is:

$$\frac{dV_T}{dt} - R_T\frac{di_L}{dt} - \frac{i_L}{C} - L\frac{d^2 i_L}{dt^2} = 0$$

Multiply both sides of the equation by C, and if the source V_T is a constant such that its time derivative is zero, the resulting second-order differential equation is:

$$LC\frac{d^2 i_L}{dt^2} + R_T C\frac{di_L}{dt} + i_L = 0 \qquad (4.30)$$

Equation 4.30 contains no forcing function (its right side is zero), which indicates that the long-term steady-state solution for i_L will be zero. In other words, the transient solution for i_L is also its complete solution.

To solve equations 4.29 and 4.30 it is first necessary to identify the *dimensionless damping ratio* ζ and the *natural frequency* ω_n. Notice that the left sides of both equations are identical, as they are for any variable in the circuit. Thus, ω_n and ζ can be found from either differential equation by comparing it to equation 4.25. The result is:

$$\frac{1}{\omega_n^2} = LC \qquad \text{and} \qquad \frac{2\zeta}{\omega_n} = R_T C$$

These two equations can be solved to yield:

$$\omega_n = \frac{1}{\sqrt{LC}} \qquad \text{and} \qquad \zeta = \frac{R_T}{2}\sqrt{\frac{C}{L}} \qquad (4.31)$$

The type of transient response for i_L and v_C depends upon ζ only. As always, when ζ is greater than, equal to, or less than 1, the transient responses $(i_L)_{\text{TR}}$ and $(v_C)_{\text{TR}}$ are *overdamped*, *critically damped*, or *underdamped*, respectively. These three types of responses are described in detail later in this section. The complete solutions are:

$$v_C(t) = (v_C)_{TR} + (v_C)_{SS} = (v_C)_{TR} + V_T$$

and

$$i_L(t) = (i_L)_{TR} + (i_L)_{SS} = (i_L)_{TR} + C\frac{dV_T}{dt}$$

Note that when V_T is a constant, $(i_L)_{SS} = 0$ and $i_L(t) = (i_L)_{TR}(t)$.

Transient Response

The transient response $x_{TR}(t)$ is found by setting the right side of the governing differential equation equal to zero. That is:

$$\frac{1}{\omega_n^2}\frac{d^2x_{TR}}{dt^2} + \frac{2\zeta}{\omega_n}\frac{dx_{TR}}{dt} + x_{TR} = 0 \qquad (4.32)$$

Just as in first-order systems, the solution of this equation has an exponential form:

$$x_{TR}(t) = \alpha e^{st} \qquad \text{Assumed transient response} \qquad (4.33)$$

Substitution into the differential equation yields the *characteristic equation*:

$$\frac{s^2}{\omega_n^2} + \frac{2\zeta}{\omega_n}s + 1 = 0 \qquad (4.34)$$

which, in turn, yields two **characteristic roots** s_1 and s_2. Specific values of s_1 and s_2 are found from the quadratic formula applied to the characteristic equation.

$$s_{1,2} = -\zeta\omega_n \pm \frac{1}{2}\sqrt{(2\zeta\omega_n)^2 - 4\omega_n^2} = -\omega_n\left(\zeta \pm \sqrt{\zeta^2 - 1}\right) \qquad (4.35)$$

The roots s_1 and s_2 are associated with the three distinct possible responses: overdamped ($\zeta > 1$), critically damped ($\zeta = 1$), and underdamped ($\zeta < 1$). The details of each of these responses are presented below.

1.   Overdamped Response ($\zeta > 1$)

Two distinct, negative, and real roots: (s_1, s_2). The transient response is overdamped when $\zeta > 1$ and the roots are $s_{1,2} = \omega_n\left(-\zeta \pm \sqrt{\zeta^2 - 1}\right)$. The general form of the solution is

$$x_{TR}(t) = \alpha_1 e^{s_1 t} + \alpha_2 e^{s_2 t} = e^{-\zeta\omega_n t}\left[\alpha_1 e^{\left(\omega_n\sqrt{\zeta^2-1}\right)t} + \alpha_2 e^{\left(-\omega_n\sqrt{\zeta^2-1}\right)t}\right]$$

$$= \alpha_1 e^{-t/\tau_1} + \alpha_2 e^{-t/\tau_2} \qquad (4.36)$$

$$\tau_1 = \frac{1}{\omega_n\left(\zeta - \sqrt{\zeta^2 - 1}\right)} \qquad \tau_2 = \frac{1}{\omega_n\left(\zeta + \sqrt{\zeta^2 - 1}\right)}$$

Thus, an overdamped response is the sum of two first-order responses, as shown in Figure 4.37.

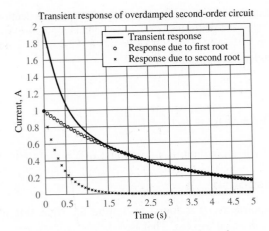

Figure 4.37 Transient response of overdamped
second-order system for $\alpha_1 = \alpha_2 = 1$; $\zeta = 1.5$; $\omega_n = 1$

Automotive Suspension

The mechanical model described in the previous sidebar can represent an automotive suspension system. The mass m, spring k, and damper b model the vehicle mass, the suspension struts (or coils), and the shock absorbers, respectively. The differential equation is:

$$\frac{m}{k}\frac{d^2 x_{body}(t)}{dt^2} + \frac{b}{k}\frac{d x_{body}(t)}{dt}$$
$$+ x_{body}(t)$$
$$= \frac{1}{k}x_{road}(t) + \frac{b}{k}\frac{d x_{road}(t)}{dt}$$

2. Critically Damped Response ($\zeta = 1$)

Two identical, negative, and real roots: (s_1, s_2). The transient response is critically damped when $\zeta = 1$. The argument of the square root in equation 4.35 is zero, such that $s_{1,2} = -\zeta\omega_n = -\omega_n$. The general form of the solution is:

$$x_{TR}(t) = \alpha_1 e^{s_1 t} + \alpha_2 t e^{s_2 t} = e^{-\omega_n t}(\alpha_1 + \alpha_2 t) = e^{-t/\tau}(\alpha_1 + \alpha_2 t)$$
$$\tau = \frac{1}{\omega_n} \tag{4.37}$$

Note that a critically damped response is the sum of a first-order exponential term plus a similar term multiplied by t. These two components and the complete response are shown in Figure 4.38.

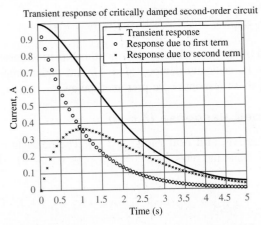

Figure 4.38 Transient response of a critically damped
second-order system for $\alpha_1 = \alpha_2 = 1$; $\zeta = 1$; $\omega_n = 1$

$m = 1{,}500$ kg
$k = 20{,}000$ N/m
$b_{new} = 15{,}000$ N-s/m
$b_{old} = 5{,}000$ N-s/m

(Continued)

3. Underdamped Response ($\zeta < 1$)

Two complex conjugate roots: (s_1, s_2). The transient response is underdamped when $\zeta < 1$. The argument of the square root in equation 4.35 is negative, such that $s_{1,2} = \omega_n\left(-\zeta \pm j\sqrt{1-\zeta^2}\right)$. The following complex exponentials appear in the general form of the response:

$$e^{\omega_n\left(-\zeta+j\sqrt{1-\zeta^2}\right)t} \qquad e^{\omega_n\left(-\zeta-j\sqrt{1-\zeta^2}\right)t} \tag{4.38}$$

Euler's formula can be used to express the complex exponentials in terms of sinusoids. The result is:

$$x_{TR}(t) = e^{-\zeta\omega_n t}[\alpha_1 \sin(\omega_d t) + \alpha_2 \cos(\omega_d t)] \tag{4.39}$$

where $\omega_d = \omega_n\sqrt{1-\zeta^2}$ is the **damped natural frequency**. Note that ω_d is the frequency of oscillation and is related to the period T of oscillation by $\omega_d T = 2\pi$. Also note that ω_d approaches the natural frequency ω_n as ζ approaches zero. The oscillation is *damped* by the decaying exponential $e^{-\zeta\omega_n t}$, which has a time constant $\tau = 1/\zeta\omega_n$, as shown in Figure 4.39. As ζ increases toward 1 (more damping), τ decreases and the oscillations decay more quickly. In the limit $\zeta \to 0$, the response is a pure sinusoid.

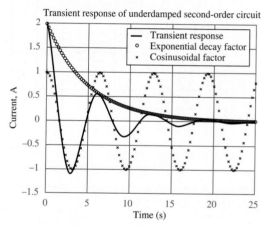

Figure 4.39 Transient response of an underdamped second-order system for $\alpha_1 = \alpha_2 = 1$; $\zeta = 0.2$; $\omega_n = 1$

Long-Term Steady-State Response

For switched DC sources, the forcing function F in equation 4.40 is a constant. The result is a constant long-term ($t \to \infty$) steady-state response x_{SS}.

$$\frac{1}{\omega_n^2}\frac{d^2 x_{SS}}{dt^2} + \frac{2\zeta}{\omega_n}\frac{dx_{SS}}{dt} + x_{SS} = K_S F \tag{4.40}$$

Since x_{SS} must also be a constant the solution for x_{SS} is:

$$x_{SS} = x(\infty) = K_S F \qquad t \to \infty \tag{4.41}$$

Complete Response

As with first-order systems, the complete response is the sum of the transient and long-term steady-state responses. The complete mathematical solutions for the overdamped, critically damped, and underdamped cases are shown in the highlighted Focus on Problem Solving section. In each of these cases, the initial conditions on the storage elements must be used to solve for the unknown constants α_1 and α_2. The required procedure uses the two initial conditions to evaluate $x(t)$ and

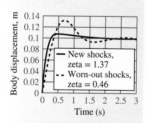

dx/dt at $t = 0^+$. The details of the procedure vary slightly in each of the three cases and are illustrated in the example problems.

One particularly useful complete solution is the *unit-step response* brought about by letting $K_S f(t)$ (see equation 4.25) be a *unit step*, which equals 0 for $t < 0$ and 1 for $t > 0$. To illustrate, assume a dimensionless damping coefficient $\zeta = 0.1$ and an underdamped period of oscillation $T = 2\pi$, such that the damped natural frequency is $\omega_d = 1$. The corresponding *unit-step response*, shown in Figure 4.40, asymptotically approaches the *long-term DC steady-state* value of 1 dictated by the unit-step input.

Also, as seen in the underdamped transient response, the magnitude of the oscillations *decays exponentially* over time. The time constant for the surrounding envelope (see dashed lines in Figure 4.40) is $\omega_d \tau = \sqrt{1 - \zeta^2}/\zeta \approx 10$, such that by $\omega_d t = 5\tau$ the oscillations are within 1 percent of the long-term DC steady-state value.

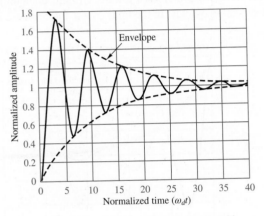

Figure 4.40 Second-order unit-step response with $K_S = 1$, $\omega_d = 1$, and $\zeta = 0.1$

Note that the rate of decay of the oscillations is governed by ζ. Figure 4.41 shows that as ζ increases the *overshoot* of the long-term DC steady-state response decreases until, when $\zeta = 1$ (critically damped), the response no longer oscillates and the overshoot is zero. The response for $\zeta > 1$ (overdamped) has no oscillations and zero overshoot.

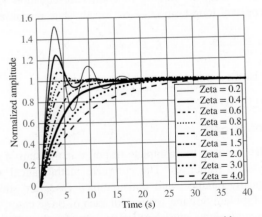

Figure 4.41 Second-order unit-step response with $K_S = 1$, $\omega_d = 1$, and ζ ranging from 0.2 to 4.0

FOCUS ON PROBLEM SOLVING

SECOND-ORDER TRANSIENT RESPONSE

1. Solve for the DC steady-state responses of $x(t)$ before and long after the switch is thrown at $(t = 0)$. These responses are designated $x(0^-)$ and $x(\infty)$.

2. Use the continuity requirements for capacitor voltage and inductor current $[v_C(0^+) = v_C(0^-)$ and $i_L(0^+) = i_L(0^-)]$ to identify the initial conditions $x(0^+)$ and $dx(0^+)/dt$.

3. Apply KVL and/or KCL to find two first-order differential equations, one of which may be the constitutive i-v relation for one of the state variables. Manipulate these equations to find a second-order differential equation in standard form (equation 4.25) and in one state variable.

4. Compare the coefficients of the differential equation to the standard form to write two equations involving the natural frequency ω_n and the dimensionless damping ratio ζ. Solve for ω_n and ζ.

5. Use the value of ζ to determine whether the transient response for $x(t)$ is overdamped, critically damped, or underdamped.

   Overdamped case ($\zeta > 1$):

$$x(t) = x_{TR}(t) + x_{SS} = e^{-\zeta\omega_n t}\left(\alpha_1 e^{\omega_n t\sqrt{\zeta^2-1}} + \alpha_2 e^{-\omega_n t\sqrt{\zeta^2-1}}\right) + x(\infty) \qquad t \geq 0$$

   Critically damped case ($\zeta = 1$):

$$x(t) = x_{TR}(t) + x_{SS} = e^{-\omega_n t}(\alpha_1 + \alpha_2 t) + x(\infty) \qquad t \geq 0$$

   Underdamped case ($\zeta < 1$):

$$x(t) = x_{TR}(t) + x_{SS} = e^{-\zeta\omega_n t}[\alpha_1 \sin(\omega_d t) + \alpha_2 \cos(\omega_d t)] + x(\infty) \qquad t \geq 0$$

$$\omega_d = \omega_n \sqrt{1 - \zeta^2}$$

6. Use the initial conditions on the state variables to solve for the constants α_1 and α_2.

   • Set $t = 0^+$ in the complete solution to find $x(0^+)$ in terms of α_1 and α_2.

   • Differentiate the complete solution, and set $t = 0^+$ to find $dx(0^+)/dt$ in terms of α_1 and α_2.

   • Use the results from step 3 to relate $x(t)$ and dx/dt at $t = 0^+$ to the state variables $v_C(0^+)$ and $i_L(0^+)$.

   Compare the two pairs of equations for $x(0^+)$ and $dx(0^+)/dt$ in step 6. Solve for α_1 and α_2. Whew! Voila!! Done!!!

FOCUS ON PROBLEM SOLVING

ROOTS OF SECOND-ORDER SYSTEMS

The general form of the roots s_1 and s_2 is
$$s_{1,2} = -\zeta\omega_n \pm \omega_n\sqrt{\zeta^2 - 1}.$$
The nature of these roots depends upon the argument of the square root.

Case 1: **Distinct, negative, real roots**. This case occurs when $\zeta > 1$ since the term under the square root sign is positive. The result is
$$s_{1,2} = -\omega_n\left[\zeta \pm \sqrt{\zeta^2 - 1}\right]$$ and a second-order **overdamped response**.

Case 2: **Identical, negative, real roots**. This case occurs when $\zeta = 1$ since the term under the square root is zero. The result is a repeated root $s = -\zeta\omega_n = -\omega_n$ and a second-order **critically damped response**.

Case 3: **Complex conjugate roots**. This case holds when $\zeta < 1$ since the term under the square root is negative. The result is a pair of complex conjugate roots $s_{1,2} = -\omega_n\left[\zeta \pm j\sqrt{1 - \zeta^2}\right]$ and a second-order **underdamped response**.

EXAMPLE 4.13   Transient Response of Second-Order Circuit

Problem

Find the transient response of $i_L(t)$ in the circuit shown in Figure 4.42.

Solution

Known Quantities: v_S; R_1; R_2; C; L.

Find: The transient response of $i_L(t)$ for the circuit in Figure 4.42.

Schematics, Diagrams, Circuits, and Given Data: $R_1 = 8$ kΩ; $R_2 = 8$ kΩ; $C = 10\ \mu$F; $L = 1$ H.

Assumptions: None

Analysis: To compute the transient response of the circuit, set the source equal to zero (short-circuit) and observe that the two resistors are in parallel and can be replaced by a single resistor $R = R_1 \| R_2$. Apply KCL to the resulting parallel *RLC* circuit, observing that the capacitor voltage is the top node voltage in the circuit:

$$\frac{v_C}{R} + C\frac{dv_C}{dt} + i_L = 0$$

This equation is a first-order differential equation in the two state variables. To obtain a second-order differential equation in one state variable use:

$$v_C = v_L = L\frac{di_L}{dt}$$

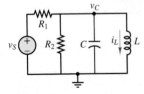

Figure 4.42 Circuit for Example 4.13

to substitute for v_C and find:

$$LC\frac{d^2 i_L}{dt^2} + \frac{L}{R}\frac{di_L}{dt} + i_L = 0$$

This equation has the form of equation 4.25, with $K_S = 1$, $\omega_n^2 = 1/LC$, and $2\zeta/\omega_n = L/R$. The roots of the differential equation can be computed from equation 4.35

$$s_{1,2} = -\zeta\omega_n \pm \frac{1}{2}\sqrt{(2\zeta\omega_n)^2 - 4\omega_n^2} = -\zeta\omega_n \pm \omega_n\sqrt{\zeta^2 - 1}$$

It is always instructive to calculate the values of the three parameters first. By inspection, $K_S = 1$, $\omega_n = 1/\sqrt{LC} = 1/\sqrt{10^{-5}} = 316.2\,\text{rad/s}$, and $\zeta = L\omega_n/2R = 0.04$. Since $\zeta < 1$, the response is *underdamped*, and the roots have the form: $s_{1,2} = -\zeta\omega_n \pm j\omega_n\sqrt{1-\zeta^2}$. Substituting numerical values, find $s_{1,2} = -12.5 \pm j316.0$, such that the transient response of any variable in the circuit has the form:

$$x_{TR}(t) = \alpha_1 e^{\left(-\zeta\omega_n + j\omega_n\sqrt{1-\zeta^2}\right)t} + \alpha_2 e^{\left(-\zeta\omega_n - j\omega_n\sqrt{1-\zeta^2}\right)t}$$

$$= \alpha_1 e^{(-12.5+j316.0)t} + \alpha_2 e^{(-12.5-j316.0)t}$$

The constants α_1 and α_2 can only be determined once the complete response is known and the initial conditions have been determined.

Comments: Note that once the second-order differential equation is expressed in standard form and the values of the three parameters are identified, the task of writing the transient solution with the aid of the Focus on Problem Solving box is straightforward.

EXAMPLE 4.14   Complete Response of Overdamped Second-Order Circuit

Problem

Determine the complete response for the loop current i shown in Figure 4.43.

Solution

Known Quantities: V_S; R; C; L.

Find: The complete response for the loop current i in the circuit of Figure 4.43.

Schematics, Diagrams, Circuits, and Given Data: $V_S = 25$ V; $R = 5$ kΩ; $C = 1$ μF; $L = 1$ H.

Assumptions: The capacitor has been charged (through a separate circuit, not shown) prior to the switch closing, such that $v_C(0) = 5$ V.

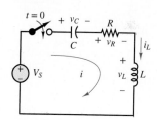

Figure 4.43 Circuit for Example 4.14.

Analysis:

Step 1: Steady-state response. While the switch is open, the current i in the circuit must be zero. Thus, the initial condition on the inductor current is also zero; that is, $i_L(0) = 0$. The initial voltage $v_C(0)$ across the capacitor cannot be deduced from the circuit itself. Instead, it must be given. In this problem, $v_C(0) = 5$ V.

After the switch has been closed for a long time and the transient response has decayed away, the remaining long-term DC steady-state value $i(\infty)$ can be found by treating the capacitor as an open-circuit and the inductor as a short-circuit. By observation, $i(\infty) = 0$ because the capacitor acts as an open-circuit in the single-loop circuit.

Step 2: Initial conditions. Two initial conditions are needed to solve a second-order circuit. These two initial conditions *always* rely on the two continuity conditions on the current through an inductor and the voltage across a capacitor, which must be continuous at all times. That is, $i_L(0^-) = i_L(0^+) = 0$ and $v_C(0^-) = v_C(0^+) = 5$ V. To solve for the two unknown constants in the complete solution it is necessary to apply the initial conditions on the two state variables to two equations involving those constants. Those two equations are $i(0^+) = i_L(0^+) = 0$ and $di(0^+)/dt = di_L(0^+)/dt$. The latter equation can be evaluated by applying KVL at $t = 0^+$:

$$V_S - v_C(0^+) - Ri_L(0^+) - v_L(0^+) = 0$$

$$V_S - v_C(0^+) - Ri_L(0^+) - L\frac{di_L(0^+)}{dt} = 0$$

$$\frac{di_L(0^+)}{dt} = \frac{V_S}{L} - \frac{v_C(0^+)}{L} = 25 - 5 = 20 \text{ A/s}$$

Step 3: Differential equation. Apply KVL around the loop to find a first-order differential equation in the two state variables v_C and i_L:

$$V_S - v_C - Ri_L(t) - L\frac{di_L(t)}{dt} = 0 \qquad t \geq 0$$

To obtain a second-order differential equation in i_L alone requires two additional steps. First, differentiate both sides of the first-order equation to find:

$$\frac{dV_S}{dt} - \frac{dv_C}{dt} - R\frac{di_L}{dt} - L\frac{d^2 i_L}{dt^2} = 0$$

Next, note that i_L is also the current through the capacitor and write the constitutive i-v relation for the capacitor as:

$$i_L = i_C = C\frac{dv_C}{dt} \qquad \text{or} \qquad \frac{dv_C}{dt} = \frac{i_C}{C} = \frac{i_L}{C}$$

Plug this result into the second-order differential equation, multiply both sides by C, and rearrange terms to obtain:

$$LC\frac{d^2 i_L}{dt^2} + RC\frac{di_L}{dt} + i_L = C\frac{dV_S}{dt} = 0 \qquad t \geq 0$$

Note that the right-hand side (forcing function) of this differential equation is zero when V_S is a DC source. This result is expected because the capacitor acts as a DC open-circuit, thus forcing the long-term DC steady-state value of i_L to be zero.

It is worthwhile to note that finding the second-order differential equation for any other variable x is now a simple matter because its right-hand side will be identical to that for i_L. The result is:

$$LC\frac{d^2 x}{dt^2} + RC\frac{dx}{dt} + x = K_S f(t) \qquad t \geq 0$$

where the forcing function $f(t)$ depends upon the independent sources in the circuit and the DC gain K_S depends upon the specific variable x. For circuits where all independent sources are direct current, the left-hand side is the long-term DC steady-state value $x(\infty)$. Thus:

$$LC\frac{d^2 x}{dt^2} + RC\frac{dx}{dt} + x = x(\infty) \qquad t \geq 0$$

Step 4: Solve for ω_n and ζ. Compare the second-order differential equation to the standard form of equation 4.25 to find:

$$\omega_n = \sqrt{\frac{1}{LC}} = 1,000 \text{ rad/s}$$

$$\zeta = RC\frac{\omega_n}{2} = \frac{R}{2}\sqrt{\frac{C}{L}} = 2.5$$

Thus, this second-order circuit response is overdamped.

Step 5: Write the complete solution. We know the circuit is overdamped, so the complete solution has the form:

$$x = x_{TR} + x_{SS} = \alpha_1 e^{\left(-\zeta\omega_n + \omega_n\sqrt{\zeta^2-1}\right)t}$$
$$+ \alpha_2 e^{\left(-\zeta\omega_n - \omega_n\sqrt{\zeta^2-1}\right)t} + x_{SS} \qquad t \geq 0$$

and since $x_{SS} = x(\infty) = 0$, the complete solution is identical to the transient solution:

$$i_L(t) = \alpha_1 e^{\left(-\zeta\omega_n + \omega_n\sqrt{\zeta^2-1}\right)t} + \alpha_2 e^{\left(-\zeta\omega_n - \omega_n\sqrt{\zeta^2-1}\right)t} \qquad t \geq 0$$

Step 6: Solve for the constants α_1 and α_2. Finally, use the initial conditions to evaluate the constants α_1 and α_2. The first initial condition yields

$$i_L(0^+) = \alpha_1 e^0 + \alpha_2 e^0$$
$$\alpha_1 = -\alpha_2$$

The second initial condition is evaluated as follows:

$$\frac{di_L(t)}{dt} = \left(-\zeta\omega_n + \omega_n\sqrt{\zeta^2-1}\right)\alpha_1 e^{\left(-\zeta\omega_n + \omega_n\sqrt{\zeta^2-1}\right)t}$$
$$+ \left(-\zeta\omega_n - \omega_n\sqrt{\zeta^2-1}\right)\alpha_2 e^{\left(-\zeta\omega_n - \omega_n\sqrt{\zeta^2-1}\right)t}$$
$$\frac{di_L(0^+)}{dt} = \left(-\zeta\omega_n + \omega_n\sqrt{\zeta^2-1}\right)\alpha_1 e^0 + \left(-\zeta\omega_n - \omega_n\sqrt{\zeta^2-1}\right)\alpha_2 e^0$$

Substituting $\alpha_1 = -\alpha_2$, we get

$$\frac{di_L(0^+)}{dt} = \left(-\zeta\omega_n + \omega_n\sqrt{\zeta^2-1}\right)\alpha_1 - \left(-\zeta\omega_n - \omega_n\sqrt{\zeta^2-1}\right)\alpha_1$$
$$= 2\left(\omega_n\sqrt{\zeta^2-1}\right)\alpha_1 = 20$$
$$\alpha_1 = \frac{20}{2\left(\omega_n\sqrt{\zeta^2-1}\right)} = 4.36 \times 10^{-3} \text{ A}$$
$$\alpha_2 = -\alpha_1 = -4.36 \times 10^{-3} \text{ A}$$

We can finally write the complete solution:

$$i_L(t) = 4.36 \times 10^{-3} e^{-208.7t} - 4.36 \times 10^{-3} e^{-4,791.3t} \text{ A} \qquad t \geq 0$$

A plot of the complete solution and of its components is given in Figure 4.44.

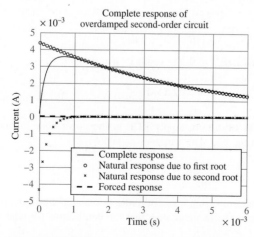

Figure 4.44 Complete response of overdamped second-order circuit

LO>

EXAMPLE 4.15   Complete Response of Critically Damped Second-Order Circuit

Problem

Determine the complete response for the voltage v_C shown in Figure 4.45.

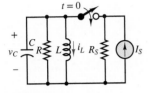

Figure 4.45 Circuit for Example 4.15.

Solution

Known Quantities: I_S; R; R_S; C; L.

Find: The complete response of the differential equation in v_C describing the circuit in Figure 4.45.

Schematics, Diagrams, Circuits, and Given Data: $I_S = 5$ A; $R = R_S = 500$ Ω; $C = 2$ μF; $L = 500$ mH.

Assumptions: None.

Analysis:

Step 1: Steady-state response. With the switch open for a long time, any energy stored in the capacitor and inductor has had time to be dissipated by the resistor; thus, the currents and voltages in the circuit are zero: $i_L(0^-) = 0$, $v_C(0^-) = v(0^-) = 0$.

After the switch has been closed for a long time and all the transients have died, the capacitor becomes an open-circuit, and the inductor behaves as a short-circuit. With the inductor behaving as a short-circuit, all the source current will flow through the inductor, and $i_L(\infty) = I_S = 5$ A. On the other hand, the current through the resistor is zero, and therefore $v_C(\infty) = v(\infty) = 0$ V.

Step 2: Initial conditions. Two initial conditions are needed to solve a second-order circuit. These two initial conditions always rely on two continuity conditions: the current through an inductor and the voltage across a capacitor are continuous. That is, $i_L(0^-) = i_L(0^+) = 0$ A and $v_C(0^-) = v_C(0^+) = 0$ V. Since the differential equation is in the variable v_C, the two needed initial conditions are $v_C(0^+)$ and $dv_C(0^+)/dt$. These can be found by applying KCL at $t = 0^+$:

$$I_S - \frac{v_C(0^+)}{R_S} - i_L(0^+) - \frac{v_C(0^+)}{R} - C\frac{dv_C(0^+)}{dt} = 0$$

Since $v_C(0^+) = 0$ and $i_L(0^+) = 0$, the result is: $dv_C(0^+)/dt$:

$$\frac{dv_C(0^+)}{dt} = \frac{I_S}{C} = \frac{5}{2 \times 10^{-6}} = 2.5 \times 10^6 \frac{V}{s}$$

Step 3: Differential equation. Apply KCL at the upper node to find a first-order differential equation in the two state variables v_C and i_L:

$$I_S - \frac{v_C}{R_S} - i_L - \frac{v_C}{R} - C\frac{dv_C}{dt} = 0 \qquad t \geq 0$$

Differentiate both sides to obtain a second-order differential equation. Then, note that v_C is also the voltage across the inductor such that the constitutive i-v relation for the inductor can be written as:

$$v_C = v_L = L\frac{di_L}{dt}$$

Use this relation to substitute for di_L/dt to obtain a second-order differential equation involving v_C only. Finally, multiply both sides by L and rearrange terms to find:

$$LC\frac{d^2v_C}{dt^2} + \frac{L(R_S + R)}{R_S R}\frac{dv_C}{dt} + v_C = L\frac{dI_S}{dt} = 0 \qquad t \geq 0$$

Note that the right-hand side (the forcing function) of this differential equation is zero when I_S is a DC source. This result is expected because the inductor acts as a DC short-circuit, thus forcing the long-term DC steady-state value of v_C to be zero.

It is worthwhile to note that finding the second-order differential equation for any other variable x is now a simple matter because its right-hand side will be identical to that for i_L. The result is:

$$LC\frac{d^2x}{dt^2} + \frac{L(R_S + R)}{R_S R}\frac{dx}{dt} + x = K_S f(t) \qquad t \geq 0$$

where the forcing function $f(t)$ depends upon the independent sources in the circuit and the DC gain K_S depends upon the specific variable x. For circuits where all independent sources are direct current, the left-hand side is the long-term DC steady-state value $x(\infty)$. Thus:

$$LC\frac{d^2x}{dt^2} + \frac{L(R_S + R)}{R_S R}\frac{dx}{dt} + x = x(\infty) \qquad t \geq 0$$

Step 4: Solve for ω_n and ζ. Compare the second-order differential equation to the standard form and observe:

$$\omega_n = \sqrt{\frac{1}{LC}} = 1{,}000 \text{ rad/s}$$

$$\zeta = \frac{L}{R_{eq}}\frac{\omega_n}{2} = \frac{1}{2R_{eq}}\sqrt{\frac{L}{C}} = 1$$

$$R_{eq} = \frac{RR_S}{R + R_S} = 250 \ \Omega$$

Thus, the second-order circuit is critically damped.

Step 5: Write the complete solution. The complete solution for the critically damped ($\zeta = 1$) case is:

$$x = x_{TR} + x_{SS} = \alpha_1 e^{-\zeta\omega_n t} + \alpha_2 t e^{-\zeta\omega_n t} + x(\infty) \qquad t \geq 0$$

and, since $v_{C_{ss}} = v_C(\infty) = 0$, the complete solution is identical to the transient solution:

$$v_C(t) = \alpha_1 e^{-\zeta\omega_n t} + \alpha_2 t e^{-\zeta\omega_n t} \qquad t \geq 0$$

Step 6: Solve for the constants α_1 and α_2. Solve for the initial conditions to evaluate the constants α_1 and α_2. The first initial condition yields:

$$v_C(0^+) = \alpha_1 e^0 + \alpha_2 \cdot 0 \cdot e^0 = 0 \qquad \text{or} \qquad \alpha_1 = 0$$

The second initial condition is:

$$\frac{dv_C(t)}{dt} = (-\zeta\omega_n)\alpha_1 e^{-\zeta\omega_n t} + (-\zeta\omega_n)\alpha_2 t e^{-\zeta\omega_n t} + \alpha_2 e^{-\zeta\omega_n t}$$

$$\frac{dv_C(0^+)}{dt} = (-\zeta\omega_n)\alpha_1 e^0 + \alpha_2 e^0 = \alpha_2 \qquad \text{or} \qquad \alpha_2 = 2.5 \times 10^6 \text{ V}$$

Finally, the complete solution is:

$$v_C(t) = 2.5 \times 10^6 t e^{-500t} \text{ V} \qquad t \geq 0$$

A plot of the complete solution and of its components is given in Figure 5.46.

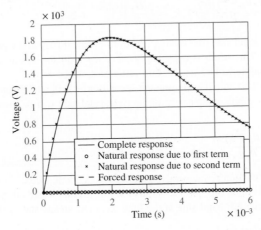

Figure 4.46 Complete response of critically damped second-order circuit

EXAMPLE 4.16   Complete Response of Underdamped Second-Order Circuit

Problem

Determine the complete response for the current i_L shown in Figure 4.47.

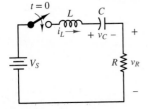

Figure 4.47 Circuit for Example 4.16.

Solution

Known Quantities: V_S; R; C; L.

Find: The complete response for the current i_L shown in Figure 4.47.

Schematics, Diagrams, Circuits, and Given Data: $V_S = 12$ V; $R = 200$ Ω; $C = 10$ μF; $L = 0.5$ H.

Assumptions: The capacitor has an initial charge such that $v_C(0^-) = v_C(0^+) = 2$ V.

Analysis:

Step 1: Steady-state response. The inductor current must be zero when the switch is open, so $i_L(0^-) = 0$. Also, the initial charge on the capacitor is $v_C(0^-) = 2$ V. After the switch has been closed for a long time, the capacitor and inductor act as open- and short-circuits, respectively, such that the loop current is zero and the battery voltage appears across the capacitor: $i(\infty) = 0$ A, $v_C(\infty) = V_S = 12$ V.

Step 2: Initial conditions. Two initial conditions are needed to solve a second-order circuit. These two initial conditions always rely on two continuity conditions: the current through an inductor and the voltage across a capacitor are continuous. That is, $i_L(0^-) = i_L(0^+) = 0$ A and $v_C(0^-) = v_C(0^+) = 2$ V. Since the differential equation is in the variable i_L, the two needed initial conditions are $i_L(0^+)$ and $di_L(0^+)/dt$. The second initial condition can be found by applying KVL at $t = 0^+$:

$$V_S - v_C(0^+) - Ri_L(0^+) - v_L(0^+) = 0$$

$$V_S - v_C(0^+) - Ri_L(0^+) - L\frac{di_L(0^+)}{dt} = 0$$

$$\frac{di_L(0^+)}{dt} = \frac{V_S}{L} - \frac{v_C(0^+)}{L} = \frac{12}{0.5} - \frac{2}{0.5} = 20 \text{ A/s}$$

Step 3: Differential equation. Apply KVL around the loop to find a first-order differential equation in the two state variables v_C and i_L:

$$V_S - L\frac{di_L}{dt} - v_C - Ri_L = 0 \qquad t \geq 0$$

To obtain a second-order differential equation in i_L alone requires two additional steps. First, differentiate both sides of the first-order equation to find:

$$\frac{dV_S}{dt} - L\frac{d^2 i_L}{dt^2} - \frac{dv_C}{dt} - R\frac{di_L}{dt} = 0$$

Next, note that i_L is also the current through the capacitor and write the constitutive i-v relation for the capacitor as:

$$i_L = i_C = C\frac{dv_C}{dt} \qquad \text{or} \qquad \frac{dv_C}{dt} = \frac{i_C}{C} = \frac{i_L}{C}$$

Plug this result into the second-order differential equation, and multiply both sides by C to obtain:

$$LC\frac{d^2 i_L}{dt^2} + RC\frac{di_L}{dt} + i_L = C\frac{dV_S}{dt} = 0 \qquad t \geq 0$$

Note that the right-hand side (the forcing function) of this differential equation is zero when V_S is a DC source. This result is expected because the capacitor acts as a DC open-circuit, thus forcing the long-term DC steady-state value of i_L to be zero.

It is worthwhile to note that finding the second-order differential equation for any other variable x is now a simple matter because its right-hand side will be identical to that for i_L. The result is:

$$LC\frac{d^2 x}{dt^2} + RC\frac{dx}{dt} + x = K_S f(t) \qquad t \geq 0$$

where the forcing function $f(t)$ depends upon the independent sources in the circuit and the DC gain K_S depends upon the specific variable x. For circuits where all independent sources are direct current, the left-hand side is the long-term DC steady-state value $x(\infty)$. Thus:

$$LC\frac{d^2 x}{dt^2} + RC\frac{dx}{dt} + x = x(\infty) \qquad t \geq 0$$

Step 4: Solve for ω_n and ζ. If we now compare the second-order differential equations to the standard form of equation 4.50, we can make the following observations:

$$\omega_n = \sqrt{\frac{1}{LC}} = 447 \text{ rad/s}$$

$$\zeta = RC\frac{\omega_n}{2} = \frac{R}{2}\sqrt{\frac{C}{L}} = 0.447$$

Thus, the second-order circuit is underdamped.

Step 5: Write the complete solution. Knowing that the circuit is underdamped ($\zeta < 1$), we write the complete solution for this case as

$$x(t) = x_{\text{TR}}(t) + x_{\text{SS}}(t) = \alpha_1 e^{\left(-\zeta\omega_n + j\omega_n\sqrt{1-\zeta^2}\right)t}$$

$$+ \alpha_2 e^{\left(-\zeta\omega_n - j\omega_n\sqrt{1-\zeta^2}\right)t} + x(\infty) \qquad t \geq 0$$

and since $x_{SS} = i_L(\infty) = 0$, the complete solution is identical to the transient solution:

$$i_L(t) = \alpha_1 e^{\left(-\zeta\omega_n + j\omega_n\sqrt{1-\zeta^2}\right)t} + \alpha_2 e^{\left(-\zeta\omega_n - j\omega_n\sqrt{1-\zeta^2}\right)t} \qquad t \geq 0$$

Step 6: Solve for the constants α_1 and α_2. Finally, we solve for the initial conditions to evaluate the constants α_1 and α_2. The first initial condition yields

$$i_L(0^+) = \alpha_1 e^0 + \alpha_2 e^0 = 0$$

$$\alpha_1 = -\alpha_2$$

The second initial condition is evaluated as follows:

$$\frac{di_L(t)}{dt} = \left(-\zeta\omega_n + j\omega_n\sqrt{1-\zeta^2}\right)\alpha_1 e^{\left(-\zeta\omega_n + j\omega_n\sqrt{1-\zeta^2}\right)t}$$

$$+ \left(-\zeta\omega_n - j\omega_n\sqrt{1-\zeta^2}\right)\alpha_2 e^{\left(-\zeta\omega_n - j\omega_n\sqrt{1-\zeta^2}\right)t}$$

$$\frac{di_L(0^+)}{dt} = \left(-\zeta\omega_n + j\omega_n\sqrt{1-\zeta^2}\right)\alpha_1 e^0 + \left(-\zeta\omega_n - j\omega_n\sqrt{1-\zeta^2}\right)\alpha_2 e^0$$

Substituting $\alpha_1 = -\alpha_2$, we get

$$\frac{di_L(0^+)}{dt} = \left(-\zeta\omega_n + j\omega_n\sqrt{1-\zeta^2}\right)\alpha_1 - \left(-\zeta\omega_n - j\omega_n\sqrt{1-\zeta^2}\right)\alpha_1 = 20 \text{ A/s}$$

$$2\left(j\omega_n\sqrt{1-\zeta^2}\right)\alpha_1 = 20$$

$$\alpha_1 = \frac{10}{j\omega_n\sqrt{1-\zeta^2}} = -j\frac{10}{\omega_n\sqrt{1-\zeta^2}} = -j0.025 \text{ A}$$

$$\alpha_2 = -\alpha_1 = j0.025 \text{ A}$$

We can finally write the complete solution:

$$i_L(t) = -j0.025 e^{(-200+j400)t} + j0.025 e^{(-200-j400)t} \qquad t \geq 0$$

$$= 0.025 e^{-200t}(-je^{j400t} + je^{-j400t}) = 0.025 e^{-200t} \times 2\sin 400t \text{ A}$$

$$= 0.05 e^{-200t}\sin 400t \text{ A} \qquad t \geq 0$$

In the above equation, we have used Euler's formula to obtain the final expression. A plot of the complete solution and of its components is given in Figure 4.48.

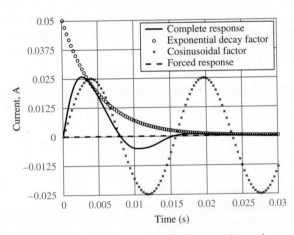

Figure 4.48 Complete response of underdamped second-order circuit

EXAMPLE 4.17   Analysis of Nonseries, Nonparallel *RLC* Circuit

Problem

Assume the circuit shown in Figure 4.49 is in DC steady state for $t < 0$. The switch closes at $t = 0$. Find the differential equations for the voltage v_C across the capacitor and the current i_L through the inductor for $t > 0$.

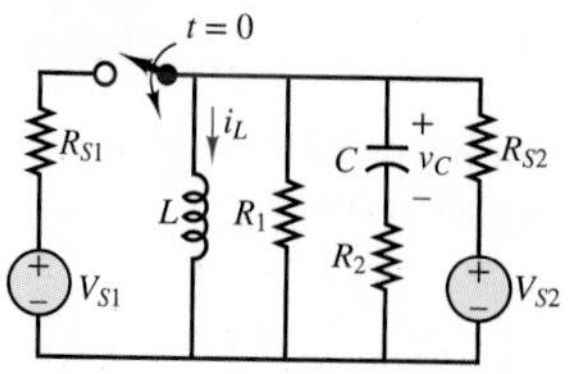

Figure 4.49 Circuit for Example 4.17.

Solution

Known Quantities: V_{S1}; R_{S1}; V_{S2}; R_{S2}; R_1; R_2; L; C.

Find: For $t > 0$, find the differential equations for the voltage v_C across the capacitor and the current i_L through the inductor shown in Figure 4.49.

Assumptions: DC steady state for $t < 0$.

Analysis: The critical difference between the circuit in this example and those in the previous examples is that the capacitor and inductor are neither in series nor in parallel. As will be shown below, it will first be necessary to find two first-order differential equations in the state variables v_C and i_L to find the second-order differential equations in each state variable.

Step 1: Steady-state response for $t < 0$. With the switch open V_{S1} and R_{S1} are disconnected from the rest of the circuit. Assuming a DC steady state the inductor acts as a short-circuit and the capacitor acts as an open-circuit. The result is that the voltage across R_1 is zero and the current through R_2 is zero. Thus, the voltage across R_{S2} is V_{S2}. Apply KCL at the top node and KVL around a loop containing the inductor and capacitor to find the following values for i_L and v_C:

$$i_L(0^-) = \frac{V_{S2}}{R_{S2}} \quad \text{and} \quad v_C(0^-) = 0$$

Step 2: Initial conditions at $t = 0$. The initial conditions on the state variables i_L and v_C are found by applying the continuity requirement.

$$i_L(0^+) = i_L(0^-) = \frac{V_{S2}}{R_{S2}} \quad \text{and} \quad v_C(0^+) = v_C(0^-) = 0$$

Step 3: Simplification of $t > 0$ circuit. With the switch closed, the first step to simplify the circuit is to divide it into a source and load. Choose the least complicated two-terminal network that contains the inductor and the capacitor as the load. As always, everything else is the source network. Figure 4.50 shows the circuit rearranged into a source and load.

The two Thévenin sources on the left can be transformed into Norton sources as shown in Figure 4.51. The resulting parallel current sources can be replaced by a single equivalent

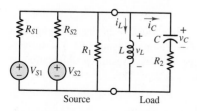

Figure 4.50 Circuit simplification, step 1.

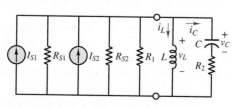

Figure 4.51 Circuit simplification, step 2.

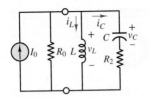

Figure 4.52 Simplified equivalent circuit

current source. Likewise, the parallel resistors can be replaced by a single equivalent resistor. The resulting simplified circuit is shown in Figure 4.52, where:

$$I_0 = I_{S1} + I_{S2} \qquad \text{and} \qquad R_0 = R_{S1} \| R_{S2} \| R_1$$

Step 4: Derivation of the differential equations. In general, when the inductor and capacitor are neither in parallel nor in series, it is necessary to apply KVL and/or KCL twice to obtain two first-order differential equations in i_L and v_C. When doing so, be careful that all the circuit elements are accounted for in the two equations. For the circuit shown in Figure 4.52, apply KCL at the top node to find:

$$I_0 - \frac{v_L}{R_0} - i_L - i_C = 0$$

Also apply KVL around the right-most loop containing the inductor and capacitor to find:

$$v_L - v_C - i_C R_2 = 0$$

The next step is to eliminate the nonstate variables v_L and i_C by substitution using the inductor and capacitor constitutive i-v relations.

$$v_L = L \frac{di_L}{dt} \qquad \text{and} \qquad i_C = C \frac{dv_C}{dt}$$

The result is two first-order differential equations in the state variables.

$$I_0 = \frac{L}{R_0} \frac{di_L}{dt} + i_L + C \frac{dv_C}{dt}$$

$$0 = L \frac{di_L}{dt} - v_C - R_2 C \frac{dv_C}{dt}$$

The next step is to combine these two first-order differential equations so as to eliminate one of the state variables. The result will be a second-order differential equation in the remaining state variable. This step may require some patience and clever manipulations.

For example, multiply the first equation above by R_2 and add the result to the second equation to yield:

$$I_0 R_2 = L \frac{R_0 + R_2}{R_0} \frac{di_L}{dt} + i_L R_2 - v_C$$

Differentiate both sides of this equation to find:

$$\frac{dv_C}{dt} = L \frac{R_0 + R_2}{R_0} \frac{d^2 i_L}{dt^2} + R_2 \frac{di_L}{dt}$$

Substitute that result into the first of the two original first-order differential equations above to yield:

$$LC \frac{R_0 + R_2}{R_0} \frac{d^2 i_L}{dt^2} + \left(R_2 C + \frac{L}{R_0} \right) \frac{di_L}{dt} + i_L = I_0$$

The resulting second-order differential equation in i_L is reasonable in that the long-term DC steady-state value for i_L is I_0, as can be seen by observation of Figure 4.52. The second-order differential equation for v_C must have the same left side. Again, by observation of Figure 4.52, the long-term DC steady-state value for v_C is zero. Thus:

$$LC \frac{R_0 + R_2}{R_0} \frac{d^2 v_C}{dt^2} + \left(R_2 C + \frac{L}{R_0} \right) \frac{dv_C}{dt} + v_C = 0$$

Step 5: Solution for ω_n and ζ. Compare either of the two second-order differential equations to equation 4.25 to write:

$$\frac{1}{\omega_n^2} = LC\frac{R_0 + R_2}{R_0} \quad \text{and} \quad \frac{2\zeta}{\omega_n} = \left(R_2 C + \frac{L}{R_0}\right)$$

These two equations can be solved for ω_n and ζ to yield:

$$\omega_n = \sqrt{\frac{1}{LC}}\sqrt{\frac{R_0}{R_0 + R_2}} \quad \text{and} \quad \zeta = \frac{\omega_n}{2}\left(R_2 C + \frac{L}{R_0}\right)$$

Step 6: The complete solution. The form (overdamped, critically damped, underdamped) of the transient solution depends upon the value of ζ, which itself depends upon the values of the various circuit elements. The complete solution is the sum of the transient solution and the long-term DC steady-state value. Regardless of the form of the transient solution, the complete solution will contain two unknown constants.

Step 7: The unknown constants. To demonstrate the process of solving for the unknown constants consider the generic complete solution for the inductor current i_L.

$$i_L(t) = (i_L)_{\text{TR}}(t) + (i_L)_{\text{SS}}$$

The two unknown constants are part of the transient solution $(i_L)_{\text{TR}}(t)$. To solve for the unknown constants it is necessary to derive two linearly independent algebraic equations for them. The first such equation is found directly from the initial condition on i_L. That is:

$$\frac{V_{S2}}{R_{S2}} = i_L(0^+) = (i_L)_{\text{TR}}(0^+) + (i_L)_{\text{SS}}$$

The second equation for the unknown constants is found by evaluating the *derivative* of i_L at $t = 0^+$. That is:

$$\frac{di_L(t)}{dt}\bigg|_{t=0^+} = \frac{di_{L_{\text{TR}}}(t)}{dt}\bigg|_{t=0^+} + \frac{di_{L_{\text{SS}}}(t)}{dt}\bigg|_{t=0^+}$$

The derivative of i_L at $t = 0^+$ can be evaluated using the initial conditions on both state variables i_L and v_C. In general, to do so requires the use of one or both of the two first-order differential equations found by applying KVL and/or KCL. From before, those two first-order differential equations were combined to produce:

$$I_0 R_2 = L\frac{R_0 + R_2}{R_0}\frac{di_L}{dt} + i_L R_2 - v_C$$

After rearranging terms and evaluating at $t = 0^+$ yields:

$$\frac{di_L(t)}{dt}\bigg|_{t=0^+} = \frac{(I_0 R_2 - i_L R_2 + v_C)R_0}{L(R_0 + R_2)}\bigg|_{t=0^+}$$

The initial conditions on the two state variables for this example were found above. Solve the two linearly independent algebraic equations for the two unknown constants. Done!

Comments: Recall that the values of the unknown constants and the long-term steady state are, in general, different for different variables. However, all variables in a circuit share the same natural frequency ω_n and dimensionless damping coefficient ζ. That is, the left side of the second-order differential equation is the same for all variables. Also, keep in mind that the initial conditions for any variable and its derivative must be related to the initial conditions on the state variables since they are the only variables guaranteed to be continuous across the transient event.

CHECK YOUR UNDERSTANDING

For what value of R in Example 4.13 will the circuit response become critically damped?

Answer: $R = 158.1\ \Omega$

CHECK YOUR UNDERSTANDING

Obtain a differential equation for v_C in the circuit of Example 4.14.

Answer: $LC\dfrac{d^2 v_C}{dt^2} + RC\dfrac{dv_C}{dt} + v_C = V_s$

CHECK YOUR UNDERSTANDING

Obtain a differential equation for i_L in the circuit of Example 4.15.

Answer: $LC\dfrac{d^2 i_L(t)}{dt^2} + \dfrac{L}{R_{eq}}\dfrac{di_L(t)}{dt} + i_L(t) = I_s \qquad t \geq 0$

CHECK YOUR UNDERSTANDING

If the inductance in Example 4.16 is reduced to one-half of its original value (from 0.5 to 0.25 H), for what range of values of R will the circuit be underdamped?

Answer: $R \leq 316\ \Omega$

Conclusion

Chapter 4 has focused on the solution of first- and second-order differential equations for the case of DC switched transients, and it has presented a number of analogies between electric circuits and other physical systems, such as thermal, hydraulic, and mechanical.

While many other forms of excitation exist, turning a DC supply on and off is a very common occurrence in electrical, electronic, and electromechanical systems. Further, the methods discussed in this chapter can be readily extended to the solution of more general problems.

A thorough study of this chapter should result in the acquisition of the following learning objectives:

1. *Write differential equations for circuits containing inductors and capacitors.* This process involves the application of KVL and/or KCL to produce first-order differential equations and the use of constitutive i-v relationships for inductors and capacitors to produce differential equations in the state variables.

2. *Determine the DC steady-state solution of circuits containing inductors and capacitors.* The DC steady-state solution of any differential equation can be easily obtained by setting the derivative terms equal to zero. Alternatively, the DC steady-state response for any circuit variable can be acquired directly from the circuit since an inductor acts as a short-circuit and a capacitor acts as an open-circuit under DC conditions.

3. *Write the differential equation of first-order circuits in standard form, and determine the complete solution of first-order circuits excited by switched DC sources.* First-order systems are most commonly described by way of two constants: the DC gain and the time constant. You have learned how to recognize these constants, how to compute the initial and final conditions, and how to write the complete solution of all first-order circuits almost by inspection.

4. *Write the differential equation of second-order circuits in standard form, and determine the complete solution of second-order circuits excited by switched DC sources.* Second-order circuits are described by three constants: the DC gain, the natural frequency, and the dimensionless damping coefficient. While the method for obtaining the complete solution for a second-order circuit is logically the same as that used for a first-order circuit, the details are more involved.

5. *Understand analogies between electric circuits and hydraulic, thermal, and mechanical systems.* Many physical systems in nature exhibit the same first- and second-order characteristics as the electric circuits studied in this chapter. This chapter introduced thermal, hydraulic, and mechanical system analogies.

HOMEWORK PROBLEMS

Section 4.2: Elements of Transient Problem Solving

4.1 Write the differential equations for $t > 0$ for i_L and v_3 in Figure P4.21. How are they related?

4.2 Write the differential equation for $t > 0$ for v_C in Figure P4.23.

4.3 Write the differential equation for $t > 0$ for i_C in Figure P4.27.

4.4 Write the differential equation for $t > 0$ for i_L in Figure P4.29.

4.5 Write the differential equation for $t > 0$ for v_C in Figure P4.32.

4.6 Write the differential equations for $t > 0$ for i_C and v_3 in Figure P4.34. How are they related?

4.7 Write the differential equation for $t > 0$ for v_C in Figure P4.41. Assume $R_1 = 5\ \Omega$, $R_2 = 4\ \Omega$, $R_3 = 3\ \Omega$, $R_4 = 6\ \Omega$, and $C_1 = C_2 = 4\ \text{F}$.

4.8 Write the differential equation for $t > 0$ for i_C in Figure P4.47. Assume $V_S = 9\ \text{V}$, $R_S = 5\ \text{k}\Omega$, $R_1 = 10\ \text{k}\Omega$, and $R_2 = R_3 = 20\ \text{k}\Omega$.

4.9 Write the differential equation for $t > 0$ for i_L in Figure P4.49.

4.10 Write the differential equations for $t > 0$ for i_L and v_1 in Figure P4.52. How are they related? Assume $L_1 = 1\ \text{H}$ and $L_2 = 5\ \text{H}$.

4.11 Determine the initial and final conditions on i_L and v_3 in Figure P4.21.

4.12 Determine the initial and final conditions on v_C in Figure P4.23.

4.13 Determine the initial and final conditions on i_C in Figure P4.27.

4.14 Determine the initial and final conditions on i_L in Figure P4.29.

4.15 Determine the initial and final conditions on v_C in Figure P4.32.

4.16 Determine the initial and final conditions on i_C and v_3 in Figure P4.34.

4.17 Determine the initial and final conditions on v_C in Figure P4.41.

4.18 Determine the initial and final conditions on i_C in Figure P4.47. Assume $V_S = 9\ \text{V}$, $R_S = 5\ \text{k}\Omega$, $R_1 = 10\ \text{k}\Omega$, and $R_2 = R_3 = 20\ \text{k}\Omega$.

4.19 Determine the initial and final conditions on i_L in Figure P4.49. Assume $L_1 = 1\ \text{H}$ and $L_2 = 5\ \text{H}$.

4.20 Determine the initial and final conditions on i_L and v_1 in Figure P4.52.

Section 4.5: First-Order Transient Analysis

4.21 At $t = 0^-$, just before the switch is opened, the current through the inductor in Figure P4.21 is $i_L = 140$ mA. Is this value the same as that for DC steady state? Was the circuit in steady state just before the switch was opened? Assume $V_s = 10$ V, $R_1 = 1$ kΩ, $R_2 = 5$ kΩ, $R_3 = 2$ kΩ, and $L = 1$ mH.

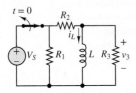

Figure P4.21

4.22 For $t < 0$, the circuit shown in Figure P4.22 is at DC steady state. The switch is thrown at $t = 0$.

$V_{S1} = 35$ V	$V_{S2} = 130$ V
$C = 11\ \mu F$	$R_1 = 17$ kΩ
$R_2 = 7$ kΩ	$R_3 = 23$ kΩ

Determine the initial current through R_3 just after the switch is thrown at $t = 0^+$.

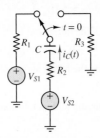

Figure P4.22

4.23 Determine the current i_C through the capacitor just before and just after the switch is closed in Figure P4.23. Assume steady-state conditions for $t < 0$. $V_1 = 15$ V, $R_1 = 0.5$ kΩ, $R_2 = 2$ kΩ, and $C = 0.4\ \mu F$.

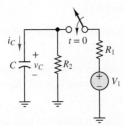

Figure P4.23

4.24 Determine the current i_C through the capacitor just before and just after the switch is closed in Figure P4.23. Assume steady-state conditions for $t < 0$. $V_1 = 10$ V, $R_1 = 200$ mΩ, $R_2 = 5$ kΩ, and $C = 300\ \mu F$.

4.25 Just before the switch is opened at $t = 0$ in Figure P4.21, the current through the inductor is $i_L = 1.5$ mA. Determine the voltage v_3 across R_3 immediately after the switch is opened. Assume $V_S = 12$ V, $R_1 = 6$ kΩ, $R_2 = 6$ kΩ, $R_3 = 3$ kΩ, and $L = 0.9$ mH.

4.26 Assume that steady-state conditions exist in the circuit shown in Figure P4.26 for $t < 0$. Determine the current through the inductor immediately before and immediately after the switch is thrown, that is, at $t = 0^-$ and at $t = 0^+$. Assume $L = 0.5$ H, $R_1 = 100$ kΩ, $R_S = 5$ Ω, and $V_S = 24$ V.

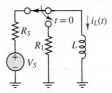

Figure P4.26

4.27 Assume that steady-state conditions exist in the circuit shown in Figure P4.27 for $t < 0$ and that $V_1 = 15$ V, $R_1 = 100$ Ω, $R_2 = 1.2$ kΩ, $R_3 = 400$ Ω, $C = 4.0\ \mu F$. Determine the current i_C through the capacitor at $t = 0^+$, just after the switch is closed.

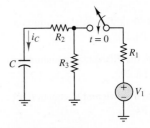

Figure P4.27

4.28 Assume that steady-state conditions exist in the circuit shown in Figure P4.28 at $t < 0$. Also assume:

$V_1 = 12$ V	$V_2 = 5$ V
$L = 3$ H	$R_1 = R_2 = 2$ Ω
$R_3 = 4$ Ω	$R_3 = 29$ kΩ

Find the Norton equivalent network seen by the inductor. Use it to determine the time constant of the circuit for $t > 0$.

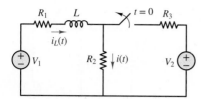

Figure P4.28

4.29   Assume that steady-state conditions exist in the circuit shown in Figure P4.29 at $t < 0$. Also assume:

$$V_{S1} = 9 \text{ V} \qquad V_{S2} = 12 \text{ V}$$
$$L = 120 \text{ mH} \qquad R_1 = 2.2 \text{ }\Omega$$
$$R_2 = 4.7 \text{ }\Omega \qquad R_3 = 18 \text{ k}\Omega$$

Find the Norton equivalent network seen by the inductor. Use it to determine the time constant of the circuit for $t > 0$.

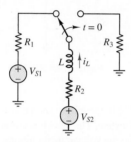

Figure P4.29

4.30   Find the Thévenin equivalent network seen by the capacitor in Figure P4.30 for $t > 0$. Use it to determine the time constant of the circuit for $t > 0$. $R_1 = 3 \text{ }\Omega$, $R_2 = 1 \text{ }\Omega$, $R_3 = 4 \text{ }\Omega$, $C = 0.2 \text{ F}$, $I_S = 3 \text{ A}$, $V_c(0) = 0$.

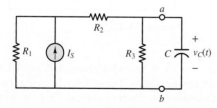

Figure P4.30

4.31   The switch shown in Figure P4.31 is closed at $t = 0$. Find the Thévenin equivalent network seen by the capacitor for $t > 0$, and use it to determine the time constant of the circuit for $t > 0$. $R_S = 8 \text{ k}\Omega$, $V_S = 40 \text{ V}$, $C = 350 \text{ }\mu\text{F}$, and $R = 24 \text{ k}\Omega$.

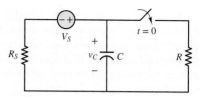

Figure P4.31

4.32   Determine the voltage v_C across the capacitor shown in Figure P4.32 for $t > 0$. The voltage across the capacitor just before the switch is thrown is $v_C(0^-) = -7 \text{ V}$. Assume:

$$I_o = 17 \text{ mA} \qquad C = 0.55 \text{ }\mu\text{F}$$
$$R_1 = 7 \text{ k}\Omega \qquad R_2 = 3.3 \text{ k}\Omega$$

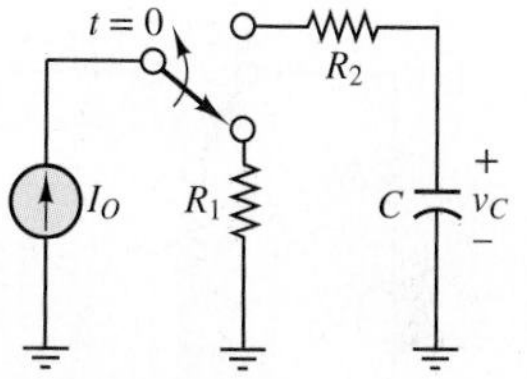

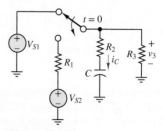

Figure P4.32

4.33   For $t < 0$, the circuit shown in Figure P4.29 is at steady state. The switch is thrown at $t = 0$. Determine the current i_L through the inductor for $t > 0$. Assume:

$$V_{S1} = 9 \text{ V} \qquad V_{S2} = 12 \text{ V}$$
$$L = 120 \text{ mH} \qquad R_1 = 2.2 \text{ }\Omega$$
$$R_2 = 4.7 \text{ }\Omega \qquad R_3 = 18 \text{ k}\Omega$$

4.34   For $t < 0$, the circuit shown in Figure P4.34 is at steady state. The switch is thrown at $t = 0$. Assume:

$$V_{S1} = 17 \text{ V} \qquad V_{S2} = 11 \text{ V}$$
$$R_1 = 14 \text{ k}\Omega \qquad R_2 = 13 \text{ k}\Omega$$
$$R_3 = 14 \text{ k}\Omega \qquad C = 70 \text{ nF}$$

Determine the

a. Current i_C through the capacitor for $t > 0$.

b. Voltage v_3 across R_3 for $t > 0$.

c. Time required for i_C and v_3 to change by 98 percent of their initial values at $t = 0^+$.

Figure P4.34

4.35   The circuit in Figure P4.35 is a simple model of an automotive ignition system. The switch models the "points" that switch electric power to the cylinder when the fuel-air mixture is compressed. R is the resistance across the gap between the electrodes of the spark plug.

$$V_G = 12 \text{ V} \qquad R_G = 0.37 \text{ }\Omega$$
$$R = 1.7 \text{ k}\Omega$$

Determine the value of L and R_1 so that the voltage across the spark plug gap just after the switch is changed is 23 kV and so that this voltage will change exponentially with a time constant $\tau = 13$ ms.

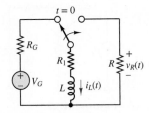

Figure P4.35

4.36   The inductor L in the circuit shown in Figure P4.36 is the coil of a relay. When the current i_L through the coil is equal to or greater than 2 mA, the relay is activated. Assume steady-state conditions at $t < 0$. If

$$V_S = 12 \text{ V}$$
$$L = 10.9 \text{ mH}$$
$$R_1 = 3.1 \text{ k}\Omega$$

Determine R_2 so that the relay activates 2.3 after the switch is thrown.

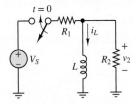

Figure P4.36

4.37   Determine the current i_C through the capacitor in Figure P4.37 for all time. Assume DC steady-state conditions for $t < 0$. Also assume: $V_1 = 10$ V, $C = 200$ μF, $R_1 = 300$ mΩ, and $R_2 = R_3 = 1.2$ kΩ.

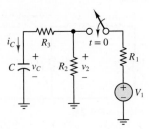

Figure P4.37

4.38   Determine the voltage v_L across the inductor in Figure P4.38 for all time. Assume DC steady-state conditions for $t < 0$. Also assume: $V_s = 15$ V, $L = 200$ mH, $R_S = 1$ Ω, and $R_1 = 20$ kΩ.

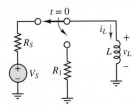

Figure P4.38

4.39   For $t < 0$, the circuit shown in Figure P4.39 is at DC steady state. The switch is closed at $t = 0$. Determine the voltage v_C for all time. Assume: $R_1 = R_3 = 3$ Ω, $R_2 = 6$ Ω, $V_1 = 15$ V, and $C = 0.5$ F.

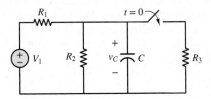

Figure P4.39

4.40   For $t < 0$, the circuit shown in Figure P4.39 is at DC steady state. The switch is opened at $t = 0$. Determine the current i_L through the inductor for all time. Assume:

$$V_S = 12 \text{ V} \qquad L = 100 \text{ mH}$$
$$R_1 = 400 \text{ }\Omega \qquad R_2 = 400 \text{ }\Omega$$
$$R_3 = 600 \text{ }\Omega$$

4.41   For the circuit shown in Figure P4.41, assume that switch S_1 is always held open and that switch S_2 is open until being closed at $t = 0$. Assume DC steady-state conditions for $t < 0$. Also assume $R_1 = 5\ \Omega$, $R_2 = 4\ \Omega$, $R_3 = 3\ \Omega$, $R_4 = 6\ \Omega$, and $C_1 = C_2 = 4$ F.

a. Find the capacitor voltage v_C at $t = 0^+$.

b. Find the time constant τ for $t > 0$.

c. Find v_C for all time and sketch the function.

d. Evaluate the ratio v_C to $v_C\,(\infty)$ at each of the following times: $t = 0,\ \tau,\ 2\tau,\ 5\tau,\ 10\tau$.

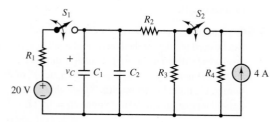

Figure P4.41

4.42   For the circuit shown in Figure P4.41, assume that switches S_1 and S_2 have been held open and closed, respectively, for a long time prior to $t = 0$. Then, simultaneously at $t = 0$, S_1 closes and S_2 opens. Also assume $R_1 = 5\ \Omega$, $R_2 = 4\ \Omega$, $R_3 = 3\ \Omega$, $R_4 = 6\ \Omega$, and $C_1 = C_2 = 4$ F.

a. Find the capacitor voltage v_C at $t = 0^+$.

b. Find the time constant τ for $t > 0$.

c. Find v_C for all time and sketch the function.

d. Evaluate the ratio v_C to $v_C\,(\infty)$ at each of the following times: $t = 0,\ \tau,\ 2\tau,\ 5\tau,\ 10\tau$.

4.43   For the circuit shown in Figure P4.41, assume that switch S_2 is always held open and that switch S_1 is closed until being opened at $t = 0$. Subsequently, S_1 closes at $t = 3\tau$ and remains closed. Also assume DC steady-state conditions for $t < 0$ and $R_1 = 5\ \Omega$, $R_2 = 4\ \Omega$, $R_3 = 3\ \Omega$, $R_4 = 6\ \Omega$, $C_1 = C_2 = 4$ F.

a. Find the capacitor voltage v_C at $t = 0$.

b. Find v_C for $0 < t < 3\tau$.

c. Use part b to find the capacitor voltage v_C at $t = 3\tau$, and use it to find v_C for $t > 3\tau$.

d. Compare the two time constants for $0 < t < 3\tau$ and $t > 3\tau$.

e. Sketch v_C for all time.

4.44   For the circuit shown in Figure P4.41, assume that switches S_1 and S_2 have been held open for a long time prior to $t = 0$ but then close at $t = 0$. Also assume

$R_1 = 5\ \Omega$, $R_2 = 4\ \Omega$, $R_3 = 3\ \Omega$, $R_4 = 6\ \Omega$, and $C_1 = C_2 = 4$ F.

a. Find the capacitor voltage v_C at $t = 0$.

b. Find the time constant τ for $t > 0$.

c. Find v_C and sketch the function.

d. Evaluate the ratio v_C to $v_C\,(\infty)$ at each of the following times: $t = 0,\ \tau,\ 2\tau,\ 5\tau,\ 10\tau$.

4.45   For the circuit shown in Figure P4.41, assume that switches S_1 and S_2 have been held closed for a long time prior to $t = 0$. S_1 then opens at $t = 0$; however, S_2 does not open until $t = 48$ s. Also assume $R_1 = 5\ \Omega$, $R_2 = 4\ \Omega$, $R_3 = 3\ \Omega$, $R_4 = 6\ \Omega$, and $C_1 = C_2 = 4$ F.

a. Find the capacitor voltage v_C at $t = 0$.

b. Find the time constant τ for $0 < t < 48$ s.

c. Find v_C for $0 < t < 48$ s.

d. Find τ for $t > 48$ s.

e. Find v_C for $t > 48$ s.

f. Sketch v_C for all time.

4.46   For the circuit shown in Figure P4.41, assume that switches S_1 and S_2 have been held closed for a long time prior to $t = 0$. S_2 then opens at $t = 0$; however, S_1 does not open until $t = 96$ s. Also assume $R_1 = 5\ \Omega$, $R_2 = 4\ \Omega$, $R_3 = 3\ \Omega$, $R_4 = 6\ \Omega$, and $C_1 = C_2 = 4$ F.

a. Find the capacitor voltage v_C at $t = 0$.

b. Find the time constant for $0 < t < 96$ s.

c. Find v_C for $0 < t < 96$ s.

d. Find the time constant for $t > 96$ s.

e. Use part c to find the capacitor voltage v_C at $t = 96$ s, and use it to find v_C for $t > 96$ s

f. Sketch v_C for all time.

4.47   For the circuit in Figure P4.47, determine the value of resistors R_1 and R_2, knowing that the time constant before the switch opens is 1.5 ms, and it is 10 ms after the switch opens. Assume: $R_S = 15$ kΩ, $R_3 = 30$ kΩ, and $C = 1\ \mu$F.

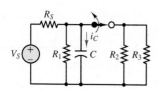

Figure P4.47

4.48   For the circuit in Figure P4.47, assume $V_S = 100$ V, $R_S = 4$ kΩ, $R_1 = 2$ kΩ, $R_2 = R_3 = 6$ kΩ, $C = 1\ \mu$F, and the circuit is in a steady-state condition before the switch opens. Find the value of v_C 2.666 ms after the switch opens.

4.49   In the circuit in Figure P4.49, how long after the switch is thrown at $t = 0$ will $i_L = 5$ A? Plot $i_L(t)$.

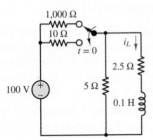

Figure P4.49

4.50   Refer to Figure P4.49 and assume that the switch takes 5 ms to move from one contact to the other. Also assume that during this time neither switch position has electrical contact. Find:

a. $i_L(t)$ for $0 < t < 5$ ms.

b. The maximum voltage between the contacts during the 5-ms duration of the switching.

Hint: This problem requires solving both a turnoff and a turn-on transient problem.

4.51   The circuit in Figure P4.51 includes a voltage-controlled switch. The switch closes or opens when the voltage across the capacitor reaches the value v_M^c or v_M^o, respectively. If $v_M^o = 1$ V and the period of the capacitor voltage waveform is 200 ms, find v_M^c.

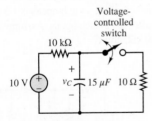

Figure P4.51

4.52   At $t = 0$ the switch in the circuit in Figure P4.52 closes. Assume that $L_1 = 1$ H, $L_2 = 5$ H, and that the circuit is in DC steady state for $t < 0$. Find $i_L(t)$ for all time.

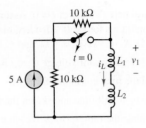

Figure P4.52

4.53   Repeat Problem P4.52, but find $v_1(t)$ for all time, instead of $i_L(t)$.

4.54   The analogy between electrical and thermal systems can be used to analyze the behavior of a pot heating on an electric stove. The heating element is modeled as shown in Figure P4.54. Find the "heat capacity" of the burner, C_S, if the burner reaches 90 percent of the desired temperature in 10 s. Assume $R_S = 1.5 \, \Omega$.

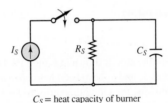

C_S = heat capacity of burner

R_S = heat loss of burner

Figure P4.54

4.55   The burner and pot of Problem 4.54 can be modeled as shown in Figure P4.55. R_0 models the thermal loss between the burner and the pot. The pot is modeled by a thermal capacitance C_P in parallel with a thermal resistance R_P.

a. Find the final temperature of the water in the pot— that is, find V_0 as $t \to \infty$ if $I_S = 75$ A, $C_P = 80$ F, $R_0 = 0.8 \, \Omega$, $R_P = 2.5 \, \Omega$, and the burner is the same as in Problem 4.54.

b. How long will it take for the water to reach 80 percent of its final temperature?

Hint: Neglect C_S since $C_S \ll C_P$.

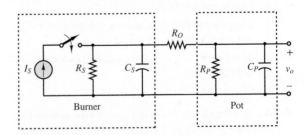

Figure P4.55

4.56 The circuit in Figure P4.56 is used as a variable delay in a burglar alarm. The alarm is a siren with an internal resistance of 1 kΩ. The alarm will not sound until the current i_0 exceeds 100 μA. Use a graphical or numerical solution to find the range of the variable resistor R for which the delay is between 1 and 2 s. Assume the capacitor is initially uncharged.

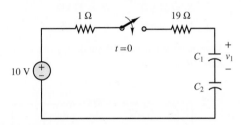

Figure P4.56

4.57 For $t > 0$, find the voltage v_1 across C_1 shown in Figure P4.57. Let $C_1 = 5$ μF and $C_2 = 10$ μF. Assume the capacitors are initially uncharged.

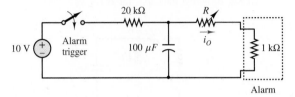

Figure P4.57

4.58 The switch shown in Figure P4.58 opens at $t = 0$. It closes at $t = 10$ s.

a. What is the time constant for $0 < t < 10$ s?

b. What is the time constant for $t > 10$ s?

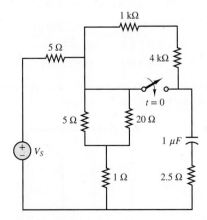

Figure P4.58

4.59 The circuit in Figure P4.59 models the charging circuit of an electronic camera flash. The flash should be charged to $v_C \geq 7.425$ V for each use. Assume $C = 1.5$ mF, $R_1 = 1$ kΩ, and $R_2 = 1$ Ω.

a. How long does it take the flash to recharge after taking a picture?

b. The shutter button stays closed for $1/30$ s. How much energy is delivered to the flash bulb R_2 in that interval? Assume the capacitor is fully charged.

c. If the shutter button is pressed 3 s after a flash, how much energy is delivered to the bulb R_2?

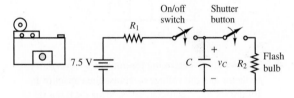

Figure P4.59

4.60 The ideal current source $i_s(t)$ in Figure P4.60 switches levels as shown. Determine and sketch the voltage $v_o(t)$ across the inductor for $0 < t < 2$ s. Assume the inductor current is zero before $t = 0$, $R = 500$ Ω, and $L = 50$ H.

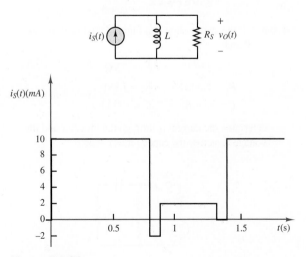

Figure P4.60

Section 4.6: Second-Order Transient Analysis

4.61   In the circuit shown in Figure P4.61:

$$V_{S1} = 15 \text{ V} \qquad V_{S2} = 9 \text{ V}$$
$$R_{S1} = 130 \text{ }\Omega \qquad R_{S2} = 290 \text{ }\Omega$$
$$R_1 = 1.1 \text{ k}\Omega \qquad R_2 = 700 \text{ }\Omega$$
$$L = 17 \text{ mH} \qquad C = 0.35 \text{ }\mu\text{F}$$

Assume that DC steady-state conditions exist for $t < 0$. Determine the voltage v_C across the capacitor and the current i_L through the inductor as $t \to \infty$.

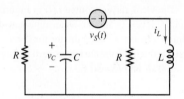

Figure P4.61

4.62   For $t > 0$, determine the current i_L through the inductor and the voltage v_C across the capacitor in Figure P4.62. Assume $v_S = -1$ V for $t < 0$ but is reversed to $v_S = 1$ V for $t > 0$. Also assume $R = 10 \text{ }\Omega$, $L = 5$ mH, $C = 100 \text{ }\mu\text{F}$, and that the circuit was in DC steady state prior to when the source was reversed.

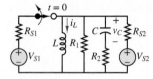

Figure P4.62

4.63   If the switch shown in Figure P4.63 is closed at $t = 0$ and:

$$V_S = 170 \text{ V} \qquad R_S = 7 \text{ k}\Omega$$
$$R_1 = 2.3 \text{ k}\Omega \qquad R_2 = 7 \text{ k}\Omega$$
$$L = 30 \text{ mH} \qquad C = 130 \text{ }\mu\text{F}$$

determine the current i_L through the inductor and the voltage v_C across the capacitor as $t \to \infty$.

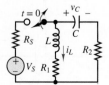

Figure P4.63

4.64   If the switch in the circuit shown in Figure P4.64 is closed at $t = 0$ and:

$$V_S = 12 \text{ V} \qquad C = 130 \text{ }\mu\text{F}$$
$$R_1 = 2.3 \text{ k}\Omega \qquad R_2 = 7 \text{ k}\Omega$$
$$L = 30 \text{ mH}$$

determine the current i_L through the inductor and the voltage v_C across the capacitor as $t \to \infty$.

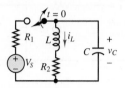

Figure P4.64

4.65   If the switch shown in Figure P4.65 is thrown at $t = 0$ and:

$$V_S = 12 \text{ V} \qquad R_S = 100 \text{ }\Omega$$
$$R_1 = 31 \text{ k}\Omega \qquad R_2 = 22 \text{ k}\Omega$$
$$L = 0.9 \text{ mH} \qquad C = 0.5 \text{ }\mu\text{F}$$

determine the current i_1 through R_1 and the voltage v_2 across R_2 as $t \to \infty$.

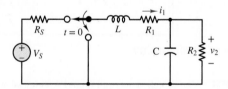

Figure P4.65

4.66   For $t < 0$, the circuit shown in Figure P4.66 is at DC steady state and the voltage across the capacitor is $+7$ V. The switch is thrown at $t = 0$, and:

$$V_S = 12 \text{ V} \qquad C = 3,300 \text{ }\mu\text{F}$$
$$R_1 = 9.1 \text{ k}\Omega \qquad R_2 = 4.3 \text{ k}\Omega$$
$$R_3 = 4.3 \text{ k}\Omega \qquad L = 16 \text{ mH}$$

Determine the initial current i_L through the inductor and the current i_2 through R_2 at $t = 0^+$.

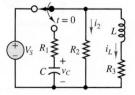

Figure P4.66

4.67   For $t < 0$, the circuit shown in Figure P4.67 is in DC steady state. Determine at $t = 0^+$, immediately after the switch is opened, the current i_L through the inductor and the voltage v_C across the capacitor.

$$V_{S1} = 15 \text{ V} \qquad V_{S2} = 9 \text{ V}$$
$$R_{S1} = 130 \text{ }\Omega \qquad R_{S2} = 290 \text{ }\Omega$$
$$R_1 = 1.1 \text{ k}\Omega \qquad R_2 = 700 \text{ }\Omega$$
$$L = 17 \text{ mH} \qquad C = 0.35 \text{ }\mu\text{F}$$

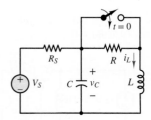

Figure P4.67

4.68   For $t < 0$, the circuit shown in Figure P4.68 is in DC steady state. The switch is closed at $t = 0$. Determine the current i_L through the inductor for $t > 0$. Assume $R = 3 \text{ k}\Omega$, $R_S = 600 \text{ }\Omega$, $V_S = 2 \text{ V}$, $C = 2 \text{ mF}$, and $L = 1 \text{ mH}$.

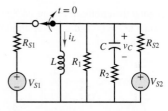

Figure P4.68

4.69   Assume the switch in the circuit in Figure P4.69 has been closed for a very long time. It is suddenly opened at $t = 0$ and then reclosed at $t = 5$ s. Determine the inductor current i_L and the voltage v across the 2-Ω resistor for $t \geq 0$.

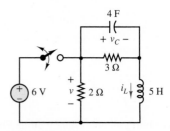

Figure P4.69

4.70   Determine if the circuit in Figure P4.70 is overdamped or underdamped for $t < 0$ and $t > 0$. Also find the capacitance that results in critical damping in both intervals. Assume $V_S = 15 \text{ V}$, $R = 200 \text{ }\Omega$, $L = 20 \text{ mH}$, and $C = 0.1 \text{ }\mu\text{F}$.

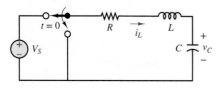

Figure P4.70

4.71   For $t < 0$, assume the capacitor in Figure P4.70 is completely discharged. If the switch is thrown at $t = 0$, find the:

a. Initial capacitor voltage v_C at $t = 0^+$.

b. Capacitor voltage v_C at $t = 20 \text{ }\mu\text{s}$.

c. Capacitor voltage v_C as $t \to \infty$.

d. Maximum capacitor voltage.

4.72   Assume the switch in the circuit in Figure P4.69 has been open for a very long time. It is suddenly closed at $t = 0$ and then reopened at $t = 5$ s. Determine the inductor current i_L, the capacitor voltage v_C, and the voltage v across the 2-Ω resistor for $t \geq 0$.

4.73   Assume that the circuit shown in Figure P4.70 is underdamped, and for $t < 0$, the circuit is in DC steady state with $v_C = V_S$. After the switch is thrown at $t = 0$, the first two zero crossings of the capacitor voltage v_C occur at $t = 5\pi/3$ μs and $t = 5\pi$ μs. At $t = 20\pi/3$ μs, the capacitor voltage v_C peaks at $0.6 \, V_S$. If $C = 1.6 \, \mu$F, what are the values of R and L?

4.74   Given the information provided in Problem 4.73, what are the values of R and L so that the peak at $20\pi/3$ μs is $v_C = 0.7 \, V_S$? Assume $C = 1.6 \, \mu$F.

4.75   Determine i_L for $t > 0$ in Figure P4.75, assuming $i(0) = 2.5$ A and $v_C(0) = 10$ V.

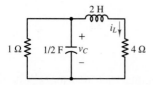

Figure P4.75

4.76 Find the maximum value of v_C for $t > 0$ in Figure P4.76, assuming DC steady state at $t = 0^-$.

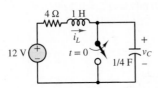

Figure P4.76

4.77 For $t > 0$, determine the time t at which $i = 2.5$ A in Figure P4.77, assuming DC steady state at $t = 0^-$.

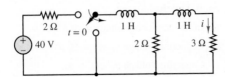

Figure P4.77

4.78 For $t > 0$, determine the time t at which $i = 6$ A in Figure P4.78, assuming DC steady state at $t = 0^-$.

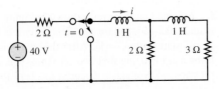

Figure P4.78

4.79 For $t > 0$, determine the time t at which $v = 7.5$ V in Figure P4.79, assuming DC steady state at $t = 0^-$.

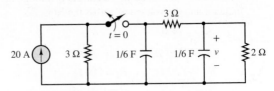

Figure P4.79

4.80 Assume the circuit in Figure P4.80 is in DC steady state at $t = 0^-$ and $L = 3$ H. Find the maximum value of v_C for $t > 0$.

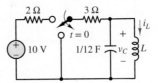

Figure P4.80

4.81 Assume the circuit in Figure P4.80 is in DC steady state at $t = 0^-$. Find the value of the inductance L that makes the circuit critically damped for $t > 0$. Find the maximum value of v for $t > 0$.

4.82 For $t > 0$, determine v in Figure P4.82, assuming DC steady state at $t = 0^-$.

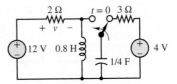

Figure P4.82

C H A P T E R

5

FREQUENCY RESPONSE AND SYSTEM CONCEPTS

requency-dependent phenomena are commonly encountered in engineering problems. For example, structures vibrate at a characteristic frequency when excited by wind forces (some high-rise buildings experience perceptible oscillation!). The propeller on a ship excites the shaft at a vibration frequency related to the engine's speed of rotation and to the number of blades on the propeller. An internal combustion engine is excited periodically by the combustion events in the individual cylinder, at a frequency determined by the firing of the cylinders. Wind blowing across a pipe excites a resonant vibration that is perceived as sound (wind instruments operate on this principle). Filters of all types depend upon frequency. In this respect, electric circuits are no different from other dynamic systems. A large body of knowledge has been developed related to the frequency response of electric circuits, most of it based on the ideas of phasors and impedance. The ideas developed in this chapter are applied, by analogy, to the analysis of other physical systems to illustrate the generality of the concepts.

In this chapter, quantities often involve angles. Unless indicated otherwise, angles are given in units of radians.

Learning Objectives

Students will learn to...

1. Understand the physical significance of frequency domain analysis, and compute the frequency response of circuits using AC circuit analysis tools. *Section 5.1.*

2. Analyze simple first- and second-order electrical filters, and determine their frequency response and filtering properties. *Section 5.2.*

3. Compute the frequency response of a circuit and its graphical representation in the form of a Bode plot. *Section 5.3.*

5.1 SINUSOIDAL FREQUENCY RESPONSE

The **sinusoidal frequency response** (or, simply, **frequency response**) of a circuit provides a measure of how the circuit responds to sinusoidal inputs of arbitrary frequency. In other words, for a given input signal with a particular amplitude, phase, and frequency, the frequency response of a circuit permits the computation of a particular output signal. For example, suppose you wanted to determine how the load voltage $\mathbf{V}_o$ or current $\mathbf{I}_o$ varied in response to different frequencies in the circuit of Figure 5.1. An analogy could be made, for example, with how an earphone (the load) responds to the audio signal generated by an MP3 player (the source) when an amplifier (the circuit) is placed between the two.[1] In the circuit of Figure 5.1, the signal source circuitry is represented by a Thévenin source. The impedances are, in general, a function of frequency. The amplifier circuit is represented by the idealized connection of two impedances $\mathbf{Z}_1$ and $\mathbf{Z}_2$, and the load is represented by an additional impedance $\mathbf{Z}_o$. The following statement provides a general definition of the frequency response of such a system:

> The frequency response of a circuit is a measure of the variation of a load-related voltage or current as a function of the frequency of the excitation signal.

Frequency Response Functions

A frequency response function is the ratio of a chosen *output* to a chosen *input*. In circuit analysis, the chosen input is often an independent voltage or current source. The chosen output can be any voltage or current elsewhere in the circuit. By convention, frequency response functions are represented by either **G** or **H**, where **G** is a

---

[1] The circuitry in a high-fidelity audio system is far more complex than the circuits discussed in this chapter. However, from the standpoint of intuition and everyday experience, the audio analogy provides a useful example. The audio spectrum terms *bass*, *midrange*, and *treble* are well known, but not well understood. The material presented in this chapter provides a technical basis for understanding these concepts.

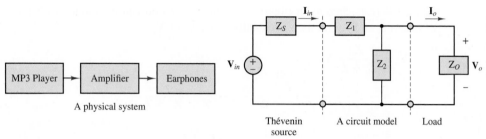

Figure 5.1 A circuit model

dimensionless *gain* and **H** is either an impedance or conductance. Four distinct types of frequency response function are defined as follows:

$$\mathbf{G}_V(j\omega) = \frac{\mathbf{V}_o(j\omega)}{\mathbf{V}_{\text{in}}(j\omega)} \qquad \mathbf{G}_I(j\omega) = \frac{\mathbf{I}_o(j\omega)}{\mathbf{I}_{\text{in}}(j\omega)}$$

$$\mathbf{H}_Z(j\omega) = \frac{\mathbf{V}_o(j\omega)}{\mathbf{I}_{\text{in}}(j\omega)} \qquad \mathbf{H}_Y(j\omega) = \frac{\mathbf{I}_o(j\omega)}{\mathbf{V}_{\text{in}}(j\omega)} \tag{5.1}$$

In many cases the inputs $\mathbf{V}_{\text{in}}$ and $\mathbf{I}_{\text{in}}$ are chosen to be independent voltage and current sources, respectively. The outputs $\mathbf{V}_o$ and $\mathbf{I}_o$ are freely chosen and, as such, represent the load in a circuit.

The above frequency response functions are related. For example, if $\mathbf{G}_V(j\omega)$ and $\mathbf{G}_I(j\omega)$ are known, the other two can be derived directly:

$$\mathbf{H}_Z(j\omega) = \frac{\mathbf{V}_o(j\omega)}{\mathbf{I}_{\text{in}}(j\omega)} = \frac{\mathbf{I}_o(j\omega)}{\mathbf{I}_{\text{in}}(j\omega)} \mathbf{Z}_o(j\omega) = \mathbf{G}_I(j\omega)\mathbf{Z}_o(j\omega) \tag{5.2}$$

$$\mathbf{H}_Y(j\omega) = \frac{\mathbf{I}_o(j\omega)}{\mathbf{V}_{\text{in}}(j\omega)} = \frac{\mathbf{V}_o(j\omega)}{\mathbf{Z}_o(j\omega)} \frac{1}{\mathbf{V}_{\text{in}}(j\omega)} = \frac{\mathbf{G}_V(j\omega)}{\mathbf{Z}_o(j\omega)} \tag{5.3}$$

Frequency response functions are important because they express the frequency response as a single function that relates an output (load) voltage or current to a given input.

Circuit Simplification

In general, the first step in determining the details of a chosen frequency response function is to divide the circuit into a load (in accord with the chosen output) and a source. Consider again the circuit shown in Figure 5.1. The network attached to the load can be replaced by its Thévenin equivalent as shown in Figure 5.2. Once

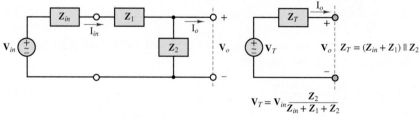

Figure 5.2 Thévenin equivalent source network

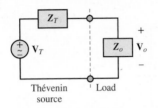

Figure 5.3 Equivalent circuit from the perspective of the load

the load is reattached as in Figure 5.3, voltage division can be applied to express $\mathbf{V}_o$ in terms of $\mathbf{V}_T$, and then eventually in terms of $\mathbf{V}_{\text{in}}$.

$$\mathbf{V}_o = \frac{\mathbf{Z}_o}{\mathbf{Z}_o + \mathbf{Z}_T}\mathbf{V}_T$$

$$= \frac{\mathbf{Z}_o}{\mathbf{Z}_o + (\mathbf{Z}_{\text{in}} + \mathbf{Z}_1)\mathbf{Z}_2/(\mathbf{Z}_{\text{in}} + \mathbf{Z}_1 + \mathbf{Z}_2)} \cdot \frac{\mathbf{Z}_2}{\mathbf{Z}_{\text{in}} + \mathbf{Z}_1 + \mathbf{Z}_2}\mathbf{V}_{\text{in}} \qquad (5.4)$$

$$= \frac{\mathbf{Z}_o\mathbf{Z}_2}{\mathbf{Z}_o(\mathbf{Z}_{\text{in}} + \mathbf{Z}_1 + \mathbf{Z}_2) + (\mathbf{Z}_{\text{in}} + \mathbf{Z}_1)\mathbf{Z}_2}\mathbf{V}_{\text{in}}$$

The gain, $\mathbf{G}_V(j\omega)$, is a dimensionless complex quantity, given by:

$$\mathbf{G}_V(j\omega) = \frac{\mathbf{V}_o}{\mathbf{V}_{\text{in}}}(j\omega) = \frac{\mathbf{Z}_o\mathbf{Z}_2}{\mathbf{Z}_o(\mathbf{Z}_{\text{in}} + \mathbf{Z}_1 + \mathbf{Z}_2) + (\mathbf{Z}_{\text{in}} + \mathbf{Z}_1)\mathbf{Z}_2} \qquad (5.5)$$

Thus, the gain is known if the circuit element impedances are known.

LO ⊳ $\mathbf{V}_o(j\omega)$ is a phase-shifted and amplitude-scaled version of $\mathbf{V}_{\text{in}}(j\omega)$.

If the phasor source voltage and the frequency response of the circuit are known, the phasor load voltage can be computed as follows:

$$\mathbf{V}_o(j\omega) = \mathbf{G}_V(j\omega) \cdot \mathbf{V}_{\text{in}}(j\omega) \qquad (5.6)$$

$$V_o e^{j\phi_o} = |\mathbf{G}_V|e^{j\angle\mathbf{G}_v} \cdot V_{\text{in}} e^{j\phi_{\text{in}}} \qquad (5.7)$$

such that

$$V_o = |\mathbf{G}_V| \cdot V_{\text{in}} \qquad (5.8)$$

and

$$\phi_o = \angle\mathbf{G}_v + \phi_{\text{in}} \qquad (5.9)$$

At any given angular frequency ω, the load voltage is a sinusoid with the same frequency as the source voltage.

First- and Second-Order Archetypes

Whenever possible, the first step toward deriving a frequency response function is to use Thévenin's or Norton's theorem to simplify the circuit. If the circuit is first-order, or second-order with the storage elements in series or parallel, it can be simplified to one of the four archetypes shown in Figures 5.4 to 5.7.

In the first-order circuit of Figure 5.4, the loop current $\mathbf{I}_C$ is related to the Thévenin source voltage $\mathbf{V}_T$ by the generalized Ohm's law:

$$\mathbf{I}_C = \frac{\mathbf{V}_T}{R_T + \mathbf{Z}_C} \qquad (5.10)$$

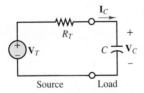

Figure 5.4 Simplified first-order equivalent circuit with one capacitor

Multiply the numerator and denominator by $(j\omega)C$ and divide both sides by $\mathbf{V}_T$ to find the frequency response function:

$$\mathbf{H}_Y(j\omega) = \frac{\mathbf{I}_C}{\mathbf{V}_T} = \frac{(j\omega)C}{1 + (j\omega)\tau} \qquad (5.11)$$

where $\tau = R_T C$.

It is now a simple matter to find the frequency response function relating $\mathbf{V}_C$ to $\mathbf{V}_T$:

$$\mathbf{G}_V(j\omega) = \frac{\mathbf{V}_C}{\mathbf{V}_T} = \frac{\mathbf{I}_C}{\mathbf{V}_T}\mathbf{Z}_C = \frac{(j\omega)C}{1+(j\omega)\tau}\frac{1}{(j\omega)C} = \frac{1}{1+(j\omega)\tau} \qquad (5.12)$$

where again $\tau = R_TC$. Notice that the denominator is the same as in $\mathbf{H}_Y$. This is a common result because the denominator expresses the basic dynamics of the circuit. The numerator expresses differences in the circuit variables. It is a useful exercise to derive $\mathbf{G}_V$ directly from voltage division. Try it!

A similar approach can be taken to find the frequency response relating the voltage $\mathbf{V}_L$ to the Norton source current $\mathbf{I}_N$ in the first-order circuit of Figure 5.5. Apply the generalized Ohm's law to write:

$$\mathbf{V}_L = \mathbf{I}_N(R_N\|\mathbf{Z}_L) = \mathbf{I}_N\frac{R_N(j\omega)L}{R_N+(j\omega)L} \qquad (5.13)$$

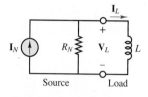

Figure 5.5 Simplified first-order circuit with one inductor

The frequency response function is found by dividing both sides by $\mathbf{I}_N$ and then dividing the numerator and denominator by R_N.

$$\mathbf{H}_Z(j\omega) = \frac{\mathbf{V}_L}{\mathbf{I}_N} = \frac{(j\omega)L}{1+(j\omega)\tau} \qquad (5.14)$$

where, in this case, $\tau = L/R_N$.

Again, it is a simple matter to find the frequency response function relating $\mathbf{I}_L$ to $\mathbf{I}_N$:

$$\mathbf{G}_I(j\omega) = \frac{\mathbf{I}_L}{\mathbf{I}_N} = \frac{\mathbf{V}_L}{\mathbf{I}_N}\frac{1}{\mathbf{Z}_L} = \frac{1}{1+(j\omega)\tau} \qquad (5.15)$$

where again $\tau = L/R_N$. Also, once more the denominator is the same as in $\mathbf{H}_Z$. It is a useful exercise to derive $\mathbf{G}_I$ directly from current division. Try it!

Second-order circuits are handled in much the same way. Consider the series LC circuit of Figure 5.6. The common loop current $\mathbf{I}_L$ is related to the Thévenin source voltage $\mathbf{V}_T$ by the generalized Ohm's law:

$$\mathbf{V}_T = \mathbf{I}_L(R_T+\mathbf{Z}_C+\mathbf{Z}_L) \qquad (5.16)$$

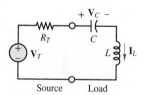

Figure 5.6 Simplified second-order circuit with one capacitor and one inductor in series

Divide both sides by $\mathbf{I}_L$, invert both sides, and multiply the resulting numerator and denominator by $j\omega C$ to find:

$$\begin{aligned}\mathbf{H}_Y(j\omega) = \frac{\mathbf{I}_L}{\mathbf{V}_T} &= \frac{(j\omega)C}{1+(j\omega)R_TC+(j\omega)^2LC} \\ &= \frac{(j\omega)C}{1+(j\omega)\tau+(j\omega/\omega_n)^2}\end{aligned} \qquad (5.17)$$

where $\tau = R_TC$ and $\omega_n^2 = 1/LC$. This latter term is the same natural frequency often found in second-order transient circuits.

The voltage gain $\mathbf{G}_V$ for the second-order series LC circuit can be found using the result for $\mathbf{H}_Y$.

$$\mathbf{G}_V(j\omega) = \frac{\mathbf{V}_C}{\mathbf{V}_T} = \frac{\mathbf{I}_C}{\mathbf{V}_T}\mathbf{Z}_C \qquad (5.18)$$

Of course, $\mathbf{I}_C = \mathbf{I}_L$ for the series loop, so:

$$\mathbf{G}_V(j\omega) = \frac{\mathbf{I}_L}{\mathbf{V}_T}\mathbf{Z}_C = \mathbf{H}_Y\mathbf{Z}_C = \frac{1}{1+(j\omega)\tau+(j\omega/\omega_n)^2} \qquad (5.19)$$

where again $\tau = R_TC$ and $\omega_n^2 = 1/LC$.

Finally, Figure 5.7 shows a second-order parallel LC circuit. The common voltage $\mathbf{V}_C$ is related to the Norton source current $\mathbf{I}_N$ by the generalized Ohm's law:

$$\mathbf{V}_C = \mathbf{I}_N(R_N \| \mathbf{Z}_L \| \mathbf{Z}_C) \tag{5.20}$$

Divide both sides by $\mathbf{I}_N$ to obtain:

$$\mathbf{H}_Z(j\omega) = \frac{\mathbf{V}_C}{\mathbf{I}_N} = \frac{1}{1/R_N + 1/j\omega L + j\omega C} \tag{5.21}$$

Multiply the numerator and denominator by $j\omega L$ to obtain:

$$\mathbf{H}_Z(j\omega) = \frac{(j\omega)L}{1 + (j\omega)L/R_N + (j\omega)^2 LC} = \frac{(j\omega)L}{1 + (j\omega)\tau + (j\omega/\omega_n)^2} \tag{5.22}$$

where $\tau = L/R_N$ and $\omega_n^2 = 1/LC$ is the same natural frequency often found in the second-order series LC circuit.

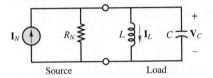

Figure 5.7 Simplified second-order circuit with one capacitor and one inductor in parallel

The current gain $\mathbf{G}_I$ for the second-order parallel LC circuit can be found using the result for $\mathbf{H}_Z$.

$$\mathbf{G}_I(j\omega) = \frac{\mathbf{I}_L}{\mathbf{I}_N} = \frac{\mathbf{V}_L}{\mathbf{I}_N}\frac{1}{\mathbf{Z}_L} \tag{5.23}$$

Of course, $\mathbf{V}_L = \mathbf{V}_C$, so:

$$\mathbf{G}_I(j\omega) = \frac{\mathbf{V}_L}{\mathbf{I}_N}\frac{1}{\mathbf{Z}_L} = \frac{\mathbf{H}_Z}{\mathbf{Z}_L} = \frac{1}{1 + (j\omega)\tau + (j\omega/\omega_n)^2} \tag{5.24}$$

where again $\tau = L/R_N$ and $\omega_n^2 = 1/LC$.

Poles and Zeros

By definition, a frequency response function is the ratio of an output to an input. Consequently, the development of any specific frequency response function will, in general, also result in a ratio. The numerator and denominator of any frequency response function can always be expressed as the product of four distinct terms. One of these terms is simply a constant. The other three terms are known as *zeros* or *poles*, depending upon whether they appear in the numerator or denominator, respectively. Each term has its own name, as indicated in the following list.

1. K A constant

2. $(j\omega)$ Pole or zero at the origin

3. $(1 + j\omega\tau)$ Simple pole or zero

4. $[1 + j\omega\tau + (j\omega/\omega_n)^2]$ Quadratic (complex) pole or zero

A simple pole or zero may also take the form $(1 + j\omega/\omega_0)$, where $\omega_0 = 1/\tau$.

The first- and second-order frequency response functions developed in the previous section are good examples of the standard form in which the numerator and denominator are expressed as products of these four terms. These same terms will appear repeatedly later in this chapter when filters and Bode plots are discussed.

EXAMPLE 5.1 Computing the Frequency Response Using Thévenin's Theorem

Problem

Compute the frequency response $\mathbf{G}_V(j\omega) = \mathbf{V}_o/\mathbf{V}_S$ for Figure 5.8.

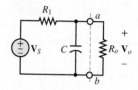

Figure 5.8 Circuit for Example 5.1.

Solution

Known Quantities: $R_1 = 10$ kΩ; $C = 10$ μF; $R_o = 10$ kΩ.

Find: The frequency response $\mathbf{G}_V(j\omega) = \mathbf{V}_o/\mathbf{V}_S$.

Assumptions: None.

Analysis: With R_o as the load resistance, the approach is to use Thévenin's theorem to determine the equivalent network of the source network; that is, the equivalent network of everything to the left of terminals a and b. The Thévenin equivalent impedance $\mathbf{Z}_T$ of the source network is:

$$\mathbf{Z}_T = (R_1 \| \mathbf{Z}_C)$$

The Thévenin (open-circuit) voltage $\mathbf{V}_T$ across terminals a and b is found from voltage division:

$$\mathbf{V}_T = \mathbf{V}_S \frac{\mathbf{Z}_C}{R_1 + \mathbf{Z}_C}$$

After reattaching the load to the Thévenin source, the voltage $\mathbf{V}_o$ across the load can be found by applying voltage division once more:

$$\mathbf{V}_o = \frac{R_o}{\mathbf{Z}_T + R_o} \mathbf{V}_T$$

$$= \frac{R_o}{R_1 \mathbf{Z}_C/(R_1 + \mathbf{Z}_C) + R_o} \frac{\mathbf{Z}_C}{R_1 + \mathbf{Z}_C} \mathbf{V}_S$$

Thus:

$$\mathbf{G}_V(j\omega) = \frac{\mathbf{V}_o}{\mathbf{V}_S}(j\omega) = \frac{R_o \mathbf{Z}_C}{R_o(R_1 + \mathbf{Z}_C) + R_1 \mathbf{Z}_C}$$

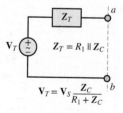

Figure 5.9 Equivalent circuit for Example 5.1.

The impedances of the circuit elements are $R_1 = 10^3$ Ω, $\mathbf{Z}_C = 1/(j\omega \times 10^{-5})\Omega$, and $R_o = 10^4$ Ω. The resulting frequency response is:

$$\mathbf{G}_V(j\omega) = \frac{\dfrac{10^4}{j\omega \times 10^{-5}}}{10^4 \left(10^3 + \dfrac{1}{j\omega \times 10^{-5}}\right) + \dfrac{10^3}{j\omega \times 10^{-5}}} = \frac{100}{110 + j\omega}$$

$$= \frac{100}{\sqrt{110^2 + \omega^2}\, e^{j\tan^{-1}(\omega/110)}} = \frac{100}{\sqrt{110^2 + \omega^2}} \angle -\arctan\left(\frac{\omega}{110}\right)$$

Comments: The use of equivalent circuit ideas is often helpful in deriving frequency response functions because it naturally forces us to identify source and load quantities. However, it is certainly not the only method of solution. For example, node analysis would have yielded the same results just as easily, by recognizing that the top node voltage is equal to the load voltage and by solving directly for $\mathbf{V}_o$ as a function of $\mathbf{V}_S$, without going through the intermediate step of computing the Thévenin equivalent source circuit.

EXAMPLE 5.2   Computing the Frequency Response

Problem

Compute the frequency response $\mathbf{H}_Z(j\omega) = \mathbf{V}_o/\mathbf{I}_S$ for Figure 5.10.

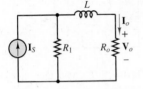

Figure 5.10 Circuit for Example 5.2.

Solution

Known Quantities: $R_1 = 1\ \text{k}\Omega$; $L = 2\ \text{mH}$; $R_o = 4\ \text{k}\Omega$.

Find: The frequency response $\mathbf{H}_Z(j\omega) = \mathbf{V}_o/\mathbf{I}_S$.

Assumptions: None.

Analysis: While it is possible to find the Thévenin or Norton equivalent network of everything attached to R_o and proceed as in the previous example to find the frequency response function, it is also possible to apply current division to find $\mathbf{I}_o$ and then apply Ohm's law to find $\mathbf{V}_o$ and thus the frequency response function.

Apply current division to write:

$$\mathbf{I}_o = \frac{R_1}{R_1 + R_o + j\omega L}\mathbf{I}_S$$

Factor out $R_1 + R_o$ in the denominator to find:

$$\mathbf{I}_o = \frac{R_1}{R_1 + R_o}\frac{1}{1 + j\omega L/(R_1 + R_o)}\mathbf{I}_S$$

Then:

$$\frac{\mathbf{V}_o}{\mathbf{I}_S}(j\omega) = \mathbf{H}_Z(j\omega) = \frac{I_o R_o}{I_S}$$

$$= \frac{R_1 R_o}{R_1 + R_o}\frac{1}{1 + j\omega L/(R_1 + R_o)}$$

Substitute numerical values to obtain:

$$\mathbf{H}_Z(j\omega) = \frac{(10^3)(4 \times 10^3)}{10^3 + 4 \times 10^3}\frac{1}{1 + (j\omega)(2 \times 10^{-3})/(5 \times 10^3)}$$

$$= (0.8 \times 10^3)\frac{1}{1 + j0.4 \times 10^{-6}\omega}$$

Comments: The units of $\mathbf{H}_Z(j\omega)$ should be ohms. Verify that they are!

CHECK YOUR UNDERSTANDING

Refer to Example 5.1 and compute the magnitude and phase of $\mathbf{G}_V$ at the frequencies $\omega = 10$, 100, and 1,000 rad/s.

Answers: Magnitude = 0.9054, 0.6727, and 0.0994; phase (degrees) = −5.1944, −42.2737, and −83.7227

CHECK YOUR UNDERSTANDING

Refer to Example 5.2 and compute the magnitude and phase of $\mathbf{H}_Z$ at the frequencies $\omega = 1$, 10, and 100 Mrad/s.

Answer: Magnitude = 742.78 Ω, 194.03 Ω, and 19.99 Ω; phase (degrees) = −21.8°, −75.96°, and −88.57°

5.2   LOW- AND HIGH-PASS FILTERS

There are many practical applications that involve filters of one kind or another. Modern sunglasses filter out eye-damaging ultraviolet radiation and reduce the intensity of sunlight reaching the eyes. The suspension system of an automobile filters out road noise and reduces the impact of potholes on passengers. An analogous concept applies to electric circuits: It is possible to *attenuate* (i.e., reduce in amplitude) or altogether eliminate signals of unwanted frequencies, such as those that may be caused by electromagnetic interference (EMI).

Low-Pass Filters

Figure 5.11 depicts a simple *RC* **filter** and denotes its input and output voltages, respectively, by $\mathbf{V}_i$ and $\mathbf{V}_o$. The frequency response for the filter may be obtained by considering the function

$$\mathbf{H}(j\omega) = \frac{\mathbf{V}_o}{\mathbf{V}_i}(j\omega) \tag{5.25}$$

and noting that the output voltage may be expressed as a function of the input voltage by means of a voltage divider, as follows:

$$\mathbf{V}_o(j\omega) = \mathbf{V}_i(j\omega)\frac{1/j\omega C}{R + 1/j\omega C} = \mathbf{V}_i(j\omega)\frac{1}{1 + j\omega RC} \tag{5.26}$$

Thus, the frequency response of the *RC* filter is

$$\frac{\mathbf{V}_o}{\mathbf{V}_i}(j\omega) = \frac{1}{1 + j\omega CR} \tag{5.27}$$

An immediate observation upon studying this frequency response is that if the signal frequency ω is zero, the value of the frequency response function

RC low-pass filter. The circuit preserves lower frequencies while attenuating the frequencies above the cutoff frequency $\omega_0 = 1/RC$. The voltages $\mathbf{V}_i$ and $\mathbf{V}_o$ are the filter input and output voltages, respectively.

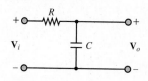

Figure 5.11 A simple *RC* filter

is 1. That is, the filter is passing all the input. Why? To answer this question, we note that at $\omega = 0$, the impedance of the capacitor, $1/j\omega C$, becomes infinite. Thus, the capacitor acts as an open-circuit, and the output voltage equals the input:

$$\mathbf{V}_o(j\omega = 0) = \mathbf{V}_i(j\omega = 0) \tag{5.28}$$

Since a signal at sinusoidal frequency equal to zero is a DC signal, this filter circuit does not in any way affect DC voltages and currents. As the signal frequency increases, the magnitude of the frequency response decreases since the denominator increases with ω. More precisely, equations 5.29 to 5.32 describe the magnitude and phase of the frequency response of the RC filter:

$$
\begin{aligned}
\mathbf{H}(j\omega) = \frac{\mathbf{V}_o}{\mathbf{V}_i}(j\omega) &= \frac{1}{1 + j\omega CR} \\
&= \frac{1}{\sqrt{1 + (\omega CR)^2}} \frac{e^{j0}}{e^{j\arctan(\omega CR/1)}} \\
&= \frac{1}{\sqrt{1 + (\omega CR)^2}} \cdot e^{-j\arctan(\omega CR)}
\end{aligned} \tag{5.29}
$$

or

$$\mathbf{H}(j\omega) = |\mathbf{H}(j\omega)| \, e^{j\angle\mathbf{H}(j\omega)} \tag{5.30}$$

with

$$|\mathbf{H}(j\omega)| = \frac{1}{\sqrt{1 + (\omega CR)^2}} = \frac{1}{\sqrt{1 + (\omega/\omega_0)^2}} \tag{5.31}$$

and

$$\angle\mathbf{H}(j\omega) = -\arctan(\omega CR) = -\arctan\frac{\omega}{\omega_0} \tag{5.32}$$

with

$$\omega_0 = \frac{1}{RC} \tag{5.33}$$

The simplest way to envision the effect of the filter is to think of the phasor voltage $\mathbf{V}_i = V_i e^{j\phi_i}$ scaled by a factor of $|\mathbf{H}|$ and shifted by a phase angle $\angle\mathbf{H}$ by the filter *at each frequency*, so that the resultant output is given by the phasor $V_o e^{j\phi_o}$, with

$$
\begin{aligned}
V_o &= |\mathbf{H}| \cdot V_i \\
\phi_o &= \angle\mathbf{H} + \phi_i
\end{aligned} \tag{5.34}
$$

and where $|\mathbf{H}|$ and $\angle\mathbf{H}$ are functions of frequency. The frequency ω_0 is called the **cutoff frequency** of the filter and, as will presently be shown, gives an indication of the filtering characteristics of the circuit.

It is customary to represent $\mathbf{H}(j\omega)$ in two separate plots, representing $|\mathbf{H}|$ and $\angle\mathbf{H}$ as functions of ω. These are shown in Figure 5.12 in normalized form,

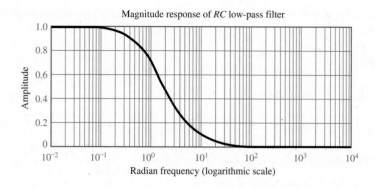

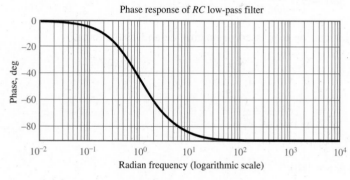

Figure 5.12 Magnitude and phase response plots for RC filter

that is, with $|\mathbf{H}|$ and $\angle\mathbf{H}$ plotted versus ω/ω_0, corresponding to a cutoff frequency $\omega_0 = 1$ rad/s. Note that, in the plot, the frequency axis has been scaled logarithmically. This is a common practice in electrical engineering because it enables viewing a very broad range of frequencies on the same plot without excessively compressing the low-frequency end of the plot. The frequency response plots of Figure 5.12 are commonly employed to describe the frequency response of a circuit since they can provide a clear idea at a glance of the effect of a filter on an excitation signal. This type of filter is called a **low-pass filter**. The cutoff frequency $\omega = 1/RC$ has a special significance in that it represents—approximately—the point where the filter begins to filter out the higher-frequency signals. The value of $|\mathbf{H}(j\omega)|$ at the cutoff frequency is $1/\sqrt{2} = 0.707$. Note how the cutoff frequency depends exclusively on the values of R and C. Therefore, one can adjust the filter response as desired simply by selecting appropriate values for C and R, and therefore one can choose the desired filtering characteristics.

Practical low-pass filters are often much more complex than simple RC combinations. The synthesis of such advanced filter networks is beyond the scope of this book; however, the implementation of some commonly used filters is discussed in Chapter 7, in connection with the operational amplifier.

High-Pass Filters

Just as a low-pass filter preserves low-frequency signals and attenuates those at higher frequencies, a **high-pass filter** attenuates low-frequency signals and preserves those

RC high-pass filter. The circuit preserves higher frequencies while attenuating the frequencies below the cutoff frequency $\omega_0 = 1/RC$.

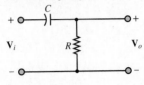

Figure 5.13 High-pass filter

at frequencies above a *cutoff frequency*. Consider the high-pass filter circuit shown in Figure 5.13. The frequency response is defined as:

$$\mathbf{H}(j\omega) = \frac{\mathbf{V}_o}{\mathbf{V}_i}(j\omega)$$

Voltage division yields:

$$\mathbf{V}_o(j\omega) = \mathbf{V}_i(j\omega)\frac{R}{R + 1/j\omega C} = \mathbf{V}_i(j\omega)\frac{j\omega CR}{1 + j\omega CR} \tag{5.35}$$

Thus, the frequency response of the filter is

$$\frac{\mathbf{V}_o}{\mathbf{V}_i}(j\omega) = \frac{j\omega CR}{1 + j\omega CR} \tag{5.36}$$

which can be expressed in magnitude-and-phase form by

$$\mathbf{H}(j\omega) = \frac{\mathbf{V}_o}{\mathbf{V}_i}(j\omega) = \frac{j\omega CR}{1 + j\omega CR} = \frac{\omega CR e^{j\pi/2}}{\sqrt{1 + (\omega CR)^2}\, e^{j\arctan(\omega CR/1)}}$$

$$= \frac{\omega CR}{\sqrt{1 + (\omega CR)^2}} \cdot e^{j[\pi/2 - \arctan(\omega CR)]}$$

or   (5.37)

$$\mathbf{H}(j\omega) = |\mathbf{H}|e^{j\angle\mathbf{H}}$$

with

$$|\mathbf{H}(j\omega)| = \frac{\omega CR}{\sqrt{1 + (\omega CR)^2}}$$

$$\angle\mathbf{H}(j\omega) = 90° - \arctan(\omega CR) \tag{5.38}$$

You can verify by inspection that the amplitude response of the high-pass filter will be zero at $\omega = 0$ and will asymptotically approach 1 as ω approaches infinity while the phase shift is $\pi/2$ at $\omega = 0$ and tends to zero for increasing ω. Amplitude-and-phase response curves for the high-pass filter are shown in Figure 5.14. These plots have been normalized to have the filter cutoff frequency $\omega_0 = 1$ rad/s. Note that, once again, it is possible to define a cutoff frequency at $\omega_0 = 1/RC$ in the same way as was done for the low-pass filter.

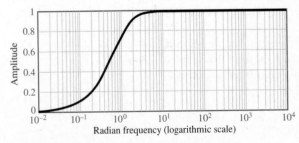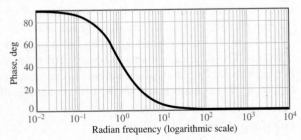

Figure 5.14 Frequency response of a high-pass filter

EXAMPLE 5.3 Frequency Response of RC Low-Pass Filter

Problem

Compute the response of the *RC* filter of Figure 5.11 to sinusoidal inputs at the frequencies of 60 and 10,000 Hz.

Solution

Known Quantities: $R = 1$ kΩ; $C = 0.47$ μF; $v_i(t) = 5 \cos(\omega t)$ V.

Find: The output voltage $v_o(t)$ at each frequency.

Assumptions: None.

Analysis: In this problem, the input signal voltage and the frequency response of the circuit (equation 5.29) are known, and the output voltage must be found at two different frequencies. If the voltages are represented in phasor form, the frequency response can be used for calculation:

$$\frac{\mathbf{V}_o}{\mathbf{V}_i}(j\omega) = \mathbf{H}_V(j\omega) = \frac{1}{1 + j\omega CR}$$

$$\mathbf{V}_o(j\omega) = \mathbf{H}_V(j\omega)\,\mathbf{V}_i(j\omega) = \frac{1}{1 + j\omega CR}\mathbf{V}_i(j\omega)$$

The cutoff frequency of the filter is $\omega_0 = 1/RC = 2{,}128$ rad/s such that the expression for the frequency response in the form of equations 5.31 and 5.32 is:

$$\mathbf{H}_V(j\omega) = \frac{1}{1 + j\omega/\omega_0} \qquad |\mathbf{H}_V(j\omega)| = \frac{1}{\sqrt{1 + (\omega/\omega_0)^2}} \qquad \angle\mathbf{H}(j\omega) = -\arctan\left(\frac{\omega}{\omega_0}\right)$$

Next, recognize that at $\omega = 60$ Hz $= 120\pi$ rad/s, the ratio $\omega/\omega_0 = 0.177$. At $\omega = 10{,}000$ Hz $= 20{,}000\pi$, $\omega/\omega_0 = 29.5$. Thus, the output voltage at each frequency can be computed as follows:

$$\mathbf{V}_o(\omega = 2\pi 60) = \frac{1}{1 + j0.177}\mathbf{V}_i(\omega = 2\pi 60) = (0.985 \times 5)\angle{-0.175}\ \text{V}$$

$$\mathbf{V}_o(\omega = 2\pi 10{,}000) = \frac{1}{1 + j29.5}\mathbf{V}_i(\omega = 2\pi 10{,}000) = (0.0339 \times 5)\angle{-1.537}\ \text{V}$$

And finally we write the time-domain response for each frequency:

$$v_o(t) = 4.923 \cos{(2\pi 60 t - 0.175)}\ \text{V} \qquad \text{at } \omega = 2\pi 60 \text{ rad/s}$$
$$v_o(t) = 0.169 \cos{(2\pi 10{,}000 t - 1.537)}\ \text{V} \qquad \text{at } \omega = 2\pi 10{,}000 \text{ rad/s}$$

The magnitude and phase responses of the filter are plotted in Figure 5.15. It should be evident from these plots that only the low-frequency components of the signal are passed by the filter. This low-pass filter would pass only the *bass range* of the audio spectrum.

Magnitude response of *RC* filter of Example 5.3

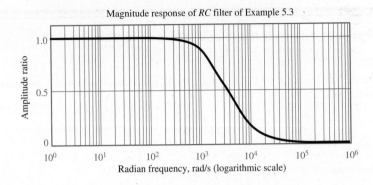

Phase response of *RC* filter of Example 5.3

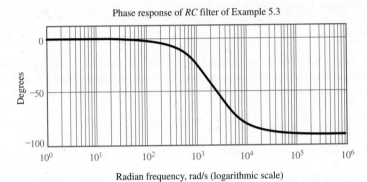

Figure 5.15 Response of *RC* filter of Example 5.3

EXAMPLE 5.4   A Realistic *RC* Low-Pass Filter Application

Problem

Determine the frequency response function $\mathbf{V}_o/\mathbf{V}_S$ and its frequency response from the network shown in Figure 5.16.

Solution

Known Quantities: $R_S = 50 \ \Omega$; $R_1 = 200 \ \Omega$; $R_o = 500 \ \Omega$; $C = 10 \ \mu F$.

Find: The frequency response function $\mathbf{V}_o/\mathbf{V}_S$, its frequency response, and the output voltage $v_o(t)$ at given frequencies.

Assumptions: None.

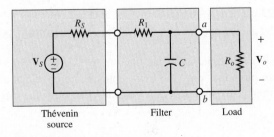

Figure 5.16 *RC* filter inserted in a circuit

Analysis: Figure 5.17 represents a more realistic filtering circuit, in that an *RC* low-pass filter is inserted between the source and load circuits. The Thévenin equivalent impedance see by the load is:

$$\mathbf{Z}_T = \mathbf{Z}_C \| (R_1 + R_S) = \frac{1/j\omega C(R_1 + R_S)}{R_1 + R_S + 1/j\omega C}$$

Multiply the numerator and denominator by $j\omega C$ to obtain:

$$\mathbf{Z}_T = \frac{R_1 + R_S}{1 + (j\omega)(R_1 + R_S)C}$$

Apply voltage division to find the Thévenin (open-circuit) voltage $\mathbf{V}_T$ across terminals a and b.

$$\mathbf{V}_T = \mathbf{V}_S \frac{1/j\omega C}{1/j\omega C + R_1 + R_S}$$

Again, multiply the numerator and denominator by $j\omega C$ to obtain:

$$\mathbf{V}_T = \mathbf{V}_S \frac{1}{1 + (j\omega)(R_1 + R_S)C}$$

Next, apply voltage division to find $\mathbf{V}_o$.

$$\mathbf{V}_o = \mathbf{V}_T \frac{R_o}{R_o + \mathbf{Z}_T}$$

Substitute for $\mathbf{V}_T$ and $\mathbf{Z}_T$, and multiply the numerator and denominator by $[1 + (j\omega)(R_1 + R_S)C]$ to obtain:

$$\mathbf{V}_o = \mathbf{V}_S \frac{R_o}{R_o[1 + (j\omega)(R_1 + R_S)C] + (R_1 + R_S)}$$

Finally, divide both sides by $\mathbf{V}_S$ and factor $(R_o + R_1 + R_S)$ out of the denominator to find:

$$\mathbf{H}_V(j\omega) = \frac{\mathbf{V}_o}{\mathbf{V}_S} = \frac{K}{1 + (j\omega)\tau}$$

where

$$K = \frac{R_o}{R_o + R_1 + R_S}$$

and

$$\tau = [R_o \| (R_1 + R_S)]C = \frac{R_o(R_1 + R_S)C}{R_o + R_1 + R_S}$$

Plug in values for the resistances and capacitance to find:

$$\mathbf{H}(j\omega) = \frac{0.667}{1 + j(\omega/600)}$$

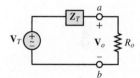

Figure 5.17 Equivalent circuit representation of Figure 5.16

Comments: Notice that the time constant τ equals the capacitance times the Thévenin equivalent resistance seen by the capacitor. Thus, the effect of placing the *RC* low-pass filter in the midst of the circuit is to shift the filter's cutoff frequency from $1/R_1C$ to $1/R_TC$. Also, note that the low-frequency amplitude of the frequency response function is simply K, which is $|\mathbf{V}_o/\mathbf{V}_S|$ when the capacitor is replaced with an open-circuit.

EXAMPLE 5.5   Low-Pass Filter Attenuation

Problem

The frequency response of a particular low-pass filter is described by the following frequency response function. At what frequency has the magnitude of the response fallen to 10 percent of its maximum?

$$\mathbf{H}(j\omega) = \frac{K}{(j\omega/\omega_1 + 1)(j\omega/\omega_2 + 1)}$$

Solution

Known Quantities: Frequency response function of a filter.

Find: Frequency $\omega_{10\%}$ at which the response amplitude equals 10 percent of its maximum.

Schematics, Diagrams, Circuits, and Given Data: $\omega_1 = 100$; $\omega_2 = 1,000$.

Assumptions: None.

Analysis: The maximum amplitude of the frequency response function is K, which occurs as $\omega \to 0$. As frequency increases, the magnitude of the frequency response function decreases monotonically, which explains why the frequency response function describes a "low-pass" filter. At low frequencies, the input is "passed" to the output; however, at higher frequencies the output is a filtered (reduced) version of the input. To solve this problem, set the amplitude of the frequency response function equal to $0.1K$ and solve for ω, as follows:

$$|\mathbf{H}(j\omega)| = \left| \frac{K}{(j\omega/\omega_1 + 1)(j\omega/\omega_2 + 1)} \right| = 0.1K$$

$$\frac{1}{\sqrt{(1 - \omega^2/\omega_1\omega_2)^2 + \omega^2(1/\omega_1 + 1/\omega_2)^2}} = 0.1$$

To simplify this expression introduce the dummy variable $\Omega = \omega^2$, and then invert and square both sides to obtain a quadratic equation in Ω:

$$\left(1 - \frac{\Omega}{\omega_1\omega_2}\right)^2 + \Omega\left(\frac{1}{\omega_1} + \frac{1}{\omega_2}\right)^2 = 100$$

$$\Omega^2 + \left[(\omega_1\omega_2)^2\left(\frac{1}{\omega_1} + \frac{1}{\omega_2}\right)^2 - 2\omega_1\omega_2\right]\Omega - 99(\omega_1\omega_2)^2 = 0$$

Plug in values for ω_1 and ω_2 and use the quadratic formula to solve for the two roots $\Omega = -1.6208 \times 10^6$ and $\Omega = 0.6108 \times 10^6$. Only the positive root has a physical meaning; thus, the solution is $\omega = \sqrt{\Omega} = 782$ rad/s. Figure 5.18(a) depicts the magnitude response of the filter. At a frequency roughly equal to 800 rad/s, the magnitude response is approximately 0.1. The phase response is shown in Figure 5.18(b).

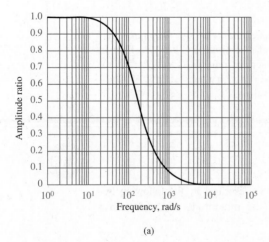

(a)

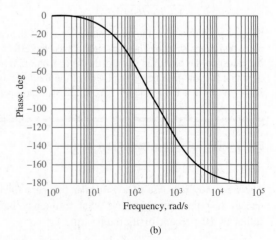

(b)

Figure 5.18 Frequency response of filter of Example 5.5; (a) magnitude response; (b) phase response

EXAMPLE 5.6   Frequency Response of *RC* High-Pass Filter

Problem

Compute the response of the *RC* high-pass filter depicted in Figure 5.19. Evaluate the response of the filter at $\omega_1 = 2\pi \times 100$ and $\omega_2 = 2\pi \times 10,000$ rad/s.

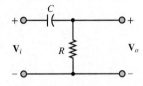

Solution

Known Quantities: $R = 200 \ \Omega$; $C = 0.199 \ \mu$F.

Find: The frequency response $\mathbf{H}_V(j\omega)$.

Assumptions: None.

Figure 5.19 High-pass *RC* filter

Analysis: The cutoff frequency of the high-pass filter is $\omega_0 = 1/RC = 25.126$ kHz $= 2\pi \times 4,000$ rad/s, which is roughly halfway between ω_1 and ω_2. The frequency response function for the circuit is given by equation 5.36:

$$\mathbf{H}_V(j\omega) = \frac{\mathbf{V}_o}{\mathbf{V}_i}(j\omega) = \frac{j\omega CR}{1 + j\omega CR}$$

$$= \frac{\omega/\omega_0}{\sqrt{1 + (\omega/\omega_0)^2}} \angle \left[\frac{\pi}{2} - \arctan\left(\frac{\omega}{\omega_0}\right)\right]$$

The frequency response function can now be evaluated at ω_1 and ω_2:

$$\mathbf{H}_V(\omega = 2\pi \times 100) = \frac{100/4,000}{\sqrt{1 + (100/4,000)^2}} \angle \left[\frac{\pi}{2} - \arctan\left(\frac{100}{4,000}\right)\right] = 0.025 \angle 1.546$$

$$\mathbf{H}_V(\omega = 2\pi \times 10,000) = \frac{10,000/4,000}{\sqrt{1 + (10,000/4,000)^2}} \angle \left[\frac{\pi}{2} - \arctan\left(\frac{10,000}{4,000}\right)\right]$$

$$= 0.929 \angle 0.38$$

These results indicate that the output is very small (2.5 percent) compared to the input at $\omega_1 \ll \omega_0$ while at $\omega_2 \gg \omega_0$ the output is comparable (92.9 percent) to the input. In general, the input is "passed" to the output at high frequencies ($\omega \gg \omega_0$) while at low frequencies ($\omega \ll \omega_0$) the output is a filtered (reduced) version of the input. The complete frequency response (amplitude and phase) is shown in Figure 5.20.

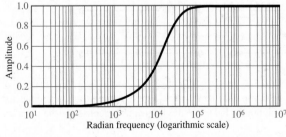

Figure 5.20 Response of high-pass *RC* filter of Example 5.6

Comments: With $\omega_0 = 2\pi \times 4,000$ (that is, 4,000 Hz), this filter would pass only the *treble range* of the audio frequency spectrum.

CHECK YOUR UNDERSTANDING

A simple *RC* low-pass filter is constructed using a 10-μF capacitor and a 2.2-kΩ resistor. Over what range of frequencies will the output of the filter be within 1 percent of the input signal amplitude (i.e., when will $V_o \geq 0.99V_S$)?

Answer: $0 \leq \omega \leq 6.48$ rad/s

CHECK YOUR UNDERSTANDING

In Figure 5.16, let $|\mathbf{V}_S| = 1$ V with an internal resistance $R_S = 50\ \Omega$. Assume $R = 1$ kΩ and $C = 0.47$. Determine the cutoff frequency ω_0 for a load resistance $R_o = 470\ \Omega$.

Answer: $\omega_0 = 6,553.3$ rad/s

CHECK YOUR UNDERSTANDING

Use the phase response plot of Figure 5.18(b) to determine at which frequency the phase shift in the output signal (relative to the input signal) is equal to $-90°$.

Answer: $\omega = 300$ rad/s (approximately)

CHECK YOUR UNDERSTANDING

Determine the cutoff frequency for each of the four "prototype" filters shown below. Which are high pass and which are low pass?

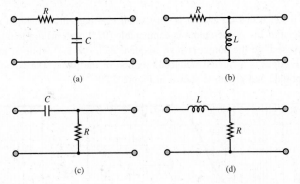

(a)   (b)

(c)   (d)

Show that it is possible to obtain a high-pass filter response simply by substituting an inductor for the capacitor in the circuit of Figure 5.11. Derive the frequency response for the circuit.

$$|H(j\omega)| = \frac{\omega L/R}{\sqrt{1 + (\omega L/R)^2}} \qquad \angle H(j\omega) = 90° + \arctan\left(\frac{-\omega L}{R}\right)$$

Answers: (a) $\omega_0 = \dfrac{1}{RC}$ (low); (b) $\omega_0 = \dfrac{R}{L}$ (high); (c) $\omega_0 = \dfrac{1}{RC}$ (high); (d) $\omega_0 = \dfrac{R}{L}$ (low)

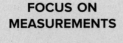

FOCUS ON MEASUREMENTS

Wheatstone Bridge Filter

Problem:

The Wheatstone bridge circuit of Example 1.15 and Focus on Measurements box, "Wheatstone Bridge and Force Measurements," in Chapter 1 is used in a number of instrumentation applications, including the measurement of force. Figure 5.21 depicts the bridge circuit. When undesired noise and interference are present in a measurement, it is often appropriate to use a low-pass filter to reduce the effect of the noise. The capacitor that is connected to the output terminals of the bridge in Figure 5.21 constitutes an effective and simple low-pass filter, in conjunction with the bridge resistance. Assume that the average resistance of each leg of the bridge is 350 Ω (a standard value for strain gauges) and that we desire to measure a sinusoidal force at a frequency of 30 Hz. From prior measurements, it has been determined that a filter with a cutoff frequency of 300 Hz is sufficient to reduce the effects of noise. Choose a capacitor that matches this filtering requirement.

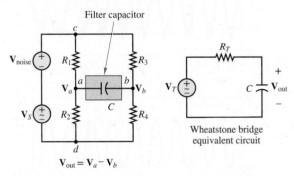

Filter capacitor

$$\mathbf{V}_{out} = \mathbf{V}_a - \mathbf{V}_b$$

Figure 5.21 Wheatstone bridge with equivalent circuit and simple capacitive filter

Solution:

By evaluating the Thévenin equivalent circuit for the Wheatstone bridge, calculating the desired value for the filter capacitor becomes relatively simple, as illustrated on the right side of Figure 5.21. The Thévenin resistance for the bridge circuit may be computed by short-circuiting the two voltage sources and removing the capacitor placed across the load terminals:

$$R_T = R_1 \| R_2 + R_3 \| R_4 = 350 \| 350 + 350 \| 350 = 350 \ \Omega$$

Since the required cutoff frequency is 300 Hz, the capacitor value can be computed from the expression

$$\omega_0 = \frac{1}{R_T C} = 2\pi \times 300$$

or

$$C = \frac{1}{R_T \omega_0} = \frac{1}{350 \times 2\pi \times 300} = 1.51 \ \mu F$$

The frequency response of the bridge circuit is of the same form as equation 5.18:

$$\frac{\mathbf{V}_{out}}{\mathbf{V}_T}(j\omega) = \frac{1}{1 + j\omega C R_T}$$

(Continued)

(Concluded)

This response can be evaluated at the frequency of 30 Hz to verify that the attenuation and phase shift at the desired signal frequency are minimal:

$$\frac{\mathbf{V}_{out}}{\mathbf{V}_T}(j\omega = j2\pi \times 30) = \frac{1}{1 + j2\pi \times 30 \times 1.51 \times 10^{-6} \times 350}$$

$$= 0.9951\angle(-5.7°)$$

Figure 5.22 depicts the appearance of a 30-Hz sinusoidal signal before and after the addition of the capacitor to the circuit.

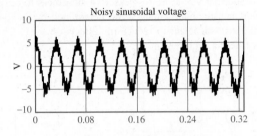

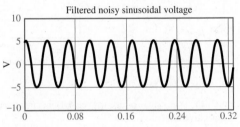

Figure 5.22 Unfiltered and filtered bridge output

5.3 BANDPASS FILTERS, RESONANCE, AND QUALITY FACTOR

Using the same principles and procedures as before, it is possible to derive a **bandpass filter** response for particular types of circuits. Such a filter passes the input to the output at frequencies *within* a certain range. The analysis of a simple *second-order* (i.e., two energy storage elements) bandpass filter is similar to that of low- and high-pass filters. Consider the circuit shown in Figure 5.23 and the designated frequency response function:

$$\mathbf{H}(j\omega) = \frac{\mathbf{V}_o}{\mathbf{V}_i}(j\omega)$$

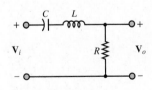

Figure 5.23 *RLC* bandpass filter

Apply voltage division to find:

$$\mathbf{V}_o(j\omega) = \mathbf{V}_i(j\omega)\frac{R}{R + 1/j\omega C + j\omega L}$$

$$= \mathbf{V}_i(j\omega)\frac{j\omega CR}{1 + j\omega CR + (j\omega)^2 LC}$$

(5.39)

Thus, the frequency response function is:

$$\frac{\mathbf{V}_o}{\mathbf{V}_i}(j\omega) = \frac{j\omega CR}{1 + j\omega CR + (j\omega)^2 LC} \tag{5.40}$$

Equation 5.36 can often be factored into the form

$$\frac{\mathbf{V}_o}{\mathbf{V}_i}(j\omega) = \frac{jA\omega}{(j\omega/\omega_1 + 1)(j\omega/\omega_2 + 1)} \tag{5.41}$$

where ω_1 and ω_2 are the two frequencies that determine the **passband** (or **bandwidth**) of the filter—that is, the frequency range over which the filter "passes" the input signal—and A is a constant that results from the factoring. An immediate observation we can make is that if the signal frequency ω is zero, the response of the filter is equal to zero since at $\omega = 0$ the impedance of the capacitor $1/j\omega C$ becomes infinite. Thus, the capacitor acts as an open-circuit, and the output voltage equals zero. Further, we note that the filter output in response to an input signal at sinusoidal frequency approaching infinity is again equal to zero. This result can be verified by considering that as ω approaches infinity, the impedance of the inductor becomes infinite, that is, an open-circuit. Thus, the filter cannot pass signals at very high frequencies. In an intermediate band of frequencies, the bandpass filter circuit will provide a variable attenuation of the input signal, dependent on the frequency of the excitation. This may be verified by taking a closer look at equation 5.39:

$$\begin{aligned}
\mathbf{H}(j\omega) = \frac{\mathbf{V}_o}{\mathbf{V}_i}(j\omega) &= \frac{jA\omega}{(j\omega/\omega_1 + 1)(j\omega/\omega_2 + 1)} \\[2mm]
&= \frac{A\omega\, e^{j\pi/2}}{\sqrt{1 + (\omega/\omega_1)^2}\sqrt{1 + (\omega/\omega_2)^2}\, e^{j\arctan(\omega/\omega_1)} e^{j\arctan(\omega/\omega_2)}} \\[2mm]
&= \frac{A\omega}{\sqrt{[1 + (\omega/\omega_1)^2][1 + (\omega/\omega_2)^2]}}\, e^{j[\pi/2 - \arctan(\omega/\omega_1) - \arctan(\omega/\omega_2)]}
\end{aligned} \tag{5.42}$$

Equation 5.36 is of the form $\mathbf{H}(j\omega) = |H| e^{j\angle H}$, with

$$|\mathbf{H}(j\omega)| = \frac{A\omega}{\sqrt{\left[1 + (\omega/\omega_1)^2\right]\left[1 + (\omega/\omega_2)^2\right]}}$$

and

$$\angle H(j\omega) = \frac{\pi}{2} - \arctan\frac{\omega}{\omega_1} - \arctan\frac{\omega}{\omega_2} \tag{5.43}$$

The magnitude and phase plots for the frequency response of the bandpass filter of Figure 5.23 are shown in Figure 5.24. These plots have been normalized to have the filter passband centered at the frequency $\omega = 1$ rad/s.

The frequency response plots of Figure 5.24 suggest that, in some sense, the bandpass filter acts as a combination of a high-pass and a low-pass filter. As illustrated in the previous cases, it should be evident that one can adjust the filter response as desired simply by selecting appropriate values for L, C, and R.

Resonance and Bandwidth

The response of second-order filters can be explained more generally by rewriting the frequency response function of the second-order bandpass filter of Figure 5.23

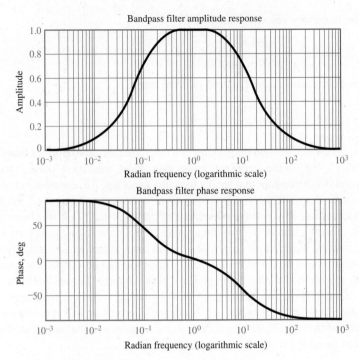

Figure 5.24 Frequency response of *RLC* bandpass filter

in the following forms:

$$\frac{\mathbf{V}_o}{\mathbf{V}_i}(j\omega) = \frac{j\omega CR}{LC(j\omega)^2 + j\omega CR + 1}$$

$$= \frac{(2\zeta/\omega_n)j\omega}{(j\omega/\omega_n)^2 + (2\zeta/\omega_n)j\omega + 1} \quad (5.44)$$

$$= \frac{(1/Q\omega_n)j\omega}{(j\omega/\omega_n)^2 + (1/Q\omega_n)j\omega + 1}$$

with the following definitions:[2]

$$\omega_n = \sqrt{\frac{1}{LC}} = \text{natural or resonant frequency}$$

$$Q = \frac{1}{2\zeta} = \frac{1}{\omega_n CR} = \omega_n \frac{L}{R} = \frac{1}{R}\sqrt{\frac{L}{C}} = \text{quality factor} \quad (5.45)$$

$$\zeta = \frac{1}{2Q} = \frac{R}{2}\sqrt{\frac{C}{L}} = \text{damping ratio}$$

---

[2] If you have already studied the section on second-order transient response in Chapter 4, you will recognize the parameters ζ and ω_n.

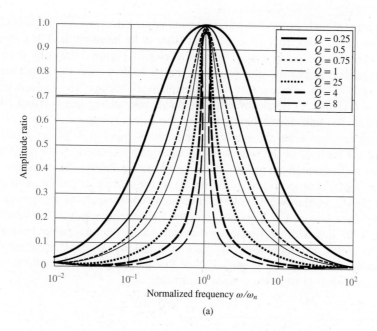

(a)

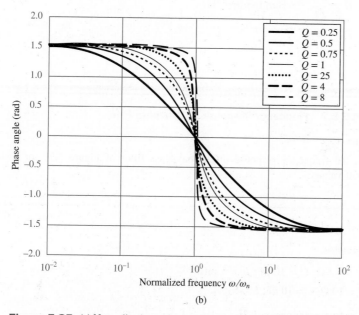

(b)

Figure 5.25 (a) Normalized magnitude response of second-order bandpass filter; (b) normalized phase response of second-order bandpass filter

Figure 5.25 depicts the normalized frequency response (magnitude and phase) of the second-order bandpass filter for $\omega_n = 1$ and various values of Q (and ζ). The peak displayed in the frequency response around the frequency ω_n is called a *resonant peak*, and ω_n is the **resonant frequency**. Note that as the **quality factor** Q increases, the sharpness of the resonance increases and the filter becomes increasingly *selective* (i.e., it has the ability to filter out most frequency components of the input signals except for a narrow band around the resonant frequency). One measure of the selectivity of a bandpass filter is its **bandwidth**. The concept of bandwidth can be easily visualized in the plot of Figure 5.25(a) by drawing a horizontal line across the plot (we have chosen to draw it at the amplitude ratio value of 0.707 for reasons that will be explained shortly). The frequency range between (magnitude) frequency response points intersecting this horizontal line is defined as the **half-power bandwidth** of the filter. The name *half-power* stems from the fact that when the amplitude response is equal to 0.707 (or $1/\sqrt{2}$), the voltage (or current) at the output of the filter has decreased by the same factor, relative to the maximum value (at the resonant frequency). Since power in an electric signal is proportional to the square of the voltage or current, a drop by a factor $1/\sqrt{2}$ in the output voltage or current corresponds to the power being reduced by a factor of $\frac{1}{2}$. Thus, we term the frequencies at which the intersection of the 0.707 line with the frequency response occurs the **half-power frequencies**. Another useful definition of bandwidth B is as follows. We shall make use of this definition in the following examples. Note that a high-Q filter has a narrow bandwidth and a low-Q filter has a wide bandwidth.

$$B = \frac{\omega_n}{Q} \qquad \text{Bandwidth}$$

(5.46)

EXAMPLE 5.7   **Frequency Response of Bandpass Filter**

Problem

Compute the frequency response of the bandpass filter of Figure 5.23 for two sets of component values.

Solution

Known Quantities:

   a. $R = 1\ \text{k}\Omega$; $C = 10\ \mu\text{F}$; $L = 5\ \text{mH}$.

   b. $R = 10\ \Omega$; $C = 10\ \mu\text{F}$; $L = 5\ \text{mH}$.

Find: The frequency response $\mathbf{H}_V(j\omega)$.

Assumptions: None.

Analysis: We write the frequency response of the bandpass filter as in equation 5.38:

$$\mathbf{H}_V(j\omega) = \frac{\mathbf{V}_o}{\mathbf{V}_i}(j\omega) = \frac{j\omega CR}{1 + j\omega CR + (j\omega)^2 LC}$$

$$= \frac{\omega CR}{\sqrt{(1 - \omega^2 LC)^2 + (\omega CR)^2}} \angle \left[\frac{\pi}{2} - \arctan\left(\frac{\omega CR}{1 - \omega^2 LC} \right) \right]$$

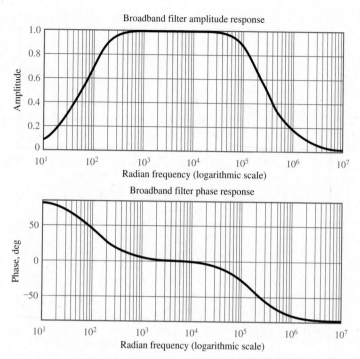

Figure 5.26 Frequency response of broadband bandpass filter of Example 5.7

We can now evaluate the response for two different values of the series resistance. The frequency response plots for case a (large series resistance) are shown in Figure 5.26. Those for case b (small series resistance) are shown in Figure 5.27. Let us calculate some quantities for each case. Since L and C are the same in both cases, the *resonant frequency* of the two circuits will be the same:

$$\omega_n = \frac{1}{\sqrt{LC}} = 4.47 \times 10^3 \text{ rad/s}$$

On the other hand, the *quality factor Q* will be substantially different:

a. $Q_a = \dfrac{1}{\omega_n CR} \approx 0.02235$

b. $Q_b = \dfrac{1}{\omega_n CR} \approx 2.235$

From these values of Q we can calculate the approximate bandwidth of the two filters:

$$B_a = \frac{\omega_n}{Q_a} \approx 200{,}000 \text{ rad/s} \qquad \text{case a}$$

$$B_b = \frac{\omega_n}{Q_b} \approx 2{,}000 \text{ rad/s} \qquad \text{case b}$$

The frequency response plots in Figures 5.26 and 5.27 confirm these observations.

Comments: It should be apparent that while at the higher and lower frequencies most of the amplitude of the input signal is filtered from the output, at the midband frequency

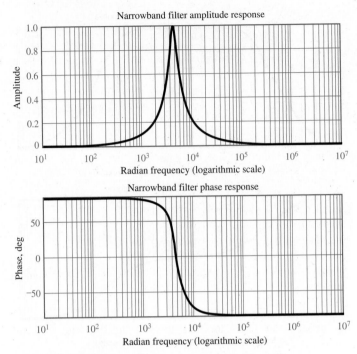

Figure 5.27 Frequency response of narrowband bandpass filter of Example 5.7

(4,500 rad/s) most of the input signal amplitude passes through the filter. The first bandpass filter analyzed in this example would "pass" the *midband range* of the audio spectrum while the second would pass only a very narrow band of frequencies around the **center frequency** of 4,500 rad/s. Such narrowband filters find application in **tuning circuits**, such as those employed in conventional AM radios (although at frequencies much higher than that of the present example). In a tuning circuit, a narrowband filter is used to tune in a frequency associated with the **carrier wave** of a radio station (e.g., for a station found at a setting of AM 820, the carrier wave transmitted by the radio station is at a frequency of 820 kHz). By using a variable capacitor, it is possible to tune in a range of carrier frequencies and therefore select the preferred station. Other circuits are then used to decode the actual speech or music signal modulated on the carrier wave.

CHECK YOUR UNDERSTANDING

Compute the frequencies ω_1 and ω_2 for the bandpass filter of Example 5.7 (with $R = 1$ kΩ) by equating the magnitude of the bandpass filter frequency response to $1/\sqrt{2}$. The result is a quadratic equation in ω, which can be solved for two frequencies, known as the **half-power frequencies**.

Answer: $\omega_1 = 99.95$ rad/s; $\omega_2 = 200.1$ krad/s

5.4   BODE PLOTS

Frequency response plots of linear systems are often displayed in the form of logarithmic plots, called **Bode plots** after the mathematician Hendrik W. Bode, where the horizontal axis represents frequency on a logarithmic scale (base 10) and the vertical axis represents either the amplitude or phase of the frequency response function. In Bode plots the amplitude is expressed in units of **decibels (dB)**, where

$$\left|\frac{A_o}{A_i}\right|_{dB} = 20\log_{10}\left|\frac{A_o}{A_i}\right| = 20\log_{10}\frac{|A_o|}{|A_i|} \tag{5.47}$$

While logarithmic plots may at first seem a daunting complication, they have two significant advantages:

1. The product of terms in a frequency response function becomes a sum of terms because $\log(ab/c) = \log(a) + \log(b) - \log(c)$. The advantage here is that Bode (logarithmic) plots can be constructed from the sum of individual plots of individual terms. Moreover, as was discussed in Section 5.1, there are only four distinct types of terms present in any frequency response function:

   a. A constant K.

   b. Poles or zeros "at the origin" $(j\omega)$.

   c. Simple poles or zeros $(1 + j\omega\tau)$ or $(1 + j\omega/\omega_0)$.

   d. Quadratic poles or zeros $[1 + j\omega\tau + (j\omega/\omega_n)^2]$.

2. The individual Bode plots of these four distinct terms are all well approximated by linear segments, which are readily summed to form the overall Bode plot of more complicated frequency response functions.

RC Low-Pass Filter Bode Plots

Consider, for example, the *RC* low-pass filter of Example 5.3 (Figure 5.11). The frequency response function is:

$$\frac{\mathbf{V}_o}{\mathbf{V}_i}(j\omega) = \frac{1}{1 + j\omega/\omega_0} = \frac{1}{\sqrt{1 + (\omega/\omega_0)^2}}\angle -\tan^{-1}\left(\frac{\omega}{\omega_0}\right) \tag{5.48}$$

where the circuit time constant is $\tau = RC = 1/\omega_0$ and ω_0 is the cutoff, or half-power, frequency of the filter. This frequency response function has a constant of value $K = 1$ and a simple pole with cutoff frequency $\omega_0 = 1/\tau = 1/RC$.

Figure 5.28 shows the Bode magnitude and phase plots for the filter. The normalized frequency on the horizontal axis is $\omega\tau$. The magnitude plot is obtained from the logarithmic form of the absolute value of the frequency response function:

$$\left|\frac{\mathbf{V}_o}{\mathbf{V}_i}\right|_{dB} = 20\log_{10}\frac{|K|}{|1 + j\omega\tau|} = 20\log_{10}\frac{|K|}{|1 + j\omega/\omega_0|} \tag{5.49}$$

When $\omega \ll \omega_0$, the imaginary part of the simple pole is much smaller than its real part, such that $|1 + j\omega/\omega_0| \approx 1$. Then:

$$\left|\frac{\mathbf{V}_o}{\mathbf{V}_i}\right|_{dB} \approx 20\log_{10}K - 20\log_{10}1 = 20\log_{10}K \quad (dB) \tag{5.50}$$

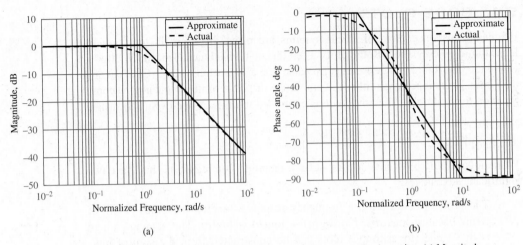

Figure 5.28 Bode plots for a low-pass RC filter; the frequency variable is normalized to ω/ω_0. (a) Magnitude response; (b) phase angle response

Thus, at very low frequencies ($\omega \ll \omega_0$), equation 5.49 is well approximated by a straight line of zero slope, which is the *low-frequency asymptote* of the Bode magnitude plot.

When $\omega \gg \omega_0$, the imaginary part of the simple pole is much larger than its real part, such that $|1 + j\omega/\omega_0| \approx |j\omega/\omega_0| = (\omega/\omega_0)$. Then:

$$\left|\frac{\mathbf{V}_o}{\mathbf{V}_i}\right|_{\text{dB}} \approx 20 \log_{10} K - 20 \log_{10} \frac{\omega}{\omega_0}$$
$$\approx 20 \log_{10} K - 20 \log_{10} \omega + 20 \log_{10} \omega_0 \tag{5.51}$$

Thus, at very high frequencies ($\omega \gg \omega_0$), equation 5.49 is well approximated by a straight line of -20 dB per **decade** slope that intercepts the log ω axis at log ω_0. This line is the *high-frequency asymptote* of the Bode magnitude plot. A decade represents a factor of 10 change in frequency. Thus, a one-decade increase in ω is equivalent to a unity change in log ω.

Finally, when $\omega = \omega_0$, the real and imaginary parts of the simple pole are equal, such that $|1 + j\omega/\omega_0| = |1 + j| = \sqrt{2}$. Then equation 5.49 becomes:

$$20 \log_{10} \frac{|K|}{|1 + j\omega/\omega_0|} = 20 \log_{10} K - 20 \log_2 1/2 = 20 \log_{10} K - 3\,\text{dB} \quad (5.52)$$

Thus, the Bode magnitude plot of a first-order low-pass filter is approximated by two straight lines intersecting at ω_0. Figure 5.28(a) clearly shows the approximation. The actual Bode magnitude plot is 3 dB lower than the approximate plot at $\omega = \omega_0$, the cutoff frequency.

The phase angle of the frequency response function $\angle(\mathbf{V}_o/\mathbf{V}_i) = -\tan^{-1}(\omega/\omega_0)$ has the following properties:

$$-\tan^{-1}\left(\frac{\omega}{\omega_0}\right) = \begin{cases} 0 & \text{when } \omega \to 0 \\ -\dfrac{\pi}{4} & \text{when } \omega = \omega_0 \\ -\dfrac{\pi}{2} & \text{when } \omega \to \infty \end{cases}$$

As a first approximation, the phase angle can be represented by three straight lines:

1. For $\omega < 0.1\omega_0$, $\angle(\mathbf{V}_o/\mathbf{V}_i) \approx 0$.

2. For $0.1\omega_0$ and $10\omega_0$, $\angle(\mathbf{V}_o/\mathbf{V}_i) \approx -(\pi/4)\log(10\omega/\omega_0)$.

3. For $\omega > 10\omega_0$, $\angle(\mathbf{V}_o/\mathbf{V}_i) \approx -\pi/2$.

These straight-line approximations are illustrated in Figure 5.28(b).

Table 5.1 lists the differences between the actual and approximate Bode magnitude and phase plots. Note that the maximum difference in magnitude is 3 dB at the cutoff frequency; thus, the cutoff is often called the **3-dB frequency** or the *half-power frequency*.

Table 5.1 **Correction factors for asymptotic approximation of first-order filter**

ω/ω_0	Magnitude response error (dB)	Phase response error (deg)
0.1	0	−5.7
0.5	−1	4.9
1	−3	0
2	−1	−4.9
10	0	+5.7

RC High-Pass Filter Bode Plots

The case of an *RC* high-pass filter (see Figure 5.13) is analyzed in the same manner as was done for the *RC* low-pass filter. The frequency response function is:

$$
\begin{aligned}
\frac{\mathbf{V}_o}{\mathbf{V}_o} &= \frac{j\omega CR}{1 + j\omega CR} = \frac{j(\omega/\omega_0)}{1 + j(\omega/\omega_0)} \\
&= \frac{(\omega/\omega_0)\angle(\pi/2)}{\sqrt{1 + (\omega/\omega_0)^2}\,\angle\arctan(\omega/\omega_0)} \\
&= \frac{\omega/\omega_0}{\sqrt{1 + (\omega/\omega_0)^2}}\angle\left(\frac{\pi}{2} - \arctan\frac{\omega}{\omega_0}\right)
\end{aligned}
\tag{5.53}
$$

Figure 5.29 depicts the Bode plots for equation 5.53, where the horizontal axis indicates the normalized frequency ω/ω_0. Straight-line asymptotic approximations may again be determined easily at low and high frequencies. The results are very similar to those for the first-order low-pass filter. For $\omega < \omega_0$, the Bode magnitude approximation intercepts the origin ($\omega = 1$) with a slope of +20 dB/decade. For $\omega > \omega_0$, the Bode magnitude approximation is 0 dB with zero slope. The straight-line approximations of the Bode phase plot are:

1. For $\omega < 0.1\omega_0$, $\angle(\mathbf{V}_o/\mathbf{V}_i) \approx \pi/2$.

2. For $0.1\omega_0$ and $10\omega_0$, $\angle(\mathbf{V}_o/\mathbf{V}_i) \approx -(\pi/4)\log(10\omega/\omega_0)$.

3. For $\omega > 10\omega_0$, $\angle(\mathbf{V}_o/\mathbf{V}_i) \approx 0$.

These straight-line approximations are illustrated in Figure 5.29(b).

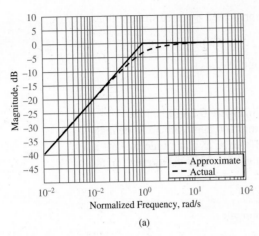
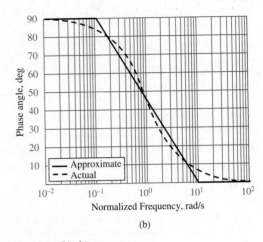

(a)   (b)

Figure 5.29 Bode plots for *RC* high-pass filter; (a) magnitude response; (b) phase response

Bode Plots of Higher-Order Filters

Bode plots of high-order systems may be obtained by combining Bode plots of factors of the higher-order frequency response function. Let, for example,

$$\mathbf{H}(j\omega) = \mathbf{H}_1(j\omega)\mathbf{H}_2(j\omega)\mathbf{H}_3(j\omega) \tag{5.54}$$

which can be expressed, in logarithmic form, as

$$|\mathbf{H}(j\omega)|_{dB} = |\mathbf{H}_1(j\omega)|_{dB} + |\mathbf{H}_2(j\omega)|_{dB} + |\mathbf{H}_3(j\omega)|_{dB} \tag{5.55}$$

and

$$\angle\mathbf{H}(j\omega) = \angle\mathbf{H}_1(j\omega) + \angle\mathbf{H}_2(j\omega) + \angle\mathbf{H}_3(j\omega) \tag{5.56}$$

Consider as an example the frequency response function

$$\mathbf{H}(j\omega) = \frac{j\omega + 5}{(j\omega + 10)(j\omega + 100)} \tag{5.57}$$

The first step in computing the asymptotic approximation consists of factoring each term in the expression so that it appears in the form $a_i(j\omega/\omega_i + 1)$, where the frequency ω_i corresponds to the appropriate 3-dB frequency, ω_1, ω_2, or ω_3. For example, the function of equation 5.57 is rewritten as:

$$\mathbf{H}(j\omega) = \frac{5(j\omega/5 + 1)}{10(j\omega/10 + 1)100(j\omega/100 + 1)}$$

$$= \frac{0.005(j\omega/5 + 1)}{(j\omega/10 + 1)(j\omega/100 + 1)} = \frac{K(j\omega/\omega_1 + 1)}{(j\omega/\omega_2 + 1)(j\omega/\omega_3 + 1)} \tag{5.58}$$

Equation 5.58 can now be expressed in logarithmic form:

$$|\mathbf{H}(j\omega)|_{dB} = |0.005|_{dB} + \left|\frac{j\omega}{5} + 1\right|_{dB} - \left|\frac{j\omega}{10} + 1\right|_{dB} - \left|\frac{j\omega}{100} + 1\right|_{dB}$$

$$\angle\mathbf{H}(j\omega) = \angle 0.005 + \angle\left(\frac{j\omega}{5} + 1\right) - \angle\left(\frac{j\omega}{10} + 1\right) - \angle\left(\frac{j\omega}{100} + 1\right) \tag{5.59}$$

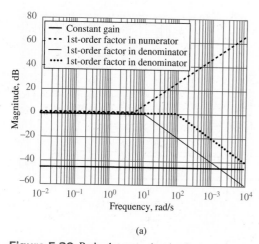

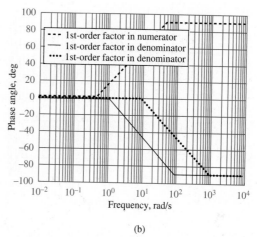

(a)

(b)

Figure 5.30 Bode plot approximation for a second-order frequency response function; (a) straight-line approximation of magnitude response; (b) straight-line approximation of phase angle response

Each of the terms in the logarithmic magnitude expression can be plotted individually. The constant corresponds to the value -46 dB, plotted in Figure 5.30(a) as a line of zero slope. The numerator term, with a 3-dB frequency $\omega_1 = 5$, is expressed in the form of the first-order Bode plot of Figure 5.28(a), except for the fact that the slope of the line leaving the zero axis at $\omega_1 = 5$ is $+20$ dB/decade; each of the two denominator factors is similarly plotted as lines of slope -20 dB/decade, departing the zero axis at $\omega_2 = 10$ and $\omega_3 = 100$. You see that the individual factors are very easy to plot by *inspection* once the frequency response function has been normalized in the form of equation 5.55.

If we now consider the phase response portion of equation 5.59, we recognize that the first term, the phase angle of the constant, is always zero. The numerator first-order term, on the other hand, can be approximated as shown in Figure 5.28(b), that is, by drawing a straight line starting at $0.1\omega_1 = 0.5$, with slope $+\pi/4$ rad/ decade (*positive because this is a numerator factor*) and ending at $10\omega_1 = 50$, where the asymptote $+\pi/2$ is reached. The two denominator terms have similar behavior, except for the fact that the slope is $-\pi/4$ and that the straight line with slope $-\pi/4$ rad/decade extends between the frequencies $0.1\omega_2$ and $10\omega_2$, and $0.1\omega_3$ and $10\omega_3$, respectively.

Figure 5.30 depicts the asymptotic approximations of the individual factors in equation 5.59, with the magnitude factors shown in Figure 5.30(a) and the phase factors in Figure 5.30(b). When all the asymptotic approximations are combined, the complete frequency response approximation is obtained. Figure 5.31 depicts the results of the asymptotic Bode approximation when compared with the actual frequency response functions.

You can see that once a frequency response function is factored into the appropriate form, it is relatively easy to sketch a good approximation of the Bode plot, even for higher-order frequency response functions. Examples 5.8 and 5.9 illustrate some additional details. The methodology is summarized in the box below.

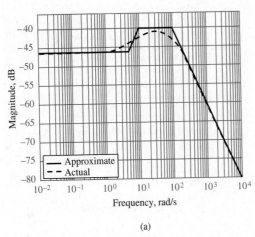

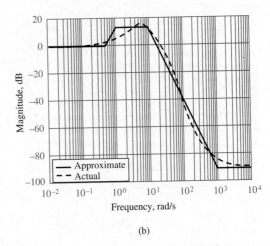

(a)

(b)

Figure 5.31 Comparison of Bode plot approximation with the actual frequency response function; (a) magnitude response of second-order frequency response function; (b) phase angle response of second-order frequency response function.

FOCUS ON PROBLEM SOLVING

BODE PLOTS

This box illustrates the Bode plot asymptotic approximation construction procedure. The method assumes that there are no complex conjugate factors in the response and that both the numerator and denominator can be factored into first-order terms with real roots.

1. Express the frequency response function in factored form, resulting in an expression similar to equation 5.43:

$$H(j\omega) = \frac{K(j\omega/\omega_1 + 1)\cdots(j\omega/\omega_m + 1)}{(j\omega/\omega_{m+1} + 1)\cdots(j\omega/\omega_n + 1)}$$

2. Select the appropriate frequency range for the semilogarithmic plot, extending at least a decade below the lowest 3-dB frequency and a decade above the highest 3-dB frequency.

3. Sketch the magnitude and phase response asymptotic approximations for each of the first-order factors, using the techniques illustrated in Figures 5.28 to 5.31.

4. Add, graphically, the individual terms to obtain a composite response.

5. If desired, apply the correction factors of Table 5.1.

EXAMPLE 5.8   Bode Plot Approximation

Problem

Sketch the asymptotic approximation of the Bode plot for the frequency response function

$$\mathbf{H}(j\omega) = \frac{0.1j\omega + 20}{2\times10^{-5}(j\omega)^3 + 0.1002(j\omega)^2 + j\omega}$$

Solution

Known Quantities: Frequency response function of a circuit.

Find: Bode plot approximation of given frequency response function.

Assumptions: None

Analysis: Following the Focus on Problem Solving box, "Bode Plots," we first factor the function into the standard form

$$\mathbf{H}(j\omega) = \frac{K(j\omega/\omega_1 + 1) \cdots (j\omega/\omega_m + 1)}{(j\omega/\omega_{m+1} + 1) \cdots (j\omega/\omega_n + 1)}$$

After a little algebra, we can obtain the following frequency response function in standard form:

$$\mathbf{H}(j\omega) = \frac{20(j\omega/200 + 1)}{j\omega(j\omega/10 + 1)(j\omega/5{,}000 + 1)}$$

We immediately notice that there is a factor of $j\omega$ in the denominator; this term needs to be treated somewhat differently. The Bode plot of the function $1/j\omega$ can be expressed in logarithmic form as follows:

$$\left|\frac{1}{j\omega}\right|_{\text{dB}} = -20\log_{10}\frac{\omega}{1}$$

$$\angle\frac{1}{j\omega} = 0 - \frac{\pi}{2} = -\frac{\pi}{2}$$

That is, the magnitude of the denominator factor $j\omega$ is represented by a line with slope of -20 dB/decade intersecting the frequency (horizontal) axis at $\omega = 1$. Its phase response is a constant equal to $-\pi/2$.

Now we can sketch the magnitude and phase response of each of the individual first-order factors, as shown in Figure 5.32(a) and (b). The composite asymptotic approximations of the magnitude and phase responses are shown in Figure 5.33(a) and (b).

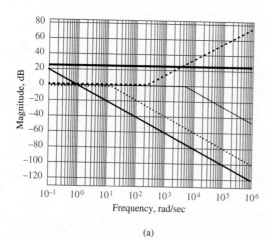

(a)

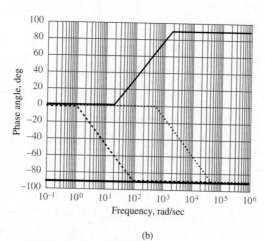

(b)

Figure 5.32 Approximate (asymptotic) frequency response of individual first-order terms; (a) straight-line approximation of magnitude response; (b) straight-line approximation of phase angle response.

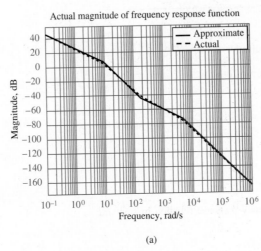

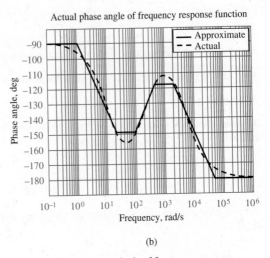

(a)   (b)

Figure 5.33 Comparison of approximate and exact frequency response; (a) actual magnitude of frequency response function; (b) actual phase angle of frequency response function.

Comments: A computer program can be used to generate the Bode plot approximation shown in Figures 5.32 and 5.33. Note that the only real effort in generating the asymptotic approximation lies in the factoring of the frequency response function.

EXAMPLE 5.9   Bode Plot Approximation

Problem

Sketch the asymptotic approximation of the Bode plot for the frequency response function

$$\mathbf{H}(j\omega) = \frac{10^{-3}(j\omega)^2 + 0.1j\omega}{[1/(9 \times 10^4)](j\omega)^2 + (3,030/90,000)j\omega + 1}$$

Solution

Known Quantities: Frequency response function of a circuit.

Find: Bode plot approximation of given frequency response function.

Assumptions: None

Analysis: Following the Focus on Problem Solving box, "Bode Plots," factor the function into standard form.

$$\mathbf{H}(j\omega) = \frac{K(j\omega/\omega_1 + 1) \cdots (j\omega/\omega_m + 1)}{(j\omega/\omega_{m+1} + 1) \cdots (j\omega/\omega_n + 1)}$$

After a little algebra, frequency response in standard form can be found as:

$$\mathbf{H}(j\omega) = \frac{0.1 j\omega(j\omega/100 + 1)}{(j\omega/30 + 1)(j\omega/3,000 + 1)}$$

Notice the $j\omega$ factor in the numerator. This factor can be expressed in logarithmic form as:

$$|j\omega|_{\text{dB}} = 20 \log_{10} \frac{\omega}{1}$$

$$\angle j\omega = \frac{\pi}{2}$$

That is, the magnitude of $j\omega$ is represented by a line with slope $+20$ dB/decade intersecting the frequency (horizontal) axis at $\omega = 1$. The phase of the factor $j\omega$ is a constant and equal to $\pi/2$.

The magnitude and phase of each individual first-order factor can be sketched as shown in Figure 5.34(a) and (b). The composite asymptotic approximations of the magnitude and phase responses are shown in Figure 5.35(a) and (b).

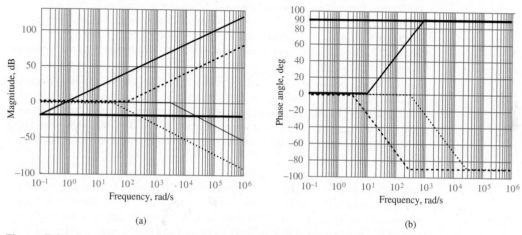

(a)   (b)

Figure 5.34 Approximate (asymptotic) frequency response of individual first-order terms; (a) straight-line approximation of magnitude response; (b) straight-line approximation of phase angle response.

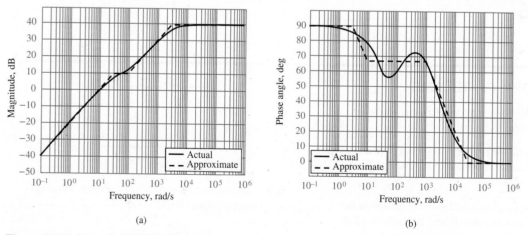

(a)   (b)

Figure 5.35 Comparison of approximate and exact frequency responses; (a) actual magnitude of frequency response function; (b) actual phase angle of frequency response function.

Comments: Bode plots can be generated using Matlab™. Circuit simulation programs, such as B2Spice, can also generate Bode plots.

Conclusion

Chapter 5 focuses on the frequency response of linear circuits, and it is a natural extension of the material covered in Chapter 3. The concepts of the spectrum of a signal, obtained through the Fourier series representation for periodic signals, and of the frequency response of a filter are very useful ideas that extend well beyond electrical engineering. For example, civil, mechanical, and aeronautical engineering students who study the vibrations of structures and machinery will find that the same methods are employed in those fields.

Upon completing this chapter, you should have mastered the following learning objectives:

1. *Understand the physical significance of frequency domain analysis, and compute the frequency response of circuits by using AC circuit analysis tools.* You had already acquired the necessary tools (phasor analysis and impedance) to compute the frequency response of circuits in Chapter 3; in the material presented in Section 5.1, these tools are put to use to determine the frequency response functions of linear circuits.

2. *Analyze simple first- and second-order electrical filters, and determine their frequency response and filtering properties.* With the concept of *frequency response* firmly in hand, now you can analyze the behavior of electrical filters and study the frequency response characteristics of the most common types, that is, low-pass, high-pass, and bandpass filters. Filters are very useful devices and are explored in greater depth in Chapter 7.

3. *Compute the frequency response of a circuit and its graphical representation in the form of a Bode plot.* Graphical approximations of Bode plots can be very useful to develop a quick understanding of the frequency response characteristics of a linear system, almost by inspection. Bode plots find use in the discipline of automatic control systems, a subject that is likely to be encountered by most engineering majors.

HOMEWORK PROBLEMS

Section 5.1: Sinusoidal Frequency Response

5.1 a. Determine the frequency response $V_{out}(j\omega)/V_{in}(j\omega)$ for the circuit of Figure P5.1. Assume $L = 0.5$ H and $R = 200$ kΩ.

b. Plot the magnitude and phase of the circuit for frequencies between 10 and 10^7 rad/s on graph paper, with a linear scale for frequency.

c. Repeat part b, using semilog paper. (Place the frequency on the logarithmic axis.)

d. Plot the magnitude response on semilog paper with magnitude in decibels.

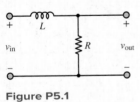

Figure P5.1

5.2 Repeat the instructions of Problem 5.1 for the circuit of Figure P5.2.

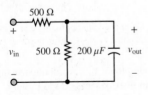

Figure P5.2

5.3   Repeat the instructions of Problem 5.1 for the circuit of Figure P5.3.

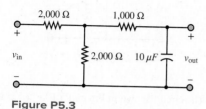

Figure P5.3

5.4   Repeat Problem 5.1 for the circuit of Figure P5.4. $R_1 = 300\ \Omega$, $R_2 = R_3 = 500\ \Omega$, $L = 4$ H, $C_1 = 40\ \mu$F, $C_2 = 160\ \mu$F.

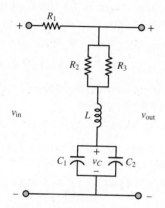

Figure P5.4

5.5   Determine the frequency response of the circuit of Figure P5.5, and generate frequency response plots. $R_1 = 20\ \text{k}\Omega$, $R_2 = 100\ \text{k}\Omega$, $L = 1$ H, $C = 100\ \mu$F.

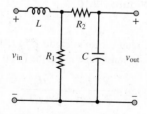

Figure P5.5

5.6   In the circuit shown in Figure P5.6, where $C = 0.5\ \mu$F and $R = 2\ \text{k}\Omega$,

a. Determine how the input impedance $\mathbf{Z}(j\omega) = \mathbf{V}_i(j\omega)/\mathbf{I}_i(j\omega)$ behaves at extremely high and low frequencies.

b. Find an expression for the impedance.

c. Show that this expression can be manipulated into the form $\mathbf{Z}(j\omega) = R[1 + j(1/\omega RC)]$.

d. Determine the frequency $\omega = \omega_C$ for which the imaginary part of the expression in part c is equal to 1.

e. Estimate (without computing it) the magnitude and phase angle of $\mathbf{Z}(j\omega)$ at $\omega = 10$ rad/s and $\omega = 10^5$ rad/s.

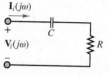

Figure P5.6

5.7   In the circuit shown in Figure P5.7, where $L = 2$ mH and $R = 2\ \text{k}\Omega$,

a. Determine how the input impedance $\mathbf{Z}(j\omega) = \mathbf{V}_i(j\omega)/\mathbf{I}_i(j\omega)$ behaves at extremely high and low frequencies.

b. Find an expression for the impedance.

c. Show that this expression can be manipulated into the form $\mathbf{Z}(j\omega) = R[1 + j(\omega L/R)]$.

d. Determine the frequency $\omega = \omega_C$ for which the imaginary part of the expression in part c is equal to 1.

e. Estimate (without computing it) the magnitude and phase angle of $\mathbf{Z}(j\omega)$ at $\omega = 10^5$ rad/s, 10^6 rad/s, and 10^7 rad/s.

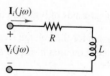

Figure P5.7

5.8   In the circuit shown in Figure P5.8, if

$$L = 190\ \text{mH} \qquad R_1 = 2.3\ \text{k}\Omega$$
$$C = 55\ \text{nF} \qquad R_2 = 1.1\ \text{k}\Omega$$

a. Determine how the input impedance behaves at extremely high or low frequencies.

b. Find an expression for the input impedance in the form

$$\mathbf{Z}(j\omega) = Z_o \left[\frac{1 + jf_1(\omega)}{1 + jf_2(\omega)} \right]$$

$$Z_o = R_1 + \frac{L}{R_2 C}$$

$$f_1(\omega) = \frac{\omega^2 R_1 LC - R_1 - R_2}{\omega(R_1 R_2 C + L)}$$

$$f_2(\omega) = \frac{\omega^2 LC - 1}{\omega C R_2}$$

c. Determine the four frequencies at which $f_1(\omega) = +1$ or -1 and $f_2(\omega) = +1$ or -1.

d. Plot the impedance (magnitude and phase) versus frequency.

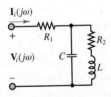

Figure P5.8

5.9 In the circuit of Figure P5.9:

$$R_1 = 1.3 \text{ k}\Omega \qquad R_2 = 1.9 \text{ k}\Omega$$
$$C = 0.5182 \ \mu F$$

Determine:

a. How the voltage frequency response function

$$\mathbf{H}_V(j\omega) = \frac{\mathbf{V}_o(j\omega)}{\mathbf{V}_i(j\omega)}$$

behaves at extremes of high and low frequencies.

b. An expression for the voltage frequency response function and show that it can be manipulated into the form

$$\mathbf{H}_v(j\omega) = \frac{H_o}{1 + jf(\omega)}$$

where

$$H_o = \frac{R_2}{R_1 + R_2} \qquad f(\omega) = \frac{\omega R_1 R_2 C}{R_1 + R_2}$$

c. The frequency at which $f(\omega) = 1$ and the value of H_o in decibels.

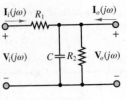

Figure P5.9

5.10 The circuit shown in Figure P5.10 is a second-order circuit because it has two reactive components (L and C). A complete solution will not be attempted. However, determine:

a. The behavior of the voltage frequency response at extremely high and low frequencies.

b. The output voltage $\mathbf{V}_o$ if the input voltage has a frequency where:

$$\mathbf{V}_i = 7.07\angle\frac{\pi}{4} \text{ V} \qquad R_1 = 2.2 \text{ k}\Omega$$
$$R_2 = 3.8 \text{ k}\Omega \qquad X_c = 5 \text{ k}\Omega \qquad X_L = 1.25 \text{ k}\Omega$$

c. The output voltage if the frequency of the input voltage doubles so that

$$X_C = 2.5 \text{ k}\Omega \qquad X_L = 2.5 \text{ k}\Omega$$

d. The output voltage if the frequency of the input voltage again doubles so that

$$X_C = 1.25 \text{ k}\Omega \qquad X_L = 5 \text{ k}\Omega$$

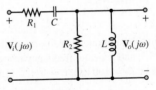

Figure P5.10

5.11 In the circuit shown in Figure P5.11, determine the frequency response function in the form:

$$\mathbf{H}_v(j\omega) = \frac{\mathbf{V}_o(j\omega)}{\mathbf{V}_i(j\omega)} = \frac{H_{vo}}{1 \pm jf(\omega)}$$

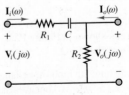

Figure P5.11

5.12   The circuit shown in Figure P5.12 has

$$R_1 = 100 \ \Omega \qquad R_o = 100 \ \Omega$$
$$R_2 = 50 \ \Omega \qquad C = 80 \ nF$$

Determine the frequency response $V_o(j\omega)/V_{in}(j\omega)$.

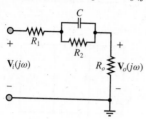

Figure P5.12

5.13

a. Determine the frequency response
   $V_{out}(j\omega)/V_{in}(j\omega)$ for the circuit of Figure P5.13.

b. Plot the magnitude and phase of the circuit for
   frequencies between 1 and 100 rad/s on graph
   paper, with a linear scale for frequency.

c. Repeat part b, using semilog paper. (Place the
   frequency on the logarithmic axis.)

d. Plot the magnitude response on semilog paper with
   magnitude in dB.

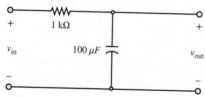

Figure P5.13

5.14   Consider the circuit shown in Figure P5.14.

a. Sketch the amplitude response of $Y = I/V_S$.

b. Sketch the amplitude response of V_1/V_S.

c. Sketch the amplitude response of V_2/V_S.

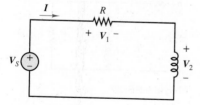

Figure P5.14

Section 5.2: Filters

5.15   Using a 15-kΩ resistance, design an RC high-pass
filter with a breakpoint at 200 kHz.

5.16   Using a 500-Ω resistance, design an RC low-pass
filter that would attenuate a 120-Hz sinusoidal voltage
by 20 dB with respect to the DC gain.

5.17   In an RLC circuit, assume ω_1 and ω_2 such that
$I(j\omega_1) = I(j\omega_2) = I_{max}/\sqrt{2}$ and $\Delta\omega$ such that
$\Delta\omega = \omega_2 - \omega_1$. In other words, $\Delta\omega$ is the width
of the current curve where the current has fallen to
$1/\sqrt{2} = 0.707$ of its maximum value at the
resonance frequency. At these frequencies, the
power dissipated in a resistance becomes one-half
of the dissipated power at the resonance frequency
(they are called the half-power points). In an
RLC circuit with a high quality factor, show that
$Q = \omega_0/\Delta\omega$.

5.18   In an RLC circuit with a high quality factor:

a. Show that the impedance at the resonance
   frequency becomes a value of Q times the
   inductive resistance at the resonance frequency.

b. Determine the impedance at the resonance
   frequency, assuming $L = 280$ mH, $C = 0.1 \ \mu F$,
   $R = 25 \ \Omega$.

5.19   At what frequency is the phase shift introduced by
the circuit of Example 5.7 equal to $-10°$?

5.20   At what frequency is the output of the circuit of
Example 5.7 attenuated by 10 percent (that is,
$V_o = 0.9V_S$)?

5.21   At what frequency is the output of the circuit of
Example 5.11 attenuated by 10 percent (that is,
$V_o = 0.9V_S$)?

5.22   At what frequency is the phase shift introduced by
the circuit of Example 5.11 equal to $20°$?

5.23   Consider the circuit shown in Figure P5.23.
Determine the resonant frequency and the bandwidth
for the circuit.

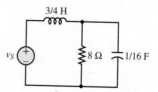

Figure P5.23

5.24   Are the filters shown in Figure P5.24
low-pass, high-pass, bandpass, or bandstop
(notch) filters?

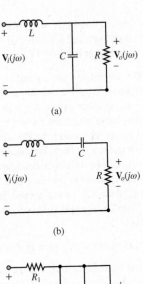

(a)

(b)

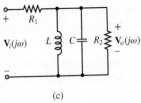

(c)

Figure P5.24

5.25   Determine if each of the circuits shown in Figure P5.25 is a low-pass, high-pass, bandpass, or bandstop (notch) filter.

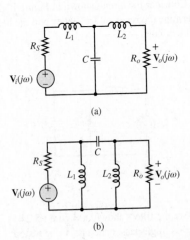

(a)

(b)

Figure P5.25

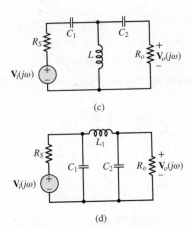

(c)

(d)

Figure P5.25 (*continued*)

5.26   For the filter circuit shown in Figure P5.26:

a. Determine if this is a low-pass, high-pass, bandpass, or bandstop filter.

b. Determine the frequency response $\mathbf{V}_o(j\omega)/\mathbf{V}_i(j\omega)$ assuming $L = 10$ mH, $C = 1$ nF, $R_1 = 50$ Ω, $R_2 = 2.5$ kΩ.

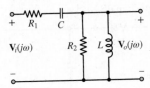

Figure P5.26

5.27   In the filter circuit shown in Figure P5.27: $L = 10$ H, $C = 1$ nF, $R_S = 20$ Ω, $R_c = 100$ Ω, $R_o = 5$ kΩ. Determine the frequency response $\mathbf{V}_o(j\omega)/\mathbf{V}_i(j\omega)$. What type of filter does this frequency response represent?

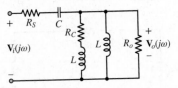

Figure P5.27

5.28   In the filter circuit shown in Figure P5.28: $L = 0.1$ mH, $C = 8$ nF, $R_S = 300$ Ω, $R_C = 10$ Ω, $R_o = 500$ Ω. Determine the frequency response $\mathbf{V}_o(j\omega)/\mathbf{V}_i(j\omega)$. What type of filter does this frequency response represent?

5.29   In the filter circuit shown in Figure P5.29:

$$R_S = 5 \text{ k}\Omega \qquad C = 56 \text{ nF}$$
$$R_o = 100 \text{ k}\Omega \qquad L = 9 \text{ μH}$$

Determine:

a. The voltage frequency response

$$\mathbf{H}_v(j\omega) = \frac{\mathbf{V}_o(j\omega)}{\mathbf{V}_i(j\omega)}$$

b. The resonant frequency.

c. The half-power frequencies.

d. The bandwidth and Q.

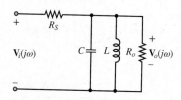

Figure P5.29

5.30   In the filter circuit shown in Figure P5.30:

$$R_S = 5 \text{ k}\Omega \qquad C = 0.5 \text{ nF}$$
$$R_o = 100 \text{ k}\Omega \qquad L = 1 \text{ mH}$$

Determine:

a. The voltage frequency response

$$\mathbf{H}_v(j\omega) = \frac{\mathbf{V}_o(j\omega)}{\mathbf{V}_i(j\omega)}$$

b. The resonant frequency.

c. The half-power frequencies.

d. The bandwidth and Q.

5.31   In the filter circuit shown in Figure P5.31:

$$R_S = 500 \text{ Ω} \qquad R_o = 5 \text{ k}\Omega$$
$$R_C = 4 \text{ k}\Omega \qquad L = 1 \text{ mH}$$
$$C = 5 \text{ pF}$$

Determine the frequency response $\mathbf{H}(j\omega)$, where:

$$\mathbf{H}(j\omega) = \frac{\mathbf{V}_o(j\omega)}{\mathbf{V}_i(j\omega)}$$

What type of filter does this frequency response represent?

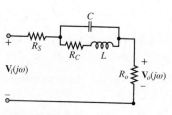

Figure P5.31

5.32   In the notch filter circuit shown in Figure P5.32, derive the voltage frequency response $\mathbf{H}(j\omega)$ in standard form, where:

$$\mathbf{H}(j\omega) = \frac{\mathbf{V}_o(j\omega)}{\mathbf{V}_i(j\omega)}$$

Assume:

$$R_S = 500 \text{ Ω} \qquad R_o = 5 \text{ k}\Omega$$
$$C = 5 \text{ pF} \qquad L = 1 \text{ mH}$$

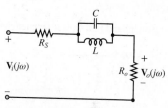

Figure P5.32

5.33   In the notch filter circuit shown in Figure P5.32, derive the voltage frequency response $\mathbf{H}(j\omega)$ in standard form, where:

$$\mathbf{H}(j\omega) = \frac{\mathbf{V}_o(j\omega)}{\mathbf{V}_i(j\omega)}$$

Assume:

$$R_S = 500 \text{ Ω} \qquad R_o = 5 \text{ k}\Omega$$
$$\omega_r = 12.1278 \text{ Mrad/s} \qquad C = 68 \text{ nF}$$
$$L = 0.1 \text{ μH}$$

Also, determine the half-power frequencies, bandwidth, and Q.

5.34   In the notch filter circuit shown in Figure P5.32, derive the voltage frequency response $\mathbf{H}(j\omega)$ in standard form, where:

$$\mathbf{H}(j\omega) = \frac{\mathbf{V}_o(j\omega)}{\mathbf{V}_i(j\omega)}$$

Assume:

$$R_S = 4.4 \text{ k}\Omega \qquad R_o = 60 \text{ k}\Omega \qquad \omega_r = 25 \text{ Mrad/s}$$

$$C = 0.8 \text{ nF} \qquad L = 2 \text{ } \mu\text{H}$$

Also, determine the half-power frequencies, bandwidth, and Q.

5.35   In the bandstop (notch) filter shown in Figure P5.35:

$$L = 0.4 \text{ mH} \qquad R_c = 100 \text{ } \Omega$$

$$C = 1 \text{ pF} \qquad R_S = R_o = 3.8 \text{ k}\Omega$$

Determine:

a. An expression for the voltage frequency response:

$$\mathbf{H}_v(j\omega) = \frac{\mathbf{V}_o(j\omega)}{\mathbf{V}_i(j\omega)} = H_o \frac{1 + jf_1(\omega)}{1 + jf_2(\omega)}$$

b. The magnitude of the frequency response at very high and very low frequencies and at the resonant frequency.

c. The magnitude of the frequency response at the resonant frequency.

d. The resonant and half-power frequencies.

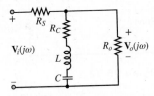

Figure P5.35

5.36   In the filter circuit shown in Figure P5.29, assume:

$$R_S = 5 \text{ k}\Omega \qquad C = 5 \text{ nF}$$

$$R_o = 50 \text{ k}\Omega \qquad L = 2 \text{ mH}$$

Determine:

a. An expression for the voltage frequency response function

$$\mathbf{H}_V(j\omega) = \frac{\mathbf{V}_o(j\omega)}{\mathbf{V}_i(j\omega)}$$

b. The resonant frequency.

c. The half-power frequencies.

d. The bandwidth and Q.

5.37   Many stereo speakers are two-way speaker systems; that is, they have a woofer for low-frequency sounds and a tweeter for high-frequency sounds. To

get the proper separation of frequencies going to the woofer and to the tweeter, *crossover* circuitry is used. A crossover circuit is effectively a bandpass, high-pass, or low-pass filter. The system model is shown in Figure P5.37. The function of the *crossover* circuitry is to channel frequencies below a given crossover frequency, f_c, into the woofer and frequencies higher than f_c into the tweeter. Assume an ideal amplifier such that $R_S = 0$ and that the desired crossover frequency is 1,200 Hz. Find C and L when $R_1 = R_2 = 8 \text{ } \Omega$. [*Hint:* Set the break frequency of the network seen by the amplifier equal to the desired crossover frequency.]

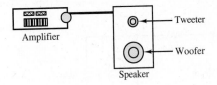

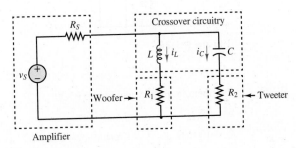

Figure P5.37

Section 5.3: Bode Plots

5.38   Determine the frequency response $\mathbf{V}_{out}(\omega)/\mathbf{V}_S(\omega)$ for the network in Figure P5.38. Generate the Bode magnitude and phase plots when $R_S = R_o = 5 \text{ k}\Omega$, $L = 10 \text{ } \mu\text{H}$, and $C = 0.1 \text{ } \mu\text{F}$.

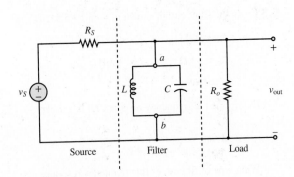

Figure P5.38

5.39 Refer to Problem 5.37 but assume that $L = 2$ mH, $C = 125$ μF, and $R_S = R_1 = R_2 = 4$ Ω in Figure P5.37.

   a. Determine the impedance seen by the amplifier as a function of frequency. At what frequency is maximum power transferred by the amplifier?

   b. Generate the Bode magnitude and phase plots of the currents through the woofer and tweeter.

5.40 For the notch filter shown in Figure P5.40 assume that $R_S = R_0 = 500$ Ω, $L = 10$ mH, and $C = 0.1$ μF.

   a. Determine the frequency response $\mathbf{V}_{out}(j\omega)/\mathbf{V}_S(j\omega)$.

   b. Generate the associated Bode magnitude and phase plots.

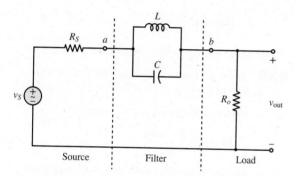

Source   Filter   Load

Figure P5.40

5.41 It is very common to see interference caused by power lines, at a frequency of 60 Hz. This problem outlines the design of the notch filter shown in Figure P5.41 to reject a band of frequencies around 60 Hz.

   a. Determine the impedance $\mathbf{Z}_{ab}(j\omega)$ between nodes a and b for the filter of Figure P5.41. r_L represents the resistance of a practical inductor.

   b. For what value of C will the center frequency of $\mathbf{Z}_{ab}(j\omega)$ equal 60 Hz when $L = 100$ mH and $r_L = 5$ Ω?

   c. Would the "sharpness," or selectivity, of the filter increase or decrease if r_L were increased?

   d. Assume that the filter is used to eliminate the 60-Hz noise from a 1 kHz sine wave. Evaluate the frequency response $\mathbf{V}_o/\mathbf{V}_{in}(j\omega)$ at both frequencies when:

$$v_g(t) = \sin(2\pi 1{,}000t) \text{ V} \qquad r_g = 50 \text{ }\Omega$$
$$v_n(t) = 3\sin(2\pi 60 t) \qquad R_o = 300 \text{ }\Omega$$

Assume $L = 100$ mH and $r_L = 5$ Ω. Use the value of C found in part b.

   e. Generate the Bode magnitude and phase plots for $\mathbf{V}_o/\mathbf{V}_{in}$. Mark the plots at 60 Hz and 1,000 Hz.

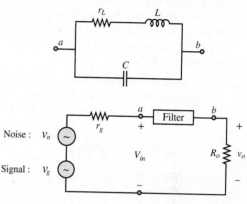

Figure P5.41

5.42 The circuit of Figure P5.42 is representative of an amplifier-speaker connection. The crossover filter allows low-frequency signals to pass to the woofer. The filter's topography is known as a π network.

   a. Find the frequency response $\mathbf{V}_o(j\omega)/\mathbf{V}_S(j\omega)$.

   b. If $C_1 = C_2 = C$, $R_S = R_o = 600$ Ω, and $1/\sqrt{LC} = R/L = 1/RC = 2{,}000\pi$, generate the Bode magnitude and phase plots in the range 100 Hz $\leq f \leq$ 10 kHz.

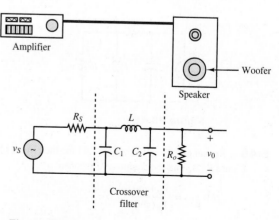

Figure P5.42

5.43   For the circuit shown in Figure P5.43:

a. Determine the frequency response:

$$\mathbf{H}(j\omega) = \frac{\mathbf{V}_{out}(j\omega)}{\mathbf{V}_{in}(j\omega)}$$

b. Sketch, by hand, the associated Bode magnitude and phase plots. List all the steps in constructing the plot. Clearly show the break frequencies on the frequency axis. (*Hint:* Use the Matlab™ command "roots" or a calculator to quickly determine the polynomial roots.)

c. Use the Matlab™ command "Bode" to generate the same plots. Verify your sketch. Assume $R_1 = R_2 = 2$ kΩ, $L = 2$ H, $C_1 = C_2 = 2$ mF.

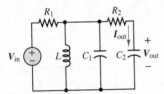

Figure P5.43

5.44   Repeat all parts of Problem 5.43 for the frequency response:

$$\mathbf{H}(j\omega) = \frac{\mathbf{I}_{out}(j\omega)}{\mathbf{V}_{in}(j\omega)}$$

Use the same component values as in Problem 5.43.

5.45   Repeat all parts of Problem 5.43 for the circuit of Figure P5.45 and the frequency response:

$$\mathbf{H}(j\omega) = \frac{\mathbf{V}_{out}(j\omega)}{\mathbf{I}_{in}(j\omega)}$$

Let $R_1 = R_2 = 1$ kΩ, $C = 1$ μF, $L = 1$ H.

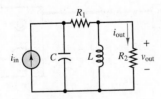

Figure P5.45

5.46   Repeat all parts of Problem 5.43 for the circuit of Figure P5.45 and the frequency response:

$$\mathbf{H}(j\omega) = \frac{\mathbf{I}_{out}(j\omega)}{\mathbf{I}_{in}(j\omega)}$$

Use the same values as in Problem 5.45.

5.47   For the circuit of Figure P5.47 determine the frequency response $\mathbf{H}(j\omega) = \mathbf{V}_{out}/\mathbf{I}_{in}$. Assume $R_1 = R_2 = 2$ kΩ, $C_1 = C_2 = 1$ mF.

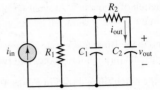

Figure P5.47

5.48   Repeat all parts of Problem 5.43 for the circuit of Figure P5.47 and the frequency response:

$$\mathbf{H}(j\omega) = \frac{\mathbf{I}_{out}(j\omega)}{\mathbf{I}_{in}(j\omega)}$$

Use the same component values as in problem 5.47.

5.49   Refer to Figure P5.4 and assume $R_1 = 300$ Ω, $R_2 = R_3 = 500$ Ω, $L = 4$ H, $C_1 = 40$ μF, $C_2 = 160$ μF.

a. Determine the frequency response:

$$\mathbf{H}(j\omega) = \frac{\mathbf{V}_{out}(j\omega)}{\mathbf{V}_{in}(j\omega)}$$

b. Sketch, by hand, the associated Bode magnitude and phase plots. List all the steps in constructing the plot. Clearly show the break frequencies on the frequency axis. (*Hint:* Use the Matlab™ command "roots" or a calculator to quickly determine the polynomial roots.)

c. Use the Matlab™ command "Bode" to generate the same plots. Verify your sketch.

5.50   Refer to Figure P5.4 and the parameter values listed in Problem 5.49.

a. Determine for the frequency response:

$$\mathbf{H}(j\omega) = \frac{\mathbf{V}_C(j\omega)}{\mathbf{V}_{in}(j\omega)}$$

b. Repeat parts b and c of Problem 5.49 for this frequency response.

5.51   Refer to Figure P5.5 and repeat the instructions of parts b and c of Problem 5.49. Assume $R_1 = 20$ kΩ, $R_2 = 100$ kΩ, $L = 1$ H, $C = 100$ μF.

5.52   Assume in a certain frequency range that the ratio of output amplitude to input amplitude is proportional to $1/\omega^3$. What is the slope of the Bode magnitude plot in this frequency range, expressed in dB/decade?

5.53 Assume that the amplitude of an output voltage depends on frequency according to:

$$\mathbf{V}(j\omega) = \frac{A\omega + B}{\sqrt{C + D\omega^2}}$$

Find:

a. The break frequency.

b. The slope (in dB/decade) of the Bode magnitude plot above the break frequency.

c. The slope (in dB/decade) of the Bode plot below the break frequency.

d. The high-frequency limit of $\mathbf{V}(j\omega)$.

5.54 Determine the equivalent impedance $\mathbf{Z}_{eq}$ in standard form as defined in Figure P5.54(a). Choose the Bode plot from Figure P5.54(b) that best describes the behavior of the impedance as a function of frequency. Describe how to find the resonant and cutoff frequencies, and the magnitude of the impedance for those ranges where it is constant. Label the Bode plot to indicate which feature you are discussing.

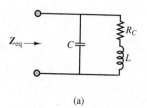

(a)

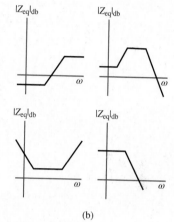

(b)

Figure P5.54

Design Credits: Mini DVI cable adapter isolated on white background: Robert Lehmann/Alamy Stock Photo; Balance scale: Alex Slobodkin/E+/Getty Images; Icon for "Focus on measurements" weighing scales: Media Bakery.

C H A P T E R

6

AC POWER

The basic concepts underlying simple AC power and the generation and distribution of electric power are extensions of those previously developed in Chapter 3, namely, phasors and impedance. Together, they pave the way for the material on electric machines in Chapter 12. The principal new concepts introduced in this chapter are average and complex power, and how they are computed for complex loads. The concept of the power factor is introduced as is the method for correcting (adjusting) it. A brief discussion of ideal transformers and maximum power transfer is provided, followed by an introduction to three-phase power, electrical safety, and finally a discussion of electric power generation and distribution.

In this chapter, quantities often involve angles. Unless indicated otherwise, angles are given in units of radians.

Learning Objectives

Students will learn to...

1. Understand the meaning of instantaneous and average power, use AC power notation, compute average power, and compute the power factor of a complex load. *Section 6.1.*

2. Use complex power notation; compute apparent, real, and reactive power for complex loads; and draw a power triangle. *Section 6.2.*

3. Compute the capacitance required to correct the power factor of a complex load. *Section 6.3.*

4. Analyze an ideal transformer; compute primary and secondary currents, voltages, and turns ratios; calculate reflected sources and impedances across ideal transformers; and understand maximum power transfer. *Section 6.4.*

5. Use three-phase AC power notation; and compute load currents and voltages for balanced wye and delta loads. *Section 6.5.*

6. Understand the basic principles of residential electrical wiring and of electrical safety. *Sections 6.6 and 6.7.*

6.1 INSTANTANEOUS AND AVERAGE POWER

When a linear electric circuit is excited by a sinusoidal source, all voltages and currents in the circuit are also sinusoids of the same frequency as the source. Figure 6.1 depicts the general form of a linear AC circuit. The most general expressions for the voltage and current delivered to an arbitrary load are as follows:

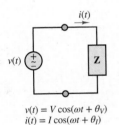

$$v(t) = V\cos(\omega t + \theta_V)$$
$$i(t) = I\cos(\omega t + \theta_I)$$

(6.1)

where V and I are the peak amplitudes of the sinusoidal voltage and current, respectively, and θ_V and θ_I are their phase angles. Two such waveforms are plotted in Figure 6.2, with unit amplitude, angular frequency 150 rad/s, and phase angles $\theta_V = 0$ and $\theta_I = \pi/3$. Notice that the current *leads* the voltage; or equivalently, the voltage *lags* the current. Keep in mind that all phase angles are relative to some reference, which is usually chosen to be the phase angle of a source. The reference phase angle is freely chosen and therefore usually set to zero for simplicity. Also keep in mind that a phase angle represents a *time delay* of one sinusoid relative to its reference sinusoid.

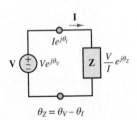

$\theta_Z = \theta_V - \theta_I$

Figure 6.1 Time and frequency domain representations of an AC circuit. The phase angle of the load is $\theta_Z = \theta_V - \theta_I$.

The **instantaneous power** dissipated by any element is the product of its instantaneous voltage and current.

$$p(t) = v(t)i(t) = VI\cos(\omega t + \theta_V)\cos(\omega t + \theta_I)$$

(6.2)

This expression is further simplified with the aid of the trigonometric identity:

$$2\cos(x)\cos(y) = \cos(x+y) + \cos(x-y)$$

(6.3)

Let $x = \omega t + \theta_V$ and $y = \omega t + \theta_I$ to yield:

$$p(t) = \frac{VI}{2}[\cos(2\omega t + \theta_V + \theta_I) + \cos(\theta_V - \theta_I)]$$

$$= \frac{VI}{2}[\cos(2\omega t + \theta_V + \theta_I) + \cos(\theta_Z)]$$

(6.4)

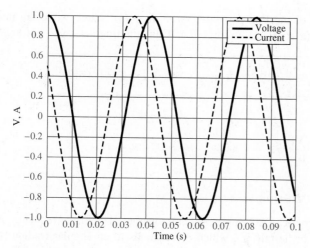

Figure 6.2 Current and voltage waveforms with unit amplitude and a phase shift of 60°

Equation 6.4 illustrates that the total instantaneous power dissipated by an element is equal to the sum of a constant $\frac{1}{2}VI\cos(\theta_Z)$ and a sinusoidal $\frac{1}{2}VI\cos(2\omega t + \theta_V + \theta_I)$, which oscillates at twice the frequency of the source. Since the time average of a sinusoid is zero over one period or over a sufficiently long interval, the constant $\frac{1}{2}VI\cos(\theta_Z)$ is the time averaged power dissipated by a complex load $\mathbf{Z}$, where θ_Z is the phase angle of that load.

Figure 6.3 shows the instantaneous and average power corresponding to the voltage and current signals of Figure 6.2. These observations can be confirmed mathematically by noting that the time average of the instantaneous power is defined by:

$$P_{\text{avg}} \equiv \frac{1}{T}\int_{t_0}^{t_0+T} p(t)\, dt \tag{6.5}$$

where T is one period of $p(t)$. Use equation 6.4 to substitute for $p(t)$ and yield:

$$P_{\text{avg}} = \frac{1}{T}\int_{t_0}^{t_0+T} \frac{VI}{2}\left[\cos(2\omega t + \theta_V + \theta_I) + \cos(\theta_Z)\right] dt$$

$$= \frac{VI}{2T}\int_{t_0}^{t_0+T} \left[\cos(2\omega t + \theta_V + \theta_I) + \cos(\theta_Z)\right] dt \tag{6.6}$$

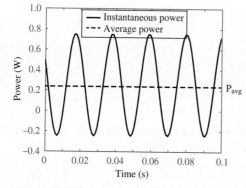

Figure 6.3 Instantaneous and average power corresponding to the signals in Figure 6.2

The integral of the first part $\cos(2\omega t + \theta_V + \theta_I)$ is zero while the integral of the second part (a constant) is $T\cos(\theta_Z)$. Thus, the time averaged power P_{avg} is:

$$P_{avg} = \frac{VI}{2}\cos(\theta_Z) = \frac{1}{2}\frac{V^2}{|\mathbf{Z}|}\cos(\theta_Z) = \frac{1}{2}I^2|\mathbf{Z}|\cos(\theta_Z) \qquad (6.7)$$

where

$$|\mathbf{Z}| = \frac{|\mathbf{V}|}{|\mathbf{I}|} = \frac{V}{I} \quad \text{and} \quad \theta_Z = \theta_V - \theta_I \qquad (6.8)$$

Effective Values

In North America, AC power systems operate at a fixed frequency of 60 cycles per second, or hertz (Hz), which corresponds to an angular (radian) frequency ω given by:

$$\omega = 2\pi \cdot 60 = 377 \text{ rad/s} \qquad \text{AC power frequency} \qquad (6.9)$$

In Europe and most other parts of the world, the AC power frequency is 50 Hz.

Unless indicated otherwise, the angular (radian) frequency ω is assumed to be 377 rad/s throughout this chapter.

It is customary in AC power analysis to employ the *effective* or *root-mean-square* (rms) amplitude (see Section 3.2) rather than the peak amplitude for AC voltages and currents. In the case of a sinusoidal waveform, the effective voltage $\tilde{V} \equiv V_{rms}$ is related to the peak voltage V by:

$$\tilde{V} = V_{rms} = \frac{V}{\sqrt{2}} \qquad (6.10)$$

Likewise, the effective current $\tilde{I} \equiv I_{rms}$ is related to the peak current I by:

$$\tilde{I} = I_{rms} = \frac{I}{\sqrt{2}} \qquad (6.11)$$

The rms, or effective, value of an AC source is the DC value that produces the same average power to be dissipated by a common resistor.

The average power can be expressed in terms of effective voltage and current by plugging $V = \sqrt{2}\tilde{V}$ and $I = \sqrt{2}\tilde{I}$ into equation 6.7 to find:

$$P_{avg} = \tilde{V}\tilde{I}\cos(\theta_Z) = \frac{\tilde{V}^2}{|\mathbf{Z}|}\cos(\theta_Z) = \tilde{I}^2|\mathbf{Z}|\cos(\theta_Z) \qquad (6.12)$$

Voltage and current phasors are also represented with effective amplitudes by the notation:

$$\tilde{\mathbf{V}} = \tilde{V}e^{j\theta_V} = \tilde{V}\angle\theta_V \tag{6.13}$$

and

$$\tilde{\mathbf{I}} = \tilde{I}e^{j\theta_I} = \tilde{I}\angle\theta_I \tag{6.14}$$

It is critical to pay close attention to the *mathematical notation* that was first introduced in Chapter 3, namely that complex quantities, such as $\mathbf{V}$, $\mathbf{I}$, and $\mathbf{Z}$ are boldface. On the other hand, scalar quantities, such as V, I, $\tilde{V}$, and $\tilde{I}$ are italic. The relationship between these quantities is $V = |\mathbf{V}|$ and $\tilde{V} = |\tilde{\mathbf{V}}|$.

Impedance Triangle

Figure 6.4 illustrates the concept of the **impedance triangle**, which is an important graphical representation of impedance as a vector in the complex plane. Basic trigonometry yields:

$$R = |\mathbf{Z}|\cos\theta \tag{6.15}$$

$$X = |\mathbf{Z}|\sin\theta \tag{6.16}$$

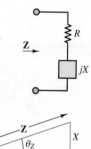

where R is the *resistance* and X is the *reactance*. Notice that both R and P_{avg} are proportional to $\cos(\theta_Z)$, which suggests that a triangle similar to (i.e., the same shape as) the impedance triangle could be constructed with P_{avg} as one leg of a right triangle. In fact, such a triangle is known as a *power triangle*. The similarity of these two types of triangles is a powerful concept for problem solving, as is shown in Section 6.2.

Figure 6.4 Impedance triangle

Power Factor

The phase angle θ_Z of the load impedance plays a very important role in AC power circuits. From equation 6.12, the average power dissipated by an AC load is proportional to $\cos(\theta_Z)$. For this reason, $\cos(\theta_Z)$ is known as the **power factor (pf)**. For purely resistive loads:

$$\theta_Z = 0 \quad \rightarrow \quad \text{pf} = 1 \qquad \text{Resistive load} \tag{6.17}$$

For purely inductive or capacitive loads:

$$\theta_Z = +\pi/2 \quad \rightarrow \quad \text{pf} = 0 \qquad \text{Inductive load} \tag{6.18}$$

$$\theta_Z = -\pi/2 \quad \rightarrow \quad \text{pf} = 0 \qquad \text{Capacitive load} \tag{6.19}$$

For loads with nonzero resistive (real) and reactive (imaginary) parts:

$$0 < |\theta_Z| < \pi/2 \quad \rightarrow \quad 0 < \text{pf} < 1 \qquad \text{Complex load} \tag{6.20}$$

Using the definition $\text{pf} = \cos(\theta_Z)$ the average power can be expressed as:

$$P_{\text{avg}} = \tilde{V}\tilde{I}\text{pf} \tag{6.21}$$

Thus, average power dissipated by a resistor is:

$$\left(P_{\text{avg}}\right)_R = \tilde{V}_R\tilde{I}_R\text{pf}_R = \tilde{V}_R\tilde{I}_R \tag{6.22}$$

because $\text{pf}_R = 1$. By contrast, the average power dissipated by a capacitor or inductor is:

$$\left(P_{\text{avg}}\right)_X = \tilde{V}_X\tilde{I}_X\text{pf}_X = 0 \tag{6.23}$$

because $\text{pf}_X = 0$, where the subscript X indicates a reactive element (i.e., either a capacitor or inductor). It is important to note that although capacitors and inductors are *lossless* (i.e., they store and release energy but do not dissipate energy), they do influence power dissipation in a circuit by affecting the voltage across and the current through resistors in the circuit.

When θ_Z is positive, the load is *inductive* and the power factor is said to be *lagging*; when θ_Z is negative, the load is capacitive and the power factor is said to be *leading*. It is important to keep in mind that $\text{pf} = \cos(\theta_Z) = \cos(-\theta_Z)$ because the cosine is an even function. Thus, while it may be important to know whether a load is inductive or capacitive, the value of the power factor only indicates the extent to which a load is inductive or capacitive. To know whether a load is inductive or capacitive, one must know whether the power factor is leading or lagging.

EXAMPLE 6.1   Computing Average and Instantaneous AC Power

Problem

Compute the average and instantaneous power dissipated by the load of Figure 6.5.

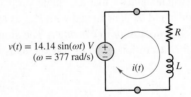

$v(t) = 14.14 \sin(\omega t)$ V
$(\omega = 377$ rad/s$)$

Figure 6.5 Circuit for Example 6.1.

Solution

Known Quantities: Source voltage and frequency, load resistance and inductance values.

Find: P_{avg} and $p(t)$ for the RL load.

Schematics, Diagrams, Circuits, and Given Data: $v(t) = 14.14 \sin(377t)$ V; $R = 4\ \Omega$; $L = 8\,\text{mH}$.

Assumptions: None.

Analysis: The source voltage is expressed in terms of $\sin(377t)$. By convention, all time-domain sinusoids should be expressed as cosines. To convert $\sin(377t)$ to $\cos(377t + \theta_V)$ recall that a sine equals a cosine shifted forward in time (to the right) by $\pi/2$ rad; that is, $\sin(377t) = \cos(377t - \pi/2)$. Thus, at the angular frequency $\omega = 377$ rad/s the source voltage is:

$$\tilde{\mathbf{V}} = 10\angle\left(-\frac{\pi}{2}\right) \text{V rms}$$

where 14.14 V = 10 V rms.

The equivalent impedance of the load is:

$$\mathbf{Z} = R + j\omega L \approx 4 + j3 = 5\angle(36.9°) = 5\angle(0.644 \text{ rad})\ \Omega$$

The current in the loop is:

$$\tilde{\mathbf{I}} = \frac{\tilde{\mathbf{V}}}{\mathbf{Z}} \approx \frac{10\angle(-\pi/2)}{5\angle(0.644)} \approx 2\angle(-2.215) \text{ A rms}$$

It is instructive to compute the average power dissipated in the circuit in two ways:

1. The most straightforward and brute force approach is to compute:

$$P_{\text{avg}} = \tilde{V}\tilde{I}\cos(\theta_Z) = 10 \times 2 \times \cos(0.644) \approx 16 \text{ W}$$

2. Another approach is to realize that the average power dissipated by the inductor is zero. Thus, the total average power dissipated equals the average power dissipated by the resistor. Thus:

$$\left(P_{\text{avg}}\right)_R = \tilde{I}^2 R\,\text{pf}_R = \tilde{I}^2 R \approx (2)^2 \times 4 = 16 \text{ W}$$

The instantaneous power is given by:

$$p(t) = v(t) \times i(t) = \sqrt{2} \times 10 \sin(377t) \times \sqrt{2} \times 2 \cos(377t - 2.215)\,\text{W}$$

The instantaneous voltage and current waveforms and the instantaneous and average power are plotted in Figure 6.6.

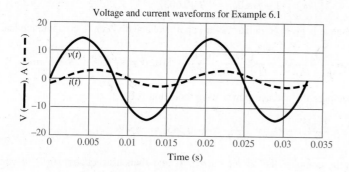

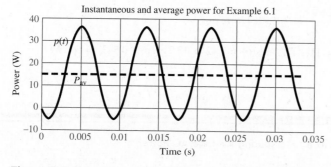

Figure 6.6 Voltage, current and power waveforms for Example 6.1.

Comment: It is standard procedure in electrical engineering practice to use rms values in power calculations. Also, note that the instantaneous power can be negative at times even though the average power is positive. This result reflects the fact that although the average power of an inductor is identically zero, the instantaneous power of an inductor can be positive or negative as the inductor charges or discharges with the sinusoidal source.

EXAMPLE 6.2   Computing Average AC Power

Problem

Compute the average power dissipated by the load of Figure 6.7.

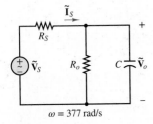

$\omega = 377$ rad/s

Figure 6.7 Circuit for
Example 6.2.

Solution

Known Quantities: Source voltage, internal resistance, load resistance, capacitance, and frequency.

Find: P_{avg} for the $R_o \| C$ load.

Schematics, Diagrams, Circuits, and Given Data: $\tilde{\mathbf{V}}_S = 110\angle 0°$ V rms; $R_S = 2\ \Omega$; $R_o = 16\ \Omega$; $C = 100\ \mu\text{F}$; $\omega = 377$ rad/s.

Assumptions: None.

Analysis: First, compute the impedance of the load at the angular frequency $\omega = 377$ rad/s:

$$\mathbf{Z}_o = R_o \| \frac{1}{j\omega C} = \frac{R_o}{1 + j\omega C R_o} = \frac{16}{1 + j0.6032} = 13.7\angle(-0.543)\ \Omega$$

where the angle is given in radians. Next, apply voltage division to compute the load voltage:

$$\tilde{\mathbf{V}}_o = \frac{\mathbf{Z}_o}{R_S + \mathbf{Z}_o} \tilde{\mathbf{V}}_S = \frac{13.7\angle(-0.543)}{2 + 13.7\angle(-0.543)} 110\angle 0 = 97.6\angle(-0.067)\ \text{V rms}$$

Finally, compute the average power using equation 6.12:

$$P_{\text{avg}} = \frac{|\tilde{\mathbf{V}}_o|^2}{|\mathbf{Z}_o|} \cos(\theta_Z) = \frac{97.6^2}{13.7} \cos(-0.543) = 595\ \text{W}$$

Alternatively, compute the source current $\tilde{\mathbf{I}}_S$ and then use equation 6.12 to compute the average power:

$$\tilde{\mathbf{I}}_S = \frac{\tilde{\mathbf{V}}_o}{\mathbf{Z}_o} = 7.12\angle 0.476\ \text{A rms}$$

$$P_{\text{avg}} = |\tilde{\mathbf{I}}_S|^2 |\mathbf{Z}_o| \cos(\theta) = 7.12^2 \times 13.7 \times \cos(-0.543) = 595\ \text{W}$$

EXAMPLE 6.3   Computing Average AC Power

Problem

Compute the average power dissipated by the load of Figure 6.8.

Solution

Known Quantities: Source voltage, internal resistance, load resistance, capacitance and inductance values, and frequency.

Find: P_{avg} for the complex load.

Schematics, Diagrams, Circuits, and Given Data: $\tilde{\mathbf{V}}_s = 110\angle0$ V rms; $R = 10$ Ω; $L = 0.05$ H; $C = 470$ μF; $\omega = 377$ rad/s. Figure 6.8.

Assumptions: None.

Analysis: First, compute the impedance of the load $\mathbf{Z}_o$ at the angular frequency $\omega = 377$ rad/s:

$$\mathbf{Z}_o = (R + j\omega L) \| \frac{1}{j\omega C} = \frac{(R + j\omega L)/j\omega C}{R + j\omega L + 1/j\omega C}$$

$$= \frac{R + j\omega L}{1 - \omega^2 LC + j\omega CR} = 1.16 - j7.18$$

$$= 7.27\angle(-1.41)\,\Omega$$

Note that the equivalent load impedance at $\omega = 377$ rad/s has a negative imaginary part, which is a feature of a *capacitive load*, as shown in Figure 6.9. The average power is:

$$P_{\text{avg}} = \frac{|\tilde{\mathbf{V}}_s|^2}{|\mathbf{Z}_o|}\cos(\theta) = \frac{110^2}{7.27}\cos(-1.41) = 266 \text{ W}$$

Comment: At $\omega = 377$ rad/s, the capacitance has a larger impact on the total equivalent impedance than the inductance. At lower frequencies, where the impedance of the capacitor is large compared to $R + j\omega L$, the parallel equivalent impedance will be inductive. It is instructive to determine the frequencies when the parallel equivalent impedance has a zero imaginary part.

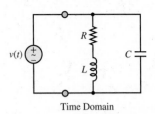

Time Domain

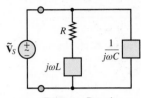

Frequency Domain

Figure 6.8 Circuit for Example 6.3

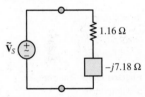

Figure 6.9 Equivalent circuit for Example 6.3

CHECK YOUR UNDERSTANDING

Consider the circuit shown in Figure 6.10. Find the impedance of the load "seen" by the voltage source, and compute the average power dissipated by the load. The constant 155.6 multiplying the cosine function is always the peak amplitude, not the rms amplitude.

Answer: $\mathbf{Z} = 4.8e^{-j33.5°}$ Ω; $P_{\text{avg}} = 2,103.4$ W

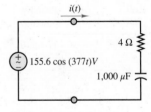

Figure 6.10 Circuit for Check Your Understanding question.

CHECK YOUR UNDERSTANDING

For Example 6.2, compute the average power dissipated by the internal source resistance R_S.

Answer: 101.46 W.

6.2 COMPLEX POWER

The computation of AC power is simplified by defining a **complex power S**, where:

$$\boxed{\mathbf{S} = \tilde{\mathbf{V}}\tilde{\mathbf{I}}^* \qquad \text{Complex power}}$$

(6.24)

where the asterisk denotes the complex conjugate (see Appendix A). Note that the effect of taking the complex conjugate of a phasor is to multiply its phase angle by -1. In other words, $\angle\mathbf{S} = \angle\tilde{\mathbf{V}} + \angle\tilde{\mathbf{I}}^* = \angle\tilde{\mathbf{V}} - \angle\tilde{\mathbf{I}} = \theta_Z$. The definition of complex power leads to:

$$
\begin{aligned}
\mathbf{S} &= \tilde{V}\tilde{I}\,\cos(\theta_Z) + j\tilde{V}\tilde{I}\,\sin(\theta_Z) \\
&= \tilde{I}^2|\mathbf{Z}|\,\cos(\theta_Z) + j\tilde{I}^2|\mathbf{Z}|\,\sin(\theta_Z) \\
&= \tilde{I}^2 R + j\tilde{I}^2 X = \tilde{I}^2\mathbf{Z}
\end{aligned} \tag{6.25}
$$

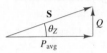

where $R = |\mathbf{Z}|\cos(\theta_Z)$ and $X = |\mathbf{Z}|\sin(\theta_Z)$ are the resistance and reactance of the impedance triangle shown in Figure 6.11. The real and imaginary parts of $\mathbf{S}$ are the **real power** $P_{\text{avg}} = \tilde{V}\tilde{I}\cos(\theta_Z)$ and the **reactive power** $Q = \tilde{V}\tilde{I}\sin(\theta_Z)$, respectively, such that:

$$
\mathbf{S} = P_{\text{avg}} + jQ \tag{6.26}
$$

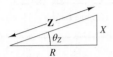

Figure 6.11 The impedance triangle

Figure 6.12 The complex power triangle

$|\mathbf{S}| = \sqrt{P_{\text{avg}}^2 + Q^2} = \tilde{V}\cdot\tilde{I}$

$P_{\text{avg}} = \tilde{V}\tilde{I}\cos\theta$

$Q = \tilde{V}\tilde{I}\sin\theta$

The magnitude $|\mathbf{S}|$ of the complex power is called the **apparent power** S and is measured in units of **volt-amperes (VA)**. The units of Q are **volt-amperes reactive**, or **VAR**.

The relationship between S, P, and Q is summarized by a *power triangle* as shown in Figure 6.12. It is important to note that the impedance and power triangles are similar; that is, they have the same shape. This result is helpful in problem solving. Table 6.1 shows the general expressions for calculating $P = P_{\text{avg}}$ and Q.

The complex power can also be expressed as:

$$
\mathbf{S} = \tilde{I}^2\mathbf{Z} = \tilde{I}^2 R + j\tilde{I}^2 X \tag{6.27}
$$

Furthermore, since $\tilde{V} = \tilde{I}|\mathbf{Z}|$ and $|\mathbf{Z}|^2 = \mathbf{Z}\mathbf{Z}^*$, the complex power can be re-expressed as:

$$
\begin{aligned}
\mathbf{S} &= \tilde{I}^2\mathbf{Z} = \frac{\tilde{I}^2\mathbf{Z}\mathbf{Z}^*}{\mathbf{Z}^*} \\
&= \frac{\tilde{V}^2}{\mathbf{Z}^*}
\end{aligned} \tag{6.28}
$$

Table 6.1 Real and reactive power

Real power P_{avg}	Reactive power Q
$\tilde{V}\tilde{I}\cos(\theta)$	$\tilde{V}\tilde{I}\sin(\theta)$
$\tilde{I}^2 R$	$\tilde{I}^2 X$

As previously stated, capacitors and inductors (reactive loads) do not dissipate energy themselves; they are lossless elements. However, they do influence power dissipation in a circuit by affecting the voltage across and current through resistors, which do dissipate energy. This influence is now quantified by the reactive power, Q, which is due entirely to capacitance and inductance in a circuit. It is worth noting that $Q = 0$, pf $= 1$, and therefore $P = S$ in purely resistive networks. It is also important to realize that P represents the real work done (per unit time) by a circuit. For example, the real power P of an electric motor represents the work done (per unit time) by the motor to perform some useful task. From the perspective of the utility company that provides the electric power for the motor and of the owner of the motor who has to pay the utility bill, it would be best if all the apparent power S provided by the utility company was converted to useful power P. (Why?) However, all electric motors have some inductance (e.g., coils of wire) such that $Q \neq 0$, pf < 1, and $P < S$. It is possible to *correct* the effect of a motor's inductance by adding capacitance in parallel with the motor so as to decrease Q and thereby decrease the apparent power S that must be provided for a given P required by the task.

FOCUS ON PROBLEM SOLVING

COMPLEX POWER COMPUTATION

1. Use AC circuit analysis methods to compute (as phasors) the voltage across and current through the load. Convert peak amplitudes to effective (rms) values.

$$\tilde{\mathbf{V}} = \tilde{V}\angle\theta_V \quad \text{and} \quad \tilde{\mathbf{I}} = \tilde{I}\angle\theta_I$$

2. Compute $\theta_Z = \theta_V - \theta_I$ and the power factor pf $= \cos(\theta_Z)$. Draw the impedance triangle, as shown in Figure 6.11.

3. Use one of the two following methods to compute P_{avg} and Q.

   - Compute the complex power $\mathbf{S} = \tilde{\mathbf{V}}\tilde{\mathbf{I}}^*$ such that $P = P_{avg} = \text{Re}(\mathbf{S})$, $Q = \text{Im}(\mathbf{S})$, and $S = |\mathbf{S}|$. The effect of taking the complex conjugate of a phasor is to multiply its phase angle by -1, such that $\angle\mathbf{S} = \angle\tilde{\mathbf{V}} - \angle\tilde{\mathbf{I}} = \theta_Z$.

   - Compute the apparent power $S = |\mathbf{S}| = \tilde{V}\tilde{I}$ such that $P = P_{avg} = S \cdot$ pf and $Q = S\sin(\theta_Z)$.

4. Draw the power triangle, as shown in Figure 6.12, and confirm that $S^2 = P^2 + Q^2$ and that $\tan(\theta_Z) = Q/P$.

5. If Q is negative, the load is capacitive and the power factor is *leading*; if Q is positive, the load is inductive and the power factor is *lagging*.

EXAMPLE 6.4   Complex Power Calculations

Problem

Compute the complex power for the load $\mathbf{Z}_o$ of Figure 6.13.

Solution

Known Quantities: Source, load voltage, and current.

Find: $\mathbf{S} = P_{avg} + jQ$ for the complex load.

Schematics, Diagrams, Circuits, and Given Data: $v(t) = 100\cos(\omega t + 0.262)$ V; $i(t) = 2\cos(\omega t - 0.262)$ A; $\omega = 377$ rad/s.

Assumptions: All angles are given in units of radians unless indicated otherwise.

Analysis: First, realize that the constants multiplying the cosine functions are always peak, not rms, values. These functions can be converted to phasor quantities with rms amplitudes as follows:

$$\tilde{\mathbf{V}} = \frac{100}{\sqrt{2}}\angle 0.262 \text{ V} \qquad \tilde{\mathbf{I}} = \frac{2}{\sqrt{2}}\angle(-0.262) \text{ A}$$

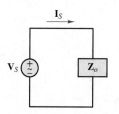

Figure 6.13 Circuit for Example 6.4.

Compute the phase angle of the load, and the real and reactive power, using the definitions of equation 6.12:

$$\theta_Z = \angle(\tilde{\mathbf{V}}) - \angle(\tilde{\mathbf{I}}) = 0.524 \text{ rad}$$

$$P_{\text{avg}} = |\tilde{\mathbf{V}}||\tilde{\mathbf{I}}| \cos(\theta_Z) = \frac{200}{2} \cos(0.524) = 86.6 \text{ W}$$

$$Q = |\tilde{\mathbf{V}}||\tilde{\mathbf{I}}| \sin(\theta_Z) = \frac{200}{2} \sin(0.524) = 50 \text{ VAR}$$

Apply the definition of complex power (equation 6.24) to repeat the same calculation:

$$\mathbf{S} = \tilde{\mathbf{V}}\tilde{\mathbf{I}}^* = \frac{100}{\sqrt{2}}\angle 0.262 \times \frac{2}{\sqrt{2}}\angle-(-0.262) = 100\angle 0.524$$
$$= (86.6 + j50)\,\text{VA}$$

Therefore

$$P_{\text{avg}} = 86.6 \text{ W} \qquad Q = 50 \text{ VAR}$$

Comments: Note how the definition of complex power yields both quantities at one time.

EXAMPLE 6.5   Real and Reactive Power Calculations

Problem

Compute the complex power for the load of Figure 6.14.

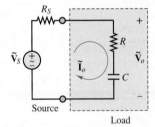

Figure 6.14 Circuit for Example 6.5.

Solution

Known Quantities: Source voltage and resistance; load impedance.

Find: $\mathbf{S} = P + jQ$ for the complex load.

Schematics, Diagrams, Circuits, and Given Data: $\tilde{\mathbf{V}}_S = 110\angle 0°$ V; $R_S = 2\ \Omega$; $R = 5\ \Omega$; $C = 2,000\ \mu\text{F}$; $\omega = 377$ rad/s.

Assumptions: All amplitudes are effective (rms) values. All angles are given in units of radians unless indicated otherwise.

Analysis: The load impedance is:

$$\mathbf{Z}_o = R + \frac{1}{j\omega C} = (5 - j1.326)\,\Omega = 5.173\angle(-0.259)\,\Omega$$

Next, apply voltage division and the generalized Ohm's law to compute the load voltage and current:

$$\tilde{\mathbf{V}}_o = \frac{\mathbf{Z}_o}{R_S + \mathbf{Z}_o}\tilde{\mathbf{V}}_S = \frac{5 - j1.326}{7 - j1.326} \times 110 = 79.86\angle(-0.072)\,\text{V}$$

$$\tilde{\mathbf{I}}_o = \frac{\tilde{\mathbf{V}}_o}{\mathbf{Z}_o} = \frac{79.86\angle(-0.072)}{5.173\angle(-0.259)} = 15.44\angle 0.187\,\text{A}$$

Finally, compute the complex power, as defined in equation 6.24:

$$\mathbf{S} = \tilde{\mathbf{V}}_o\tilde{\mathbf{I}}_o^* = 79.9\angle(-0.072) \times 15.44\angle(-0.187) = 1,233\angle(-0.259)$$
$$= (1,192 - j316)\,\text{VA}$$

Therefore:

$$P = 1,192 \text{ W} \qquad Q = -316 \text{ VAR}$$

Comment: Is the reactive power capacitive or inductive? Since $Q < 0$, the reactive power is capacitive!

EXAMPLE 6.6   Real Power Transfer for Complex Loads

Problem

Compute the complex power for the load between terminals a and b of Figure 6.15. Repeat the computation with the inductor removed from the load, and compare the real power for the two cases.

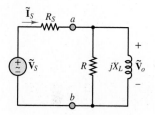

Figure 6.15 Circuit for Example 6.6.

Solution

Known Quantities: Source voltage and resistance; load impedance.

Find:

1. $\mathbf{S}_1 = P_1 + jQ_1$ for the complex load.
2. $\mathbf{S}_2 = P_2 + jQ_2$ for the real load.
3. For each case, compute the ratio of the real power dissipated by the load to the overall real power dissipated by the circuit.

Schematics, Diagrams, Circuits, and Given Data: $\tilde{\mathbf{V}}_S = 110\angle 0° \text{ V}$; $R_S = 4 \text{ }\Omega$; $R = 10 \text{ }\Omega$; $jX_L = j6 \text{ }\Omega$.

Assumptions: All amplitudes are effective (rms) values. All angles are given in units of radians unless indicated otherwise.

Analysis:

1. With the inductor included in the load, its impedance $\mathbf{Z}_o$ is:

$$\mathbf{Z}_o = R \| j\omega L = \frac{10 \times j6}{10 + j6} = 5.145\angle 1.03 \text{ }\Omega$$

Apply voltage division to compute the load voltage $\tilde{\mathbf{V}}_o$ and the generalized Ohm's law to compute the current $\tilde{\mathbf{I}}_o = \tilde{\mathbf{I}}_S$.

$$\tilde{\mathbf{V}}_o = \frac{\mathbf{Z}_o}{R_S + \mathbf{Z}_o} \tilde{\mathbf{V}}_S = \frac{5.145\angle 1.03}{4 + 5.145\angle 1.03} \times 110 = 70.9\angle 0.444 \text{ V}$$

$$\tilde{\mathbf{I}}_o = \frac{\tilde{\mathbf{V}}_o}{\mathbf{Z}_o} = \frac{70.9\angle 0.444}{5.145\angle 1.03} = 13.8\angle(-0.586) \text{ A}$$

Finally, compute the complex power, as defined in equation 6.24:

$$\mathbf{S}_1 = \tilde{\mathbf{V}}_o \tilde{\mathbf{I}}_o^* = 70.9\angle 0.444 \times 13.8\angle 0.586 = 978\angle 1.03 \text{ VA}$$
$$= (503 + j839) \text{ VA}$$

Therefore:

$$P_1 = 503 \text{ W} \qquad Q_1 = 839 \text{ VAR}$$

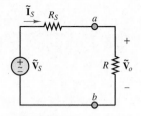

Figure 6.16 Circuit for Example 6.6 with inductor removed.

2. With the inductor excluded from the load (Figure 6.16), its impedance is:

$$\mathbf{Z}_o = R = 10\ \Omega$$

Compute the load voltage and current:

$$\tilde{\mathbf{V}}_o = \frac{\mathbf{Z}_o}{R_S + \mathbf{Z}_o}\tilde{\mathbf{V}}_S = \frac{10}{4+10}\times 110 = 78.57\angle 0\ \text{V}$$

$$\tilde{\mathbf{I}}_o = \frac{\tilde{\mathbf{V}}_o}{\mathbf{Z}_o} = \frac{78.57\angle 0}{10} = 7.857\angle 0\ \text{A}$$

Finally, compute the complex power, as defined in equation 6.24:

$$\mathbf{S}_2 = \tilde{\mathbf{V}}_o\tilde{\mathbf{I}}_o^* = 78.57\angle 0 \times 7.857\angle 0 = 617\angle 0 = (617 + j0)\,\text{VA}$$

Therefore:

$$P_2 = 617\ \text{W} \qquad Q_2 = 0\ \text{VAR}$$

3. To compute the overall real power P_{total} dissipated by the circuit, it is necessary to include the impact of the line resistance R_S and compute for each case:

$$\mathbf{S}_{\text{total}} = \tilde{\mathbf{V}}_S\tilde{\mathbf{I}}_S^* = P_{\text{total}} + jQ_{\text{total}}$$

For case 1:

$$\tilde{\mathbf{I}}_S = \frac{\tilde{\mathbf{V}}_S}{\mathbf{Z}_{\text{total}}} = \frac{\tilde{\mathbf{V}}_S}{R_S + \mathbf{Z}_o} = \frac{110}{4 + 5.145\angle 1.03} = 13.8\angle(-0.586)\,\text{A}$$

$$\mathbf{S}_{1_{\text{total}}} = \tilde{\mathbf{V}}_S\tilde{\mathbf{I}}_S^* = 110 \times 13.8\angle(+0.586) = (1{,}264 + j838)\,\text{VA} = P_{1_{\text{total}}} + jQ_{1_{\text{total}}}$$

The percent real power transfer is:

$$100 \times \frac{P_1}{P_{1_{\text{total}}}} = \frac{503}{1{,}264} = 39.8\%$$

For case 2:

$$\tilde{\mathbf{I}}_S = \frac{\tilde{\mathbf{V}}_S}{\mathbf{Z}_{\text{total}}} = \frac{\tilde{\mathbf{V}}_S}{R_S + R} = \frac{110}{4 + 10} = 7.857\angle 0\ \text{A}$$

$$\mathbf{S}_{2_{\text{total}}} = \tilde{\mathbf{V}}_S\tilde{\mathbf{I}}_S^* = 110 \times 7.857 = (864 + j0)\,\text{VA} = P_{2_{\text{total}}} + jQ_{2_{\text{total}}}$$

The percent real power transfer is:

$$100 \times \frac{P_2}{P_{2_{\text{total}}}} = \frac{617}{864} = 71.4\%$$

Comments: If it were possible to eliminate the reactive part of the impedance, the percentage of real power transferred from the source to the load would be increased significantly. The procedure to accomplish this goal is called *power factor correction*.

EXAMPLE 6.7   Complex Power and Power Triangle

Problem

Find the reactive and real power for the load of Figure 6.17. Draw the associated power triangle.

Solution

Known Quantities: Source voltage; load impedance.

Find: $\mathbf{S} = P_{\text{avg}} + jQ$ for the complex load.

Schematics, Diagrams, Circuits, and Given Data: $\tilde{\mathbf{V}}_S = 60\angle 0 \, \text{V}$; $R = 3 \, \Omega$; $jX_L = j9 \, \Omega$; $jX_C = -j5 \, \Omega$.

Assumptions: All amplitudes are effective (rms) values. All angles are given in units of radians.

Analysis: First, compute the load current:

$$\tilde{\mathbf{I}}_o = \frac{\tilde{\mathbf{V}}_o}{\mathbf{Z}_o} = \frac{60\angle 0}{3 + j9 - j5} = \frac{60\angle 0}{5\angle 0.9273} = 12\angle(-0.9273)\,\text{A}$$

Next, compute the complex power, as defined in equation 6.24:

$$\mathbf{S} = \tilde{\mathbf{V}}_o \tilde{\mathbf{I}}_o^* = 60\angle 0 \times 12\angle 0.9273 = 720\angle 0.9273 = (432 + j576)\,\text{VA}$$

Therefore:

$$P = 432 \, \text{W} \qquad Q = 576 \, \text{VAR}$$

The total reactive power must be the sum of the reactive powers in each of the elements, such that $Q = Q_C + Q_L$. Compute these two quantities as follows:

$$Q_C = |\tilde{\mathbf{I}}_o|^2 \times X_C = (144)(-5) = -720 \, \text{VAR}$$
$$Q_L = |\tilde{\mathbf{I}}_o|^2 \times X_L = (144)(9) = 1{,}296 \, \text{VAR}$$

and

$$Q = Q_L + Q_C = 576 \, \text{VAR}$$

Comments: The power triangle corresponding to this circuit is drawn in Figure 6.18. The vector diagram shows how the complex power $\mathbf{S}$ results from the vector addition of the three components P, Q_C, and Q_L.

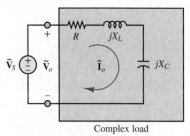

Figure 6.17 Circuit for Example 6.7.

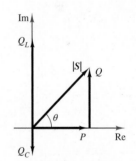

Note: $S = P + jQ_C + jQ_L$

Figure 6.18 Power triangle for Example 6.7.

CHECK YOUR UNDERSTANDING

Compute the real and reactive power for the load of Example 6.2.

Answer: $P_{\text{avg}} = 595 \, \text{W}; \; Q = -359 \, \text{VAR}$

CHECK YOUR UNDERSTANDING

Compute the real and reactive power for the load of Figure 6.10.

Answer: $P_{\text{avg}} = 2.1 \, \text{kW}; \; Q = 1.39 \, \text{kVAR}$

CHECK YOUR UNDERSTANDING

Refer to Example 6.6, and compute the percent of real power transfer for the case where the inductance of the load is one-half of the original value.

Answer: 29.3%

CHECK YOUR UNDERSTANDING

Compute the power factor for the load of Example 6.7 with and without the inductor in the circuit.

Answer: pf = 0.6, lagging (with L in circuit); pf = 0.5145, leading (without L)

6.3   POWER FACTOR CORRECTION

A power factor close to unity signifies an efficient transfer of energy from the AC source to the load while a small power factor corresponds to inefficient use of energy, as illustrated in Example 6.6. If a load requires a given real power P, the current required by the load will be minimized when the power factor is maximized, that is, when pf $= \cos(\theta_Z) \rightarrow 1$. When pf < 1, it is possible to increase it (i.e., *correct* it) by adding, as appropriate, reactance (e.g., capacitance) to the load. When pf is leading, inductance must be added; when pf is lagging, capacitance must be added.

> If $\theta_Z > 0$, then $Q > 0$, the load is inductive, the load current *lags* the load voltage, and the power factor pf is lagging. Alternatively, if $\theta_Z < 0$, then $Q < 0$, the load is capacitive, the load current *leads* the load voltage, and the power factor pf is leading.

Table 6.2 illustrates and summarizes these concepts. For simplicity, the phase angle of the voltage phasor $\tilde{\mathbf{V}}$ shown in the table is zero and acts as a reference angle for the current phasor.

In practice, the load designed for a useful industrial task is often inductive because of the presence of electric motors. The power factor of an inductive load can be *corrected* by adding capacitance in parallel with the load. This procedure is called *power factor correction*.

The measurement and correction of the power factor for the load are an extremely important aspect of any industrial engineering application that requires the use of substantial quantities of electric power. In particular, industrial plants, construction sites, heavy machinery, and other heavy users of electric power must be aware of the power factor that their loads present to the electric utility company. As was already observed, a low power factor results in greater current draw from the electric utility and greater line losses. Thus, computations related to the power factor of complex loads are of great utility to any practicing engineer.

Table 6.2 **Important facts related to complex power**

	Resistive load	**Capacitive load**	**Inductive load**
Ohm's law	$\tilde{V} = \tilde{I}Z$	$\tilde{V} = \tilde{I}Z$	$\tilde{V} = \tilde{I}Z$
Complex impedance	$Z = R$	$Z = R + jX$ $X < 0$	$Z = R + jX$ $X > 0$
Phase angle	$\theta = 0$	$\theta < 0$	$\theta > 0$
Complex plane sketch			
Explanation	The current is in phase with the voltage.	The current "leads" the voltage.	The current "lags" the voltage.
Power factor	Unity	Leading, < 1	Lagging, < 1
Reactive power	0	Negative	Positive

FOCUS ON PROBLEM SOLVING

POWER FACTOR CORRECTION

1. Follow the steps outlined in the Focus on Problem Solving box "Complex Power Computation" to find the initial phase angle of the load θ_{Z_i}, power factor pf_i, real power P_i, and reactive power Q_i. If both P_i and either pf or θ_Z are given, compute Q directly using $Q = P \tan(\theta_Z)$. An initial power triangle is helpful for visualizing this information.

2. For a lagging power factor, augment the load with a parallel capacitor such that:

$$\Delta Q = Q_C = \frac{\tilde{V}^2}{|Z_C|}\sin(\theta_Z) = -\omega C \tilde{V}^2$$

3. Express the final reactive power Q_f as:

$$Q_f = Q_i + \Delta Q$$

4. The real power is unchanged by the addition of the capacitor in parallel. Thus, $P_f = P_i$ and the final (corrected) phase angle of the augmented load is:

$$\theta_{Z_f} = \tan^{-1}\left(\frac{Q_f}{P_f}\right)$$

It is helpful to draw a final power triangle to visualize the effect of the parallel capacitor.

5. The final corrected power factor is:

$$pf_f = \cos(\theta_{Z_f})$$

EXAMPLE 6.8 **Power Factor Correction**

Problem

Calculate the power factor for the circuit of Figure 6.19. Correct it to unity by adding a capacitor in parallel with the load.

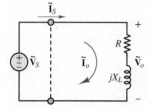

Figure 6.19 Circuit for Example 6.8.

Solution

Known Quantities: Source voltage; load impedance.

Find:

1. $\mathbf{S} = P + jQ$ for the complex load.
2. Value of parallel capacitance that results in pf = 1.

Schematics, Diagrams, Circuits, and Given Data: $\tilde{\mathbf{V}}_S = 117\angle 0\,\text{V rms}$; $R = 50\,\Omega$; $jX_L = j86.7\,\Omega$; $\omega = 377\,\text{rad/s}$.

Assumptions: All amplitudes are effective (rms) values. All angles are given in units of radians unless indicated otherwise.

Analysis:

1. First, compute the load impedance:

$$\mathbf{Z}_o = R + jX_L = 50 + j86.7 = 100\angle 1.047\,\Omega$$

Next, compute the load current $\tilde{\mathbf{I}}_o = \tilde{\mathbf{I}}_S$:

$$\tilde{\mathbf{I}}_o = \frac{\tilde{\mathbf{V}}_o}{\mathbf{Z}_o} = \frac{117\angle 0}{50 + j86.7} = \frac{117\angle 0}{100\angle 1.047} = 1.17\angle(-1.047)\,\text{A}$$

The complex power, as defined in equation 6.24, is:

$$\mathbf{S} = \tilde{\mathbf{V}}_o\tilde{\mathbf{I}}_o^* = 117\angle 0 \times 1.17\angle 1.047 = 137\angle 1.047 = (68.4 + j118.5)\,\text{VA}$$

Therefore:

$$P = 68.4\,\text{W} \qquad Q = 118.5\,\text{VAR}$$

The power triangle corresponding to this circuit is drawn in Figure 6.20. The vector diagram shows how the complex power $\mathbf{S}$ results from the vector addition of the two components P and Q.

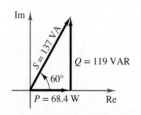

Figure 6.20 Power triangle for Example 6.8.

2. To correct the power factor to unity it is necessary to subtract 118.5 VAR. This goal can be accomplished by adding in parallel a capacitor with $Q_C = -118.5$ VAR. The required capacitance is found by:

$$X_C = \frac{|\tilde{\mathbf{V}}_o|^2}{Q_C} = -\frac{(117)^2}{118.5} = -115\,\Omega$$

The reactance X_C is related to the capacitance by:

$$jX_C = \frac{1}{j\omega C} = -\frac{j}{\omega C}$$

Thus, the result is:

$$C = -\frac{1}{\omega X_C} = -\frac{1}{377(-115)} = 23.1\,\mu\text{F}$$

3. The total current required of the source is $\tilde{\mathbf{I}}_S = \tilde{\mathbf{I}}_o + \tilde{\mathbf{I}}_c$, where:

$$\tilde{\mathbf{I}}_c = \frac{\tilde{\mathbf{V}}_S}{\mathbf{Z}_c} = (j\omega C)(117\angle 0) = (377)(23.1\,\mu\text{F})(117)\angle(\pi/2) \approx 1.02\angle 90° \text{ A}$$

Notice that $|\tilde{\mathbf{I}}_c| = |\tilde{\mathbf{V}}_S|/|X_c| \approx 117/115 \approx 1.02$ A. The total current is computed by phasor addition to be:

$$\tilde{\mathbf{I}}_S \approx 1.17\angle(-1.047) + 1.02\angle(\pi/2) \approx 0.585\angle 0 \text{ A}$$

The corrected power factor pf $= 1$ implies that the impedance of the load is now purely real; that is, $\theta_Z = 0$. Thus, the source current must now be *in phase* with the source voltage; and it is.

Comments: Notice that the magnitude of the source current is reduced by increasing the power factor. The power factor correction, which is a very common procedure in electric power systems, is illustrated in Figure 6.21.

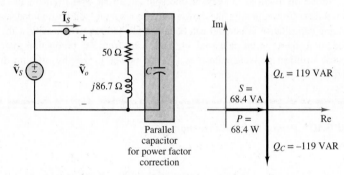

Figure 6.21 Power factor correction

EXAMPLE 6.9   **Can a Series Capacitor Be Used for Power Factor Correction?**

Problem

The circuit of Figure 6.22 suggests the use of a series capacitor for power factor correction. Why is this approach *not* a feasible alternative to the parallel capacitor approach demonstrated in Example 6.8?

Solution

Known Quantities: Source voltage; load impedance.

Find: Load (source) current.

Schematics, Diagrams, Circuits, and Given Data: $\tilde{\mathbf{V}}_S = 117\angle 0$ V; $R = 50\ \Omega$; $jX_L = j86.7\ \Omega$; $jX_C = -j86.7\ \Omega$.

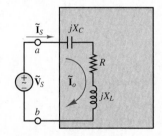

Figure 6.22 Circuit for Example 6.9.

Assumptions: All amplitudes are effective (rms) values. All angles are given in units of radians unless indicated otherwise.

Analysis: First, compute the impedance of the load between terminals a and b:

$$\mathbf{Z}_o = R + jX_L + jX_C = 50 + j86.7 - j86.7 = 50\ \Omega$$

Notice that the reactance of the capacitor was chosen so as to make the total load purely resistive. Thus, $\theta_Z = 0$ and the corrected power factor is pf = 1. So far, so good.

Next, compute the current through the series load:

$$\tilde{\mathbf{I}}_o = \tilde{\mathbf{I}}_S = \frac{\tilde{\mathbf{V}}_S}{\mathbf{Z}_o} = \frac{117\angle 0}{50} = 2.34 \text{ A}$$

The corrected power factor pf = 1 implies that the impedance of the load is now purely real; that is, $\theta_Z = 0$. Thus, the source current must now be *in phase* with the source voltage; and it is.

The problem with this approach to power factor correction is revealed by computing the initial current through the load, prior to the addition of the capacitor.

$$(\tilde{\mathbf{I}}_o)_{\text{initial}} = \frac{\tilde{\mathbf{V}}_S}{R + jX_L} = \frac{117\angle 0}{50 + j86.7} \approx 1.17\angle(-\pi/3)\,\text{A}$$

Comments: Notice the twofold increase in the source current as a result of the additional capacitor in series. Consequently, the power required by the source doubled as well. In practice, adding capacitance in parallel can be accomplished relatively easily with one large bank located somewhere on an industrial site and away from the production motors themselves. Electric utilities motivate industries to raise power factors by offering discounted rates ($/kWh).

EXAMPLE 6.10   Power Factor Correction

Problem

A capacitor is used to correct the power factor of the 100 kW and lagging pf = 0.7 load of Figure 6.23. Determine the reactive power of the load alone, and compute the capacitance required for a corrected power factor pf = 1.

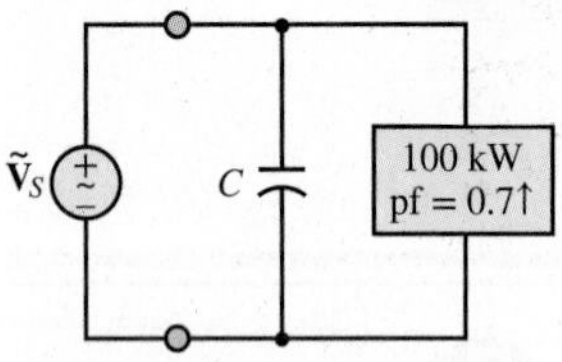

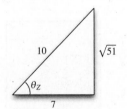

Figure 6.23 Circuit for Example 6.10.

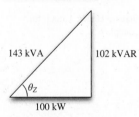

Figure 6.24 Relative dimensions of power triangle

Figure 6.25 Power triangle for Example 6.10.

Solution

Known Quantities: Source voltage; load power and power factor.

Find:

1. The reactive power Q of the load alone.
2. The capacitance C required for a corrected power factor pf = 1.

Schematics, Diagrams, Circuits, and Given Data: $\tilde{\mathbf{V}}_S = 480\angle 0 \text{ V rms}$; $P = 10^5 \text{ W}$; pf = 0.7 lagging for the load; $\omega = 377$ rad/s.

Assumptions: All amplitudes are effective (rms) values. All angles are given in units of radians unless indicated otherwise.

Analysis:

1. For the load alone, pf = 0.7 lagging or $\cos(\theta_Z) = 7/10$, and the power triangle has the *shape* shown in Figure 6.24. The real power is given as $P = 100$ kW, so the reactive power of the load can be computed using the relative triangle dimensions to be:

$$Q = P\tan(\theta_Z) = (100 \text{ kW})(\sqrt{51}/7) = 102 \text{ kVAR}$$

Since the power factor is lagging, the reactive power is positive as indicated in Table 6.2 and shown in the power triangle of Figure 6.25.

2. To set the corrected power factor to pf $= 1$ the capacitance must contribute -102 kVAR of reactive power. That is:

$$Q_C = \Delta Q = Q_{final} - Q_{initial} = 0 - 102 \text{ kVAR} = -102 \text{ kVAR}$$

Since the voltage across capacitor $\tilde{\mathbf{V}}_C$ equals the source voltage $\tilde{\mathbf{V}}_S$, the reactive power of the capacitor is:

$$Q_C = \frac{|\tilde{\mathbf{V}}_C|^2}{|X_C|} \sin(-90°) = -(\omega C)|\tilde{\mathbf{V}}_S|^2 = -(377)(480^2)C$$

Thus, to correct the power factor to pf $= 1$ (zero total reactive power), the capacitor must satisfy:

$$Q_C = -(377)(480^2)C = -102 \text{ kVAR}$$

or

$$C = \frac{102 \text{ kVAR}}{(377)(480^2)} = 1,175 \ \mu\text{F}$$

Use trigonometry and/or the Pythagorean theorem to show that the apparent power $|S| = 143$ kVA, as indicated in Figure 6.25.

Comments: Note that it is not necessary to know the load impedance to perform power factor correction; however, it is a useful exercise to compute the equivalent impedance seen by $\tilde{\mathbf{V}}_S$ and check that $\cos(\theta_Z) = 0.7$.

EXAMPLE 6.11   Power Factor Correction

Problem

Figure 6.26 shows a second load added to the circuit of Figure 6.23. Determine the capacitance required for an overall corrected power factor pf $= 1$. Draw the phasor diagram showing the relationship between $\tilde{\mathbf{I}}_C$, $\tilde{\mathbf{I}}_1$, and $\tilde{\mathbf{I}}_2$.

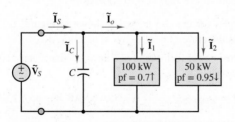

Figure 6.26 Circuit for Example 6.11.

Solution

Known Quantities: Source voltage; load power and power factor.

Find:

1. The total reactive power of loads 1 and 2.

2. The capacitance C required for an overall power factor pf $= 1$.

3. $\tilde{\mathbf{I}}_C$, $\tilde{\mathbf{I}}_1$, and $\tilde{\mathbf{I}}_2$, and construct a phasor diagram of these currents.

Schematics, Diagrams, Circuits, and Given Data: $\tilde{\mathbf{V}}_S = 480\angle 0$ V rms; $P_1 = 100$ kW; $\text{pf}_1 = 0.7$ lagging; $P_2 = 50$ kW; $\text{pf}_2 = 0.95$ leading; $\omega = 377$ rad/s.

Assumptions: All amplitudes are effective (rms) values. All angles are given in units of radians unless indicated otherwise.

Analysis:

1. Compute $\tilde{\mathbf{I}}_1$ and $\tilde{\mathbf{I}}_2$ using the relation $P = |\tilde{\mathbf{V}}||\tilde{\mathbf{I}}|\text{pf}$.

$$P_1 = |\tilde{\mathbf{V}}_S||\tilde{\mathbf{I}}_1|\cos(\theta_1) \quad \rightarrow \quad |\tilde{\mathbf{I}}_1| = \frac{P_1}{|\tilde{\mathbf{V}}_S|\cos(\theta_1)} \approx 298 \text{ A}$$

and

$$\angle\tilde{\mathbf{V}}_S = \angle\tilde{\mathbf{I}}_1 + \theta_{Z_1} \quad \rightarrow \quad \angle\tilde{\mathbf{I}}_1 = \angle\tilde{\mathbf{V}}_S - \theta_{Z_1} = 0 - \cos^{-1}(0.7) \approx -0.795 \text{ rad}$$

It is important to keep in mind that although inverse trigonometric functions are double-valued [e.g., $\cos^{-1}(0.7) \approx \pm 0.795$ rad], the power factor for load 1 is lagging such that $\theta_{Z_1} = +0.795$ rad is the correct choice.

Similarly, for load 2:

$$P_2 = |\tilde{\mathbf{V}}_S||\tilde{\mathbf{I}}_2|\cos(\theta_2) \quad \rightarrow \quad |\tilde{\mathbf{I}}_2| = \frac{P_2}{|\tilde{\mathbf{V}}_S|\cos(\theta_2)} \approx 110 \text{ A}$$

and

$$\angle\tilde{\mathbf{V}}_S = \angle\tilde{\mathbf{I}}_2 + \theta_{Z_2} \quad \rightarrow \quad \angle\tilde{\mathbf{I}}_2 = \angle\tilde{\mathbf{V}}_S - \theta_{Z_2} = 0 - \cos^{-1}(0.95) \approx +0.318 \text{ rad}$$

The power factor for load 2 is leading such that $\theta_{Z_2} = -0.318$ rad is the correct choice.

Now use the given data and the relation $Q = P\tan(\theta_Z)$ to compute the reactive power for each load.

$$Q_1 = P_1 \tan(+0.795 \text{ rad}) \approx +102 \text{ kVAR}$$

and

$$Q_2 = P_2 \tan(-0.318 \text{ rad}) \approx -16.4 \text{ kVAR}$$

The power triangles for the two loads are shown in Figures 6.27 and 6.28. The total reactive power is therefore $Q = Q_1 + Q_2 \approx 85.6$ kVAR.

2. To set the corrected power factor to $\text{pf} = 1$ the capacitance must contribute -85.6 kVAR of reactive power. That is:

$$Q_C = \Delta Q = Q_{\text{final}} - Q_{\text{initial}} = 0 - 85.6 \text{ kVAR} = -85.6 \text{ kVAR}$$

For a capacitor alone its reactive power is:

$$Q_C = \frac{|\tilde{\mathbf{V}}_C|^2}{X_C} = -(\omega C)|\tilde{\mathbf{V}}_S|^2 = -(377)(480^2)C$$

Thus, to correct the power factor to $\text{pf} = 1$ (zero total reactive power), the capacitor must satisfy:

$$Q_C = -(377)(480^2)C = -85.6 \text{ kVAR}$$

or

$$C = \frac{85.6 \text{ kVAR}}{(377)(480^2)} \approx 985 \,\mu\text{F}$$

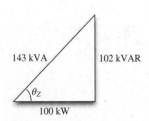

Figure 6.27 Power triangle for load 1

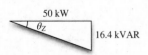

Figure 6.28 Power triangle for load 2

3. To compute the capacitor current it is not possible to use $P = |\tilde{\mathbf{V}}||\tilde{\mathbf{I}}|\text{pf}$ because $P = 0$ and pf = 0 for a capacitor. Instead, the generalized Ohm's law provides an alternative approach.

$$\tilde{\mathbf{V}}_C = \tilde{\mathbf{I}}_C \mathbf{Z}_C \quad \rightarrow \quad |\tilde{\mathbf{I}}_C| = \frac{|\tilde{\mathbf{V}}_C|}{|\mathbf{Z}_C|} = \omega C|\tilde{\mathbf{V}}_C| \approx 178.2 \text{ A}$$

where $\tilde{\mathbf{V}}_C = \tilde{\mathbf{V}}_S$. The phase angle of $\tilde{\mathbf{I}}_C$ is:

$$\angle\tilde{\mathbf{I}}_C = \angle\tilde{\mathbf{V}}_C - \theta_{Z_C} = 0 - (-\pi/2) = +\pi/2 \text{ rad}$$

The current phasor diagram can now be drawn as shown in Figure 6.29.

Comment: The power triangle suggests that the capacitor current can also be calculated using the relation $Q_C = |\tilde{\mathbf{V}}_C||\tilde{\mathbf{I}}_C| \sin(\theta_C)$, where $\theta_C = -\pi/2$ and $Q_C = |\tilde{\mathbf{V}}_C|^2/X_C = -(\omega C)|\tilde{\mathbf{V}}_C|^2$. Try it!

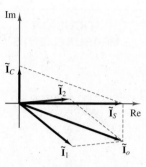

Figure 6.29 Phasor diagram for Example 6.11.

CHECK YOUR UNDERSTANDING

Two cases of the voltage across and the current through a load are given below. Determine the power factor of the load, and whether it is leading or lagging, for each case.

  a. $v(t) = 540 \cos(\omega t + 15°)$ V, $i(t) = 2 \cos(\omega t + 47°)$ A
  b. $v(t) = 155 \cos(\omega t - 15°)$ V, $i(t) = 2 \cos(\omega t - 22°)$ A

CHECK YOUR UNDERSTANDING

Determine if a load is capacitive or inductive, given the following facts:

  a. pf = 0.87, leading
  b. pf = 0.42, leading
  c. $v(t) = 42 \cos(\omega t)$ V, $i(t) = 4.2 \sin(\omega t)$ A [*Hint:* $\sin(\omega t)$ lags $\cos(\omega t)$.]
  d. $v(t) = 10.4 \cos(\omega t - 22°)$ V, $i(t) = 0.4 \cos(\omega t - 22°)$ A

CHECK YOUR UNDERSTANDING

Compute the power factor for an inductive load with $L = 100$ mH in series with $R = 0.4 \ \Omega$. Assume $\omega = 377$ rad/s.

FOCUS ON MEASUREMENTS

The Wattmeter

The instrument used to measure power is called a wattmeter. The external part of a wattmeter consists of four connections and a metering mechanism that displays the amount of real power dissipated by a circuit. The external and internal appearance of a wattmeter is depicted in Figure 6.30. Inside the wattmeter are two coils: a current-sensing coil and a voltage-sensing coil. In this example, we assume for simplicity that the impedance of the current-sensing coil $\mathbf{Z}_I$ is zero and that the impedance of the voltage-sensing coil $\mathbf{Z}_V$ is infinite. In practice, this will not necessarily be true; some correction mechanism will be required to account for the impedance of the sensing coils.

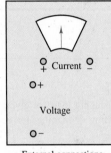
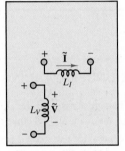

External connections   Wattmeter coils (inside)

Figure 6.30 Wattmeter connections.

A wattmeter should be connected as shown in Figure 6.31 to provide both current and voltage measurements. We see that the current-sensing coil is placed in series with the load and that the voltage-sensing coil is placed in parallel with the load. In this manner, the wattmeter is seeing the current through and the voltage across the load. Remember that the power dissipated by a circuit element is related to these two quantities. The wattmeter, then, is constructed to provide a readout of the real power absorbed by the load: $P = \mathrm{Re}(S) = \mathrm{Re}(\tilde{\mathbf{V}}\tilde{\mathbf{I}}^*)$.

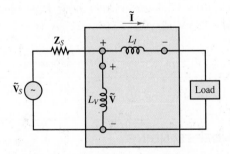

Figure 6.31 How to connect a wattmeter in a circuit.

(Continued)

Problem:

1. For the circuit shown in Figure 6.32, show the connections of a wattmeter between the ideal voltage source and the load and find the power dissipated by the load.

2. Show the connections that will determine the power dissipated by R_2. What should the meter read?

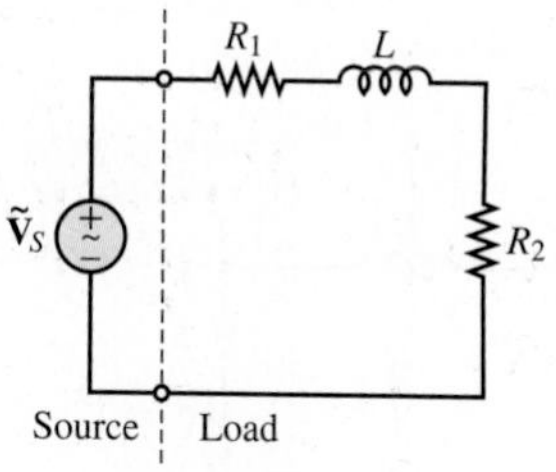

$v_S(t) = 156 \cos(377t)$
$R_1 = 10 \ \Omega$
$R_2 = 5 \ \Omega$
$L = 20 \ \text{mH}$

Figure 6.32 Circuit for power measurement example.

Solution:

1. To measure the power dissipated by the load, we must know the current through and the voltage across the entire load circuit. This means that the wattmeter must be connected as shown in Figure 6.33. The wattmeter should read

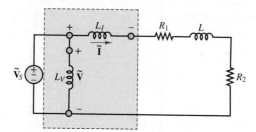

Figure 6.33 Circuit with wattmeter inserted.

$$P = \text{Re}[\tilde{\mathbf{V}}_S \tilde{\mathbf{I}}^*] = \text{Re}\left[\left(\frac{156}{\sqrt{2}} \angle 0 \right) \left(\frac{(156/\sqrt{2}) \angle 0}{R_1 + R_2 + j\omega L} \right)^* \right]$$

$$= \text{Re}\left[110 \angle 0° \left(\frac{110 \angle 0}{15 + j7.54} \right)^* \right]$$

$$= \text{Re}\left[110 \angle 0° \left(\frac{110 \angle 0}{16.79 \angle 0.466} \right)^* \right] = \text{Re}\left[\frac{110^2}{16.79 \angle (-0.466)} \right]$$

$$= \text{Re}[720.67 \angle 0.466]$$

$$= 643.88 \ \text{W}$$

(*Continued*)

(*Concluded*)

2. To measure the power dissipated by R_2 alone, we must measure the current through R_2 and the voltage across R_2 *alone*. The connection is shown in Figure 6.34. The meter will read

$$P = |\tilde{\mathbf{I}}^2| R_2 = \left[\frac{110}{(15^2 + 7.54^2)^{1/2}} \right]^2 \times 5 = \frac{110^2}{15^2 + 7.54^2} \times 5$$

$$= 215 \text{ W}$$

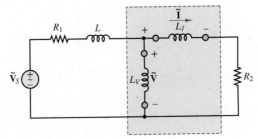

Figure 6.34 Circuit with wattmeter inserted to measure only the power dissipated by R_2

Power Factor

Problem:

A capacitor is being used to correct the power factor of a load to unity, as shown in Figure 6.35. The capacitor value is varied, and measurements of the total current are taken. Explain how it is possible to zero in on the capacitance value necessary to bring the power factor to unity just by monitoring the current $\tilde{\mathbf{I}}_S$.

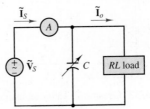

Figure 6.35 Circuit for illustration of power factor correction.

Solution:

The current through the load is

$$\tilde{\mathbf{I}}_o = \frac{\tilde{V}_S \angle 0°}{R + j\omega L} = \frac{\tilde{V}_S}{R^2 + \omega^2 L^2}(R - j\omega L)$$

$$= \frac{\tilde{V}_S R}{R^2 + \omega^2 L^2} - j\frac{\tilde{V}_S \omega L}{R^2 + \omega^2 L^2}$$

(*Continued*)

(*Concluded*)

The current through the capacitor is

$$\tilde{\mathbf{I}}_C = \frac{\tilde{V}_S \angle 0°}{1/j\omega C} = j\tilde{V}_S \omega C$$

The source current to be measured is

$$\tilde{\mathbf{I}}_S = \tilde{\mathbf{I}}_o + \tilde{\mathbf{I}}_C = \frac{\tilde{V}_S R}{R^2 + \omega^2 L^2} + j\left(\tilde{V}_S \omega C - \frac{\tilde{V}_S \omega L}{R^2 + \omega^2 L^2} \right)$$

The magnitude of the source current is

$$\tilde{I}_S = \sqrt{\left(\frac{\tilde{V}_S R}{R^2 + \omega^2 L^2} \right)^2 + \left(\tilde{V}_S \omega C - \frac{\tilde{V}_S \omega L}{R^2 + \omega^2 L^2} \right)^2}$$

We know that when the load is a pure resistance, then the current and voltage are in phase, the power factor is 1, and all the power delivered by the source is dissipated by the load as real power. This corresponds to equating the imaginary part of the expression for the source current to zero or, equivalently, to the following expression:

$$\frac{\tilde{V}_S \omega L}{R^2 + \omega^2 L^2} = \tilde{V}_S \omega C$$

in the expression for $|\tilde{I}_S|$. Thus, the magnitude of the source current is actually a minimum when the power factor is unity! It is therefore possible to "tune" a load to a unity pf by observing the readout of the ammeter while changing the value of the capacitor and selecting the capacitor value that corresponds to the lowest source current value.

6.4 TRANSFORMERS

Two separate AC circuits are often interfaced by a **transformer**, which acts as a magnetic coupling and *transforms* the voltage and current at the interface (e.g., by matching the high-voltage, low-current output of one circuit to the low-voltage, high-current input required by the other). Transformers play a major role in electric power engineering and are a necessary part of the electric power distribution network. The objective of this section is to introduce the ideal transformer and the concepts of impedance reflection and impedance matching. The operations of practical transformers, and more advanced models, are discussed in Chapter 12.

The Ideal Transformer

The ideal transformer consists of two coils coupled to each other by a magnetic medium. There is no conducting electrical connection between the coils. The input side of a transformer is known as the **primary** while the output side is known as the **secondary**. The number of turns in the primary and secondary coils are designated n_1 and n_2, respectively. The **turns ratio** N is defined by:

$$N = \frac{n_2}{n_1} \tag{6.29}$$

Figure 6.36 illustrates the convention by which voltages and currents are usually assigned at a transformer. The solid black dots in Figure 6.36 are used to mark coil terminals that have the same polarity.

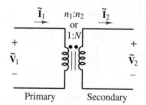

Figure 6.36 Ideal transformer

Recall from Faraday's law that each coil experiences *self-induction* in that a time-varying current through a coil produces a time-varying magnetic flux through the coil itself, which, in turn, induces a potential difference opposing the time-varying magnetic flux. The net effect of this self-induction is expressed by the inductance L of a coil. However, when two coils are present, as in a transformer, both coils also experience *mutual induction* in that some of the time-varying magnetic flux due to one coil passes through the other coil and induces another opposing potential difference. The net effect of the mutual induction is expressed by the mutual inductance M of the two coils. Both L and M contribute to the behavior of a transformer.

Notice the emphasis on *time variations* in the previous paragraph. One result of Faraday's law is that a leave in current through a coil, which generates a constant magnetic field, induces no opposing reaction within the coil itself (no self-induction) nor within any nearby coil (no mutual induction). Instead, a coil acts as a short-circuit in the presence of leave in current and transformers perform no useful function in DC circuits. See Chapter 12 for further discussion of Faraday's law as it relates to electromechanics.

As depicted in Figure 6.36, the relationships between primary and secondary currents and voltages in an ideal transformer are:

$$\begin{aligned} \tilde{\mathbf{V}}_2 &= N\tilde{\mathbf{V}}_1 \\ \tilde{\mathbf{I}}_2 &= \frac{\tilde{\mathbf{I}}_1}{N} \end{aligned} \qquad \text{Ideal transformer} \tag{6.30}$$

When $N > 1$, $|\tilde{\mathbf{V}}_2| > |\tilde{\mathbf{V}}_1|$ and a transformer is called a **step-up transformer**. When $N < 1$, $|\tilde{\mathbf{V}}_2| < |\tilde{\mathbf{V}}_1|$ and a transformer is called a **step-down transformer**. Either side of an ideal transformer can be used as the primary; thus, to produce a step-up transformer from a step-down transformer one only need exchange the primary and secondary connections. (Exchanging the primary and secondary by mistake can lead to significant dangers in a laboratory experiment!) Finally, when $N = 1$, a transformer is called an **isolation transformer**, which can be used to electrically couple or isolate two circuits and adjust the output and input impedances at the interface of two circuits.

A comparison of the complex power at the primary and secondary terminals of an ideal transformer reveals that they are the same:

$$\mathbf{S}_1 = \tilde{\mathbf{I}}_1^* \tilde{\mathbf{V}}_1 = N\tilde{\mathbf{I}}_2^* \frac{\tilde{\mathbf{V}}_2}{N} = \tilde{\mathbf{I}}_2^* \tilde{\mathbf{V}}_2 = \mathbf{S}_2 \tag{6.31}$$

That is, **ideal transformers conserve power**.

As shown in Figure 6.37, the secondary coil of many practical transformers is *center-tapped*, which splits the secondary voltage into two equal halves. This type of transformer is found at the entry of a power line into a household, where a high-voltage primary is transformed to 240 V as well as split into two 120-V lines. Referring to Figure 6.37, $\tilde{\mathbf{V}}_2$ and $\tilde{\mathbf{V}}_3$ would both provide 120 V for common household appliances while $(\tilde{\mathbf{V}}_2 + \tilde{\mathbf{V}}_3)$ would provide 240 V for higher-powered devices, such as clothes dryers and electric ranges.

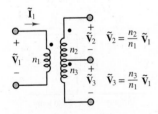

Figure 6.37 Center-tapped transformer

Impedance Reflection

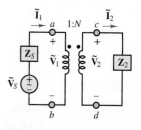

Figure 6.38 Operation of an ideal transformer

Transformers are commonly used to couple one AC circuit to another, as depicted in Figure 6.38, where an AC Thévenin source network is connected to a load $\mathbf{Z}_2$ by means of a transformer.

The equivalent impedance seen by the Thévenin source is that of the entire network to the right of terminals a and b. Applying the definition of equivalent impedance and using the ideal transformer relations from equation 6.30, the result is:

$$
\begin{aligned}
\mathbf{Z}_1 \equiv \frac{\tilde{\mathbf{V}}_1}{\tilde{\mathbf{I}}_1} &= \frac{\tilde{\mathbf{V}}_2}{N}\frac{1}{N\tilde{\mathbf{I}}_2} \\
&= \frac{1}{N^2}\frac{\tilde{\mathbf{V}}_2}{\tilde{\mathbf{I}}_2} \\
&= \frac{1}{N^2}\mathbf{Z}_2
\end{aligned}
\qquad (6.32)
$$

Thus, the equivalent impedance seen by the AC Thévenin source is the load impedance $\mathbf{Z}_2$ reduced by the factor $1/N^2$.

Likewise, the equivalent network seen by $\mathbf{Z}_2$ is the Thévenin equivalent of the entire network to the left of terminals c and d. When $\mathbf{Z}_2$ is replaced by an open-circuit, $\tilde{\mathbf{I}}_2 = 0$ and the Thévenin (open-circuit) voltage is:

$$
\tilde{\mathbf{V}}_T = (\tilde{\mathbf{V}}_2)_{\text{OC}} = N\tilde{\mathbf{V}}_1
\qquad (6.33)
$$

However, since $\tilde{\mathbf{I}}_1 = N\tilde{\mathbf{I}}_2 = 0$, the voltage drop across $\mathbf{Z}_S$ is zero such that $\tilde{\mathbf{V}}_1 = \tilde{\mathbf{V}}_S$ with the result:

$$
\tilde{\mathbf{V}}_T = (\tilde{\mathbf{V}}_2)_{\text{OC}} = N\tilde{\mathbf{V}}_S
\qquad (6.34)
$$

When $\mathbf{Z}_2$ is replaced by a short-circuit, $\tilde{\mathbf{V}}_2 = 0$ and the short-circuit current is:

$$
(\mathbf{I}_2)_{\text{SC}} = \frac{\mathbf{I}_1}{N}
\qquad (6.35)
$$

However, since $\tilde{\mathbf{V}}_1 = \tilde{\mathbf{V}}_2/N = 0$, the voltage drop across $\mathbf{Z}_S$ is $\tilde{\mathbf{V}}_S$ such that $\tilde{\mathbf{I}}_1 = \tilde{\mathbf{V}}_S/\mathbf{Z}_S$ with the result:

$$
(\tilde{\mathbf{I}}_2)_{\text{SC}} = \frac{1}{N}\frac{\tilde{\mathbf{V}}_S}{\mathbf{Z}_S}
\qquad (6.36)
$$

Thus, the Thévenin equivalent impedance seen by $\mathbf{Z}_2$ is:

$$
\mathbf{Z}_T = \frac{(\tilde{\mathbf{V}}_2)_{\text{OC}}}{(\tilde{\mathbf{I}}_2)_{\text{SC}}} = N\tilde{\mathbf{V}}_S\frac{N\mathbf{Z}_S}{\tilde{\mathbf{V}}_S} = N^2\mathbf{Z}_S
\qquad (6.37)
$$

Thus, the equivalent impedance seen by $\mathbf{Z}_2$ is the source impedance $\mathbf{Z}_S$ multiplied by N^2.

Figure 6.39 summarizes and illustrates these effects, which are known as **impedance reflection** across a transformer and which play an important role in power transfer.

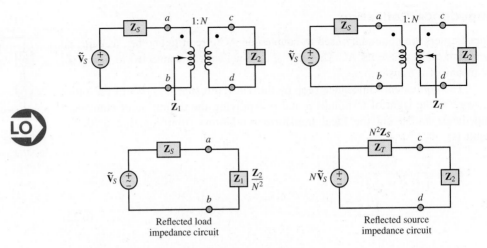

Figure 6.39 Impedance reflection across a transformer

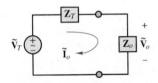

Figure 6.40 The maximum power transfer problem in AC circuits

Maximum Power Transfer

Recall that in resistive DC circuits, maximum power is transferred to a load when the load equals the Thévenin equivalent resistance of the source network. For AC circuits, the analogous maximum power transfer condition is known as **impedance matching**.

Consider the general form of an AC circuit, shown in Figure 6.40, and assume that the source impedance $\mathbf{Z}_T$ is:

$$\mathbf{Z}_T = R_T + jX_T \tag{6.38}$$

What value of the load $\mathbf{Z}_o$ results in the maximum *real power* transfer to the load itself? The real power absorbed by the load is:

$$P_o = \tilde{V}_o \tilde{I}_o \cos\theta_{Z_o} = \mathrm{Re}(\tilde{\mathbf{V}}_o \tilde{\mathbf{I}}_o^*) \tag{6.39}$$

Apply voltage division and the generalized Ohm's law to find:

$$\tilde{\mathbf{V}}_o = \frac{\mathbf{Z}_o}{\mathbf{Z}_T + \mathbf{Z}_o}\tilde{\mathbf{V}}_T \qquad \tilde{\mathbf{I}}_o = \frac{\tilde{\mathbf{V}}_T}{\mathbf{Z}_T + \mathbf{Z}_o} \tag{6.40}$$

Let $\mathbf{Z}_o = R_o + jX_o = |\mathbf{Z}_o|\cos\theta_{Z_o} + j|\mathbf{Z}_o|\sin\theta_{Z_o}$, and since $\tilde{V}_o = |\tilde{\mathbf{V}}_o|$ and $\tilde{I}_o = |\tilde{\mathbf{I}}_o|$, the real power absorbed by the load can be expressed as:

$$P_o = \frac{|\mathbf{Z}_o|}{|\mathbf{Z}_T + \mathbf{Z}_o|}\tilde{V}_T \times \frac{1}{|\mathbf{Z}_T + \mathbf{Z}_o|}\tilde{V}_T \times \frac{R_o}{|\mathbf{Z}_o|} \tag{6.41}$$

Or, after simplification:

$$P_o = \frac{R_o}{|\mathbf{Z}_T + \mathbf{Z}_o|^2}\tilde{V}_T^2 = \frac{R_o}{(R_T + R_o)^2 + (X_T + X_o)^2}\tilde{V}_T^2 \tag{6.42}$$

The condition for the maximum value of P_o can be found by solving:

$$dP_o = \frac{\partial P_o}{\partial R_o}dR_o + \frac{\partial P_o}{\partial X_o}dX_o = 0 \tag{6.43}$$

or

$$\frac{\partial P_o}{\partial R_o} = 0 \qquad \text{and} \qquad \frac{\partial P_o}{\partial X_o} = 0 \tag{6.44}$$

Both of these conditions are satisfied when $R_o = R_T$ and $X_o = -X_T$. That is, the condition for maximum real power transfer to a load is $\mathbf{Z}_o = \mathbf{Z}_T^*$.

$$\mathbf{Z}_o = \mathbf{Z}_T^* \qquad \text{Maximum power transfer} \tag{6.45}$$

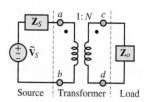

Maximum power is transferred to the load when its impedance equals the complex conjugate of the Thévenin equivalent impedance of the source. When this condition is satisfied, the load and source impedances are *matched*.

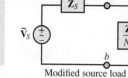

Modified source load

Figure 6.41 Maximum power transfer in an AC circuit with a transformer

In some cases, it may not be possible to match the load to the source because of practical limitations. In these situations, it may be possible to use a transformer as the interface between the source and the load to achieve maximum power transfer. Figure 6.41 illustrates how the reflected load impedance, as seen by the source, is equal to $\mathbf{Z}_o/N^2$, such that the condition for maximum power transfer is:

$$\frac{\mathbf{Z}_o}{N^2} = \mathbf{Z}_S^*$$

$$R_o = N^2 R_S \tag{6.46}$$

$$X_o = -N^2 X_S$$

EXAMPLE 6.12   Ideal Transformer Turns Ratio

Problem

We require a transformer (see Figure 6.42) to output 500 mA at 24 V from a 120 V rms input line source. The primary has $n_1 = 3,000$ turns. How many turns are required in the secondary? What is the primary current?

Solution

Known Quantities: Primary and secondary voltages; secondary current; number of turns in the primary coil.

Find: n_2 and $\tilde{\mathbf{I}}_1$.

Schematics, Diagrams, Circuits, and Given Data: $\tilde{\mathbf{V}}_1 = 120$ V; $\tilde{\mathbf{V}}_2 = 24$ V; $\tilde{\mathbf{I}}_2 = 500$ mA; $n_1 = 3,000$ turns.

Assumptions: All amplitudes are effective (rms) values. All angles are given in units of radians.

Analysis: Use equation 6.30 to compute the number of turns in the secondary coil:

$$\frac{\tilde{\mathbf{V}}_1}{n_1} = \frac{\tilde{\mathbf{V}}_2}{n_2} \qquad n_2 = n_1 \frac{\tilde{\mathbf{V}}_2}{\tilde{\mathbf{V}}_1} = 3,000 \times \frac{24}{120} = 600 \text{ turns}$$

Figure 6.42 Example 6.12

Again, use equations 6.29 and 6.30 to compute the primary current:

$$n_1\tilde{\mathbf{I}}_1 = n_2\tilde{\mathbf{I}}_2 \qquad \tilde{\mathbf{I}}_1 = \frac{n_2}{n_1}\tilde{\mathbf{I}}_2 = \frac{600}{3,000} \times 500 = 100 \text{ mA}$$

Comment: Note that since the transformer does not affect the phase of the voltages and currents, it was possible to solve the problem using only the rms amplitudes.

EXAMPLE 6.13   Center-Tapped Transformer

Problem

An ideal center-tapped power transformer (Figure 6.43) has a 4,800-V primary and a 240-V secondary. The center-tap is located such that $\tilde{V}_2 = \tilde{V}_3 = 120$ V. Three resistive loads are attached to the secondary terminals. Compute the current in the primary assuming that R_2, R_3, and R_4 each absorb P_2, P_3, and P_4, respectively. Also compute the current through each load and the resistance of each load.

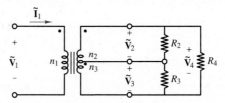

Figure 6.43 Example 6.13

Solution

Known Quantities: Primary and secondary voltages; load power ratings.

Find: $\tilde{I}_{\text{primary}} = |\tilde{\mathbf{I}}|$

Schematics, Diagrams, Circuits, and Given Data: $\tilde{\mathbf{V}}_1 = 4,800$ V rms; $\tilde{\mathbf{V}}_2 = 120$ V rms; $\tilde{\mathbf{V}}_3 = 120$ V rms; $P_2 = 5,000$ W; $P_3 = 1,000$ W; $P_4 = 1,500$ W.

Assumptions: All amplitudes are effective (rms) values. All angles are given in units of radians unless indicated otherwise. The transformer is ideal.

Analysis: Power is conserved for an ideal transformer; thus:

$$\mathbf{S}_{\text{primary}} = \mathbf{S}_{\text{secondary}}$$

Since each load is purely resistive, $\theta_Z = 0$ and pf $= \cos\theta_Z = 1$ such that:

$$|\mathbf{S}|_{\text{secondary}} = P_{\text{secondary}} = P_2 + P_3 + P_4 = 7,500 \text{ W}$$

Since $|\mathbf{S}|_{\text{primary}} = |\mathbf{S}|_{\text{secondary}}$:

$$\tilde{V}_{\text{primary}} \times \tilde{I}_{\text{primary}} = 7,500 \text{ W}$$

Thus:

$$\tilde{I}_{\text{primary}} = \frac{7,500 \text{ W}}{4,800 \text{ V rms}} = 1.5625 \text{ A rms}$$

The current through each resistor is simply:

$$\tilde{I}_2 = \frac{P_2}{\tilde{V}_2} = \frac{5{,}000 \text{ W}}{120 \text{ V rms}} \approx 41.7 \text{ A rms}$$

$$\tilde{I}_3 = \frac{P_3}{\tilde{V}_3} = \frac{1{,}000 \text{ W}}{120 \text{ V rms}} \approx 8.3 \text{ A rms}$$

$$\tilde{I}_4 = \frac{P_4}{\tilde{V}_2 + \tilde{V}_3} = \frac{1{,}500 \text{ W}}{240 \text{ V rms}} = 6.25 \text{ A rms}$$

The resistor values are:

$$\tilde{R}_2 = \frac{P_2}{\tilde{I}_2^2} = 2.88 \ \Omega$$

$$\tilde{R}_3 = \frac{P_3}{\tilde{I}_3^2} = 14.4 \ \Omega$$

$$\tilde{R}_4 = \frac{P_4}{\tilde{I}_4^2} = 38.4 \ \Omega$$

Comments: The calculations in this example were particularly straightforward because the load was purely resistive, such that $\theta_Z = 0$, the power triangle is flat, and the apparent power S equals the real power P. When the load is complex, $\theta_Z > 0$, the power triangle is not flat, and the apparent power S equals $P \cos \theta_Z$. Then, the calculations are more complicated.

Also, KCL can be used to determine the current drawn from/to the outside and center taps. Try it!

EXAMPLE 6.14   Use of Transformers to Improve Power Line Efficiency

Problem

Figure 6.44 illustrates the use of transformers in electric power transmission lines. The line voltage is transformed before and after being transmitted over long distances. This example illustrates the efficiency gained through the use of transformers. For the sake of simplicity, ideal transformers and simple resistive models for the generator, transmission line, and load have been assumed.

Solution

Known Quantities: Values of circuit elements.

Find: Calculate the power transfer efficiency for the two circuits of Figure 6.44.

Schematics, Diagrams, Circuits, and Given Data: Step-up transformer turns ratio is N, step-down transformer turns ratio is $M = 1/N$. All transformers are ideal.

Assumptions: None.

Analysis: Since the load and source currents are equal in Figure 6.44(a), the power transmission efficiency is:

$$\eta = \frac{P_{\text{load}}}{P_{\text{source}}} = \frac{\tilde{V}_{\text{load}} \tilde{I}_{\text{load}}}{\tilde{V}_{\text{source}} \tilde{I}_{\text{load}}} = \frac{\tilde{V}_{\text{load}}}{\tilde{V}_{\text{source}}} = \frac{R_{\text{load}}}{R_{\text{source}} + R_{\text{line}} + R_{\text{load}}}$$

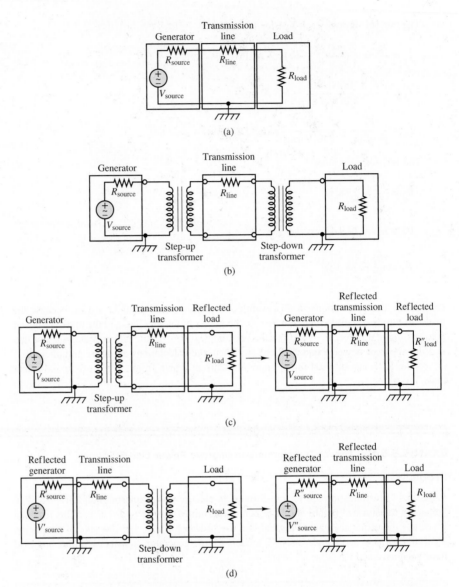

Figure 6.44 Electric power transmission: (a) direct power transmission; (b) power transmission with transformers; (c) equivalent circuit seen by generator; (d) equivalent circuit seen by load

In Figure 6.44(b), transformers are introduced between each of the three portions of the overall circuit. The equivalent load resistance seen by the transmission line (or "reflected" by the step-down transformer) is found from equation 6.32 to be:

$$R'_{\text{load}} = \frac{1}{M^2} R_{\text{load}} = N^2 R_{\text{load}}$$

Now, the step-up transformer sees the equivalent impedance $R'_{\text{load}} + R_{\text{line}}$. The resistance seen by the generator (or "reflected" by the step-up transformer) is:

$$R''_{\text{load}} = \frac{1}{N^2} (R'_{\text{load}} + R_{\text{line}}) = R_{\text{load}} + \frac{1}{N^2} R_{\text{line}}$$

These transformations are depicted in Figure 6.44(c). The effect of the two transformers is to reduce the line resistance seen by the source by N^2. The source current is:

$$\tilde{I}_{\text{source}} = \frac{V_{\text{source}}}{R_{\text{source}} + R''_{\text{load}}} = \frac{\tilde{V}_{\text{source}}}{R_{\text{source}} + (1/N^2)R_{\text{line}} + R_{\text{load}}}$$

The source power is:

$$P_{\text{source}} = \frac{\tilde{V}^2_{\text{source}}}{R_{\text{source}} + (1/N^2)R_{\text{line}} + R_{\text{load}}}$$

The same process can be repeated starting from the left and reflecting the source circuit to the right of the step-up transformer:

$$\tilde{V}'_{\text{source}} = N\tilde{V}_{\text{source}} \quad\text{and}\quad R'_{\text{source}} = N^2 R_{\text{source}}$$

Now the circuit to the left of the step-down transformer comprises the series combination of $\tilde{V}'_{\text{source}}$, R'_{source}, and R_{line}, which can be reflected to the right of the step-down transformer to obtain $\tilde{V}''_{\text{source}} = M\tilde{V}'_{\text{source}} = \tilde{V}_{\text{source}}$, $R'_{\text{source}} = M^2 R'_{\text{source}} = R_{\text{source}}$, $R'_{\text{line}} = M^2 R_{\text{line}}$, and R_{load} in series. These transformations are depicted in Figure 6.44(d). Thus, the load voltage, current, and power are:

$$\tilde{I}_{\text{load}} = \frac{\tilde{V}_{\text{source}}}{R_{\text{source}} + (1/N^2)R_{\text{line}} + R_{\text{load}}}$$

$$\tilde{V}_{\text{load}} = \tilde{V}_{\text{source}} \frac{R_{\text{load}}}{R_{\text{source}} + (1/N^2)R_{\text{line}} + R_{\text{load}}}$$

$$P_{\text{load}} = \tilde{I}_{\text{load}}\tilde{V}_{\text{load}} = \frac{\tilde{V}^2_{\text{source}} R_{\text{load}}}{\left[R_{\text{source}} + (1/N^2)R_{\text{line}} + R_{\text{load}}\right]^2}$$

Finally, the power efficiency can be computed as the ratio of the load to source power:

$$\eta = \frac{P_{\text{load}}}{P_{\text{source}}} = \frac{\tilde{V}^2_{\text{source}} R_{\text{load}}}{\left[R_{\text{source}} + (1/N^2)R_{\text{line}} + R_{\text{load}}\right]^2} \frac{R_{\text{source}} + (1/N^2)R_{\text{line}} + R_{\text{load}}}{\tilde{V}^2_{\text{source}}}$$

$$= \frac{R_{\text{load}}}{R_{\text{source}} + (1/N^2)R_{\text{line}} + R_{\text{load}}}$$

Notice that the power transmission efficiency calculated for Figure 6.44(a) was improved by reducing the effect of the line resistance by a factor of $1/N^2$.

EXAMPLE 6.15   Maximum Power Transfer Through a Transformer

Problem

Find the transformer turns ratio N and the load reactance X_o that results in maximum power transfer in the transformer shown in Figure 6.45.

Solution

Known Quantities: Source voltage, frequency, and impedance; load resistance.

Find: Transformer turns ratio and load reactance.

Schematics, Diagrams, Circuits, and Given Data: $\tilde{V}_S = 240\angle 0\text{ V rms}$; $R_S = 10\ \Omega$; $L_S = 0.1\text{ H}$; $R_o = 400\ \Omega$; $\omega = 377\text{ rad/s}$.

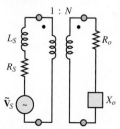

Figure 6.45 Circuit for Example 6.15.

Assumptions: All amplitudes are effective (rms) values. All angles are given in units of radians. The transformer is ideal.

Analysis: The requirements for maximum power transfer, as given by equation 6.46, are $R_o = N^2 R_S$ and $X_o = -N^2 X_S = -N^2 (\omega \times 0.1)$. Thus:

$$N^2 = \frac{R_o}{R_S} = \frac{400}{10} = 40 \qquad N = \sqrt{40} = 6.325$$

$$X_o = -40 \times 37.7 = -1,508 \ \Omega$$

Thus, the load reactance should be a capacitor with value:

$$C = -\frac{1}{X_o \omega} = -\frac{1}{(-1,508)(377)} = 1.76 \ \mu F$$

CHECK YOUR UNDERSTANDING

With reference to Example 6.12, compute the number of primary turns required if $n_2 = 600$ but the transformer is required to deliver 1 A. What is the primary current now?

Answers: $n_1 = 3,000;\ I_1 = 200$ mA

CHECK YOUR UNDERSTANDING

If the transformer of Example 6.12 has 300 turns in the secondary coil, how many turns will the primary require?

Answer: $n_2 = 6,000$

CHECK YOUR UNDERSTANDING

Assume that the generator produces a source voltage of 480 V rms, and that $N = 300$. Further assume that the source impedance is 2 Ω, the line impedance is also 2 Ω, and that the load impedance is 8 Ω. Calculate the efficiency improvement for the circuit of Figure 6.37(b) over the circuit of Figure 6.37(a).

Answer: 80% vs. 67%

CHECK YOUR UNDERSTANDING

The transformer shown in Figure 6.46 is ideal. Assume that $\mathbf{Z}_S = 1,800 \ \Omega$ and $\mathbf{Z}_o = 8 \ \Omega$ to find the turns ratio N that will ensure maximum power transfer to the load.

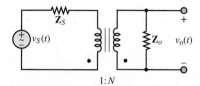

Figure 6.46 Figure for Check Your Understanding.

Now assume that $N = 5.4$ and $\mathbf{Z}_o = 2 + j10 \ \Omega$. Find the source impedance $\mathbf{Z}_S$ that will ensure maximum power transfer to the load.

Answers: $N = 0.0667; \ \mathbf{Z}_S = 0.0686 - j0.3429 \ \Omega$

6.5 THREE-PHASE POWER

The material presented so far in this chapter has dealt exclusively with **single-phase AC power**, which implies a single sinusoidal source. However, most of the AC power used today is generated and distributed as **three-phase power**, which implies three sinusoidal sources, each out of phase with the other. The primary benefit is efficiency: The weight of the conductors and other components in a three-phase system is much lower than that in a single-phase system delivering the same amount of power. Further, while the power produced by a single-phase system has a pulsating nature (recall the results of section 6.1), a three-phase system can deliver a steady, constant supply of power. For example, later in this section it will be shown that a three-phase generator producing three **balanced voltages**—that is, voltages of equal amplitude and frequency displaced in phase by 120°—has the property of delivering constant instantaneous power.

The change to three-phase AC power systems from the early DC system proposed by Edison was due to a number of reasons: the efficiency resulting from transforming voltages up and down to minimize transmission losses over long distances, the ability to deliver constant power, a more efficient use of conductors, and the ability to provide starting torque for industrial motors.

Consider a three-phase source connected in a **wye (Y) configuration**, as shown in Figure 6.47. Each of the three voltages is 120° out of phase with the

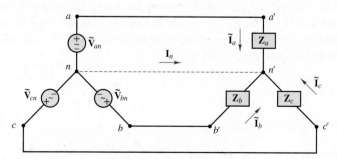

Figure 6.47 Balanced three-phase AC circuit

others, such that:

$$\begin{aligned} \tilde{\mathbf{V}}_{an} &= \tilde{V}_{an}\angle 0° \\ \tilde{\mathbf{V}}_{bn} &= \tilde{V}_{bn}\angle-(120°) \\ \tilde{\mathbf{V}}_{cn} &= \tilde{V}_{cn}\angle(-240°) = \tilde{V}_{cn}\angle 120° \end{aligned} \qquad \text{Phase voltages} \tag{6.47}$$

If the three-phase source is *balanced*, then:

$$\tilde{\mathbf{V}}_{an} + \tilde{\mathbf{V}}_{bn} + \tilde{\mathbf{V}}_{cn} = 0 \qquad \text{Balanced phase voltages} \tag{6.48}$$

For three balanced **phase voltages** each separated by 120°, the phase amplitudes are also equal:

$$\tilde{V}_{an} = \tilde{V}_{bn} = \tilde{V}_{cn} = \tilde{V} \tag{6.49}$$

The result is the so-called **positive abc sequence**, as shown in Figure 6.48. In the wye configuration, the three phase voltages share a common *neutral node*, denoted by *n*.

It is also possible to define **line voltages** as the potential differences between lines *aa′* and *bb′*, lines *aa′* and *cc′*, and lines *bb′* and *cc′*. Each line voltage is related to the phase voltages by:

$$\begin{aligned} \tilde{\mathbf{V}}_{ab} &= \tilde{\mathbf{V}}_{an} - \tilde{\mathbf{V}}_{bn} = \sqrt{3}\tilde{V}\angle 30° \\ \tilde{\mathbf{V}}_{bc} &= \tilde{\mathbf{V}}_{bn} - \tilde{\mathbf{V}}_{cn} = \sqrt{3}\tilde{V}\angle(-90°) \\ \tilde{\mathbf{V}}_{ca} &= \tilde{\mathbf{V}}_{cn} - \tilde{\mathbf{V}}_{an} = \sqrt{3}\tilde{V}\angle 150° \end{aligned} \qquad \text{Line voltages} \tag{6.50}$$

It is instructive to note that the circuit of Figure 6.47 can be redrawn as shown in Figure 6.49, where it is clear that the three branches are in parallel.

When $\mathbf{Z}_a = \mathbf{Z}_b = \mathbf{Z}_c = \mathbf{Z}$, the wye load configuration is also balanced. When both the source and load networks are balanced, KCL requires that the current $\tilde{\mathbf{I}}_n$ in the neutral line $n - n'$ be identically zero.

$$\tilde{\mathbf{I}}_n = \tilde{\mathbf{I}}_a + \tilde{\mathbf{I}}_b + \tilde{\mathbf{I}}_c = \frac{\tilde{\mathbf{V}}_{an} + \tilde{\mathbf{V}}_{bn} + \tilde{\mathbf{V}}_{cn}}{\mathbf{Z}} = 0 \tag{6.51}$$

Another important characteristic of a balanced three-phase power system is illustrated by a simplified version of Figure 6.49, where the balanced load impedances are replaced by three equal resistors R. Since $\theta_R = 0$, the instantaneous power $p(t)$ delivered to each resistor is given by equation 6.4 [with $\theta_V = \theta_I$ and with the freely chosen reference $(\theta_V)_a = 0$] to be:

$$p_a(t) = \frac{\tilde{V}^2}{R}(1 + \cos 2\omega t)$$

$$p_b(t) = \frac{\tilde{V}^2}{R}[1 + \cos(2\omega t - 120°)] \tag{6.52}$$

$$p_c(t) = \frac{\tilde{V}^2}{R}[1 + \cos(2\omega t + 120°)]$$

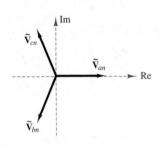

Figure 6.48 Positive, or *abc*, sequence for balanced three-phase voltages

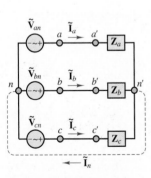

Figure 6.49 Balanced three-phase AC circuit (redrawn)

The total instantaneous power $p(t)$ delivered to the total load is the sum:

$$p(t) = p_a(t) + p_b(t) + p_c(t)$$

$$= \frac{\tilde{V}^2}{R}[3 + \cos 2\omega t + \cos(2\omega t - 120°) + \cos(2\omega t + 120°)] \qquad (6.53)$$

$$= \frac{3\tilde{V}^2}{R} = \text{constant!}$$

It is worthwhile to verify that the sum of the three cosine terms is identically zero. (*Hint:* Consider the phasor sum of $e^{j(2\omega t)}$, $e^{j(2\omega t - \pi/3)}$, and $e^{j(2\omega t + \pi/3)}$.)

Thus, with the simplified balanced resistive load, the total power delivered to the load by the balanced three-phase source is constant. This is an extremely important result, for a very practical reason: Delivering power in a steady fashion (as opposed to the pulsating nature of single-phase power) reduces "wear and tear" on the source and load.

It is also possible to connect three AC sources in a **delta (Δ) configuration**, as shown in Figure 6.50 although it is rarely used in practice.

Balanced Wye Loads

These results for purely resistive loads can be generalized for any arbitrary balanced complex load. Consider again in Figure 6.47, where now the balanced load consists of three complex impedances:

$$\mathbf{Z}_a = \mathbf{Z}_b = \mathbf{Z}_c = \mathbf{Z}_y = |\mathbf{Z}_y|\angle\theta \qquad (6.54)$$

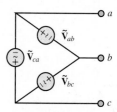

A delta-connected three-phase generator with line voltages V_{ab}, V_{bc}, V_{ca}

Figure 6.50 Delta configuration

Because of the common neutral line $n - n'$, each impedance sees the corresponding phase voltage across itself. Therefore, since $\tilde{V}_{an} = \tilde{V}_{bn} = \tilde{V}_{cn}$, it is also true that $\tilde{I}_a = \tilde{I}_b = \tilde{I}_c$ and the phase angles of the currents will differ by $\pm 120°$. Consequently, it is possible to compute the power for each phase from the phase voltage and the associated line current. Denote the complex power for each phase by **S**, where:

$$\mathbf{S} = P + jQ$$

$$= \tilde{V}\tilde{I}\cos\theta + j\tilde{V}\tilde{I}\sin\theta \qquad (6.55)$$

The total real power delivered to the balanced wye load is $3P$, and the total reactive power is $3Q$. The total complex power $\mathbf{S}_T$ is:

$$\mathbf{S}_T = P_T + jQ_T = 3P + j3Q$$

$$= \sqrt{(3P)^2 + (3Q)^2}\angle\theta \qquad (6.56)$$

The apparent power $|\mathbf{S}_T|$ is:

$$|\mathbf{S}_T| = 3\sqrt{(\tilde{V}\tilde{I})^2\cos^2\theta + (\tilde{V}\tilde{I})^2\sin^2\theta}$$

$$= 3\tilde{V}\tilde{I} \qquad (6.57)$$

such that:

$$P_T = |\mathbf{S}_T|\cos\theta$$

$$Q_T = |\mathbf{S}_T|\sin\theta \qquad (6.58)$$

Balanced Delta Loads

It is also possible to assemble a balanced load in a delta configuration. A wye generator and a delta load are shown in Figure 6.51.

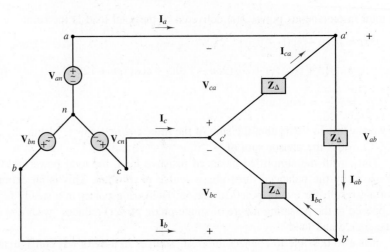

Figure 6.51 Balanced wye generators with balanced delta load

Note immediately that each impedance $\mathbf{Z}_\Delta$ sees a corresponding line voltage, rather than a phase voltage. For example, the voltage across $\mathbf{Z}_{c'a'}$ is $\tilde{\mathbf{V}}_{ca}$. Thus, the three load currents are:

$$
\begin{aligned}
\tilde{\mathbf{I}}_{ab} &= \frac{\tilde{\mathbf{V}}_{ab}}{\mathbf{Z}_\Delta} = \frac{\sqrt{3}\tilde{V}\angle(\pi/6)}{|\mathbf{Z}_\Delta|\angle\theta} \\
\tilde{\mathbf{I}}_{bc} &= \frac{\tilde{\mathbf{V}}_{bc}}{\mathbf{Z}_\Delta} = \frac{\sqrt{3}\tilde{V}\angle(-\pi/2)}{|\mathbf{Z}_\Delta|\angle\theta} \\
\tilde{\mathbf{I}}_{ca} &= \frac{\tilde{\mathbf{V}}_{ca}}{\mathbf{Z}_\Delta} = \frac{\sqrt{3}\tilde{V}\angle(5\pi/6)}{|\mathbf{Z}_\Delta|\angle\theta}
\end{aligned}
\tag{6.59}
$$

The relationship between a delta load and a wye load can be illustrated by determining the delta load $\mathbf{Z}_\Delta$ that would draw the same amount of current as a wye load $\mathbf{Z}_y$, assuming a given source voltage. Consider the circuits shown in Figures 6.47 and 6.51. For instance, the line current drawn in phase a by a wye load is:

$$
(\tilde{\mathbf{I}}_a)_y = \frac{\tilde{\mathbf{V}}_{an}}{\mathbf{Z}} = \frac{\tilde{V}}{|\mathbf{Z}_y|}\angle(-\theta)
\tag{6.60}
$$

The current drawn by a delta load is:

$$
\begin{aligned}
(\tilde{\mathbf{I}}_a)_\Delta &= \tilde{\mathbf{I}}_{ab} - \tilde{\mathbf{I}}_{ca} \\
&= \frac{\tilde{\mathbf{V}}_{ab}}{\mathbf{Z}_\Delta} - \frac{\tilde{\mathbf{V}}_{ca}}{\mathbf{Z}_\Delta} \\
&= \frac{1}{\mathbf{Z}_\Delta}(\tilde{\mathbf{V}}_{an} - \tilde{\mathbf{V}}_{bn} - \tilde{\mathbf{V}}_{cn} + \tilde{\mathbf{V}}_{an}) \\
&= \frac{1}{\mathbf{Z}_\Delta}(2\tilde{\mathbf{V}}_{an} - \tilde{\mathbf{V}}_{bn} - \tilde{\mathbf{V}}_{cn}) \\
&= \frac{3\tilde{\mathbf{V}}_{an}}{\mathbf{Z}_\Delta} = \frac{3\tilde{V}}{|\mathbf{Z}_\Delta|}\angle(-\theta)
\end{aligned}
\tag{6.61}
$$

The two currents $(\tilde{\mathbf{I}}_a)_\Delta$ and $(\tilde{\mathbf{I}}_a)_y$ are equal if:

$$\mathbf{Z}_\Delta = 3\mathbf{Z}_y \tag{6.62}$$

This result also implies that a delta load will draw three times as much current and absorb three times as much power as a wye load with the same branch impedance.

EXAMPLE 6.16   Per-Phase Solution of Balanced Wye-Wye Circuit

Problem

Compute the power delivered to the load by the three-phase generator in the circuit shown in Figure 6.52.

Solution

Known Quantities: Source voltage, line resistance, load impedance.

Find: Power delivered to the load P_{load}.

Schematics, Diagrams, Circuits, and Given Data: $\tilde{\mathbf{V}}_{an} = 480\angle 0$ V rms; $\tilde{\mathbf{V}}_{bn} = 480\angle(-2\pi/3)$ V rms; $\tilde{\mathbf{V}}_{cn} = 480\angle(2\pi/3)$ V rms; $R_{\text{line}} = 2\ \Omega$; $R_{\text{neutral}} = 10\ \Omega$; $\mathbf{Z}_y = R_o + jX_o = 2 + j4 = 4.47\angle 1.107\ \Omega$.

Assumptions: All amplitudes are effective (rms) values. All angles are given in units of radians unless indicated otherwise.

Analysis: Since the circuit is balanced, $\tilde{\mathbf{V}}_{n-n'} = 0$ and the current through the neutral line is zero. As a result, each phase has the structure shown in Figure 6.53. For example, the real power absorbed by the load in phase a is:

$$P_a = |\tilde{\mathbf{I}}_a|^2 R_o$$

where

$$|\tilde{\mathbf{I}}_a| = \left|\frac{\tilde{\mathbf{V}}_a}{\mathbf{Z}_y + R_{\text{line}}}\right| = \left|\frac{480\angle 0}{2 + j4 + 2}\right| = \left|\frac{480\angle 0}{5.66\angle(\pi/4)}\right| = 84.85 \text{ A rms}$$

and $P_a = (84.85\,\text{A})^2 (2\ \Omega) = 14.4\,\text{kW}$. Since the circuit is balanced, the results for phases b and c are identical, such that:

$$P_{\text{load}} = 3P_a = 43.2 \text{ kW}$$

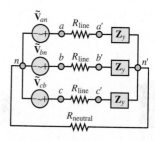

Figure 6.52 Circuit for Example 6.52.

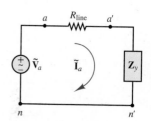

Figure 6.53 One phase of the three-phase circuit

EXAMPLE 6.17   Parallel Wye-Delta Load Circuit

Problem

Compute the power delivered to the wye-delta load by the three-phase generator in the circuit shown in Figure 6.54.

Solution

Known Quantities: Source voltage, line resistance, load impedance.

Find: Power delivered to the load P_{load}.

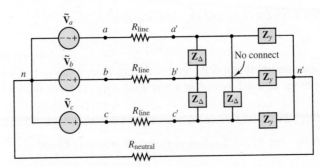

Figure 6.54 AC circuit with delta and wye loads

Schematics, Diagrams, Circuits, and Given Data: $\tilde{\mathbf{V}}_{an} = 480\angle 0$ V rms; $\tilde{\mathbf{V}}_{bn} = 480\angle(-2\pi/3)$ V rms; $\tilde{\mathbf{V}}_{cn} = 480\angle(2\pi/3)$ V rms; $\mathbf{Z}_y = 2 + j4 = 4.47\angle 1.107$ Ω; $\mathbf{Z}_\Delta = 5 - j2 = 5.4\angle(-0.381)$ Ω; $R_{line} = 2$ Ω; $R_{neutral} = 10$ Ω.

Assumptions: All amplitudes are effective (rms) values. All angles are given in units of radians.

Analysis: First, convert the balanced delta load to an equivalent wye load, according to equation 6.62. Figure 6.55 illustrates the effect of this conversion.

$$\mathbf{Z}_{\Delta-y} = \frac{\mathbf{Z}_\Delta}{3} = 1.667 - j0.667 = 1.8\angle(-0.381)\,\Omega$$

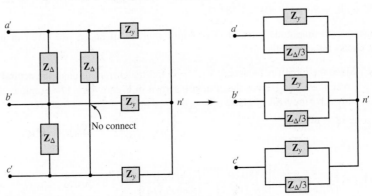

Figure 6.55 Conversion of delta load to equivalent wye load

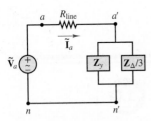

Figure 6.56 Per-phase circuit

Since the circuit is balanced, $\tilde{\mathbf{V}}_{n-n'} = 0$ and the current through the neutral line is zero. The resulting per-phase circuit is shown in Figure 6.56. For example, the real power absorbed by the load in phase a is:

$$P_a = |\tilde{\mathbf{I}}_a|^2 R_a = |\tilde{\mathbf{I}}_a|^2 \,\text{Re}\,(\mathbf{Z}_a)$$

where

$$\mathbf{Z}_a = \mathbf{Z}_y \,\|\, \mathbf{Z}_{\Delta-y} = \frac{\mathbf{Z}_y \times \mathbf{Z}_{\Delta-y}}{\mathbf{Z}_y + \mathbf{Z}_{\Delta-y}} = 1.62 - j0.018 = 1.62\angle(-0.011)\,\Omega$$

The load current $|\tilde{\mathbf{I}}_a|$ is:

$$|\tilde{\mathbf{I}}_a| = \left|\frac{\tilde{\mathbf{V}}_a}{\mathbf{Z}_o + R_{line}}\right| = \left|\frac{480\angle 0}{1.62 + j0.018 + 2}\right| = 132.6 \text{ A rms}$$

Thus, $P_a = (132.6)^2 \times \text{Re}\,(\mathbf{Z}_o) = 28.5$ kW. Since the circuit is balanced, the results for phases b and c are identical, such that:

$$P_{load} = 3P_a = 85.5 \text{ kW}$$

CHECK YOUR UNDERSTANDING

Find the power lost in the line resistance shown in Example 6.16.

Compute the complex power $\mathbf{S}_o$ delivered to the balanced load of Example 6.16 if the lines have zero resistance and $\mathbf{Z}_y = 1 + j3\ \Omega$.

Show that the voltage across each branch of the wye load is equal to the corresponding phase voltage (e.g., the voltage across $\mathbf{Z}_a$ is $\tilde{\mathbf{V}}_a$).

Prove that the sum of the instantaneous powers absorbed by the three branches in a balanced wye-connected load is constant and equal to $3\tilde{V}\tilde{I}\cos\theta$.

6.6   RESIDENTIAL WIRING; GROUNDING AND SAFETY

Common residential electric power service consists of a three-wire AC system supplied by the local power company. The three wires originate from a utility pole and consist of a neutral wire, which is connected to earth ground, and two "hot" wires. Each of the hot lines supplies 120 V rms to the residential circuits; the two lines are 180° out of phase, for reasons that will become apparent during the course of this discussion. The phasor line voltages, shown in Figure 6.57, are usually referred to by means of a subscript convention derived from the color of the insulation on the different wires: W for white (neutral), B for black (hot), and R for red (hot). This convention is adhered to uniformly.

The voltages across the hot lines are given by

$$\tilde{\mathbf{V}}_B - \tilde{\mathbf{V}}_R = \tilde{\mathbf{V}}_{BR} = \tilde{\mathbf{V}}_B - (-\tilde{\mathbf{V}}_B) = 2\tilde{\mathbf{V}}_B = 240\angle 0° \tag{6.63}$$

Appliances such as electric stoves, air conditioners, and heaters are powered by the 240 V rms arrangement. On the other hand, lighting and all the electric outlets in the house used for small appliances are powered by a single 120 V rms line.

The use of 240 V rms service for appliances that require a substantial amount of power to operate is dictated by power transfer considerations. Consider the two circuits shown in Figure 6.58. In delivering the necessary power to a load, a lower line loss will be incurred with the 240 V rms wiring since the power loss in the lines (the I^2R **loss**, as it is commonly referred to) is directly related to the current required by the load. In an effort to minimize line losses, the size of the wires is increased for the lower-voltage case. This typically reduces the wire resistance by a factor of 2. In the top circuit, assuming $R_S/2 = 0.01\ \Omega$, the current required by the 10-kW load is approximately 83.3 A while in the bottom circuit, with $R_S = 0.02\ \Omega$, it is approximately one-half as much (41.7 A). (You should be able to verify that the approximate I^2R losses are 69.4 W in the top circuit and 34.7 W in the bottom circuit.) Limiting the I^2R losses is important from the viewpoint of efficiency, besides reducing the amount of heat generated in the wiring for safety considerations. Figure 6.59 shows some typical wiring configurations for a home. Note that several circuits are wired and fused separately.

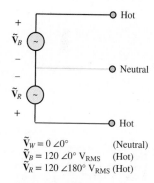

$\tilde{\mathbf{V}}_W = 0\angle 0°$ (Neutral)
$\tilde{\mathbf{V}}_B = 120\angle 0°\ \mathrm{V_{RMS}}$ (Hot)
$\tilde{\mathbf{V}}_R = 120\angle 180°\ \mathrm{V_{RMS}}$ (Hot)

Figure 6.57 Line voltage convention for residential circuits

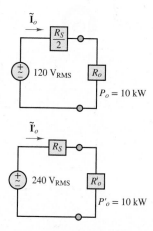

Figure 6.58 Line losses in 120- and 240-VAC circuits

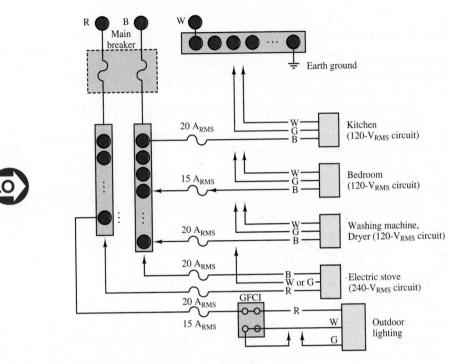

Figure 6.59 A typical residential wiring arrangement

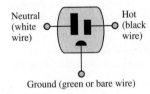

Figure 6.60 A three-wire outlet

Today, most homes have three wire connections to their outlets. The outlets appear as sketched in Figure 6.60. Then why are both the ground and neutral connections needed in an outlet? The answer to this question is *safety:* The ground connection is used to connect the chassis of the appliance to earth ground. Without this provision, the appliance chassis could be at any potential with respect to ground, possibly even at the hot wire's potential if a segment of the hot wire were to lose some insulation and come in contact with the inside of the chassis! Poorly grounded appliances can thus be a significant hazard. Figure 6.61 illustrates schematically how even though the chassis is intended to be insulated from the electric circuit, an unintended connection (represented by the dashed line) may occur, for example, because of corrosion or a loose mechanical connection. A path to ground might be provided by the body of a person touching the chassis with a hand. In the figure, such an undesired ground loop current is indicated by I_G. In this case, the ground current I_G would pass directly through the body to ground and could be harmful.

In some cases the danger posed by such undesired ground loops can be great, leading to death by electric shock. Figure 6.62 describes the effects of electric currents on an average male when the point of contact is dry skin. Particularly hazardous conditions are liable to occur whenever the natural resistance to current provided by the skin breaks down, as would happen in the presence of water. Thus, the danger presented to humans by unsafe electric circuits is very much dependent on the particular conditions—whenever water or moisture is present, the natural electrical resistance of dry skin, or of dry shoe soles, decreases dramatically, and even relatively low voltages can lead to fatal currents. Proper grounding procedures, such as those required by the National Electrical Code, help prevent fatalities due to electric shock.

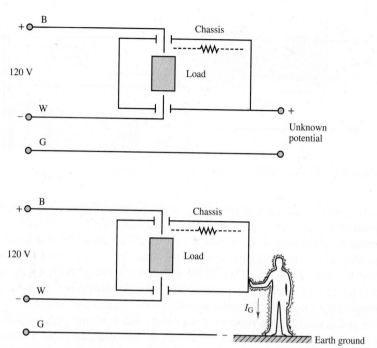

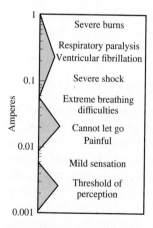

Figure 6.61 Unintended connection

The **ground fault circuit interrupter**, labeled **GFCI** in Figure 6.59, is a special safety circuit used primarily with outdoor circuits and in bathrooms, where the risk of death by electric shock is greatest. Its application is best described by an example.

Consider the case of an outdoor pool surrounded by a metal fence, which uses an existing light pole for a post, as shown in Figure 6.63. The light pole and the metal fence can be considered as forming a chassis. If the fence were not

Figure 6.62 Physiological effects of electric currents

Figure 6.63 Outdoor pool

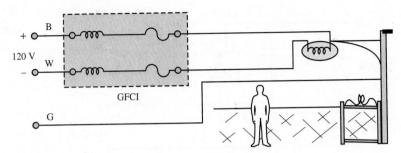

Figure 6.64 Use of a GFCI in a potentially hazardous setting

properly grounded all the way around the pool and if the light fixture were poorly insulated from the pole, a path to ground could easily be created by an unaware swimmer reaching, say, for the metal gate. A GFCI provides protection from potentially lethal ground loops, such as this one, by sensing both the hot-wire (B) and the neutral (W) currents. If the difference between the hot-wire current I_B and the neutral current I_W is more than a few milliamperes, then the GFCI disconnects the circuit nearly instantaneously. Any significant difference between the hot and neutral (return-path) currents means that a second path to ground has been created (by the unfortunate swimmer, in this example) and a potentially dangerous condition has arisen. Figure 6.64 illustrates the idea. GFCIs are typically resettable circuit breakers, so that one does not need to replace a fuse every time the GFCI circuit is enabled.

CHECK YOUR UNDERSTANDING

Use the circuits of Figure 6.58 to show that the I^2R losses will be higher for a 120-V service appliance than a 240-V service appliance if both have the same power usage rating.

Answer: The 120 V rms circuit has double the losses of the 240 V rms circuit for the same power rating.

6.7   POWER GENERATION AND DISTRIBUTION

We now conclude the discussion of power systems with a brief description of the various elements of a power system. Electric power originates from a variety of sources; in Chapter 13, electric generators are introduced as a means of producing electric power from a variety of energy conversion processes. In general, electric power may be obtained from hydroelectric, thermoelectric, geothermal, wind, solar, and nuclear sources. The choice of a given source is typically dictated by the power requirement for the given application, and by economic and environmental factors. In this section, the structure of an AC power network, from the power-generating station to the residential circuits discussed in Section 6.6, is briefly outlined.

A typical generator will produce electric power at 18 kV rms, as shown in the diagram of Figure 6.65. To minimize losses along the conductors, the output of the generators is processed through a step-up transformer to achieve line voltages

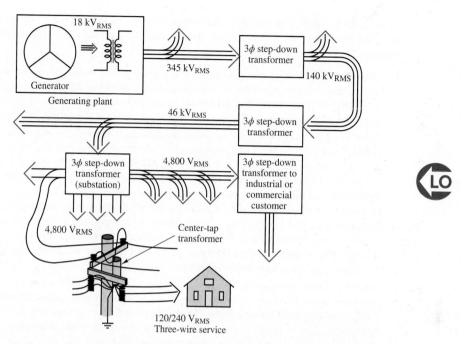

Figure 6.65 Structure of an AC power distribution network

of hundreds of kilovolts (345 kV rms, in Figure 6.65). Without this transformation, the majority of the power generated would be lost in the **transmission lines** that carry the electric current from the power station.

The local electric company operates a power-generating plant that is capable of supplying several hundred megavolt-amperes (MVA) on a three-phase basis. For this reason, the power company uses a three-phase step-up transformer at the generation plant to increase the line voltage to around 345 kV rms. One can immediately see that at the rated power of the generator (in megavolt-amperes) there will be a significant reduction of current beyond the step-up transformer.

Beyond the generation plant, an electric power network distributes energy to several **substations**. This network is usually referred to as the **power grid**. At the substations, the voltage is stepped down to a lower level (10 to 150 kV rms, typically). Some very large loads (e.g., an industrial plant)may be served directly from the power grid although most loads are supplied by individual substations in the power grid. At the local substations (one of which you may have seen in your own neighborhood), the voltage is stepped down further by a three-phase step-down transformer to 4,800 V. These substations distribute the energy to residential and industrial customers. To further reduce the line voltage to levels that are safe for residential use, step-down transformers are mounted on utility poles. These drop the voltage to the 120/240-V three-wire single-phase residential service discussed in Section 6.6. Industrial and commercial customers receive 460- and/or 208-V three-phase service.

Conclusion

Chapter 6 introduces the essential elements that permit the analysis of AC power systems. AC power is essential to all industrial activities and to the conveniences we are accustomed to in residential life. Virtually all engineers will be exposed to AC power systems in their careers, and the material presented in this chapter provides all the necessary tools to understand the analysis of AC power circuits. Upon completing this chapter, you should have mastered the following learning objectives:

1. *Understand the meaning of instantaneous and average power, master AC power notation, and compute average power for AC circuits. Compute the power factor of a complex load.* The power dissipated by a load in an AC circuit consists of the sum of an average and a fluctuating component. In practice, the average power is the quantity of interest.

2. *Learn complex power notation; compute apparent, real, and reactive power for complex loads. Draw the power triangle, and compute the capacitor size required to perform power factor correction on a load.* AC power can best be analyzed with the aid of complex notation. Complex power S is defined as the product of the phasor load voltage and the complex conjugate of the load current. The real part of S is the real power actually consumed by a load (that for which the user is charged); the imaginary part of S is called the reactive power and corresponds to energy stored in the circuit—it cannot be directly used for practical purposes. Reactive power is quantified by a quantity called the *power factor*, and it can be minimized through a procedure called *power factor correction*.

3. *Analyze the ideal transformer; compute primary and secondary currents and voltages and turns ratios. Calculate reflected sources and impedances across ideal transformers. Understand maximum power transfer.* Transformers find many applications in electrical engineering. One of the most common is in power transmission and distribution, where the electric power generated at electric power plants is stepped "up" and "down" before and after transmission, to improve the overall efficiency of electric power distribution.

4. *Learn three-phase AC power notation; compute load currents and voltages for balanced wye and delta loads.* AC power is generated and distributed in three-phase form. Residential services are typically single-phase (making use of only one branch of the three-phase lines) while industrial applications are often served directly by three-phase power.

5. *Understand the basic principles of residential electrical wiring, of electrical safety, and of the generation and distribution of AC power.*

HOMEWORK PROBLEMS

Section 6.1: Instantaneous and Average Power

6.1 The heating element in a soldering iron has a resistance of 20 Ω. Find the average power dissipated in the soldering iron if it is connected to a voltage source of 90 V rms.

6.2 A coffeemaker has a rated power of 1,000 W at 240 V rms. Find the resistance of the heating element.

6.3 A current source $i(t)$ is connected to a 50-Ω resistor. Find the average power delivered to the resistor, given that $i(t)$ is

a. $7 \cos 100t$　A

b. $7 \cos (100t - 30°)$　A

c. $7 \cos 100t - 3 \cos (100t - 60°)$　A

d. $7 \cos 100t - 3$　A

6.4 Find the rms value of each of the following periodic currents:

a. $\cos 200t + 3 \cos 200t$

b. $\cos 10t + 2 \sin 10t$

c. $\cos 50t + 1$

d. $\cos 30t + \cos(30t + \pi/6)$

6.5 A current of 2.5 A through a neon light advertisement is supplied by a 115 V rms voltage source. The current lags the voltage by 30°. Find the impedance of the light, the real power dissipated by it, and its power factor.

6.6 Compute the average power dissipated by the load seen by the voltage source in Figure P6.6. Let $\omega = 377$ rad/s, $\tilde{\mathbf{V}}_s = 50\angle0$, $R = 10$ Ω, $L = 0.08$ H, and $C = 200$ μF.

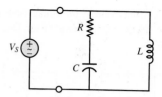

Figure P6.6

6.7   A drilling machine is driven by a single-phase induction machine connected to a 110 V rms supply. Assume that the machining operation requires 1 kW, that the tool machine has 90 percent efficiency, and that the supply current is 14 A rms with a power factor of 0.8. Find the AC machine efficiency.

6.8   Given the waveform of a voltage source shown in Figure P6.8, find:

a. The steady DC voltage that would cause the same heating effect across a resistance.

b. The average current supplied to a 10-Ω resistor connected across the voltage source.

c. The average power supplied to a 1-Ω resistor connected across the voltage source.

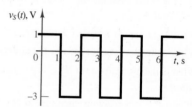

Figure P6.8

6.9   A current source $i(t)$ is connected to a 100-Ω resistor. Find the average power delivered to the resistor, given that $i(t)$ is:

a. $4 \cos(100t)$ A

b. $4 \cos(100t - 50°)$ A

c. $4 \cos(100t - 3) \cos(100t - 50°)$ A

d. $4 \cos(100t - 3)$ A

6.10   Find the rms value of each of the following periodic currents:

$\cos(377t) + \cos(377t)$ A

$\cos(2t) + \sin(2t)$ A

$\cos(377t) + 1$ A

$\cos(2t) + \cos(2t + 135°)$ A

$\cos(2t) + \cos(3t)$ A

Section 6.2: Complex Power

6.11   A current of 10 A rms results when a single-phase circuit is placed across a 220 V rms source. The current lags the voltage by 60°. Find the power dissipated by the circuit and the power factor.

6.12   A network is supplied by a 120 V rms, 60-Hz voltage source. An ammeter and a wattmeter indicate that 12 A rms is drawn from the source and 800 W are consumed by the network. Determine:

a. The network power factor.

b. The network phase angle.

c. The network impedance.

d. The equivalent resistance and reactance of the network.

6.13   For the following numeric values, determine the average power, P, the reactive power, Q, and the complex power, S, of the circuit shown in Figure P6.13. *Note:* Phasor quantities are rms.

a. $v_S(t) = 650 \cos(377t)$ V
   $i_o(t) = 20 \cos(377t - 10°)$ A

b. $\tilde{\mathbf{V}}_S = 460\angle 0°$ V rms
   $\tilde{\mathbf{I}}_o = 14.14\angle -45°$ A rms

c. $\tilde{\mathbf{V}}_S = 100\angle 0°$ V rms
   $\tilde{\mathbf{I}}_o = 8.6\angle -86°$ A rms

d. $\tilde{\mathbf{V}}_S = 208\angle -30°$ V rms
   $\tilde{\mathbf{I}}_o = 2.3\angle -63°$ A rms

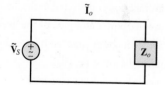

Figure P6.13

6.14   For the circuit of Figure P6.13, determine the power factor for the load $\mathbf{Z}_o$ and determine whether it is leading or lagging for the following conditions:

a. $v_S(t) = 679 \cos(\omega t + 15°)$ V
   $i_o(t) = 20 \cos(\omega t + 47°)$ A

b. $v_S(t) = 163 \cos(\omega t + 15°)$ V
   $i_o(t) = 20 \cos(\omega t - 22°)$ A

c. $v_S(t) = 294 \cos(\omega t)$ V
   $i_o(t) = 1.7 \cos(\omega t + 175°)$ A

d. $\mathbf{Z}_o = (48 + j16)$ Ω

6.15   For the circuit of Figure P6.13, determine whether the load is capacitive or inductive, assuming:

a. pf = 0.87 (leading)

b. pf = 0.42 (leading)

c. $v_S(t) = 42 \cos(\omega t)$ V
$i_L(t) = 4.2 \sin(\omega t)$ A

d. $v_S(t) = 10.4 \cos(\omega t - 12°)$ V
$i_L(t) = 0.4 \cos(\omega t - 12°)$ A

6.16   For the circuit shown in Figure P6.16, assume $C = 265$ μF, $L = 25.55$ mH, and $R = 10$ Ω. Find the instantaneous real and reactive power if:

a. $v_S(t) = 120 \cos(377t)$ V (i.e., the frequency is 60 Hz)

b. $v_S(t) = 650 \cos(314t)$ V (i.e., the frequency is 50 Hz)

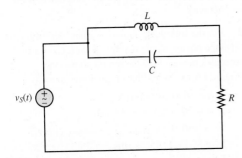

Figure P6.16

6.17   A load impedance, $\mathbf{Z}_o = 10 + j3$ Ω, is connected to a source with line resistance equal to 1 Ω, as shown in Figure P6.17. Calculate the following values:

a. The average power delivered to the load.

b. The average power absorbed by the line.

c. The apparent power supplied by the generator.

d. The power factor of the load.

e. The power factor of line plus load.

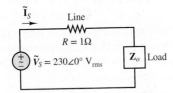

Figure P6.17

Section 6.3: Power Factor Correction

6.18   A single-phase motor draws 220 W at a power factor of 0.8 lagging when connected across a 240 V rms, 60-Hz source. A capacitor is connected in parallel with the load to produce a unity power factor. Determine the required capacitance.

6.19   The networks seen by the voltage sources in Figure P6.19 have unity power factor. Determine C_P and C_S.

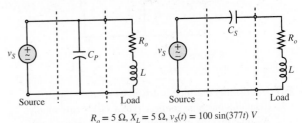

$R_o = 5$ Ω, $X_L = 5$ Ω, $v_S(t) = 100 \sin(377t)$ V

Figure P6.19

6.20   A 1,000-W electric motor is connected to a 120 V_{rms}, 60-Hz source. The power factor seen by the source is 0.8, lagging. To correct the pf to 0.95 lagging, a capacitor is placed in parallel with the motor. Calculate the current drawn from the source with and without the capacitor connected. Determine the value of the capacitor required to make the correction.

6.21   The motor inside a blender can be modeled as a resistance in series with an inductance, as shown in Figure P6.21. The wall socket source is modeled as an ideal 120 V rms voltage source in series with a 2-Ω output resistance. Assume the source frequency is $\omega = 377$ rad/s.

a. What is the power factor of the motor?

b. What is the power factor seen by the voltage source?

c. What is the average power, P_{AV}, consumed by the motor?

d. What value of capacitor when placed in parallel with the motor will change the power factor seen by the voltage source to 0.9 lagging?

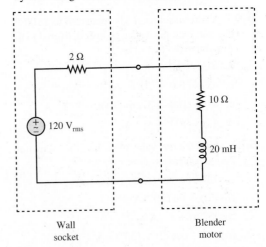

Figure P6.21

6.22   For the circuit shown in Figure P6.22, find:

a. The Thévenin equivalent network seen by the load.

b. The power dissipated by the load resistor.

c. The load impedance that would result in maximum power transfer to the load.

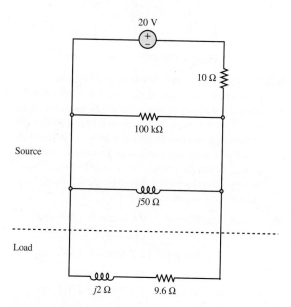

Figure P6.22

6.23   For the following numerical values, determine the capacitance to be placed in parallel with the load $\mathbf{Z}_o$ shown in Figure P6.13 that will result in a unity power factor seen by the voltage source. Assume $\omega = 377$ rad/s.

a. $\tilde{\mathbf{V}}_s = 300\angle 0$ V rms, $\tilde{\mathbf{I}}_o = 80\angle(-0.15\pi)$ A rms

b. $\tilde{\mathbf{V}}_s = 100\angle 0$ V rms, $\tilde{\mathbf{I}}_o = 30\angle(-\pi/4)$ A rms

c. $\tilde{\mathbf{V}}_s = 12\angle(-\pi/4)$ V rms, $\tilde{\mathbf{I}}_o = 3\angle(-\pi/2)$ A rms

6.24   For the circuit of Figure P6.13, determine the power factor of the load for each case listed below. Is it leading or lagging?

a. $v_s(t) = 50 \cos(\omega t)$ V
    $i_o(t) = 20 \sin(\omega t + 1.2)$ A

b. $v_s(t) = 110 \cos(\omega t + 0.1)$ V
    $i_o(t) = 10 \cos(\omega t - 0.1)$ A

c. $\mathbf{Z}_o = (20 + j5)$ Ω

d. $\mathbf{Z}_o = (20 - j5)$ Ω

6.25   For the circuit of Figure P6.13, determine whether the load $\mathbf{Z}_o$ is capacitive or inductive, if:

a. its power factor is pf = 0.76 lagging.

b. its power factor is pf = 0.5 (leading).

c. $v_s(t) = 10 \cos(\omega t)$ V, $i_o(t) = \cos(\omega t)$ A.

d. $v_s(t) = 100 \cos(\omega t)$ V, $i_o(t) = 12 \cos(\omega t + \pi/4)$ A.

6.26   Find the real and reactive power supplied by the voltage source shown in Figure P6.26 for $\omega = 5$ rad/s and $\omega = 15$ rad/s. Let $v_S = 15 \cos(\omega t)$ V, $R = 5$ Ω, $C = 0.1$ F, $L_1 = 1$ H, $L_2 = 2$ H.

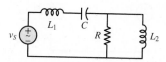

Figure P6.26

6.27   In Figure P6.27, assume $\tilde{\mathbf{V}}_{S1} = 10\angle-\pi/4$ V rms, $\tilde{\mathbf{V}}_{S2} = 12\angle 0.8$ V rms, $R_1 = 2$ Ω, $R_2 = 3$ Ω, $X_L = 4$ Ω, and $X_C = -4$ Ω. Find:

a. The amplitude of the current supplied by each source.

b. The total real power supplied by each source.

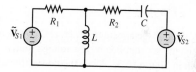

Figure P6.27

6.28   For the circuit shown in Figure P6.28, assume $f = 60$ Hz, $\tilde{\mathbf{V}}_S = 90\angle 0$ V rms, $R = 25$ Ω, $X_L = 70$ Ω, and $X_C = -8$ Ω. Calculate:

a. The capacitance C and the inductance L.

b. The power factor seen by the voltage source.

c. The new capacitance required to correct that power factor to unity.

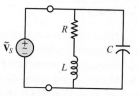

Figure P6.28

6.29 The load $\mathbf{Z}_o$ shown in Figure P6.29 consists of a 20-Ω resistor in series with a 0.01-H inductor. Assuming $f = 60$ Hz, $R = 0.5\ \Omega$, $\tilde{\mathbf{V}}_S = 100\angle 0$ V rms. Calculate:

a. The apparent power supplied by the voltage source.

b. The apparent power delivered to the load.

c. The power factor of the load.

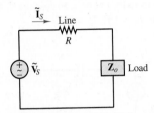

Figure P6.29

6.30 Calculate the real and reactive power of the load between terminals a and b in Figure P6.30. Assume $f = 60$ Hz, $\tilde{\mathbf{V}}_S = 70\angle 0$ V rms, $R_S = 2\ \Omega$, $R_o = 18\ \Omega$, and $X_L = 5\ \Omega$.

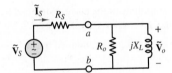

Figure P6.30

6.31 Calculate the apparent power, real power, and reactive power supplied by the voltage source shown in Figure P6.31. Draw the power triangle. Assume $f = 60$ Hz, $\tilde{\mathbf{V}}_S = 70\angle 0$ V rms, $R = 18\ \Omega$, $C = 50\ \mu$F, and $L = 0.001$ H.

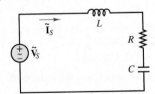

Figure P6.31

6.32 Refer to Problem 6.31 and determine the capacitance needed in parallel with the voltage source to correct the power factor seen by the source to 0.95. Draw the power triangle.

6.33 A single-phase motor is modeled as a resistor R in series with an inductor L as shown in Figure P6.33. The capacitor corrects the power factor between terminals a and b to unity. Assume the meters shown are ideal and $f = 50$ Hz, $V = 220$ V rms, $I = 20$ A rms, and $I_1 = 25$ A rms. Find the capacitor value.

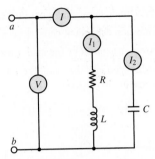

Figure P6.33

6.34 Suppose that the electricity in your home has gone out on a hot, humid summer day and the power company will not be able to fix the problem for several days. The freezer in the basement contains $300 worth of food that you cannot afford to let spoil. You would also like to keep one window air conditioner running, as well as run the refrigerator in your kitchen. When these appliances are on, they draw the following currents (all values are rms):

Air conditioner:	9.6 A rms @ 120 V rms
	pf = 0.90 lagging
Freezer:	4.2 A rms @ 120 V rms
	pf = 0.87 lagging
Refrigerator:	3.5 A rms @ 120 V rms
	pf = 0.80 lagging

In the worst-case scenario, how much power must an emergency generator supply?

6.35 The French TGV high-speed train absorbs 11 MW at 300 km/h (186 mi/h). The power supply module shown in Figure P6.35 consists of two 25-kV rms single-phase AC power stations connected at the same overhead line, one at each end of the module. For the return circuits, the rail is used. The train is also designed to operate at a low speed with 1.5-kV DC in railway stations or under the old electrification lines. The natural (average) power factor in the AC operation is 0.8. Assume that the equivalent specific resistance of the overhead line is 0.2 Ω/km and that the rail resistance can be neglected. Find:

a. A simple circuit model for the system.

b. The locomotive's current in the condition of a 10 percent voltage drop.

c. The reactive power supplied by the power stations.

d. The supplied real power, overhead line losses, and maximum distance between two power stations supplied in the condition of a 10 percent voltage drop when the train is located at the half-distance between the stations.

e. Overhead line losses in the condition of a 10 percent voltage drop when the train is located at the half-distance between the stations, assuming pf = 1. (The French TGV is designed with a state-of-the-art power compensation system.)

f. The maximum distance between the two power stations supplied in the condition of a 10 percent voltage drop when the train is located at the half-distance between the stations, assuming the DC (1.5-kV) operation at one-quarter power.

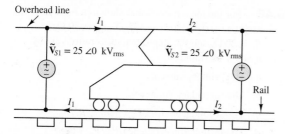

Figure P6.35

6.36   An industrial assembly hall is continuously lit by one hundred 40-W mercury vapor lamps in parallel and supplied by a 120 V rms, 60-Hz source. The power factor seen by the source is 0.65, which is so low that a 25 percent penalty is applied at billing. If the average price of 1 kWh is $0.05 and the average cost of a capacitor is $50 per mF, compute how long it will take before the billing penalty equals the cost of the capacitor needed to correct the power factor to 0.85.

6.37   Refer to Problem 6.36 and assume that each lamp is now available with a compensating capacitor in parallel with the original lamp. Find:

a. The compensating capacitor value for unity power factor seen by the source.

b. The maximum number of additional lamps that can be installed without exceeding the original current supplied by the source when using uncompensated lamps.

6.38   The voltage and current supplied by a source to a load are:

$$\tilde{\mathbf{V}}_s = 7\angle 0.873 \text{ V rms} \qquad \tilde{\mathbf{I}}_s = 13\angle(-0.349) \text{ A rms}$$

Determine:

a. The real power consumed as work and dissipated as heat in the load.

b. The reactive power stored in the load.

c. The impedance angle of the load and its power factor.

6.39   Determine the real power dissipated and the reactive power stored in each of the impedances shown in Figure P6.39. Assume:

$$\tilde{\mathbf{V}}_{s1} = \frac{170}{\sqrt{2}}\angle 0 \text{ V rms}$$

$$\tilde{\mathbf{V}}_{s2} = \frac{170}{\sqrt{2}}\angle \pi/2 \text{ V rms}$$

$$\omega = 377 \text{ rad/s}$$

$$\mathbf{Z}_1 = 0.7\angle\frac{\pi}{6} \, \Omega$$

$$\mathbf{Z}_2 = 1.5\angle 0.105 \, \Omega$$

$$\mathbf{Z}_3 = 0.3 + j0.4 \, \Omega$$

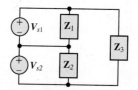

Figure P6.39

6.40   The following are supplied by a source to a load:

$$\tilde{\mathbf{V}}_s = 170\angle(-0.157) \text{ V rms} \qquad \tilde{\mathbf{I}}_s = 13\angle 0.28 \text{ A rms}$$

Determine:

a. The real power consumed as work and dissipated as heat in the load.

b. The reactive power stored in the load.

c. The impedance angle of the load and its power factor.

Section 6.4: Transformers

6.41   A center-tapped transformer has the schematic representation shown in Figure P6.41. The primary-side voltage is stepped down to two secondary-side voltages. Assume that each secondary supplies a 7-kW resistive load and that the primary is connected to 100 V rms. Find:

a. The primary power.

b. The primary current.

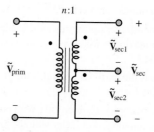

Figure P6.41

6.42   A center-tapped transformer has the schematic representation shown in Figure P6.41. The primary-side voltage is stepped down to a secondary-side voltage $\tilde{\mathbf{V}}_{sec}$ by a ratio of n:1. On the secondary side, $\tilde{\mathbf{V}}_{sec1} = \tilde{\mathbf{V}}_{sec2} = \frac{1}{2}\tilde{\mathbf{V}}_{sec}$.

a. $\tilde{\mathbf{V}}_{prim} = 220\angle 0°$ V rms and $n = 11$, find $\tilde{\mathbf{V}}_{sec}$, $\tilde{\mathbf{V}}_{sec1}$, and $\tilde{\mathbf{V}}_{sec2}$.

b.   What must n be if $\tilde{\mathbf{V}}_{prim} = 110\angle 0°$ V rms and we desire $|\tilde{\mathbf{V}}_{sec2}|$ to be 5 V rms?

6.43   For the circuit shown in Figure P6.43, assume that $\tilde{V}_g = 80\angle 0$ V rms, $R_g = 2\ \Omega$, and $R_o = 12\ \Omega$. Assume an ideal transformer. Find:

a.   The equivalent resistance seen by the voltage source.

b.   The power P_{source} supplied by the voltage source.

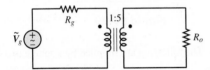

Figure P6.43

6.44   Refer to Problem 6.43 and find:

a.   The power P_{load} consumed by R_o.

b.   The installation efficiency P_{load}/P_{source}.

c.   The load R_o that results in maximum power transfer to the load.

6.45   An ideal transformer is rated to deliver 460 kVA at 380 V rms to a customer, as shown in Figure P6.45.

a.   How much current can the transformer supply to the customer?

b.   If the customer's load is purely resistive (i.e., if pf = 1), what is the maximum power that the customer can receive?

c.   If the customer's power factor is 0.8 lagging, what is the maximum usable power the customer can receive?

d.   What is the maximum power if the pf is 0.7 lagging?

e.   If the customer requires 300 kW to operate, what is the minimum power factor with the given size transformer?

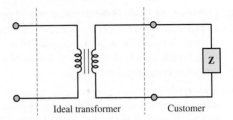

Figure P6.45

6.46   For the ideal transformer shown in Figure P6.46, assume $v_{in}(t) = 240\cos(377t)$ V, $R_{in} = 50\ \Omega$, $R_o = 20\ \Omega$, and the step-down turns ratio is set by $n = 3$. Determine:

a.   The primary current i_{in}.

b.   The secondary voltage v_o.

c.   The secondary power $P_o = i_o^2 R_o = v_o^2/R_o$.

d.   The installation efficiency P_{in}/P_o, where $P_{in} = i_{in}^2 R_{in}$.

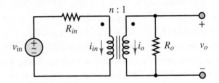

Figure P6.46

6.47   For Figure P6.47, assume the transformer is ideal. Find the step-down turns ratio $M = n$ that provides maximum power transfer to R_o. Let $R_{in} = 1,200\ \Omega$, $R_o = 100\ \Omega$, and $v_{in}(t) = V_{pk}\cos(\omega t)$.

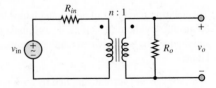

Figure P6.47

6.48   Consider the 8-Ω resistor shown in Figure P6.48 to be the load. Assume $\tilde{V}_g = 110\angle 0$ V rms and a variable turns ratio n. What value of n results in maximum power (a) dissipated by the load? (b) supplied by the voltage source? What value of n results in maximum power transfer efficiency from source to load?

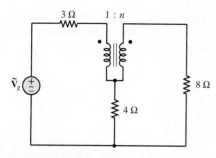

Figure P6.48

6.49   Assume the transformer shown in Figure P6.49 delivers 70 A rms at 90 V rms to a resistive load. What is the power transfer efficiency between voltage source and load? Let $R_s = 2\ \Omega$, $X_{C_1} = -10\ \Omega$, $X_{C_2} = -5\ \Omega$.

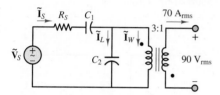

Figure P6.49

6.50   A method for determining the equivalent network of a non-ideal transformer consists of two tests: the open-circuit test and the short-circuit test. The open-circuit test, shown in Figure P6.50(a), is usually done by applying rated voltage to the primary side of the transformer while leaving the secondary side open.

The current into the primary side is measured, as is the power dissipated. The short-circuit test, shown in Figure P6.50(b), is performed by increasing the primary voltage until rated current is going into the transformer while the secondary side is short-circuited. The current into the transformer, the applied voltage, and the power dissipated are measured.

The equivalent circuit of a transformer is shown in Figure P6.50(c), where r_w and L_w represent the winding resistance and inductance, respectively, and r_c and L_c represent the losses in the core of the transformer and the inductance of the core. The ideal transformer is also included in the model.

With the open-circuit test, we may assume that $\tilde{\mathbf{I}}_{primary} = \tilde{\mathbf{I}}_{secondary} = 0$. Then all the current that is measured is directed through the parallel combination of r_c and L_c. We also assume that $|r_c||j\omega L_c|$ is much greater than $r_w + j\omega L_w$. Using these assumptions and the open-circuit test data, we can find the resistance r_c and the inductance L_c.

In the short-circuit test, we assume that $\tilde{\mathbf{V}}_{secondary}$ is zero, so that the voltage on the primary side of the ideal transformer is also zero, causing no current

through the $r_c||L_c$ parallel combination. Using this assumption with the short-circuit test data, we are able to find the resistance r_w and inductance L_w.

The following test data was measured by the meters indicated in Figure P6.50(a) and (b):

Open-circuit test:	$\tilde{\mathbf{V}} = 241$ V rms
	$\tilde{\mathbf{I}} = 0.95$ A rms
	$P = 32$ W
Short-circuit test:	$\tilde{\mathbf{V}} = 5$ V rms
	$\tilde{\mathbf{I}} = 5.25$ A rms
	$P = 26$ W

Both tests were made at $\omega = 377$ rad/s. Use the data to determine the equivalent network of the non-ideal transformer.

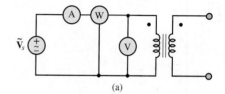

(a)

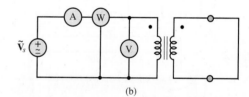

(b)

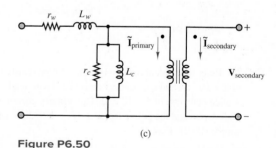

(c)

Figure P6.50

6.51   Use the methods outlined in Problem 6.50 and the following data to find the equivalent network of a nonideal transformer.

Open-circuit test:	$\tilde{\mathbf{V}} = 4{,}600$ V rms
	$\tilde{\mathbf{I}} = 0.7$ A rms
	$P = 200$ W
Short-circuit test:	$P = 50$ W
	$\tilde{\mathbf{V}} = 5.2$ V rms

The transformer is rated at 460 kVA, and tests are performed at 60 Hz.

6.52 A method of thermal treatment for a steel pipe is to heat the pipe by the Joule effect, caused when a current is directed through the pipe. In most cases, a low-voltage, high-current transformer is used to deliver the current through the pipe. In this problem, we consider a single-phase transformer at 220 V rms, which delivers 1.2 V rms. Because of the pipe's resistance variation with temperature, a secondary voltage regulation is needed in the range of 10 percent, as shown in Figure P6.52. The voltage regulation is obtained with five different slots in the primary winding (high-voltage regulation). Assuming that the secondary coil has two turns, find the number of turns for each slot.

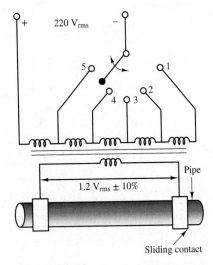

Figure P6.52

6.53 Refer to Problem 6.52 and assume a pipe resistance of 2×10^{-4} Ω and a secondary resistance (wire leads + slide contacts) of 5×10^{-5} Ω. The primary current is 28.8 A rms and the power factor seen by the 220 V rms source is 0.91. Find:

a. The slot number.

b. The secondary reactance.

c. The power transfer efficiency.

6.54 A single-phase transformer used for street lighting (high-pressure sodium discharge lamps) converts 6 kV rms to 230 V rms with an efficiency of 0.95. Assuming the power factor seen by the high voltage source is 0.8 and the primary apparent power is 30 kVA, find:

a. The secondary current.

b. The transformer turns ratio N.

6.55 The transformer shown in Figure P6.55 has several sets of windings on the secondary side. The windings have the following turns ratios:

a. $: N = 1/15$

b. $: N = 1/4$

c. $: N = 1/12$

d. $: N = 1/18$

If $\tilde{\mathbf{V}}_{primary} = 120\angle0°$ V rms, find and draw the connections that will allow you to produce the following secondary voltages:

a. $24.67\angle0°$ V rms

b. $36.67\angle0°$ V rms

c. $18\angle0°$ V rms

d. $54.67\angle180°$ V rms

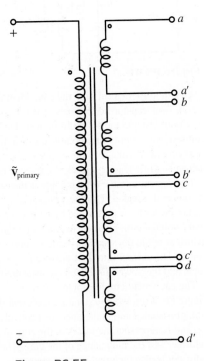

Figure P6.55

6.56 The circuit in Figure P6.56 shows the use of ideal transformers for impedance matching. You have a limited choice of turns ratios among available transformers. Suppose you can find transformers with turns ratios of 2:1, 7:2, 120:1, 3:2, and 6:1. If $\mathbf{Z}_o$ is $475\angle-25°$ Ω and $\mathbf{Z}_{ab}$ must be $267\angle-25°$, find the combination of transformers that will provide this impedance. (You may assume that polarities are easily reversed on these transformers.)

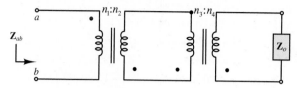

Figure P6.56

6.57   Before cable TV was generally available, TV networks broadcast their signals wirelessly. Large antennas were often installed atop residential homes to improve the reception of these signals. The impedance of the wire connecting the roof antenna to the TV set was typically 300 Ω, as shown in Figure P6.57(a). However, a typical TV had a 75-Ω impedance connection, as shown in Figure P6.57(b). To achieve maximum power transfer from the antenna to the television set, an ideal transformer was placed between the antenna and the TV, as shown in Figure P6.57(c). What is the turns ratio, $N = 1/n$, needed to obtain maximum power transfer?

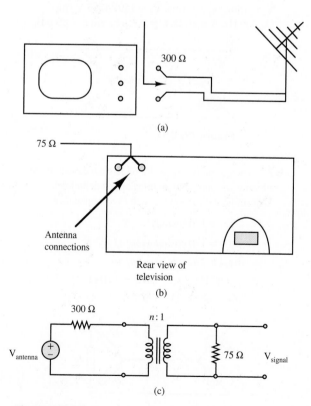

Figure P6.57

Section 6.5: Three-Phase Power

6.58   The magnitude of the phase voltage of a balanced three-phase wye system is 208 V rms. Express each phase and line voltage in both polar and rectangular coordinates.

6.59   The phase currents in a four-wire wye-connected load, such as that shown in Figure 6.49, are:

$$\tilde{\mathbf{I}}_{an} = 22\angle 0 \text{ A rms} \quad \tilde{\mathbf{I}}_{bn} = 10\angle\frac{2\pi}{3} \text{ A rms} \quad \tilde{\mathbf{I}}_{cn} = 15\angle\frac{\pi}{4}$$

Determine the current in the neutral wire.

6.60   Each voltage source shown in Figure P6.60 has a relative phase difference of $2\pi/3$.

a. Find $\tilde{\mathbf{V}}_{RW}$, $\tilde{\mathbf{V}}_{WB}$, and $\tilde{\mathbf{V}}_{BR}$, where $\tilde{\mathbf{V}}_{RW} = \tilde{\mathbf{V}}_R - \tilde{\mathbf{V}}_W$, $\tilde{\mathbf{V}}_{WB} = \tilde{\mathbf{V}}_W - \tilde{\mathbf{V}}_B$, and $\tilde{\mathbf{V}}_{BR} = \tilde{\mathbf{V}}_B - \tilde{\mathbf{V}}_R$.

b. Compare the results of part a with the calculations:

$$\tilde{\mathbf{V}}_{RW} = \tilde{\mathbf{V}}_R\sqrt{3}\angle(-\pi/6)$$
$$\tilde{\mathbf{V}}_{WB} = \tilde{\mathbf{V}}_W\sqrt{3}\angle(-\pi/6)$$
$$\tilde{\mathbf{V}}_{BR} = \tilde{\mathbf{V}}_B\sqrt{3}\angle(-\pi/6)$$

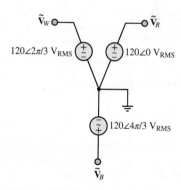

Figure P6.60

6.61   For the three-phase network shown in Figure P6.61, find the current in each wire and the real power consumed by the wye network. Let $\tilde{\mathbf{V}}_R = 110\angle 0$ V rms, $\tilde{\mathbf{V}}_W = 110\angle 2\pi/3$ V rms, and $\tilde{\mathbf{V}}_B = 110\angle 4\pi/3$ V rms. $R = 50 \Omega$, $L = 120$ mH, $C = 133 \mu F$, $f = 60$ Hz.

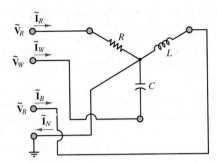

Figure P6.61

6.62 For the three-phase network shown in Figure P6.62, find the current in each wire and the real power consumed by the wye network. Let $\tilde{\mathbf{V}}_R = 170\angle 0$ V rms, $\tilde{\mathbf{V}}_W = 170\angle 2\pi/3$ V rms, and $\tilde{\mathbf{V}}_B = 170\angle 4\pi/3$ V rms.

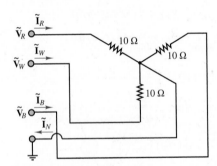

Figure P6.62

6.63 A three-phase steel-treatment electric oven has a phase resistance of 10 Ω and is connected at three-phase 380 V rms AC. Compute

  a. The current through the resistors in wye and delta connections.

  b. The power of the oven in wye and delta connections.

6.64 A naval in-board synchronous generator has an apparent power of 50 kVA and supplies a three-phase network of 380 V rms. Compute the phase currents, the real power, and the reactive power if:

  a. The power factor is 0.85.

  b. The power factor is 1.

6.65 The three-phase circuit shown in Figure P6.65 has a balanced wye source but an unbalanced wye load.

$$v_{s1} = 170\cos(\omega t) \text{ V}$$
$$v_{s2} = 170\cos(\omega t + 2\pi/3) \text{ V}$$
$$v_{s3} = 170\cos(\omega t - 2\pi/3) \text{ V}$$

$f = 60$ Hz $\mathbf{Z}_1 = 0.5\angle 20°$ Ω

$\mathbf{Z}_2 = 0.35\angle 0°$ Ω $\mathbf{Z}_3 = 1.7\angle(-90°)$ Ω

Determine the current through $\mathbf{Z}_1$, using the following methods:

  a. Mesh analysis.

  b. Superposition.

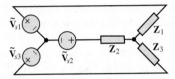

Figure P6.65

6.66 Determine the current through R shown in Figure P6.66. Assume: $\tilde{\mathbf{V}}_1 = 150\angle 0$ V rms, $\tilde{\mathbf{V}}_2 = 150\angle 2\pi/3$ V rms, $\tilde{\mathbf{V}}_3 = 150\angle 4\pi/3$ V rms, $f = 300$ Hz, $R = 80$ Ω, $C = 0.3$ μF, and $L = 80$ mH.

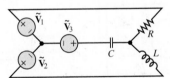

Figure P6.66

6.67 The circuit of Figure P6.67 has a balanced three-phase wye source but an unbalanced delta load. Determine the current through each impedance.

$$v_1(t) = 170\cos(\omega t) \qquad \text{V}$$
$$v_2(t) = 170\cos(\omega t + 2\pi/3) \qquad \text{V}$$
$$v_3(t) = 170\cos(\omega t - 2\pi/3) \qquad \text{V}$$

$f = 60$ Hz $\mathbf{Z}_1 = 3\angle 0$ Ω

$\mathbf{Z}_2 = 7\angle \pi/2$ Ω $\mathbf{Z}_3 = -j11$ Ω

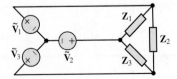

Figure P6.67

6.68 If we model each winding of a three-phase motor like the circuit shown in Figure P6.68(a) and connect the windings as shown in Figure P6.68(b), we have the three-phase circuit shown in Figure P6.68(c). The motor can be constructed so that $R_1 = R_2 = R_3$ and $L_1 = L_2 = L_3$, as is the usual case. If we connect the motor as shown in Figure P6.68(c), find the currents $\tilde{\mathbf{I}}_R$, $\tilde{\mathbf{I}}_W$, $\tilde{\mathbf{I}}_B$, and $\tilde{\mathbf{I}}_N$, assuming that the resistances are 40 Ω each and each inductance is 5 mH. The frequency of each source is 60 Hz.

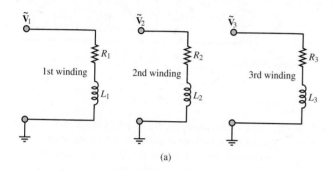

(a)

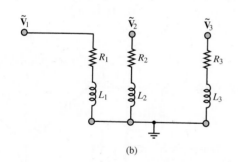

(b)

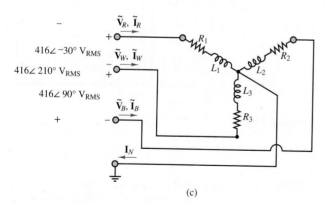

(c)

Figure P6.68

6.69 With reference to the motor of Problem 6.67,

a. How much power (in watts) is delivered to the motor?

b. What is the motor's power factor?

c. Why is it common in industrial practice *not* to connect the ground lead to motors of this type?

6.70 In general, a three-phase induction motor is designed for wye connection operation. However, for short-time operation, a delta connection can be used at the nominal wye voltage. Find the ratio between the power delivered to the same motor in the wye and delta connections.

6.71 A residential four-wire system supplies power at 240 V rms to the following single-phase appliances: On the first phase, there are ten 60-W bulbs. On the second phase, there is a 1-kW vacuum cleaner with a power factor of 0.9. On the third phase, there are ten 23-W compact fluorescent lamps with a power factor of 0.61. Find:

a. The current in the neutral wire.

b. The real, reactive, and apparent power for each phase.

6.72 The electric power company is concerned with the loading of its transformers. Since it is responsible for a large number of customers, it must be certain that it can supply the demands of *all* customers. The power company's transformers will deliver rated kVA to the secondary load. However, if the demand increased to a point where greater than rated current were required, the secondary voltage would have to drop below rated value. Also, the current would increase, and with it the I^2R losses (due to winding resistance), possibly causing the transformer to overheat. Unreasonable current demand could be caused, for example, by excessively low power factors at the load.

The customer, on the other hand, is not greatly concerned with an inefficient power factor, provided that sufficient power reaches the load. To make the customer more aware of power factor considerations, the power company may install a penalty on the customer's bill. A typical penalty–power factor chart is shown in Table 6.3. Power factors below 0.7 are not permitted. A 25 percent penalty will be applied to any billing after two consecutive months in which the customer's power factor has remained below 0.7.

Table 6.3

Power factor	Penalty
0.850 and higher	None
0.8 to 0.849	1%
0.75 to 0.799	2%
0.7 to 0.749	3%

Courtesy of Detroit Edison.

The wye-wye circuit shown in Figure P6.72 is representative of a three-phase motor load.

a. Find the total power supplied to the motor.

b. Find the power converted to mechanical energy if the motor is 80 percent efficient.

c. Find the power factor.

d. Does the company risk facing a power factor penalty on its next bill if all the motors in the factory are similar to this one?

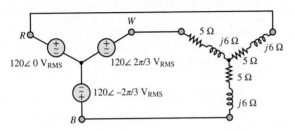

Figure P6.72

6.73 To correct the power factor problems of the motor in Problem 6.72, the company has decided to install capacitors as shown in Figure P6.73.

a. What capacitance must be installed to achieve a unity power factor if the line frequency is 60 Hz?

b. Repeat part a if the power factor is to be 0.85 lagging.

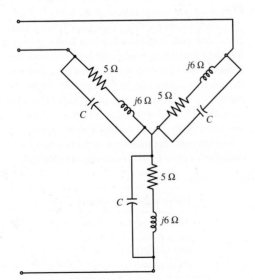

Figure P6.73

6.74 Find the apparent power and the real power delivered to the load in the Y-Δ circuit shown in Figure P6.74. What is the power factor?

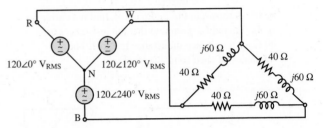

Figure P6.74

6.75 The circuit shown in Figure P6.75 is a Y-Δ-Y connected three-phase circuit. The primaries of the transformers are wye-connected, the secondaries are delta-connected, and the load is wye-connected. Find the currents $\tilde{\mathbf{I}}_{RP}$, $\tilde{\mathbf{I}}_{WP}$, $\tilde{\mathbf{I}}_{BP}$, $\tilde{\mathbf{I}}_A$, $\tilde{\mathbf{I}}_B$, and $\tilde{\mathbf{I}}_C$.

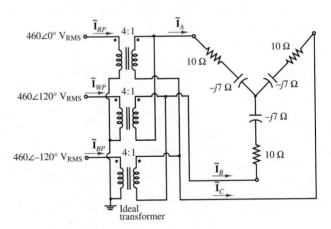

Figure P6.75

6.76 A three-phase motor is modeled by the wye-connected circuit shown in Figure P6.76. At $t = t_1$, a line fuse is blown (modeled by the switch). Find the line currents $\tilde{\mathbf{I}}_R$, $\tilde{\mathbf{I}}_W$, and $\tilde{\mathbf{I}}_B$ and the power dissipated by the motor in the following conditions:

a. $t \ll t_1$

b. $t \gg t_1$

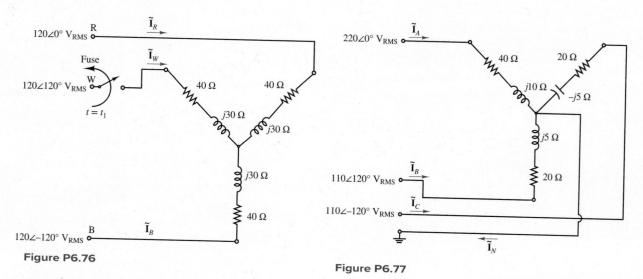

Figure P6.76

Figure P6.77

6.77 For the circuit shown in Figure P6.77, find the currents $\tilde{\mathbf{I}}_A$, $\tilde{\mathbf{I}}_B$, $\tilde{\mathbf{I}}_C$ and $\tilde{\mathbf{I}}_N$, and the real power dissipated by the load.

Design Credits: Mini DVI cable adapter isolated on white background: Robert Lehmann/Alamy Stock Photo; Balance scale: Alex Slobodkin/E+/Getty Images; Icon for "Focus on measurements" weighing scales: Media Bakery.

PART II
ELECTRONICS

Circuit Board: Arthur S. Aubry/Stockbyte/Getty Images

Buckeye on salt flats: Giorgio Rizzoni

C H A P T E R

7

OPERATIONAL AMPLIFIERS

A mplification and switching are the two fundamental operations carried out by diodes and transistors, which are themselves the two fundamental electronic components. Of course, many specialized electronic devices have been developed from diodes and transistors. One of these is the *operational amplifier*, or op-amp, the mastery of which is essential to any practical application of electronics. This chapter presents the general features of an ideal amplifier and the specific features of the operational amplifier and various popular and powerful circuits based upon it. The effects of feedback in amplifier circuits are discussed as well as the gain and frequency response of the operational amplifier. The models presented in this chapter are based on concepts that have already been explored at length in earlier chapters, namely, Thévenin and Norton equivalent circuits and frequency response. The chapter is designed to provide both a thorough analytical and practical understanding of the operational amplifier so that a student can successfully use it in practical amplifier circuits found in many engineering applications.

> ## (LO) Learning Objectives
>
> *Students will learn to...*
> 1. Understand the properties of ideal amplifiers and the concepts of gain, input impedance, output impedance, and feedback. *Section 7.1.*
> 2. Understand the difference between open-loop and closed-loop op-amp configurations; and compute the gain of (or complete the design of) simple inverting, non-inverting, summing, and differential amplifiers using ideal op-amp analysis. Analyze more advanced op-amp circuits, using ideal op-amp analysis; and identify important performance parameters in op-amp data sheets. *Section 7.2.*
> 3. Analyze and design simple active filters. Analyze and design ideal integrator and differentiator circuits. *Sections 7.3–7.4.*
> 4. Understand the principal physical limitations of an op-amp. *Section 7.5.*

Figure 7.1 Typical digital audio player (*Jim Kearns*)

7.1 IDEAL AMPLIFIERS

Amplifiers are an essential aspect of many electronic applications. Perhaps the most familiar use of an amplifier is to convert the low-voltage, low-power signal from a digital audio player (e.g., iPhone or MP3 player) to a level suitable for driving a pair of earbuds or headphones, as shown in Figure 7.1. Amplifiers have important applications in practically every field of engineering because the vast majority of transducers and sensors used for measurement produce electrical signals, which are then amplified, filtered, sampled, and processed by analog and digital electronic instrumentation. For example, mechanical engineers use thermistors, accelerometers, and strain gauges to convert temperature, acceleration, and strain into electrical signals. These signals must be amplified prior to transmission and then filtered (a function carried out by amplifiers) prior to sampling the data in preparation for producing a digital version of the original analog signal. Other, less obvious, functions such as impedance isolation are also performed by amplifiers. It should now be clear that amplifiers do more than simply produce an enlarged replica of a signal although that function is certainly very important. This chapter explores the general features of amplifiers and focuses on the characteristics and applications of a particularly important integrated-circuit amplifier, the **operational amplifier**.

Ideal Amplifier Characteristics

The simplest model for an amplifier is depicted in Figure 7.2, where a signal v_S is amplified by a constant factor G, called the *voltage gain* of the amplifier. Ideally, the *input impedance* of the amplifier is infinite such that $v_{\text{in}} = v_S$; if its *output impedance* is zero, v_o will be determined by the amplifier independent of R such that:

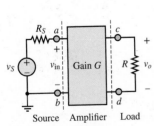

Source   Amplifier   Load

Figure 7.2 Amplifier between source and load

$$v_o = Gv_{\text{in}} = Gv_S \qquad \text{Ideal amplifier} \qquad (7.1)$$

Note that the input seen by the amplifier is a Thévenin source (v_S in series with R_S), while the output seen by the amplifier is a single equivalent resistance R.

A more realistic (but still quite simple) amplifier model is shown in Figure 7.3. In this figure the concepts of input and output impedance of the amplifier are incorporated as single resistances R_{in} and R_{out}, respectively. That is, from the perspective of the load R the amplifier acts as a Thévenin source ($A\upsilon_{\text{in}}$ in series with R_{out}), while from the perspective of the external source (υ_S in series with R_S) the amplifier acts as an equivalent resistance R_{in}. The constant A is the multiplier associated with the dependent (controlled) voltage source and is known as the *open-loop gain*.[1]

Using the amplifier model of Figure 7.3 and applying voltage division, the input voltage to the amplifier is now:

$$\upsilon_{ab} = \upsilon_{\text{in}} = \frac{R_{\text{in}}}{R_S + R_{\text{in}}}\upsilon_S \tag{7.2}$$

The output voltage of the amplifier can also be found by applying voltage division, where:

$$\upsilon_o = A\upsilon_{\text{in}}\frac{R}{R_{\text{out}} + R} \tag{7.3}$$

Substitute for υ_{in} and divide both sides by υ_S to obtain:

$$\frac{\upsilon_o}{\upsilon_S} = A\frac{R_{\text{in}}}{R_{\text{in}} + R_S}\frac{R}{R + R_{\text{out}}} \tag{7.4}$$

which is the overall voltage gain from υ_S to υ_o. The voltage gain G of the amplifier itself is:

$$G \equiv \frac{\upsilon_o}{\upsilon_{\text{in}}} = A\frac{R}{R_{\text{out}} + R} \tag{7.5}$$

For this model, the voltage gain G is dependent upon the external resistance R, which means that the amplifier performs differently for different loads. Moreover, the input voltage υ_{in} to the amplifier is a modified version of υ_S. Neither of these results seem desirable. Rather, it stands to reason that the gain of a "quality" amplifier would be independent of its load and would not impact its source signal. These attributes are achieved when $R_{\text{out}} \ll R$ and $R_{\text{in}} \gg R_S$. In the limit that $R_{\text{out}} \to 0$:

$$\lim_{R_{\text{out}} \to 0} \frac{R}{R + R_{\text{out}}} = 1 \tag{7.6}$$

such that:

$$G \equiv \frac{\upsilon_o}{\upsilon_{\text{in}}} \approx A \qquad \text{when} \quad R_{\text{out}} \to 0 \tag{7.7}$$

Also, in the limit that $R_{\text{in}} \to \infty$:

$$\lim_{R_{\text{in}} \to \infty} \frac{R_{\text{in}}}{R_{\text{in}} + R_S} = 1 \tag{7.8}$$

such that

$$\upsilon_{\text{in}} \approx \upsilon_S \qquad \text{when} \quad R_{\text{in}} \to \infty \tag{7.9}$$

In general, a "quality" voltage amplifier will have a very small output impedance and a very large input impedance.

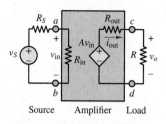

Figure 7.3 Simple voltage amplifier model

---

[1] The voltage gain G and the open-loop gain A may also be designated as A_V and $A_{V_{\text{OL}}}$, respectively. Electrical conductance is also designated as G; as always, it is important to correctly interpret a symbol from the context in which it is used. Happily, conductance G is rarely used in engineering work. Its inverse, resistance R, is preferred instead.

Input and Output Impedance

In general, the input impedance R_{in} and the output impedance R_{out} of an amplifier are defined as:

$$R_{in} = \frac{v_{in}}{i_{in}} \qquad \text{and} \qquad R_{out} = \frac{v_{OC}}{i_{SC}} \tag{7.10}$$

where v_{OC} is the open-circuit voltage and i_{SC} is the short-circuit current at the output of the amplifier. An ideal voltage amplifier has zero output impedance and infinite input impedance so that the amplifier does not suffer from loading effects at its input or output terminals. In practice, voltage amplifiers are designed to have large input impedance and small output impedance.

It is a worthwhile exercise to show that an ideal *current amplifier* has zero input impedance and infinite output impedance. Also, as was suggested in Chapter 6, an ideal *power amplifier* is designed so that its input impedance matches its source network and its output impedance matches its load impedance.

Feedback

Feedback, which is the process of using the output of an amplifier to reinforce or inhibit its input, plays a critical role in many amplifier applications. Without feedback an amplifier is said to be in *open-loop* mode; with feedback an amplifier is said to be in *closed-loop* mode. The output of the amplifier model shown in Figure 7.3 does not affect its input (because there is no path from output to input), so feedback is not present, and the model is open loop. As suggested earlier, the most basic characteristic of an amplifier is its *gain*, which is simply the ratio of the output to the input. The open-loop gain A of a practical amplifier (e.g., an operational amplifier) is usually very large, whereas the closed-loop gain G is a reduced version of the open-loop gain. The relationship between A and G is developed and explored in the rest of this chapter.

There are two types of feedback possible in closed-loop mode: *positive feedback*, which tends to reinforce the amplifier input, and *negative feedback*, which tends to inhibit the amplifier input. Both positive and negative feedback have useful applications; however, negative feedback is by far the most common type of feedback found in applications. In general, negative feedback causes the large open-loop gain A of an amplifier to be exchanged for a smaller closed-loop gain G. While this exchange may seem undesirable at first glance, several key benefits accompany the exchange. These benefits to the amplifier are:

1. Decreased sensitivity to variations in circuit and environmental parameters, most notably temperature.

2. Increased bandwidth.

3. Increased linearity.

4. Increased signal-to-noise ratio.

In addition, negative feedback is implemented by establishing one or more paths from the output to the input of the amplifier. The impedance of each feedback path can be adjusted to produce improved input and output impedances of the overall amplifier circuit. These input and output impedances are key characteristics for understanding the *loading effects* of other circuits attached to an amplifier.

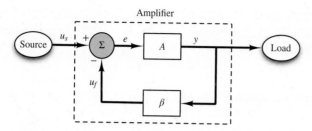

Figure 7.4 Signal-flow diagram of generic amplifier

Figure 7.4 shows a *signal-flow diagram* of an amplifier situated between a source and a load. The arrows indicate the direction of signal flow. The signals shown are u_s, u_f, e, and y. The output signal of each rectangle is a multiple of its input signal, where the two constants, A and β, are both positive such that:

$$y = Ae \quad \text{and} \quad u_f = \beta y \tag{7.11}$$

The circle sums its inputs, u_s and u_f, to produce one output, e. The polarity signs ($\pm$) indicate that u_s and u_f make positive and negative contributions to the sum, respectively. That is:

$$e = u_s - u_f = u_s - \beta y \tag{7.12}$$

Because the feedback signal u_f makes a negative contribution to the sum, the signal flow diagram of Figure 7.4 is said to employ negative feedback.

Equations 7.11 and 7.12 can be combined to yield:

$$y = Ae = A(u_s - u_f) = A(u_s - \beta y) \tag{7.13}$$

which can be rearranged to solve for y. Then, the closed-loop gain of the amplifier is:

$$G \equiv \frac{y}{u_s} = \frac{A}{1 + A\beta} \tag{7.14}$$

The quantity $A\beta$ is known as the *loop gain*. Implicit in the derivation of equation 7.14 is that the behavior of the blocks within the amplifier is not affected by the other blocks nor by the external source and load. In other words, the blocks are *ideal* such that *loading effects* are zero.

Two important observations can be made at this point:

1. The closed-loop gain G depends upon β, which is known as the *feedback factor*.

2. Since $A\beta$ is positive, the closed-loop gain G is smaller than the open-loop gain A.

Furthermore, for most practical amplifiers, $A\beta$ is quite large such that:

$$G \approx \frac{1}{\beta} \tag{7.15}$$

This result is particularly important (and probably surprising!) because it indicates that the closed-loop gain G of the amplifier is largely *independent* of the open-loop gain A, as long as $A\beta \gg 1$, and that G is, in turn, determined largely by the feedback factor, β.

When $A\beta \gg 1$, the closed-loop gain G of an amplifier is determined largely by the feedback factor, β.

Furthermore, equation 7.14 can be used to find the ratio of the two inputs, u_s and u_f.

$$\frac{u_f}{u_s} = \frac{y}{u_s}\frac{u_f}{y} = \frac{A}{1+A\beta}\beta = \frac{A\beta}{1+A\beta} \tag{7.16}$$

Thus, when $A\beta \gg 1$, another important result is:

$$\frac{u_f}{u_s} \rightarrow 1 \qquad \text{or} \qquad u_s - u_f \rightarrow 0 \tag{7.17}$$

This result indicates that when the loop gain $A\beta$ is large, the *difference* between the input signal u_s and the feedback signal u_f is driven toward zero.

When $A\beta \gg 1$, the *difference* between the input signal u_s and the feedback signal u_f is driven toward zero.

Both of the results of equations 7.15 and 7.17 will show up repeatedly in the analysis of operational amplifier circuits in closed-loop mode.

Benefits of Negative Feedback

As mentioned in the previous section, negative feedback provides several benefits in exchange for a reduced gain. For example, take the derivative of both sides of equation 7.14 to find:

$$dG = \frac{dA}{1+A\beta} - \frac{A\beta \, dA}{(1+A\beta)^2} = \frac{dA}{(1+A\beta)^2} \tag{7.18}$$

Divide the left side by G and the right side by $A/(1+A\beta)$ to obtain:

$$\frac{dG}{G} = \frac{1}{1+A\beta}\frac{dA}{A} \tag{7.19}$$

When $A\beta \gg 1$, this result indicates that the percentage change in G due to a percentage change in A is relatively small. In other words, the closed-loop gain G is relatively insensitive to changes in the open-loop gain A.

When $A\beta \gg 1$, the closed-loop gain G is relatively insensitive to changes in the open-loop gain A.

For any amplifier, the open-loop gain A is a function of frequency. For example, the open-loop gain $A(\omega)$ of an op-amp is characterized by a simple pole such that:

$$A(\omega) = \frac{A_0}{1 + j\omega/\omega_o} \tag{7.20}$$

where ω_o is its 3-dB break frequency. The Bode magnitude characteristic plot is shown in Figure 7.5. Equation 7.20 can be substituted into equation 7.14 to obtain:

$$G(\omega) = \frac{A(\omega)}{1 + A(\omega)\beta} = \frac{A_0/(1 + j\omega/\omega_o)}{1 + A_0\beta(1 + j\omega/\omega_o)} \qquad (7.21)$$

Multiply the numerator and denominator on the right side of equation 7.21 by $1 + j\omega/\omega_o$ and then factor out $1 + A_0\beta$ from the denominator to obtain:

$$G(\omega) = \frac{A_0}{1 + A_0\beta}\frac{1}{1 + j\omega/\omega_g} = G_o\frac{1}{1 + j\omega/\omega_g} \qquad (7.22)$$

where $\omega_g = \omega_o(1 + A_0\beta)$. Thus, the closed-loop 3-dB break frequency is $(1 + A_0\beta)$ *larger* than the open-loop 3-dB break frequency.

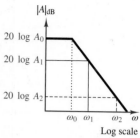

Figure 7.5 Typical amplifier Bode magnitude characteristic

> The closed-loop 3-dB break frequency is $(1 + A_0\beta)$ larger than the open-loop 3-dB break frequency.

Likewise, if the amplifier is characterized by a simple zero, its 3-dB break frequency will be $(1 + A_0\beta)$ *smaller* than the open-loop 3-dB break frequency. It is a worthwhile exercise to work out this result.

Similar analyses can be performed to show the increased linearity and increased signal-to-noise ratio resulting from negative feedback. All these benefits are acquired at the expense of amplifier gain. Finally, all of the features of a generic amplifier with negative feedback outlined in this section also occur in closed-loop amplifiers constructed using operational amplifiers and other basic components.

7.2   THE OPERATIONAL AMPLIFIER

An **operational amplifier** (op-amp) is an **integrated circuit (IC)** that contains a large number of microscopic electrical and electronic components integrated on a single silicon wafer. An op-amp can be used in conjunction with other common components to create circuits that perform amplification and filtering, as well as mathematical operations, such as addition, subtraction, multiplication, differentiation, and integration, on electrical signals. Op-amps are found in most measurement and instrumentation applications, serving as a versatile building block for many applications.

The behavior of an op-amp is well described by fairly simple models, which permit an understanding of its effects and applications without delving into its internal details. Its simplicity and versatility make the op-amp an appealing electronic device with which to begin understanding electronics and integrated circuits. The lower right portion of Figure 7.6 shows a standard single op-amp IC chip pin layout. It has two input pins (2 and 3) and one output pin (6). Also notice the two DC power supply pins (4 and 7) that provide external power to the chip and thus *enable* the op-amp. Operational amplifiers are *active* devices; that is, they need an external power source to function. Pin 4 is held at a low DC voltage V_S^-, while pin 7 is held at a high DC voltage V_S^+. These two DC voltages are well below and above, respectively, the op-amp's reference voltage and bound the output of the op-amp.

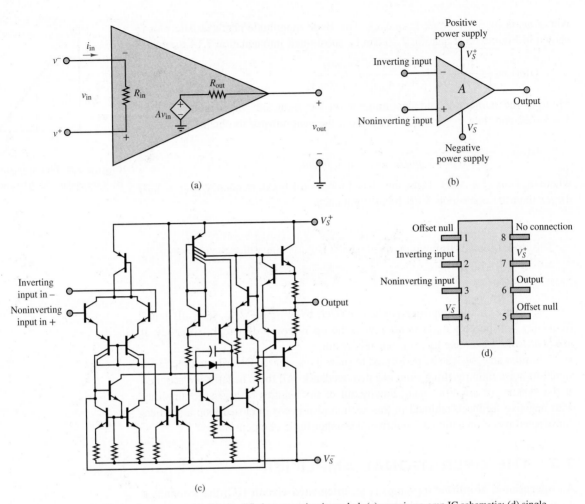

Figure 7.6 (a) Small-signal op-amp model; (b) simplified op-amp circuit symbol; (c) generic op-amp IC schematic; (d) single op-amp IC chip pin layout

The upper left portion of Figure 7.6 shows the so-called small-signal, low-frequency model of an op-amp, which is exactly the same amplifier model shown in Figure 7.3. For this model, the input impedance is R_{in} and the output impedance is R_{out}. The op-amp itself is a *differential amplifier* because its output is a function of the difference between two input voltages, v^+ and v^-, which are known as the *noninverting* and *inverting* inputs, respectively. Notice that the value of the internal dependent voltage source is $A(v^+ - v^-)$, where A is the *open-loop gain* of the op-amp. In a practical op-amp, A is quite large by design, typically on the order of 10^5 to 10^7. As discussed in the previous section, this large open-loop gain can be exchanged, by design, for a smaller *closed-loop gain G* to acquire various beneficial characteristics for an amplifier circuit, of which the op-amp is just one component.[2]

---

[2]The operational amplifier of Figure 7.6 is a *voltage amplifier*; another type of operational amplifier, called a *current* or *transconductance amplifier*, is described in the homework problems.

The Ideal Op-Amp

Practical op-amps have a large open-loop gain A, as noted above. The input impedance R_{in} is also large, typically on the order of 10^6 to 10^{12} Ω, while the output impedance R_{out} is small, typically on the order of 10^0 or 10^1 Ω. In the ideal case, the open-loop gain and the input impedance would be infinite, while the output impedance would be zero. When the output impedance is zero, the output voltage of an ideal op-amp is simply:

$$v_{out} = A(v^+ - v^-) = A\Delta v \tag{7.23}$$

But is this relationship practical when the open-loop gain A approaches infinity? The implication for a practical op-amp is that one of the two following possibilities will hold:

1. In the case that $\Delta v \neq 0$, the output voltage *saturates* near either the positive or negative DC power supply value, V_S^+ or V_S^-, as shown in Figure 7.7. These external DC power supply *rails* enable a practical op-amp to function but also bound the op-amp output voltage v_{out}. This case applies to all practical applications of an op-amp used in *open-loop mode*; that is, when there is no feedback from v_{out} to v^-.

2. In the case that $\Delta v = 0$, the output voltage is not determined by the op-amp itself but by whatever other circuitry is attached to it. Recall from Section 7.1 that when $A\beta \gg 1$ the closed-loop gain of an amplifier is approximately equal to $1/\beta$ and largely independent of A itself. Thus, this case applies to all practical applications of an op-amp in *closed-loop mode*; that is, when negative feedback is present from v_{out} to v^-.

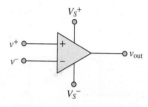

Figure 7.7 Ideal operational amplifier

By far the most prominent open-loop mode application of an op-amp is the *comparator*. Many of the practical applications of the op-amp in closed-loop mode are explored in this chapter.

Notice in Figure 7.7 that the letter "A" does not appear within the triangle symbol, thus implying that the open-loop gain is infinite. Also implied by the ideal op-amp symbol is that the current into or out of either input terminal is zero. This result is a consequence of the infinite input impedance of an ideal op-amp and is known as the *first golden rule* of ideal op-amps:

$$i^+ = i^- = 0 \qquad \text{First golden rule} \tag{7.24}$$

Also recall from the discussion of negative feedback in Section 7.1 that when $A\beta \gg 1$ the difference between the two amplifier inputs, u_s and u_f, approaches zero. In the context of ideal op-amps, where $A \rightarrow \infty$, the difference between the two amplifier inputs, v^+ and v^-, will be zero exactly *as long as there is a feedback path from v_{out} to v^-*.

$$v^+ = v^- \qquad \text{Second golden rule; negative feedback required} \tag{7.25}$$

The Golden Rules of Ideal Op-Amps:

1. $i^+ = i^- = 0$.
2. $v^+ = v^-$ (when negative feedback is present).

Amplifier Archetypes

There are three fundamental amplifiers that utilize the operational amplifier and employ *negative feedback*. They are:

- The inverting amplifier.

- The noninverting amplifier.

- The isolation buffer (or voltage follower).

These archetypes have many important applications and are the building blocks for other important amplifiers. Understanding and recognizing these archetypes is an essential first step in the study of amplifiers based upon the op-amp. It is worth emphasizing that the op-amp is rarely used as a stand-alone amplifier; rather it is used along with other components to form specialized amplifiers.

The Inverting Amplifier

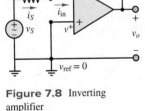

Figure 7.8 Inverting amplifier

Figure 7.8 shows a basic *inverting amplifier* circuit. The name derives from the fact that the input signal v_S "sees" the inverting terminal $(-)$ and that, as is shown below, the output signal v_o is an inverted (negative) version of the input signal. The goal of the following analysis is to determine the relationship between the output and the input signals. To begin, assume the op-amp is ideal and apply KCL at the inverting node marked v^-.

$$i_S = i_F + i_{\text{in}} \tag{7.26}$$

However, the first golden rule of ideal op-amps states that $i_{\text{in}} = i^- = 0$. Thus, $i_S = i_F$ such that R_S and R_F form a *virtual* series connection. Ohm's law can be applied to each resistor to yield:

$$i_S = \frac{v_S - v^-}{R_S} \qquad i_F = \frac{v^- - v_o}{R_F} \tag{7.27}$$

These expressions can be simplified by noting that $v^+ = 0$ and then applying the second golden rule of ideal op-amps to realize $v^- = v^+ = 0$. Thus:

$$i_S = i_F$$

or $\tag{7.28}$

$$\frac{v_S}{R_S} = \frac{-v_o}{R_F}$$

Cross-multiply to find the closed-loop gain G:

$$\boxed{G = \frac{v_o}{v_S} = -\frac{R_F}{R_S} \qquad \text{Inverting amplifier}} \tag{7.29}$$

Note that the magnitude of G can be greater or less than 1.

An alternate approach is to apply voltage division across the virtual series connection of R_S and R_F.

$$\frac{v_S - v_o}{v_S - 0} = \frac{R_S + R_F}{R_S}$$

or $\tag{7.30}$

$$1 - \frac{v_o}{v_S} = 1 + \frac{R_F}{R_S}$$

Subtract 1 from each side of this expression to find the same result as equation 7.29.

Notice that the closed-loop gain G of an inverting amplifier is determined solely by the choice of resistors. This result was derived for an ideal op-amp. For a practical op-amp the result is only slightly different as long as the open-loop gain A is large. It is important to remember that this result depends upon both golden rules of ideal op-amps and that, in particular, the second golden rule is valid only when negative feedback is present.

> As long as the open-loop gain A is large, the presence of negative feedback from the output to the inverting input drives the voltage difference between the two input terminals to zero.

The input impedance of the inverting amplifier is simply:

$$R_{in} = \frac{v_{in}}{i_{in}} = \frac{v_S - 0}{i_S} = R_S \tag{7.31}$$

Notice the important role played by the virtual ground at the inverting terminal in making this calculation so easy. This result also reveals a shortcoming of the inverting amplifier. In general, an ideal amplifier would have an infinite input impedance so as to not load the source network. It is tempting to correct this problem by choosing R_S to be very large; however, in so doing, the closed-loop gain (equation 7.29) will be reduced. Thus, it is not possible to design an inverting amplifier to have a large gain and also a large input impedance. Alas, there is no such thing as a free lunch!

The Noninverting Amplifier

Figure 7.9 shows a basic *noninverting amplifier* circuit. The name derives from the fact that the input signal v_S "sees" the noninverting terminal (+) and that, as is shown below, the output signal v_o is a noninverted (positive) version of the input signal. The goal of the following analysis is to determine the relationship between the output and the input signals. As was done for the inverting amplifier circuit, assume the op-amp is ideal and apply KCL at the inverting node marked v^-.

$$i_F = i_1 + i_{in} \tag{7.32}$$

However, the first golden rule of ideal op-amps states that $i_{in} = i^- = i^+ = 0$. Thus, $i_F = i_1$ such that R_F and R_1 form a *virtual* series connection. Ohm's law can be applied to each resistor to yield:

$$i_1 = \frac{v^- - 0}{R_1} \qquad i_F = \frac{v_o - v^-}{R_F}$$

or   (7.33)

$$\frac{v^-}{R_1} = \frac{v_o - v^-}{R_F}$$

Since there is negative feedback present, the second golden rule of ideal op-amps can be applied such that $v^- = v^+$. Notice that because $i_{in} = 0$, the voltage drop across

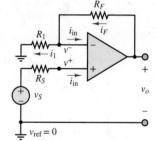

Figure 7.9 Noninverting amplifier

R_S is zero with the result that $v^- = v^+ = v_S$. Substitute this result into equation 7.33 and rearrange terms to yield the closed-loop gain G:

$$G = \frac{v_o}{v_S} = 1 + \frac{R_F}{R_1} \qquad \text{Noninverting amplifier} \tag{7.34}$$

Note that $G \geq 1$.

An alternate approach is to apply voltage division across the virtual series connection of R_1 and R_F.

$$\frac{v_o - 0}{v^- - 0} = \frac{R_1 + R_F}{R_1} \tag{7.35}$$

Since $v^- = v_S$:

$$\frac{v_o}{v_S} = 1 + \frac{R_F}{R_1} \tag{7.36}$$

which is the same result as that found in equation 7.34.

Notice that the closed-loop gain G of a noninverting amplifier is also determined solely by the choice of resistors. This result was derived for an ideal op-amp. For a practical op-amp the result is only slightly different as long as the open-loop gain A is large. It is important to remember that this result depended upon both golden rules of ideal op-amps and that, in particular, the second golden rule is valid only when negative feedback is present.

As long as the open-loop gain A is large, the presence of negative feedback from the output to the inverting input drives the voltage difference between the two input terminals to zero.

The input impedance of the noninverting amplifier is simply:

$$R_{\text{in}} = \frac{v_{\text{in}}}{i_{\text{in}}} = \frac{v_S - 0}{i_{\text{in}}} \rightarrow \infty \tag{7.37}$$

In practice, the input impedance of a noninverting amplifier is very large due to the very large input impedance of the op-amp, which limits i_{in} to very small values. Notice that the closed-loop gain of the noninverting amplifier is independent of its input impedance. Thus, the noninverting amplifier does not suffer from a trade-off between gain and input impedance, as does the inverting amplifier. However, the gain of a noninverting amplifier is limited to values greater than one, whereas the gain of the inverting amplifier can take on any value. Alas, again there is no such thing as a free lunch!

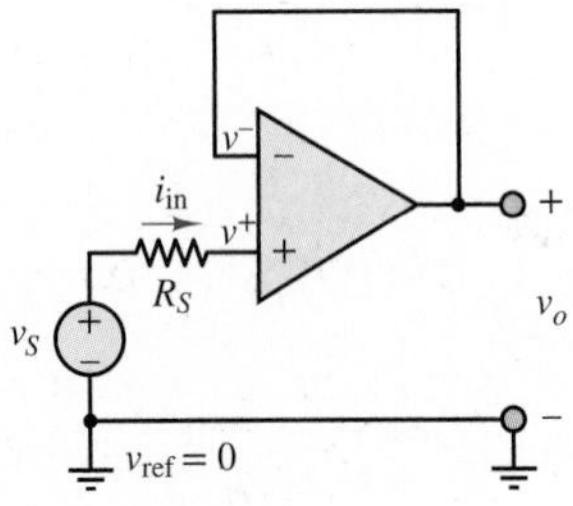

Isolation Buffer or Voltage Follower

Figure 7.10 shows an *isolation buffer*, which is also known as a *voltage follower*. Notice that the input signal v_S "sees" the noninverting terminal (+) such that the output signal v_o should be a noninverted (positive) version of v_S. The analysis of

Figure 7.10 Isolation buffer or voltage follower

this circuit is as simple as the circuit itself. Assume that the op-amp is ideal. Since negative feedback is present, both golden rules are valid. That is:

$$i^+ = i^- = 0 \qquad \text{and} \qquad v^+ = v^- \tag{7.38}$$

By observation, $v^+ = v_S$ and $v^- = v_{\text{out}}$ with the result that the closed-loop gain G is:

$$G = \frac{v_o}{v_S} = 1 \qquad \text{Isolation buffer (voltage follower)} \tag{7.39}$$

The reason this circuit is called a voltage follower should now be obvious; the output voltage v_o "follows" (equals) the input voltage v_S. On the other hand, the reason this circuit is also known as an isolation buffer is not obvious. However, since $i^+ = 0$, the ideal op-amp is said to possess an infinite *input resistance* or *input impedance* such that the voltage source experiences no loading from the op-amp. Yet the circuit still reproduces v_S at the output. Any loading effects at the output are experienced by the op-amp rather than the voltage source, such that the source is *isolated* or *buffered* from the output.

The input impedance of an isolation buffer is simply:

$$R_{\text{in}} = \frac{v_{\text{in}}}{i_{\text{in}}} = \frac{v_S - 0}{i_{\text{in}}} \rightarrow \infty \tag{7.40}$$

In practice, the input impedance of an isolation buffer is very large due to the very large input impedance of the op-amp, which limits i_{in} to very small values. The closed-loop gain is fixed at unity as long as the open-loop gain A is large such that v^- will be driven to v^+ by negative feedback.

Application of Thévenin's Theorem

Notice in Figures 7.8 and 7.9 that the input source is represented as a Thévenin source. The implication is that the previous results for inverting and noninverting amplifiers can be applied to any case where the input source of the amplifier circuit is linear and can be simplified to an equivalent Thévenin source. In other words, R_S and v_S are the Thévenin equivalent resistance and the open-circuit voltage, respectively, of any arbitrary linear input circuit.

For example, consider the inverting amplifier circuit shown in Figure 7.11. It does not have the same form as the archetype of Figure 7.8. However, the voltage source v_{in} "sees" the inverting terminal; therefore, the output voltage v_o will be an inverted version of v_{in}. The circuit is an inverting amplifier. To solve for v_o replace the entire linear circuit to the left of terminals a and b with its Thévenin equivalent.

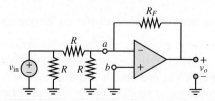

Figure 7.11 Inverting amplifier before simplification to archetype

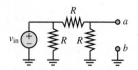

Figure 7.12 Source network detached at terminals a and b

Figure 7.12 shows the source circuit detached at terminals a and b. To find the Thévenin equivalent resistance of the input circuit set the voltage source to zero and replace it with a short-circuit. Then:

$$R_T = R_{ab} = R \| R = \frac{R}{2} \tag{7.41}$$

The Thévenin (open-circuit) voltage across terminals a and b can be found from voltage division:

$$V_T = V_{ab} = \frac{R}{R + R} v_{in} = \frac{v_{in}}{2} \tag{7.42}$$

The Thévenin equivalent source network attached to the rest of the amplifier circuit is shown in Figure 7.13. Notice that the simplified amplifier is now identical in form to the inverting amplifier archetype of Figure 7.8. Thus, using equation 7.29:

$$\frac{v_o}{V_T} = -\frac{R_F}{R_T} = -\frac{2R_F}{R} \tag{7.43}$$

To obtain the closed-loop gain G of the original amplifier circuit, write:

$$\begin{aligned} G = \frac{v_o}{v_{in}} &= \frac{v_o}{V_T} \cdot \frac{V_T}{v_{in}} \\ &= -\frac{2R_F}{R} \cdot \frac{1}{2} \\ &= -\frac{R_F}{R} \end{aligned} \tag{7.44}$$

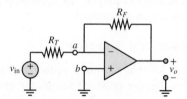

Figure 7.13 Inverting amplifier after simplification to archetype

Figure 7.13 generalizes Figure 7.8 by representing explicitly the source network as the Thévenin equivalent network of any linear input source network. The same approach can be taken to generalize the noninverting amplifier and isolation buffer circuits shown in Figures 7.9 and 7.10, respectively, where v_S and R_S are now the Thévenin (open-circuit) voltage and the Thévenin equivalent resistance, respectively, of the input source network.

Multiple Sources and the Principle of Superposition

There are many situations that call for an amplifier to accommodate multiple input source networks. The analysis of these amplifiers can be accomplished using basic principles, such as KCL, KVL, and Ohm's law. However, it is often useful to apply the principle of superposition to simplify the overall amplifier circuit into multiple component amplifiers, each with only one independent source still turned on. Thévenin's theorem can often be used to transform these component amplifiers into one of the amplifier archetypes: the inverting amplifier, the noninverting amplifier, and the isolation buffer. Two important examples of amplifiers with multiple input sources are the summing amplifier and the differential amplifier.

The Summing Amplifier

A useful op-amp circuit that is based on the inverting amplifier is the **op-amp summer**, or **summing amplifier**, as shown in Figure 7.14. Assume the op-amp is ideal. The first golden rule of op-amps states that $i^+ = i^- = 0$. Thus, when KCL is applied at the inverting node, the result is:

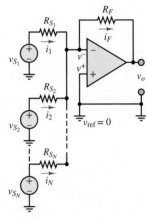

Figure 7.14 Summing amplifier

$$\sum_{n=1}^{N} i_n = i_1 + i_2 + \cdots + i_N = i_F \qquad (7.45)$$

Since negative feedback is present, the second golden rule is also valid such that $v^- = v^+ = 0$. Ohm's law can then be applied at each resistor to obtain:

$$i_n = \frac{v_{S_n} - 0}{R_{S_n}} \qquad n = 1, 2, \ldots, N$$

and

$$i_F = \frac{0 - v_o}{R_F}$$

$$(7.46)$$

The results of equation 7.46 can be plugged into equation 7.45 to find:

$$\sum_{n=1}^{N} \frac{v_{S_n}}{R_{S_n}} = -\frac{v_o}{R_F}$$

or

$$(7.47)$$

$$v_o = -\sum_{n=1}^{N} \frac{R_F}{R_{S_n}} v_{S_n}$$

The output of the summing amplifier is the weighted sum of N input signal sources, where the weighting factor for each source v_{S_n} equals the ratio of the feedback resistance R_F to the source resistance R_{S_n}. Notice that if $R_S = R_{S_1} = R_{S_2} = \cdots = R_{S_N}$, then:

$$\boxed{v_o = -\frac{R_F}{R_S} \sum_{n=1}^{N} v_{S_n} \qquad \text{Summing amplifier}} \qquad (7.48)$$

The summing amplifier can also be analyzed using the principle of superposition. Consider turning off all the voltage sources except v_{S_1}. The result is that the voltage drop across the resistors $R_2, \ldots, R_N$ is zero since a zero voltage source is equivalent to a short-circuit. Thus, for this case, $i_2 = i_3 = \cdots = i_N = 0$ as shown

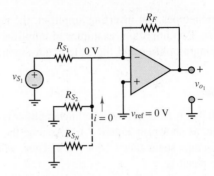

Figure 7.15 Summing amplifier with only one source turned on

in Figure 7.15. Assume the op-amp is ideal such that $i^+ = i^- = 0$. Then KCL applied at the inverting terminal node yields simply:

$$i_1 = i_F \tag{7.49}$$

Again, because negative feedback is present, the second golden rule is valid such that $v^- = v^+ = 0$. Ohm's law can then be applied to R_{S_1} and R_F to obtain:

$$i_1 = \frac{v_{S_1} - 0}{R_{S_1}} \quad \text{and} \quad i_F = \frac{0 - v_{o_1}}{R_F} \tag{7.50}$$

Plug these two results into equation 7.49 and rearrange to yield:

$$v_{o_1} = -\frac{R_F}{R_{S_1}} v_{S_1} \tag{7.51}$$

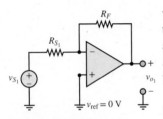

Figure 7.16 Equivalent inverting amplifier circuit for summing amplifier with only one source turned on

where v_{o_1} is the component of v_o due to the voltage source v_{S_1}. It is worth noting that this result is equivalent to what would be obtained for the inverting amplifier archetype shown in Figure 7.16. This equivalence is due to the fact that the currents $i_2, i_3, \ldots, i_N$ are all zero such that R_{S_1} and R_F are still in a virtual series connection as in the inverting amplifier archetype.

Since the Thévenin source pairs v_{S_n} and R_{S_n} in Figure 7.14 are all in parallel, the component of v_o due to v_{S_n} is:

$$v_{o_n} = -\frac{R_F}{R_{S_n}} v_{S_n} \qquad n = 1, 2, \ldots, N \tag{7.52}$$

Summing all these component contributions yields:

$$v_o = -\sum_{n=1}^{N} \frac{R_F}{R_{S_n}} v_{S_n} \tag{7.53}$$

which is the same result as that found in equation 7.47.

The Differential Amplifier

A useful op-amp circuit that is based on the inverting and noninverting amplifier archetypes is the **differential amplifier** shown in Figure 7.17. This amplifier finds frequent use in situations where the difference between two signals needs to be amplified. The two sources v_1 and v_2 may be independent of each other or may originate from the same process, as they do in the Focus on Measurements box, "Electrocardiogram (EKG) Amplifier."

The analysis of the differential amplifier can be accomplished by applying basic principles (e.g., KCL, Ohm's law) or by applying the principle of superposition.

Both approaches will assume an ideal op-amp, and since negative feedback is present, both golden rules are valid. The former approach begins by noting that $i^+ = i^- = 0$ such that R_1 and R_F are in a virtual series connection as are R_2 and R_3. Thus, the voltage at the noninverting terminal v^+ can be computed from voltage division.

$$v^+ = \frac{R_3}{R_3 + R_2}v_2 = \frac{R_3/R_2}{1 + (R_3/R_2)}v_2 \tag{7.54}$$

Likewise, voltage division along the other virtual series connection yields:

$$i = \frac{v_1 - v_o}{R_1 + R_F} = \frac{v^- - v_o}{R_F} \tag{7.55}$$

Solving for v^- yields:

$$v^- = \frac{R_F v_1 + R_1 v_o}{R_1 + R_F} = \frac{(R_F/R_1)v_1 + v_o}{1 + (R_F/R_1)} \tag{7.56}$$

The second golden rule is $v^+ = v^-$ such that:

$$\frac{R_3/R_2}{1 + (R_3/R_2)}v_2 = \frac{(R_F/R_1)v_1 + v_o}{1 + (R_F/R_1)}$$

or $\tag{7.57}$

$$v_o = \frac{1 + (R_F/R_1)}{1 + (R_3/R_2)}\frac{R_3}{R_2}v_2 - \frac{R_F}{R_1}v_1$$

In this form the expression for v_o is too complicated to leave much of an impression. However, it is greatly simplified by choosing the resistor values to satisfy:

$$\frac{R_F}{R_1} = \frac{R_3}{R_2} \tag{7.58}$$

such that:

$$v_o = \frac{R_F}{R_1}(v_2 - v_1) \qquad \text{Differential amplifier} \tag{7.59}$$

Figure 7.18 shows one particular version of a differential amplifier where equation 7.58 is satisfied by setting $R_3 = R_F$ and $R_2 = R_1$.

The circuit in Figure 7.17 can also be analyzed using the principle of superposition. The op-amp is still assumed to be ideal, and since negative feedback is present, both golden rules are valid. To begin, set $v_2 = 0$ and find the component of v_o due to v_1 as shown in Figure 7.19. Since $i^+ = 0$, there can be no voltage drop across R_2 and R_3 with the result that $v^+ = 0$. Thus, the circuit is an inverting amplifier with the output given by equation 7.29 as:

$$v_{o_1} = -\frac{R_F}{R_1}v_1 \tag{7.60}$$

Now set $v_1 = 0$ and find the component of V_o due to v_2 as shown in Figure 7.20. Since $i^+ = 0$, v_2, R_2, and R_3 are in a virtual series connection. Apply voltage division to yield:

$$v^+ = \frac{R_3}{R_3 + R_2}v_2 \tag{7.61}$$

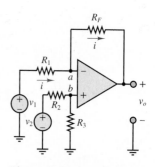

Figure 7.17 Amplifier with input sources at the inverting and noninverting terminals

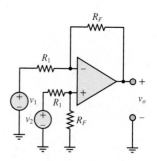

Figure 7.18 Differential amplifier

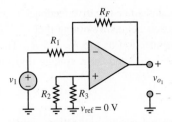

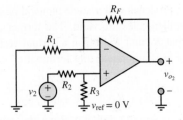

Figure 7.19 Inverting amplifier appears when $v_2 = 0$

Figure 7.20 Amplifier of Figure 7.17 but with $v_1 = 0$

Thus, the equivalent circuit is a noninverting amplifier as shown in Figure 7.21 with the output given by equation 7.34 as:

$$v_{o_2} = \left(1 + \frac{R_F}{R_1}\right) v^+ = \left(1 + \frac{R_F}{R_1}\right) \frac{R_3}{R_3 + R_2} v_2 \tag{7.62}$$

Finally, apply the principle of superposition to obtain:

$$v_o = v_{o_1} + v_{o_2} = -\frac{R_F}{R_1} v_1 + \left(1 + \frac{R_F}{R_1}\right) \frac{R_3}{R_3 + R_2} v_2 \tag{7.63}$$

As before, this expression is greatly simplified by choosing the resistor values such that:

$$\frac{R_F}{R_1} = \frac{R_3}{R_2} \tag{7.64}$$

The result is (of course!) equation 7.59.

Both of the solution methods shown above are completely valid. Neither is particularly easier than the other although tastes do vary! However, the principle of superposition has the added appeal of determining the individual contributions of each voltage source and therefore allows for a quick recalculation of the solution when only one of the voltage sources is changed.

It is important to realize that if the linear source network seen by either terminal is more complicated than those shown in Figure 7.17 it is possible to simplify them using Thévenin's theorem. For example, the source network seen by the noninverting terminal is shown in Figure 7.22, where:

$$v_T = \frac{R_3}{R_3 + R_2} v_2 \quad \text{and} \quad R_T = R_2 \| R_3 \tag{7.65}$$

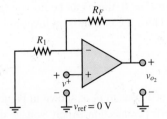

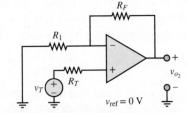

Figure 7.21 Noninverting amplifier appears when $v_1 = 0$

Figure 7.22 Figure 7.20 but with the network seen by the noninverting terminal replaced with its Thévenin equivalent network

Common and Differential Modes

It is often necessary to amplify the difference between two signals that may both be corrupted by noise or interference. The two input signals v_1 and v_2 can be decomposed be into two parts: the **common mode** (CM) and the **differential mode** (DM). These two modes are defined mathematically as:

$$v_{CM} = \frac{v_1 + v_2}{2} \quad \text{and} \quad v_{DM} = v_2 - v_1 \tag{7.66}$$

where the common mode v_{CM} is the average value of v_1 and v_2.

$$v_1 = v_{CM} - \frac{v_{DM}}{2} \quad \text{and} \quad v_2 = v_{CM} + \frac{v_{DM}}{2} \tag{7.67}$$

With these definitions, the output of an ideal differential amplifier is simply:

$$v_o = \frac{R_F}{R_1}(v_2 - v_1) = \frac{R_F}{R_1}v_{DM} \tag{7.68}$$

In other words, the common mode of the two input signals is *rejected* by the differential amplifier. In many situations, the noise and interference of one input is identical to (or nearly the same as) that of the other input. Thus, a differential amplifier can be used to eliminate noise and interference that is common to both inputs. In practice, the output of a differential amplifier is given by:

$$v_o = A_{DM}(v_2 - v_1) + A_{CM}\left(\frac{v_2 + v_1}{2}\right) \tag{7.69}$$

where A_{DM} and A_{CM} are the differential-mode and common-mode gains, respectively. In the ideal case, $A_{CM} = 0$, such as for the circuit of Figure 7.18 when the op-amp is ideal and the external resistors satisfy equation 7.58 exactly. The extent to which a practical differential amplifier rejects the common mode is known as the **common-mode rejection ratio (CMRR)**:

$$CMRR = 20 \log \left| \frac{A_{DM}}{A_{CM}} \right| \quad \text{(in dB)} \tag{7.70}$$

For example, op-amps themselves are differential amplifiers. A particular op-amp known as the 741 has a typical CMRR of 90 dB. The Focus on Measurements box, "Electrocardiogram (EKG) Amplifier," provides a realistic look at a common application of a differential amplifier.

Table 7.1 summarizes the basic op-amp circuits presented in this section.

Table 7.1 **Summary of basic amplifiers**

Configuration	Circuit diagram	Output voltage (ideal op-amp)
Inverting amplifier	Figure 7.8	$-\dfrac{R_F}{R_S}v_S$
Noninverting amplifier	Figure 7.9	$\left(1 + \dfrac{R_F}{R_S}\right)v_S$
Isolation buffer	Figure 7.10	v_S
Summing amplifier	Figure 7.14	$-\dfrac{R_F}{R_S}\displaystyle\sum_{n=1}^{N} v_{S_n}$
Differential amplifier	Figure 7.18	$\dfrac{R_F}{R_1}(v_2 - v_1)$

FOCUS ON MEASUREMENTS

Electrocardiogram (EKG) Amplifier

This example illustrates the principle behind a two-lead electrocardiogram (EKG) measurement. The desired cardiac waveform is given by the difference between the potentials measured by two electrodes suitably placed on the patient's chest, as shown in Figure 7.23. A healthy, noise-free EKG waveform $v_1 - v_2$ is shown in Figure 7.24.

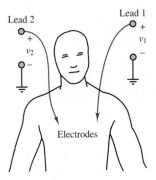

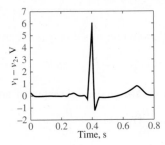

Figure 7.24 EKG waveform

Figure 7.23 Two-lead electrocardiogram

Unfortunately, noise present on the 60-Hz, 110-V AC line used to power the equipment may appear in the EKG itself, due to capacitive coupling. Ambient electromagnetic interference can also interact with the closed-loop formed by the lead wires to generate another source of noise. Other sources of noise include changes at the electrode-skin interface due to respiration, muscle contractions, and other displacements. In addition, different DC offsets due to the electrodes complicate the signals. The signal processing associated with an actual EKG involves instrumentation amplifiers (see Example 7.2) and active filters (see Section 7.3). In this example, the focus is limited to the role of a differential amplifier in rejecting common-mode 60-Hz noise found in a typical EKG. With that limitation in mind, assume that the EKG signals v_1 and v_2 indicated in Figure 7.23 are represented by:

Lead 1:

$$v_1(t) + v_n(t) = v_1(t) + V_n \cos(377t + \phi_n)$$

Lead 2:

$$v_2(t) + v_n(t) = v_2(t) + V_n \cos(377t + \phi_n)$$

As shown in Figure 7.25, the interference signal $V_n = \cos(377t + \phi_n)$ is approximately the same at both leads because the electrodes are designed to be identical and are used in close proximity to each other. If the resistors of the differential amplifier are properly matched, the voltage output will be:

$$v_o = \frac{R_2}{R_1} [(v_1 + v_n) - (v_2 + v_n)] = \frac{R_2}{R_1} (v_1 - v_2)$$

Thus, common-mode 60-Hz noise is eliminated, or greatly reduced, while the desired EKG waveform is amplified. Great! In practice, the common-mode rejection ratio is not infinite but can be made quite large to satisfy the design specifications required for a proper diagnosis.

(Continued)

(*Concluded*)

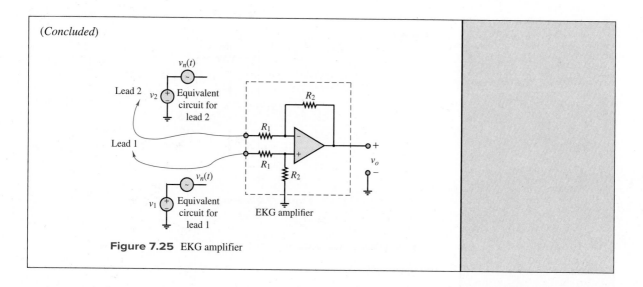

Figure 7.25 EKG amplifier

Sensor Calibration Circuit

In many practical instances, the output of a sensor is related to the physical variable we wish to measure in a form that requires some signal conditioning. The most desirable form of a sensor output is one in which the electrical output of the sensor (e.g., voltage) is related to the physical variable by a constant factor. Such a relationship is depicted in Figure 7.26(a), where k is the calibration constant relating voltage to temperature. Note that k is a positive number, and that the *calibration curve* passes through the (0, 0) point. On the other hand, the sensor characteristic of Figure 7.26(b) is best described by the following equation:

$$v_{\text{sensor}} = -\beta T + V_0$$

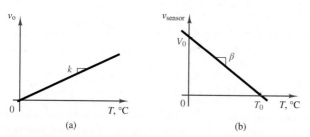

Figure 7.26 Sensor calibration curves

It is possible to modify the sensor calibration curve of Figure 7.26(b) to the more desirable one of Figure 7.26(a) by means of the simple circuit displayed in Figure 7.27. This circuit provides the desired calibration constant k by a simple gain adjustment, while the zero (or bias) offset is adjusted by means of a potentiometer connected to the voltage supplies. The detailed operation of the circuit is described in the following paragraphs.

(*Continued*)

(*Concluded*)

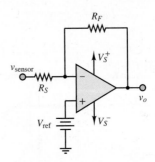

Figure 7.27 Sensor calibration circuit

This nonideal characteristic is described by:

$$v_{\text{sensor}} = -\beta T + V_0$$

When $V_{\text{ref}} = 0$, the sensor voltage input sees an inverting amplifier such that:

$$(v_o)_{\text{sensor}} = -\frac{R_F}{R_S} v_{\text{sensor}}$$

Likewise, when $v_{\text{sensor}} = 0$, the battery voltage sees a noninverting amplifier such that:

$$(v_o)_{\text{ref}} = 1 + \frac{R_F}{R_S} V_{\text{ref}}$$

Thus, the total output of the op-amp circuit of Figure 7.27 may be determined from the principle of superposition:

$$v_o = (v_o)_{\text{sensor}} + (v_o)_{\text{ref}}$$

$$= -\frac{R_F}{F_S} v_{\text{sensor}} + \left(1 + \frac{R_F}{R_S}\right) V_{\text{ref}}$$

$$= -\frac{R_F}{R_S}(-\beta T + V_0) + \left(1 + \frac{R_F}{R_S}\right) V_{\text{ref}}$$

The requirement for a linear response such as that shown in Figure 7.26(a) is $v_o = kT$, where k is the constant slope of the linear response. This requirement is satisfied by suitable choices of R_F, R_S, and V_{ref} such that:

$$kT = -\frac{R_F}{R_S}(-\beta T + V_0) + \left(1 + \frac{R_F}{R_S}\right) V_{\text{ref}}$$

For this equation to hold, the coefficients of T on both sides must be equal and the sum of the constant terms on the right side must equal zero. That is:

$$k = \frac{R_F}{R_S}\beta$$

and

$$\frac{R_F}{R_S} V_0 = \left(1 + \frac{R_F}{R_S}\right) V_{\text{ref}}$$

or

$$V_{\text{ref}} = \frac{R_F}{R_S + R_F} V_0$$

It is worth noting that $V_{ref} \approx V_0$ when $R_F \gg R_S$. Thus, when this condition holds, the appropriate battery voltage for the sensor calibration circuit can be determined directly from the sensor calibration curve of Figure 7.26(b). One should also attempt to pick a large enough value of R_S such that the sensor is not loaded by the calibration circuit.

It is also worth noting that the effect of the inverting aspect of the amplifier is to invert (change the sign of) the slope, while the effect of the reference battery voltage is to raise or lower the inverted calibration curve so that it passes through the origin. For this reason, the sensor calibration circuit is known more generally as a *level shifter*. See Example 7.3 for further discussion.

EXAMPLE 7.1   Inverting Amplifier Circuit

Problem

Determine the voltage gain and output voltage for the inverting amplifier circuit of Figure 7.8. What will the uncertainty in the gain be if 5 and 10 percent tolerance resistors are used, respectively?

Solution

Known Quantities: Feedback and source resistances, source voltage.

Find: $G = v_{out}/v_{in}$; maximum percent change in G for 5 and 10 percent tolerance resistors.

Schematics, Diagrams, Circuits, and Given Data: $R_S = 1\ k\Omega$; $R_F = 10\ k\Omega$; $v_S(t) = A\ \cos(\omega t)$; $A = 0.015\ V$; $\omega = 50\ rad/s$.

Assumptions: The amplifier behaves ideally; that is, the input current into the op-amp is zero, and negative feedback forces $v^+ = v^-$.

Analysis: Using equation 7.29, the output voltage is:

$$v_o(t) = G \times v_S(t) = -\frac{R_F}{R_S} \times v_S(t) = -10 \times 0.015\ \cos(\omega t) = -0.15\ \cos(\omega t)$$

The input and output waveforms are sketched in Figure 7.28.

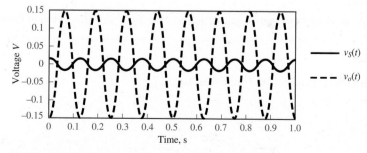

Figure 7.28 Input and output signal waveforms for Example 7.1.

The nominal gain of the amplifier is $G_{\text{nom}} = -10$. If 5 percent tolerance resistors are employed, the worst-case error will occur at the extremes:

$$G_{\text{min}} = -\frac{R_{F\text{min}}}{R_{S\text{max}}} = -\frac{9{,}500}{1{,}050} = -9.05 \qquad G_{\text{max}} = -\frac{R_{F\text{max}}}{R_{S\text{min}}} = -\frac{10{,}500}{950} = -11.05$$

The percentage error is therefore computed as

$$100 \times \frac{|G_{\text{nom}}| - |G_{\text{min}}|}{|G_{\text{nom}}|} = 100 \times \frac{10 - 9.05}{10} = 9.5\%$$

$$100 \times \frac{|G_{\text{nom}}| - |G_{\text{max}}|}{|G_{\text{nom}}|} = 100 \times \frac{10 - 11.05}{10} = -10.5\%$$

Thus, the amplifier gain could vary by as much as ± 10 percent (approximately) when 5 percent resistors are used. If 10 percent resistors were used, we would calculate a percent error of approximately ± 20 percent, as shown below.

$$G_{\text{min}} = -\frac{R_{F\text{min}}}{R_{S\text{max}}} = -\frac{9{,}000}{1{,}100} = -8.18 \qquad G_{\text{max}} = -\frac{R_{F\text{max}}}{R_{S\text{min}}} = -\frac{11{,}000}{900} = -12.2$$

$$100 \times \frac{|G_{\text{nom}}| - |G_{\text{min}}|}{|G_{\text{nom}}|} = 100 \times \frac{10 - 8.18}{10} = 18.2\%$$

$$100 \times \frac{|G_{\text{nom}}| - |G_{\text{max}}|}{|G_{\text{nom}}|} = 100 \times \frac{10 - 12.2}{10} = -22.2\%$$

Comments: Note that the worst-case percent error in the closed-loop gain G is approximately double the resistor tolerance. This result can be calculated by assuming a resistor tolerance x and noting that the worst case is:

$$|\Delta G| = \frac{R_F(1 + x)}{R_S(1 - x)} - \frac{R_F}{R_S}$$

Let $G_{\text{nom}} = -R_F/R_S$ such that:

$$\frac{|\Delta G|}{|G_{\text{nom}}|} = \frac{1 + x}{1 - x} - 1 = \frac{2x}{1 - x}$$

$$= 2x(1 + x + x^2 + \cdots) \approx 2x \qquad (x \ll 1)$$

EXAMPLE 7.2   Instrumentation Amplifier

Problem

Determine the closed-loop voltage gain of the instrumentation amplifier circuit of Figure 7.29.

Solution

Known Quantities: Feedback and source resistances.

Find:

$$G = \frac{v_o}{v_1 - v_2}$$

Assumptions: Assume ideal op-amps.

Analysis: Often, to provide impedance isolation between bridge transducers and the differential amplifier stage, the signals v_1 and v_2 are amplified separately. This technique gives rise to the **instrumentation amplifier (IA)**, shown in Figure 7.29.

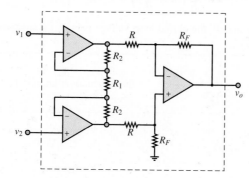

Figure 7.29 Instrumentation amplifier

Because the instrumentation amplifier has widespread application—and to ensure the best possible match between resistors—the entire circuit of Figure 7.29 is often packaged as a single integrated circuit. The advantage of this configuration is that resistors R_1 and R_2 can be matched much more precisely in an integrated circuit than would be possible by using discrete components.

Consider the input circuit first. Thanks to the symmetry of the circuit, we can represent one-half of the circuit as illustrated in Figure 7.30(a), depicting the lower half of the first *stage* of the instrumentation amplifier. We next recognize that the circuit of Figure 7.30(a) is a noninverting amplifier (see Figure 7.9), such that the closed-loop voltage gain is (equation 7.34):

$$A = 1 + \frac{R_2}{R_1/2} = 1 + \frac{2R_2}{R_1}$$

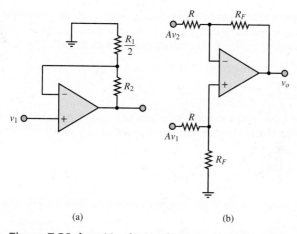

(a)   (b)

Figure 7.30 Input (a) and output (b) stages of instrumentation amplifier

Each of the two inputs v_1 and v_2 is therefore an input to the second *stage* of the instrumentation amplifier, shown in Figure 7.30(b). We recognize the second stage to be a differential amplifier (see Figure 7.18), and can therefore write the output voltage using equation 7.59:

$$v_o = \frac{R_F}{R}(Av_1 - Av_2) = \frac{R_F}{R}\left(1 + \frac{2R_2}{R_1}\right)(v_1 - v_2)$$

from which we can compute the closed-loop voltage gain of the instrumentation amplifier:

$$G = \frac{v_o}{v_1 - v_2} = \frac{R_F}{R}\left(1 + \frac{2R_2}{R_1}\right) \qquad \text{Instrumentation amplifier}$$

EXAMPLE 7.3   Level Shifter

Problem

The level shifter of Figure 7.31 has the ability to add or subtract a DC offset to or from a signal. Analyze the circuit, and design it so that it can remove a 1.8-V DC offset from a sensor signal.

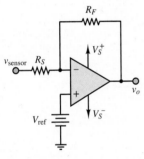

Figure 7.31 Level shifter

Solution

Known Quantities: Sensor (input) voltage; feedback and source resistors.

Find: Value of V_{ref} required to remove DC bias.

Schematics, Diagrams, Circuits, and Given Data: $v_{\text{sensor}}(t) = 1.8 + 0.1\cos(\omega t)$; $R_F = 220$ kΩ; $R_S = 10$ kΩ.

Assumptions: Assume an ideal op-amp.

Analysis: The output voltage can be computed quite easily using the principle of superposition. When the reference voltage source V_{ref} is set to zero and replaced by a short-circuit, the sensor input voltage v_{sensor} sees an inverting amplifier such that:

$$\frac{v_{o_1}}{v_{\text{sensor}}} = -\frac{R_F}{R_S}$$

When the sensor input voltage source is set to zero and replaced by a short-circuit, the reference voltage source (the battery) sees a noninverting amplifier such that:

$$\frac{v_{o_2}}{V_{\text{ref}}} = 1 + \frac{R_F}{R_S}$$

Thus, the total output voltage is the sum of contributions from the two sources:

$$v_o = v_{o_1} + v_{o_2} = -\frac{R_F}{R_S}v_{\text{sensor}} + \left(1 + \frac{R_F}{R_S}\right)V_{\text{ref}}$$

Substitute the expression for v_{sensor} into the previous equation to find:

$$v_o = -\frac{R_F}{R_S}[1.8 + 0.1\cos(\omega t)] + \left(1 + \frac{R_F}{R_S}\right)V_{\text{ref}}$$

$$= -\frac{R_F}{R_S}[0.1\cos(\omega t)] - \frac{R_F}{R_S}(1.8) + \left(1 + \frac{R_F}{R_S}\right)V_{\text{ref}}$$

To remove the DC offset, require:

$$-\frac{R_F}{R_S}(1.8) + \left(1 + \frac{R_F}{R_S}\right)V_{\text{ref}} = 0$$

or

$$V_{\text{ref}} = 1.8\frac{R_F/R_S}{1 + R_F/R_S} = 1.722\,\text{V}$$

Comments: The presence of a precision voltage source in the circuit is undesirable because it may add considerable expense to the circuit design and, in the case of a battery, it is not adjustable. The circuit of Figure 7.32 illustrates how an adjustable voltage reference can be produced from the DC supplies already used by the op-amp, two fixed resistors R, and a potentiometer R_p. The fixed resistors are included to guarantee a minimum resistance R from the wiper to either power supply at all times and thus prevent possible overheating of the potentiometer. An expression for V_{ref} is obtained from voltage division:

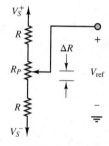

$$\frac{V_{\text{ref}} - V_S^-}{V_S^+ - V_S^-} = \frac{R + \Delta R}{2R + R_p}$$

If the voltage supplies are symmetric, as is usually the case, $V_S^+ = -V_S^-$ such that:

$$\frac{V_{\text{ref}} + V_S^+}{2V_S^+} = \frac{R + \Delta R}{2R + R_p}$$

Figure 7.32 Adjustable voltage reference for Example 7.3.

Rearrange terms to find:

$$V_{\text{ref}} = \frac{2\Delta R - R_p}{2R + R_p}V_S^+$$

The value of V_{ref} is determined by the position of the wiper ΔR. Also, when $R_p \gg R$, the range of V_{ref} is approximately $\pm V_S^+$.

EXAMPLE 7.4   Temperature Control Using Op-Amps

Problem

Op-amps often serve as building blocks in analog control systems. The objective of this example is to illustrate the use of op-amps in a temperature control circuit. Figure 7.33(a) depicts a system for which the temperature is to be maintained constant at 20°C in a variable temperature environment. The system temperature is measured via a thermocouple. Heat is added to the system by a coil of resistance R_{coil}. The heat flux is $q_{\text{in}} = i^2 R_{\text{coil}}$, where i is the current provided by a power amplifier. The system is insulated on three sides. The fourth side is not insulated such that heat is transferred across the boundary by convection, which is represented by an equivalent thermal resistance R_t. The system has mass m, specific heat c, and thermal capacitance $C_t = mc$ (see the Make the Connection boxes "Thermal Capacitance" and "Thermal System Dynamics" in Chapter 4).

Solution

Known Quantities: Sensor (input) voltage; feedback and source resistors, thermal system component values.

Find: Select desired value of proportional gain K_P to achieve automatic temperature control.

Schematics, Diagrams, Circuits, and Given Data: $R_{coil} = 5\ \Omega$; $R_t = 2°C/W$; $C_t = 50\ J/°C$; $\alpha = 1\ V/°C$. Figure 7.33(a) to (e).

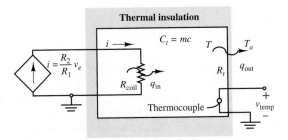

Figure 7.33(a) Thermal system

Assumptions: Assume ideal op-amps.

Analysis: Conservation of energy requires that:

$$q_{in} - q_{out} = \frac{dE_{stored}}{dt}$$

where q_{in} represents the heat added to the system by the electrical heater, q_{out} represents the heat lost from the system through convection to the surrounding air, and E_{stored} represents the energy stored in the system due to its thermal capacitance. The system temperature T is measured by a thermocouple whose output voltage is proportional to temperature: $v_{temp} = \alpha T$. Further, assume that the power amplifier is modeled by a *voltage-controlled current source* (VCCS) such that:

$$i = K_v K_p v_e = \frac{R_2}{R_1} v_e = \frac{R_2}{R_1}(v_{ref} - v_{temp}) = \frac{R_2}{R_1}\alpha(T_{ref} - T)$$

where v_e is the error or difference between the reference voltage and the measured voltage. The negative feedback system shown in Figure 7.33(b) tends to drive v_e to zero. When v_e

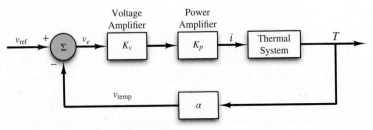

Figure 7.33(b) Block diagram of control system

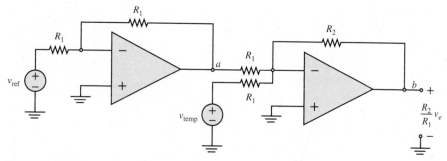

Figure 7.33(c) Circuit for generating proportional gain of error voltage

is positive, $v_{ref} > v_{temp}$ and the system calls for heating; on the other hand, when v_e is negative, $v_{ref} < v_{temp}$ and the system calls for cooling. The power amplifier outputs a positive current for a positive v_e. Thus, the block diagram shown in Figure 7.33(b) corresponds to an *automatic control system* that increases or decreases the heating coil current to maintain the system temperature at the desired (reference) value. The *proportional gain K_p* of the power amplifier determines *how much* to increase coil current and allows the user to optimize the response of the system for a specific design requirement. For example, a system specification could require that the automatic temperature control system be designed so as to maintain the temperature to within 1 degree of the reference temperature for external temperature disturbances as large as 10 degrees. The response of the system can be adjusted by varying the proportional gain.

The voltage amplifier can be realized by a two-stage amplifier using two op-amps as shown in Figure 7.33(c). The first stage is an inverting amplifier with closed-loop gain $G_1 = -1$ such that the voltage at node a is $-v_{ref}$. The second stage is a summing amplifier with a closed-loop gain of $G_2 = -R_2/R_1$ for each input. Thus, the output voltage at node b is:

$$v_b = -\frac{R_2}{R_1}(v_a + v_{temp}) = \frac{R_2}{R_1}(v_{ref} - v_{temp}) = \frac{R_2}{R_1}(v_e)$$

The coefficient R_2/R_1 is the voltage gain K_v. In other words, *selecting the feedback resistor R_2 is equivalent to choosing K_v.*

The thermal system itself is described by the conservation of energy equation given above. The rate of energy added to the system by the heating coil is simply $i^2 R_{coil}$. The rate of energy subtracted from the system by convective heat transfer is defined as $(T - T_a)/R_t$, where R_t is a lumped parameter called the *thermal resistance*. Small values of R_t correspond to large values of the convective heat transfer coefficient, and vice versa. Finally, the net rate at which energy is stored in the system is proportional to the rate at which the system temperature T is changing, where the constant of proportionality C_t is known as the *thermal capacitance*. With these definitions in place, the conservation of energy equation can be rewritten as:

$$i^2 R_{coil} - \frac{T - T_a}{R_t} = C_t \frac{dT}{dt}$$

or

$$R_t C_t \frac{dT}{dt} + T = R_t R_{coil} i^2 + T_a$$

where $i = K_p K_v v_e = K_p K_v \alpha(T_{ref} - T)$. Notice this equation is, in general, a nonlinear first-order ordinary differential equation. The time constant is $\tau = R_t C_t = 2°C/W \times 50 \ J/°C = 100 \ s$.

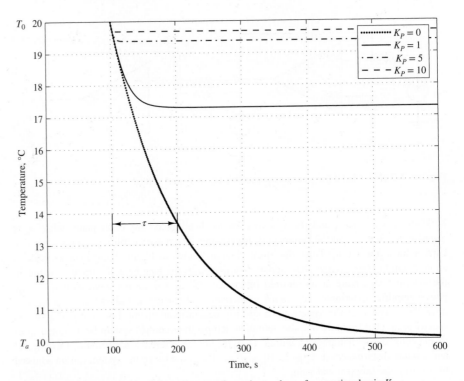

Figure 7.33(d) Response of thermal system for various values of proportional gain K_p

When $K_p = 0$, no current is supplied to the heating coil and the thermal system response is simply its own natural response; that is, no automatic control is active when $K_p = 0$ and the system response is the *open-loop* response. In that case, the governing differential equation is:

$$\tau \frac{dT}{dt} + T = T_a$$

The solution is (see Chapter 4):

$$T = (T_0 - T_a)e^{-t/\tau} + T_a$$

where T_0 is the initial value of the system temperature. For example, assume $T_0 = 20°C$ and $T_a = 10°C$. The time constant τ is $R_t C_t = 100$ s, using the data given for R_t and C_t. Thus:

$$T = (10°C)e^{-t/100} + 10°C \qquad \text{Open-loop response}$$

When the gain K_p is increased to 1, v_e increases as soon as the temperature drops below the reference value. The transduction constant of the thermocouple was given as $\alpha = 1$ such that the voltage v_{temp} is numerically equal to the system temperature. Figure 7.33(d) shows the temperature response for values of K_p ranging from 1 to 10. As the gain increases, the error between the desired and actual temperatures decreases very quickly. Observe that the error is less than 1 degree (recall the design specification) for $K_p = 5$. To better understand the workings of the complete control system, it is helpful to observe the heater current, which is an amplified version of the error voltage. Figure 7.33(e) shows that when $K_p = 1$ the current increases to a final value of roughly 2.7 A; when $K_p = 5$ and 10, the current increases more rapidly, and eventually settles to values of 3 and 3.1 A, respectively.

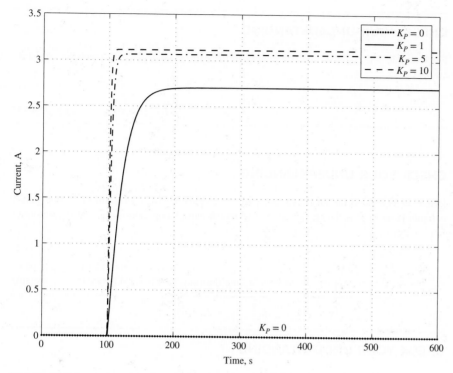

Figure 7.33(e) Power amplifier output current for various proportional gain K_p

The steady-state value of the current is reached in about 17 s for $K_p = 5$, and in about 8 s for $K_p = 10$.

Comments: As K_p increases, the system's speed of response increases; however, the system's *steady-state error* also increases. The design specifications anticipate this effect by setting a tolerance of 1°C.

CHECK YOUR UNDERSTANDING

Consider an ideal inverting amplifier (see Figure 7.8) with a nominal closed-loop gain of $-1,000$. The impact of a nonideal op-amp with a finite, but large, open-loop gain A on the closed-loop gain can be derived by assuming that the voltage v^- at the inverting terminal is only approximately equal to the voltage $v^+ = 0$ at the noninverting terminal. Under this assumption, $v_{out} = -Av^-$. The first golden rule still applies such that $i_{in} = 0$ and R_S is virtually in series with R_F. Use this information to find an expression for the closed-loop gain as a function of the open-loop gain A. Compute the closed-loop gain when A equals 10^7, 10^6, 10^5, 10^4. How large is the open-loop gain when the closed-loop gain is less than 0.1 percent away from its nominal value?

CHECK YOUR UNDERSTANDING

For Example 7.1, calculate the uncertainty in the gain when 1 percent "precision" resistors are used.

Answer: $+1.98$ to -2.02 percent

CHECK YOUR UNDERSTANDING

Derive an expression for the closed-loop gain of an isolation buffer when the open-loop gain A is finite. How large is the open-loop gain when the closed-loop gain is only 0.1 percent away from unity?

Answer: The expression for the closed-loop gain is $v_{out}/v_{in} = 1 + 1/A$; thus A should equal 10^4 for 0.1 percent accuracy.

CHECK YOUR UNDERSTANDING

For Example 7.3, find the value ΔR that removes the DC bias from the sensor signal. Assume the supply voltages are symmetric at ±15 V and a 10-kΩ potentiometer is tied to two 10-kΩ fixed resistors as in Figure 7.32. What is the range of V_{ref} when a 10-kΩ potentiometer is tied to two 10-kΩ fixed resistors?

Answers: $\Delta R = 6.722$ kΩ; V_{ref} is between ±0.714 V

CHECK YOUR UNDERSTANDING

How much steady-state power, in watts, will be input to the thermal system of Example 7.4 to maintain its temperature in the face of a 10°C ambient temperature drop for values of K_P of 1, 5, and 10?

Answers: $K_P = 1$: 36.5 W; $K_P = 5$: 45 W; $K_P = 10$: 48 W

CHECK YOUR UNDERSTANDING

With reference to the Focus on Measurements box, "Sensor Calibration Circuit," find numerical values of R_F/R_S and V_{ref} if the temperature sensor has $\beta = 0.235$ and $V_0 = 0.7$ V and the desired relationship is $v_{out} = 10\,T$.

Answers: $R_F/R_S = 42.55$; $V_{ref} = 0.684$ V

7.3 ACTIVE FILTERS

The range of useful applications of an operational amplifier is greatly expanded if energy storage elements are introduced into the design; the frequency-dependent properties of these elements, studied in Chapters 3 and 5, will prove useful in the design of various types of op-amp circuits. In particular, it will be shown that it is possible to shape the frequency response of an operational amplifier by appropriate use of complex impedances in the input and feedback circuits. The class of filters one can obtain by means of op-amp designs is called **active filters** because op-amps can provide amplification (gain) in addition to the filtering effects already studied in Chapter 5 for passive circuits (i.e., circuits comprising exclusively resistors, capacitors, and inductors).

The easiest way to see how the frequency response of an op-amp can be shaped (almost) arbitrarily is to replace the resistors R_F and R_S in Figures 7.8 and 7.9 with impedances $\mathbf{Z}_F$ and $\mathbf{Z}_S$, as shown in Figure 7.34. It is a straightforward matter to show that in the case of the inverting amplifier, the expression for the closed-loop gain is given by

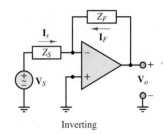

Inverting

$$\frac{\mathbf{V}_o}{\mathbf{V}_S}(j\omega) = -\frac{\mathbf{Z}_F}{\mathbf{Z}_S} \tag{7.71}$$

whereas for the noninverting case, the gain is

$$\frac{\mathbf{V}_o}{\mathbf{V}_S}(j\omega) = 1 + \frac{\mathbf{Z}_F}{\mathbf{Z}_S} \tag{7.72}$$

where $\mathbf{Z}_F$ and $\mathbf{Z}_S$ can be arbitrarily complex impedance functions and where $\mathbf{V}_S$, $\mathbf{V}_o$, $\mathbf{I}_F$, and $\mathbf{I}_S$ are all phasors. Thus, it is possible to shape the frequency response of an ideal op-amp filter simply by selecting suitable ratios of feedback impedance to source impedance. By connecting a circuit similar to the low-pass filters studied in Chapter 5 in the feedback loop of an op-amp, the same filtering effect can be achieved and, in addition, the signal can be amplified.

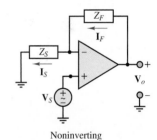

Noninverting

Figure 7.34 Op-amp circuits employing complex impedances

The simplest op-amp low-pass filter is shown in Figure 7.35. Its analysis is quite simple if we take advantage of the fact that the closed-loop gain, as a function of frequency, is given by

$$\mathbf{G}_{\mathrm{LP}}(j\omega) = -\frac{\mathbf{Z}_F}{\mathbf{Z}_S} \tag{7.73}$$

where

$$\mathbf{Z}_F = R_F \| \frac{1}{j\omega C_F} = \frac{R_F}{1 + j\omega C_F R_F} \tag{7.74}$$

and

$$\mathbf{Z}_S = R_S \tag{7.75}$$

Note the similarity between $\mathbf{Z}_F$ and the low-pass characteristic of the passive RC circuit! The closed-loop gain $\mathbf{G}_{\mathrm{LP}}(j\omega)$ is then computed to be

$$\mathbf{G}_{\mathrm{LP}}(j\omega) = -\frac{\mathbf{Z}_F}{\mathbf{Z}_S} = -\frac{R_F/R_S}{1 + j\omega C_F R_F} \qquad \text{Low-pass filter} \tag{7.76}$$

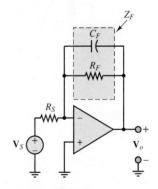

Figure 7.35 Active low-pass filter

This expression can be factored into two terms. The first is an amplification factor analogous to the amplification that would be obtained with a simple inverting amplifier (i.e., the same circuit as that of Figure 7.35 with the capacitor removed); the second is a low-pass filter, with a cutoff frequency dictated by the parallel

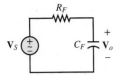

Figure 7.36 Passive low-pass filter

combination of R_F and C_F in the feedback loop. The filtering effect is completely analogous to what would be attained by the passive circuit shown in Figure 7.36. However, the op-amp filter also provides amplification by a factor of R_F/R_S.

It should be apparent that the response of this op-amp filter is just an amplified version of that of the passive filter. Figure 7.37 depicts the amplitude response of the active low-pass filter (in the figure, $R_F/R_S = 10$ and $1/R_FC_F = 1$) in two different graphs; the first plots the amplitude ratio $\mathbf{V}_o(j\omega)$ versus radian frequency ω on a logarithmic scale, while the second plots the amplitude ratio $20 \log \mathbf{V}_S(j\omega)$ (in units of decibels), also versus ω on a logarithmic scale. Recall from Chapter 5 that decibel frequency response plots are often encountered. Note that in the decibel plot, the slope of the filter response for frequencies significantly higher than the cutoff frequency,

$$\omega_0 = \frac{1}{R_F C_F} \tag{7.77}$$

is -20 dB/decade, while the slope for frequencies significantly lower than this cutoff frequency is equal to zero. The value of the response at the cutoff frequency is found to be, in units of decibel,

$$|\mathbf{G}_{\text{LP}}(j\omega_0)|_{\text{dB}} = 20 \log_{10} \frac{R_F}{R_S} - 20 \log \sqrt{2} \tag{7.78}$$

where

$$-20 \log_{10} \sqrt{2} = -3 \text{ dB} \tag{7.79}$$

Thus, ω_0 is also called the *3-dB frequency*.

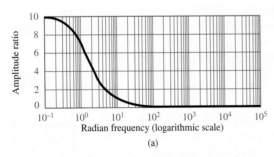

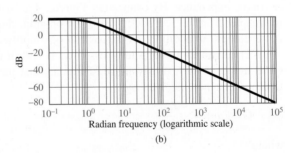

Figure 7.37 Normalized response of active low-pass filter: (a) amplitude ratio response; (b) dB response

Among the advantages of such active low-pass filters is the ease with which the gain and the bandwidth can be adjusted by controlling the ratios R_F/R_S and $1/R_FC_F$, respectively.

It is also possible to construct other types of filters by suitably connecting resistors and energy storage elements to an op-amp. For example, a high-pass active filter can easily be obtained by using the circuit shown in Figure 7.38. The impedance of the input path is:

$$\mathbf{Z}_S = R_S + \frac{1}{j\omega C_S} \tag{7.80}$$

The impedance of the feedback path is:

$$\mathbf{Z}_F = R_F \tag{7.81}$$

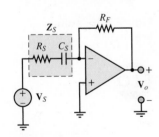

Figure 7.38 Active high-pass filter

The closed-loop gain for this inverting amplifier is:

$$\mathbf{G}_{\text{HP}}(j\omega) = -\frac{\mathbf{Z}_F}{\mathbf{Z}_S} = -\frac{j\omega C_S R_F}{1 + j\omega R_S C_S} \qquad \text{High-pass filter} \qquad (7.82)$$

Note that $G \to 0$ as $\omega \to 0$. Also note that as $\omega \to \infty$, the closed-loop gain G approaches a constant:

$$\lim_{\omega \to \infty} \mathbf{G}_{\text{HP}}(j\omega) = -\frac{R_F}{R_S} \qquad (7.83)$$

That is, above a certain frequency range, the circuit acts as a linear amplifier. This is exactly the behavior one would expect of a high-pass filter. The high-pass response is depicted in Figure 7.39, in both linear and decibel plots (in the figure, $R_F/R_S = 10$ and $1/R_S C = 1$). Note that in the decibel plot, the slope of the filter response for frequencies significantly lower than $\omega = 1/R_S C_S = 1$ is +20 dB/decade, while the slope for frequencies significantly higher than this cutoff (or 3 dB) frequency is equal to zero.

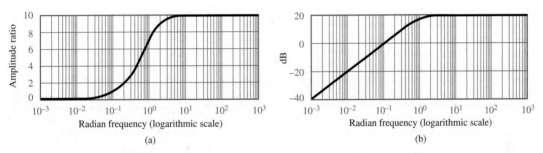

Figure 7.39 Normalized response of active high-pass filter: (a) amplitude ratio response; (b) dB response

As a final example of active filters, let us look at a simple active bandpass filter configuration. This type of response may be realized simply by combining the high- and low-pass filters we examined earlier. The circuit is shown in Figure 7.40.

The analysis of the bandpass circuit follows the same structure used in previous examples. First we evaluate the feedback and input impedances:

$$\mathbf{Z}_F = R_F \| \frac{1}{j\omega C_F} = \frac{R_F}{1 + j\omega C_F R_F} \qquad (7.84)$$

$$\mathbf{Z}_S = R_S + \frac{1}{j\omega C_S} = \frac{1 + j\omega C_S R_S}{j\omega C_S} \qquad (7.85)$$

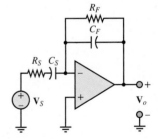

Figure 7.40 Active bandpass filter

Next we compute the closed-loop frequency response of the op-amp, as follows:

$$\mathbf{G}_{\text{BP}}(j\omega) = -\frac{\mathbf{Z}_F}{\mathbf{Z}_S} = -\frac{j\omega C_S R_F}{(1 + j\omega C_F R_F)(1 + j\omega C_S R_S)} \qquad \begin{matrix}\text{Bandpass} \\ \text{filter}\end{matrix} \qquad (7.86)$$

The form of the op-amp response we just obtained should not be a surprise. It is very similar (although not identical) to the product of the low-pass and high-pass responses of equations 7.76 and 7.82. In particular, the denominator of $\mathbf{G}_{\text{BP}}(j\omega)$ is exactly the product of the denominators of $\mathbf{G}_{\text{LP}}(j\omega)$ and $\mathbf{G}_{\text{HP}}(j\omega)$. It is particularly enlightening to rewrite $\mathbf{G}_{LP}(j\omega)$ in a slightly different form, after making the observation that each RC product corresponds to some "critical" frequency:

$$\omega_1 = \frac{1}{R_F C_S} \qquad \omega_{LP} = \frac{1}{R_F C_F} \qquad \omega_{HP} = \frac{1}{R_S C_S} \qquad (7.87)$$

It is easy to verify that for the case where

$$\omega_{HP} > \omega_{LP} \qquad (7.88)$$

the response of the op-amp filter may be represented as shown in Figure 7.41 in both linear and decibel plots (in the figure, $\omega_1 = 1$, $\omega_{HP} = 1,000$, and $\omega_{LP} = 10$). The decibel plot is very revealing, for it shows that, in effect, the bandpass response is the graphical superposition of the low-pass and high-pass responses shown earlier. The two 3-dB (or cutoff) frequencies are the same as in $\mathbf{G}_{LP}(j\omega)$, $1/R_F C_F$; and in $\mathbf{G}_{HP}(j\omega)$, $1/R_S C_S$. The third frequency, $\omega_1 = 1/R_F C_S$, represents the point where the response of the filter crosses the 0-dB axis (rising slope). Since 0 dB corresponds to a gain of 1, this frequency is called the **unity gain frequency**.

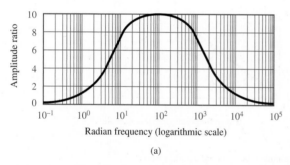

(a)

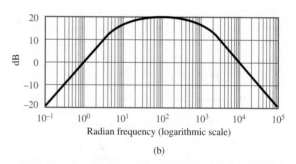

(b)

Figure 7.41 Normalized amplitude response of active bandpass filter: (a) amplitude ratio response; (b) dB response

The ideas developed thus far can be employed to construct more complex functions of frequency. In fact, most active filters one encounters in practical applications are based on circuits involving more than one or two energy storage elements. By constructing suitable functions for $\mathbf{Z}_F$ and $\mathbf{Z}_S$, it is possible to realize filters with greater frequency selectivity (i.e., sharpness of cutoff), as well as flatter bandpass or band-rejection functions (i.e., filters that either allow or reject signals in a limited band of frequencies). One remark that should be made in passing, though, pertains to the exclusive use of capacitors in the circuits analyzed thus far. One of the advantages of op-amp filters is that it is not necessary to use both capacitors and inductors to obtain a bandpass response. Suitable connections of capacitors can accomplish that task in an op-amp. This seemingly minor fact is of great importance in practice because inductors are expensive to mass-produce to close tolerances and exact specifications and are often bulkier than capacitors with equivalent energy storage capabilities. On the other hand, capacitors are easy to manufacture in a wide variety of tolerances and values, and in relatively compact packages, including in integrated-circuit form.

Example 7.5 illustrates how it is possible to construct active filters with greater frequency selectivity by adding energy storage elements to the design.

EXAMPLE 7.5   Second-Order Low-Pass Filter

Problem

Determine the closed-loop voltage gain as a function of frequency for the op-amp circuit of Figure 7.42.

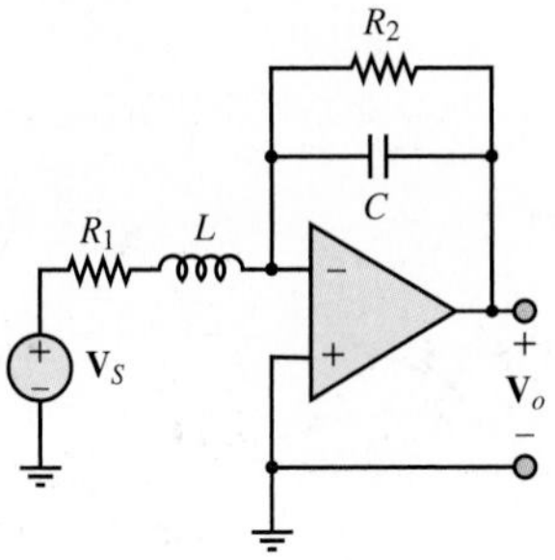

Figure 7.42 Circuit for Example 7.5.

Solution

Known Quantities: Feedback and source impedances.

Find:

$$G(j\omega) = \frac{V_o(j\omega)}{V_S(j\omega)}$$

Schematics, Diagrams, Circuits, and Given Data: $1/R_2C = R_1/L = \omega_0$.

Assumptions: Assume an ideal op-amp.

Analysis: The expression for the gain of the filter of Figure 7.42 can be determined by using equation 7.71:

$$G(j\omega) = \frac{V_o(j\omega)}{V_S(j\omega)} = -\frac{Z_F(j\omega)}{Z_S(j\omega)}$$

where

$$Z_F(j\omega) = R_2 \| \frac{1}{j\omega C} = \frac{R_2}{1 + j\omega C R_2} = \frac{R_2}{1 + j\omega/\omega_0}$$

$$Z_S(j\omega) = R_1 + j\omega L = R_1\left(1 + j\omega\frac{L}{R_1}\right) = R_1\left(1 + \frac{j\omega}{\omega_0}\right)$$

Thus, the closed-loop gain G of the filter is:

$$G(j\omega) = -\frac{R_2/(1 + j\omega/\omega_0)}{R_1(1 + j\omega/\omega_0)}$$

$$= -\frac{R_2/R_1}{(1 + j\omega/\omega_0)^2}$$

Note that it is possible to simplify the circuit in Figure 7.42 at very low and very high frequencies. The solutions for these simplified forms can be used to validate the previous expression. For example, at very low frequencies, the inductor acts like a short-circuit and the capacitor acts like an open-circuit. Using these approximations, the circuit becomes a simple inverting amplifier with closed-loop gain:

$$G(j\omega) \approx -\frac{R_2}{R_1} \qquad \omega \ll \omega_0$$

This approximation matches the complete solution when $\omega \ll \omega_0$ because $1 + j\omega/\omega_0 \approx 1$. Likewise, at very high frequencies, the inductor acts as an open-circuit and the capacitor acts as a short-circuit. Using these approximations, the circuit experiences a virtual short-circuit from its output to the inverting terminal and the source V_S sees a very large input impedance. The result is that $V_o \approx 0$ due to the virtual ground at the inverting terminal and thus the closed-loop gain is:

$$\mathbf{G}(j\omega) \approx 0 \qquad \omega \gg \omega_0$$

This approximation matches the complete solution when $\omega \gg \omega_0$ because $1 + j\omega/\omega_0 \rightarrow j\omega/\omega_0 \rightarrow \infty$ as $\omega \rightarrow \infty$. Validations are an important aspect of good problem solving because they add confidence to solutions and, if done smartly, usually expose erroneous solutions.

Comments: Note the similarity between the expression for the gain of the filter of Figure 7.42 and that given in equation 7.76 for the gain of a first-order low-pass filter. Clearly, the circuit analyzed in this example is also a second-order low-pass filter, as indicated by the quadratic term in the denominator. Figure 7.43 compares the two responses in both linear and decibel (Bode) magnitude plots. The slope of the decibel plot for the second-order filter at higher frequencies is twice that of the first-order filter (−40 versus −20 dB/decade). We should also remark that the use of an inductor in the filter design is not recommended in practice, as explained in the above section, and that we have used it in this example only because of the simplicity of the resulting gain expressions.

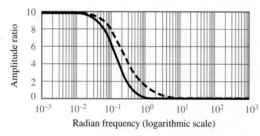
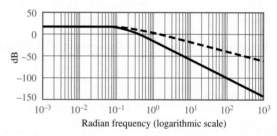

Figure 7.43 Comparison of first- and second-order active low-pass filters: (a) amplitude ratio response; (b) dB response

CHECK YOUR UNDERSTANDING

(a) Design a low-pass filter with a closed-loop gain of 100 and cutoff (3-dB) frequency equal to 800 Hz. Assume that only 0.01-μF capacitors are available. Find R_F and R_S.

(b) Repeat the design of the exercise above for a high-pass filter with a cutoff frequency of 2,000 Hz. This time, however, assume that only standard values of resistors are available (see the table of standard values in Chapter 1). Select the nearest component values, and calculate the percent error in cutoff frequency.

(c) Find the frequencies corresponding to 1-dB attenuation from the low-frequency gains of the filters of parts a and b.

(d) What is the decibel gain for the filter of Example 7.5 at the cutoff frequency ω_0? Find the 3-dB frequency for this filter in terms of the cutoff frequency ω_0, and note that the two are not the same.

7.4   INTEGRATORS AND DIFFERENTIATORS

In the preceding sections, we examined the frequency response of op-amp circuits for sinusoidal inputs. However, certain op-amp circuits containing energy storage elements reveal some of their more general properties if we analyze their response to inputs that are time varying but not necessarily sinusoidal. Among such circuits are the commonly used integrator and differentiator; the analysis of these circuits is presented in the following paragraphs.

The Ideal Integrator

Consider the circuit of Figure 7.44, where $v_S(t)$ is an arbitrary function of time (e.g., a pulse train, a triangular wave, or a square wave). The op-amp circuit shown provides an output that is proportional to the integral of $v_S(t)$. The analysis of the integrator circuit is, as always, based on the observation that

$$i_S(t) = -i_F(t) \tag{7.89}$$

where

$$i_S(t) = \frac{v_S(t)}{R_S} \tag{7.90}$$

It is also known that

$$i_F(t) = C_F \frac{dv_o(t)}{dt} \tag{7.91}$$

from the fundamental definition of the capacitor. The source voltage can then be expressed as a function of the derivative of the output voltage:

$$\frac{1}{R_S C_F} v_S(t) = -\frac{dv_o(t)}{dt} \tag{7.92}$$

Integrate both sides of equation 7.92 to obtain:

$$v_o(t) = -\frac{1}{R_S C_F} \int_{-\infty}^{t} v_S(t') dt' \qquad \text{Integrator} \tag{7.93}$$

There are numerous applications for integrators in practical circuits.

The Ideal Differentiator

Using an argument similar to that employed for the integrator, we can derive a result for the ideal differentiator circuit of Figure 7.45. The relationship between input and output is obtained by observing that

$$i_S(t) = C_S \frac{dv_S(t)}{dt} \tag{7.94}$$

and

$$i_F(t) = \frac{V_o(t)}{R_F} \tag{7.95}$$

so that the output of the differentiator circuit is proportional to the derivative of the input:

$$V_o(t) = -R_F C_S \frac{dv_S(t)}{dt} \qquad \text{Op-amp differentiator} \tag{7.96}$$

Although mathematically attractive, the differentiation property of this op-amp circuit is seldom used in practice because differentiation tends to amplify any noise that may be present in a signal.

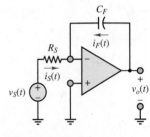

Figure 7.44 Op-amp integrator

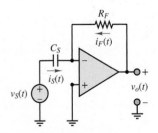

Figure 7.45 Op-amp differentiator

FOCUS ON MEASUREMENTS

Charge Amplifiers

One of the most common families of transducers for the measurement of force, pressure, and acceleration is that of piezoelectric transducers. These transducers contain a piezoelectric crystal that generates an electric charge in response to deformation. Thus, if a force is applied to the crystal (leading to a displacement), a charge is generated within the crystal. If the external force generates a displacement x_i, then the transducer will generate a charge q according to the expression:

$$q = K_P x_i$$

Figure 7.46 depicts the basic structure of the piezoelectric transducer, and a simple circuit model. The model consists of a current source in parallel with a capacitor, where the current source represents the rate of change of the charge generated in response to an external force; and the capacitance is a consequence of the structure of the transducer, which consists of a piezoelectric crystal (e.g., quartz or Rochelle salt) sandwiched between conducting electrodes (in effect, this is a parallel-plate capacitor).

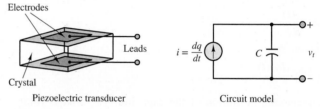

Figure 7.46 Piezoelectric transducer

Although it is possible, in principle, to employ a conventional voltage amplifier to amplify the transducer output voltage v_t, given by

$$v_t = \frac{1}{C} \int i \, dt = \frac{1}{C} \int \frac{dq}{dt} dt = \frac{q}{C} = \frac{K_P x_i}{C}$$

it is often advantageous to use a **charge amplifier**. The charge amplifier is essentially an integrator circuit, as shown in Figure 7.47, characterized by an extremely high input impedance.[3] The high impedance is essential; otherwise, the charge generated by the transducer would leak to ground through the input impedance of the amplifier.

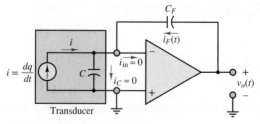

Figure 7.47 Charge amplifier

(*Continued*)

---

[3]Special op-amps are employed to achieve extremely high input impedance, through FET input circuits. See Chapter 10.

(Concluded)

Because of the high input impedance, the input current to the amplifier is negligible; further, because of the high open-loop gain of the amplifier, the inverting-terminal voltage is essentially at ground potential. Thus, *the voltage across the transducer is effectively zero.* As a consequence, to satisfy KCL, the feedback current $i_F(t)$ must be equal and opposite to the transducer current i:

$$i_F(t) = -i$$

and since

$$V_o(t) = \frac{1}{C_F} \int i_F(t)\, dt$$

it follows that the output voltage is proportional to the charge generated by the transducer, and therefore to the displacement:

$$V_o(t) = \frac{1}{C_F} \int -i\ dt = \frac{1}{C_F} \int -\frac{dq}{dt}\, dt = -\frac{q}{C_F} = -\frac{K_P x_i}{C_F}$$

Since the displacement is caused by an external force or pressure, this sensing principle is widely adopted in the measurement of force and pressure.

EXAMPLE 7.6   Integrating a Square Wave

Problem

Determine the output voltage for the integrator circuit of Figure 7.44 if the input is a square wave of amplitude $\pm A$ and period T, as shown in Figure 7.48.

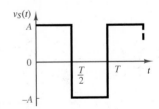

Figure 7.48 Square wave for Example 7.6.

Solution

Known Quantities: Feedback and source impedances; input waveform characteristics.

Find: $v_o(t)$.

Schematics, Diagrams, Circuits, and Given Data: $T = 10$ ms; $C_S = 1\ \mu\text{F}$; $R_F = 10\,\text{k}\Omega$.

Assumptions: The op-amp is ideal and $v_o = 0$ at $t = 0$.

Analysis: Equation 7.93 expresses the output of an integrator as:

$$v_o(t) = -\frac{1}{R_F C_S} \int_{-\infty}^{t} v_S(t')\, dt' = -\frac{1}{R_F C_S} \left[\int_{-\infty}^{0} v_S(t')\, dt' + \int_{0}^{t} v_S(t')\, dt' \right]$$

$$= v_o(0) - \frac{1}{R_F C_S} \int_{0}^{t} v_S(t')\, dt'$$

The square wave can be integrated in a piecewise fashion by observing that $v_S(t) = A$ for $0 \le t < T/2$ and $v_S(t) = -A$ for $T/2 \le t < T$. Thus, for the two half periods of the waveform:

$$v_o(t) = -\frac{1}{R_F C_S}\int_0^t v_S(t')\,dt' = -100\int_0^t A\,dt'$$

$$= -100At \qquad 0 \le t < \frac{T}{2}$$

$$v_o(t) = v_o\left(\frac{T}{2}\right) - \frac{1}{R_F C_s}\int_{T/2}^t v_S(t')\,dt' = -100A\frac{T}{2} - 100\int_{T/2}^t (-A)\,dt'$$

$$= -100A\frac{T}{2} + 100A\left(t - \frac{T}{2}\right) = -100A(T-t) \qquad \frac{T}{2} \le t < T$$

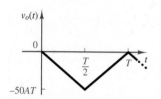

Since the waveform is periodic, the above result will repeat with period T, as shown in Figure 7.49. Note also that the average value of the output voltage is not zero.

Comments: The integral of a square wave is thus a triangular wave. This is a useful fact to remember. Note that the effect of the initial condition is very important since it determines the starting point of the triangular wave.

Figure 7.49 Output of integrator for Example 7.6.

EXAMPLE 7.7   Proportional-Integral Control With Op-Amps

Problem

The aim of this example is to illustrate the very common practice of *proportional-integral*, or *PI, control*. Consider the temperature control circuit of Example 7.4, shown again in Figure 7.50(a), where it was discovered that the proportional control implemented with the gain K_P could still give rise to a steady-state error in the final temperature of the system. This error can be eliminated by using an automatic control system that feeds back a component that is proportional to the *integral of the error voltage*, in addition to the proportional term. Figure 7.50(b) depicts the block diagram of such a PI controller. Now, the design of the control system requires selecting two gains, the *proportional gain K_P* and the *integral gain K_I*.

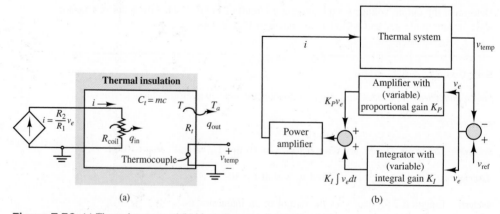

Figure 7.50 (a) Thermal system and (b) block diagram of control system

---

Solution

Known Quantities: Sensor (input) voltage; feedback and source resistors, thermal system component values.

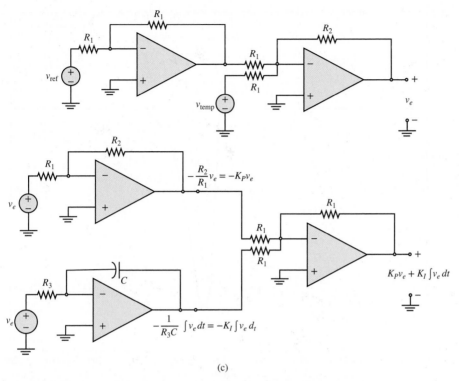

(c)

Figure 7.50(c) Circuit for generating error voltage and proportional gain

Find: Select desired value of proportional gain K_P and integral gain K_I to achieve automatic temperature control with zero steady-state error.

Schematics, Diagrams, Circuits and Given Data: $R_{coil} = 5\ \Omega$; thermal resistance $R_t = 2°C/W$; thermal capacitance $C_t = 50\ J/°C$; sensor calibration constant, $\alpha = 1\ V/°C$.

Assumptions: Assume ideal op-amps.

Analysis: The circuit of Figure 7.50(c) shows two op-amp circuits—the top circuit generates the error voltage v_e. The only difference is that in this case the circuit does not provide any gain. The bottom circuit amplifies v_e by the proportional gain $-K_P = -R_2/R_1$ and also computes the integral of v_e times the integral gain $-K_I = -1/R_3C$. These two quantities are then summed through another inverting summer circuit, which takes care of the sign change as well.

Figure 7.50(d) depicts the temperature response of the system for $K_P = 5$ (as selected in Example 7.5) and different values of K_I. Note that the steady-state error is now zero! This is a property of controllers that incorporate an integral term. Figure 7.50(e) shows the current supplied to the heater coil. Note that the response is quite fast and that the temperature deviation is minimal.

Comments: The addition of the integral term in the controller causes the system temperature to oscillate in response to the $-10°C$ temperature disturbance described in Example 7.4 (for sufficiently high values of K_I). This oscillation is a characteristic of an underdamped second-order system (see Chapter 4)—but we originally started out with a first-order thermal system! The addition of the integral term has increased the order of the system, and now it is possible for the system to display oscillatory behavior, that is, to have complex conjugate roots (poles). To those familiar with thermal systems, this behavior should cause a raised

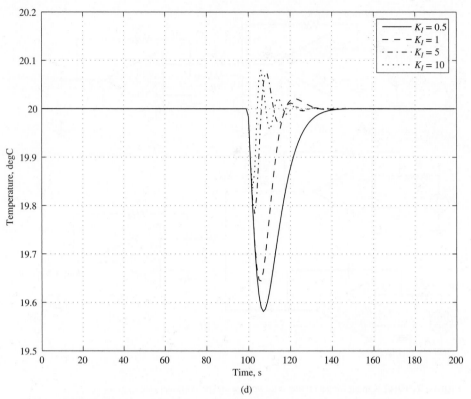

(d)

Figure 7.50(d) Response of thermal system for various values of integral gain, K_I $(K_P = 5)$

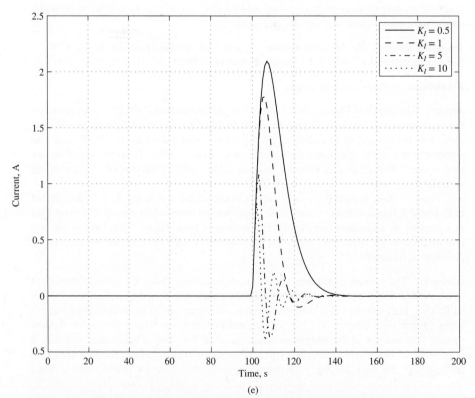

(e)

Figure 7.50(e) Power amplifier current system for various values of integral gain K_I $(K_P = 5)$

eyebrow! It is well known that thermal systems cannot display underdamped behavior (that is, there is no thermal system property analogous to inductance). The introduction of the integral gain can, in fact, cause temperature oscillations as if an artificial "thermal inductor" were introduced in the system.

CHECK YOUR UNDERSTANDING

Plot the frequency response of an ideal integrator in the form of a Bode plot. Determine the slope of the straight-line segments in dB/decade. Assume $R_S C_F = 10$ s.

Answer: -20 dB/decade

CHECK YOUR UNDERSTANDING

Plot the frequency response of an ideal differentiator in the form of a Bode plot. Determine the slope of the straight-line segments in dB/decade. Assume $R_F C_S = 100$ s.

Verify that, if the triangular wave of Example 7.6 is the input to the ideal differentiator of Figure 7.45, the resulting output is a square wave.

Answer: $+20$ dB/decade

EXAMPLE 7.8   Using Cascaded Amplifiers to Simulate a Differential Equation

Problem

Derive the differential equation corresponding to the circuit shown in Figure 7.51.

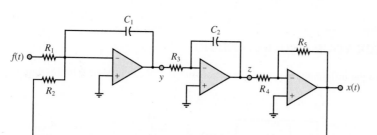

Figure 7.51 Analog computer simulation of unknown system

Solution

Known Quantities: Resistor and capacitor values.

Find: Differential equation in $x(t)$.

Schematics, Diagrams, Circuits, and Given Data: $R_1 = 0.4$ MΩ; $R_2 = R_3 = R_5 = 1$ MΩ; $R_4 = 2.5$ kΩ; $C_1 = C_2 = 1$ μF.

Assumptions: Assume ideal op-amps.

Analysis: Begin the analysis from the right-hand side of the circuit to determine the intermediate variable z as a function of x:

$$x = -\frac{R_5}{R_4}z = -400z$$

Moving to the left, next determine the relationship between y and z:

$$z = -\frac{1}{R_3 C_2}\int y(t')\,dt' \quad \text{or} \quad y = -\frac{dz}{dt}$$

Finally, determine y as a function of x and f:

$$y = -\frac{1}{R_2 C_1}\int x(t')\,dt' - \frac{1}{R_1 C_1}\int f(t')\,dt' = -\int[x(t') + 2.5f(t')]\,dt'$$

or

$$\frac{dy}{dt} = -x - 2.5f$$

Substitute the expressions into one another and eliminate y and z to obtain:

$$x = -400z$$

$$\frac{dx}{dt} = -400\frac{dz}{dt} = 400y$$

$$\frac{d^2x}{dt^2} = 400\frac{dy}{dt} = 400(-x - 2.5f)$$

and

$$\frac{d^2x}{dt^2} + 400x = -1{,}000f$$

Comments: Note that the summing and integrating functions have been combined into a single block in the first amplifier.

CHECK YOUR UNDERSTANDING

Derive the differential equation corresponding to the circuit shown in the figure.

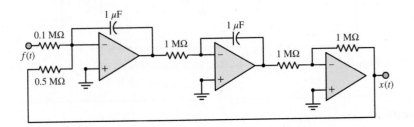

7.5 PHYSICAL LIMITATIONS OF OPERATIONAL AMPLIFIERS

In nearly all the discussion and examples so far, the op-amp has been treated as an ideal device, characterized by infinite input impedance, zero output resistance, and infinite open-loop voltage gain. Although this model is adequate to represent its behavior in a large number of applications, practical op-amps are not ideal but exhibit limitations that should be considered in the design of instrumentation. In particular, in dealing with relatively large voltages and currents, and in the presence of high-frequency signals, it is important to be aware of the nonideal properties of the op-amp.

Voltage Supply Limits

As indicated in Figure 7.6, operational amplifiers (and all amplifiers, in general) are powered by external DC voltage supplies V_S^+ and V_S^-, which are usually symmetric and on the order of ± 10 to ± 20 V. Some op-amps are especially designed to operate from a single voltage supply, but for the sake of simplicity from here on we shall consider only symmetric supplies. The effect of limiting supply voltages is that amplifiers are capable of amplifying signals *only within the range of their supply voltages*; it would be physically impossible for an amplifier to generate a voltage greater than V_S^+ or less than V_S^-. This limitation may be stated as follows:

$$V_S^- < v_o < V_S^+ \qquad \text{Voltage supply limitation} \tag{7.97}$$

For most op-amps, the limit is actually approximately 1.5 V less than the supply voltages. How does this practically affect the performance of an amplifier circuit? An example will best illustrate the idea.

Note how the voltage supply limit actually causes the peaks of the sine wave to be clipped in an abrupt fashion. This type of hard nonlinearity changes the characteristics of the signal quite radically and could lead to significant errors if not taken into account. Just to give an intuitive idea of how such clipping can affect a signal, have you ever wondered why rock guitar has a characteristic sound that is very different from the sound of classical or jazz guitar? The reason is that the "rock sound" is obtained by overamplifying the signal, attempting to exceed the voltage supply limits, and causing clipping similar in quality to the distortion introduced by voltage supply limits in an op-amp. This clipping broadens the spectral content of each tone and causes the sound to be distorted.

One of the circuits most directly affected by supply voltage limitations is the op-amp integrator.

Frequency Response Limits

Another property of all amplifiers that may pose severe limitations to the op-amp is their finite bandwidth. We have so far assumed, in our ideal op-amp model, that the open-loop gain is a very large constant. In reality, A is a function of frequency and is characterized by a low-pass response. For a typical op-amp,

$$A(j\omega) = \frac{A_0}{1 + j\omega/\omega_0} \qquad \text{Finite bandwidth limitation} \tag{7.98}$$

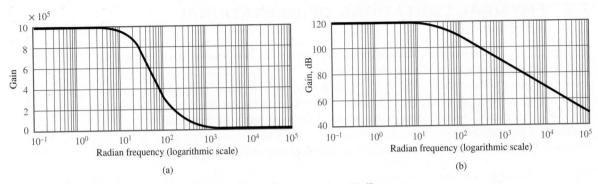

Figure 7.52 Open-loop gain of practical op-amp (a) amplitude ratio response; (b) dB response

The cutoff frequency of the op-amp open-loop gain ω_0 represents approximately the point where the amplifier response starts to drop off as a function of frequency and is analogous to the cutoff frequencies of the RC and RL circuits of Chapter 5. Figure 7.52 depicts $A(j\omega)$ in both linear and decibel plots for the fairly typical values $A_0 = 10^6$ and $\omega_0 = 10\pi$. It should be apparent from Figure 7.52 that the assumption of a very large open-loop gain becomes less and less accurate for increasing frequency. Recall the initial derivation of the closed-loop gain for the inverting amplifier: In obtaining the final result $\mathbf{V}_o/\mathbf{V}_S = -R_F/R_S$, it was assumed that $A \to \infty$. This assumption is clearly inadequate at the higher frequencies.

The finite bandwidth of the practical op-amp results in a fixed **gain-bandwidth product** for any given amplifier. The effect of a constant gain-bandwidth product is that as the closed-loop gain of the amplifier is increased, its 3-dB bandwidth is proportionally reduced until, in the limit, if the amplifier were used in the open-loop mode, its gain would be equal to A_0 and its 3-dB bandwidth would be equal to ω_0. The constant gain-bandwidth product is therefore equal to the product of the open-loop gain and the open-loop bandwidth of the amplifier: $A_0\omega_0 = K$. When the amplifier is connected in a closed-loop configuration (e.g., as an inverting amplifier), its gain is typically much less than the open-loop gain and the 3-dB bandwidth of the amplifier is proportionally increased. To explain this further, Figure 7.53 depicts the case in which two different linear amplifiers (achieved through any two different negative feedback configurations) have been designed for the same op-amp. The first has closed-loop gain $G_1 = A_1$, and the second has closed-loop gain $G_2 = A_2$. The bold line in the figure indicates the open-loop frequency response, with gain A_0 and cutoff frequency ω_0. As the gain decreases from A_0 to A_1, the cutoff frequency increases from ω_0 to ω_1. As the gain decreases to A_2, the bandwidth increases to ω_2. Thus:

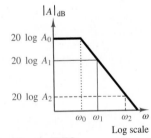

Figure 7.53

The gain-bandwidth product of any given op-amp is constant.

$$A_0 \times \omega_0 = A_1 \times \omega_1 = A_2 \times \omega_2 = K$$

(7.99)

Input Offset Voltage

Another limitation of practical op-amps results because even in the absence of any external inputs, it is possible that an **offset voltage** will be present at the input of an op-amp. This voltage is usually denoted by $\pm V_{os}$, and it is caused by mismatches in the internal circuitry of the op-amp. The offset voltage appears as a differential input voltage between the inverting and noninverting input terminals. The presence of an additional input voltage will cause a DC bias error in the amplifier output. Typical and maximum values of V_{os} are quoted in manufacturers' data sheets. The worst case effects due to the presence of offset voltages can therefore be predicted for any given application.

Input Bias Currents

Another nonideal characteristic of op-amps results from the presence of small input bias currents at the inverting and noninverting terminals. Once again, these are due to the internal construction of the input stage of an operational amplifier. Figure 7.54 illustrates the presence of nonzero input bias currents I_B going into an op-amp.

Typical values of I_{B+} and I_{B-} depend on the semiconductor technology employed in the construction of the op-amp. Op-amps with bipolar transistor input stages may see input bias currents as large as 1 μA, while for FET input devices, the input bias currents are less than 1 nA. These currents depend on the internal design of the op-amp and are not necessarily equal.

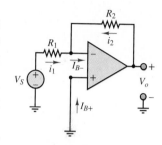

Figure 7.54 Input bias currents.

One often designates the **input offset current** I_{os} as

$$I_{os} = I_{B+} - I_{B-} \tag{7.100}$$

The latter parameter is sometimes more convenient from the standpoint of analysis.

Output Offset Adjustment

Both the offset voltage and the input offset current contribute to an output offset voltage $V_{o,os}$. Some op-amps provide a means for minimizing $V_{o,os}$. For example, the μA741 op-amp provides a connection for this procedure. Figure 7.55 shows a typical pin configuration for an op-amp in an eight-pin dual-in-line package (DIP) and the circuit used for nulling the output offset voltage. The variable resistor is adjusted until v_{out} reaches a minimum (ideally, 0 V). Nulling the output voltage in this manner removes the effect of both input offset voltage and current on the output.

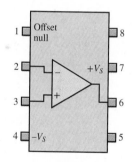

Slew Rate Limit

Another important restriction in the performance of a practical op-amp is associated with rapid changes in voltage. The op-amp can produce only a finite rate of change at its output. This limit rate is called the **slew rate**. Consider an ideal step input, where at $t = 0$ the input voltage is switched from 0 to V volts. Then we would expect the output to switch from 0 to AV volts, where A is the amplifier gain. However, v_o can change at only a finite rate; thus,

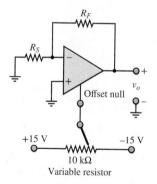

Figure 7.55 Output offset voltage adjustment

$$\left|\frac{dv_o}{dt}\right|_{max} = S_0 \qquad \text{Slew rate limitation} \tag{7.101}$$

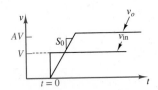

Figure 7.56 Slew rate limit in op-amps

Figure 7.56 shows the response of an op-amp to an ideal step change in input voltage. Here, S_0, the slope of v_o, represents the slew rate.

The slew rate limitation can affect sinusoidal signals, as well as signals that display abrupt changes, as does the step voltage of Figure 7.56. This may not be obvious until we examine the sinusoidal response more closely. It should be apparent that the maximum rate of change for a sinusoid occurs at the zero crossing, as shown by Figure 7.57. To evaluate the slope of the waveform at the zero crossing, let

$$v_{\text{in}}(t) = V \sin \omega t \qquad \text{such that} \qquad v_o(t) = AV \sin \omega t \tag{7.102}$$

Then:

$$\frac{dv_o}{dt} = \omega AV \cos \omega t \tag{7.103}$$

The maximum slope of the sinusoidal signal will therefore occur at $\omega t = 0,\ \pi,\ 2\pi,\ \ldots$, so that

$$\left.\frac{dv_o}{dt}\right|_{\text{max}} = \omega AV = S_0 \tag{7.104}$$

Thus, the maximum slope of a sinusoid is proportional to both the signal frequency and the amplitude. The curve shown by a dashed line in Figure 7.57 should indicate that as ω increases, so does the slope of $v(t)$ at the zero crossings. What is the direct consequence of this result, then?

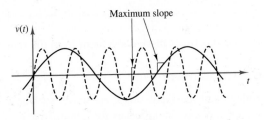

Figure 7.57 The maximum slope of a sinusoidal signal varies with the signal frequency.

Short-Circuit Output Current

Recall the model for the op-amp shown in Figure 7.3, which depicted the internal circuit of the op-amp as an equivalent input impedance R_{in} and a controlled voltage source Av_{in}. In practice, the internal source is not ideal because it cannot provide an infinite amount of current (to the load, to the feedback connection, or to both). The immediate consequence of this nonideal op-amp characteristic is that the maximum output current of the amplifier is limited by the so-called short-circuit output current I_{SC}:

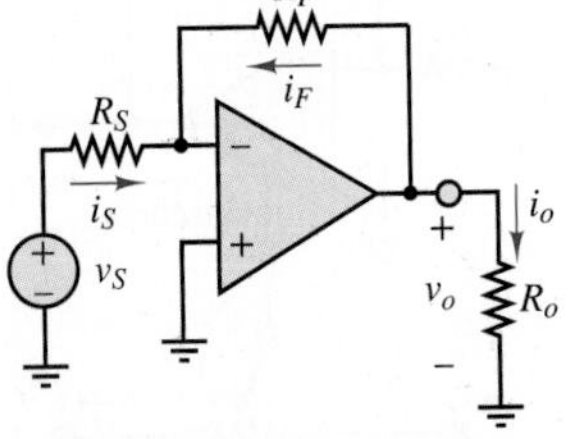

Figure 7.58 Circuit for illustration of short-circuit output current.

$$\boxed{|i_o| < I_{\text{SC}} \qquad \text{Short-circuit output current limitation}} \tag{7.105}$$

To further explain this point, consider that the op-amp needs to provide current to the feedback path (in order to "zero" the voltage differential at the input) and to whatever load resistance, R_o, may be connected to the output. Figure 7.58 illustrates this idea for the case of an inverting amplifier, where I_{SC} is the load current that would be provided to a short-circuit load ($R_o = 0$).

Common-Mode Rejection Ratio

The concepts of common-mode and differential-mode voltages as well as the common-mode rejection ratio (CMRR) were introduced in Section 7.2 and expressed mathematically by equations 7.66 to 7.70. The CMRR is an amplifier characteristic that can be found in the data sheet for any particular amplifier, such as a 741 operational amplifier.

$$\text{CMRR} = 20 \log \left| \frac{A_{\text{DM}}}{A_{\text{CM}}} \right| \qquad \text{(in dB)}$$

Practical Op-Amp Considerations

The results presented in the preceding pages suggest that operational amplifiers permit the design of rather sophisticated circuits in a few simple steps, by selecting appropriate resistor values. This is certainly true, provided that the circuit component selection satisfies certain criteria. A few important practical criteria for selecting op-amp circuit component values are summarized here.

1. Use standard resistor values. While any arbitrary value of gain can, in principle, be achieved by selecting the appropriate combination of resistors, the designer is often constrained to the use of standard 5 percent resistor values. For example, if a design requires a gain of 25, it might be tempting to select, say, 100- and 4-kΩ resistors to achieve $R_F/R_S = 25$ for the inverting amplifier shown in Figure 7.58. However, 4 kΩ is not a standard value; the closest 5 percent tolerance resistor value is 3.9 kΩ, leading to a gain of 25.64. Can you find a combination of standard 5 percent resistors whose ratio is closer to 25?

2. Ensure that the load current is reasonable. Assume the maximum output voltage in the step 1 example is 10 V. The feedback current required by your design with $R_F = 100$ kΩ and $R_S = 4$ kΩ would be $I_F = 10/100,000 = 0.1$ mA. This is a very reasonable value for an op-amp. If you tried to achieve the same gain by using, say, a 10-Ω feedback resistor and a 0.39-Ω source resistor, the feedback current would become as large as 1 A. This value is generally beyond the capabilities of a general-purpose op-amp, so very low resistor values are generally not acceptable. On the other hand, 10-kΩ and 390-Ω resistors would still lead to acceptable currents. As a general rule of thumb, avoid resistor values lower than 100 Ω in practical designs.

3. Avoid stray capacitance by avoiding excessively large resistances, which can cause unwanted signals to couple into the circuit through a mechanism known as *capacitive coupling*. Large resistances can also cause other problems. As a general rule of thumb, avoid resistor values higher than 1 MΩ in practical designs.

4. Precision designs may be warranted. If a certain design requires that the amplifier gain be set to a very accurate value, it may be appropriate to use the (more expensive) option of precision resistors: for example, 1 percent tolerance resistors are commonly available, at a premium cost. Some of the examples and homework problems explore the variability in gain due to the use of higher- and lower-tolerance resistors.

EXAMPLE 7.9   Voltage Supply Limits in an Inverting Amplifier

Problem

Compute and sketch the output voltage of the inverting amplifier of Figure 7.59.

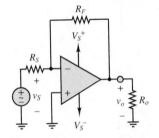

Figure 7.59 Circuit for Example 7.9.

Solution

Known Quantities: Resistor and supply voltage values; input voltage.

Find: $v_o(t)$.

Schematics, Diagrams, Circuits, and Given Data: $R_o = 1 \text{ k}\Omega$; $R_F = 10 \text{ k}\Omega$; $R_o = 1 \text{ k}\Omega$; $V_S^+ = 15 \text{ V}$; $V_S^- = -15 \text{ V}$; $v_S(t) = 2\sin(1{,}000t)$.

Assumptions: Assume a supply voltage–limited op-amp.

Analysis: For an ideal op-amp the output would be

$$v_o(t) = -\frac{R_F}{R_S}v_S(t) = -10 \times 2\sin(1{,}000t) = -20\sin(1{,}000t)$$

However, the supply voltage is limited to ±15 V, and the op-amp output voltage will therefore saturate before reaching the theoretical peak output value of ±20 V. Figure 7.60 depicts the output voltage waveform.

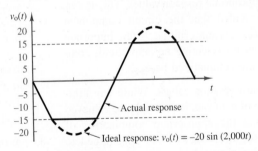

Figure 7.60 Op-amp output with voltage supply limit

Comments: In a practical op-amp, saturation would be reached at 1.5 V below the supply voltages, or at approximately ±13.5 V.

EXAMPLE 7.10   Voltage Supply Limits in an Op-Amp Integrator

Problem

Compute and sketch the output voltage of the integrator of Figure 7.44.

Solution

Known Quantities: Resistor, capacitor, and supply voltage values; input voltage.

Find: $v_o(t)$.

Schematics, Diagrams, Circuits, and Given Data: $R_S = 10 \text{ k}\Omega$; $C_F = 20 \text{ }\mu\text{F}$; $V_S^+ = 15 \text{ V}$; $V_S^- = -15 \text{ V}$; $v_S(t) = 0.5 + 0.3 \cos(10t)$.

Assumptions: Assume a supply voltage–limited op-amp. The initial condition is $v_{\text{out}}(0) = 0$.

Analysis: For an ideal op-amp integrator the output would be

$$v_{\text{out}}(t) = -\frac{1}{R_S C_F}\int_{-\infty}^{t} v_S(t')dt' = -\frac{1}{0.2}\int_{-\infty}^{t}[0.5 + 0.3\cos(10t')]\,dt'$$

$$= -2.5t + 1.5\sin(10t)$$

However, the supply voltage is limited to ± 15 V, and the integrator output voltage will therefore saturate at the lower supply voltage value of -15 V as the term $2.5t$ increases with time. Figure 7.61 depicts the output voltage waveform.

Comments: Note that the DC offset in the waveform causes the integrator output voltage to increase linearly with time. The presence of even a very small DC offset will always cause integrator saturation. One solution to this problem is to include a large feedback resistor in parallel with the capacitor.

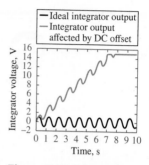

Figure 7.61 Effect of DC offset on integrator

EXAMPLE 7.11   Gain-Bandwidth Product Limit in an Op-Amp

Problem

Determine the maximum allowable closed-loop voltage gain of an op-amp if the amplifier is required to have an audio-range bandwidth of 20 kHz.

Solution

Known Quantities: Gain-bandwidth product.

Find: G_{max}.

Schematics, Diagrams, Circuits, and Given Data: $A_0 = 10^6$; $\omega_0 = 2\pi \times 5$ rad/s.

Assumptions: Assume a gain-bandwidth product limited op-amp.

Analysis: The gain-bandwidth product of the op-amp is

$$A_0 \times \omega_0 = K = 10^6 \times 2\pi \times 5 = \pi \times 10^7 \text{ rad/s}$$

The desired bandwidth is $\omega_{\text{max}} = 2\pi \times 20,000$ rad/s, and the maximum allowable gain will therefore be

$$G_{\text{max}} = \frac{K}{\omega_{\text{max}}} = \frac{\pi \times 10^7}{\pi \times 4 \times 10^4} = 250\,\frac{V}{V}$$

For any closed-loop voltage gain greater than 250, the amplifier would have reduced bandwidth.

Comments: If we desired to achieve gains greater than 250 and maintain the same bandwidth, two options would be available: (1) Use a different op-amp with greater gain-bandwidth product, or (2) connect two amplifiers in cascade, each with lower gain and greater bandwidth, such that the product of the gains would be greater than 250.

To further explore the first option, you may wish to look at the device data sheets for different op-amps and verify that op-amps can be designed (at a cost!) to have substantially greater gain-bandwidth product than the amplifier used in this example.

EXAMPLE 7.12 Increasing the Gain-Bandwidth Product by Means of
Amplifiers in Cascade

Problem

Determine the overall 3-dB bandwidth of the cascade amplifier of Figure 7.62.

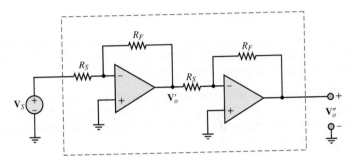

Figure 7.62 Cascade amplifier

Solution

Known Quantities: Gain-bandwidth product and gain of each amplifier.

Find: $\omega_{3\,dB}$ of cascade amplifier.

Schematics, Diagrams, Circuits, and Given Data: $A_0\,\Omega_0 = K = 4\pi \times 10^6$ for each amplifier.
$R_F/R_S = 100$ for each amplifier.

Assumptions: Assume gain-bandwidth product limited (otherwise ideal) op-amps.

Analysis: Let G_1 and ω_1 denote the gain and the 3-dB bandwidth of the first amplifier,
respectively, and G_2 and ω_2 those of the second amplifier.
      The 3-dB bandwidth of the first amplifier is

$$\omega_1 = \frac{K}{G_1} = \frac{4\pi \times 10^6}{10^2} = 4\pi \times 10^4 \frac{rad}{s}$$

The second amplifier will also have

$$\omega_2 = \frac{K}{G_2} = \frac{4\pi \times 10^6}{10^2} = 4\pi \times 10^4 \frac{rad}{s}$$

Thus, the approximate bandwidth of the cascade amplifier is $4\pi \times 10^4$, and the gain of the
cascade amplifier is $G_1 G_2 = 100 \times 100 = 10^4$ or 80 dB.
      Had we attempted to achieve the same gain with a single-stage amplifier having the
same K, we would have achieved a bandwidth of only

$$\omega_3 = \frac{K}{G_3} = \frac{4\pi \times 10^6}{10^4} = 4\pi \times 10^2 \frac{rad}{s}$$

Comments: In practice, the actual 3-dB bandwidth of the cascade amplifier is not quite as
large as that of each of the two stages because the gain of each amplifier starts decreasing
at frequencies somewhat lower than the nominal cutoff frequency.

EXAMPLE 7.13   Effect of Input Offset Voltage on an Amplifier

Problem

Determine the effect of the input offset voltage V_{os} on the output of the amplifier shown in Figure 7.63.

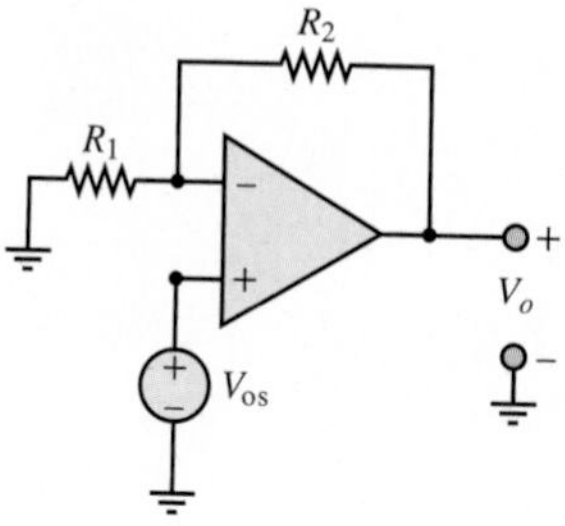

Solution

Known Quantities: Nominal closed-loop voltage gain; input offset voltage.

Find: The offset voltage component in the output voltage V_o, os.

Schematics, Diagrams, Circuits, and Given Data: $A_{nom} = 100$; $V_{os} = 1.5$ mV.

Assumptions: Assume an input offset voltage–limited (otherwise ideal) op-amp.

Analysis: The amplifier is connected in a noninverting configuration; thus its gain is

$$G_{nom} = 100 = 1 + \frac{R_F}{R_S}$$

The DC offset voltage, represented by an ideal voltage source, is represented as being directly applied to the noninverting input; thus

$$V_o, \text{os} = G_{nom} V_{os} = 100 V_{os} = 150 \text{ mV}$$

Thus, we should expect the output of the amplifier to be shifted upward by 150 mV.

Comments: The input offset voltage is not, of course, an external source, but it represents a voltage offset between the inputs of the op-amp. Figure 7.55 depicts how such an offset can be zeroed. The worst-case offset voltage is usually listed in the device data sheets. Typical values are 2 mV for the 741c general-purpose op-amp and 5 mV for the FET-input TLO81.

Figure 7.63 Op-amp input offset voltage

EXAMPLE 7.14   Effect of Input Offset Current on an Amplifier

Problem

Determine the effect of the input offset current I_{os} on the output of the amplifier of Figure 7.64.

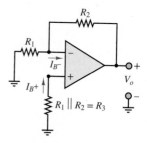

Solution

Known Quantities: Resistor values; input offset current.

Find: The offset voltage component in the output voltage $v_{out,os}$.

Schematics, Diagrams, Circuits, and Given Data: $I_{os} = 1$ μA; $R_2 = 10$ kΩ.

Assumptions: Assume an input offset current–limited (otherwise ideal) op-amp.

Analysis: We calculate the inverting and noninverting terminal voltages caused by the offset current in the absence of an external input:

$$v^+ = R_3 I_{B^+} \qquad v^- = v^+ = R_3 I_{B^+}$$

Figure 7.64 Circuit for Example 7.14.

With these values we can apply KCL at the inverting node and write

$$\frac{V_o - v^-}{R_2} - \frac{v^+}{R_1} = I_{B^-}$$

$$\frac{V_o}{R_2} - \frac{-R_3 I_{B^+}}{R_2} - \frac{-R_3 I_{B^+}}{R_1} = I_{B^-}$$

$$V_o = R_2\left[-I_{B^+}R_3\left(\frac{1}{R_2} + \frac{1}{R_1}\right) + I_{B^-}\right] = -R_2 I_{os}$$

Thus, we should expect the output of the amplifier to be shifted downward by $R_2 I_{os}$, or $10^4 \times 10^{-6} = 10$ mV for the data given in this example.

Comments: Usually, the worst-case input offset currents (or input bias currents) are listed in the device data sheets. Values can range from 100 pA (for CMOS op-amps, for example, LMC6061) to around 200 nA for a low-cost general-purpose amplifier (for example, μA741c).

EXAMPLE 7.15   Effect of Slew Rate Limit on an Amplifier

Problem

Determine the effect of the slew rate limit S_0 on the output of an inverting amplifier for a sinusoidal input voltage of known amplitude and frequency.

Solution

Known Quantities: Slew rate limit S_0; amplitude and frequency of sinusoidal input voltage; amplifier closed-loop gain.

Find: Sketch the theoretically correct output and the actual output of the amplifier in the same graph.

Schematics, Diagrams, Circuits, and Given Data: $S_0 = 1$ V/μs; $v_S = \sin(2\pi \times 10^5 t)$; $G = 10$.

Assumptions: Assume the op-amp is slew rate–limited, but otherwise ideal.

Analysis: Given the closed-loop voltage gain of 10, compute the theoretical output voltage to be:

$$v_o = -10\sin(2\pi \times 10^5 t)$$

The maximum slope of the output voltage is then computed as follows:

$$\left|\frac{dv_o}{dt}\right|_{max} = G\omega = 10 \times 2\pi \times 10^5 = 6.28 \frac{\text{V}}{\mu\text{s}}$$

Clearly, the value calculated above far exceeds the slew rate limit. Figure 7.65 depicts the approximate appearance of the waveforms that one would measure in an experiment.

Comments: Note that in this example the slew rate limit has been exceeded severely, and the output waveform is visibly distorted, to the point that it has effectively become a triangular wave. The effect of the slew rate limit is not always necessarily so dramatic and visible; thus one needs to pay attention to the specifications of a given op-amp. The slew rate limit is listed in the device data sheets. Typical values can range from 13 V/μs, for the TLO81, to around 0.5 V/μs for a low-cost general-purpose amplifier (for example, μA741c).

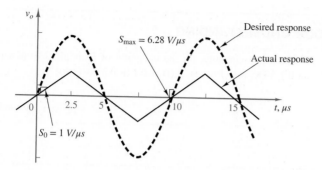

Figure 7.65 Distortion introduced by slew rate limit

EXAMPLE 7.16   Effect of Short-Circuit Current Limit on an Amplifier

Problem

Determine the effect of the short-circuit limit I_{SC} on the output of an inverting amplifier for a sinusoidal input voltage of known amplitude.

Solution

Known Quantities: Short-circuit current limit I_{SC}; amplitude of sinusoidal input voltage; amplifier closed-loop gain.

Find: Compute the minimum allowable load resistance value $R_{o_{\min}}$, and sketch the theoretical and actual output voltage waveforms for resistances smaller than $R_{o_{\min}}$.

Schematics, Diagrams, Circuits, and Given Data: $I_{SC} = 50$ mA; $v_S = 0.05 \sin(\omega t)$; $G = 100$.

Assumptions: Assume the op-amp is short-circuit current–limited, but otherwise ideal.

Analysis: Given the closed-loop voltage gain of 100, compute the theoretical output voltage to be:

$$v_o(t) = -Gv_S(t) = -5\sin(\omega t)$$

To assess the effect of the short-circuit current limit, calculate the peak value of the output voltage since this is the condition that will require the maximum output current from the op-amp:

$$V_{o_{peak}} = 5 \text{ V}$$
$$I_{SC} = 50 \text{ mA}$$
$$R_{o_{\min}} = \frac{V_{o_{peak}}}{I_{SC}} = \frac{5 \text{ V}}{50 \text{ mA}} = 100 \text{ } \Omega$$

For any load resistance less than 100 Ω, the required load current will be greater than I_{SC}. For example, if we chose a 75-Ω load resistor, we would find that

$$V_{o_{peak}} = I_{SC} \times R_o = 3.75 \text{ V}$$

That is, the output voltage cannot reach the theoretically correct 5-V peak and would be "compressed" to reach a peak voltage of only 3.75 V. This effect is depicted in Figure 7.66.

Comments: The short-circuit current limit is listed in the device data sheets. Typical values for a low-cost general-purpose amplifier (say, the 741c) are in the tens of milliamperes.

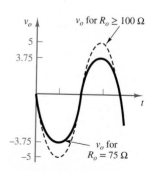

Figure 7.66 Distortion introduced by short-circuit current limit

CHECK YOUR UNDERSTANDING

How long will it take (approximately) for the integrator of Example 7.10 to saturate if the input signal has a 0.1-VDC bias [that is, $v_S(t) = 0.1 + 0.3 \cos(10t)$]?

Answer: Approximately 30 s

CHECK YOUR UNDERSTANDING

What is the maximum gain that could be achieved by the op-amp of Example 7.11 if the desired bandwidth is 100 kHz?

Answer: $A_{max} = 50$

CHECK YOUR UNDERSTANDING

In Example 7.12, the closed-loop gain of each amplifier was assumed constant at frequencies below the cutoff frequency. In practice, this is only approximately true, since the open-loop gain A of each op-amp begins to decrease with frequency at frequencies usually much lower than the closed-loop gain cutoff frequency. The frequency response for the open-loop gain of an op-amp is well approximated by:

$$A(j\omega) = \frac{A_0}{1 + j\omega/\omega_0}$$

Use this expression to find an expression for the closed-loop gain of the cascade amplifier. (*Hint:* The combined gain is equal to the product of the individual closed-loop gains.) What is the actual gain in decibels at the cutoff frequency ω_0 for the cascade amplifier?

What is the 3-dB bandwidth of the cascade amplifier of Example 7.12? [*Hint:* The gain of the cascade amplifier is the product of the individual op-amp frequency responses. Compute the magnitude of this product, set it equal to $(1/\sqrt{2}) \times 10,000$, and solve for ω.]

Answers: 74 dB; $\omega_{3dB} = 2\pi \times 12,800$ rad/s

CHECK YOUR UNDERSTANDING

What is the maximum gain that can be accepted in the op-amp circuit of Example 7.13 if the offset is not to exceed 50 mV?

Answer: $A_{Vmax} = 33.3$

CHECK YOUR UNDERSTANDING

Given the desired peak output amplitude (10 V), what is the maximum frequency that will not result in violating the slew rate limit for the op-amp of Example 7.15.

Answer: $f_{max} = 159$ kHz.

Conclusion

Operational amplifiers constitute the single most important building block in analog electronics. The contents of this chapter will be frequently referenced in later sections of this book. Upon completing this chapter, the following learning objectives should have been mastered:

1. *Understand the properties of ideal amplifiers and the concepts of gain, input impedance, output impedance, and feedback.* Ideal amplifiers represent fundamental building blocks of electronic instrumentation. With the concept of an ideal amplifier clearly established, one can design practical amplifiers, filters, integrators, and many other signal processing circuits. A practical op-amp closely approximates the characteristics of ideal amplifiers.

2. *Understand the difference between open-loop and closed-loop op-amp configuration; and compute the gain (or complete the design of) simple inverting, noninverting, summing, and differential amplifiers using ideal op-amp analysis. Analyze more advanced op-amp circuits, using ideal op-amp analysis, and identify important performance parameters in op-amp data sheets.* Analysis of op-amp circuits is made easy by a few simplifying assumptions, which are based on the op-amp having a very large input impedance, a very small output impedance, and a large open-loop gain. The simple inverting and noninverting amplifier configurations permit the design of very useful circuits simply by appropriately selecting and placing a few resistors.

3. *Analyze and design simple active filters. Analyze and design ideal integrator and differentiator circuits.* The use of capacitors in op-amp circuits extends the applications of this useful element to include filtering, integration, and differentiation.

4. *Understand the structure and behavior of analog computers, and design analog computer circuits to solve simple differential equations.* The properties of op-amp summing amplifiers and integrators make it possible to construct analog computers that can serve as an aid in the solution of differential equations and in the simulation of dynamic systems. While digital computer-based numerical simulations have become very popular in the last two decades, there is still a role for analog computers in some specialized applications.

5. *Understand the principal physical limitations of an op-amp.* It is important to understand that there are limitations in the performance of op-amp circuits that are not predicted by the simple op-amp models presented in the early sections of the chapter. In practical designs, issues related to voltage supply limits, bandwidth limits, offsets, slew rate limits, and output current limits are very important if one is to achieve the design performance of an op-amp circuit.

HOMEWORK PROBLEMS

Section 7.1: Ideal Amplifiers

7.1 The circuit shown in Figure P7.1 has a DC signal source, two stages of amplification, and a load. Determine, in decibels, the power gain

$$G = P_0/P_S = V_o I_o/V_S I_S, \text{ where:}$$

$R_s = 0.5\ \text{k}\Omega$	$R_{o3} = 0.7\ \text{k}\Omega$
$R_{i1} = 3.2\ \text{k}\Omega$	$R_{i2} = 2.8\ \text{k}\Omega$
$R_{o1} = 2.2\ \text{k}\Omega$	$R_{o2} = 2.2\ \text{k}\Omega$
$A_1 = 90$	$H_2 = 300\ \text{mS}$

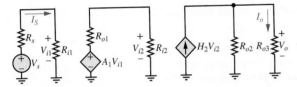

Figure P7.1

7.2 A temperature sensor in a production line under normal operating conditions produces a no-load (i.e., sensor current = 0) voltage:

$$v_s = V_{pk}\cos(\omega t) \qquad R_s = 400\ \Omega$$
$$V_{pk} = 500\ \text{mV} \qquad \omega = 6.28\ \text{krad/s}$$

The temperature is monitored on a display (the load) with a vertical line of light-emitting diodes. Normal conditions are indicated when a string of the bottommost diodes 2 cm in length is on. This requires that a voltage be supplied to the display input terminals where

$$R_o = 12\ \text{k}\Omega \qquad v_o = V_m\cos(\omega t) \qquad V_m = 6\ \text{V}$$

The signal from the sensor must be amplified. Therefore, a voltage amplifier, shown in Figure P7.2, is connected between the sensor and CRT with

$$R_i = 2\ \text{k}\Omega \qquad R_o = 3\ \text{k}\Omega$$

Determine the required no-load gain of the amplifier.

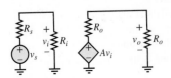

Figure P7.2

7.3 What approximations are valid for the voltages and currents shown in Figure P7.3 of an ideal operational amplifier? What conditions must be satisfied for these approximations?

Figure P7.3

7.4 What approximations are usually made about the circuit components and parameters shown in Figure P7.4 for an ideal op-amp?

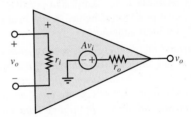

Figure P7.4

Section 7.2: The Operational Amplifier

7.5 Find v_1 in the circuits of Figure P7.5(a) and (b). In Figure P7.5(a) the 3-kΩ resistor "loads" the output; that is, v_1 is changed by attaching the 3-kΩ resistor in parallel with the lower 6-kΩ resistor. However, in Figure P7.5(b) the isolation buffer holds v_1 to $v_g/2$, regardless of the presence of the 3-kΩ resistor and its value!

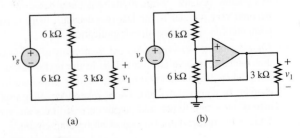

(a)   (b)

Figure P7.5

7.6   Find the current i in the circuit of Figure P7.6.

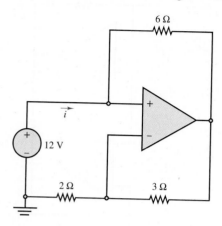

Figure P7.6

7.7   Find the voltage v_o in Figure P7.7 by (i) applying nodal analysis, and by (ii) finding the Thévenin equivalent network seen to the left of nodes a and b to form an archetypical inverting amplifier.

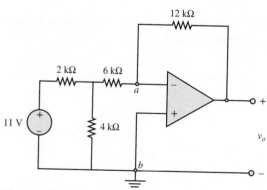

Figure P7.7

7.8   Find the Thévenin equivalent network seen between the noninverting terminal node and the reference node in Figure P7.8.

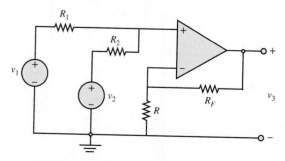

Figure P7.8

7.9   Determine an expression for the closed-loop voltage gain $G = v_o/v_1$ for the circuit of Figure P7.9. Find the input conductance i_1/v_1 seen by the voltage source. Assume the op-amp is ideal.

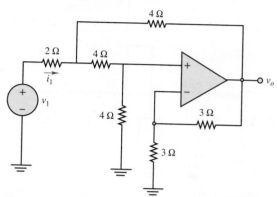

Figure P7.9

7.10   Differential amplifiers are often used in conjunction with a Wheatstone bridge, such as that shown in Figure P7.10, where each resistor is a temperature sensing element, and their change in resistance ΔR is directly proportional to their change in temperature ΔT. The constant of proportionality is the temperature coefficient $\pm\alpha$, which can be positive (PTC) or negative (NTC). Find the Thévenin equivalent network seen by the amplifier to the left of nodes a and b. Assume that $\Delta R = \pm\Delta T$ and $|\Delta R| \ll R_0$.

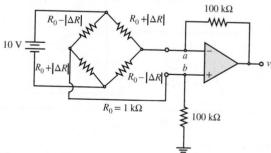

Figure P7.10

7.11   The circuit shown in Figure P7.11 is a *negative impedance converter*. Find the input impedance $\mathbf{Z}_{in}$:

$$\mathbf{Z}_{in} = \frac{V_1}{I_1}$$

when:

a. $\mathbf{Z}_o = R$

b. $\mathbf{Z}_o = \dfrac{1}{j\omega C}$

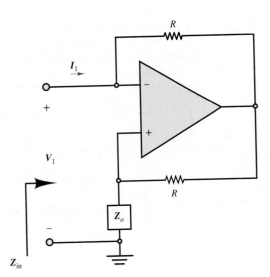

Figure P7.11

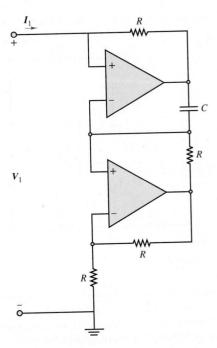

Figure P7.13

7.12   The circuit of Figure P7.12 demonstrates that op-amp feedback can create a resonant circuit without the use of an inductor. Assume $R_1 = R_2 = 1\ \Omega$, $C_1 = 2Q$ F, and $C_2 = 1/2Q$ F, where Q is the quality factor introduced in Chapter 5. Use nodal analysis to determine the voltage gain v_o/v_{in}.

7.14   In the circuit of Figure P7.14, determine the input impedance $\mathbf{Z}_{in} = V_1/I_1$.

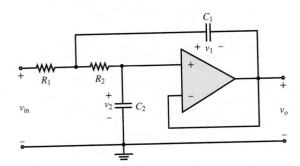

Figure P7.12

7.13   Inductors are difficult to use as components of integrated circuits due to the need for large coils of wire, which require significant space and tend to act as excellent antennas for ambient noise. As an alternative, a "solid-state inductor" can be constructed as shown in Figure P7.13.

a.   Determine the input impedance $\mathbf{Z}_{in} = V_1/I_1$.

b.   What is $\mathbf{Z}_{in}$ when $R = 1{,}000\ \Omega$ and $C = 0.02\ \mu F$?

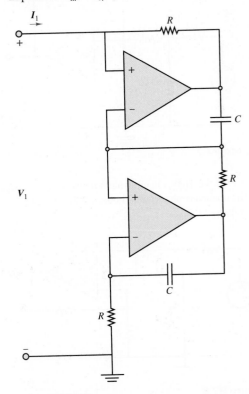

Figure P7.14

7.15   It is easy to construct a current source using an inverting amplifier configuration, as in Figure P7.15. Verify that the current I through R_o is independent of the value of R_o, assuming that the op-amp stays in its linear operating region, and find the value of I.

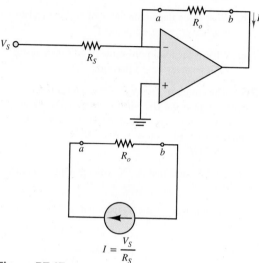

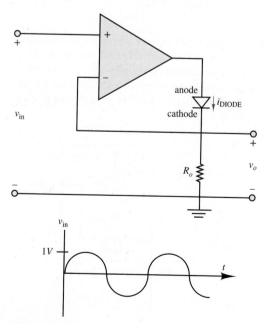

$$I = \frac{V_S}{R_S}$$

Figure P7.15

7.16   A "super diode" or "precision diode" circuit, which eliminates the diode offset voltage, is shown in Figure P7.16. The diode permits current directed from anode to cathode only, as indicated in the figure. Determine the output voltage $v_o(t)$ for the given input voltage $v_{in}(t)$.

Figure P7.16

7.17   Determine the response function $\tilde{V}_2/\tilde{V}_1$ for the circuit of Figure P7.17.

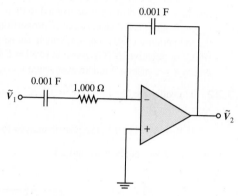

Figure P7.17

7.18   Time delays are often encountered in engineering systems. They can be approximated using Euler's definition as

$$e^{-sT} = \lim_{N \to \infty} \left[\frac{1}{sT/N + 1} \right]^N$$

If $T = 1$ and $N = 1$, then the approximation can be implemented by the circuit of Problem 7.17 (see Figure P7.17), with the addition of a unity gain inverting amplifier to eliminate the negative sign. Modify the circuit of Figure P7.17 as needed, and use it as many times as necessary to design an approximate time delay for $T = 1$ and $N = 4$ in Euler's definition of the exponential.

7.19   Show that the output voltage v_3 shown in Figure P7.8 has the form $v_3 = a_1 v_1 + a_2 v_2$, where a_1 and a_2 are constants.

7.20   For the circuit of Figure P7.20, find v_o.

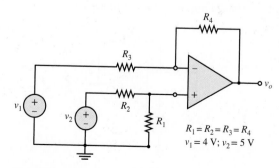

$R_1 = R_2 = R_3 = R_4$
$v_1 = 4\ \text{V};\ v_2 = 5\ \text{V}$

Figure P7.20

7.21 Differential amplifiers are often used in conjunction with a Wheatstone bridge, such as that shown in Figure P7.10, where each resistor is a temperature sensing element, and their change in resistance ΔR is directly proportional to their change in temperature ΔT. The constant of proportionality is the temperature coefficient $\pm\alpha$, which can be positive (PTC) or negative (NTC). Assume $|\Delta R| = K\Delta T$, where $K = $ constant. Find an expression for $v_o (\Delta T)$.

7.22 Consider the circuit of Figure P7.22. Assume $\omega = 1,000$ rad/s:

a. If $\mathbf{V}_1 - \mathbf{V}_2 = 1\angle 0$ V, use phasor analysis to find $|\mathbf{V}_o|$.

b. Use phasor analysis to find $\angle \mathbf{V}_o$.

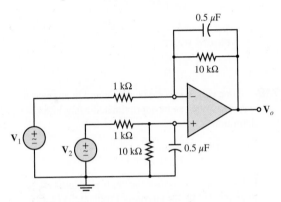

Figure P7.22

7.23 Find an expression for the voltage gain $\tilde{\mathbf{V}}_{\text{out}}/\tilde{\mathbf{V}}_{\text{in}}$ of the circuit of Figure P7.12. Assume $R_1 = 3\ \Omega$, $R_2 = 2\ \Omega$, and $C_1 = C_2 = \frac{1}{6}$ F.

7.24 In the circuit of Figure P7.24, assume $R_F = 12$ kΩ and that it is critical that the voltage gain v_o/v_S remain within ± 2 percent of the nominal gain of 20. What value of R_S is needed for the nominal gain? What are the allowed maximum and minimum values of R_S? Will a standard 5 percent tolerance resistor be adequate to satisfy this requirement? (See Table 1.3 of standard resistor values in Chapter 1.)

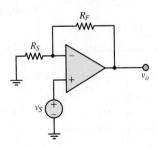

Figure P7.24

7.25 The two 5 percent tolerance resistors of an inverting amplifier (see Figure 7.8) have nominal values $R_F = 33$ kΩ and $R_S = 1.5$ kΩ.

a. What is the nominal voltage gain $G = v_o/v_S$ of the amplifier?

b. What is the maximum value of G if the resistor values can swing ± 5 percent?

c. What is the minimum value of G if the resistor values can swing ± 5 percent?

7.26 The circuit of Figure P7.26 is a *level shifter*, which adjusts the DC portion of the input voltage $v_1(t)$ while also amplifying the AC portion. Let: $v_1(t) = 10 + 10^{-3} \sin \omega t$ V, $R_F = 10$ kΩ and $V_{\text{batt}} = 20$ V.

a. Find R_S such that no DC voltage appears at the output.

b. What is $v_o(t)$, using R_S from part a?

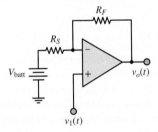

Figure P7.26

7.27 Figure P7.27 shows a simple practical amplifier that uses a 741 op-amp chip. Pin numbers are as indicated. Assume the op amp has a 2-MΩ input resistance, an open-loop gain $A = 200,000$, and an output impedance $R_o = 50\ \Omega$. Find the closed-loop gain $G = v_o/v_i$.

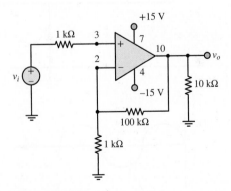

Figure P7.27

7.28 Design an inverting summing amplifier to obtain the following weighted sum of four different signal sources:

$$v_o = -(2 \sin\omega_1 t + 4 \sin\omega_2 t + 8 \sin\omega_3 t + 16 \sin\omega_4 t)$$

Assume that $R_F = 5\ \text{k}\Omega$, and determine the required source resistors.

7.29 The amplifier shown in Figure P7.29 has a signal source (v_s in series with R_s) and load R_o separated by an amplification stage built upon the Motorola MC1741C op-amp. Assume:

$$R_s = 2.2\ \text{k}\Omega \qquad R_1 = 1\ \text{k}\Omega$$
$$R_F = 8.7\ \text{k}\Omega \qquad R_o = 20\ \Omega$$

The op-amp itself has a 2-MΩ input resistance, a 75-Ω output resistance, and a 200K open-loop gain. To a first approximation, the op-amp would be modeled as ideal. A better model would include the effects of the parameters listed above.

a. Assume the op-amp is not ideal, and derive an expression for the input resistance $r_i = v_i/i_i$ of the overall amplifier, where $v_i = v_s - i_i R_s$.

b. Determine the value of that input resistance, and compare it to the input resistance derived for an ideal op-amp.

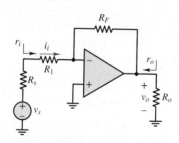

Figure P7.29

7.30 In the circuit shown in Figure P7.30, assume $R_1 = 40\ \text{k}\Omega$, $R_2 = 2\ \text{k}\Omega$, $R_F = 150\ \text{k}\Omega$, and $v_s = 0.01 + 0.005 \cos(\omega t)$ V. Determine an expression for the output voltage v_o and its value. Assume an ideal op-amp.

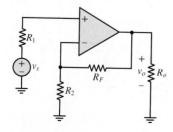

Figure P7.30

7.31 For the circuit shown in Figure P7.31, assume $v_S = 0.3 + 0.2 \cos(\omega t)$, $R_s = 4\ \Omega$, and $R_o = 15\ \Omega$. Determine the output voltage V_o for an ideal op-amp and also for a Motorola MC1741C op-amp with characteristics as given in Problem P7.29.

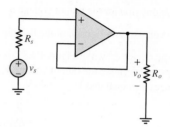

Figure P7.31

7.32 For the circuit shown in Figure P7.32, assume:

$$v_{S1} = 2.9 \times 10^{-3} \cos(\omega t)\ \text{V}$$
$$v_{S2} = 3.1 \times 10^{-3} \cos(\omega t)\ \text{V}$$
$$R_1 = 1\ \text{k}\Omega \qquad R_2 = 3.3\ \text{k}\Omega$$
$$R_3 = 10\ \text{k}\Omega \qquad R_4 = 18\ \text{k}\Omega$$

Determine an expression and value for the output voltage v_o.

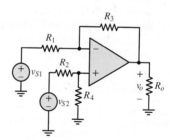

Figure P7.32

7.33 For the circuit shown in Figure P7.33, assume $v_{S1} = -2$ V, $v_{S2} = 2 \sin(2\pi \cdot 2{,}000t)$ V, $R_1 = 100\ \text{k}\Omega$, $R_2 = 50\ \text{k}\Omega$, and $R_F = 150\ \text{k}\Omega$. Determine the output voltage v_o.

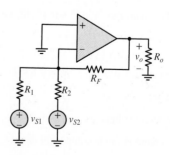

Figure P7.33

7.34   For the circuit shown in Figure P7.33, assume: $v_{S1} = v_{S2} = 5$ mV, $R_1 = 50\ \Omega$, $R_2 = 2$ kΩ, and $R_F = 2$ kΩ. The nonideal MC1741C op-amp has a 2-mΩ input resistance, a 75-Ω output resistance, and an open-loop gain of 200K. Determine:

a. An expression for the output voltage v_o.

b. The voltage gain for each of the two input signals.

7.35   In the circuit shown in Figure P7.35, assume ideal op-amps to determine the output voltage V_o. All resistances are equal and $V_{in} = 4\angle 0$ V.

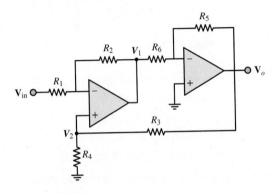

Figure P7.35

7.36   In the circuit shown in Figure P7.36, assume $V_2 = 8\angle 0$ V and find the input voltage V_{in} such that $V_o = 0$. Assume ideal op-amps.

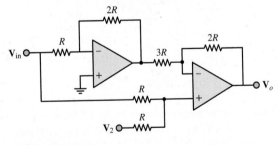

Figure P7.36

7.37   In the circuit shown in Figure P7.32, assume:

$$v_{S1} = 1.3\ \text{V} \qquad v_{S2} = 1.9\ \text{V}$$
$$R_1 = R_2 = 4.7\ \text{k}\Omega$$
$$R_3 = R_4 = 10\ \text{k}\Omega \qquad R_o = 1.8\ \text{k}\Omega$$

Determine:

a. The output voltage v_o.

b. The common-mode component of v_o.

c. The differential-mode component of v_o.

7.38   In the circuit shown in Figure P7.38, determine the output voltage V_o. Let $R_1 = 10$ kΩ, $R_2 = 10$ kΩ, $R_3 = 15$ kΩ, $R_4 = 10$ kΩ, $R_F = 50$ kΩ, and $V_{in} = 6$ V.

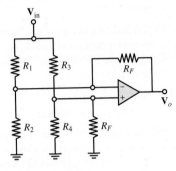

Figure P7.38

7.39   A linear potentiometer R_P is used to sense and generate a voltage v_y proportional to the y-coordinate of an xy inkjet printer head. A reference signal v_R is supplied by the software controlling the printer. The difference between these voltages is amplified to drive a motor. The motor changes the position of the printer head until that difference equals zero. For proper operation, the motor voltage must be 10 times the difference between the signal and reference voltage. For rotation in the proper direction, the motor voltage must be negative with respect to v_y. In addition, i_P must be negligibly small to avoid loading the pot and causing an erroneous signal voltage.

a. Design an op-amp circuit that satisfies these specifications. Redraw Figure P7.39, replacing the dotted line box with your amplifier circuit. Be sure to indicate component values.

b. Mark the pin numbers on your redrawn figure for an 8-pin single μA741C op-amp chip.

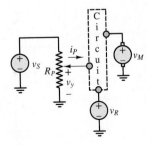

Figure P7.39

7.40 Compute the current I_{batt} delivered by the battery in Figure P7.40. Assume: $R_{S1} = 30$ kΩ, $R_{S2} = 30$ kΩ, $R_{f1} = 100$ kΩ, $R_{f2} = 60$ kΩ, $R_1 = 5$ kΩ, $R_2 = 7$ kΩ, and $V_{batt} = 3$ V.

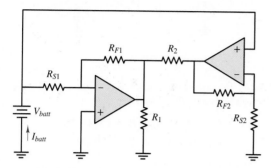

Figure P7.40

7.41 Figure P7.41 shows a simple voltage-to-current converter. Show that the current I_o through the light-emitting diode (LED), and therefore its brightness, is proportional to the source voltage V_s as long as $V_s > 0$. The LED permits current in the direction shown only.

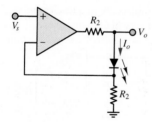

Figure P7.41

7.42 Figure P7.42 shows a simple current-to-voltage converter. Show that the voltage V_o is proportional to the current generated by the cadmium sulfide (CdS) solar cell. Also show that the transimpedance of the circuit V_o/I_s is $-R$!

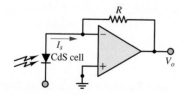

Figure P7.42

7.43 An op-amp voltmeter circuit as in Figure P7.43 is required to measure a maximum input of $V_S = 15$ mV. The op-amp input current is $I_B = 0.25$ μA. The ammeter is designed for full-scale deflection when

$I_m = 80$ μA and $r_m = 8$ kΩ. Determine suitable values for R_3 and R_4 so that the full-scale deflection of the ammeter corresponds to $V_S = 15$ mV. What is the significance of I_B for a nonideal op-amp?

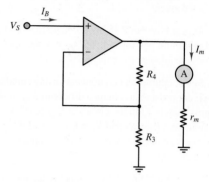

Figure P7.43

7.44 Find an expression for the voltage gain v_o/v_s in Figure P7.44. What is the gain when $R_{S1} = R_{S2} = 2.5$ kΩ and $R_{F1} = R_{F2} = 9.0$ kΩ?

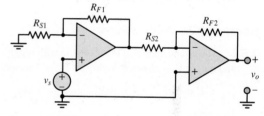

Figure P7.44

7.45 Select appropriate components using standard 5 percent resistors to obtain a voltage gain $v_o/v_s \approx 80$ for the circuit of Figure P7.44. How closely can you approximate the desired gain? Compute the expected error.

7.46 Repeat Problem 7.45 but compute the maximum and minimum possible voltage gains if the resistor values are allowed to swing ± 5 percent.

7.47 The circuit shown in Figure P7.47 can function as a precision ammeter. Assume that the voltmeter has a range of 0 to 10 V and an internal resistance of 20 kΩ. The full-scale reading of the ammeter is intended to be 1 mA. Find the resistance R such that the voltmeter reading is 10 V when $i_{in} = 1$ mA.

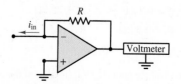

Figure P7.47

7.48   Select appropriate components using standard 5 percent resistors to obtain a voltage gain $v_o/v_s = 20$ for the circuit of Figure P7.30. How closely can you approximate the desired gain? Compute the expected error.

7.49   Repeat Problem 7.48 but compute the maximum and minimum possible voltage gains if the resistor values are allowed to swing ±5 percent.

7.50   Select appropriate components using standard 1 percent resistors to obtain a differential gain of approximately 15 in the circuit of Figure P7.32. Assume that $R_3 = R_4$ and $R_1 = R_2$. How closely can you approximate the desired gain? Compute the expected error.

7.51   Repeat Problem 7.50 but compute the maximum and minimum possible voltage gains if the resistor values are allowed to swing ±1 percent. Also compute the maximum common-mode output for the same allowed ±1 percent swing. Pick the nominal resistor values so that $R_3 = R_4$ and $R_1 = R_2$.

Section 7.3: Active Filters

7.52   The circuit shown in Figure P7.52 with input V_s and output V_o is an active high-pass filter. Assume:

$$C = 1\ \mu F \qquad R = 10\ k\Omega \qquad R_o = 1\ k\Omega$$

Determine:

a. The voltage gain $|V_o/V_s|$ (in dB) in the passband.

b. The cutoff frequency.

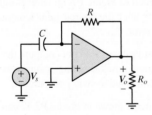

Figure P7.52

7.53   The op-amp circuit shown in Figure P7.53 is used as a high-pass filter. Assume:

$$C = 0.2\ \mu F \qquad R_o = 222\ \Omega$$
$$R_1 = 1.5\ k\Omega \qquad R_2 = 5.5\ k\Omega$$

Determine:

a. The voltage gain $|V_o/V_s|$, (in dB), in the passband.

b. The cutoff frequency.

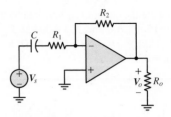

Figure P7.53

7.54   The op-amp circuit shown in Figure P7.53 is used as a high-pass filter. Assume:

$$C = 200\ pF \qquad R_o = 1\ k\Omega$$
$$R_1 = 10\ k\Omega \qquad R_2 = 220\ k\Omega$$

Determine:

a. The voltage gain $|V_o/V_s|$, (in dB), in the passband.

b. The cutoff frequency.

7.55   The circuit shown in Figure P7.55 is an active filter. Assume:

$$C = 120\ pF \qquad R_o = 180\ k\Omega$$
$$R_1 = 3\ k\Omega \qquad R_2 = 50\ k\Omega$$

Determine the break frequencies and $|V_o/V_i|$ at very low and at very high frequencies.

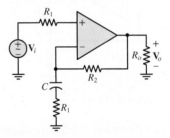

Figure P7.55

7.56   The circuit shown in Figure P7.56 is an active filter. Assume:

$$C = 15\ nF \qquad R_o = 4\ k\Omega$$
$$R_1 = 1.2\ k\Omega \qquad R_2 = 5.6\ k\Omega$$
$$R_3 = 62\ k\Omega$$

Determine:

a. An expression for the voltage gain in standard form:
$$\mathbf{H}_v(j\omega) = \frac{\mathbf{V}_o(j\omega)}{\mathbf{V}_i(j\omega)}$$

b. The break frequencies.

c. The passband gain.

d. The Bode magnitude and phase plots of $\mathbf{V}_o/\mathbf{V}_i$.

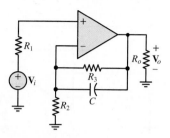

Figure P7.56

7.57 The op-amp circuit shown in Figure P7.57 is used as a low-pass filter. Assume:

$$C = 0.8\,\mu\text{F} \qquad R_o = 1\,\text{k}\Omega$$
$$R_1 = 5\,\text{k}\Omega \qquad R_2 = 15\,\text{k}\Omega$$

Determine:

a. An expression in standard form for the voltage gain V_o/V_s.

b. The gain, in dB, in the passband and at the cutoff frequency.

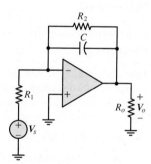

Figure P7.57

7.58 The op-amp circuit shown in Figure P7.57 is a low-pass filter. Assume:

$$R_1 = 2.2\,\text{k}\Omega \qquad R_2 = 68\,\text{k}\Omega$$
$$C = 0.47\,\text{nF} \qquad R_o = 1\,\text{k}\Omega$$

Determine:

a. An expression in standard form for the voltage gain V_o/V_s.

b. The gain, in dB, in the passband and at the cutoff frequency.

7.59 The circuit shown in Figure P7.59 is a bandpass filter. Assume:

$$R_1 = R_2 = 10\,\text{k}\Omega \qquad R_o = 4.7\,\text{k}\Omega$$
$$C_1 = C_2 = 0.1\,\mu\text{F}$$

Determine:

a. The voltage gain $|V_o/V_i|$ in the passband.

b. The resonant frequency.

c. The break frequencies.

d. The quality factor Q.

e. The Bode magnitude and phase plots of V_o/V_i.

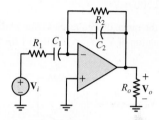

Figure P7.59

7.60 The op-amp circuit shown in Figure P7.57 is a low-pass filter. Assume:

$$R_1 = 12\,\text{k}\Omega \qquad R_2 = 4.7\,\text{k}\Omega \qquad R_o = 3.3\,\text{k}\Omega$$
$$C = 0.7\,\text{nF}$$

Determine:

a. An expression in standard form for the voltage gain V_o/V_s.

b. The gain, in dB, in the passband and at the cutoff frequency.

c. Would such small resistance values cause a practical op-amp to behave in a significantly nonideal manner?

7.61 The circuit shown in Figure P7.59 is a bandpass filter. Assume:

$$R_1 = 2.2\,\text{k}\Omega \qquad R_2 = 100\,\text{k}\Omega$$
$$C_1 = 2.2\,\mu\text{F} \qquad C_2 = 1\,\text{nF}$$

Determine the passband gain.

7.62 Derive the frequency response function V_o/V_{in} for the circuit shown in Figure P7.62.

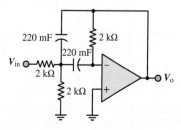

Figure P7.62

7.63 The circuit shown in Figure P7.63 can be used as a low-pass filter.

a. Derive the frequency response $\mathbf{V}_o/\mathbf{V}_{in}$ of the circuit.

b. If $R_1 = R_2 = 100\ k\Omega$ and $C = 0.1\ \mu F$, compute the attenuation, in dB, of $\mathbf{V}_o/\mathbf{V}_{in}$ at $\omega = 1,000$ rad/s.

c. Compute the amplitude and phase of $\mathbf{V}_o/\mathbf{V}_{in}$ at $\omega = 2,500$ rad/s.

d. Find the range of frequencies over which the attenuation of $\mathbf{V}_o/\mathbf{V}_{in}$ is less than 1 dB.

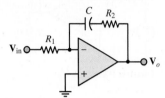

Figure P7.63

7.64 Determine a symbolic expression in standard form for the voltage gain $\mathbf{V}_o/\mathbf{V}_{in}$ in Figure P7.64. What kind of a filter does the voltage gain represent?

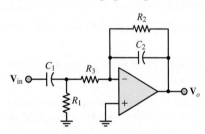

Figure P7.64

7.65 For the circuit of Figure P7.65, sketch the amplitude response of $\mathbf{V}_2/\mathbf{V}_1$, indicating the half-power frequencies. Assume the op-amp is ideal.

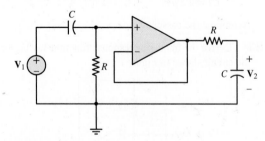

Figure P7.65

7.66 Determine a symbolic expression for the voltage gain $\mathbf{V}_o/\mathbf{V}_{S1}$ of Figure P7.66. What kind of a filter does the gain represent?

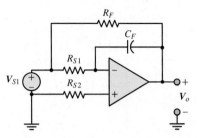

Figure P7.66

7.67 Determine a symbolic expression for the voltage gain $\mathbf{V}_o/\mathbf{V}_S$ of Figure P7.67. What kind of a filter does the gain represent?

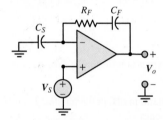

Figure P7.67

Section 7.4: Integrators and Differentiators

7.68 The circuit shown in Figure P7.68(a) produces an output voltage v_o which is either the integral or the derivative of the source voltage v_s shown in Figure P7.68(b) multiplied by some gain. Assume:

$$C = 1.5\ \mu F \qquad R = 5\ k\Omega \qquad R_o = 1.5\ k\Omega$$

For the given source voltage, determine the output voltage as a function of time and plot it.

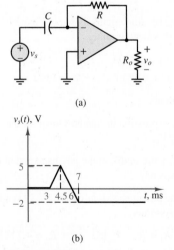

Figure P7.68

7.69 The circuit shown in Figure P7.69(a) produces an output voltage v_o which is either the integral or the derivative of the source voltage v_s shown in Figure P7.69(b) multiplied by some gain. Assume:

$$C = 0.5 \; \mu F \qquad R = 8 \; k\Omega \qquad R_o = 2 \; k\Omega$$

For the given source voltage, determine the output voltage as a function of time and plot it.

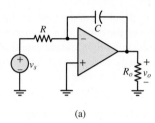

(a)

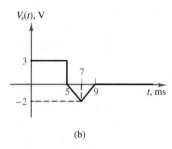

(b)

Figure P7.69

7.70 The circuit shown in Figure P7.70 is an integrator. The capacitor is initially uncharged, and the source voltage is

$$v_{in}(t) = 10^{-2} \, V + \sin(2,000 \, \pi t) \; V$$

a. At $t = 0$, the switch S_1 is closed. How long does it take before clipping occurs at the output if $R_s = 10 \; k\Omega$ and $C_F = 0.008 \; \mu F$?

b. At what times does the integration of the DC input cause the op-amp to saturate fully?

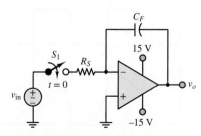

Figure P7.70

7.71 A practical integrator is shown in Figure 7.35. Note that the resistor in parallel with the feedback capacitor provides a path for the capacitor to discharge DC voltage. Usually, the time constant $R_F C_F$ is chosen to be large enough not to interfere with the integration.

a. If $R_S = 10 \; k\Omega$, $R_F = 2 \; M\Omega$, $C_F = 0.008 \; \mu F$, and $v_S(t) = 10 \; V + \sin(2,000 \pi t) \; V$, find $v_o(t)$, using phasor analysis.

b. Repeat part a if $R_F = 200 \; k\Omega$ and if $R_F = 20 \; k\Omega$.

c. Compare the time constants $R_F C_F$ with the period of the waveform for parts a and b. What can you say about the time constant and the ability of the circuit to integrate?

7.72 The circuit of Figure 7.40 is a practical differentiator. Assume an ideal op-amp, and $v_S(t) = 10^4 \sin(2,000 \pi t) \; mV$, $C_S = 100 \; \mu F$, $C_F = 0.008 \; \mu F$, $R_F = 2 \; M\Omega$, and $R_S = 10 \; k\Omega$.

a. Determine the voltage gain V_o/V_S.

b. Sum the DC and AC components of $v_o(t)$ to find the total output voltage.

7.73 Derive the differential equation in $x(t)$ for the circuit of Figure P7.73.

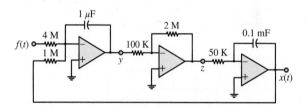

Figure P7.73

7.74 Construct a circuit corresponding to the following differential equation:

$$\frac{d^2 x}{dt^2} + 100 \frac{dx}{dt} + 10x = -5f(t)$$

Section 7.5: Physical Limitations of Operational Amplifiers

7.75 Consider the noninverting amplifier of Figure 7.9. Find the error introduced in v_o when the op-amp has an input offset voltage of 2 mV. Assume the input bias currents are zero and $R_1 = R_F = 4.7 \; k\Omega$. Assume that the offset voltage appears as shown in Figure 7.63.

7.76 In the circuit shown in Figure P7.76, sketch the output voltage $v_o(t)$ for the two input voltages $v_1(t)$ and $v_2(t)$. Assume $R_1 = 120$ kΩ, $R_2 = 150$ kΩ, and $C = 2$ nF. Also assume the op-amp slew rate limit is $S_0 = 1.0$ V/μs and the capacitor is initially uncharged.

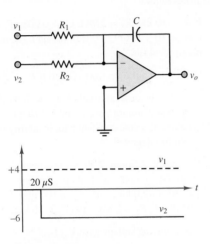

Figure P7.76

7.77 Consider a standard inverting amplifier, as shown in Figure P7.77. Assume that the offset voltage can be neglected and that the two input bias currents are equal. Find the value of R_x that eliminates the error in the output voltage due to the bias currents.

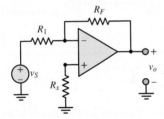

Figure P7.77

7.78 Determine the effect of the slew rate limit $S_0 = 0.5$ V/μs on the output of the isolation buffer for each of the following sinusoidal input voltages:

a. $v_{in} = 0.8 \sin(2\pi \cdot 6,000t)$ V

b. $v_{in} = 0.9 \sin(2\pi \cdot 7,500t)$ V

c. $v_{in} = 0.9 \sin(2\pi \cdot 15,000t)$ V

7.79 In the circuit shown in Figure P7.79, derive the output voltage $v_o(t)$ as a function of $v_{in}(t)$. Compute the effect of the slew rate on the maximum slope of the output voltage. Assume v_{in} is zero but then undergoes a step increase of amplitude ΔV.

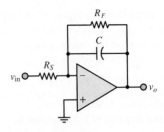

Figure P7.79

7.80 Determine the effect of the slew rate limit $S_0 = 0.5$ V/μs on the output of a noninverting amplifier with closed-loop voltage gain $G = 10$ for a symmetric square wave v_{in}. Sketch the output waveform for each following case:

a. v_{in} switches between ± 0.5 V and $f = 500$ Hz.

b. v_{in} switches between ± 1.25 V and $f = 5$ kHz.

c. v_{in} switches between ± 0.5 V and $f = 25$ kHz.

7.81 Consider a differential amplifier. We desire the common-mode output to be less than 1 percent of the differential-mode output. Find the minimum decibel common-mode rejection ratio to fulfill this requirement if the differential-mode gain $A_{dm} = 1,000$. Let

$$v_1 = \sin(2,000\pi t) + 0.1 \sin(120\pi t) \text{ V}$$
$$v_2 = \sin(2,000\pi t + 180°) + 0.1 \sin(120\pi t) \text{ V}$$
$$v_o = A_{dm}(v_1 - v_2) + A_{cm}\frac{v_1 + v_2}{2}$$

7.82 Square wave testing can be used with operational amplifiers to estimate the *slew rate*, which is defined as the maximum rate at which the output can change (in volts per microsecond). Input and output waveforms for a noninverting op-amp circuit are shown in Figure P7.82. As indicated, the rise time t_R of the output waveform is defined as the time it takes for that waveform to increase from 10 percent to 90 percent of its final value, or

$$t_R \triangleq t_B - t_A = -\tau(\ln 0.1 - \ln 0.9) = 2.2\tau$$

where τ is the circuit time constant. Estimate the slew rate for the op-amp.

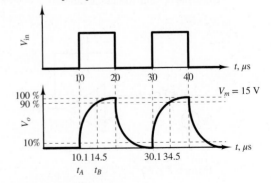

Figure P7.82

7.83 The nonideal op-amp used in the inverting amplifier of Figure 7.8 has an open-loop voltage gain $A = 250 \times 10^3$. Assume that v^- is small but nonzero. The input terminal currents i_{in} can still be assumed zero. Apply equation 7.23 to find:

$$\frac{v_o}{v_S} = \frac{-R_F/R_S}{1 + (1/A)[(R_F + R_S)/R_S]}$$

a. If $R_S = 10$ kΩ and $R_F = 1$ MΩ, find the closed-loop voltage gain $G = v_o/v_S$.

b. Repeat part a for $R_F = 10$ MΩ.

c. Repeat part a for $R_F = 100$ MΩ.

d. Evaluate G as $A \rightarrow \infty$ for parts a to c.

7.84 The nonideal op-amp used in the noninverting amplifier of Figure P7.84 has an open-loop voltage gain $A = 250 \times 10^3$. Assume $v_{in} = v^+ + \Delta v$, where Δv is small but nonzero, as suggested in equation 7.23. The input terminal currents i_{in} can still be assumed zero. Find:

a. The closed-loop gain v_o/v_{in} for $R_F = R_S = 7.5$ kΩ;

b. The closed-loop gain v_o/v_{in} for $R_F = R_S = 7.5$ kΩ.

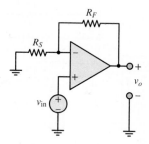

Figure P7.84

7.85 Given the unity-gain bandwidth for an ideal op-amp equal to 5.0 MHz, find the voltage gain at a frequency of $f = 500$ kHz.

7.86 The open-loop gain A of real (nonideal) op-amps is very large at low frequencies but decreases markedly as frequency increases. As a result, the closed-loop gain of op-amp circuits can be strongly dependent on frequency. Determine the relationship between a finite and frequency-dependent open-loop gain $A(\omega)$ and the closed-loop gain $G(\omega)$ of an inverting amplifier as a function of frequency. Plot G versus ω. Notice that $-R_F/R_S$ is the low-frequency closed-loop gain.

7.87 A sinusoidal sound (pressure) wave $p(t)$ impinges upon a condenser microphone of sensitivity S (mV/kPa). The voltage output of the microphone v_s is amplified by two cascaded inverting amplifiers to produce an amplified signal v_0. Determine the peak amplitude of the sound wave (in dB) if $v_0 = 5$ V_{RMS}. Estimate the maximum peak magnitude of the sound wave in order that v_0 not contain any saturation effects of the op-amps.

7.88 For the circuit shown in Figure P7.88, assume a nonideal op-amp and:

$$v_{S1} = 2.8 + 0.01 \cos(\omega t) \quad V$$
$$v_{S2} = 3.5 - 0.007 \cos(\omega t) \quad V$$
$$A_1 = -13 \quad A_2 = 10 \quad \omega = 4 \text{ krad/s}$$

where A_1 and A_2 are the open-loop voltage gains associated with inputs v_{S1} and v_{S2}, respectively.

Determine:

a. Common- and differential-mode input signals.

b. Common- and differential-mode closed-loop gains.

c. Common- and differential-mode components of the output voltage.

d. Total output voltage.

e. Common-mode rejection ratio.

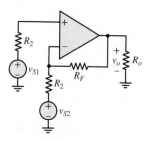

Figure P7.88

7.89 For the circuit shown in Figure P7.88, assume a nonideal op-amp and:

$$v_{S1} = 3.5 + 0.01 \cos(\omega t) \quad V$$
$$v_{S2} = 3.5 - 0.01 \cos(\omega t) \quad V$$
$$A_c = 10 \text{ dB} \quad A_d = 20 \text{ dB} \quad \omega = 4 \text{ krad/s}$$

where A_c and A_d are the common- and differential-mode open-loop voltage gains, respectively. Determine:

a. Common- and differential-mode input voltages.

b. The voltage gains for v_{S1} and v_{S2}.

c. The common-mode component and differential-mode component of the output voltage.

d. The common-mode rejection ratio (CMRR), in dB.

7.90 In the circuit shown in Figure P7.90, the two voltage sources are temperature sensors with T = temperature (Kelvin) and

$$v_{S1} = kT_1 \quad v_{S2} = kT_2$$

where

$$k = 120 \ \mu V/K$$
$$R_1 = R_3 = R_4 = 5 \ k\Omega$$
$$R_2 = 3 \ k\Omega \qquad R_o = 600 \ \Omega$$

If

$$T_1 = 310 \ K \qquad T_2 = 335 \ K$$

determine

a. The voltage gains for the two input voltages.

b. The common-mode and differential-mode input voltages.

c. The common-mode and differential-mode gains.

d. The common-mode component and the differential-mode component of the output voltage.

e. The common-mode rejection ratio (CMRR), in dB.

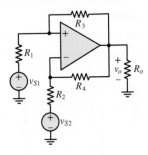

Figure P7.90

7.91   In the differential amplifier shown in Figure P7.90,

$$v_{S1} = 13 \ mV \qquad v_{S2} = 9 \ mV$$
$$v_o = v_{oc} + v_{od}$$
$$v_{oc} = 33 \ mV \qquad \text{(common-mode output voltage)}$$
$$v_{od} = 18 \ V \qquad \text{(difference-mode output voltage)}$$

Determine

a. The common-mode gain.

b. The differential-mode gain.

c. The common-mode rejection ratio, in dB.

7.92   The ideal charge amplifier discussed in the Focus on Measurements box, "Charge Amplifiers," will saturate in the presence of any DC offsets. Figure P7.92 presents a practical charge amplifier in which the user is provided with a choice of three time constants—RC_F, $10RC_F$, and $100RC_F$—which can be selected by means of a switch. Assume that $R = 0.1 \ M\Omega$, and $C_F = 0.1 \ \mu F$. Analyze the frequency response of the practical charge amplifier for each time constant, and determine the lowest input signal frequency that can be amplified without excessive distortion for each case. Can this circuit amplify a DC signal?

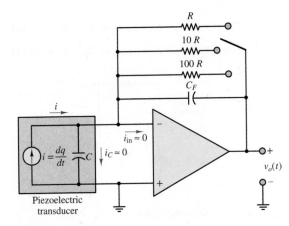

Figure P7.92

Design Credits: Mini DVI cable adapter isolated on white background: Robert Lehmann/Alamy Stock Photo; Balance scale: Alex Slobodkin/E+/Getty Images; Icon for "Focus on measurements" weighing scales: Media Bakery.

C H A P T E R

8

SEMICONDUCTORS
AND DIODES

M arvelous advances have taken place in the field of **solid state electronics** ever since the invention of the *diode* and *transistor*. Modern analog and digital electronic systems are possible because these discrete electronic elements have been integrated into complex devices and systems. Although discrete electronic elements have been replaced in many applications by *integrated circuits* (e.g., operational amplifiers), it is nonetheless important to understand how these elements function. The aim of Part II of this textbook is to explore the behavior and applications of diodes, transistors, and other *electronic devices*.

This chapter explains the workings of the semiconductor diode, a device that finds use in many practical circuits used in electric power systems and in high- and low-power electronic circuits. While the *i-v* characteristic of a diode is inherently nonlinear, simple *linear models* can be used to approximate the diode characteristic and thus produce *linear circuits* that can be analyzed using the analytical tools developed in earlier chapters.

Learning Objectives

Students will learn to...

1. Understand the basic principles underlying the physics of semiconductor devices in general and of the *pn* junction in particular. Become familiar with the diode equation and *i-v* characteristic. *Sections 8.1–8.2.*
2. Use various circuit models of the semiconductor diode in simple circuits. These are divided into two classes: large-signal models, useful to study rectifier circuits, and small-signal models, useful in signal processing applications. *Section 8.2.*
3. Study practical full-wave rectifier circuits and learn to analyze and determine the practical specifications of a rectifier by using large-signal diode models. *Section 8.3.*
4. Understand the basic operation of Zener diodes as voltage references, and use simple circuit models to analyze elementary voltage regulators. *Section 8.4.*
5. Use the diode models presented in Section 8.2 to analyze the operation of various practical diode circuits in signal processing applications. *Section 8.5.*

8.1 ELECTRICAL CONDUCTION IN SEMICONDUCTOR DEVICES

Elemental[1] or intrinsic **semiconductors** are those elements, specifically silicon and germanium, from group IV of the periodic table whose conductivity is much weaker than that of a typical conductor but significantly stronger than that of a typical insulator. For example, typical conductivities of copper (a good conductor) and glass (a common insulator) are 5.96×10^7 S/m and 10^{-13} S/m, respectively. By comparison, silicon and germanium, both semiconductors, have conductivities on the order of 10^{-3} and 10^0, respectively. Another important property of silicon and germanium is that their conductivities *increase* with temperature, whereas the conductivity of most conductors (e.g., metals) decreases with temperature. It is important to note that most of the group IV elements are *not* semiconductors; tin and lead are metals whose conductivity is large and decreases with temperature.

Conducting materials have enough weakly bonded electrons in the outer conduction band that a modest electric field can easily produce a significant current. By contrast, the outer-band electrons in a semiconducting material are held by **covalent bonds** such that much stronger electric fields are needed to liberate them. Figure 8.1 depicts the lattice arrangement for a pure silicon (Si) matrix. At sufficiently high temperatures, thermal energy causes the atoms in the lattice to vibrate; when sufficient kinetic energy is present, some of the valence electrons break their bonds with the lattice structure and become available as conduction electrons. These **free electrons** enable current flow in the semiconductor. As the temperature increases, more valence electrons are liberated, which explains why the conductivity of a semiconductor increases with temperature.

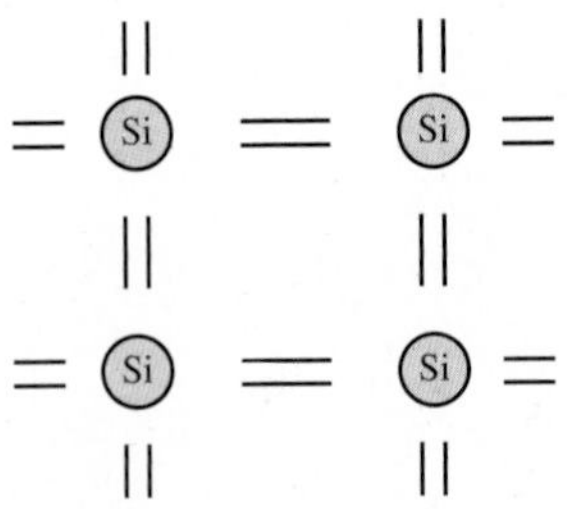

Figure 8.1 Lattice structure of silicon, with four valence electrons

However, the free valence electrons are not the only charge carriers present in a semiconductor. Whenever a free electron is liberated from the lattice, a

[1]Semiconductors can also be made of more than one element, in which case the elements are not necessarily from group IV.

corresponding net positive charge or **hole** within the lattice is also created as depicted by Figure 8.2. Holes act as positive charge carriers within a semiconducting material but with a different **mobility**—the ease with which charge carriers move through the lattice—than free electrons. Free electrons move far more easily around the lattice than holes. These two charge carriers also move in opposite directions when subjected to an external electric field, as illustrated in Figure 8.3.

Occasionally, a free electron traveling in the immediate neighborhood of a hole will recombine with it to form a covalent bond. The result is two lost charge carriers. This additional phenomenon of **recombination** is proportional to the number of free electrons and holes and reduces the number of charge carriers in a semiconductor. However, in spite of recombination, at any given temperature a number of free electrons and holes will be available for conduction. The number of available charge carriers is called the **intrinsic concentration** n_i. The most commonly reported expression for n_i is:

$$n_i \propto T^{1.5} e^{-E_g/2kT} \tag{8.1}$$

where T is temperature in K; E_g is the bandgap energy, which for silicon is 1.12 eV; and k is Boltzmann's constant 8.62×10^{-5} eV/K. At $T = 300$ K, n_i is approximately 1.5×10^{10} carriers/cm^3. Note the strong dependence on temperature.[2]

As noted above, pure semiconductors are not particularly good conductors. To enhance its concentration of charge carriers and thus its conductivity, a semiconductor can be **doped**, whereby either *trivalent* (group III) or *pentavalent* (group V) impurities are added to the crystalline structure of the semiconductor. Trivalent impurities, such as boron and gallium, add holes to the semiconductor's lattice and are known as *acceptors*; pentavalent impurities, such as phosphorus and arsenic, add free electrons, as depicted in Figure 8.4, and are known as *donors*.

Free electrons are the *majority charge carrier*, and holes are the *minority charge carrier* in semiconductors doped with donor elements. These materials are called ***n*-type semiconductors**. Likewise, holes are the majority charge carrier, and free electrons are the minority charge carrier in semiconductors doped with acceptor elements. These materials are called ***p*-type semiconductors**. In thermal equilibrium, the concentration of free electrons n (negative) is related to the concentration of holes p (positive) by:

$$pn = n_i^2 \tag{8.2}$$

In a doped semiconductor, the concentration of donated atoms is usually much greater than the intrinsic concentration of the semiconductor. In this case, the concentration of majority charge carriers is approximately the same as the concentration of donated atoms, which is determined by the doping process and is not a function of temperature. However, the concentration of minority charge carriers is determined by temperature and is usually much less than the intrinsic concentration of the semiconductor. For example, in an *n*-type material, the concentration of free electrons n_n is approximately equal to the concentration of donor atoms n_D. Since $p_n n_n = n_i^2$, the result is:

$$n_n \approx n_D \gg n_i \quad \text{and} \quad p_n = \frac{n_i^2}{n_n} \approx \frac{n_i^2}{n_D} \ll n_i \quad \text{n-type} \tag{8.3}$$

---

[2]Another reported relation [A.B. Sproul and M.A. Green, *J. Appl. Phys.* 70, 846 (1991)] is $n_i \propto T^2 e^{-E_G/2kT}$ with a value at 300 K of approximately 1.0×10^{10} carriers/cm^3.

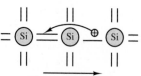

⊕ = Hole   Electron jumps to fill hole

The net effect is a hole moving to the right

A vacancy (or hole) is created whenever a free electron leaves the structure.
This "hole" can move around the lattice if other electrons replace the free electron.

Figure 8.2 Free electrons and "holes" in the lattice structure

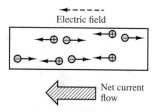

Electric field

Net current flow

An external electric field forces holes to migrate to the left and free electrons to the right. The net current flow is to the left.

Figure 8.3 Current in a semiconductor

An additional free electron is created when Si is "doped" with a group V element.

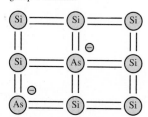

Figure 8.4 Doped semiconductor

Likewise, in a *p*-type material where n_A is the concentration of acceptor atoms:

$$p_p \approx n_A \gg n_i \quad \text{and} \quad n_p = \frac{n_i^2}{p_p} \approx \frac{n_i^2}{n_A} \ll n_i \quad p\text{-type} \tag{8.4}$$

In the previous two equations, the subscripts *i*, *n*, and *p* indicate whether the material is intrinsic (pure) semiconductor, *n*-type, or *p*-type, respectively.

It is important to keep in mind that doped *n*- and *p*-type materials are *electrically neutral* because the donor and acceptor elements have equal numbers of protons and electrons. The material type simply indicates the nature of the *mobile* majority charge carriers present in the conduction band of the material lattice.

8.2   THE *pn* JUNCTION AND THE SEMICONDUCTOR DIODE

A simple section of *n*- or *p*-type material is not particularly useful for the construction of electronic circuits. However, when sections of *n*- and *p*-type material are brought in contact to form a **pn junction**, a **diode** is formed. Diodes have a number of interesting and useful properties that are due entirely to the nature of the *pn* junction.

Figure 8.5 depicts an idealized *pn* junction. The difference in concentrations of free electrons in the *n*-type material compared to the *p*-type material results in a *diffusion* of free electrons from right to left across the junction. Likewise, the difference in concentration of holes on either side of the junction results in diffusion of holes from left to right across the junction. In both cases, the **diffusion current** I_d is directed left to right because a positive current is defined as either positive holes moving left to right or negative free electrons moving right to left.

As free electrons leave the *n*-type material and enter the *p*-type material they tend to recombine with holes. Likewise, as holes leave the *p*-type material and enter the *n*-type material they tend to recombine with free electrons. Once free electrons and holes recombine they are no longer mobile, but held in place in the material lattice by covalent bonds. At first, most of the recombinations occur close to the junction. However, as time passes, more and more of the mobile charges near the junction have recombined such that diffusing mobile charges must travel further from the junction to encounter a partner with which to recombine. Thus, this diffusion process results in recombinations on both sides of the junction and, as the process continues, an expanding **depletion region** wherein virtually no mobile charge carriers remain. This region is *electrically charged* because mobile charge carriers that have recombined to form the region have no electrical counterpart in the lattice where they have become fixed. In Figure 8.5 this result is

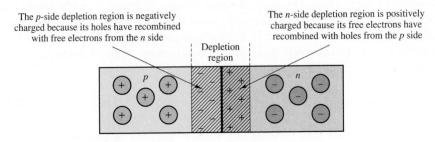

Figure 8.5 A *pn* junction

depicted by the negatively charged p-type region to the left of the junction and the positively charged n-type region to the right of the junction.

Once the depletion region begins to form, the resulting net charge separation produces an electric field pointing from the positively charged n-type to the negatively charged p-type portions of the depletion region. This electric field slows the ongoing diffusion of majority charge carriers by establishing a **potential barrier** or **contact potential** across the depletion region. This potential depends upon the semiconductor material (about 0.6 to 0.7 V for silicon) and is also known as the *offset voltage* V_γ.

In addition to the diffusion current associated with majority charge carriers, an oppositely directed **drift current** I_S associated with minority charge carriers is established across the depletion region. Specifically, free electrons and holes are thermally generated in the p- and n-type materials, respectively. Any of these minority carriers that manage to reach the depletion region are swept across it by the electric field. Note that both components of the drift current contribute to a positive current from right to left because a positive current is defined as either positive holes moving right to left or negative free electrons moving left to right.

Figure 8.6 depicts the presence of both a diffusion current and drift current across the depletion region. Its equilibrium width is reached when the average net drift current exactly offsets the average net diffusion current. Recall that the magnitude of the diffusion current is largely determined by the concentration of the donor and acceptor elements while the magnitude of the drift current is highly temperature dependent. Thus, the equilibrium width of the depletion region depends upon both temperature and the doping process.

Now consider the case shown in Figure 8.7(a) where a battery has been connected across a pn junction in the **reverse-biased** direction. Assume that suitable contacts between the battery and the p- and n-type materials are established. The reverse-bias orientation of the battery widens the depletion region and increases the potential barrier across it such that the majority carrier diffusion current decreases. On the other hand, the minority carrier drift current increases such that there is now a small (on the order of nano-amperes) nonzero current I_0 directed from the n- to p-type region. I_0 is small because it is comprised of minority carriers. Thus, when reverse-biased, the diode current i_D is:

$$i_D = -I_0 = I_S \qquad \text{Reverse-biased diode current} \qquad (8.5)$$

where I_S is known as the **reverse saturation current**.

When the pn junction is forward-biased as in Figure 8.7(b), the depletion region is narrowed and the potential barrier across it is lowered such that the majority carrier diffusion current increases. As the forward-biased diode voltage v_D is increased the diffusion current I_d increases exponentially:

$$I_d = I_0 e^{q_e v_D / kT} = I_0 e^{v_D / V_T} \qquad (8.6)$$

where $q_e = 1.6 \times 10^{-19}$ C is the elementary charge, T is the material temperature (in K), and $V_T = kT/q_e$ is the **thermal voltage**. At room temperature, $V_T \approx 25$ mV. The net diode current under forward bias is:

$$\boxed{i_D = I_d - I_0 = I_0(e^{v_D / V_T} - 1) \qquad \text{Diode equation}} \qquad (8.7)$$

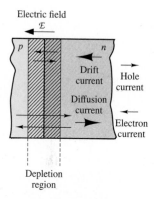

Electric field

Figure 8.6 Drift and diffusion currents in a *pn* junction

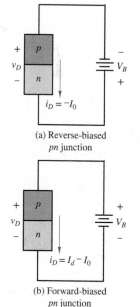

(a) Reverse-biased *pn* junction

(b) Forward-biased *pn* junction

Figure 8.7 Forward- and reverse-biased *pn* junctions

Figure 8.8 depicts the diode *i-v* characteristic described by the diode equation for a fairly typical silicon diode for $v_D > 0$. Since I_0 is typically very small (10^{-9} to 10^{-15} A), the diode equation is often approximated by:

$$i_D = I_0 e^{v_D/V_T} \tag{8.8}$$

This expression is a good approximation for a silicon diode at room temperature when v_D is greater than a few tenths of a volt.

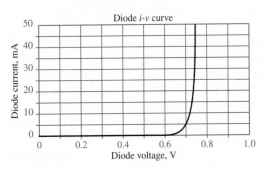

Figure 8.8 Typical diode *i-v* characteristic curve

The ability of the *pn* junction to conduct significant current only in the forward-biased direction allows it to function in electric circuits much like a check valve functions in mechanical circuits. A generic *pn* junction and the diode circuit symbol are shown in Figure 8.9. Notice that the triangle shape suggests the direction of forward-biased current. Positive current i_D passes from the **anode** to the **cathode**, where the term *cathode* always refers to the source of electrons (negative charge carriers) whether used in reference to a diode or battery.[3]

Figure 8.10 shows the complete *i-v* characteristic of a diode. Note that the diode current is approximately zero when $v_D < 0$ unless v_D is sufficiently large and negative (reverse-biased) such that **reverse breakdown** occurs. When $v_D < -V_Z$, the diode conducts current in the *reverse-biased direction*. Two effects contribute to this reverse-biased current: the *Zener effect* and *avalanche breakdown*. In silicon diodes, the Zener effect tends to dominate when $V_Z < 5.6$ V while avalanche breakdown tends to dominate at larger, more negative diode voltages.

The root causes of these two effects, while similar, are not the same. The Zener effect is significant when the depletion region is designed to be heavily doped but very thin such that for a given potential difference v_D, the electric field is large enough to sever covalent bonds in the depletion region and generate pairs of free electrons and holes, which are then swept away by the electric field, thus creating a current. Avalanche breakdown occurs when the potential difference v_D is large enough that the kinetic energy of minority charge carriers is sufficient to break covalent bonds during collisions. These collisions may liberate free electrons and holes, which, again, are swept away by the electric field. The process by which energy is imparted to new charge carriers is called *impact ionization*. These new charge carriers may also have enough energy to energize other low-energy electrons, such that a sufficiently large reverse-biased diode voltage may initiate an avalanche of liberated charge carriers.

In **Zener breakdown** the high concentration of charge carriers provides the means for a substantial reverse-biased current to be sustained, at a nearly constant

The triangle in the circuit symbol for the diode indicates the direction of current flow when the diode is forward-biased.

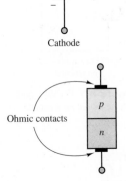

Figure 8.9 Diode circuit symbol

[3]The positive terminal of a battery is referred to as the cathode because internally it is the source of negative ions traveling toward the negative terminal.

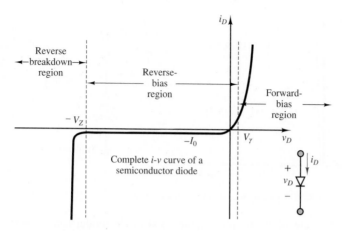

Figure 8.10 The diode i-v characteristic

Hydraulic Check Valves

The operation of a diode can be understood intuitively by reference to a very common hydraulic device that finds application whenever one wishes to restrict the flow of a fluid to a single direction and to prevent (check) reverse flow. Hydraulic check valves perform this task in a number of ways. A few examples are illustrated here.

The first figure below depicts a swing check valve. In this design, flow from left to right is permitted, as the greater fluid pressure on the left side of the valve forces the swing door to open. If flow were to reverse, the reversal of fluid pressure (greater pressure on the right) would cause the swing door to shut.

In this design, flow from left to right is permitted, as the greater fluid pressure on the left side of the valve forces the swing "door" to open. If flow were to reverse,

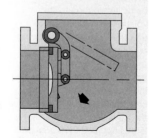

Swing check valve

(Continued)

reverse-biased voltage, the **Zener voltage** V_Z. This effect is very useful in applications where one would like to regulate (hold constant) the voltage across a load. It should also be noted that a typical silicon diode is not designed for use in reverse breakdown, where even a modest current at a large V_Z will likely generate more power than the diode can dissipate through heat transfer. The result could be a melted or burned diode!

8.3   LARGE-SIGNAL MODELS FOR THE SEMICONDUCTOR DIODE

From the viewpoint of a *user* of electronic circuits (as opposed to a *designer*), it is often sufficient to characterize a device in terms of its i-v characteristic, using either load-line analysis or appropriate circuit models to determine the operating currents and voltages. This section shows how it is possible to use the i-v characteristics of the semiconductor diode to construct simple yet useful *circuit models*. Depending on the desired level of detail, it is possible to construct *large-signal models* of the diode, which describe the gross behavior of the device in the presence of relatively large voltages and currents; or *small-signal models*, which are capable of describing the behavior of the diode in finer detail and, in particular, the response of the diode to small changes in the average diode voltage and current. From the user's standpoint, these circuit models greatly simplify the analysis of diode circuits and make it possible to effectively analyze relatively "difficult" circuits simply by using the familiar circuit analysis tools of Chapter 2. The first two major divisions of this section describe different diode models and the assumptions under which they are obtained, to provide the knowledge you will need to select and use the appropriate model for a given application.

Ideal Diode Model

The simplest large-signal diode model is the **ideal diode**, which approximates a diode as a simple on/off device (like a check valve in hydraulic circuits). The circuit symbol for an ideal diode, its i-v approximation, and the i-v characteristic of a typical diode are shown in Figure 8.11. An ideal diode behaves as an open-circuit when reverse-biased ($v_D < 0$) and as a short-circuit when forward-biased

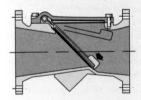
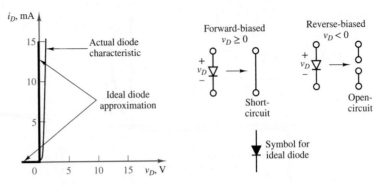

Figure 8.11 Large-signal on/off ideal diode model

($v_D \geq 0$). In spite of its simplicity, the ideal diode model can be very useful in circuit analysis.

Ideal diodes are represented by the solid black triangle symbol shown in Figure 8.11.

A general method for analyzing diode circuits is illustrated using the circuit shown in Figure 8.12, which contains a 1.5-V battery, an ideal diode, and a 1-kΩ resistor. The method is simply to assume that the ideal diode is forward-biased ($v_D \geq 0$) and thus equivalent to a short-circuit, as indicated in Figure 8.13. Under this assumption, $v_D = 0$ such that the loop current is $i_D = 1.5\,\text{V}/1\,\text{k}\Omega = 1.5\,\text{mA}$. Since the resulting direction of the current and the diode voltage are consistent with the assumption of a conducting diode ($v_D \geq 0$, $i_D > 0$), the assumption is correct. If the assumption had resulted in diode current and voltage that contradict the assumption, then the assumption would have been deemed incorrect, and the opposite assumption of a nonconducting diode could be tested, and presumably found to be true.

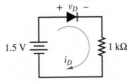

Figure 8.12 Circuit containing ideal diode

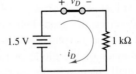

Figure 8.13 Circuit of Figure 8.12, assuming that the ideal diode conducts

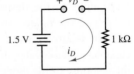

Figure 8.14 Circuit of Figure 8.12, assuming that the ideal diode does not conduct

To test the opposite assumption, assume the ideal diode is reverse-biased ($v_D < 0$) and thus equivalent to an open-circuit, as shown in Figure 8.14. Since the loop does not form a closed path, the current i_D must be zero and thus Ohm's law requires the voltage across the resistor to also be zero. Then, KVL requires that $v_D = 1.5\,\text{V}$. However, this result contradicts the assumption that the ideal diode is reverse-biased. Thus, the assumption is deemed incorrect.

The method can be applied to more complicated circuits involving multiple diodes by simply testing all the possible combinations of forward- and reverse-biased assumptions for the diodes. In such cases, it is helpful to consider which combinations are more likely to yield a correct solution and test those combinations first. With practice, such educated guesses should become more and more effective in reducing the number of tests necessary for any particular problem. Remember that it is only necessary to find one set of assumptions that does not result in a contradiction.

Offset Diode Model

While the ideal diode model is useful in approximating the large-scale characteristics of a physical diode, it does not account for the diode offset voltage. A better model is the **offset diode model**, which consists of an ideal diode in series with a battery, as shown in Figure 8.15, where the battery voltage equals the offset voltage (for silicon diodes $V_\gamma = 0.6$ V unless otherwise indicated). The effect of the battery is to shift the characteristic of the ideal diode to the right on the voltage axis, as shown in Figure 8.16.

The behavior of a diode in the offset diode model is described as follows:

$$v_D \geq 0.6 \text{ V} \qquad \text{Diode} \rightarrow 0.6\text{-V battery}$$
$$v_D < 0.6 \text{ V} \qquad \text{Diode} \rightarrow \text{open-circuit}$$

Offset diode model   (8.9)

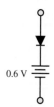

Figure 8.15 Offset diode as an extension of the ideal diode model

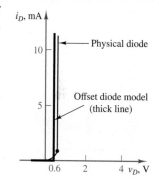

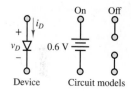

Figure 8.16 Offset diode states

LO

FOCUS ON PROBLEM SOLVING

DETERMINING THE CONDUCTION STATE OF IDEAL DIODES

1. Assume a diode conduction state (forward- or reverse-biased) for each diode.

2. Replace each diode with an ideal diode model (short-circuit if forward-biased, open-circuit if reverse-biased).

3. Solve for the diode currents and voltages, using linear circuit analysis.

4. If the entire solution is consistent with the assumptions, then the initial assumptions were correct; if not, at least one of the initial diode conduction state assumptions is wrong. Change at least one of the assumed diode conduction states, and solve the new circuit. Continue to iterate this process until a solution is found that is consistent with the assumptions. Be careful to keep track of the conduction state combinations that have been tested.

EXAMPLE 8.1   Determining the Conduction State of an Ideal Diode

Problem

Determine whether the ideal diode of Figure 8.17 is conducting.

Solution

Known Quantities: $V_S = 12$ V; $V_B = 11$ V; $R_1 = 5$ Ω; $R_2 = 10$ Ω; $R_3 = 10$ Ω.

Find: The conduction state of the diode.

Assumptions: Use the ideal diode model.

LO

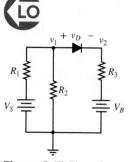

Figure 8.17 Figure for Example 8.1.

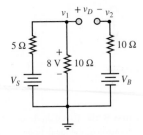

Figure 8.18 Circuit with ideal diode assumed "off".

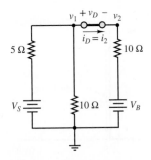

Figure 8.19 Circuit with ideal diode assumed "on".

Analysis: Assume initially that the ideal diode does not conduct, and replace it with an open-circuit, as shown in Figure 8.18. The voltage across R_2 can then be computed by using the voltage divider rule:

$$v_1 = \frac{R_2}{R_1 + R_2} V_S = \frac{10}{5 + 10} 12 = 8 \text{ V}$$

Applying KVL to the right-hand-side mesh (and observing that no current flows in the circuit since the diode is assumed off), we obtain

$$v_1 = v_D + V_B \quad \text{or} \quad v_D = 8 - 11 = -3 \text{ V}$$

The result indicates that the diode is reverse-biased and confirms the initial assumption. Thus, the diode is not conducting.

As further illustration, assume that the diode conducts. In this case, the diode is replaced with a short-circuit, as shown in Figure 8.19. The resulting circuit can be solved by node analysis, noting that $v_1 = v_2$ because of the short-circuit.

$$\frac{V_S - v_1}{R_1} = \frac{v_1 - 0}{R_2} + \frac{v_2 - V_B}{R_3}$$

$$\frac{V_S}{R_1} + \frac{V_B}{R_3} = \frac{v_1}{R_1} + \frac{v_1}{R_2} + \frac{v_2}{R_3}$$

$$\frac{12}{5} + \frac{11}{10} = \left(\frac{1}{5} + \frac{1}{10} + \frac{1}{10}\right) v_1$$

$$v_1 = 2.5(2.4 + 1.1) = 8.75 \text{ V}$$

With $v_1 = v_2 = 8.75$ V, the current through the branch containing the diode is:

$$i_D = \frac{v_1 - V_B}{R_3} = \frac{8.75 - 11}{10} = -0.225 \text{ A}$$

However, this negative current violates the forward-biased assumption about the diode. Thus, the forward-biased conducting assumption is incorrect and the diode must not be conducting.

EXAMPLE 8.2   Determining the Conduction State of an Ideal Diode

Problem

Determine whether the ideal diode of Figure 8.20 is conducting.

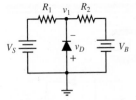

Figure 8.20 Circuit for Example 8.2.

Solution

Known Quantities: $V_S = 12$ V; $V_B = 11$ V; $R_1 = 5 \ \Omega$; $R_2 = 4 \ \Omega$.

Find: The conduction state of the diode.

Assumptions: Use the ideal diode model.

Analysis: Assume initially that the ideal diode does not conduct, and replace it with an open-circuit, as shown in Figure 8.21. The current through the resulting series loop is:

$$i = \frac{V_S - V_B}{R_1 + R_2} = \frac{1}{9} \text{ A}$$

The voltage at node v_1 is

$$\frac{12 - v_1}{5} = \frac{v_1 - 11}{4}$$

$$v_1 = 11.44 \text{ V}$$

The result indicates that the diode is strongly reverse-biased, since $v_D = 0 - v_1 = -11.44$ V, and is in accord with the initial assumption. Thus, the diode is not conducting.

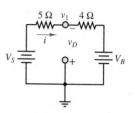

Figure 8.21 Circuit with ideal diode assumed "off".

EXAMPLE 8.3   Using the Offset Diode Model

Problem

Use the offset diode model to determine the value of v_1 for which diode D_1 first conducts in the circuit of Figure 8.22.

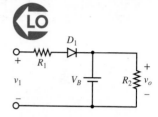

Figure 8.22 Circuit for Example 8.3.

Solution

Known Quantities: $V_B = 2$ V; $R_1 = 1$ kΩ; $R_2 = 500$ Ω; $V_\gamma = 0.6$ V.

Find: The lowest value of v_1 for which diode D_1 conducts.

Assumptions: Use the offset diode model.

Analysis: Start by replacing the diode with the offset diode model, as shown in Figure 8.23. If v_1 is negative, the diode will certainly be off. The point at which the diode turns on as v_1 is increased can be determined by analyzing the circuit with the diode assumed to be off. In a laboratory experiment, v_1 could be progressively increased until the diode conducts, that is, until the current through R_1 is nonzero. With the diode off, KVL yields:

$$v_1 = v_{D1} + 0.6 + 2 \qquad \text{or} \qquad v_{D1} = v_1 - 2.6$$

Thus, the condition required for the diode to conduct is:

$$v_1 \geq 2.6 \text{ V} \qquad \text{Diode "on" condition}$$

Comments: The same solution method can be used for problems involving the ideal diode model or the offset diode model.

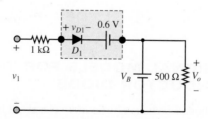

Figure 8.23 Circuit including diode offset model.

CHECK YOUR UNDERSTANDING

If the resistor R_2 is replaced with an open-circuit in the circuit of Figure 8.17, will the diode conduct?

CHECK YOUR UNDERSTANDING

Repeat the analysis of Example 8.2, assuming that the diode is conducting, and show that this assumption leads to inconsistent results.

CHECK YOUR UNDERSTANDING

Determine which of the diodes conduct in the circuit shown below for each of the following voltages. Treat the diodes as ideal.

a. $v_1 = 0$ V; $v_2 = 0$ V
b. $v_1 = 5$ V; $v_2 = 5$ V
c. $v_1 = 0$ V; $v_2 = 5$ V
d. $v_1 = 5$ V; $v_2 = 0$ V

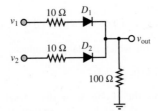

CHECK YOUR UNDERSTANDING

Determine which of the diodes conduct in the circuit shown below. Each diode has an offset voltage of 0.6 V.

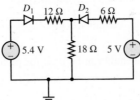

8.4   SMALL-SIGNAL MODELS FOR THE SEMICONDUCTOR DIODE

As one examines the diode *i-v* characteristic more closely, it becomes apparent that the short-circuit approximation is not adequate to represent the *small-signal behavior* of the diode. The term *small-signal behavior* usually signifies the response of the diode to small time-varying signals that may be superimposed on the average diode current and voltage. Figure 8.8 provides a more detailed view of a silicon diode *i-v* curve. Clearly, the short-circuit approximation is not very accurate when a diode's behavior is viewed on a finer scale. To a first-order approximation, however, the *i-v* characteristic is linear for voltages greater than the offset voltage. Thus, it seems reasonable to model a conducting diode as a resistor. Load-line analysis can be exploited to determine the diode **small-signal resistance**, which is related to the slope of its *i-v* characteristic.

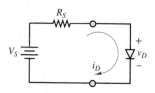

Figure 8.24 Diode circuit used to illustrate load-line analysis

Consider the circuit of Figure 8.24, which represents the Thévenin equivalent circuit of an arbitrary linear resistive circuit connected to a diode. KVL yields the *governing equation*:

$$V_S = i_D R_S + v_D \tag{8.10}$$

The *constitutive relation* for the diode is:

$$i_D = I_0(e^{v_D/V_T} - 1) \tag{8.11}$$

These two equations in two unknowns cannot be solved analytically since one of the equations is *transcendental*; that is, it contains the unknown v_D in exponential form. Transcendental equations of this type can be solved graphically or numerically. Only a graphical solution is considered here.

Consider a plot of the two preceding equations in the i_D–v_D plane. The diode equation gives rise to the familiar curve of Figure 8.8. The *load-line equation*, obtained from KVL, is the equation of a line with slope $-1/R_S$, open-circuit voltage V_S, and short-circuit current V_S/R_S. That is:

$$i_D = \frac{V_S - v_D}{R_S} = -\frac{1}{R_S}v_D + \frac{V_S}{R_S} \qquad \text{Load-line equation} \tag{8.12}$$

The superposition of these two curves gives rise to the plot of Figure 8.25, where the solution to the two equations is found graphically to be the pair of values (I_Q, V_Q). The intersection of the two curves is called the **quiescent (operating) point**, or **Q point**. The voltage $v_D = V_Q$ and the current $i_D = I_Q$ are the actual diode voltage and current when the diode is connected as in the circuit of Figure 8.24. Note that this method is also useful for circuits containing a larger number of elements, where the diode is treated as the load and provided that Thévenin's theorem can be used to simplify the remaining source network.

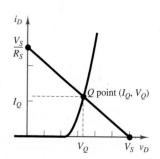

Figure 8.25 Graphical solution of equations 8.13 and 8.14

Piecewise Linear Diode Model

The graphical solution of diode circuits can be somewhat tedious, and its accuracy is limited by the resolution of the graph. However, it does provide insight into the **piecewise linear diode model** in which the diode is treated as an open-circuit in the "off" state and as a linear resistor in series with a battery of value V_γ in the "on" state. Figure 8.26 provides a graphical illustration of this model. The straight line that approximates the on state of the diode is chosen to be tangent to the operating point *Q*. Thus, in the neighborhood of the *Q* point, the diode in this model acts as a linear small-signal resistance, with slope given by $1/r_D$, where:

$$\frac{1}{r_D} = \frac{\partial i_D}{\partial v_D}\bigg|_{(I_Q, V_Q)} \qquad \text{Diode incremental resistance} \tag{8.13}$$

The diode offset voltage is defined as the intersection of the tangent line at *Q* with the voltage axis. Thus, rather than represent the diode as a short-circuit in its

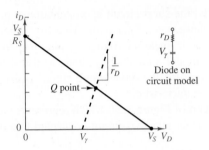

Figure 8.26 Piecewise linear diode model

forward-biased state, it can be treated as a linear resistor r_D. The piecewise linear model offers the convenience of a linear representation once the state of the diode is established, and of a more accurate model than either the ideal or the offset diode model. This model is very useful in illustrating the performance of diodes in real-world applications.

FOCUS ON PROBLEM SOLVING

DETERMINING THE OPERATING POINT OF A DIODE

1. Reduce the circuit to a Thévenin or Norton equivalent circuit with the diode as the load.

2. Determine the diode load line equation (equation 8.12).

3. Solve numerically two simultaneous equations in two unknowns (the load-line equations and the diode equation) for the diode current and voltage.

   or

4. Solve graphically by finding the intersection of the diode curve (e.g., from a data sheet) with the load-line curve. The intersection of the two curves is the diode operating point Q.

EXAMPLE 8.4   Using Load-Line Analysis and Diode Curves to Determine the Operating Point of a Diode

Problem

Determine the operating point of the 1N914 diode in the circuit of Figure 8.27, and compute the total power output of the 12-V battery.

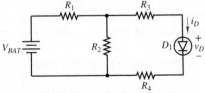

Figure 8.27 Circuit for Example 8.4.

Solution

Known Quantities: $V_{BAT} = 12$ V; $R_1 = 50\ \Omega$; $R_2 = 10\ \Omega$; $R_3 = 20\ \Omega$; $R_4 = 20\ \Omega$.

Find: The diode operating voltage and current and the power supplied by the battery.

Assumptions: Use the diode nonlinear model, as described by its i-v curve (Figure 8.28).

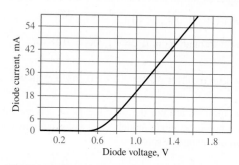

Figure 8.28 The 1N914 diode i-v curve

Analysis: Consider the diode in Figure 8.27 to be the load and everything else attached to it as its source network. Replace the source network with its Thévenin equivalent (Figure 8.29) and determine the load line as shown in Figure 8.30. The Thévenin equivalent resistance and the Thévenin (open-circuit) voltage seen by the diode are:

$$R_S = R_3 + R_4 + (R_1 \| R_2) = 20 + 20 + (10 \| 50) = 48.33\ \Omega$$

$$V_S = \frac{R_2}{R_1 + R_2} V_{BAT} = \frac{10}{60} 12 = 2\ \text{V}$$

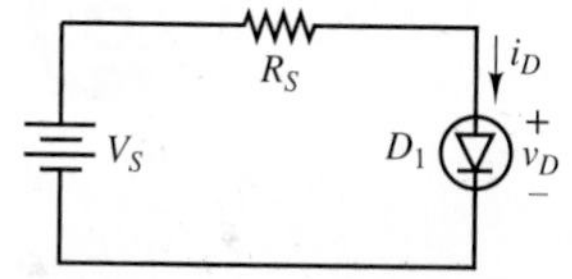

Figure 8.29 Equivalent circuit with diode as the load.

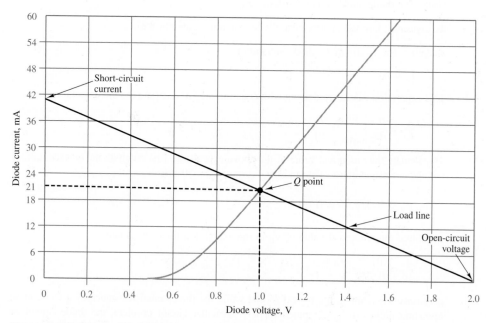

Figure 8.30 Superposition of load line and diode i-v characteristic

The short-circuit current is $V_S/R_S = 41\,\text{mA}$. The intersection of the diode curve and the load line is the *quiescent* or *operating point* Q of the diode, which is given by the values $V_Q = 1.0\,\text{V}$ and $I_Q = 21\,\text{mA}$.

To determine the battery power output, observe that the power supplied by the battery is $P_B = 12 \times I_B$ and that I_B is equal to the current through R_1. Upon further inspection, the battery current must, by KCL, be equal to the sum of the currents through R_2 and the diode. The current through the diode is I_Q. To determine the current through R_2, observe that the voltage across R_2 is equal to the sum of the voltages across R_3, R_4, and D_1:

$$V_{R2} = I_Q(R_3 + R_4) + V_Q = 0.021 \times 40 + 1 = 1.84\,\text{V}$$

and therefore the current through R_2 is $I_{R2} = V_{R2}/R_2 = 0.184\,\text{A}$.

Finally:

$$P_B = 12 \times I_B = 12 \times (0.021 + 0.184) = 12 \times 0.205 = 2.46\,\text{W}$$

Comments: Graphical solutions are not the only means of solving the nonlinear equations that result from using a nonlinear diode model. The same equations could be solved numerically by using a nonlinear equation solver.

EXAMPLE 8.5   Computing the Incremental (Small-Signal) Resistance of a Diode

Problem

Determine the incremental resistance of a diode, using the diode equation.

Solution

Known Quantities: $I_0 = 10^{-14}\,\text{A}$; $V_T = 25\,\text{mV}$ (at $T = 300\,\text{K}$); $I_Q = 50\,\text{mA}$.

Find: The diode small-signal resistance r_D.

Assumptions: Use the approximate diode equation (equation 8.8).

Analysis: The approximate diode equation is:

$$i_D = I_0 e^{v_D/V_T}$$

This expression can be used along with equation 8.13 to compute the incremental resistance:

$$\frac{1}{r_D} = \left.\frac{\partial i_D}{\partial v_D}\right|_{(I_Q, V_Q)} = \frac{I_0}{V_T} e^{v_Q/V_T}$$

To calculate the numerical value of the above expression, first compute the quiescent diode voltage corresponding to the quiescent current $I_Q = 50\,\text{mA}$:

$$V_Q = V_T \log_e \frac{I_Q}{I_0} = 0.731\,\text{V}$$

Substitute the numerical value of V_Q in the expression for r_D to obtain:

$$\frac{1}{r_D} = \frac{10^{-14}}{0.025} e^{0.731/0.025} = 2\,\text{S} \qquad \text{or} \qquad r_D = 0.5\,\Omega$$

Comments: It is important to realize that while the incremental resistance of a diode at an operating point can be computed for any particular circuit problem, the diode cannot be treated simply as a resistor. The small-signal resistance of the diode is used in the piecewise linear diode model to account for the fact that there is a dependence between diode voltage and current (i.e., the diode *i-v* characteristic is not a vertical line for voltages above the offset voltage—see Figure 8.26). It is also important to realize that the incremental resistance of a diode will change if the operating point changes since the incremental resistance is, after all, the slope of the *i-v* characteristic at the operating point.

EXAMPLE 8.6 Using the Piecewise Linear Diode Model

Problem

Determine the load voltage v_o in the *rectifier* circuit of Figure 8.31, using a piecewise linear approximation.

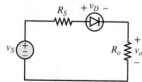

Figure 8.31 Circuit for Example 8.6.

Solution

Known Quantities: $v_S(t) = 10 \cos \omega t$; $V_\gamma = 0.6$ V; $r_D = 0.5\ \Omega$; $R_S = 1\ \Omega$; $R_o = 10\ \Omega$.

Find: The load voltage v_o.

Assumptions: Use the piecewise linear diode model (Figure 8.26).

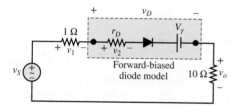

Figure 8.32 Piecewise linear model of forward-biased diode inserted in circuit of Figure 8.31

Analysis: Use KVL to determine the requirement for diode conduction. For the circuit in this example, KVL yields:

$$v_S = v_1 + v_D + v_o = v_1 + v_2 + V_\gamma + v_o \qquad \text{Conducting forward-biased diode}$$
$$v_S = v_D \qquad\qquad\qquad\qquad\qquad\qquad\qquad \text{Nonconducting diode}$$

Observe that when v_S is negative, the diode will be off; it will act as an open-circuit; the voltages v_1, v_2, and v_o will be zero; and $v_D = v_S$. At the onset of conduction the diode is forward-biased but the diode current is still zero. Under this condition v_1, v_2, and v_o are zero (Ohm's law) and the ideal diode forward voltage drop is zero (as always) such that $v_D = v_S = V_\gamma = 0.6$ V. Thus, the condition for conduction is:

$$v_D = v_S = V_\gamma = 0.6 \text{ V} \qquad \text{Onset of conduction}$$

Once the diode conducts, the difference between v_S and V_γ is divided among the three series resistors according to the voltage division rule. Thus:

$$v_o = \begin{cases} 0 & v_S < V_\gamma = 0.6 \text{ V} \\[2mm] \dfrac{R_o}{R_S + r_D + R_o}(v_S - V_\gamma) = 8.7 \cos \omega t - 0.52 & v_S \geq 0.6 \text{ V} \end{cases}$$

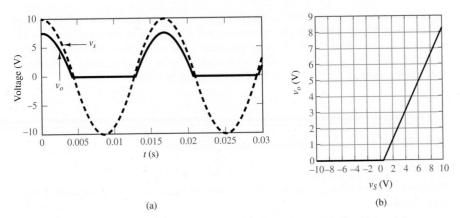

(a)   (b)

Figure 8.33 (a) Source voltage and rectified load voltage; (b) voltage transfer characteristic

The source and load voltage are plotted in Figure 8.33(a). The *transfer characteristic* of the circuit is shown as a plot of v_o versus v_S in Figure 8.33(b).

CHECK YOUR UNDERSTANDING

Use load-line analysis to determine the operating point Q of the diode circuit shown below. The diode has the *i-v* characteristic shown in Figure 8.30. Graph the load line using the short-circuit current V_S/R_S as the ordinate intercept and $-1/R_S$ as the slope of the load line.

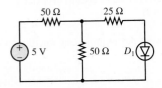

Answer: $V_o = 1.11$ V, $I_o = 27.7$ mA

CHECK YOUR UNDERSTANDING

Compute the incremental resistance of the diode of Example 8.5 if the current through the diode, I_Q, is 250 mA.

Answer: $r_D = 0.1$ Ω

CHECK YOUR UNDERSTANDING

Consider the *half-wave rectifier* circuit shown below where $v_i = 18 \cos t$ V and $R = 4\ \Omega$. Sketch the output voltage waveform if the piecewise linear diode model is used to represent the diode, with $V_\gamma = 0.6$ V and $r_D = 1\ \Omega$. What is the peak value of the output waveform?

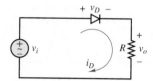

8.5   RECTIFIER CIRCUITS

The need for converting one form of electric energy to another arises frequently in practice. The most readily available form of electric power is alternating current, as generated and delivered by electric power utilities. However, DC power is frequently required for applications ranging from the control of electric motors to the operation of consumer electronic circuits, such as MP3 players, tablet computers, and smartphones. An important part of the process of converting an AC signal to direct current is *rectification*, which is the process of converting an electrical signal so that all its parts have the same sign. Of particular interest is the process of converting an AC signal (e.g., a typical 120 V rms line voltage) with zero average (DC) value to a signal with a nonzero DC value. For example, power supplies use rectification to produce a DC output from the readily available AC line voltage. The basic principle of rectification is well illustrated using ideal diodes, particularly when the magnitude of the AC voltage is large compared to the diode offset voltage V_γ.

This section introduces the following three types of rectifier circuits:

- Half-wave rectifier
- Full-wave rectifier
- Bridge rectifier

The Half-Wave Rectifier

Consider the circuit of Figure 8.34, where an AC source v_i is connected to an ideal diode and a resistive load in a series loop. The diode will conduct only when it is forward-biased ($v_D \geq 0$), which occurs during the positive half-cycle of the sinusoidal voltage. During that interval the ideal diode acts as a short-circuit such that $v_o = v_i$ and $i_D = v_i/R$. During the negative half-cycle of the sinusoid the ideal diode is reverse-biased ($v_D < 0$) and acts as an open-circuit. The loop current i_D is then zero, and, by Ohm's law, the output voltage v_o is also zero. The input voltage v_i and the resulting output voltage v_o are shown in Figure 8.35, where the

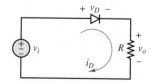

Figure 8.34 Ideal diode acting as a half-wave rectifier

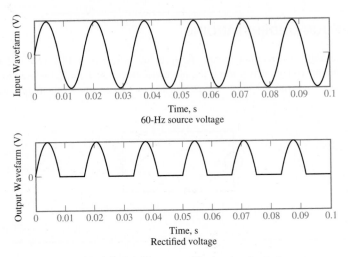

Figure 8.35 Ideal diode half-wave rectifier input and output

frequency is assumed to be $\omega = 2\pi f = 2\pi(60\,\text{Hz})$. Notice that although the input voltage has a zero average (DC) value, the rectified output voltage v_o has a non-zero average (DC) value, which is computed, in general, as:

$$(v_o)_{\text{avg}} = \frac{1}{T}\int_0^T v_o(t)\,dt = \frac{\omega}{2\pi}\int_0^{2\pi/\omega} v_o(t)\,dt \tag{8.14}$$

where T is the period of the output waveform. For example, assume $v_i = 120\sqrt{2}\,\sin(\omega t)$ V. Then:

$$(v_o)_{\text{avg}} = \frac{\omega}{2\pi}\left[\int_0^{\pi/\omega} 120\sqrt{2}\,\sin(\omega t)\,dt + \int_{\pi/\omega}^{2\pi/\omega} 0\,dt\right] \tag{8.15}$$

$$= \frac{120\sqrt{2}}{\pi} = 54.0\,\text{V}$$

The circuit of Figure 8.34 is known as a **half-wave rectifier**, because only the positive half of the input waveform appears across the output. This result is not particularly satisfying nor efficient since half of the input waveform is lost. Luckily, it is possible to do better using a *full-wave rectifier*.

The Full-Wave Rectifier

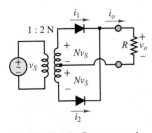

Figure 8.36 Center-tapped AC transformer and a full-wave rectifier with two ideal divides

The half-wave rectifier is not an efficient AC-DC converter because, by not conducting current during the negative half-cycle of an AC waveform, half of the available energy is not utilized. The **full-wave rectifier** shown in Figure 8.36 offers a substantial improvement. The first section of the full-wave rectifier circuit includes an AC source and a center-tapped transformer with 1:2N turns ratio. The purpose of the transformer is to step up ($N > 1$) or step down ($N < 1$) the primary voltage v_S prior to rectification. The voltage amplitude across each half of the secondary coil is Nv_S. In addition to scaling the source voltage, the transformer isolates the rectifier circuit from the AC source voltage since there is no direct electrical connection between the input and output of a transformer.

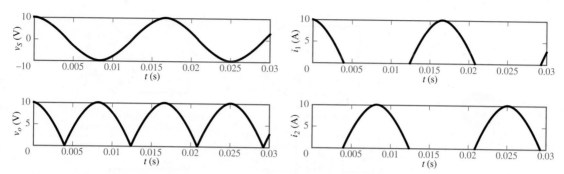

Figure 8.37 Full-wave rectifier current and voltage waveforms ($R = 1\,\Omega$)

In most applications, the amplitude of the secondary voltage (the input voltage to the rectifier) is much larger than the offset voltage of the diodes. When this condition is true, the diodes can be approximated as ideal without significantly compromising the result of the analysis. The key to the operation of the full-wave rectifier is to note that as the sign of v_S periodically alternates between positive and negative, the two diodes alternate in turns between forward- and reverse-biased states. For instance, during the positive half-cycle of v_S, the top diode is forward-biased while the bottom diode is reverse-biased. Alternately, during the negative half-cycle of v_S, the top diode is reverse-biased while the bottom diode is forward-biased. Therefore, the output current i_o satisfies the following two relations:

$$i_o = i_1 = N\frac{v_S}{R} \qquad v_S \geq 0 \tag{8.16}$$

$$i_o = i_2 = -N\frac{v_S}{R} \qquad v_S \leq 0 \tag{8.17}$$

Wow! The direction of i_o does not alternate! It is always positive as shown.

The source voltage, the output voltage, and the currents i_1 and i_2 are shown in Figure 8.37 for a load resistance $R = 1\,\Omega$ and $N = 1$. Notice that the output voltage is exactly the superposition of the output of two half-wave rectifiers $180°$ out of phase. Thus, the DC output of the full-wave rectifier should be twice that of the half-wave rectifier. This observation can be confirmed by computing the DC value of the full-wave rectifier output.

$$\begin{aligned}(v_o)_{\text{avg}} &= \frac{1}{T}\int_0^T v_o(t)\,dt = \frac{\omega}{2\pi}\int_0^{2\pi/\omega} v_o(t)\,dt \\[2mm] &= \frac{\omega}{2\pi}\left[\int_0^{\pi/\omega} |v_o(t)|\,dt + \int_{\pi/\omega}^{2\pi/\omega} |v_o(t)|\,dt\right] \\[2mm] &= 2\frac{\omega}{2\pi}\left[\int_0^{\pi/\omega} |v_o(t)|\,dt\right]\end{aligned} \tag{8.18}$$

Keep in mind that this result is approximate because the impact of the diode offset voltage was ignored by assuming ideal diodes. When the offset voltage is included, there will be periods typically brief when both diodes are reverse-biased and the output voltage is zero. The net effect is to reduce the output waveform shown in Figure 8.37 by V_γ. However, those portions of the adjusted waveform that would otherwise be negative (between 0 and $-V_\gamma$) are, in fact, zero because both diodes are reverse-biased for the brief periods when $-V_\gamma < v_S < V_\gamma$.

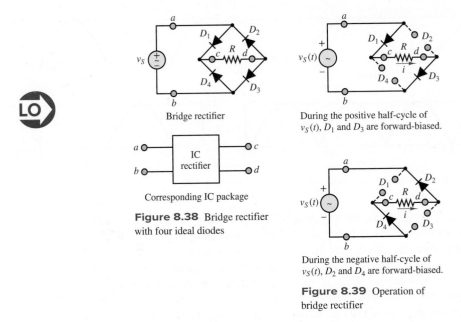

Bridge rectifier

During the positive half-cycle of
$v_S(t)$, D_1 and D_3 are forward-biased.

Corresponding IC package

Figure 8.38 Bridge rectifier
with four ideal diodes

During the negative half-cycle of
$v_S(t)$, D_2 and D_4 are forward-biased.

Figure 8.39 Operation of
bridge rectifier

The Bridge Rectifier

Another rectifier circuit commonly available "off the shelf" as a single *integrated circuit* is the *bridge rectifier*, which employs four diodes in the bridge configuration shown in Figure 8.38.

The analysis of the bridge rectifier is best understood by observing that as the sign of v_S periodically alternates between positive and negative, *pairs* of the four bridge diodes alternate in turns between forward- and reverse-biased states, as shown in Figure 8.39. During the positive half-cycle of v_S, diodes D_1 and D_3 are forward-biased while diodes D_2 and D_4 are reverse-biased. Alternately, during the negative half-cycle of v_S, diodes D_1 and D_3 are reverse-biased while diodes D_2 and D_4 are forward-biased. It is important to note that the current i through R is directed from node c to node d during both half-cycles.

The input and rectified output waveforms are shown in Figure 8.40(a) and (b) for the case of ideal diodes and a 30-V peak AC source input. If each diode is assumed to have an offset voltage $V_\gamma = 0.6$ V, the effect is to reduce the output waveform by $2V_\gamma = 1.2$ V, as shown in Figure 8.40(c). The $2V_\gamma$ reduction occurs during both half-cycles. During the positive half-cycle of v_S, the path from node a to node b contains two forward-biased diodes D_1 and D_3. Alternately, during the negative half-cycle of v_S, the path from node b to node a also contains two forward-biased diodes D_2 and D_4. Each of these forward-biased diodes requires a "toll" of V_γ.

As with the full-wave rectifier, no portion of the rectified output waveform is negative even when reduced by $2V_\gamma$. Instead, during those periods when $-2V_\gamma < v_S < 2V_\gamma$, all four diodes are reverse-biased and the rectified output waveform is zero.

In most practical applications of rectifier circuits, the signal waveform to be rectified is the 60-Hz, 110 V rms line voltage. As shown in Figures 8.37 and 8.40, the fundamental frequency of the rectified output waveform is twice that of the

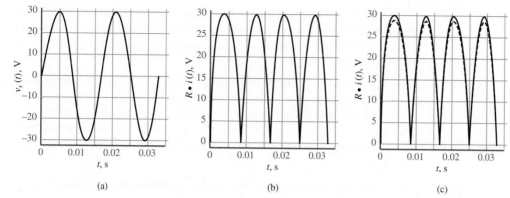

Figure 8.40 Dashed line represents diode offset model, solid line diode ideal model. (a) Unrectified source voltage; (b) rectified load voltage (ideal diodes); (c) rectified load voltage (ideal and offset diodes).

input waveform. Thus, for a 60-Hz input waveform, the fundamental ripple frequency is 120 Hz or 754 rad/s. A low-pass filter is required such that:

$$\omega_0 \ll \omega_{\text{ripple}} \tag{8.19}$$

Figure 8.41 shows the resulting waveforms.

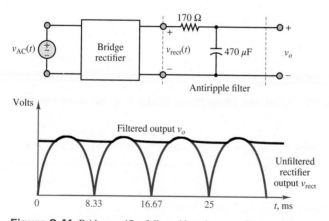

Figure 8.41 Bridge rectifier followed by a low-pass filter, and the resulting waveforms

DC Power Supplies

The rectification of an AC input waveform is just one of four fundamental steps needed to convert an AC input to a practical DC output. In a typical **DC power supply** these steps are, in order:

Step 1: Scale (step up or step down) the amplitude of the AC input waveform. This step is commonly accomplished by a transformer although high-frequency switch-mode circuits can also provide scaling of a DC output.

Step 2: Rectify the scaled AC input waveform. This step may be accomplished by a full-wave or bridge rectifier. Rectification can also be accomplished by more exotic devices, such as gate turnoff thyristors (GTOs) and insulated-gate bipolar transistors (IGBTs).

Step 3: Filter the rectified output waveform to remove remaining AC components known as *ripple*. This step can be accomplished by an *RC* low-pass (antiripple) filter in a simple DC power supply, as shown in Figure 8.41, or by more sophisticated active low-pass filters.

Step 4: Regulate the filtered DC output voltage to maintain the desired DC value for a large range of loads. The Zener diode provides a very inexpensive and simple form of voltage regulation. Linear voltage regulators, which have very good noise characteristics, and switched-mode regulators, which have very high energy efficiency, are available as integrated circuits (e.g., the 78xx linear series).

These steps are represented in the generic depiction of a DC power supply shown in Figure 8.42.

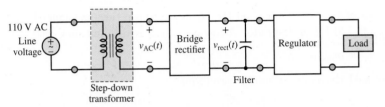

Figure 8.42 DC power supply

EXAMPLE 8.7   Using the Offset Diode Model in a Half-Wave Rectifier

Problem

Compute and plot the rectified load voltage v_R in the circuit of Figure 8.43.

Solution

Known Quantities: $v_S(t) = 3 \cos \omega t$; $V_\gamma = 0.6 \, \text{V}$.

Find: An analytical expression for the load voltage.

Assumptions: Use the offset diode model.

Analysis: Replace the diode with the offset diode model, as shown in the lower half of Figure 8.43, and use the method developed earlier for ideal diode problems.

First, assume that the diode is reverse-biased and replace it with an open-circuit, as shown in Figure 8.44(a). Since the current through R is zero, the diode voltage v_D is found from KVL to be:

$$v_D = v_S \quad \text{when } v_S < V_\gamma \quad \text{Reverse-bias condition}$$

When the source voltage is greater than $V_\gamma = 0.6 \, \text{V}$, the diode is forward-biased such that it behaves as a short-circuit in series with a small offset voltage drop, as shown in Figure 8.44(b). The loop current i and the voltage v_R across R are given by:

$$i = \frac{v_S - V_\gamma}{R} \qquad v_R = iR = v_S - V_\gamma$$

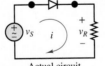

Actual circuit

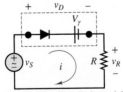

Circuit with offset diode model

Figure 8.43 Circuit for Example 8.7.

Thus, the half-wave rectifier circuit behavior is summarized by:

$$v_R = \begin{cases} 0 & v_S < 0.6 \text{ V} \\ v_S - 0.6 & v_S \geq 0.6 \text{ V} \end{cases}$$

The resulting rectified waveform $v_R(t)$ is plotted along with $v_S(t)$ in Figure 8.45. The effect of the offset voltage is to lower the positive portion of the rectified waveform by V_γ. The period T^+ during which the rectified waveform is positive is slightly shorter than half the period T of the input waveform. For ideal diodes, the maximum amplitude of the rectified waveform equals the amplitude of the input waveform and $T^+ = T/2$.

Comments: The rectified waveform is shifted downward by an amount equal to the offset voltage V_γ. The shift is visible in the case of this example because V_γ is a substantial fraction of the source voltage. If the source voltage had peak values of tens or hundreds of volts, such a shift would be negligible, and an ideal diode model would serve just as well.

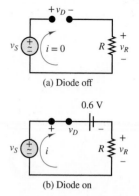

(a) Diode off

(b) Diode on

Figure 8.44 Circuit for Example 8.7 with diode "off" and "on".

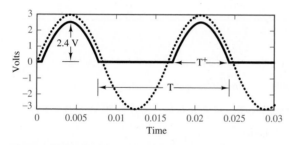

Figure 8.45 Source waveform (···) and rectified waveform (—) for the circuit of Figure 8.43

EXAMPLE 8.8   Half-Wave Rectifiers

Problem

A half-wave rectifier, similar to that in Figure 8.34, is used to provide a DC supply to a 50-Ω load. If the AC source voltage is 20 V rms, find the peak and average current in the load. Assume an ideal diode.

Solution

Known Quantities: Value of circuit elements and source voltage.

Find: Peak and average values of load current in half-wave rectifier circuit.

Schematics, Diagrams, Circuits, and Given Data: $v_S = 20$ V rms, $R = 50\ \Omega$.

Assumptions: Ideal diode.

Analysis: According to the ideal diode model, the peak load voltage is equal to the peak sinusoidal source voltage. Thus, the peak load current is

$$i_{\text{peak}} = \frac{v_{\text{peak}}}{R} = \frac{\sqrt{2}\, v_{\text{rms}}}{R} = 0.567 \text{ A}$$

To compute the average current, integrate the half-wave rectified sinusoid:

$$\langle i \rangle = \frac{1}{T}\int_0^T i(t)\,dt = \frac{1}{T}\left[\int_0^{T/2} \frac{v_{\text{peak}}}{R}\sin(\omega t)\,dt + \int_{T/2}^T 0\,dt\right]$$

$$= \frac{v_{\text{peak}}}{\pi R} = \frac{\sqrt{2}\,v_{\text{rms}}}{\pi R} = 0.18\ \text{A}$$

EXAMPLE 8.9   Bridge Rectifier

Problem

A bridge rectifier, similar to that in Figure 8.38, is used to produce a 50-V, 5-A DC supply. What is the resistance of the load R that will result in a 5-A DC output current? What is the required source voltage v_S (in V rms) to achieve the desired DC output voltage? Assume ideal diodes.

Solution

Known Quantities: Value of circuit elements and source voltage.

Find: Source voltage v_S (in V rms) and the load resistance R.

Schematics, Diagrams, Circuits, and Given Data: $\langle v_o \rangle = 50\ \text{V}$; $\langle i_o \rangle = 5\ \text{A}$.

Assumptions: Ideal diodes.

Analysis: The load resistance that will result in an average direct current of 5 A is:

$$R = \frac{\langle v_o \rangle}{\langle i_o \rangle} = \frac{50}{5} = 10\ \Omega$$

which is the lowest value of R for which the DC supply will be able to provide the required current. To compute the required source voltage, we observe that the average load voltage can be found from the expression

$$\langle v_o \rangle = R\langle i_o \rangle = \frac{R}{T}\int_0^T i(t)\,dt = \frac{R}{T}\left[\int_0^{T/2} \frac{v_{\text{peak}}}{R}\sin(\omega t)\,dt\right] \times 2$$

$$= \frac{2v_{\text{peak}}}{\pi} = \frac{2\sqrt{2}\,v_{\text{rms}}}{\pi} = 50\ \text{V}$$

Hence:

$$v_{\text{rms}} = \frac{50\pi}{2\sqrt{2}} = 55.5\ \text{V}$$

CHECK YOUR UNDERSTANDING

Compute the DC value of the rectified waveform for the circuit of Figure 8.34 for $v_i = 52\cos\omega t$ V.

CHECK YOUR UNDERSTANDING

In Example 8.8, what is the peak current if an offset diode model is used with $V_\gamma = 0.6\,\text{V}$?

Answer: 0.554 A

CHECK YOUR UNDERSTANDING

Show that the DC output voltage of the full-wave rectifier of Figure 8.36 is $2Nv_{\text{Speak}}/\pi$.

CHECK YOUR UNDERSTANDING

Compute the peak voltage output of the bridge rectifier of Figure 8.38, assuming diodes with 0.6-V offset voltage and a 110 V rms AC supply.

Answer: 154.36 V

Conclusion

This chapter introduces the topic of electronic devices by presenting the semiconductor diode. Upon completing this chapter, you should have mastered the following learning objectives:

1. *Understand the basic principles underlying the physics of semiconductor devices in general and of the pn junction in particular. Become familiar with the diode equation and i-v characteristic.* Semiconductors have conductive properties that fall between those of conductors and insulators. These properties make the materials useful in the construction of many electronic devices that exhibit nonlinear *i-v* characteristics. Of these devices, the diode is one of the most commonly employed.

2. *Use various circuit models of the semiconductor diode in simple circuits. These are divided into two classes: the large-signal models useful to study rectifier circuits and the small-signal models useful in signal processing applications.* The semiconductor diode acts as a one-way current valve, permitting the flow of current only when it is biased in the forward direction. The behavior of the diode is described by an exponential equation, but it is possible to approximate the operation of the diode by means of simple circuit models. The simplest (ideal) model treats the diode either as a short-circuit (when it is forward-biased) or as an open-circuit (when it is reverse-biased). The ideal model can be extended to include an offset voltage, which represents the contact potential at the diode *pn* junction. A further model, useful for small-signal circuits, includes a resistance that models the forward resistance of the diode. With the aid of these models it is possible to analyze diode circuits by using the DC and AC circuit analysis methods of earlier chapters.

3. *Study practical full-wave rectifier circuits, and learn to analyze and determine the practical specifications of a rectifier by using large-signal diode models.* One of the most important properties of the diode is its ability to rectify AC voltages and currents. Diode rectifiers can be of the half-wave and full-wave types. Full-wave rectifiers can be constructed in a two-diode configuration or in a four-diode bridge configuration. Diode rectification is an essential element of DC power supplies.

Another important part of a DC power supply is the filtering, or smoothing, that is usually accomplished by using capacitors.

4. *Understand the basic operation of Zener diodes as voltage references, and use simple circuit models to analyze elementary voltage regulators.* In addition to rectification and filtering, the power supply requires output voltage regulation. Zener diodes can be used to provide a voltage reference that is useful in voltage regulators.

5. *Use the diode models presented in Section 8.2 to analyze the operation of various practical diode circuits in signal processing applications.* In addition to power supply applications, diodes find use in many signal processing and signal conditioning circuits. Of these, the diode peak detector, the diode limiter, and the diode clamp are explored in this chapter.

6. *Understand the basic principle of operation of photodiodes, including solar cells, photosensors, and light-emitting diodes.* Semiconductor material properties can also be affected by light intensity. Certain types of diodes, known as *photodiodes*, find applications in light detectors, solar cells, or light-emitting diodes.

HOMEWORK PROBLEMS

Section 8.1: Electrical Conduction in Semiconductor Devices;

Section 8.2: The *pn* Junction and the Semiconductor Diode

8.1 In a semiconductor material, the net charge is zero. This requires the density of positive charges to be equal to the density of negative charges. Both charge carriers (free electrons and holes) and ionized dopant atoms have a charge equal to the magnitude of one electronic charge. Therefore the charge neutrality equation (CNE) is:

$$p_o + N_d^+ - n_o - N_a^- = 0$$

where

n_o = equilibrium negative carrier density

p_o = equilibrium positive carrier density

N_a^- = ionized acceptor density

N_d^+ = ionized donor density

The carrier product equation (CPE) states that as a semiconductor is doped, the product of the charge carrier densities remains constant:

$$n_o p_o = \text{const}$$

For intrinsic silicon at $T = 300$ K:

$$\text{Const} = n_{io} p_{io} = n_{io}^2 = p_{io}^2$$

$$= \left(1.5 \times 10^{16} \frac{1}{\text{m}^3}\right)^2 = 2.25 \times 10^{32} \frac{1}{\text{m}^2}$$

The semiconductor material is *n*- or *p*-type depending on whether donor or acceptor doping is greater. Almost all dopant atoms are ionized at room temperature. If intrinsic silicon is doped:

$$N_A \approx N_a^- = 10^{17} \frac{1}{\text{m}^3} \qquad N_d = 0$$

Determine:

a. If this is an *n*- or *p*-type extrinsic semiconductor.

b. Which are the major and which are the minority charge carriers.

c. The density of majority and minority carriers.

8.2 If intrinsic silicon is doped, then

$$N_a \approx N_a^- = 10^{17} \frac{1}{\text{m}^3} \qquad N_d \approx N_d^+ = 5 \times 10^{18} \frac{1}{\text{m}^3}$$

Determine:

a. If this is an *n*- or *p*-type extrinsic semiconductor.

b. Which are the majority and which are the minority charge carriers.

c. The density of majority and minority carriers.

8.3 Describe the microscopic structure of semiconductor materials. What are the three most commonly used semiconductor materials?

8.4 Describe the thermal production of charge carriers in a semiconductor and how this process limits the operation of a semiconductor device.

8.5   Describe the properties of donor and acceptor dopant atoms and how they affect the densities of charge carriers in a semiconductor material.

8.6   Physically describe the behavior of the charge carriers and ionized dopant atoms in the vicinity of a semiconductor *pn* junction that causes the potential (energy) barrier that tends to prevent charge carriers from crossing the junction.

Section 8.3–8.4: Circuit Models for the Semiconductor Diode

8.7   Consider the circuit of Figure P8.7. Determine whether the diode is conducting. Assume $V_A = 12$ V, $V_B = 10$ V, and that the diode is ideal.

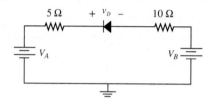

Figure P8.7

8.8   Repeat Problem 8.7 for $V_A = 12$ V and $V_B = 15$ V.

8.9   Consider the circuit of Figure P8.9. Determine whether the diode is conducting. Assume $V_A = 12$ V, $V_B = 10$ V, $V_C = 5$ V and that the diode is ideal.

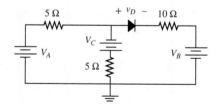

Figure P8.9

8.10   Repeat Problem 8.9 for $V_B = 15$ V.

8.11   Repeat Problem 8.9 for $V_C = 15$ V.

8.12   Repeat Problem 8.9 for $V_B = 15$ V and $V_C = 10$ V.

8.13   For the circuit of Figure P8.13, sketch $i_D(t)$ using:

a. The ideal diode model.

b. An ideal diode model with offset ($V_\gamma = 0.6$ V).

c. The piecewise linear approximation diode model with $r_D = 1$ kΩ and $V_\gamma = 0.6$ V.

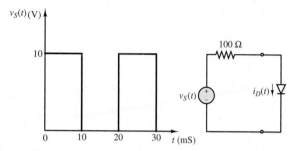

Figure P8.13

8.14   For the circuit of Figure P8.14, find the range of V_{in} for which D_1 is forward-biased. Assume an ideal diode.

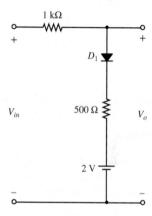

Figure P8.14

8.15   One of the more interesting applications of a diode, based on the diode equation, is an electronic thermometer. The concept is based on the empirical observation that if the current through a diode is nearly constant, the offset voltage is nearly a linear function of the diode temperature, as shown in Figure P8.15(a).

a. Show that i_D in the circuit of Figure P8.15(b) is nearly constant in the face of variations in the diode voltage v_D. To do so, compute the percent change in i_D for a given percent change in v_D. Assume that v_D changes by 10 percent from 0.6 V to 0.66 V.

b. On the basis of the graph of Figure P8.15(a), write an equation for $v_D(T°)$ of the form

$$v_D = \alpha T° + \beta$$

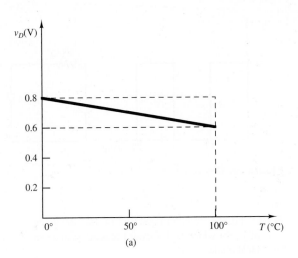

(a)

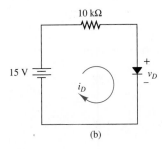

(b)

Figure P8.15

8.16 Find expressions for the voltage v_o in Figure P8.16, where D is an ideal diode, for positive and negative values of v_S. Sketch a plot of v_o versus v_S.

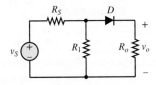

Figure P8.16

8.17 Repeat Problem 8.16, using the offset diode model.

8.18 Find the power dissipated in diode D, and the power dissipated in R in Figure P8.18. Use the exponential diode equation and assume $R = 2 \text{ k}\Omega$, $V_S = 5 \text{ V}$, $V_D = 900 \text{ mV}$, $q/KT = \frac{1}{52} \text{ mV}$, and $I_0 = 15 \text{ nA}$.

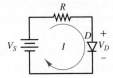

Figure P8.18

8.19 Determine the Thévenin equivalent network seen by diode D in Figure P8.19, and use it to determine the diode current i_D. Also, solve for the currents i_1 and i_2. Assume $R_1 = 5 \text{ k}\Omega$, $R_2 = 3 \text{ k}\Omega$, $V_{cc} = 10 \text{ V}$, and $V_{dd} = 15 \text{ V}$.

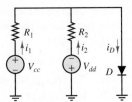

Figure P8.19

8.20 In Figure P8.20, assume a sinusoidal source $V_S = 50 \text{ V rms}$, $R = 170 \text{ }\Omega$, and $V_\gamma = 0.6 \text{ V}$. Use the offset diode model for a silicon diode to determine:

a. The maximum forward current.

b. The peak reverse voltage across the diode.

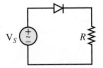

Figure P8.20

8.21 Determine voltages V_o assuming the diodes are ideal in each of the configurations shown in Figure P8.21.

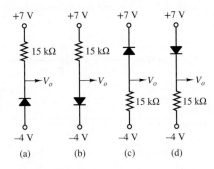

Figure P8.21

8.22 In the circuit of Figure P8.22, find the range of V_{in} for which D_1 is forward-biased. Assume ideal diodes.

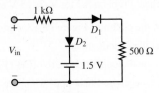

Figure P8.22

8.23 Determine which diodes are forward-biased and which are reverse-biased in the configurations shown in Figure P8.23. Assuming a 0.7-V drop across each forward-biased diode, determine v_{out}.

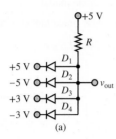

(a)

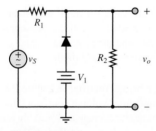

(b)   (c)

Figure P8.23

8.24 Sketch the output waveform and the voltage transfer characteristic for the circuit of Figure P8.24. Assume an ideal diode, $v_S(t) = 8 \sin(\pi t)$, $V_1 = 3$ V, $R_1 = 8$ Ω, and $R_2 = 5$ Ω.

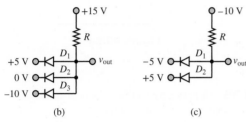

Figure P8.24

8.25 Repeat Problem 8.24, using the offset diode model with $V_\gamma = 0.55$ V.

8.26 Repeat Problem 8.24 for $v_S(t) = 1.5 \sin(2,000\pi t)$, $V_1 = 1$ V, and $R_1 = R_2 = 1$ kΩ. Use the piecewise linear model with $r_D = 200$ Ω.

8.27 The silicon diode shown in Figure P8.27 is described by:

$$i_D = I_o(e^{v_D/V_T} - 1)$$

where at $T = 300$ K

$$I_o = 250 \times 10^{-12} \text{ A} \qquad V_T = \frac{kT}{q} \approx 26 \text{ mV}$$

$$v_S = 4.2 \text{ V} + 110 \cos(\omega t) \text{ mV}$$

$$\omega = 377 \text{ rad/s} \qquad R = 7 \text{ k}\Omega$$

Determine the current i_D at the operating point Q:

a. Using the diode offset model.

b. By graphically solving the circuit characteristic (i.e., the DC load-line equation) and the device characteristic (i.e., the diode equation).

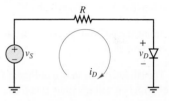

Figure P8.27

8.28 Repeat Problem 8.27 using the following data:

$$i_D = I_o(e^{v_D/V_T} - 1)$$

where at $T = 300$ K

$$I_o = 2.030 \times 10^{-15} \text{ A} \qquad V_T = \frac{kT}{q} \approx 26 \text{ mV}$$

$$v_S = 5.3 \text{ V} + 7 \cos(\omega t) \text{ mV}$$

$$\omega = 377 \text{ rad/s} \qquad R = 4.6 \text{ k}\Omega$$

8.29 A diode with the i-v characteristic shown in Figure 8.8 is connected in series with a 5-V voltage source (in the forward-bias direction) and a load resistance of 200 Ω. Determine:

a. The load current and voltage.

b. The power dissipated by the diode.

c. The load current and voltage if the load is changed to 100 Ω and 500 Ω.

8.30 A diode with the i-v characteristic shown in Figure 8.28 is connected in series with a 2-V source (in the forward-bias direction) and a 200-Ω load resistance. Determine:

a. The load current and voltage.

b. The power dissipated by the diode.

c. The load current and voltage if the load is changed to 100 Ω and 300 Ω.

8.31   The silicon diode shown in Figure P8.27 is described by:

$$i_D = I_o(e^{v_D/V_T} - 1)$$

where at $T = 300$ K

$$I_o = 250 \times 10^{-12} \text{ A} \qquad V_T = \frac{kT}{q} \approx 26 \text{ mV}$$

$$v_S = V_S + v_s = 4.2 \text{ V} + 110 \cos(\omega t) \text{ mV}$$

$$\omega = 377 \text{ rad/s} \qquad R = 7 \text{ k}\Omega$$

The DC operating (quiescent) point Q and the AC small-signal equivalent resistance at Q are:

$$I_{DQ} = 0.548 \text{ mA} \qquad V_{DQ} = 0.365 \text{ V} \qquad r_d = 47.45 \text{ }\Omega$$

Determine the AC voltage across the diode and the AC current through it.

8.32   The silicon diode shown in Figure P8.32 is in series with two voltage sources and a resistor, where:

$$R = 2.2 \text{ k}\Omega \qquad V_{S2} = 3 \text{ V} \qquad V_r = 0.7 \text{ V}$$

Determine the minimum value of V_{S1} at which the diode will be forward-biased and conduct charge.

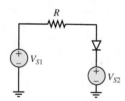

Figure P8.32

Section 8.5: Rectifier Circuits

8.33   Find the average value of the output voltage v_o shown in Figure P8.33. Assume $v_{in} = 10 \sin(\omega t)$ V, $C = 80$ nF, and $V_\gamma = 0.5$ V.

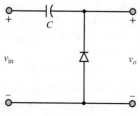

Figure P8.33

8.34   The circuit of Figure P8.34 is driven by a sinusoidal source $v_S(t) = 6 \sin(314t)$ V. Determine the average and peak diode currents, using:

a. The ideal diode model.

b. The offset diode model.

c. The piecewise linear model with resistance r_D.

Assume $R_o = 200 \text{ }\Omega$, $r_D = 25 \text{ }\Omega$, and $V_\gamma = 0.8$ V.

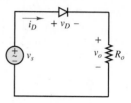

Figure P8.34

8.35   A half-wave rectifier produces an average voltage of 50 V at its output.

a. Draw a schematic diagram of the circuit.

b. Sketch the output voltage waveform.

c. Determine the peak value of the output voltage.

d. Sketch the input voltage waveform.

e. What is the rms voltage at the input?

8.36   A half-wave rectifier is used to provide a DC supply to a 80-Ω load. If the AC source voltage is 32 V rms, find the peak and average current in the load. Assume an ideal diode.

8.37   The bridge rectifier in Figure P.8.37 is driven by a sinusoidal voltage source $v_s(t) = 6 \sin(314t)$ V. Determine the average and peak forward current through each diode and $R_o = 200 \text{ }\Omega$. Assume ideal diodes.

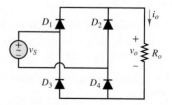

Figure P8.37

8.38   In the full-wave power supply shown in Figure P8.38 the silicon diodes are 1N4001 with a rated peak reverse voltage of 25 V.

$$n = 0.05883$$
$$C = 80\,\mu\text{F} \qquad R_o = 1\,\text{k}\Omega$$
$$v_{\text{line}} = 170\cos(377t)\ \text{V}$$

a. Determine the actual peak reverse voltage across each diode.

b. Explain why these diodes are or are not suitable for the specifications given.

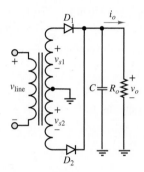

Figure P8.38

8.39   In the full-wave power supply shown in Figure P8.38,

$$n = 0.1$$
$$C = 80\,\mu\text{F} \qquad R_o = 1\,\text{k}\Omega$$
$$v_{\text{line}} = 170\cos(377t)\ \text{V}$$

The silicon diodes are 1N914 switching diodes (but used here for AC-DC conversion) with the following performance ratings:

$$P_{\text{max}} = 500\,\text{mW} \qquad \text{at } T = 25°\text{C}$$
$$V_{\text{pk-rev}} = 30\,\text{V}$$

The derating factor is 3 mW/°C for $25°\text{C} < T \le 125°\text{C}$ and 4 mW/°C for $125°\text{C} < T \le 175°\text{C}$.

a. Determine the actual peak reverse voltage across each diode.

b. Are these diodes suitable for the specifications given? Explain.

8.40   Refer to Problem 8.38 and assume a load voltage waveform as shown in Figure P8.40. Also assume:

$$|i_o|_{avg} = 60\,\text{mA} \qquad |v_o|_{avg} = 5\,\text{V} \qquad |V_{\text{ripple}}| = 5\%$$
$$v_{\text{line}} = 170\cos(\omega t)\ \text{V} \qquad \omega = 377\,\text{rad/s}$$

Determine:

a. The turns ratio n.

b. The capacitor C.

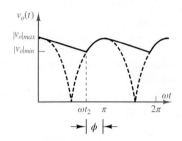

Figure P8.40

8.41   Refer to Problem 8.38. Assume:

$$|i_o|_{avg} = 600\,\text{mA} \qquad |v_o|_{avg} = 50\,\text{V}$$
$$V_r = 8\% = 4\,\text{V}$$
$$v_{\text{line}} = 170\cos(\omega t)\ \text{V} \qquad \omega = 377\,\text{rad/s}$$

Determine:

a. The turns ratio n.

b. The capacitor C.

8.42   Repeat Problem 8.37, using the diode offset model with $V_\gamma = 0.8$ V.

8.43   You have been asked to design a bridge rectifier for a power supply. A step-down transformer has already been chosen. It will supply 12 V rms to your rectifier. The bridge rectifier is shown in Figure P8.43.

a. If the diodes have an offset voltage of 0.6 V, sketch the input source voltage $v_S(t)$ and the output voltage $v_o(t)$, and state which diodes are on and which are off in the appropriate cycles of $v_S(t)$. The frequency of the source is 60 Hz.

b. If $R_o = 1{,}000\,\Omega$ and a filtering capacitor has a value of 8 μF, sketch the output voltage $v_o(t)$.

c. Repeat part b, with the capacitance equal to 100 μF.

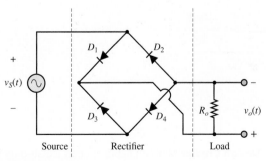

Figure P8.43

8.44 In the bridge rectifier of the power supply shown in Figure P8.44 the silicon diodes are 1N4001 with a rated peak reverse voltage of 50 V.

$$v_{line} = 170 \cos(377t) \text{ V}$$
$$n = 0.2941$$
$$C = 700 \, \mu\text{F} \qquad R_o = 2.5 \text{ k}\Omega$$

a. Determine the actual peak reverse voltage across each diode.

b. Are these diodes suitable for the specifications given? Explain.

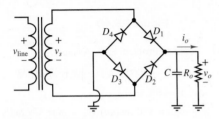

Figure P8.44

8.45 Refer to Problem 8.44. Assume the diodes have a rated peak reverse voltage of 10 V and:

$$v_{line} = 156 \cos(377t) \text{ V}$$
$$n = 0.04231 \qquad V_r = 0.2 \text{ V}$$
$$|i_o|_{avg} = 2.5 \text{ mA} \qquad |v_o|_{avg} = 5.1 \text{ V}$$

a. Determine the actual peak reverse voltage across the diodes.

b. Explain why these diodes are or are not suitable for the specifications given.

8.46 Refer to Problem 8.44. Assume:

$$|i_o|_{avg} = 650 \text{ mA} \qquad |v_o|_{avg} = 10 \text{ V}$$
$$V_r = 1 \text{ V} \qquad \omega = 377 \text{ rad/s}$$
$$v_{line} = 170 \cos(\omega t) \text{ V} \qquad \phi = 23.66°$$

Determine the value of the average and peak current through each diode.

8.47 Repeat Problem 8.37, using the piecewise linear diode model with $V_\gamma = 0.8$ V and resistance $R_D = 25 \, \Omega$.

8.48 Refer to Problem 8.44. Assume:

$$|i_o|_{avg} = 250 \text{ mA} \qquad |v_o|_{avg} = 10 \text{ V}$$
$$V_r = 2.4 \text{ V} \qquad \omega = 377 \text{ rad/s}$$
$$V_{line} = 156 \cos(\omega t) \text{ V}$$

Determine:

a. The turns ratio n.

b. The capacitor C.

CHAPTER

9

BIPOLAR JUNCTION TRANSISTORS: OPERATION, CIRCUIT MODELS, AND APPLICATIONS

Over the last half-century, transistor technology has revolutionized the manner in which power and information are transmitted and utilized within our society. The impact of this technology is difficult to overstate, and examples of it are ubiquitous. Moreover, the technology and the products that depend upon it continue to develop at an exponential rate. It is astounding to consider that the first Macintosh personal computer was introduced by Apple Computer Co. in January 1984 with 64 kB of ROM, 128 kB of RAM, a motherboard running at 8 MHz, a display with 384×256 *pixel* resolution, all for the modest price of $2,495, which is roughly equivalent to $5,500 in 2012. In the same year, IBM released its second-generation AT (advanced technology) personal computer, which featured the 16-bit, 6-MHz Intel 80286 microprocessor, a 20-MB hard drive. Less than 30 years later, the specifications of a modest desktop computer include a 64-bit, 3.0-GHz processor, a 1.3-GHz data bus with 6 GB of RAM, and a monitor resolution of 1600×900.

Of course, advances in analog and digital technology have not been limited to personal computers. In general, communication systems of all kinds have been revolutionized. Until 1983, interpersonal telecommunications were limited to land-line phone calls. The only asynchronous form of telecommunication was provided by analog telephone tape recorders and by letter and package carriers such as the U.S. Postal Service, UPS, and FedEx. While these services continue to play an important role in our society, new forms of communication, particularly real-time asynchronous communications, have exploded. On a daily, if not hourly, basis we now transmit, exchange, and broadcast digital images, video, text, and voice using handheld mobile devices. It is not unreasonable to describe these "smart" phones as pocket-sized supercomputers. According to the Pew Research Center's Internet & American Life Project, as of May 2011, roughly 35 percent of American adults owned a smartphone of one type or another. Today that number has risen to 81 percent.

Fundamentally, all this progress has relied on advances in transistor technology. Given the broad impact of this technology, it would seem essential that engineers of all stripes possess a basic understanding of transistors and how they are used to produce the two building blocks of all communication and power devices. These two building blocks are the **switch** and the **amplifier**. Chapters 9 and 10 are dedicated to revealing how transistors are utilized to produce various types of switches and amplifiers. Chapter 9 focuses on a family of transistors known as **bipolar junction transistors** (BJTs). The underlying physics is discussed in sufficient detail to provide a comfortable basis for understanding the three modes of BJT operation. Practical examples are provided to illustrate important BJT circuits and their analysis using linear circuit models.

Learning Objectives

Students will learn to...

1. Understand the basic principles of amplification and switching. *Section 9.1.*
2. Understand the physical operation of bipolar transistors; determine the operating point of a bipolar transistor circuit. *Section 9.2.*
3. Understand the large-signal model of the bipolar transistor, and apply it to simple amplifier circuits. *Section 9.3.*
4. Select the operating point of a bipolar transistor circuit; understand the principle of small-signal amplifiers. *Section 9.4.*
5. Understand the operation of a bipolar transistor as a switch, and analyze basic analog and digital gate circuits. *Section 9.5.*

9.1 AMPLIFIERS AND SWITCHES

A transistor is a three-terminal semiconductor device that can perform two functions that are fundamental to the design of electronic circuits: **amplification** and **switching**. Amplification consists of using an external power source to produce a scaled reproduction of a signal. Switching consists of using a relatively small input current or voltage to control a larger output current or voltage.

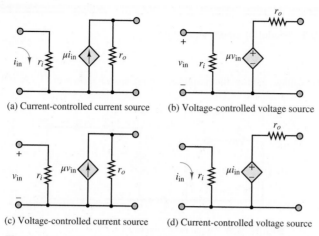

(a) Current-controlled current source   (b) Voltage-controlled voltage source

(c) Voltage-controlled current source   (d) Current-controlled voltage source

Figure 9.1 Controlled-source models of linear amplifiers

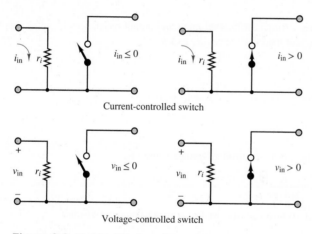

Current-controlled switch

Voltage-controlled switch

Figure 9.2 Models of ideal transistor switches

Four different linear amplifier models are shown in Figure 9.1. Controlled voltage and current sources generate an output proportional to an input current or voltage; the proportionality constant μ is called the internal *gain* of the transistor. Bipolar junction transistor (BJTs) are well modeled as current-controlled devices.[1]

Transistors are also operated in a nonlinear mode, as voltage- or current-controlled switches. Figure 9.2 depicts the idealized operation of the transistor as a switch, suggesting that the switch is closed (on) whenever a control voltage or current is greater than zero and is open (off) otherwise. More realistic conditions on transistors acting in a switch mode are discussed later in this chapter and Chapter 10.

---

[1]Another family of transistors, the field-effect transistors (FETs), are well modeled as voltage controlled devices. See Chapter 10.

EXAMPLE 9.1   Model of Linear Amplifier

Problem

Determine the voltage gain of the amplifier circuit model shown in Figure 9.3.

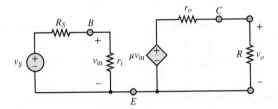

Figure 9.3 Circuit for Example 9.1.

Solution

Known Quantities: Amplifier internal input and output resistances r_i and r_o; amplifier internal gain μ; source and load resistances R_S and R.

Find: $G = \dfrac{v_o}{v_S}$

Analysis: Apply voltage division to determine v_{in}:

$$v_{\text{in}} = \frac{r_i}{r_i + R_S} v_S$$

Then, the output of the controlled voltage source is:

$$\mu v_{\text{in}} = \mu \frac{r_i}{r_i + R_S} v_S$$

and the output voltage can also be found using voltage division:

$$v_o = \mu \frac{r_i}{r_i + R_S} v_S \times \frac{R}{r_o + R}$$

Finally, the amplifier voltage gain can be computed:

$$G = \frac{v_o}{v_S} = \mu \frac{r_i}{r_i + R_S} \times \frac{R}{r_o + R}$$

Comments: Note that the voltage gain computed above is always less than the transistor internal voltage gain μ. One can easily show that if $r_i \gg R_S$ and $r_o \ll R$, then $G \approx \mu$. In general, the amplifier gain always depends on the ratio of the source R_S to input r_i resistances and the ratio of output r_o to load R resistances.

CHECK YOUR UNDERSTANDING

Repeat Example 9.1 for a current-controlled voltage source (CCVS) as shown in Figure 9.1(d). What is the amplifier voltage gain? Under what conditions would $G = \mu/R_S$?

Repeat Example 9.1 for the current-controlled current source (CCCS) of Figure 9.1(a). What is the amplifier voltage gain?

Repeat Example 9.1 for the voltage-controlled current source (VCCS) of Figure 9.1(c). What is the amplifier voltage gain?

Answers: $G = \mu \dfrac{1}{r_i} \dfrac{R}{r_i + R_S} \dfrac{r_o}{r_o + R}$; $r_i \to 0, r_o \to 0$; $G = \dfrac{1}{r_i} \dfrac{r_o R}{r_i + R_S} \dfrac{r_o}{r_o + R} \mu$; $G = \mu \dfrac{r_i}{r_i + R_S} \dfrac{r_o R}{r_o + R}$

9.2   THE BIPOLAR JUNCTION TRANSISTOR

A BJT is formed by joining three sections of alternating *p*- and *n*-type material. An *npn* transistor is a BJT with a thin, lightly doped *p*-type **base** region sandwiched between a heavily doped *n*-type **emitter** region and a large, lightly doped *n*-type **collector** region. The BJT counterpart to the *npn* is the *pnp* transistor, which utilizes the same doping scheme except that the *n* and *p* regions are swapped with respect to the *npn*. In both of these BJT types, the heavily doped emitter region is often labeled n^+ or p^+ to distinguish it from the lightly doped collector. Figure 9.4

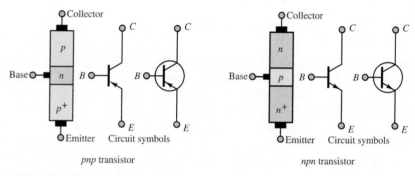

Figure 9.4 Bipolar junction transistors

illustrates the construction, symbols, and nomenclature for the two types of BJTs. Notice that there are the two *pn* junctions in a BJT: the **emitter-base junction** (EBJ) and the **collector-base junction** (CBJ). The operating mode of a BJT depends upon whether these junctions are reverse- or forward-biased, as indicated in Table 9.1.

Table 9.1 **BJT operating modes**

Mode	EBJ	CBJ	Application
Cutoff	Reverse-biased	Reverse-biased	Open switch
Active	Forward-biased	Reverse-biased	Amplifier
Saturation	Forward-biased	Forward-biased	Closed switch

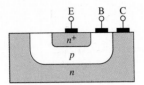

Figure 9.5 Cross section of an *npn* transistor. Notice that the collector is much larger and much more lightly doped than the emitter. The base is, in fact, very thin compared to the emitter and collector.

Although the construction of a BJT results in two opposing *pn* junctions, it is important to avoid modeling a BJT as two identical but opposing diodes. The EBJ always behaves as a true diode; however, because of the thin base region and the lightly doped collector region, the CBJ does not. Figure 9.5 depicts the basic geometry of a cross section of a BJT. The base region is shown much thicker (compared to the emitter and collector) for the sake of clarity. There are two key points to note from the figure: (1) The base is a very thin envelope around the emitter, and (2) the collector is much larger than the emitter and the base because it envelopes both and is itself relatively thick compared to the emitter. The result of this geometry is that the collector can receive large numbers of mobile charge carriers without any significant impact upon its density of charge carriers.

Cutoff Mode (EBJ Reverse-biased; CBJ Reverse-biased)

When both *pn* junctions are reverse-biased, no current is present across either junction and the path from collector to emitter can be approximated as an open-circuit. In fact, small reverse currents due to minority carriers are present across the junctions, but for most practical applications these reverse currents are negligible. In silicon-based BJTs, the offset voltage for the EBJ is the same 0.6 V presented in Chapter 8 for single silicon diodes. Thus, in cutoff mode, when $v_{BE} < 0.6$ V, the transistor acts as a switch in its off (open-circuit) condition.

Active Mode (EBJ Forward-biased; CBJ Reverse-biased)

Figure 9.6 shows a Norton source connected across the base and emitter terminals of an *npn* transistor and the resulting *i-v* characteristic of its EBJ. Notice that $i_B \approx 0$ when $v_{BE} \leq 0.6$ V, which is cutoff mode. However, when the EBJ is forward-biased such that $v_{BE} \geq 0.6$ V, current is conducted as in a typical diode. Majority carriers in the emitter and base *drift* across the EBJ under the influence of the forward-bias voltage in excess of the potential barrier of the depletion region. However, since the emitter is heavily doped while the base is lightly doped, the current I_E across the EBJ is dominated by the majority carriers from the emitter.

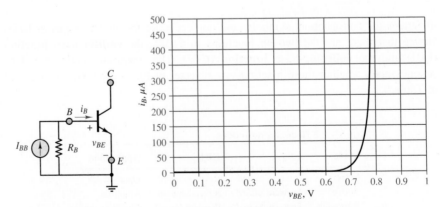

Figure 9.6 The *i-v* characteristic of the emitter-base junction of a typical *npn* transistor

The *i-v* characteristics of the EBJ for *npn* and *pnp* transistors are identical except that the abscissas are v_{BE} and v_{EB}, respectively. The discussion below is based upon the behavior of an *npn* transistor; however, the behavior of a *pnp* transistor is completely analogous to that of an *npn* transistor, except that positive and negative charge carriers are interchanged and the EBJ is forward-biased from emitter to base rather than from base to emitter.

> The behavior of a *pnp* transistor is completely analogous to that of an *npn* transistor, except that positive and negative charge carriers are interchanged and the EBJ is forward-biased from emitter to base rather than from base to emitter.

For an *npn* BJT, the majority carriers in the emitter are electrons while the majority carriers in the base are holes, as indicated in Figure 9.7. Some of these electrons recombine with holes in the base; however, since the base is lightly doped, most of these electrons remain mobile *minority carriers* in the *p*-type base. As these mobile electrons cross the EBJ, their growing concentration in the base causes them to *diffuse* toward the CBJ. The equilibrium concentration of these mobile electrons throughout the base region is a maximum at the EBJ and is given by:

$$(n_p)_{max} = (n_p)_o (e^{v_{BE}/V_T} - 1) \qquad (9.1)$$

where v_{BE} is the forward-bias voltage from base to emitter and $(n_p)_o$ is the thermal equilibrium concentration of electrons in the base. Since the base is very thin, the equilibrium concentration *gradient* across the base is nearly linear, as depicted in Figure 9.8, such that the electron diffusion *rate* from the EBJ to the CBJ can be approximated as:

$$\frac{Aq_e D_n (n_p)_{max}}{W} \qquad (9.2)$$

where A is the cross-sectional area of the EBJ, W is the width of the base (not including the width of the two bounding depletion regions), and D_n is the diffusivity of electrons in the base. It is important to note that this electron diffusion rate

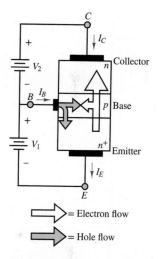

Figure 9.7 Flow of emitter electrons into the collector in an *npn* transistor

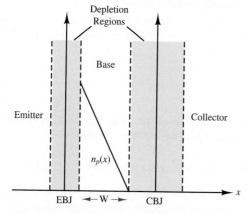

Figure 9.8 Equilibrium concentration gradient of free electrons in the *p*-type base of a forward-biased *NPN* transistor

is temperature dependent and that it represents a diffusion *current* directed from the CBJ to the EBJ because of the convention that the direction of positive current is the direction of flow of positive charge carriers. Once these diffusing electrons reach the CBJ they are swept into the collector by the reverse-bias voltage across the CBJ. Thus, the **collector current** i_C is:

$$
\begin{aligned}
i_C &= \frac{A q_e D_n (n_p)_o}{W} (e^{v_{BE}/V_T} - 1) \\
&= \frac{A q_e D_n n_i^2}{W N_A} (e^{v_{BE}/V_T} - 1) \\
&= I_S (e^{v_{BE}/V_T} - 1) \qquad \text{Ebers-Moll equation}
\end{aligned}
\tag{9.3}
$$

where N_A is the doping concentration of holes in the base and I_S is known as the **scale current** because it scales with the cross-sectional area A of the EBJ. Typical values of I_S range from 10^{-12} A to 10^{-15} A.

The **base current** i_B (from base to emitter) is comprised of those majority carriers in the base (e.g., holes for an *npn* transistor) that traverse the EBJ. Some of these carriers recombine with the majority carriers in the emitter (e.g., electrons for an *npn* transistor); however, those majority carriers lost to recombination are replaced by additional majority carriers supplied by V_1. Because the concentration of these majority carriers is proportional to $e^{v_{BE}/V_T} - 1$, the base current is proportional to the collector current i_C such that:

$$
i_B = \frac{i_C}{\beta} = \frac{i_C}{h_{FE}}
\tag{9.4}
$$

where β is known as the forward **common-emitter current gain** with typical values ranging from 20 to 200. Although β can vary significantly from one transistor to another, most practical electronic devices only require that $\beta \gg 1$. Figure 9.7 depicts the flow of charge carriers from emitter to base to collector and from base to emitter, as discussed above, for an *npn* transistor. A BJT is a *bipolar* device because its current is comprised of both electrons and holes.[2]

The parameter β is not often found in a data sheet. Instead, the forward **DC** value of β is listed as h_{FE}, which is the **large-signal current gain**. A related parameter, h_{fe}, is the **small-signal current gain**.

Finally, to satisfy KCL, the **emitter current** i_E must be the sum of the collector and base currents and, therefore, must also be proportional to e^{v_{BE}/V_T}. Thus:

$$
\begin{aligned}
i_E &= I_{ES} (e^{v_{BE}/V_T} - 1) \\
&= i_C + i_B = \frac{\beta + 1}{\beta} i_C = \frac{i_C}{\alpha}
\end{aligned}
\tag{9.5}
$$

where I_{ES} is the reverse **saturation current** and α is the **common-base current gain** with a typical value close to, but not exceeding, 1.

Saturation Mode (EBJ Forward-biased; CBJ Forward-biased)

A BJT remains in active mode as long as the CBJ is reverse-biased; that is, as long as $V_2 > 0$, electrons diffusing across the base will be swept away into the collector

---

[2]By contrast, a field-effect transistor (FET) is a *unipolar* device. See Chapter 10.

once they reach the CBJ. However, when the CBJ is forward-biased ($V_2 < 0$), these diffusing electrons are no longer swept away into the collector but instead accumulate at the CBJ such that the concentration of minority carrier electrons there is no longer zero. The magnitude of this concentration increases as V_2 decreases, such that the concentration gradient across the base decreases. The result is that the diffusion of minority carrier electrons across the base decreases; in other words, the collector current i_C decreases as the forward bias of the CBJ increases.

It is important to realize that as the concentration gradient across the base decreases and the rate of diffusion across the base decreases, the rate of increase of the concentration near the CBJ slows and the concentration gradient across the base approaches zero asymptotically. This asymptotic process expresses itself as an upper limit on the forward-bias voltage across the CBJ. Figure 9.9 defines three voltages across the terminals of an *npn* transistor. In saturation, the action of the transistor limits v_{CB} such that v_{CE} is always positive, although small, with a typical value of 0.2 V. In fact, saturation mode is often best determined by the value of v_{CE}, which has a value of approximately 0.2 V for a silicon-based BJT.

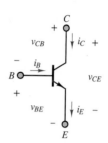

KCL: $i_E = i_B + i_C$
KVL: $v_{CE} = v_{CB} + v_{BE}$

Figure 9.9
Definition of BJT voltages and currents

> In saturation, the collector current is no longer proportional to the base current and the collector-emitter voltage V_{CE} for a silicon-based BJT is small (< 0.4 V). An increasing base current drives a BJT further into saturation, and V_{CE} approaches the saturation limit of $V_{CE\,\text{sat}} \approx 0.2$ V.

$$\boxed{V_{CE\,\text{sat}} \approx 0.2 \text{ V} \qquad \text{Saturation limit}} \tag{9.6}$$

Key BJT Characteristics

The voltages and currents shown in Figure 9.9 for an *npn* transistor are related by KCL and KVL.

$$v_{CE} = v_{CB} + v_{BE} \qquad \text{KVL} \tag{9.7}$$

$$i_E = i_C + i_B \qquad \text{KCL} \tag{9.8}$$

The BJT currents are temperature dependent because they are proportional to both n_i^2 and e^{v_{BE}/V_T}. These currents are also proportional to the cross-sectional area A of the EBJ and inversely proportional to the effective width W of the base.

The relationships between these voltages and currents are commonly represented by a graph of i_C versus v_{CE}, with i_B treated as a parameter. A typical example of such a graph is shown in Figure 9.10. The operating mode of a BJT is completely specified by these three variables. The three modes of operation are indicated in the figure. Cutoff and saturation modes occur when i_C and v_{CE} are very small, respectively.

For any fixed value of i_B, the slope of the transistor characteristic is very small in active mode. In the ideal case, this slope would be zero; however, the effective width of the base decreases with v_{CE} such that the concentration gradient of charge carriers in the base increases and, thus, the collector current increases as well. This increase in i_C with v_{CE} is known as the **Early effect** or **base-width modulation**.

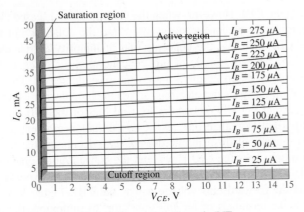

Figure 9.10 Typical characteristic lines of a BJT

It is important to realize that the operating values of i_B, i_C, and v_{CE}, and the operating mode itself, are determined by the external circuitry attached to the BJT. An important objective of this chapter is to provide a method to design the external circuitry so as to dictate and control the operating mode of a BJT. To understand the development of such a method, it is essential to keep in mind the key characteristics of the cutoff, active, and saturation modes, which are the same for both *npn* and *pnp* transistors, and which are summarized in the following box.

Cutoff mode: Both the EBJ and CBJ are reverse-biased such that all three currents i_C, i_B, and i_E are approximately zero. In cutoff mode, a BJT acts as an open switch between the collector and emitter.

Active mode: The EBJ is forward-biased while the CBJ is reverse-biased. The BJT currents are related by:

$$i_C = \beta i_B \qquad i_C = \alpha i_E$$

In active mode, these currents are largely independent of v_{CB} and the BJT acts as a linear amplifier.

Saturation mode: Both the EBJ and CBJ are forward-biased such that $v_{BE} \approx 0.7$ V and $v_{CE} \approx 0.2$ V. The collector current i_C is highly sensitive to small changes in v_{CE}, and, since v_{CE} is small, i_C is largely determined by external circuitry attached to the collector terminal. In saturation mode, the BJT approximates a closed switch between the collector and emitter.

Determining the Operating Mode of a BJT

A few simple voltage measurements permit a quick determination of the state of a transistor. Consider, for example, an *npn* transistor placed in the circuit of Figure 9.11, where:

$$R_B = 40 \text{ k}\Omega \qquad R_C = 1 \ \Omega \qquad R_E = 161 \ \Omega$$

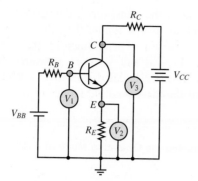

Figure 9.11 Determination of the operating mode of a BJT

and

$$V_{BB} = 4\,\text{V} \qquad V_{CC} = 12\,\text{V}$$

Assume that the measured collector, emitter, and base terminal voltages are:

$$V_B = V_1 = 2.0\,\text{V} \qquad V_E = V_2 = 1.3\,\text{V} \qquad V_C = V_3 = 4.0\,\text{V}$$

The method used in determining the state of a transistor is to assume an operating mode and then test the assumption against the known data. It is usually best to first assume cutoff mode and check whether the EBJ is reverse-biased. The voltage across the EBJ is:

$$V_{BE} = V_B - V_E = 0.7\,\text{V}$$

Thus, the EBJ is forward-biased, not reverse-biased, and the transistor is not in cutoff mode.

One can next assume either active or saturation mode and test the assumption. For this example, assume saturation mode and test whether the CBJ is forward-biased. The voltage across the CBJ is:

$$V_{BC} = V_B - V_C = -2.0\,\text{V}$$

Thus, the CBJ is reverse-biased and the transistor is in active mode. The same determination could be made by evaluating the voltage across the collector-emitter terminals.

$$V_{CE} = V_C - V_E = 2.7\,\text{V}$$

The requirement for saturation mode is $V_{CE} < 0.4$ V, which is clearly not satisfied.

Since the transistor is in active mode, it is possible to calculate the common-emitter current gain β. The base current is:

$$I_B = \frac{V_{BB} - V_B}{R_B} = \frac{4-2}{40{,}000} = 50\,\mu\text{A}$$

The collector current is:

$$I_C = \frac{V_{CC} - V_C}{R_C} = \frac{12-4}{1{,}000} = 8\,\text{mA}$$

Thus, the current amplification factor is:

$$\frac{I_C}{I_B} = \beta = 160$$

The operating point of the transistor in the given circuit can be located on a characteristic plot such as that in Figure 9.10. It is important to note that the operating mode of the transistor is determined by the attached circuitry. In this example, the values of V_B, V_C, and V_E were measured. However, for analytic problems, these values can be calculated using KCL, KVL, Ohm's law, and the known characteristics of the three possible modes of operation.

EXAMPLE 9.2   Determining the Operating Mode of a BJT

Problem

Determine the operating mode of the BJT in the circuit of Figure 9.11.

Solution

Known Quantities: Base, collector, and emitter voltages with respect to ground.

Find: Operating mode of the transistor.

Schematics, Diagrams, Circuits, and Given Data: $V_1 = V_B = 1.0$ V; $V_2 = V_E = 0.3$ V; $V_3 = V_C = 0.6$ V; $R_B = 40$ kΩ; $R_C = 1$ kΩ; $R_E = 26$ Ω.

Analysis: Compute V_{BE} and V_{BC} to determine the bias conditions of the EBJ and CBJ, which determine the mode of operation of the transistor.

$$V_{BE} = V_B - V_E = 0.7 \text{ V}$$
$$V_{BC} = V_B - V_C = 0.4 \text{ V}$$

Since both junctions are forward-biased, the transistor is in saturation mode. Also, notice that $V_{CE} = V_C - V_E = 0.3$ V is less than 0.4 V, which indicates that the BJT is operating near or in saturation.

The operating point of this transistor can be located in Figure 9.10 by calculating:

$$I_C = \frac{V_{CC} - V_C}{R_C} = \frac{12 - 0.6}{1,000} = 11.4 \text{ mA}$$

and

$$I_B = \frac{V_{BB} - V_B}{R_B} = \frac{4 - 1.0}{40,000} = 75.0 \text{ } \mu\text{A}$$

Notice that the operating point in Figure 9.10 is near the elbow in the $I_B = 75.0$ μA curve at $V_{CE} = 0.3$ V.

Comments: KCL requires $I_E = I_C + I_B$. The latter sum is 11.475 mA, whereas I_E is 0.3 V/26 Ω = 11.5 mA. The difference between these two currents is due entirely to numerical approximations. In fact, KCL is—as it must be—satisfied exactly.

It is important to notice that by only changing R_E from 26 to 161 Ω, the operating mode of the transistor is changed from saturation mode to active mode.

CHECK YOUR UNDERSTANDING

Describe the operation of a *pnp* transistor in active mode by analogy with that of the *npn* transistor.

CHECK YOUR UNDERSTANDING

For the circuit of Figure 9.11, the voltmeter readings are $V_1 = 3$ V, $V_2 = 2.4$ V, and $V_3 = 2.7$ V. Determine the operating mode of the transistor.

Answer: Saturation

9.3   BJT LARGE-SIGNAL MODEL

The i-v characteristics of a BJT indicate that it acts as a current-controlled current source (CCCS) in the cutoff and active operating modes. In those two modes, the base current dictates the behavior of the BJT. These characteristics form part of a **large-signal model** for the BJT that describes its behavior in terms of the amplitudes of the base and collector currents. Like all models, the large-signal model does not account for every characteristic of a BJT. In particular, it does not account for the Early effect nor temperature effects. However, this model does provide a useful and simple starting point for the analysis of transistor circuits.

It is worth noting that Section 9.4 introduces the **small-signal model** for a BJT that approximates the behavior of a transistor in the presence of small variations in current or voltage. In simple terms, the large- and small-signal models relate BJT variables (I, V) and $(\Delta I, \Delta V)$.

Large-Signal Model of the *npn* BJT

In cutoff mode, the *BE* junction is reverse-biased, the base and collector currents are approximately zero, and therefore the transistor acts as a *virtual* open-circuit. In practice, there is always a leakage current, denoted by I_{CEO}, through the collector, even when $V_{BE} = 0$ and $I_B = 0$.

In active mode, the *BE* junction is forward-biased, and the collector current is proportional to the base current, where the constant of proportionality is β.

$$I_C = \beta I_B \tag{9.9}$$

Since $\beta \gg 1$, this relationship indicates that the collector current is controlled by a relatively small base current.

Finally, in saturation mode, the base current is sufficiently large that the collector-emitter voltage V_{CE} reaches its saturation limit, and the collector current is no longer proportional to the base current. In fact, the collector-emitter pathway acts like a *virtual* short-circuit, except for the small potential drop $V_{CEsat} \approx 0.2$ V.

All three of these operating modes are described by the simple circuit models shown in Figure 9.12. Each of these individual models approximates one of the three operating modes indicated in Figure 9.10. Notice that the large-signal model treats the forward-biased *BE* junction as an offset diode.

Selecting an Operating Point for a BJT

The family of curves shown for the collector i-v characteristic in Figure 9.10 reflects the dependence of the collector current on the base current. For each value of the base current i_B, there exists a corresponding i_C-v_{CE} curve. Thus, by selecting the base current and collector current (or collector-emitter voltage), an operating point Q for

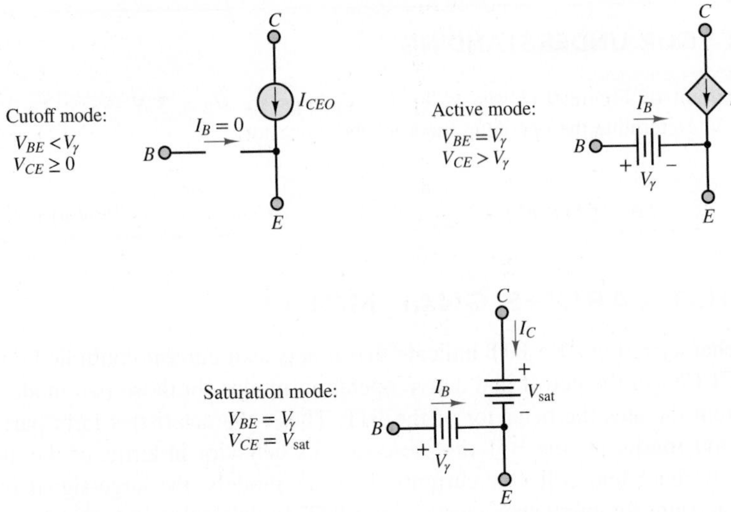

Cutoff mode:
$V_{BE} < V_\gamma$
$V_{CE} \geq 0$

$I_B = 0$

I_{CEO}

Active mode:
$V_{BE} = V_\gamma$
$V_{CE} > V_\gamma$

I_B

βI_B

V_γ

Saturation mode:
$V_{BE} = V_\gamma$
$V_{CE} = V_{sat}$

I_C

I_B

V_{sat}

V_γ

Figure 9.12 An *npn* BJT large-signal model

the transistor is determined. Q is defined in terms of the **quiescent** (or **idle**) **currents** and **voltages** that are present at the terminals of the device under DC conditions. The circuit of Figure 9.13 illustrates an ideal (not practical) **DC bias circuit**, used to set the operating point Q such that $V_{CE} \approx V_{CC}/2$. (A practical bias circuit is discussed later in this chapter.) The underlying principle is to pick R_C and R_B such that under quiescent DC conditions the BJT is maintained in active mode for all anticipated variations in I_B, I_C, and V_{CE} under operating (nonquiescent) conditions.

> By appropriate choice of I_{BB}, R_B, R_C and V_{CC}, the desired Q point may be selected.

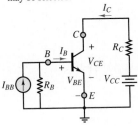

Figure 9.13 A simple ideal bias circuit for a BJT amplifier

KVL can be applied to yield the following equations:

$$I_B = I_{BB} - \frac{V_{BE}}{R_B} \tag{9.10}$$

and

$$V_{CE} = V_{CC} - I_C R_C \tag{9.11}$$

or

$$I_C = \frac{V_{CC} - V_{CE}}{R_C} \tag{9.12}$$

Note that equation 9.11 represents a *load line* for the source network of V_{CC} in series with R_C. When $V_{CE} = 0$, the collector current is $I_C = V_{CC}/R_C$; when $I_C = 0$, the collector-emitter voltage is $V_{CE} = V_{CC}$. These two conditions represent the virtual short- and open-circuit cases for the collector-emitter pathway, that is, the saturation and cutoff modes of the BJT, respectively. The load line can be super-imposed upon the plot of BJT characteristics as shown in Figure 9.14. The slope of the load line is $-1/R_C$. The operating point Q is the intersection of the load line with one of the BJT characteristic lines. The particular characteristic line is determined by the base current I_B, as given by equation 9.9. The particular load line shown in Figure 9.14 assumes $V_{CC} = 15$ V, $V_{CC}/R_C = 40$ mA, and $I_B = 150$ μA.

Once the operating point is established, the BJT is considered **biased** and prepared to operate as a linear amplifier. It is important to note that in circuit diagrams transistors are usually designated Q_1, Q_2, etc. The use of Q to denote transistors is related to the use of Q to denote an operating point, but the two uses serve two different purposes.

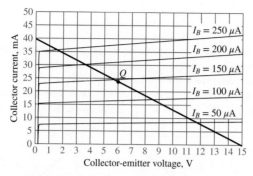

Figure 9.14 Load-line analysis of a simplified
BJT amplifier

Large-Signal Amplifier for Diode Thermometer

Problem:

A diode can be used as the temperature transducer in an electronic thermometer (see
the Focus on Measurements box, "Diode Thermometer," in Chapter 8). In this example,
a diode element again acts as a temperature transducer within the transistor amplifier
circuit shown in Figure 9.15.

Solution:

Known Quantities—Diode and transistor amplifier bias circuits; diode voltage
versus temperature response.

Find—Collector resistance and transistor output voltage versus temperature.

Schematics, Diagrams, Circuits, and Given Data—$V_{CC} = 12$ V; large-signal
$\beta = 188.5$; $V_{BE} = 0.75$ V; $R_S = 500$ Ω; $R_B = 10$ kΩ.

Assumptions—A 1N914 diode and a 2N3904 transistor are used in the circuit.

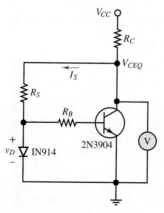

Figure 9.15 Large-signal
amplifier for diode thermometer

Analysis—The diode temperature response characteristic is shown in Figure 9.16(a).
The midrange diode thermometer output voltage is approximately 1.1 V. Thus, the

(Continued)

(*Concluded*)

operating point of the transistor amplifier should be designed so that $v_D \approx 1.1$ V to reduce the risk of distortion due to nonlinearities elsewhere in the diode temperature response characteristic. Since the collector supply $V_{CC} = 12$ V, choose the quiescent operating point Q such that V_{CEQ} is half of V_{CC}, or 6 V in this example. In this way, V_{CEQ} has roughly $V_{CC}/2$ *headroom* to vary, up or down, as the diode temperature varies.

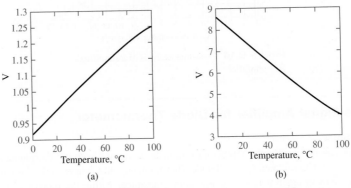

(a)   (b)

Figure 9.16 (a) Diode voltage temperature dependence; (b) amplifier output

With $v_D = 1.1$ V at the quiescent point, the quiescent base current is:

$$v_D - I_{BQ}R_B - V_{BEQ} = 0$$

$$I_{BQ} = \frac{v_D - V_{BEQ}}{R_B} = \frac{1.1 - 0.75}{10,000} = 35 \ \mu A$$

The collector current can be computed as:

$$I_{CQ} = \beta I_{BQ} = 188.5 \times 35 \ \mu A = 6.6 \ mA$$

Finally, KVL can be applied to find a relationship for the resistance R_C:

$$V_{CC} - (I_{CQ} + I_S)R_C - V_{CEQ} = 0$$

$$R_C = \frac{V_{CC} - V_{CEQ}}{I_{CQ} + I_S} = \frac{12 \ V - 6 \ V}{6.6 \ mA + (V_{CEQ} - v_D/R_S)} = \frac{6 \ V}{16.4 \ mA} = 366 \ \Omega$$

Once the circuit is designed according to these specifications, the output voltage can be determined by computing the base current as a function of the diode voltage (which is a function of temperature); from the base current, we can compute the collector current and use the collector equation to determine the output voltage $v_{\text{out}} = v_{CE}$. The result is plotted in Figure 9.16(b).

Comments—Note that the transistor amplifies the slope of the temperature by a factor of approximately 6. Observe also that the *common-emitter amplifier* used in this example causes a sign inversion in the output (the output voltage now decreases for increasing temperatures while the diode voltage increases). Finally, note that the design assumes that the impedance of the voltmeter is infinite, which is a reasonable assumption because a typical voltmeter has a very large input resistance compared to the transistor output resistance. If the output measured by the voltmeter were connected to another circuit, one would have to pay close attention to the input resistance of the second circuit to ensure that it did not load the transistor circuit and thereby affect its behavior.

EXAMPLE 9.3   LED Driver

Problem

Design a transistor switch to control an LED as shown in Figure 9.17. The LED is required to turn on and off with the on/off signal from a digital output port of a microcontroller that is in series with the transistor base.

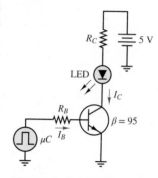

Figure 9.17 LED driver circuit

Solution

Known Quantities: Microcontroller output resistance and output signal voltage and current levels; LED offset voltage, required current, and power rating; BJT current gain and base-emitter junction offset voltage.

Find: (a) Collector resistance R_C such that the transistor is in saturation when the microcontroller outputs 5 V; (b) power dissipated by LED.

Schematics, Diagrams, Circuits, and Given Data:

Microcontroller: output resistance = $R_B = 1$ kΩ; $V_{ON} = 5$ V; $V_{OFF} = 0$ V.

Transistor: $V_{CC} = 5$ V; $V_\gamma = 0.7$ V; $\beta = 95$; $V_{CEsat} = 0.2$ V.

LED: $V_{LED} = 1.4$ V; $I_{LED} = 30$ mA; $P_{max} = 100$ mW.

Assumptions: Use the large-signal models for cutoff and saturation as shown in Figure 9.12. In saturation, $V_{CE} \approx V_{CEsat} = 0.2$ V.

Analysis: When the microcontroller output voltage is zero, the BJT is in cutoff mode since the base current is zero. When the microcontroller output voltage is $V_{ON} = 5$ V, the transistor should be in saturation mode so that the LED sees a virtual short-circuit from collector to emitter. Figure 9.18(a) depicts the equivalent base-emitter circuit when the microcontroller output voltage is $V_{ON} = 5$ V. Figure 9.18(b) depicts the collector circuit, and Figure 9.18(c), the same collector circuit with the large-signal model for the transistor in place of the BJT. Apply KVL to obtain:

$$V_{CC} = R_C I_C + V_{LED} + V_{CEsat}$$

or

$$R_C = \frac{V_{CC} - V_{LED} - V_{CEsat}}{I_C} = \frac{3.4}{I_C}$$

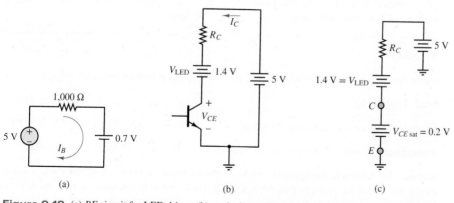

(a)   (b)   (c)

Figure 9.18 (a) *BE* circuit for LED driver; (b) equivalent collector circuit of LED driver, assuming that the BJT is in the linear active mode; (c) LED driver equivalent collector circuit, assuming that the BJT is in saturation mode

A typical LED requires at least 15 mA to be on. In this example, the LED current is specified as 30 mA to ensure that the LED is reasonably bright when on. The collector resistance R_C needed to provide this current is $\approx 113\ \Omega$.

To confirm that the transistor is in saturation when the microcontroller voltage is 5 V, the ratio I_C/I_B should be less than β. For the given specifications, the base current is:

$$I_B = \frac{V_{ON} - V_\gamma}{R_B} = \frac{4.3}{1{,}000} = 4.3\ \text{mA}$$

Thus:

$$\frac{I_C}{I_B} \approx 7$$

In active mode, the ratio $I_C/I_B = \beta = 95$. For sufficiently large values of the base current, the transistor leaves active mode and enters saturation. In saturation, the ratio I_C/I_B is no longer constant and is always less than β. Clearly, this condition is satisfied when the microcontroller output is on. For any particular transistor, the value of β can be significantly different from its typical value given in a generic data sheet. Thus, in practice, it is a good idea to make sure that $I_C/I_B \ll \beta_{typ}$. In this example, $7 \ll 95$ such that it is reasonably certain that the transistor will be in saturation for the design specification of $R_C \approx 113\ \Omega$.

The power dissipated by the LED is:

$$P_{LED} = V_{LED} I_C = 1.4 \times 0.03 = 42\ \text{mW} < 100\ \text{mW}$$

Since the power rating of the LED has not been exceeded, the design is complete.

Comments: The large-signal model of the BJT is easy to apply because the *BE* and *CE* junctions are approximated as short-circuits in series with an independent voltage source. Remember to first check the operating mode of the transistor by assuming a mode and then verifying that the assumption is not contradicted by the resulting voltages across the EBJ, CBJ, and CEJ.

EXAMPLE 9.4   Simple BJT Battery Charger (Current Source)

Problem

Design a constant-current battery charging circuit by selecting values of V_{CC}, R_1, R_2 (a potentiometer) that will cause the transistor Q_1 to act as a current-controlled current source (CCCS) with a selectable range $10\ \text{mA} \le i_C \le 100\ \text{mA}$.

Solution

Known Quantities: Transistor large-signal parameters; NiCd battery nominal voltage.

Find: V_{CC}, R_1, R_2.

Schematics, Diagrams, Circuits, and Given Data: Figure 9.19. $V_{CC} = 12$ V; $V_\gamma = 0.6$ V; $\beta = 100$.

Assumptions: Assume that the transistor can be represented by the large-signal model.

Analysis: To determine the operating mode of the transistor, assume one of the three possible modes and check for any contradictions. First, if cutoff mode is assumed, $i_B = 0$ and $i_C = 0$. Clearly, this mode is not useful for charging. Moreover, KVL requires $V_{BE} + i_B(R_1 + R_2) = V_{CC}$, or since $i_B = 0$, $V_{BE} = V_{CC} = 12$ V. Thus, the EBJ would be forward-biased if $i_B = 0$. This result is a contradiction of the cutoff mode assumption, which therefore must be incorrect.

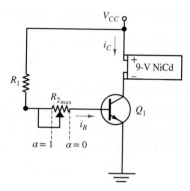

Figure 9.19 Simple battery charging circuit

Second, if the saturation mode is assumed, $V_{CE} \approx V_{CEsat} = 0.2$ V. However, KVL requires $V_{CE} + 9$ V $= V_{CC} = 12$ V, or $V_{CE} = 3$ V, which is a contradiction of the assumed saturation mode.

Thus, the transistor must be in active mode. The base and collector currents, i_B and i_C, are given by Ohm's law and $i_C = \beta i_B$, respectively.

$$i_B = \frac{V_{CC} - V_\gamma}{R_1 + R_2} \quad \text{and} \quad i_C = \beta \frac{V_{CC} - V_\gamma}{R_1 + R_2}$$

The bounds on the collector current i_C, which charges the battery, are:

$$10 \text{ mA} \leq i_C \leq 100 \text{ mA}$$

The potentiometer wiper can be set to any value in the range $0 \leq \alpha \leq 1$ such that the resistance seen by the base is $R_1 + \alpha R_{2_{max}}$. The maximum collector current is obtained when the wiper is set to the far right position $\alpha = 0$. Thus, select R_1 by setting $i_C = i_{C_{max}} = 100$ mA when $\alpha = 0$.

$$100 \text{ mA} = \beta \left(\frac{V_{CC} - V_\gamma}{R_1} \right)$$

or

$$R_1 = (V_{CC} - V_\gamma) \frac{\beta}{10^{-1}} = (12 - 0.6) \frac{100}{10^{-1}} = 11,400 \text{ } \Omega$$

If the value of R_1 is restricted to the E12 series of standard resistor values, the closest standard value is $R_1 = 12$ kΩ, which will result in a slightly lower maximum collector current. The rated value of the potentiometer $R_{2_{max}}$ is found by requiring that $i_C = i_{C_{min}} = 10$ mA when the wiper is set to the far left position $\alpha = 1$. Thus:

$$i_{C_{min}} = 10 \text{ mA} = \beta \frac{V_{CC} - V_\gamma}{R_1 + R_{2_{max}}}$$

or

$$R_{2_{max}} = \frac{\beta}{10 \text{ mA}} (V_{CC} - V_\gamma) - R_1 = 102,600 \text{ } \Omega$$

Again, if the value of $R_{2_{max}}$ is restricted to the E12 series of standard resistor values, the closest standard value is $R_{2_{max}} = 100$ kΩ, which results in a slightly higher minimum collector current.

Comments: A practical note on NiCd batteries: a standard 9-V NiCd battery is made up of eight 1.2-V cells. Thus, the actual nominal battery voltage is 9.6 V. Further, as the battery becomes fully charged, each cell rises to approximately 1.3 V, leading to a fully charged voltage of 10.4 V.

EXAMPLE 9.5 Simple BJT Motor Drive Circuit

Problem

The aim of this example is to design a BJT driver for the Lego® 9V Technic DC motor, model 43362. Figure 9.20 shows the driver circuit and a picture of the motor. The motor has a maximum (stall) current of 340 mA. The minimum current needed to start motor rotation is 20 mA. The aim of the circuit is to control the current to the motor (and therefore the motor torque, which is proportional to the current) through the potentiometer $R_{2_{\max}}$.

Solution

Known Quantities: Transistor large-signal parameters; component values.

Find: Values of R_1 and $R_{2_{\max}}$.

Schematics, Diagrams, Circuits, and Given Data: Figure 9.20. Maximum (stall) of 340 mA; minimum (start) current of 20 mA; $V_\gamma = 0.6$ V; $\beta = 40$; $V_{CC} = 12$ V.

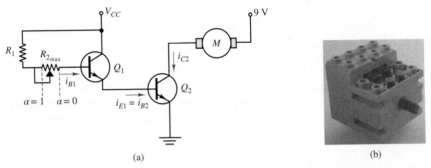

(a)   (b)

Figure 9.20 Motor drive circuit; (a) BJT driver circuit; (b) Lego® 9V Technic motor, model 43362 *(Courtesy Philippe "Philo" Hurbain)*

Assumptions: Use the large-signal model with $\beta = 40$ for each transistor.

Analysis: This example circuit is a good example of how to stage transistors to accomplish a task that is difficult or impossible to accomplish with one transistor alone. Assume that both transistors are in active mode such that $i_C = \beta i_B$ for each transistor. Once a solution is found, the voltages across the EBJ and CBJ can be checked for compatibility with the active mode assumption; however, the i-v characteristic of the motor is needed to do so. For this example, it is assumed that the active mode assumption is correct.

It is important to recognize that the emitter current from Q_1 is the base current for Q_2. Since $i_{E1} = i_{C1} + i_{B1} = (\beta + 1)i_{B1}$, $i_{B2} = i_{E1}$, and $i_{C2} = \beta i_{B2}$, the collector current i_{C2} of Q_2 is related to the base current i_{B1} of Q_1 by:

$$i_{C2} = \beta i_{B2} = \beta i_{E1} = \beta(\beta + 1)i_{B1}$$

The base current i_{B1} of Q_1 is given by Ohm's law.

$$i_{B1} = \frac{V_{CC} - 2V_\gamma}{R_1 + R_{2_{\max}}}$$

Therefore, the range of the motor current is:

$$i_{C2_{min}} \leq \beta(\beta + 1)\left(\frac{V_{CC} - 2V_\gamma}{R_1 + R_{2_{max}}}\right) \leq i_{C2_{max}}$$

The potentiometer wiper can be set to any value in the range $0 \leq \alpha \leq 1$ such that the resistance seen by the base of Q_1 is $R_1 + \alpha R_{2_{max}}$. The maximum (stall) current for the motor is obtained when the wiper is set to the far right position $\alpha = 0$. Thus, select R_1 by setting $i_{C2} = i_{C2_{max}} = 340$ mA when $\alpha = 0$.

$$i_{C2_{max}} = 0.34 \text{ A} = \beta(\beta + 1)\left(\frac{V_{CC} - 2V_\gamma}{R_1}\right)$$

or

$$R_1 = \frac{\beta(\beta + 1)}{0.34}(V_{CC} - 2V_\gamma) = 52,094 \text{ }\Omega$$

If the value of R_1 is restricted to the E12 series of standard resistor values, the closest standard value is $R_1 = 56$ kΩ, which will result in a somewhat lower maximum motor current. The rated value of the potentiometer $R_{2_{max}}$ is found by requiring that $i_{C2} = i_{C2_{min}} = 20$ mA when the wiper is set to the far left position $\alpha = 1$. Thus:

$$R_{2_{max}} = \frac{\beta(\beta + 1)}{0.02}(V_{CC} - 2V_\gamma) - R_1 = 829.6 \text{ k}\Omega$$

Again, if the value of $R_{2_{max}}$ is restricted to the E12 series of standard resistor values, the closest standard value that is still greater than 833,508 Ω is $R_{2_{max}} = 1$ MΩ, which results in a slightly lower minimum motor current. The lower minimum motor current will allow the motor to be turned off by adjusting the potentiometer. Great!

Comments: This design is simple and permits manual control of the motor current (and torque). If the motor is to be controlled by a microcontroller, the circuit should be redesigned to accept an external voltage input.

EXAMPLE 9.6   Calculating an Operating Point for a BJT Amplifier

Problem

Determine the DC operating point of the BJT amplifier in the circuit of Figure 9.21.

Solution

Known Quantities: Base and collector resistances R_B and R_C; base and collector supply voltages V_{BB} and V_{CC}; BJT characteristic curves; BE junction offset voltage.

Find: Quiescent currents I_{BQ} and I_{CQ}, and collector-emitter voltage V_{CEQ}.

Schematics, Diagrams, Circuits, and Given Data: $R_B = 62.7$ kΩ; $R_C = 375$ Ω; $V_{BB} = 10$ V; $V_{CC} = 15$ V; $V_\gamma = 0.6$ V. The BJT characteristic curves from Figure 9.14.

Assumptions: The transistor is in active mode.

Analysis: KVL provides the load-line equation for the source network V_{CC} in series with R_C.

$$V_{CE} = V_{CC} - R_C I_C = 15 - 375 I_C$$

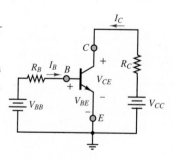

Figure 9.21 Circuit for Example 9.6.

This load line is shown in Figure 9.14. To determine the Q point it is necessary to know the base current. Applying KVL around the base circuit, and assuming that the BE junction is forward-biased, the base current is:

$$I_B = \frac{V_{BB} - V_{BE}}{R_B} = \frac{V_{BB} - V_\gamma}{R_B} = \frac{10 - 0.6}{62,700} = 150 \; \mu A$$

The intersection of the load line with the 150-μA base curve is the operating or quiescent point of the transistor, defined below by the three values:

$$V_{CEQ} = 6 \; V \qquad I_{CQ} = 25 \; mA \qquad I_{BQ} = 150 \; \mu A$$

Comments: Although this example employed two separate voltage sources V_{BB} and V_{CC}, it is possible to bias a transistor using a single voltage source. Note that the transistor dissipates power even in its quiescent state; as should be expected, most of the power is dissipated by R_C: $P_{CQ} = V_{CEQ} \times I_{CQ} = 150 \; mW$.

CHECK YOUR UNDERSTANDING

Repeat the analysis of Example 9.3 for $R_C = 400 \; \Omega$. In which mode is the transistor operating? What is the collector current?

What is the power dissipated by the LED in Example 9.3 if $R_C = 30 \; \Omega$?

Answers: Saturation; 8.5 mA; 159 mW

CHECK YOUR UNDERSTANDING

In Example 9.4, what is V_{CE} when the battery is fully charged (10.4 V)? Is this value consistent with the assumption that the transistor is in active mode?

Answer: $V_{CE} \approx 1.6 \; V \gg V_{CE_{sat}} = 0.2 \; V$; Yes

---

CHECK YOUR UNDERSTANDING

Compute the maximum and minimum possible motor currents for the circuit in Example 9.5 using the selected standard resistor values for R_1 and $R_{2_{max}}$.

Answer: $i_{C_{2_{max}}} = 316.3 \; mA$; $i_{C_{2_{min}}} = 16.8 \; mA$

CHECK YOUR UNDERSTANDING

In Example 9.6, how would the Q point change if R_B was decreased such that the base current increased to 200 μA?

Answer: $V_{CEQ} \approx 4 \; V$, $I_{CQ} \approx 31 \; mA$

9.4 A BRIEF INTRODUCTION TO SMALL-SIGNAL AMPLIFICATION

The purpose of a DC operating point Q for a BJT circuit is to *bias* the BJT so that it is prepared to act as a linear amplifier for a *relatively small* time-varying input signal.

Typically, a time-varying voltage signal ΔV_B is superimposed upon a much larger DC voltage V_{BB}, as shown in Figure 9.22, such that the base current is also a time-varying function $I_B + \Delta I_B$. The primary objective of the DC biasing is to prevent the variation in the base current ΔI_B from driving the BJT out of active mode. This objective will be achieved if the maximum variation in the base current $\Delta I_{B_{max}}$ is *small compared to the DC bias current* I_B and if I_B is picked such that the operating point of the BJT is located in active mode, far from cutoff and saturation. An example of such an operating point Q is shown in Figure 9.14. In that figure, notice that I_B would have to change by at least $\pm 100~\mu A$ from the 150-μA bias current for the BJT to leave active mode and enter either cutoff or saturation. As the base current changes, the location of Q simply moves along the load line, either up and to the left as I_B increases, or down and to the right as I_B decreases.

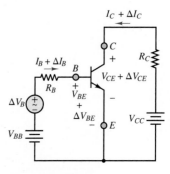

Figure 9.22 Circuit illustrating the amplification effect in a BJT

> The phrase *small-signal model* refers to the fact that the maximum variation in the amplified signal must be small compared to the DC bias conditions.

As long as the BJT remains in active mode the collector current will be roughly proportional to the base current, such that:

$$I_C + \Delta I_C = \beta(I_B + \Delta I_B) \tag{9.13}$$

Further, as seen in Figure 9.22, KVL around the collector source network yields:

$$V_{CE} + \Delta V_{CE} = V_{CC} - (I_C + \Delta I_C)R_C = V_{CC} - \beta(I_B + \Delta I_B)R_C \tag{9.14}$$

In the quiescent state (no time-varying input signal), this equation becomes:

$$V_{CE} = V_{CC} - I_C R_C = V_{CC} - \beta I_B R_C \tag{9.15}$$

Subtract equation 9.15 from equation 9.14 to obtain:

$$\Delta V_{CE} = \Delta I_C R_C = \beta \Delta I_B R_C \tag{9.16}$$

Notice that the variation in the collector-emitter voltage ΔV_{CE} is proportional to the variation in the base current, where the constant of proportionality is βR_C.

Further analysis applying KVL around the base source network yields:

$$V_{BB} + \Delta V_{BB} = (I_B + \Delta I_B)R_B + V_{BE} + \Delta V_{BE} \tag{9.17}$$

In the quiescent state, this equation becomes:

$$V_{BB} = I_B R_B + V_{BE} \tag{9.18}$$

Subtract equation 9.18 from equation 9.17 to obtain:

$$\Delta V_{BB} = \Delta I_B R_B + \Delta V_{BE} \tag{9.19}$$

or

$$\Delta I_B = \frac{\Delta V_{BB} - \Delta V_{BE}}{R_B} \tag{9.20}$$

Use this result to substitute for ΔI_B in equation 9.15 to obtain:

$$\Delta V_{CE} = \beta \frac{R_C}{R_B}(\Delta V_{BB} - \Delta V_{BE}) \tag{9.21}$$

This equation shows that the time-varying component ΔV_{BB} of the *input voltage* is amplified by a factor of $\beta R_C/R_B$ to produce a time-varying component ΔV_{CE} of the *output voltage*. Notice that the *output* of the BJT circuit in Figure 9.22 is considered to be the collector-emitter voltage.

It is important to mention that equation 9.20 shows that the expression for ΔV_{CE} is proportional to ΔV_{BB} only if ΔV_{BE} is negligibly small compared to ΔV_{BB}. Keep in mind that when the BJT is in active mode the EBJ is forward-biased such that the operating point for the EBJ diode is located along the steep portion of the curve shown in Figure 9.6. As a result, ΔV_{BE} tends to be quite small for changes in I_B. Whether ΔV_{BE} is negligible requires more analysis than is appropriate here. Besides, there are other nonideal behaviors of a BJT that prevent the amplifier from being completely linear. The key point is that if the BJT is properly biased, these nonideal effects can be kept small.

An example of the amplification process described above is illustrated in Figure 9.23, where a time-varying sinusoidal collector current $I_C + \Delta I_C$ is shown to the right of the horizontal time axis and the resulting time-varying sinusoidal collector-emitter voltage $V_{CE} + \Delta V_{CE}$ is shown below the V_{CE} axis. Notice that the base current oscillates between 110 and 190 μA, causing the collector current to correspondingly fluctuate between 15.3 and 28.6 mA. Thus, the BJT acts as a *current amplifier*.

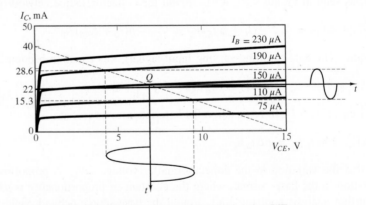

Figure 9.23 Amplification of sinusoidal oscillations in a BJT

A Practical Self-Biasing BJT Circuit

In practice, the circuit shown in Figure 9.22 can be used to bias a BJT; however, it has some weaknesses that can create serious problems in applications. In particular, variations in temperature can cause the operating point Q to shift significantly, and perhaps result in *thermal runaway*. Even if temperature effects are compensated for by other means, the Q points for two apparently identical reproductions of this circuit can be significantly different if the β values for the two BJTs are significantly different, as is often the case even in BJTs of the same type and lot.

A much better *self-biasing* circuit that automatically compensates for such parameter variations is shown in Figure 9.24. This circuit also has the added advantage of needing only one common power supply V_{CC}. Notice that V_{CC} appears across both (R_1, R_2) and (R_C, R_E) such that the circuit can be redrawn as shown in Figure 9.25(a). The Thévenin equivalent network seen by the base is shown in Figure 9.25(b), where:

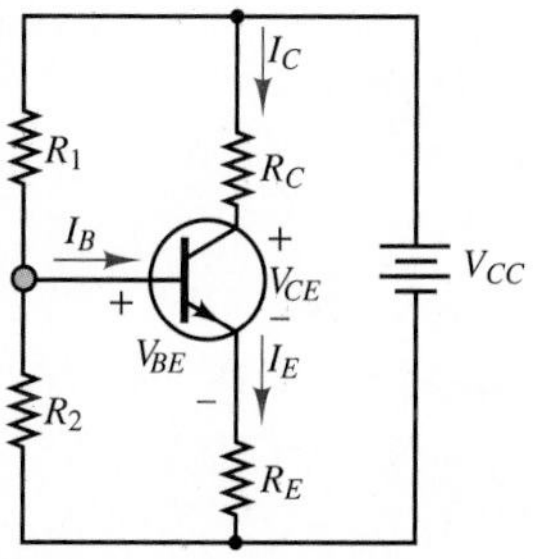

$$V_{BB} = \frac{R_2}{R_1 + R_2} V_{CC} \tag{9.22}$$

and

$$R_B = R_1 \| R_2 = \frac{R_1 R_2}{R_1 + R_2} \tag{9.23}$$

Notice that the circuit in Figure 9.25(b) closely resembles the circuit in Figure 9.22. The important difference is the presence of R_E between the emitter and the node along the bottom portion of the diagram.

Figure 9.24 Practical single power supply BJT self-bias DC circuit

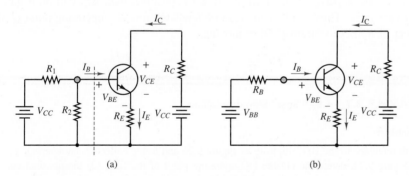

Figure 9.25   (a) DC self bias circuit; (b) equivalent circuit form.

KVL can be applied around the base and collector networks to yield:

$$V_{BB} = I_B R_B + V_{BE} + I_E R_E = [R_B + (\beta + 1) R_E] I_B + V_{BE} \tag{9.24}$$

and

$$V_{CC} = I_C R_C + V_{CE} + I_E R_E = I_C \left(R_C + \frac{\beta + 1}{\beta} R_E \right) + V_{CE} \tag{9.25}$$

where

$$I_E = I_C + I_B = (\beta + 1) I_B = \frac{\beta + 1}{\beta} I_C \tag{9.26}$$

These two equations can be solved to obtain:

$$I_B = \frac{V_{BB} - V_{BE}}{R_B + (\beta + 1)R_E} \qquad\qquad I_C = \beta I_B \tag{9.27}$$

and

$$V_{CE} = V_{CC} - I_C \left(R_C + \frac{\beta + 1}{\beta} R_E \right) \tag{9.28}$$

The latter equation is the load line for the bias circuit. Notice that the effective load resistance seen by the collector circuit is now:

$$R_C + \frac{\beta + 1}{\beta} R_E \approx R_C + R_E \qquad \beta \gg 1$$

rather than simply R_C.

The role of R_E is to provide *negative feedback* to a change in the operating point Q due to, for example, a change in temperature that, in turn, changes β of the transistor. Refer to Figure 9.25(b) for the case of a change $\Delta\beta$. The most immediate effect is a change in the collector current $\Delta I_C = \Delta\beta I_B$. In turn, this change results in a change in the emitter current $\Delta I_E = \Delta I_C + \Delta I_B$. It is here that R_E plays its part. The change in the emitter current results in a change in the voltage across R_E of $\Delta I_E R_E$, which then brings about a change in the voltage V_{BE} across the EBJ. Finally, this change in V_{BE} brings about a change in the base current due to the fact that the EBJ is a diode. At this point, it is important to realize that the change in base current *always* tends to offset the original change in the collector current because $\Delta I_C = \beta \Delta I_B$. In other words, if $\Delta\beta$ is positive, then ΔI_B will be negative, and vice versa. Thus, while a change in β tends to move the operating point Q, the effect of R_E is to restrain Q from moving.

EXAMPLE 9.7   A BJT Small-Signal Amplifier

Problem

With reference to the BJT amplifier of Figure 9.26 and to the collector characteristic curves of Figure 9.23, determine (1) the DC operating point of the BJT, (2) the nominal current gain β at the operating point, and (3) the AC voltage gain $G = \Delta V_o / \Delta V_B$.

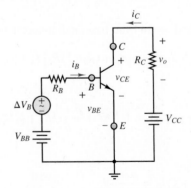

Figure 9.26 Circuit for Example 9.7.

Solution

Known Quantities: Base, collector, and emitter resistances; base and collector supply voltages; collector characteristic curves; *BE* junction offset voltage.

Find: (1) DC (quiescent) base and collector currents I_{BQ} and I_{CQ} and collector-emitter voltage V_{CEQ}, (2) $\beta = \Delta I_C / \Delta I_B$, and (3) $G = \Delta V_o / \Delta V_B$.

Schematics, Diagrams, Circuits, and Given Data: $R_B = 10$ kΩ; $R_C = 375$ Ω; $V_{BB} = 2.1$ V; $V_{CC} = 15$ V; $V_\gamma = 0.6$ V. The collector characteristic curves are shown in Figure 9.28.

Assumptions: Assume that the *BE* junction resistance is negligible compared to the base resistance. Assume that each voltage and current can be represented by the superposition of a DC (quiescent) value and an AC component, for example, $v_0 = V_{0Q} + \Delta V_0$.

Analysis:

1. *DC operating point.* If the resistance of the *BE* junction is assumed to be much smaller than R_B, any change in the voltage across the EBJ is negligible such that $v_{BE} = V_{BEQ} = V_\gamma$. Figure 9.27 shows the resulting DC equivalent base circuit. KVL yields:

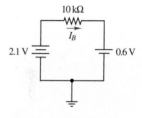

Figure 9.27 DC equivalent base circuit or Example 9.7.

$$V_{BB} = R_B I_{BQ} + V_{BEQ}$$

The quiescent base current can be computed as:

$$I_{BQ} = \frac{V_{BB} - V_{BEQ}}{R_B} = \frac{V_{BB} - V_\gamma}{R_B} = \frac{2.1 - 0.6}{10,000} = 150 \ \mu\text{A}$$

The load-line equation for the collector circuit is given by KVL as:

$$V_{CE} = V_{CC} - R_C I_C = 15 - 375 I_C$$

The load line and its intersection Q with the $I_B = 150$ μA line is shown in Figure 9.28. At the operating or quiescent point Q, $V_{CEQ} = 7.2$ V, $I_{CQ} = 22$ mA, and $I_{BQ} = 150$ μA.

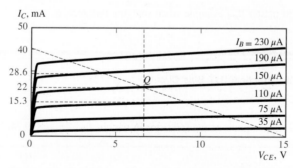

Figure 9.28 Operating point on the characteristic curve

2. *AC gain.* The current gain is determined from the characteristic curves of Figure 9.28. The collector current values corresponding to base currents of 190 and 110 μA are 28.6 and 15.3 mA, respectively. These collector current excursions ΔI_C from the Q point correspond to the effects of an oscillation ΔI_B in the base current. Thus, the current gain of the BJT amplifier can be computed as:

$$\beta = \frac{\Delta I_C}{\Delta I_B} = \frac{28.6 \times 10^{-3} - 15.3 \times 10^{-3}}{190 \times 10^{-6} - 110 \times 10^{-6}} = 166.25$$

which is the nominal current gain of the transistor.

3. *AC voltage gain.* To determine the AC voltage gain $G = \Delta V_o / \Delta V_B$, express ΔV_o as a function of ΔV_B. Observe that $v_o = -R_C i_C = -R_C I_{CQ} - R_C \Delta I_C$. Thus:

$$\Delta V_o = -R_C \, \Delta I_C = -R_C \beta \, \Delta I_B$$

The principle of superposition allows ΔI_B to be computed from the KVL equation for the base circuit.

$$\Delta V_B = R_B \, \Delta I_B + \Delta V_{BE}$$

However, due to the assumed small EBJ resistance, ΔV_{BE} is negligible. Thus:

$$\Delta I_B = \frac{\Delta V_B}{R_B}$$

Substitute this result into the expression for ΔV_o to find:

$$\Delta V_o = -R_C \beta \, \Delta I_B = -\frac{R_C \beta \Delta V_B}{R_B}$$

or

$$\frac{\Delta V_o}{\Delta V_B} = G = -\frac{R_C}{R_B} \beta = -6.23$$

Comments: The circuit examined in this example is not self-biasing, but it demonstrates most of the essential features of BJT amplifiers, which are summarized below.

- Transistor amplifier analysis is greatly simplified by applying the principle of superposition to consider the DC bias circuit and the AC equivalent circuits separately.

- Once the bias point Q has been determined, the current gain can also be determined. Its value is somewhat dependent on the location of Q.

- The AC voltage gain of the amplifier is strongly dependent on R_B and R_C. Note that the AC voltage gain ΔV_o is negative! This inversion corresponds to a $180°$ phase shift for a sinusoidal AC input.

It is important to master this example when studying this section.

EXAMPLE 9.8   Practical BJT Bias Circuit

Problem

Determine the DC bias point of the transistor in the circuit of Figure 9.24.

Solution

Known Quantities: Base, collector, and emitter resistances; collector supply voltage; nominal transistor current gain; *BE* junction offset voltage.

Find: DC (quiescent) base and collector currents I_{BQ} and I_{CQ} and collector-emitter voltage V_{CEQ}.

Schematics, Diagrams, Circuits, and Given Data: $R_1 = 100$ kΩ; $R_2 = 50$ kΩ; $R_C = 5$ kΩ; $R_E = 3$ kΩ; $V_{CC} = 15$ V; $V_\gamma = 0.7$ V, $\beta = 100$.

Analysis: We first determine the equivalent base voltage from equation 9.7,

$$V_{BB} = \frac{R_2}{R_1 + R_2} V_{CC} = \frac{50}{100 + 50} 15 = 5 \text{ V}$$

and the equivalent base resistance from equation 9.8,

$$R_B = R_1 \| R_2 = 33.3 \text{ k}\Omega$$

Now we can compute the base current from equation 9.11,

$$I_B = \frac{V_{BB} - V_{BE}}{R_B + (\beta + 1)R_E} = \frac{V_{BB} - V_\gamma}{R_B + (\beta + 1)R_E} = \frac{5 - 0.7}{33{,}000 + 101 \times 3{,}000} = 12.8 \ \mu A$$

and knowing the current gain of the transistor β, we can determine the collector current:

$$I_C = \beta I_B = 1.28 \text{ mA}$$

Finally, the collector-emitter junction voltage can be computed with reference to equation 9.12:

$$V_{CE} = V_{CC} - I_C \left(R_C + \frac{\beta + 1}{\beta} R_E \right)$$

$$= 15 - 1.28 \times 10^{-3} \left(5 \times 10^3 + \frac{101}{100} \times 3 \times 10^3 \right) = 4.78 \text{ V}$$

Thus, the Q point of the transistor is given by:

$$V_{CEQ} = 4.73 \text{ V} \qquad I_{CQ} = 1.28 \text{ mA} \qquad I_{BQ} = 12.8 \ \mu A$$

CHECK YOUR UNDERSTANDING

In Example 9.7, find the new Q point if R_C is increased to 680 Ω.

Answer: Since V_{BB} and R_B are unchanged and the change in V_{BEQ} is negligible, I_{BQ} will remain approximately equal to 150 μA. By observation, $V_{CEQ} \approx 0.5$ V is much smaller and the BJT is close to saturation. The new collector current $I_{CQ} \approx 20$ mA.

CHECK YOUR UNDERSTANDING

In the circuit of Figure 9.25, find the value of V_{BB} that yields a collector current $I_C = 6.3$ mA. What is the corresponding collector-emitter voltage? Assume that $V_{BE} = 0.6$ V, $R_B = 50$ kΩ, $R_E = 200 \ \Omega$, $R_C = 1$ kΩ, $\beta = 100$, and $V_{CC} = 14$ V.

What percentage change in collector current would result if β were changed to 150 in Example 9.8? Why does the collector current increase less than 50 percent?

Answers: $V_{BB} = 5$ V, $V_{CE} = 6.43$ V; 3.74 percent. Because R_E provides negative feedback action that will keep I_C and I_E nearly constant

9.5 GATES AND SWITCHES

In describing the properties of transistors, it was suggested that, in addition to serving as amplifiers, three-terminal devices can be used as electronic switches in which one terminal controls the flow of current between the other two. It had also been hinted in Chapter 8 that diodes can act as on/off devices as well. In this section, we discuss the operation of diodes and transistors as electronic switches, illustrating the use of these electronic devices as the switching circuits that are at the heart of **analog** and **digital gates**. Transistor switching circuits form the basis of digital logic circuits, which are discussed in greater detail in Chapter 11. The objective of this section is to discuss the internal operation of these circuits and to provide the reader interested in the internal workings of digital circuits with an adequate understanding of the basic principles.

An **electronic gate** is a device that, on the basis of one or more input signals, produces one of two or more prescribed outputs; as will be seen shortly, one can construct both digital and analog gates. A word of explanation is required, first, regarding the meaning of the words *analog* and *digital*. An analog voltage or current—or, more generally, an analog signal—is one that varies in a continuous fashion over time, in *analogy* (hence the expression *analog*) with a physical quantity. An example of an analog signal is a sensor voltage corresponding to ambient temperature on any given day, which may fluctuate between, say, 30 and 50°F. A digital signal, on the other hand, is a signal that can take only a finite number of values; in particular, a commonly encountered class of digital signals consists of **binary signals**, which can take only one of two values (for example, 1 and 0). A typical example of a binary signal would be the control signal for the furnace in a home heating system controlled by a conventional thermostat, where one can think of this signal as being on (or 1) if the temperature of the house has dropped below the thermostat setting (desired value), or off (or 0) if the house temperature is greater than or equal to the set temperature (say, 68°F). Figure 9.29 illustrates the appearance of the analog and digital signals in this furnace example.

The discussion of digital signals will be continued and expanded in Chapter 11. Digital circuits are an especially important topic because a large part of today's industrial and consumer electronics is realized in digital form.

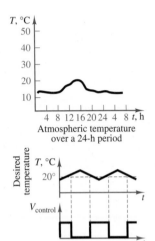

Atmospheric temperature over a 24-h period

Average temperature in a house and related digital control voltage

Figure 9.29 Illustration of analog and digital signals

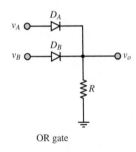

OR gate

OR gate operation

$v_A = v_B = 0$ V → Diodes are off and $v_o = 0$

$\left.\begin{array}{l} v_A = 5 \text{ V} \\ v_B = 0 \text{ V} \end{array}\right\}$ → D_A is on, D_B is off

Equivalent circuit

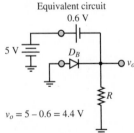

$v_o = 5 - 0.6 = 4.4$ V

Figure 9.30 Diode OR gate

Diode Gates

You will recall that a diode conducts current when it is forward-biased and otherwise acts very much as an open-circuit. Thus, the diode can serve as a switch if properly employed. The circuit of Figure 9.30 is called an **OR gate**; it operates as follows. Let voltage levels greater than, say, 2 V correspond to a "logic 1" and voltages less than 2 V represent a "logic 0." Suppose, then, that input voltages v_A and v_B can be equal to either 0 V or 5 V. If $v_A = 5$ V, diode D_A will conduct; if $v_A = 0$ V, D_A will act as an open-circuit. The same argument holds for D_B. It should be apparent, then, that the voltage across the resistor R will be 0 V, or logic 0 if both v_A and v_B are 0. If either v_A or v_B is equal to 5 V, though, the corresponding diode will conduct, and—assuming an offset model for the diode with $V_\gamma = 0.6$ V—we find that $v_o = 4.4$ V, or logic 1. Similar analysis yields an equivalent result if both v_A and v_B are equal to 5 V.

This type of gate is called an OR gate because v_o is equal to logic 1 (or "high") if either v_A *or* v_B is on while it is logic 0 (or "low") if neither v_A *nor* v_B is on. Other functions can also be implemented; however, the discussion of diode gates will be limited to this simple introduction because diode gate circuits, such as the one of Figure 9.30, are rarely, if ever, employed in practice. Most modern digital circuits employ transistors to implement switching and gate functions.

BJT Gates

In discussing large-signal models for the BJT, we observed that the i-v characteristic of this family of devices includes a *cutoff* mode, where virtually no current flows through the transistor. On the other hand, when a sufficient amount of current is injected into the base of the transistor, a bipolar transistor will reach *saturation*, and a substantial amount of collector current will flow. This behavior is quite well suited to the design of electronic gates and switches and can be visualized by superimposing a load line on the collector characteristic, as shown in Figure 9.31.

The operation of the simple **BJT switch** is illustrated in Figure 9.31, by means of load-line analysis. The load-line equation of the collector circuit is:

$$v_{CE} = V_{CC} - i_C R_C \tag{9.29}$$

and

$$v_o = v_{CE} \tag{9.30}$$

Thus, when the input voltage v_{in} is low (say, 0 V), the transistor is in cutoff mode and its currents are very small. Then:

$$v_o = v_{CE} = V_{CC} \tag{9.31}$$

such that the output is "logic high."

When v_{in} is large enough to drive the transistor into the saturation mode, a substantial amount of collector current will flow and the collector-emitter voltage will be reduced to the small saturation value $V_{CE\,sat}$, which is typically a fraction of a volt. This corresponds to the point labeled B on the load line. For the input voltage v_{in} to drive the BJT of Figure 9.31 into saturation, a base current of approximately 50 μA will be required. Suppose, then, that the voltage v_{in} could take the values 0 or 5 V. Then if $v_{in} = 0$ V, v_o will be nearly equal to V_{CC}, or, again, 5 V. If, on the other hand, $v_{in} = 5$ V and R_B is, say, equal to 89 kΩ [so that the base current required for saturation is $i_B = (v_{in} - V_\gamma)/R_B = (5 - 0.6)/89,000 \approx 50$ μA], the BJT is in saturation, and $v_o = V_{CE\,sat} \approx 0.2$ V.

Thus, whenever v_{in} corresponds to a logic high (or logic 1), v_o takes a value close to 0 V, or logic low (or 0); conversely, $v_{in} =$ "0" (logic "low") leads to $v_o =$ "1." The values of 5 and 0 V for the two logic levels 1 and 0 are quite common in practice and are the standard values used in a family of logic circuits denoted by the acronym **TTL**, which stands for **transistor-transistor logic**.[3] One of the more common TTL blocks is the **inverter** shown in Figure 9.31, so called because it "inverts" the input by providing a low output for a high input, and vice versa. This type of inverting, or "negative," logic behavior is quite typical of BJT gates (and of transistor gates in general).

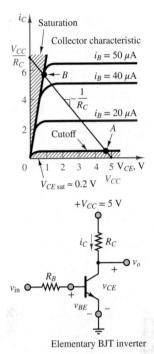

Figure 9.31 BJT switching characteristic

EXAMPLE 9.9   TTL NAND Gate

Problem

Refer to Figure 9.32 and complete the table below to determine the logic gate operation of a TTL NAND gate, which acts as an inverted AND gate (thus the prefix N in NAND, which stands for NOT).

---

[3]TTL logic values are actually quite flexible, with v_{HIGH} as low as 2.4 V and v_{LOW} as high as 0.4 V.

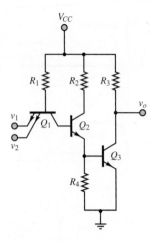

Figure 9.32 TTL NAND gate

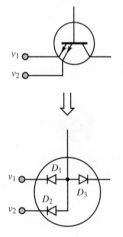

Figure 9.33 Offset diode model of transistor Q_1

v_1 (V)	v_2 (V)	State of Q_1	State of Q_2	v_o
0	0			
0	5			
5	0			
5	5			

Solution

Known Quantities: Resistor values; V_{BEon} and V_{CEsat} for each transistor.

Find: v_o for each of the four combinations of v_1 and v_2.

Schematics, Diagrams, Circuits, and Given Data: $R_1 = 5.7$ kΩ; $R_2 = 2.2$ kΩ; $R_3 = 2.2$ kΩ; $R_4 = 1.8$ kΩ; $V_{CC} = 5$ V; $V_{BEon} = V_\gamma = 0.7$ V; $V_{CEsat} = 0.2$ V.

Assumptions: Treat the *BE* and *BC* junctions of Q_1 as offset diodes. Assume that the transistors are in saturation when conducting.

Analysis: The inputs to the TTL gate, v_1 and v_2, are applied to the emitter of transistor Q_1. The transistor is designed so as to have two emitter circuits in parallel. Transistor Q_1 is modeled by the offset diode model, as shown in Figure 9.33. Consider each of the four cases.

1. $v_1 = v_2 = 0$ V. With the emitters of Q_1 connected to ground and the base of Q_1 at 5 V, the *BE* junction will clearly be forward-biased and Q_1 is on. This result means that the base current of Q_2 (equal to the collector current of Q_1) is negative, and therefore Q_2 must be off. If Q_2 is off, its emitter current must be zero, and therefore no base current can flow into Q_3, which is in turn also off. With Q_3 off, no current flows through R_3, and therefore $v_o = 5 - v_{R3} = 5$ V.

2. $v_1 = 5$ V; $v_2 = 0$ V. With reference to Figure 9.33, diode D_2 is still forward-biased, but D_1 is now reverse-biased because of the 5-V potential at v_2. Thus, the EBJ conducts current and Q_1 is on. The remainder of the analysis is the same as in case 1, and Q_2 and Q_3 are both off, leading to $v_o = 5$ V.

3. $v_1 = 0$ V; $v_2 = 5$ V. By symmetry with case 2, one emitter branch is conducting, Q_1 is on, Q_2 and Q_3 are off, and $v_o = 5$ V.

4. $v_1 = 5$ V; $v_2 = 5$ V. Here, diodes D_1 and D_2 are both reverse-biased, there is no emitter current, and Q_1 is off. Note, however, that although D_1 and D_2 are reverse-biased, D_3 is forward-biased, and a base current exists for Q_2; thus, Q_2 is on and its emitter current turns on Q_3. To determine the output voltage, assume that Q_3 is operating in saturation such that:

$$v_o \approx V_{CEsat}$$

KVL can be applied to the collector circuit to find:

$$V_{CC} = I_{C3}R_3 + V_{CE3}$$

or

$$I_{C3} = \frac{V_{CC} - V_{CE3}}{R_3} = \frac{V_{CC} - V_{CEsat}}{R_3} = \frac{5 - 0.2}{2,200} \approx 2.2 \text{ mA}$$

A reasonable question is, Can Q_2 also be in saturation? If it is, then R_2 and R_4 are virtually in series and the base voltage of Q_3 can be computed by voltage division.

$$V_{B_3} \approx \frac{R_4}{R_2 + R_4}V_{CC} = \frac{1.8}{2.2 + 1.8}5 = 2.25 \text{ V}$$

Since the emitter of Q_3 is tied directly to the reference node ($V = 0$), the voltage across the EBJ of Q_3 would also be 2.25 V. But this value is incompatible with the assumption that $V_\gamma \approx 0.7$ V for silicon-based transistors. Thus, Q_2 cannot be in saturation. But since it is on, it must be in active mode.

The results for all four cases are summarized in the table below. The output values are consistent with TTL logic; the output voltage for case 4 is sufficiently close to zero to be considered zero for logic purposes.

v_1 (V)	v_2 (V)	State of Q_2	State of Q_3	v_o (V)
0	0	Off	Off	5
0	5	Off	Off	5
5	0	Off	Off	5
5	5	On	On	0.2

Comments: While exact analysis of TTL logic gate circuits could be tedious and involved, the method demonstrated in this example—to determine whether transistors are on or off—leads to a very simple analysis. When working with logic devices, the primary interest is in logic levels rather than exact values; thus, approximations are appropriate.

CHECK YOUR UNDERSTANDING

Use the BJT switching characteristic of Figure 9.31 to find the value of R_B required to drive the transistor to saturation. Assume a base current of 50 μA when the minimum v_{in} to turn on the transistor is 2.5 V.

Answer: $R_B \leq 38$ kΩ

Conclusion

This chapter introduces the bipolar junction transistor, and by way of the simple circuit model demonstrates its operation as an amplifier and a switch. Upon completing this chapter, you should have mastered the following learning objectives:

1. *Understand the basic principles of amplification and switching.* Transistors are three-terminal electronic semiconductor devices that can serve as amplifiers and switches.
2. *Understand the physical operation of bipolar transistors; determine the operating point of a bipolar transistor circuit.* The bipolar junction transistor has four modes of operation. These can be readily identified by simple voltage measurements.
3. *Understand the large-signal model of the bipolar transistor, and apply it to simple amplifier circuits.* The large-signal model of the BJT is very easy to use, requiring only a basic understanding of DC circuit analysis, and can be readily applied to many practical situations.
4. *Select the operating point of a bipolar transistor circuit.* Biasing a bipolar transistor consists of selecting the appropriate values for the DC supply voltage(s) and for the resistors that comprise a transistor amplifier circuit. When biased in the forward active mode, the bipolar transistor acts as a current-controlled current source and can amplify small currents injected into the base by as much as a factor of 200.

5. *Understand the operation of a bipolar transistor as a switch, and analyze basic analog and digital gate circuits.* The operation of a BJT as a switch is very straightforward, and consists of designing a transistor circuit that will go from cutoff to saturation when an input voltage changes from a high to a low value, or vice versa. Transistor switches are commonly used to design digital logic gates.

HOMEWORK PROBLEMS

Section 9.2: The Bipolar Junction Transistor

9.1 For each transistor shown in Figure P9.1, determine whether the BE and BC junctions are forward- or reverse-biased, and determine the operating mode.

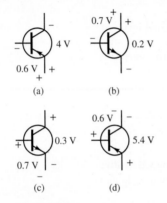

Figure P9.1

9.2 Determine the mode of operation for the following transistors:

a. npn, $V_{BE} = 0.8$ V, $V_{CE} = 0.4$ V

b. npn, $V_{CB} = 1.4$ V, $V_{CE} = 2.1$ V

c. pnp, $V_{CB} = 0.9$ V, $V_{CE} = 0.4$ V

d. npn, $V_{BE} = -1.2$ V, $V_{CB} = 0.6$ V

9.3 Given the circuit of Figure P9.3, determine the operating point of the transistor. Let $\beta = 100 \div 200$, $R_B = 100$ kΩ, $R_c = 200$ Ω, and $V_{CC} = 7$ V.

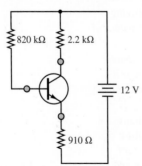

Figure P9.3

9.4 Refer to Figure 9.4 and assume that for a *pnp* transistor the emitter and base currents are $I_E = 5$ mA and $I_B = 0.2$ mA, respectively. The voltage drops across the emitter-base and collector-base junctions are $V_{EB} = 0.67$ V and $V_{CB} = 7.8$ V. Find:

a. V_{CE}.

b. The collector current.

c. The total power dissipated in the transistor, defined here as $P = V_{CE}I_C + V_{BE}I_B$.

9.5 For the circuit shown in Figure P9.5, determine the emitter current I_E and the collector-base voltage V_{CB}, as defined in Figure 9.9. Assume $V_\gamma = 0.62$ V.

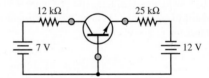

Figure P9.5

9.6 Given the circuit of Figure P9.6, determine V_{CE} and I_C. Assume $\beta = 80$, $R_1 = 15$ kΩ, $R_2 = 25$ kΩ, $R_c = 2$ kΩ, $V_{BB} = 5$ V, $V_{CC} = 10$ V, and $V_{AA} = -4$ V.

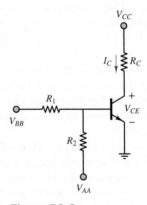

Figure P9.6

9.7 Given the circuit of Figure P9.7, determine the emitter current I_E and the collector-base voltage V_{CB}. Assume the offset voltage is $V_\gamma = 0.6$ V.

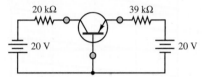

Figure P9.7

9.8 Given the circuit of Figure P9.8, determine V_{CE} and I_C. Assume $R_1 = 50$ kΩ, $R_2 = 10$ kΩ, $R_C = 600$ Ω, $R_E = 400$ Ω, $V_{BE} = 0.9$ V, $I_B = 25$ μA, $I_2 = 200$ μA, and $V_{CC} = 18$ V.

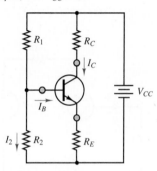

Figure P9.8

9.9 The collector characteristics for a certain transistor are shown in Figure P9.9.

a. Find the ratio I_C/I_B for $V_{CE} = 10$ V and $I_B = 100$, 200, and 600 μA.

b. If the maximum allowable collector power dissipation is $P = i_C v_{CE} = 0.5$ W for $I_B = 500$ μA, find V_{CE}.

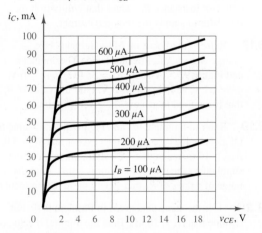

Figure P9.9

9.10 Given the circuit of Figure P9.10, determine the current I_R. Let $R_B = 30$ kΩ, $R_{C1} = 1$ kΩ, $R_{C2} = 3$ kΩ, $R = 7$ kΩ, $V_{BB_1} = 4$ V, $V_{BB_2} = 3$ V, $V_{CC} = 10$ V, $\beta_1 = 40$, and $\beta_2 = 60$.

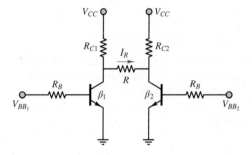

Figure P9.10

9.11 For the circuit shown in Figure P9.11, determine I_R. Let $R_B = 50$ kΩ, $R_C = 1$ kΩ, $R = 2$ kΩ, $V_{BB} = 2$ V, $V_{CC} = 12$ V, and $\beta = 120$.

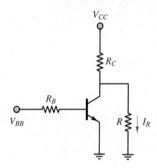

Figure P9.11

9.12 For the circuit shown in Figure P9.12, determine whether the transistor is in saturation. Let $R_B = 8$ kΩ, $R_E = 260$ Ω, $R_C = 1.1$ kΩ, $V_{CC} = 13$ V, $V_{BB} = 7$ V, and $\beta = 100$.

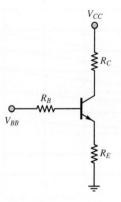

Figure P9.12

9.13 For the circuit shown in Figure P9.8, $V_{CC} = 20$ V, $R_C = 5$ kΩ, and $R_E = 1$ kΩ. Determine the operating mode of the transistor if:

a. $I_C = 1$ mA, $I_B = 20$ μA, $V_{BE} = 0.7$ V

b. $I_C = 3.2$ mA, $I_B = 0.3$ mA, $V_{BE} = 0.8$ V

c. $I_C = 3$ mA, $I_B = 1.5$ mA, $V_{BE} = 0.85$ V

9.14 For the circuit shown in Figure P9.14, find the minimum input voltage v_{in} required to saturate the transistor. Assume $V_{CC} = 5$ V, $R_C = 2$ kΩ, $R_B = 50$ kΩ, $V_{CE\,sat} = 0.1$ V, $V_{BE\,sat} = 0.6$ V, and $\beta = 50$.

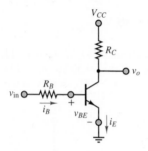

Figure P9.14

9.15 An *npn* transistor, such as that in Figure 9.9, is operated in active mode with $i_C = 60i_B$ and with junction voltages of $V_{BE} = 0.6$ V and $V_{CB} = 7.2$ V. If $|I_E| = 4$ mA, find (a) I_B and (b) V_{CE}.

9.16 Use the collector characteristics of the 2N3904 *npn* transistor shown in Figure P9.16(a) and (b) to determine I_C and V_{CE} of the transistor in Figure P9.16(c). Is the transistor in the active mode? If so, determine its value of β.

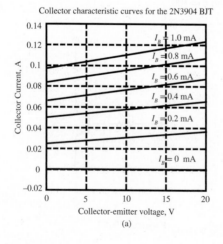

Collector characteristic curves for the 2N3904 BJT

(a)

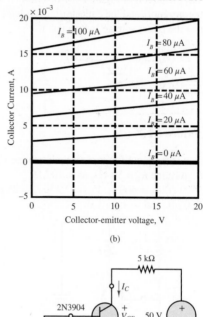

Collector characteristic curves for the 2N3904 BJT

(b)

(c)

Figure P9.16

Section 9.3: BJT Large-Signal Model

9.17 Refer to Example 9.3 and Figure 9.17. Assume that all given values are unchanged except that the application requires $I_{LED} = 10$ mA. Find the range of collector resistance R_C values that will permit the transistor to supply the required current.

9.18 Refer to the Focus on Measurements box, "Large-Signal Amplifier for Diode Thermometer," and Figure 9.15. Assume $R_B = 33$ kΩ, $V_{CEQ} = 6$ V, $v_D = 1.1$ V, and $V_{BEQ} = 0.75$ V. Find the value of R_C that is required to achieve the given Q point.

9.19 Refer to Example 9.3 and Figure 9.17. Assume that all given values are unchanged except that $R_C = 340$ Ω, $I_{LED} \geq 10$ mA, and that the maximum base current supplied by the microprocessor is 5 mA. Find the range of values of R_B that satisfy these requirements.

9.20 Use the same data given in Problem 9.19, but assume that $R_B = 10$ kΩ. Find the minimum value of β that satisfies the requirements.

9.21 Repeat Problem 9.20 for the case of a microprocessor operating on a 2.8-V supply (that is, $V_{ON} = 2.8$ V).

9.22 Consider the LED driver circuit of Figure 9.17. This circuit is now used to drive an automotive fuel injector (an electromechanical solenoid valve). The differences in the circuit are as follows: The collector resistor and the LED are replaced by the fuel injector, which can be modeled as a series RL circuit. The voltage supply for the fuel injector is 13 V (instead of 5 V). For the purposes of this problem, it is reasonable to assume $R = 12\ \Omega$ and $L \sim 0$. Assume that the maximum current that can be supplied by the microprocessor is 1 mA, that the current required to drive the fuel injector must be at least 1 A, and that the transistor saturation voltage is $V_{CEsat} = 1$ V. Find the minimum value of β required for the transistor.

9.23 Refer to Problem 9.22. Assume $\beta = 7,000$. Find the allowable range of R_B.

9.24 Given the circuit of Figure P9.8, find the minimum value of R_C such that transistor operates in active mode and dissipates less than 15 mW. Let $V_{CC} = 10$ V, $R_1 = R_2 = 40$ kΩ, $R_E = 1.5$ kΩ, $V_{BE} = 0.7$ V, $\beta = 70$, and $V_{CEsat} = 0.25$ V.

9.25 The circuit shown in Figure P9.25 is a 9-V battery charger. The purpose of the Zener diode is to provide a constant voltage across resistor R_2, such that the transistor will source a constant emitter (and therefore collector) current. Select the values of R_2, R_1, and V_{CC} such that the battery will be charged with a constant 40-mA current.

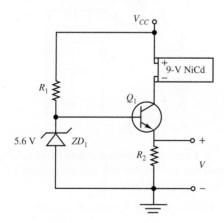

Figure P9.25

9.26 The circuit shown in Figure P9.26 is a variation of that shown in Figure P9.25. Analyze the operation of the circuit, and explain how this circuit will provide a

decreasing charging current (taper current cycle) until the NiCd battery is fully charged (10.4 V—see note in Example 9.4). Choose appropriate values of V_{CC} and R_1 that would result in a practical design. Use standard resistor values.

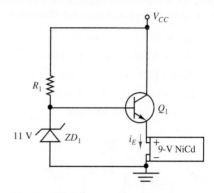

Figure P9.26

9.27 The circuit shown in Figure P9.27 is a variation of the motor driver circuit of Example 9.5. The external voltage v_{in} represents the analog output of a microcontroller and alternates between 0 and 5 V. Complete the design of the circuit by selecting the value of the base resistor R_b such that the motor will see the maximum design current when $v_{in} = 5$ V. Use the design specifications given in the example.

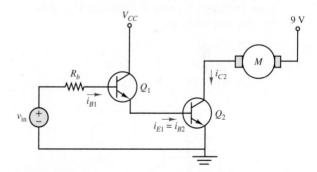

Figure P9.27

9.28 For the circuit shown in Figure 9.21, $R_C = 1$ kΩ, $V_{BB} = 5$ V, $\beta_{min} = 50$, and $V_{CC} = 10$ V. Find the range of R_B so that the transistor is in saturation.

9.29 For the circuit shown in Figure 9.21, $V_{CC} = 5$ V, $R_C = 1$ kΩ, $R_B = 10$ kΩ, and $\beta_{min} = 50$. Find the range of values of V_{BB} so that the transistor is in saturation.

9.30 For the circuit shown in Figure 9.13, $V_\gamma = 0.6$ V, $R_B = 100$ kΩ, $I_{BB} = 26\ \mu$A, $R_C = 2$ kΩ, $V_{CC} = 10$ V, and $\beta = 100$. Find I_C, I_E, V_{CE}, and V_{CB}.

Section 9.4: The Small-Signal Model and AC Amplification

9.31 The circuit shown in Figure P9.31 is a common-emitter amplifier stage. Determine the DC Thévenin equivalent of the network between the base node and the reference node. Use it to redraw the circuit.

$$V_{CC} = 20 \text{ V} \qquad \beta = 130$$
$$R_1 = 1.8 \text{ M}\Omega \qquad R_2 = 300 \text{ k}\Omega$$
$$R_C = 3 \text{ k}\Omega \qquad R_E = 1 \text{ k}\Omega$$
$$R_o = 1 \text{ k}\Omega \qquad R_S = 0.6 \text{ k}\Omega$$
$$v_S = 1 \cos(6.28 \times 10^3 t) \text{ mV}$$

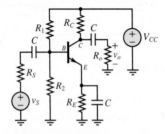

Figure P9.31

9.32 The circuit shown in Figure P9.32 is a common-collector (or emitter follower) amplifier stage implemented with an *npn* silicon transistor and a single DC supply $V_{CC} = 12$ V. Determine V_{CEQ} at the DC operating (Q) point.

$$R_o = 16 \text{ }\Omega \qquad \beta = 130$$
$$R_1 = 82 \text{ k}\Omega \qquad R_2 = 22 \text{ k}\Omega$$
$$R_S = 0.7 \text{ k}\Omega \qquad R_E = 0.5 \text{ k}\Omega$$

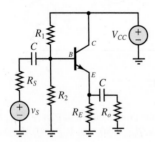

Figure P9.32

9.33 Shown in Figure P9.33 is a common-emitter amplifier stage implemented with an *npn* silicon transistor and two DC supplies $V_{CC} = 12$ V and $V_{EE} = 4$ V. Determine V_{CEQ} and the mode of operation.

$$\beta = 100 \qquad R_B = 100 \text{ k}\Omega$$
$$R_C = 3 \text{ k}\Omega \qquad R_E = 3 \text{ k}\Omega$$
$$R_o = 6 \text{ k}\Omega \qquad R_S = 0.6 \text{ k}\Omega$$
$$v_S = 1 \cos(6.28 \times 10^3 t) \text{ mV}$$

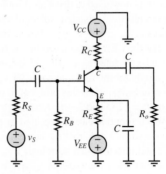

Figure P9.33

9.34 Shown in Figure P9.34 is a common-emitter amplifier stage implemented with an *npn* silicon transistor and a single DC supply $V_{CC} = 12$ V. Determine V_{CEQ} and the mode of operation.

$$\beta = 130 \qquad R_B = 325 \text{ k}\Omega$$
$$R_C = 1.9 \text{ k}\Omega \qquad R_E = 2.3 \text{ k}\Omega$$
$$R_o = 10 \text{ k}\Omega \qquad R_S = 0.5 \text{ k}\Omega$$
$$v_S = 1 \cos(6.28 \times 10^3 t) \text{ mV}$$

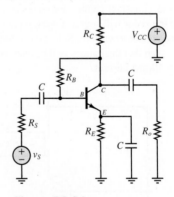

Figure P9.34

9.35 For the circuit shown in Figure P9.35 v_S is a small sine wave signal with average value of 3 V. If $\beta = 100$ and $R_B = 60$ kΩ,

a. Find the value of R_E so that I_E is 1 mA.

b. Find R_C so that V_C is 5 V.

c. For $R_o = 5$ kΩ, find the small-signal equivalent circuit of the amplifier.

d. Find the voltage gain.

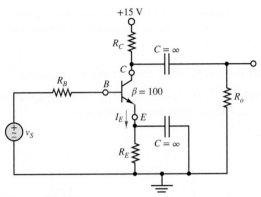

Figure P9.35

9.36 The circuit in Figure P9.36 is similar to a common collector when R_C is small. Assume $R_C = 200\ \Omega$.

a. Find the operating point Q of the transistor.

b. Find the voltage gain v_o/v_{in}.

c. Find the current gain i_o/i_{in}.

d. Find the input resistance $r_i = v_{in}/i_{in}$.

e. Find the output resistance $r_o = v_o/i_o$.

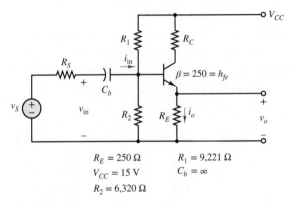

$$R_E = 250\ \Omega \qquad R_1 = 9{,}221\ \Omega$$
$$V_{CC} = 15\ V \qquad C_b = \infty$$
$$R_2 = 6{,}320\ \Omega$$

Figure P9.36

9.37 An automobile fuel injector system is depicted in Figure P9.37(a). The internal circuitry of the injector can be modeled as shown in Figure P9.37(b). The injector will inject gasoline into the intake manifold when $I_{inj} \geq 0.1\ A$. A voltage pulse train v_{signal} is shown in Figure P9.37(c). For a cold engine at start-up, the pulse width τ is determined by:

$$\tau = BIT \times K_C + VCIT$$

where

$$BIT = \text{basic injection time} = 1\ ms$$
$$K_C = \text{compensation constant of temperature}$$
$$\quad\text{of coolant } (T_C)$$
$$VCIT = \text{voltage-compensated injection time}$$

The characteristics of VCIT and K_C are shown in Figure P9.37(d). Assume the transistor Q_1 saturates at $V_{CE} = 0.3\ V$ and $V_{BE} = 0.9\ V$. Find the period of the fuel injector pulse if:

a. $V_{batt} = 13\ V$, $T_C = 100°C$

b. $V_{batt} = 8.6\ V$, $T_C = 20°C$

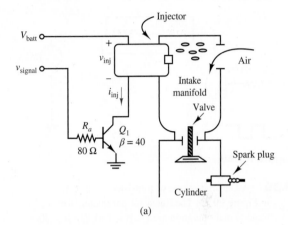

(a)

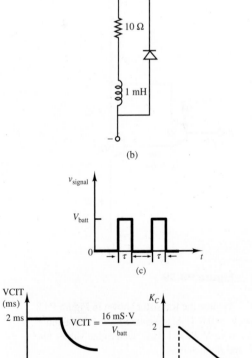

Figure P9.37

9.38 The circuit shown in Figure P9.38 is used to switch a relay that turns a light off and on under the control of a microcontroller. The relay dissipates 0.5 W at 5 VDC. It switches on at 3 VDC and off at 1.0 VDC. What is the maximum frequency with which the light can be switched? The inductance of the relay is 5 mH, and the transistor saturates at 0.2 V, $V_\gamma = 0.8$ V.

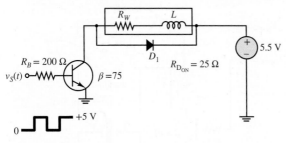

Figure P9.38

9.39 A Darlington pair of transistors is shown in Figure P9.39. The transistor parameters for large-signal operation are Q_1: $\beta = 130$; Q_2: $\beta = 70$. Calculate the overall current gain.

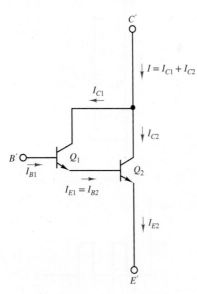

Figure P9.39

9.40 Assume the transistor shown in Figure P9.8 has $V_\gamma = 0.6$ V. Also assume $R_C = 1.5$ kΩ and $V_{CC} = 18$ V, $R_E = 1.0$ kΩ. Determine values for R_1 and R_2 such that:

a. The DC collector-emitter voltage V_{CEQ} is 5 V.

b. The DC collector current I_{CQ} will vary no more than 10 percent as β varies from 20 to 50.

c. Values of R_1 and R_2 that will permit maximum symmetrical swing in the collector current. Assume $\beta = 100$.

Section 9.5: BJT Switches and Gates

9.41 Show that the circuit of Figure P9.41 functions as an OR gate if the output is taken at v_{o1}.

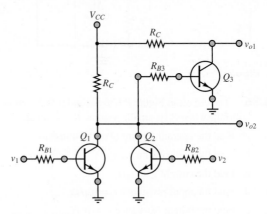

Figure P9.41

9.42 Show that the circuit of Figure P9.41 functions as a NOR gate if the output is taken at v_{o2}.

9.43 Show that the circuit of Figure P9.43 functions as an AND gate if the output is taken at v_{o1}.

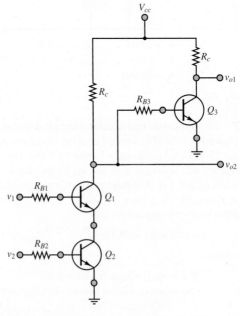

Figure P9.43

9.44   Refer to the circuit in Figure P9.14. The input voltage waveform is shown in Figure P9.44. Determine v_o assuming $\beta = 90$, $R_B = 40\ \text{k}\Omega$, $R_C = 2\ \text{k}\Omega$, and $V_{CC} = 4\ \text{V}$.

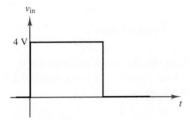

Figure P9.44

9.45   For the circuit shown in Figure P9.14, assume $\beta > 10$, and the minimum value of v_{in} for a high input is 2.0 V. Find the range for resistor R_B that guarantees the transistor is on.

9.46   Figure P9.46 shows a circuit with two transistor inverters connected in series, where

$$R_{1C} = R_{2C} = 10\ \text{k}\Omega \qquad \text{and} \qquad R_{1B} = R_{2B} = 27\ \text{k}\Omega$$

a. Find v_B, v_o, and the state of transistor Q_1 when v_{in} is low (0 V).

b. Find v_B, v_o, and the state of transistor Q_1 when v_{in} is high (5 V).

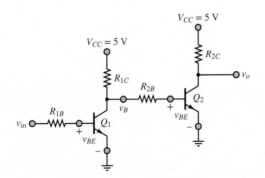

Figure P9.46

9.47   For the circuit shown in Figure P9.47, determine $v_o(t)$, where $v_{in}(t)$ is as shown in Figure P9.44. Let $\beta = 120$, $R_B = 10\ \text{k}\Omega$, $R_{C1} = R_{C2} = 1\ \text{k}\Omega$, and $V_{CC} = 4\ \text{V}$.

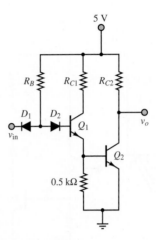

Figure P9.47

9.48   For the circuit shown in Figure P9.48, determine $v_o(t)$, where $v_{in}(t)$ is as shown in Figure P9.44. Let $\beta = 90$, $R_B = 3\ \text{k}\Omega$, $R_C = 5\ \text{k}\Omega$, and $V_{CC} = 6\ \text{V}$.

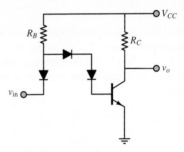

Figure P9.48

9.49   The basic circuit of a TTL gate is shown in Figure P9.49. Determine its logic function.

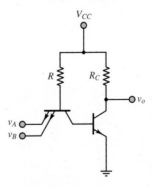

Figure P9.49

9.50   Figure P9.50 shows a three-input TTL NAND gate. Assuming that all the input voltages are high, find v_{B1}, v_{B2}, v_{B3}, v_{C2}, and v_o. Also indicate the operating mode of each transistor.

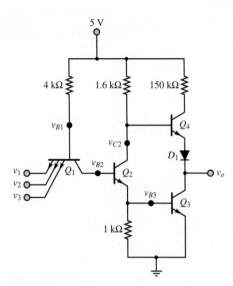

Figure P9.50

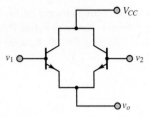

Figure P9.51

9.51 Show that two or more emitter-follower outputs connected to a common load, as shown in Figure P9.51, result in an OR operation; that is, $v_o = v_1 + v_2$. Here, the + sign represents a logical OR operation.

9.52 Verify that the circuit of Figure P9.52 is a NAND gate. Assume that a low state is 0.2 V, a high state is 5 V, and $\beta_{min} = 40$.

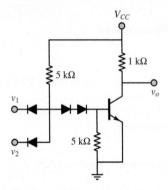

Figure P9.52

Design Credits: Mini DVI cable adapter isolated on white background: Robert Lehmann/Alamy Stock Photo; Balance scale: Alex Slobodkin/E+/Getty Images; Icon for "Focus on measurements" weighing scales: Media Bakery.

C H A P T E R

10

FIELD-EFFECT TRANSISTORS: OPERATION, CIRCUIT MODELS, AND APPLICATIONS

hapter 10 introduces the family of field-effect transistors, or FETs, in which an external electric field is used to control the conductivity of a *channel*, causing the FET to behave either as a voltage-controlled resistor or as a voltage-controlled current source. FETs are the dominant transistor family in today's integrated electronics, and although these transistors come in several different configurations, it is possible to understand the operation of the different devices by focusing principally on one type. Two large families of FETs are the JFETs (junction FETs) and the MOSFETs (metal-oxide semiconducting FETs). Both families can be further classified by a mode (enhancement or depletion) and a channel type (*n* or *p*). In this chapter, the focus is on the enhancement-mode MOSFET with either an *n*-type channel (NMOS) or *p*-type channel (PMOS). The very important CMOS technology, which combines both NMOS and PMOS, is also introduced.

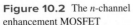

Learning Objectives

Students will learn to...
1. Understand the classification of field-effect transistors. *Section 10.1.*
2. Learn the basic operation of enhancement-mode MOSFETs by understanding their *i-v* curves and defining equations. *Section 10.2.*
3. Learn how enhancement-mode MOSFET circuits are biased. *Section 10.3.*
4. Understand the concept and operation of FET large-signal amplifiers. *Section 10.4.*
5. Understand the concept and operation of FET switches. *Section 10.5.*
6. Analyze FET switches and digital gates. *Section 10.5.*

Enhancement MOS

p channel   *n* channel

Depletion MOS

p channel   *n* channel

JFET

p channel   *n* channel

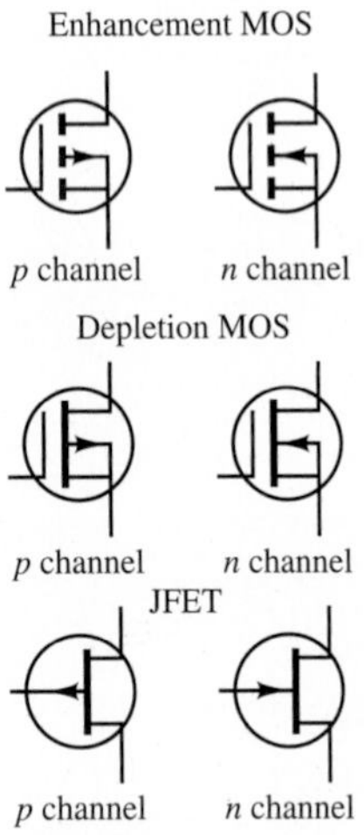

Figure 10.1 Classification of field-effect transistors

Gate

Source   Drain

n^+ *p* n^+

Bulk (substrate)

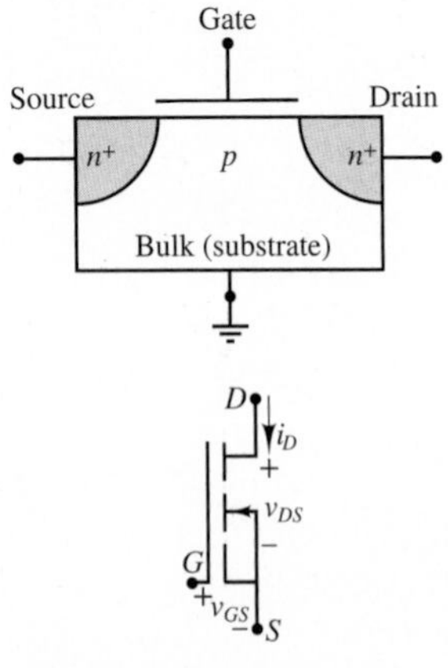

Figure 10.2 The *n*-channel enhancement MOSFET construction and circuit symbol

10.1   FIELD-EFFECT TRANSISTOR CLASSES

There are three major classes of field-effect transistors:

1. **Enhancement-mode MOSFETs**
2. **Depletion-mode MOSFETs**
3. **Junction field-effect transistors**, or **JFETs**

Each of these classes is comprised of *n* and *p*-channel devices, where the *n* or *p* designation indicates the nature of the doping in the semiconductor channel. The acronym MOSFET stands for **metal-oxide semiconductor field-effect transistor**, and although the specific materials and processes used in fabricating transistors has, of course, evolved over time, the acronym continues to be used to describe all enhancement-mode and depletion-mode FETs.

Figure 10.1 shows typical circuit symbols for the *n*- and *p*-channel devices within each of the three transistor classes. These transistors have similar behaviors and applications; for the sake of brevity, only the enhancement-mode MOSFET is discussed in detail in this chapter. All the FETs are *unipolar* devices in that current is conducted by only one type of charge carrier, either holes or electrons, unlike BJTs that conduct current using both holes and electrons. Also, whereas both FETs and BJTs are three-terminal devices, the BJTs are asymmetric devices because the collector and emitter are not interchangeable. However, in FETs, the analogous terminals, known as the drain and source, are completely symmetric and therefore interchangeable.

10.2   ENHANCEMENT-MODE MOSFETs

Figure 10.2 depicts the circuit symbol and the construction of a typical *n*-channel enhancement-mode MOSFET. The device has four regions: the **gate**, the **drain**, the **source**, and the **bulk**.[1] Each of these regions has its own conducting terminal. The bulk and source terminals are often electrically connected, in which case the bulk terminal is not shown in the circuit symbol. The gate consists of a conducting plate separated from the *p*-type bulk by a thin (10^{-9} m) insulating layer, usually silicon oxide SiO_2.[2] The drain and source regions are both composed of n^+ material.

---

[1] The bulk is also known as the **substrate**, **body**, or **base**.

[2] In the past, a metal oxide was used, which explains the terminology *metal-oxide semiconductor* MOS.

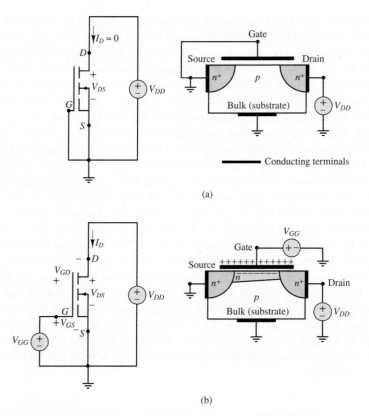

Figure 10.3 Channel formation in NMOS transistor: (a) With zero gate
voltage, the source-bulk and bulk-drain junctions are both reverse-biased, and
the channel acts as an open-circuit. (b) When a positive gate voltage is applied,
positive majority carriers in the bulk (i.e., holes) are repelled by the gate leaving
behind negatively charged atoms. Also, negative majority carriers from the
source and drain (i.e., electrons) are drawn toward the gate. The result is a
conducting n-type channel between the source and drain regions.

Consider the case when the gate and source terminals are connected to a
reference node and the drain terminal is connected to a positive voltage supply
V_{DD}, as shown in Figure 10.3(a). The bulk terminal is also connected to the
reference node, by virtue of its connection to the source terminal, and so the pn^+
junction between the bulk and drain is reverse-biased. Likewise, the voltage
across the pn^+ junction between the bulk and the source is zero, and thus that
junction is also reverse-biased. Thus, a path between drain and source consists
of two reverse-biased pn^+ junctions such that the current from drain to source is
effectively zero. In this case, the resistance from drain to source is on the order
of 10^{12} Ω.

When the voltage from gate to source is zero, the n-channel enhancement-
mode MOSFET acts as an open-circuit. Thus, enhancement-mode devices
are referred to as *normally off* and their channels as *normally open*.

Suppose now that a positive DC voltage V_{GG} is applied to the gate as shown in Figure 10.3(b). Positive majority charge carriers in the bulk (i.e., holes) are repelled in the region nearest the gate. At the same time, negative majority charge carriers in the source and drain (i.e., electrons) are drawn to the same region. The result is a narrow n-type **channel** beneath the insulating layer that separates the gate from the bulk. For a given drain voltage, the higher the gate voltage, the higher the concentration of negative charge carriers in the channel, and the higher its conductivity. The term *enhancement mode* refers to the influence of the gate voltage in enhancing the conductivity of the channel. The term *field effect* refers to the effect of the electric field from gate to bulk that is associated with the gate voltage.

Depletion-mode devices also exist, in which an externally applied field depletes the channel of charge carriers by reducing the effective channel width. Depletion-mode MOSFETs are normally on (i.e., the channel is conducting) and are turned off (i.e., the channel is not conducting) by an external gate voltage.

Both enhancement- and depletion-mode MOSFETs are available with either n- or p-type channels. Enhancement-mode devices do not have a conducting channel built in; however, one can be created by the *action* of the gate. On the other hand, depletion-mode devices do have a built-in conducting channel that can be depleted by the *action* of the gate. Depending upon the mode and channel type, FETs can be *active high* or *active low* devices, where *high* and *low* refer to the voltage of the gate relative to a common reference. Table 10.1 summarizes these results. n- and p-channel MOSFETs are referred to as **NMOS** and **PMOS** transistors, respectively.

Table 10.1

Channel Type	Mode	
	Enhancement	Depletion
n	Active high	Active low
p	Active low	Active high

Operating Regions and the Threshold Voltage V_t

When the gate-to-bulk voltage of an NMOS transistor (see Figure 10.4) is less than a threshold voltage V_t, a channel will not form between the source and drain. The result is that no current can be conducted from drain to source and the transistor is in the *cutoff region*. Typical values of V_t are between 0.3 and 1.0 V, although it can be significantly larger.

When the gate-to-bulk voltage is greater than the threshold voltage V_t at any point between the source and drain, a conducting n-type channel is formed at that point. If, as usual, the source and bulk are both connected to a common reference, then the gate-to-bulk voltage is the same as the gate-to-source voltage v_{GS}. If the drain is also connected to the same common reference such that $v_{DS} = 0$, then a channel of uniform thickness will be formed from drain to source when $v_{GS} > V_t$. It is common to introduce the *overdrive voltage* $v_{OV} = v_{GS} - V_t$, which is the gate-to-source voltage in excess of what is necessary to create a channel. Note that $v_{OV} > 0$ is another way to write $v_{GS} > V_t$.

Note that if $v_{DS} = 0$, then $v_{GD} \equiv v_{GS} - v_{DS} = v_{GS}$, the channel has a uniform thickness, and its resistance per unit channel length is also uniform. In this state, known as the *ohmic region*, the channel effectively acts as a variable resistor whose resistance is dictated by the gate voltage. In other words, for a given value of v_{GS}, the channel current i_D is proportional to v_{DS}. This linear relationship between i_D and v_{DS} is valid for small values of v_{DS}.

$$i_D \propto v_{DS} \qquad \text{when} \quad v_{DS} \ll v_{OV} \qquad \text{Ohmic region} \tag{10.1}$$

When $v_{GS} > V_t$ and the drain-to-source voltage v_{DS} is no longer small but held at a positive value V_{DD}, the channel is thinner near the drain than near the source, as depicted in Figure 10.3(b). In addition, as long as $v_{GD} > V_t$, the channel will still exist from source to drain. This condition is equivalent to the requirement that $v_{DS} < v_{OV}$. In this state, the channel resistance per unit length is no longer uniform, the channel current i_D is proportional to v_{DS}^2, and the transistor is in the *triode region*.

$$i_D \propto v_{DS}^2 \qquad \text{when} \quad v_{DS} < v_{OV} \qquad \text{Triode region} \tag{10.2}$$

It is important to realize that the ohmic region is simply one part of the triode region when $v_{DS} \ll v_{OV}$.

Eventually, if v_{DS} is continually increased, it will exceed v_{OV} such that the channel thickness at the drain goes to zero. In fact, the depletion region of the bulk-drain junction has expanded sufficiently, due to the increase in v_{DS}, to take the place of the channel. This condition is often called *channel pinch-off*. Although the channel thickness is now zero, current is still conducted in the channel because the voltage at the drain is large enough to drive mobile electrons in the channel across the depletion region. However, any increase in v_{DS} beyond v_{OV} is confined to the depletion region such that the voltage across the channel length remains constant. The result is that the channel current is independent of v_{DS} and depends only upon v_{OV}. In this state, the transistor is in the *saturation region*.

$$i_D \propto v_{OV}^2 \qquad \text{when} \quad v_{DS} > v_{OV} \qquad \text{Saturation region} \tag{10.3}$$

These operating regions and their dependence upon v_{GD} and v_{GS} are depicted in Figure 10.4.

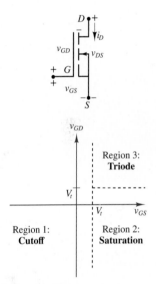

Figure 10.4 Regions of operation of NMOS transistor

Channel Current i_D and the Conductance Parameter K

The ability of the channel to conduct is dependent on various mechanisms, the effects of which are captured in a conductance parameter K, defined as:

$$K = \frac{W}{L} \frac{\mu C_{ox}}{2} \tag{10.4}$$

where W is the cross-sectional width of the channel, L is the channel length, μ is the mobility of the majority channel charge carrier (electrons in n-channel devices, holes in p-channel devices), and C_{ox} is the gate-channel capacitance due to the thin insulating oxide layer. The units of K are A/V^2.

With this definition of the conductance parameter, the relationship between i_D and v_{DS} can be expressed in the various operating regions as listed here. In the cutoff region:

$$i_D = 0 \qquad \text{when} \quad v_{GS} \ll V_t \qquad \text{Cutoff region} \tag{10.5}$$

In the triode region:

$$i_D = K(2v_{OV} - v_{DS})v_{DS} \qquad \text{when} \quad v_{DS} < v_{OV} \qquad \text{Triode region} \tag{10.6}$$

When $v_{DS} \ll v_{OV}$, this expression is approximated by

$$i_D \approx 2K v_{OV} v_{DS} \qquad \text{when} \quad v_{DS} \ll v_{OV} \qquad \text{Ohmic region} \tag{10.7}$$

which is the linear relationship characteristic of the ohmic region. In the ohmic region, the transistor acts as a voltage-controlled resistor. This property allows transistors to act as resistors in integrated-circuit (IC) designs. Other applications of a voltage-controlled resistor are found in tunable (variable-gain) amplifiers and in analog gates.

In the saturation region:

$$i_D \approx K v_{OV}^2 \qquad \text{when} \quad v_{DS} > v_{OV} \qquad \text{Saturation region} \tag{10.8}$$

This relationship is only approximate. This relationship is made more exact by accounting for the **Early effect**, which describes the effect of v_{DS} on the effective length of the channel. This effect is accounted for by incorporating the **Early voltage** V_A as:

$$i_D = K v_{OV}^2 \left(1 + \frac{v_{DS}}{V_A} \right) \qquad \text{when} \quad v_{DS} > v_{OV} \qquad \text{Saturation region} \tag{10.9}$$

When V_A is large compared to v_{DS}, as is often the case, the Early effect is small and equation 10.9 is well approximated by equation 10.8. When this condition is true, the transistor acts as a *voltage-controlled current source*.

The three regions of operation can also be identified in the characteristic curves shown in Figure 10.5, which can be generated from the circuit of Figure 10.3(b) by varying the gate and drain voltages relative to the source voltage. Notice that for $v_{GS} < V_t$ the transistor is in the cutoff region and $i_D = 0$. The boundary between the saturation and triode regions is indicated by the curve $i_D = K v_{DS}^2$, which is the locus of all points where the slope of the characteristic curve first becomes zero as v_{DS} increases. (If the Early voltage V_A is not negligible, then the slope of the characteristic lines in saturation is not zero, but some small positive constant.)

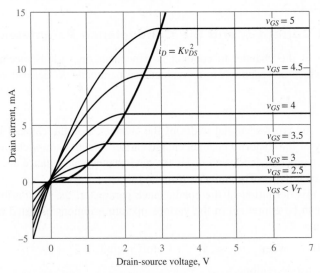

Figure 10.5 Characteristic drain curves for an NMOS transistor with $V_t = 2$ V and $K = 1.5$ mA/V^2

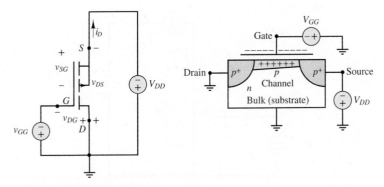

Figure 10.6 The *p*-channel enhancement-mode field-effect transistor (PMOS)

In the saturation region, the transistor drain current is nearly constant and independent of v_{DS}. In fact, its value is proportional to v_{GS}^2. Finally, in the triode region, the drain current is strongly dependent on v_{GS} and v_{DS}. As $v_{DS} \to 0$ the slope of each characteristic curve becomes approximately constant, which is the characteristic of the ohmic region.

Operation of the *P*-channel Enhancement-Mode MOSFET

The operation of a PMOS enhancement-mode transistor is very similar in concept to that of an NMOS device. Figure 10.6 depicts a test circuit and a sketch of the device construction. Note that the roles of the *n*-type and *p*-type materials are reversed and that the charge carriers in the channel are holes, not electrons. Further, the threshold voltage V_t is now negative. However, if v_{GS} is replaced with v_{SG}, v_{GD} with v_{DG}, and v_{DS} with v_{SD}, and $|V_t|$ is used in place of V_t, then the analysis of the device is completely analogous to that of an NMOS transistor. In particular, Figure 10.7 depicts the behavior of a PMOS transistor in terms of the gate-to-drain and gate-to-source voltages, in analogy with Figure 10.4. The resulting equations for the three modes of operation of the PMOS transistor are summarized below:

Cutoff region: when $v_{SG} < |V_t|$ and $v_{DG} < |V_t|$.

$$i_D = 0 \qquad \text{Cutoff region} \tag{10.10}$$

Saturation region: when $v_{SG} > |V_t|$ and $v_{DG} < |V_t|$.

$$i_D \cong K(v_{SG} - |V_t|)^2 \qquad \text{Saturation region} \tag{10.11}$$

Triode region: when $v_{SG} > |V_t|$ and $v_{DG} > |V_t|$.

$$i_D = K[2(v_{SG} - |V_t|)v_{SD} - v_{SD}^2] \qquad \text{Triode or ohmic region} \tag{10.12}$$

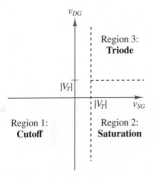

Figure 10.7 Regions of operation of PMOS transistor

EXAMPLE 10.1 Determining the Operating State of a MOSFET

Problem

Determine the operating state of the MOSFET shown in the circuit of Figure 10.8 for the given values of V_{DD} and V_{GG} if the ammeter and voltmeter shown read the following values:

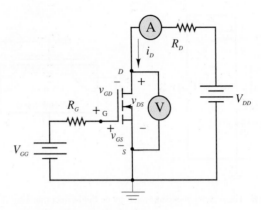

Figure 10.8 Circuit for Example 10.1.

---

a. $V_{GG} = 1$ V; $V_{DD} = 10$ V; $v_{DS} = 10$ V; $i_D = 0$ mA; $R_D = 100$ Ω.

b. $V_{GG} = 4$ V; $V_{DD} = 10$ V; $v_{DS} = 2.8$ V; $i_D = 72$ mA; $R_D = 100$ Ω.

c. $V_{GG} = 3$ V; $V_{DD} = 10$ V; $v_{DS} = 1.5$ V; $i_D = 13.5$ mA; $R_D = 630$ Ω.

Solution

Known Quantities: MOSFET drain resistance; drain and gate supply voltages; MOSFET equations.

Find: MOSFET quiescent drain current i_{DQ} and quiescent drain-source voltage v_{DSQ}.

Schematics, Diagrams, Circuits, and Given Data: $V_t = 2$ V; $K = 18$ mA/V^2.

Assumptions: None.

Analysis: First, notice that the diode indicator in Figure 10.8 points from bulk to channel. These arrows always point from p to n; thus, the channel is n-type and the transistor is an NMOS. The channel is also marked by a dashed line indicating enhancement mode.

a. Since the drain current is zero, the MOSFET is in the cutoff region. You should verify that both the conditions $v_{GS} < V_t$ and $v_{GD} < V_t$ are satisfied.

b. In this case, $v_{GS} = V_{GG} = 4$ V $> V_t$. On the other hand, $v_{GD} = v_G - v_D = 4 - 2.8 = 1.2$ V $< V_t$. Thus, the transistor is in the saturation region. We can calculate the drain current to be $i_D = K(v_{GS} - V_t)^2 = 18 \times (4 - 2)^2 = 72$ mA. Alternatively, the drain current can be calculated as:

$$i_D = \frac{V_{DD} - v_{DS}}{R_D} = \frac{10 - 2.8}{0.1\,\text{k}\Omega} = 72 \text{ mA}$$

c. In the third case, $v_{GS} = V_{GG} = v_G = 3$ V $> V_t$. The drain voltage is measured to be $v_{DS} = v_D = 1.5$ V, and therefore $v_{GD} = 3 - 1.5 = 1.5$ V $< V_t$. In this case, the MOSFET is in the ohmic, or triode, region. We can now calculate the current to be $i_D = K[2(v_{GS} - V_t)v_{DS} - v_{DS}^2] = 18 \times [2 \times (3 - 2) \times 1.5 - 1.5^2] = 13.5$ mA. The drain current can also be calculated as:

$$i_D = \frac{V_{DD} - v_{DS}}{R_D} = \frac{(10 - 1.5)\,\text{V}}{0.630 \text{ k}\Omega} = 13.5 \text{ mA}$$

CHECK YOUR UNDERSTANDING

What is the operating state of the MOSFET of Example 10.1 for the following conditions?

$$V_{GG} = \frac{10}{3} \text{ V} \qquad V_{DD} = 10 \text{ V} \qquad v_{DS} = 3.6 \text{ V} \qquad i_D = 32 \text{ mA} \qquad R_D = 200 \ \Omega$$

10.3   BIASING MOSFET CIRCUITS

Now that the basic characteristics of enhancement-mode MOSFETs and the means for identifying operating regions are known, it is time to develop systematic procedures for biasing a MOSFET. This section presents two bias circuits, which are identical to those presented for biasing BJTs. The first, illustrated in Examples 10.2 and 10.3, uses two distinct voltage supplies. This bias circuit is easier to understand, but not very practical—as was discussed in the chapter on BJTs, it is preferable to have a single DC voltage supply and to enable the circuit to regulate its bias point. These features are presented in the second bias circuit, described in Examples 10.4 and 10.5.

EXAMPLE 10.2   MOSFET Q-point Graphical Determination

Problem

Determine the Q point for the MOSFET in the circuit of Figure 10.9.

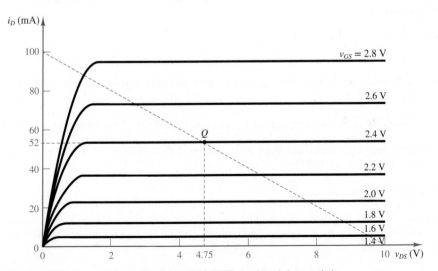

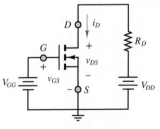

Figure 10.9 An n-channel enhancement MOSFET circuit and characteristics

---

Solution

Known Quantities: MOSFET drain resistance; drain and gate supply voltages; MOSFET drain curves.

Find: MOSFET quiescent drain current i_{DQ} and quiescent drain-source voltage v_{DSQ}.

Schematics, Diagrams, Circuits, and Given Data: $V_{GG} = 2.4$ V; $V_{DD} = 10$ V; $R_D = 100\ \Omega$.

Assumptions: Use the characteristic curves of Figure 10.9.

Analysis: First, notice that the diode indicator in Figure 10.9 points from bulk to channel. These arrows always point from p to n; thus, the channel is n-type and the transistor is an NMOS. The channel is also marked by a dashed line indicating enhancement mode.
   To determine the Q point, write the drain circuit equation and apply KVL:

$$V_{DD} = R_D i_D + v_{DS}$$
$$10 = 100\, i_D + v_{DS}$$

The resulting curve is plotted as a dashed line on the drain curves of Figure 10.9 by noting that the drain current axis intercept is equal to $V_{DD}/R_D = 100$ mA and that the drain-source voltage axis intercept is equal to $V_{DD} = 10$ V. The Q point is then given by the intersection of the load line with the $v_{GS} = 2.4$ V curve. Thus, $i_{DQ} = 52$ mA and $v_{DSQ} = 4.75$ V.

Comments: The determination of a Q point for a MOSFET is easier than for a BJT because the gate current is essentially zero.

EXAMPLE 10.3   MOSFET Q-point Calculation

Problem

Use the MOSFET characteristic curves shown in Figure 10.9 to determine the Q point for the conditions listed below.

Solution

Known Quantities: MOSFET drain resistance; drain and gate supply voltages; MOSFET equations.

Find: MOSFET quiescent drain current i_{DQ} and quiescent drain-source voltage v_{DSQ}.

Schematics, Diagrams, Circuits, and Given Data: $V_{GG} = 2.4$ V; $V_{DD} = 10$ V; $V_t = 1.4$ V; $K = 48.5$ mA/V^2; $R_D = 100\ \Omega$.

Assumptions: None.

Analysis: The gate supply V_{GG} ensures that $v_{GSQ} = V_{GG} = 2.4$ V. Thus, $v_{GSQ} > V_t$. We assume that the MOSFET is in the saturation region, and we proceed to use equation 10.8 to calculate the drain current:

$$i_{DQ} = K(v_{GS} - V_t)^2 = 48.5 \times 10^{-3}(2.4 - 1.4)^2 = 48.5\ \text{mA}$$

Applying KVL to the drain loop, we can calculate the quiescent drain-to-source voltage as:

$$v_{DSQ} = V_{DD} - R_D i_{DQ} = 10 - 100 \times 48.5 \times 10^{-3} = 5.15 \text{ V}$$

Now we can verify the assumption that the MOSFET was operating in the saturation region. Recall that the conditions required for operation in region 2 (saturation) were $v_{GS} > V_t$ and $v_{GD} < V_t$. The first condition is clearly satisfied. The second can be verified by recognizing that:

$$v_{GD} = v_{GS} + v_{SD} = v_{GS} - v_{DS} = -2.75 \text{ V}$$

Clearly, the condition $v_{GD} < V_t$ is also satisfied, and the MOSFET is indeed operating in the saturation region.

EXAMPLE 10.4   MOSFET Self-Bias Circuit

Problem

Figure 10.10(a) depicts a self-bias circuit for a MOSFET. Determine the Q point for the MOSFET by choosing R_S such that $v_{DSQ} = 8$ V.

Solution

Known Quantities: MOSFET drain and gate resistances; drain supply voltage; MOSFET parameters V_t and K; desired drain-to-source voltage v_{DSQ}.

Find: MOSFET quiescent gate-source voltage v_{GSQ}, quiescent drain current i_{DQ}, and quiescent drain-source voltage v_{DSQ}.

Schematics, Diagrams, Circuits, and Given Data: $V_{DD} = 30$ V; $R_D = 10$ kΩ; $R_1 = R_2 = 1.2$ MΩ; $V_t = 4$ V; $K = 0.2188$ mA/V²; $v_{DSQ} = 8$ V.

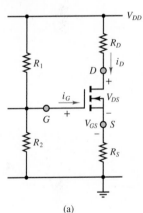

(a)

Assumptions: Assume operation in the saturation region.

Analysis: First we reduce the circuit of Figure 10.10(a) to the circuit of Figure 10.10(b), in which the voltage divider rule has been used to compute the value of the equivalent network seen by the gate.

$$V_{GG} = \frac{R_2}{R_1 + R_2} V_{DD} = 15 \text{ V} \qquad R_G = R_1 \| R_2 = 600 \text{ kΩ}$$

Let all currents be expressed in milliamps and all resistances in kilo-ohms. Applying KVL around the equivalent gate circuit of Figure 10.10(b) yields:

$$v_{GSQ} + i_{GQ} R_G + i_{DQ} R_S = V_{GG} = 15 \text{ V}$$

Since $i_{GQ} = 0$, due to the infinite input resistance of the MOSFET, the gate equation simplifies to:

$$v_{GSQ} + i_{DQ} R_S = 15 \text{ V} \tag{a}$$

The drain circuit equation is:

$$v_{DSQ} + i_{DQ} R_D + i_{DQ} R_S = V_{DD} = 30 \text{ V} \tag{b}$$

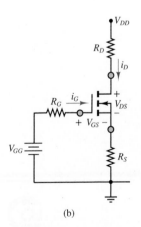

(b)

Figure 10.10 (a) Self-bias circuit; (b) equivalent circuit for part (a)

Use equation 10.8 to obtain:

$$i_{DQ} = K(v_{GS} - V_t)^2 \tag{c}$$

The third equation needed to solve for the three unknowns v_{GSQ}, i_{DQ}, and v_{DSQ} can be obtained from equation (a) as:

$$i_{DQ}R_S = V_{GG} - v_{GSQ} = \frac{V_{DD}}{2} - v_{GSQ}$$

Substitute the result into equation (b) to find:

$$V_{DD} = i_{DQ}R_D + v_{DSQ} + \frac{V_{DD}}{2} - v_{GSQ}$$

or

$$i_{DQ} = \frac{1}{R_D}\left(\frac{V_{DD}}{2} - v_{DSQ} + v_{GSQ}\right)$$

Substitute the above equation for i_{DQ} into equation (c) to obtain a quadratic equation that can be solved for v_{GSQ} since the desired value of v_{DSQ} is known:

$$\frac{1}{R_D}\left(\frac{V_{DD}}{2} - v_{DSQ} + v_{GSQ}\right) = K(v_{GSQ} - V_t)^2$$

$$K v_{GSQ}^2 - 2K V_t v_{GSQ} + K V_t^2 - \frac{1}{R_D}\left(\frac{V_{DD}}{2} - v_{DSQ}\right) - \frac{1}{R_D} v_{GSQ} = 0$$

$$v_{GSQ}^2 - \left(2V_t + \frac{1}{KR_D}\right) v_{GSQ} + V_t^2 - \frac{1}{KR_D}\left(\frac{V_{DD}}{2} - v_{DSQ}\right) = 0$$

$$v_{GSQ}^2 - 8.457 v_{GSQ} + 12.8 = 0$$

The two solutions for the above quadratic equation are:

$$v_{GSQ} = 6.48 \text{ V} \qquad \text{and} \qquad v_{GSQ} = 1.97 \text{ V}$$

Only the first of these two values is acceptable for operation in the saturation region, since the second root corresponds to a value of v_{GS} lower than the threshold voltage $V_t = 4$ V. Substitute the first value into equation (c) to compute the quiescent drain current:

$$i_{DQ} = 1.35 \text{ mA}$$

Use this value in the gate circuit equation (a) to compute the solution for the source resistance:

$$R_S = 6.32 \text{ k}\Omega$$

Comments: There are two mathematical solutions to this problem because the drain current equation is a quadratic equation. The physical constraints of the problem must be used to select the appropriate solution.

EXAMPLE 10.5   Analysis of a MOSFET Amplifier

Problem

Determine the gate and drain-source voltage and the drain current for the MOSFET amplifier of Figure 10.11.

Solution

Known Quantities: Drain, source, and gate resistors; drain supply voltage; MOSFET parameters.

Find: v_{GS}; v_{DS}; i_D.

Schematics, Diagrams, Circuits, and Given Data: $R_1 = R_2 = 1$ MΩ; $R_D = 6$ kΩ; $R_S = 6$ kΩ; $V_{DD} = 10$ V; $V_t = 1$ V; $K = 0.5$ mA/V^2.

Assumptions: The MOSFET is operating in the saturation region.

Analysis: The gate voltage is computed by applying the voltage divider rule between resistors R_1 and R_2 since the gate current is zero:

$$v_G = \frac{R_2}{R_1 + R_2}V_{DD} = \frac{1}{2}V_{DD} = 5 \text{ V}$$

Assuming saturation region operation, write:

$$v_{GS} = v_G - v_S = v_G - R_S i_D = 5 - 6i_D$$

The drain current can be computed from equation 10.8:

$$i_D = K(v_{GS} - V_t)^2 = 0.5\,(5 - 6i_D - 1)^2$$

or

$$36\,i_D^2 - 50\,i_D + 16 = 0$$

with solutions

$$i_D = 0.89 \text{ mA} \qquad \text{and} \qquad i_D = 0.5 \text{ mA}$$

To determine which of these two solutions should be chosen, compute the gate-source voltage for each one. For $i_D = 0.89$ mA, $v_{GS} = 5 - 6i_D = -0.34$ V. For $i_D = 0.5$ mA, $v_{GS} = 5 - 6i_D = 2$ V. Since v_{GS} must be greater than V_t for the MOSFET to be in the saturation region, we select the solution

$$i_D = 0.5 \text{ mA} \qquad v_{GS} = 2 \text{ V}$$

The corresponding drain voltage is therefore found to be

$$v_D = v_{DD} - R_D i_D = 10 - 6i_D = 7 \text{ V}$$

And therefore

$$v_{DS} = v_D - v_S = v_D - i_D R_S = 7 - 3 = 4 \text{ V}$$

Comments: Now that the desired voltages and current have been computed, the assumption of a saturation operating condition can be verified: $v_{GS} = 2 > V_t$ and $v_{GD} = v_{GS} - v_{DS} = 2 - 4 = -2 < V_t$. Since the inequalities are satisfied, the MOSFET is indeed operating in the saturation region.

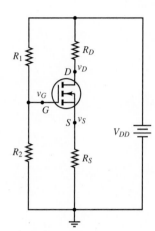

Figure 10.11 Circuit for Example 10.5.

CHECK YOUR UNDERSTANDING

Determine the operating region of the MOSFET of Example 10.2 when $v_{GS} = 3.5$ V.

Answer: The MOSFET is in the ohmic region.

CHECK YOUR UNDERSTANDING

Find the lowest value of R_D for the MOSFET of Example 10.3 that will place the MOSFET in the triode region.

Answer: $\approx 185.6\ \Omega$

CHECK YOUR UNDERSTANDING

Determine the appropriate value of R_S if we wish to move the operating point of the MOSFET of Example 10.4 to $v_{DSQ} = 12$ V. Also find the values of v_{GSQ} and i_{DQ}. Are these values unique?

Answer: The answer is unique. One of the two solutions is $v_{GS} = 2.42$ V, but this value is less than V_t, so it is not valid. The other solution is $v_{GS} = 6.03$ V with $R_S = 9.9$ kΩ and $i_D = 0.9$ mA.

10.4   MOSFET LARGE-SIGNAL AMPLIFIERS

The objective of this section is to illustrate how a MOSFET can be used as a large-signal amplifier, in applications similar to those illustrated in Chapter 9 for bipolar transistors. Equation 10.8 describes the approximate saturation region relationship between the drain current and gate-source voltage for the MOSFET in a large-signal amplifier application. Appropriate biasing, as explained in the preceding section, is used to ensure that the MOSFET is operating in saturation.

$$i_D \approx K(v_{GS} - V_t)^2 \tag{10.13}$$

MOSFET amplifiers are commonly found in one of two configurations: *common-source* and *source-follower*. Figure 10.12 depicts a basic common-source configuration. Note that when the MOSFET is in saturation, this amplifier is essentially a

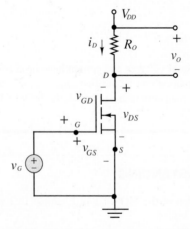

Figure 10.12 Common-source MOSFET amplifier

voltage-controlled current source (VCCS), in which the drain current is controlled by the gate voltage. Thus, the load voltage v_o across the load R_o is:

$$v_o = R_o i_D \approx R_o K(v_{GS} - V_t)^2 = R_o K(V_G - V_t)^2 \tag{10.14}$$

A source-follower amplifier is shown in Figure 10.13(a). Note that the load is now connected between the source and ground. The behavior of this circuit depends on the load current and can be analyzed for the resistive load of Figure 10.13 by observing that the load voltage is given by the expression $v_o = R_o i_D$, where

$$i_D \approx K(v_{GS} - V_t)^2 = K(v_G - V_t - v_o)^2 = K(\Delta v - R_o i_D)^2 \tag{10.15}$$

where $\Delta v = v_G - V_t$. Expand the quadratic term to obtain:

$$i_D = K(\Delta v - R_o i_D)^2 = K\Delta v^2 - 2K\Delta v R_o i_D + R_o^2 i_D^2 \tag{10.16}$$

This expression can be rearranged to yield:

$$i_D^2 - \frac{1}{R_o^2}(2K\Delta v R_o + 1)i_D + \frac{K}{R_o^2}\Delta v^2 = 0 \tag{10.17}$$

Use the quadratic equation to solve for the load current:

$$i_D = \frac{1}{2}\left(-b \pm \sqrt{b^2 - 4c}\right) \tag{10.18}$$

where

$$b = \frac{2K\Delta v R_o + 1}{R_o^2} \quad \text{and} \quad c = \frac{K\Delta v^2}{R_o^2} \tag{10.19}$$

Figure 10.13(b) depicts the drain current response of the source-follower MOSFET amplifier when the gate voltage varies between the threshold voltage and 5 V for a

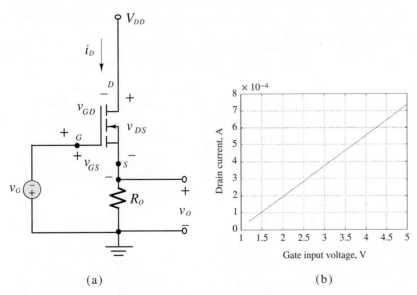

(a)   (b)

Figure 10.13 (a) Source-follower MOSFET amplifier; (b) drain current response of a common-source amplifier for a 100-Ω load when $K = 0.018$ and $V_t = 1.2$ V

100-Ω load when $K = 0.018$ and $V_t = 1.2$ V. Note that the response of this amplifier is linear in the gate voltage. This behavior is due to the fact that the source voltage increases as the drain current increases, since the source voltage is proportional to i_D.

 EXAMPLE 10.6 Using a MOSFET as a Current Source for Battery Charging

Problem

Analyze the two battery charging circuits shown in Figure 10.14(a) and (b). Use the transistor parameters to determine the range of required gate voltages v_G to provide a variable charging current up to a maximum of 0.1 A. Assume that the terminal voltage of a discharged battery is 9 V, and of a charged battery is 10.5 V.

Solution

Known Quantities: Transistor large-signal parameters, NiCd battery nominal voltage.

Find: V_{DD}, v_G, range of gate voltages leading to a maximum charging current of 0.1 A.

Schematics, Diagrams, Circuits, and Given Data: Figure 10.14(a) and (b). $V_t = 1.2$ V; $K = 18$ mA/V^2, $V_B = 9$ V.

Assumptions: Assume that the MOSFETs are operating in the saturation region.

Analysis:

a. The conditions for the MOSFET to be in the saturation region are: $v_{GS} > V_t$ and $v_{GD} < V_t$. The first condition is satisfied when $v_G \geq 1.2$ V. Assuming for the moment that both conditions are satisfied, and that V_{DD} is sufficiently large, the drain current can be calculated as:

$$i_D = K(v_{GS} - V_t)^2 = 0.018 \times (v_G - 1.2)^2 \text{ A}$$

The plot of Figure 10.14(c) depicts the battery charging (drain) current as a function of the gate voltage. The maximum charging current of 100 mA is generated at a gate voltage of approximately 3.5 V.

The requirement for the saturation region is that $v_{DS} > v_{GS} - V_t$, which is equivalent to $v_{GD} < V_t$. But be careful in interpreting this last equation. If $v_{DS} > v_{GS} - V_t$,

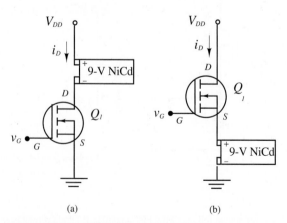

(a)   (b)

Figure 10.14 MOSFET battery charger. (a) common-source current source; (b) common-drain current source

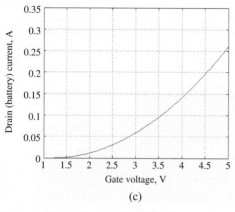

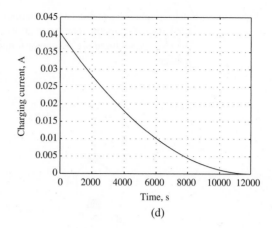

Figure 10.14 (*Continued*) MOSFET battery charger

then v_{GD} may be negative, since the drain voltage could be larger than the source voltage. Consider that $v_{GD} = v_G - v_D$ and $v_D = V_{DD} - V_B$, where V_B is the battery voltage. Then, the condition $v_{GD} < V_t$ can be rewritten as:

$$v_{GD} = v_G - v_D < V_t$$
$$v_D > v_G - V_t$$
$$V_{DD} - V_B > v_G - V_t$$

To ensure that the NMOS remains in the saturation region throughout the range of battery voltage, V_{DD} must be sufficiently large. In this case V_{DD} should be larger than 12.8 V.

b. The analysis of the second circuit is based on the observation that the voltage at the source terminal of the MOSFET is equal to the gate voltage minus the threshold voltage. If the battery is to be charged to 10.5 V, the gate voltage must be at least 11.7 V to satisfy $v_{GS} > V_t$. Assume that the battery is initially discharged (9 V), and calculate the initial charging current.

$$i_D = K(v_{GS} - V_t)^2 = K(v_G - V_B - V_t)^2 = 0.018 \times (11.7 - 9 - 1.2)^2$$
$$= 0.0405 \text{ A}$$

Also assume that during charging the battery voltage increases linearly from 9 to 10.5 V over a period of 20 min to calculate the charging current as the battery voltage increases. Note that when the battery is charged, v_{GS} is no longer larger than V_t and the transistor is cut off. A plot of the drain (charging) current as a function of time is shown in Figure 10.14(d). Note that the charging current naturally tapers to zero as the battery voltage increases.

Comments: In the circuit of part b, please note that the battery voltage is not likely to actually increase linearly. The voltage rise will begin to taper off as the battery begins to approach full charge. In practice, this means that the charging process will take longer than projected in Figure 10.14(d).

EXAMPLE 10.7   MOSFET DC Motor Drive Circuit

Problem

The aim of this example is to design a MOSFET driver for the **Lego® 9V Technic motor, model 43362**. Figure 10.15(a) and (b) show the driver circuit and a picture of the motor, respectively. The motor has a maximum (stall) current of 340 mA. The minimum current needed to start motor rotation is 20 mA. The aim of the circuit is to control the current to the motor (and therefore the motor torque, which is proportional to the current) via the gate voltage.

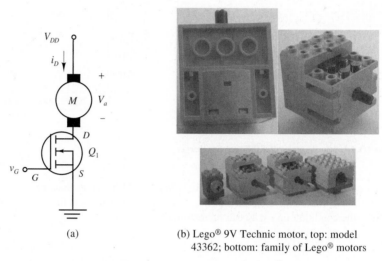

(a)

(b) Lego® 9V Technic motor, top: model 43362; bottom: family of Lego® motors

Figure 10.15 (a) MOSFET DC motor drive circuit; (b) Lego® motors (*Courtesy Philippe "Philo" Hurbain*)

Solution

Known Quantities: Transistor large-signal parameters, component values.

Find: R_1 and R_2, and the value of v_G needed to drive the motor.

Schematics, Diagrams, Circuits, and Given Data: Figure 10.15. $V_t = 1.2$ V; $K = 0.08$ A/V².

Assumptions: Assume that the MOSFET is in the saturation region.

Analysis: The conditions for the MOSFET to be in the saturation region are $v_{GS} > V_t$ and $v_{GD} < V_t$. The first condition is satisfied whenever the gate voltage is above 1.2 V. Thus the transistor will first begin to conduct when $v_G = 1.2$ V. Assuming for the moment that both conditions are satisfied, and that V_{DD} is sufficiently large, we can calculate the drain current to be:

$$i_D = K(v_{GS} - V_t)^2 = 0.08 \times (v_G - 1.2)^2 \text{ A}$$

The plot of Figure 10.15(c) depicts the DC motor (drain) current as a function of the gate voltage. The maximum current of 340 mA can be generated with a gate voltage of approximately 3.3 V. It would take approximately 1.5 V at the gate to generate the minimum required current of 20 mA.

Comments: This circuit could be quite easily implemented in practice to drive the motor with a signal from a microcontroller. In practice, instead of trying to output an analog voltage, a microcontroller is better suited to the generation of a digital (on/off) signal. For example, the gate drive signal could be a *pulse-width modulated* (PWM) 0–5 V pulse train, in which the ratio of the *on* time to the period of the waveform time is called the *duty cycle*. Figure 10.15(d) depicts the possible appearance of a digital PWM gate voltage input.

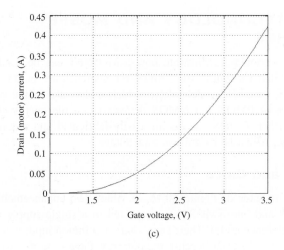

(c)

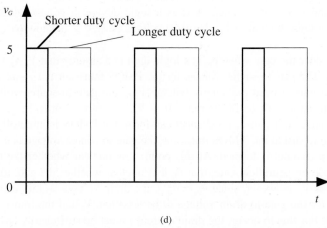

(d)

Figure 10.15 (*Continued*) (c) drain-gate voltage curve for MOSFET in saturation; (d) Pulse-width modulation (PWM) gate voltage waveforms

CHECK YOUR UNDERSTANDING

What is the maximum power dissipation of the MOSFET for each of the circuits in Example 10.6?

Answers: Part a: $P_\text{NMOS} = v_{DS} \times i_D = (v_D - v_S) \times i_D = 2.36 \times 0.1 = 236$ mW
Part b: $P_\text{NMOS} = v_{DS} \times i_D = (v_D - v_S) \times i_D = 2.36 \times 0.0405 = 95.4$ mW

CHECK YOUR UNDERSTANDING

What is the range of duty cycles needed to cover the current range of the Lego motor in Example 10.7?

Answer: 30 to 66 percent

10.5   CMOS TECHNOLOGY AND MOSFET SWITCHES

The objective of this section is to illustrate how a MOSFET can be used as an analog or a digital switch (or gate). Most MOSFET switches are based upon a **complementary MOS**, or **CMOS**, technology, which makes use of the complementary characteristics of PMOS and NMOS devices to enable energy-efficient integrated circuits. Further, CMOS circuits are easily fabricated and require only a single supply voltage, which is a significant advantage.

Digital Switches and Gates

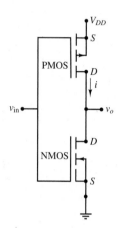

Figure 10.16 CMOS inverter

Consider the **CMOS inverter** of Figure 10.16, in which two enhancement-mode transistors, one PMOS and one NMOS, are connected to a single supply voltage (V_{DD}, relative to the reference node). Their gates share a common input voltage v_{in}. This device is known as an inverter because the output voltage $v_o \approx V_{DD}$ whenever $v_{in} \approx 0$, and vice versa. When used as a logic device, a voltage close to V_{DD} is known as a *logic high*, or a 1, whereas a voltage close 0 V is known as a *logic low*, or a 0.

Consider the case when v_{in} is a logic high and assume that $V_t \ll V_{DD}$ for both transistors. The gate-to-source voltage for the PMOS transistor is $v_{in} - v_o \approx V_{DD} - v_o$. Since V_o cannot exceed the supply voltage V_{DD}, the gate-to-source voltage for the PMOS transistor must be in the range $0 \to V_{DD}$. In other words, its gate-to-source voltage cannot be negative, no channel can form, the PMOS is in cutoff, and $i = 0$.

With regard to the NMOS transistor, the gate-to-source voltage is $v_{in} - 0 \approx V_{DD}$ such that a channel is formed. At this point, it is unclear whether the transistor is in the triode or saturation state; that is, it is unclear whether the gate-to-drain voltage exceeds V_t. However, since $v_{in} \approx V_{DD}$, the drain voltage would also need to be near V_{DD} for the gate-to-drain voltage to be less than V_t and the transistor to be in saturation. For this to occur, the drain current i must be sufficiently large such that $iR_D \approx V_{DD}$. But $i = 0$ because the PMOS transistor is in cutoff. As a result, the NMOS transistor is in the triode state, its drain current is zero, and the voltage from drain-to-source v_{out} is zero. This result can be checked for compatibility with triode current equation 10.6:

$$i_D = K(2v_{GS} - 2V_t - v_{DS})v_{DS} = 0$$

Clearly, $v_{DS} = 0$ is a solution of $i_D = 0$.

The net result is that for a logic high input voltage v_{in}, the output voltage v_{out} is a logic low. Note that the PMOS transistor is in the cutoff state and acts like an open-circuit ($i_D = 0$). On the other hand, the NMOS transistor is in the triode state with an open channel and acts like a short-circuit. These two states can be represented as ideal open and closed switches, respectively, as shown in Figure 10.17(a).

The same analysis can be applied to the case when v_{in} is a logic low. In this case, the PMOS transistor sees a large negative gate-to-source voltage and a channel is formed in the triode state. Inversely, the NMOS transistor sees a gate-to-source voltage near zero such that no channel is formed in the cutoff state. Figure 10.17(b) represents this situation in terms of ideal switches. Note that this circuit does not require the transistors to be biased. Also, note that the drain current i_D is zero in both cases such that a CMOS inverter consumes very little power.

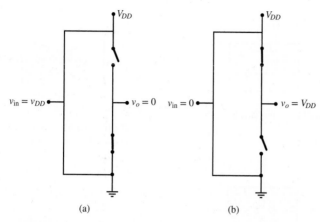

Figure 10.17 CMOS inverter approximated by ideal switches:
(a) When v_{in} is high, v_{out} is tied to ground; (b) when v_{in} is low, v_{out} is
tied to V_{DD}.

Analog Switches

A common analog gate employs a FET and takes advantage of the fact that its
current can be bidirectional in the ohmic region. Recall that a MOSFET operating in the ohmic state acts very much as a linear resistor. For example, for an
NMOS enhancement-mode transistor the conditions for the ohmic state can be
defined as:

$$v_{GS} > V_t \quad \text{and} \quad |v_{DS}| \leq \frac{1}{4}(v_{GS} - V_t) \tag{10.20}$$

As long as the NMOS satisfies these conditions, it acts as a simple linear resistor
with a channel resistance of:

$$r_{DS} = \frac{1}{2K(v_{GS} - V_t)} \tag{10.21}$$

Thus, the drain current can be simply represented as:

$$i_D \approx \frac{v_{DS}}{r_{DS}} \quad \text{for} \quad |v_{DS}| \leq \frac{1}{4}(v_{GS} - V_t) \quad \text{and} \quad v_{GS} > V_t \tag{10.22}$$

The most important feature of the MOSFET operating in the ohmic region is that
it acts as a voltage-controlled resistor, with the gate-source voltage v_{GS} controlling
the channel resistance R_{DS}. The use of the MOSFET as a switch in the ohmic region
consists of providing a gate-source voltage that can either hold the MOSFET in
the cutoff region ($v_{GS} \leq V_t$) or the ohmic region.

Consider the circuit shown in Figure 10.18, where v_G can be varied externally and v_{in} is an analog input signal source that is to be connected to the load
R_o at some appropriate time. When $v_G \leq V_t$, the FET is in the cutoff region and
acts as an open-circuit. If $v_{GS} \geq V_t$ such that the MOSFET is in the ohmic region,
then $v_G > V_t$ and the transistor acts as a linear resistance R_{DS}. If $R_{DS} \ll R_o$, then
$v_o \approx v_{in}$.

MOSFET analog switches are usually produced in integrated-circuit (IC)
form and denoted by the symbol shown in Figure 10.19, where v_G is the controlling
voltage (v_G in Figure 10.18).

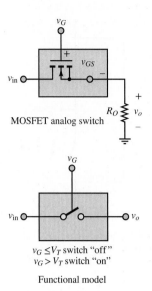

MOSFET analog switch

$v_G \leq V_T$ switch "off"
$v_G > V_T$ switch "on"

Functional model

Figure 10.18 MOSFET analog switch

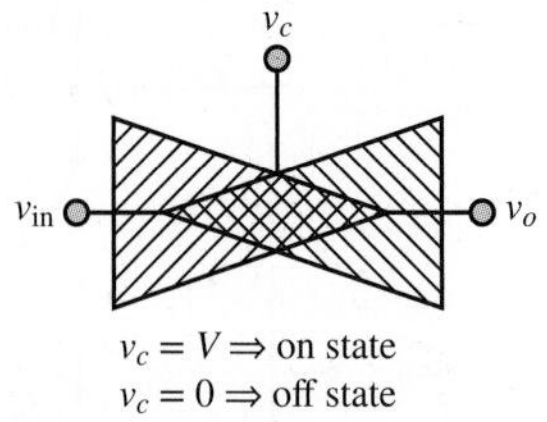

$v_c = V \Rightarrow$ on state
$v_c = 0 \Rightarrow$ off state

Figure 10.19 Symbol for a bilateral FET analog gate

FOCUS ON MEASUREMENTS

MOSFET Bidirectional Analog Gate

The variable-resistor feature of MOSFETs in the ohmic state finds application in the **analog transmission gate**. The circuit shown in Figure 10.20 depicts a circuit constructed using CMOS technology. The circuit operates on the basis of a control voltage v_C that can be either low (say, 0 V) or high ($v_C > V_t$), where V_t is the threshold voltage for the n-channel MOSFET and $-V_t$ is the threshold voltage for the p-channel MOSFET. The circuit operates in one of two modes. When the gate of Q_1 is connected to the high voltage and the gate of Q_2 is connected to the low voltage, the path between v_{in} and V_o has a relatively small resistance and the transmission gate conducts. When the gate of Q_1 is connected to the low voltage and the gate of Q_2 is connected to the high voltage, the transmission gate acts as a very large resistance and is an open-circuit for all practical purposes. A more precise analysis follows.

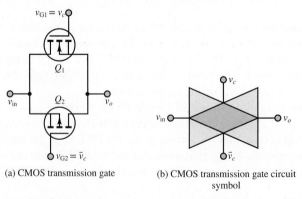

(a) CMOS transmission gate

(b) CMOS transmission gate circuit symbol

Figure 10.20 Analog transmission gate: (a) circuit; (b) circuit symbol.

(Continued)

(*Concluded*)

Let $v_C = V > V_t$ and $\bar{v}_c = 0$. Assume that the input voltage v_{in} is in the range $0 \leq v_{in} \leq V$. To determine the state of the transmission gate, we consider only the extreme cases $v_{in} = 0$ and $v_{in} = V$. When $v_{in} = 0$, $v_{GS1} = v_C - v_{in} = V - 0 = V > V_t$. Since V is above the threshold voltage, MOSFET Q_1 conducts (in the ohmic region). Further, $v_{GS2} = \bar{v}_c - v_{in} = 0 > -V_t$. Since the gate-source voltage is not more negative than the threshold voltage, Q_2 is in cutoff and does not conduct. Since one of the two possible paths between v_{in} and v_o is conducting, the transmission gate is on. Now consider the other extreme, where $v_{in} = V$. By reversing the previous argument, we can see that Q_1 is now off, since $v_{GS1} = 0 < V_t$. However, now Q_2 is in the ohmic state, because $v_{GS2} = \bar{v}_c - v_{in} = 0 - V < -V_t$. In this case, then, it is Q_2 that provides a conducting path between the input and the output of the transmission gate, and the transmission gate is also on. We have therefore concluded that when $v_C = V$ and $\bar{v}_C = 0$, the transmission gate conducts and provides a near-zero-resistance (typically tens of ohms) connection between the input and the output of the transmission gate, for values of the input ranging from 0 to V.

Let us now reverse the control voltages and set $v_C = 0$ and $\bar{v}_C = V > V_t$. It is very straightforward to show that in this case, regardless of the value of v_{in}, both Q_1 and Q_2 are always off; therefore, the transmission gate is essentially an open-circuit.

EXAMPLE 10.8   NMOS Switch

Problem

Determine the operating points of the NMOS switch of Figure 10.21 when the input signal is equal to 0 and 2.5 V, respectively.

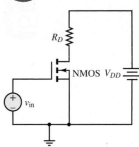

Figure 10.21 Circuit for example 10.8.

Solution

Known Quantities: Drain resistor; V_{DD}; input signal voltage.

Find: The Q point for each value of the input signal voltage.

Schematics, Diagrams, Circuits, and Given Data: $R_D = 125\ \Omega$; $V_{DD} = 10$ V; $v_{in} = 0$ V for $t < 0$; $v_{in} = 2.5$ V for $t \geq 0$.

Assumptions: Use the drain characteristic curves for the NMOS (Figure 10.22).

Analysis: Apply KVL around the drain circuit to find its load line:

$$V_{DD} = R_D i_D + v_{DS} \qquad 10 = 125 i_D + v_{DS}$$

If $i_D = 0$, then $v_{DS} = 10$ V. Likewise, if $v_{DS} = 0$, then $i_D = 10/125 = 80$ mA.

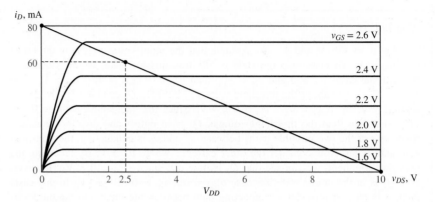

Figure 10.22 Drain curves for the NMOS of Figure 10.21

1. $t < 0$ s: When the input signal is zero, the gate voltage is zero and the NMOS is in the cutoff region. The Q point is

$$v_{GSQ} = 0 \text{ V} \qquad i_{DQ} = 0 \text{ mA} \qquad v_{DSQ} = 10 \text{ V}$$

2. $t \geq 0$ s: When the input signal is 2.5 V, the gate voltage is 2.5 V and the NMOS is in the saturation region. The Q point is

$$v_{GSQ} = 2.5 \text{ V} \qquad i_{DQ} = 60 \text{ mA} \qquad v_{DSQ} = 2.5 \text{ V}$$

This result satisfies KVL because $R_D i_D = 0.06 \times 125 = 7.5$ V

EXAMPLE 10.9   CMOS Gate

Problem

Determine the logic function implemented by the CMOS gate of Figure 10.23. Use the table below to summarize the behavior of the circuit.

v_1 (V)	v_2 (V)	State of M_1	State of M_2	State of M_3	State of M_4	v_o
0	0					
0	5					
5	0					
5	5					

Solution

Find: The logic value of v_{out} for each combination of v_1 and v_2.

Schematics, Diagrams, Circuits, and Given Data: $V_t = 1.7$ V; $V_{DD} = 5$ V.

Assumptions: Treat the MOSFETs as open-circuits when off and as linear resistors when on.

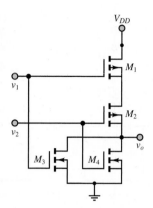

V_{DD}

M_1

v_1

M_2

v_o

M_3 M_4

The transistors in this circuit
show the substrate for
each transistor connected to
its respective source terminal.
In a true CMOS IC, the
substrates for the p-channel
transistors are connected to
5 V and the substrates of the
n-channel transistors are
connected to ground.

Figure 10.23 Circuit for Example 10.9.

Analysis: Note that the state of an NMOS transistor for a high (5 V) gate input is the same as the state of a PMOS transistor for a low (0 V) gate input; both result in the formation of a channel with the transistors in the triode (ohmic) state. In these two cases, the transistors can be represented by simple linear resistors.

On the other hand, the state of an NMOS transistor for a low (0V) gate input is the same as the state of a PMOS transistor for a high (5V) gate input; both result in no channel formation with the transistors in the cutoff state. In these two cases, the transistors can be represented by open-circuits.

a. $v_1 = v_2 = 0$ **V**: With both input voltages equal to zero, M_3 and M_4 are in cutoff and are off since $v_{GS} < V_t$ for both transistors. On the other hand, both M_1 and M_2 form channels, are on, and act as simple linear resistors. However, because both M_3 and M_4 act as open-circuits, there is no current through M_1 and M_2, which act as *pull-up* resistors; that is, with no current through M_1 and M_2, there is no voltage drop across either transistor and, thus, $v_o = V_{DD} = 5$ V, which is a logic high. This situation is depicted in Figure 10.24(a).

b. $v_1 = 0$ **V**; $v_2 = 5$ **V**: With $v_1 = 0$, M_1 forms a channel, is on, and acts as a linear resistor. However, M_3 does not form a channel, is off, and acts as an open-circuit. With $v_2 = 5$ V, M_2 does not form a channel, is off, and acts as an open-circuit, whereas M_4 forms a channel, is on, and acts as a linear resistor. This situation is depicted in Figure 10.24(b). Notice that there can be no current through M_4 because M_2 prevents M_4 from seeing the 5-V source. The result is that $v_o = 0$, which is a logic low.

c. $v_1 = 5$ **V**; $v_2 = 0$ **V**: By symmetry with case b, when the values of v_1 and v_2 are inverted, the states of the four transistors are also inverted. As a result, M_1 and M_4 are off and act as open-circuits, whereas M_2 and M_3 are on and act as linear resistors, as depicted in Figure 10.24(c). Again, an open-circuit, this time M_1, prevents M_3 from seeing the 5-V source such that there is no current through M_3. The result is that once again $v_o = 0$, which is a logic low.

d. $v_1 = v_2 = 5$ **V**: Finally, with both input voltages equal to 5 V, M_1 and M_2 do not form channels, are off, and act as open-circuits. Although M_3 and M_4 both form channels, are on, and act as linear resistors, as depicted in Figure 10.24(d), they are unable to see the 5-V supply voltage and, thus, their currents are zero. Therefore, $v_o = 0$, which is a logic low. Notice that this situation is the inverse of that in case a.

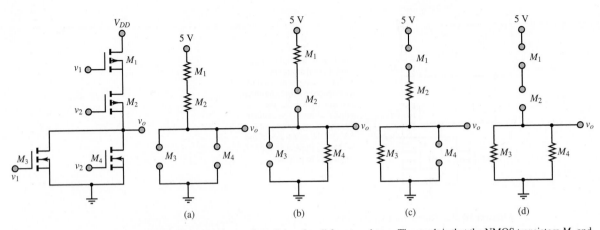

Figure 10.24 When $v_1 = v_2 = 0$, the gate-to-source voltage is low for all four transistors. The result is that the NMOS transistors M_3 and M_4 are off, while the PMOS transistors M_1 and M_2 are on.

These results are summarized in the table below.

v_1 (V)	v_2 (V)	M_1	M_2	M_3	M_4	v_o (V)
0	0	On	On	Off	Off	5
0	5	On	Off	Off	On	0
5	0	Off	On	On	Off	0
5	5	Off	Off	On	On	0

Columns v_1, v_2, and v_o represent a two-variable *truth table* when 0 V and 5 V are interpreted as FALSE and TRUE conditions, respectively. The results indicate that the output variable v_o is TRUE if and only if both input variables are FALSE. Otherwise, the output is FALSE. Such a truth table describes a two-input NOR gate.

CHECK YOUR UNDERSTANDING

What value of R_D would ensure a drain-to-source voltage v_{DS} of 5 V in the circuit of Example 10.8?

Answer: 83.3 Ω

CHECK YOUR UNDERSTANDING

Analyze the CMOS gate of Figure 10.25, and find the output voltages for the following conditions: (a) $v_1 = 0$, $v_2 = 0$; (b) $v_1 = 5$ V, $v_2 = 0$; (c) $v_1 = 0$, $v_2 = 5$ V; (d) $v_1 = 5$ V, $v_2 = 5$ V. Identify the logic function accomplished by the circuit.

Answer:

v_1 (V)	v_2 (V)	v_o (V)
0	0	5
5	0	5
0	5	5
5	5	0

NAND gate

CHECK YOUR UNDERSTANDING

Show that the CMOS bidirectional gate described in the Focus on Measurements box, "MOSFET Bidirectional Analog Gate," is off for all values of υ_{in} between 0 and V whenever $\upsilon_C = 0$ and $\bar{\upsilon}_C = V > V_t$.

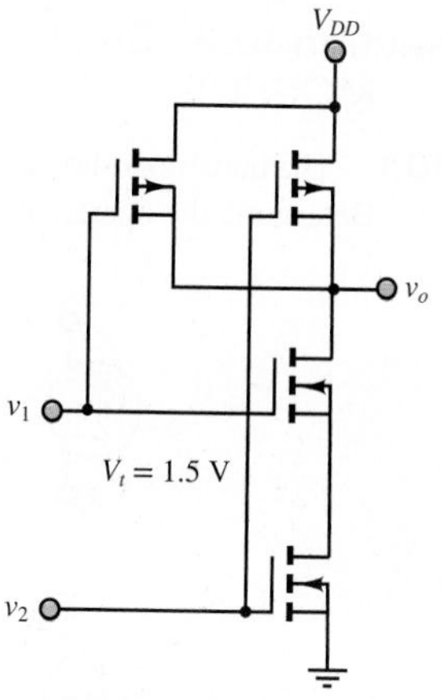

Figure 10.25 CMOS gate

Conclusion

This chapter has introduced field-effect transistors, focusing primarily on metal-oxide semiconductor enhancement-mode n-channel devices to explain the operation of FETs as amplifiers. A brief introduction to p-channel devices is used as the basis to introduce CMOS technology and to present analog and digital switches and logic gate applications of MOSFETs. Upon completing this chapter, you should have mastered the following learning objectives:

1. *Understand the classification of field-effect transistors.* FETs include three major families; the enhancement-mode family is the most commonly used and is the one explored in this chapter. Depletion-mode and junction FETs are only mentioned briefly.
2. *Learn the basic operation of enhancement-mode MOSFETs by understanding their i-υ curves and defining equations.* MOSFETs can be described by the *i-υ* drain characteristic curves and by a set of nonlinear equations linking the drain current to the gate-to-source and drain-to-source voltages. MOSFETs can operate in one of four regions: *cutoff*, in which the transistor does not conduct current; *triode*, in which the transistor can act as a voltage-controlled resistor under certain conditions; *saturation*, in which the transistor acts as a voltage-controlled current source and can be used as an amplifier; and *breakdown* when the limits of operation are exceeded.
3. *Learn how enhancement-mode MOSFET circuits are biased.* MOSFET circuits can be biased to operate around a certain operating point, known as the Q point, when appropriate supply voltages and resistors are selected.
4. *Understand the concept and operation of FET large-signal amplifiers.* Once a MOSFET circuit is properly biased in the saturation region, it can serve as an amplifier by virtue of its voltage-controlled current source property: small changes in the gate-to-source voltages are translated to proportional changes in drain current.
5. *Understand the concept and operation of FET switches.* MOSFETs can serve as analog and digital switches: by controlling the gate voltage, a MOSFET can be turned on and off (digital switch), or its resistance can be modulated (analog switch).
6. *Analyze FET switches and digital gates.* These devices find application in CMOS circuits as digital logic gates and analog transmission gates.

HOMEWORK PROBLEMS

Section 10.2: Enhancement-Mode MOSFETs

10.1   The transistors shown in Figure P10.1 have $|V_t| = 3$ V. Determine the operating region.

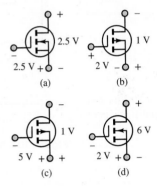

(a)   (b)

(c)   (d)

Figure P10.1

10.2   The three terminals of an n-channel enhancement-mode MOSFET are at potentials of 4, 5, and 10 V with respect to ground. Draw the circuit symbol, with the appropriate voltages at each terminal, if the device is operating

a.   In the ohmic region.

b.   In the saturation region.

10.3   An enhancement-type NMOS transistor with $V_t = 2$ V has its source grounded and a 3-VDC source connected to the gate. Determine the operating state if

a. $v_D = 0.5$ V

b. $v_D = 1$ V

c. $v_D = 5$ V

10.4   In the circuit shown in Figure P10.4, the PMOS transistor has $|V_t| = 2$ V and $K = 10$ mA/V^2. Find R and v_S for $i_S = 0.4$ mA.

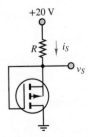

Figure P10.4

10.5   An enhancement-type NMOS transistor has $V_t = 2.5$ V and $i_D = 0.8$ mA when $v_{GS} = v_{DS} = 4$ V. Find the value of i_D for $v_{GS} = 5$ V.

10.6   The NMOS transistor shown in Figure P10.6 has $V_t = 1.5$ V and $K = 0.4$ mA/V^2. If v_G is a pulse with 0 to 5 V, find the voltage levels of the pulse signal at the drain.

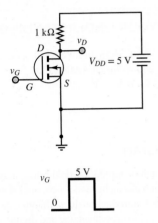

Figure P10.6

10.7   In the circuit shown in Figure P10.7, a drain voltage of 0.1 V is established. Find the current i_D for $V_t = 1$ V and $k = 0.5$ mA/V^2.

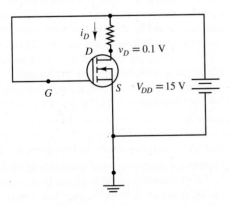

Figure P10.7

Section 10.3: Biasing MOSFET Circuits

10.8   An n-channel enhancement-mode MOSFET, shown in Figure P10.8, is operated in the ohmic region. Size the resistors so that the quiescent drain current $I_{DQ} = 4$ mA. Let $V_{DD} = 15$ V, $K = 0.3$ mA/V^2, and $V_t = 3.3$ V.

Let $V_{DD} = 18$ V, $K = 0.3$ mA/V^2, $V_t = 3$ V, $R_1 = 5.5$ MΩ, $R_2 = 4.5$ MΩ, $R_D = 2$ kΩ, and $R_S = 1$ kΩ.

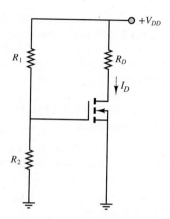

Figure P10.8

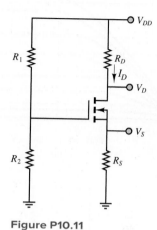

Figure P10.11

10.9   Compute the power dissipated by the circuit in Figure P10.9. Let $V_{DD} = V_{SS} = 15$ V, $R_1 = R_2 = 90$ kΩ, $R_D = 0.1$ kΩ, $V_t = 3.5$ V, $K = 0.816$ mA/V^2.

10.12   In the circuit shown in Figure P10.12, the MOSFET operates in the saturation region, for $I_S = 0.5$ mA and $V_S = 3$ V. This enhancement-type PMOS has $V_t = -1$ V, and $K = 0.5$ mA/V^2. Find:

a. R_S.

b. The largest allowable value of R_S for the MOSFET to remain in the saturation region.

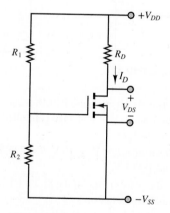

Figure P10.9

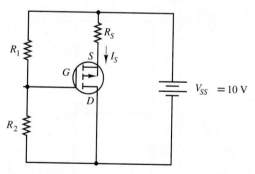

Figure P10.12

10.10   Find the operating region of the enhancement-type NMOS transistor shown in Figure P10.8. Let $V_{DD} = 20$ V, $K = 0.2$ mA/V^2, $V_t = 4$ V, $R_1 = 4$ MΩ, $R_2 = 3$ MΩ, and $R_D = 3$ kΩ.

10.11   Find the operating region of the enhancement-type NMOS transistor shown in Figure P10.11.

10.13   The i-v characteristic of an n-channel enhancement MOSFET is shown in Figure P10.13(a); a standard amplifier circuit based on the n-channel MOSFET is shown in Figure P10.13(b). Determine the quiescent current I_{DQ} and drain-to-source voltage V_{DS} when $V_{DD} = 10$ V and $R_D = 5$ Ω. In what region is the transistor operating?

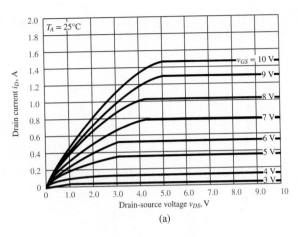

(a)

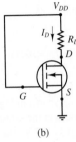

(b)

Figure P10.13

10.14 Given the enhancement-type NMOS transistor and drain characteristic shown in Figure P10.14, compute R_S and V_{DD}. Let $R_1 = 200$ kΩ and $R_2 = 100$ kΩ.

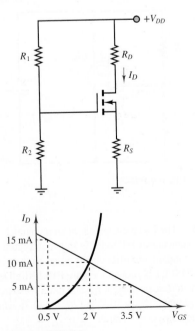

Figure P10.14

10.15 Given the enhancement-type NMOS transistor shown in Figure P10.8, compute R_1, R_2 and R_D. Let $I_D = 2$ mA, $V_t = 4$ V, $V_{DS} = 8$ V, $V_{DD} = 16$ V, $K = 0.375$ mA/V^2, and total dissipated power $P_T = 35$ mW.

10.16 Given the enhancement-type NMOS transistor shown in Figure P10.11, compute R_1, R_2, R_S and R_D. Let $I_D = 4$ mA, $V_D = 9$ V, $V_{DS} = 4.5$ V, $V_{DD} = 18$ V, $V_t = 4$ V, $K = 0.625$ mA/V^2, and maximum total dissipated power $P_{T,\max} = 75$ mW.

Section 10.4: MOSFET Large-Signal Amplifiers

10.17 The power MOSFET circuit of Figure P10.17 is configured as a voltage-controlled current source (VCCS). Let $K = 1.5$ A/V^2 and $V_t = 3$ V.

a. If $V_G = 5$ V, find the range of R for which the VCCS will operate.

b. If $R = 1$ Ω, determine the range of V_G for which the VCCS will operate.

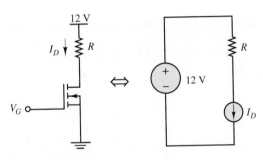

Figure P10.17

10.18 The circuit of Figure P10.18 is called a *source follower* and acts as a voltage-controlled current source (VCCS).

a. Determine I_S if $V_G = 10$ V, $R = 2$ Ω, $K = 0.5$ A/V^2 and $V_t = 4$ V.

b. If the power rating of the MOSFET is 50 W, how small can R be?

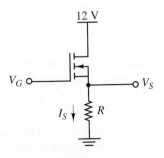

Figure P10.18

10.19 The circuit of Figure P10.19 is a class A amplifier.

  a. Determine the output current for the given biased audio tone input $v_G = 10 + 0.1 \cos(500t)$ V. Let $K = 2$ mA/V^2 and $V_t = 3$ V.

  b. Determine the output voltage v_o.

  c. Determine the voltage gain of the $\cos(500t)$ signal.

  d. Determine the DC power consumption of the resistor and the MOSFET.

10.21 Sometimes it is necessary to discharge batteries before recharging. To do this, an electronic load can be used. A high-power electronic load is shown in Figure P10.21, for the battery discharge application. With $K = 4$ A/V^2, $V_t = 3$ V, and $V_G = 8$ V, determine the discharging current I_D and the required MOSFET power rating.

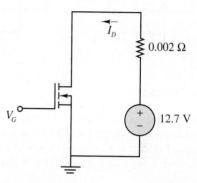

Figure P10.21

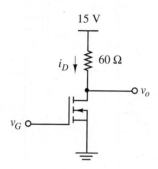

Figure P10.19

10.20 The circuit of Figure P10.20 is a source-follower amplifier. Let $K = 30$ mA/V^2, $V_t = 4$ V, and $v_G = 9 + 0.1 \cos(500t)$ V.

  a. Determine the source current i_s.

  b. Determine the output voltage v_o.

  c. Determine the voltage gain for the $\cos(500t)$ signal.

  d. Determine the DC power consumption of the MOSFET and the 4-Ω resistor.

10.22 A precision voltage source can be created by driving the drain of a MOSFET. Figure P10.22 shows a circuit that will accomplish this function. With $I_{\text{Ref}} = 0.01$ A, determine the output V_G. Let $K = 0.006$ A/V^2 and $V_t = 1.5$ V.

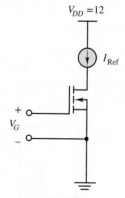

Figure P10.22

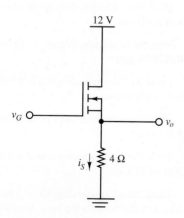

Figure P10.20

10.23 To allow more current in a MOSFET amplifier, several MOSFETs can be connected in parallel. Determine the currents I_D and I_S in the circuits of Figure P10.23. Let $K = 0.2$ A/V^2, $V_t = 3$ V, and $V_G = V_{DD}$.

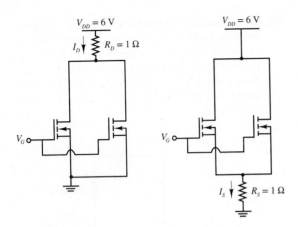

Figure P10.23

10.24 A push-pull amplifier can be constructed from matched n and p-channel MOSFETs, as shown in Figure P10.24. Let $K_n = K_p = 0.5$ A/V^2, $V_{tn} = +3$ V, $V_{tp} = -3$ V, and $v_{in} = 0.8 \cos(1,000t)$ V. Determine v_o and i_o.

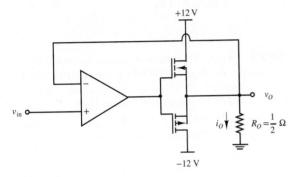

Figure P10.24

10.25 Show that the NMOS shown in Figure P10.25 cannot be in triode mode. Determine its i-v characteristic to show that it acts as a VCCS.

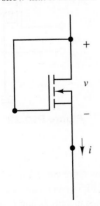

Figure P10.25

10.26 Determine v_o and i_o for the two-stage amplifier shown in the circuit of Figure P10.26, with identical MOSFETs having $K = 1$ A/V^2 and $V_t = 3$ V, for

a. $v_G = 4$ V.

b. $v_G = 5$ V.

c. $v_G = 4 + 0.1 \cos(750t)$.

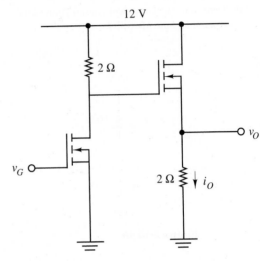

Figure P10.26

Section 10.5: CMOS Technology and MOSFET Switches

10.27 For the CMOS NOR gate of Figure 10.23 identify the state of each transistor for $v_1 = v_2 = 5$ V. Assume $V_{DD} = 5$ V.

10.28 Repeat Problem 10.27 for $v_1 = 5$ V and $v_2 = 0$ V.

10.29 Draw the schematic diagram of a two-input CMOS OR gate.

10.30 Draw the schematic diagram of a two-input CMOS AND gate.

10.31 Draw the schematic diagram of a two-input CMOS NOR gate.

10.32 Draw the schematic diagram of a two-input CMOS NAND gate.

10.33 Draw the schematic diagram of a three-input CMOS OR gate.

10.34 Draw the schematic diagram of a three-input CMOS AND gate.

10.35 Draw the schematic diagram of a three-input CMOS gate that realizes the logic function $\overline{A(B + C)}$.

10.36   Show that the circuit of Figure P10.36 functions as a logic inverter.

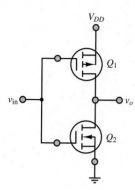

Figure P10.36

10.37   Show that the circuit of Figure P10.37 functions as a NOR gate.

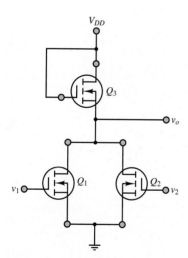

Figure P10.37

10.38   Show that the circuit of Figure P10.38 functions as a NAND gate.

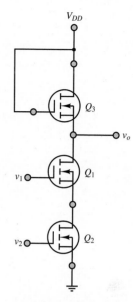

Figure P10.38

10.39   Determine the logic function implemented by the CMOS gate of Figure P10.39. Use a table to summarize the behavior of the circuit.

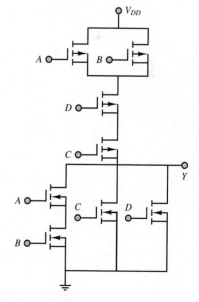

Figure P10.39

C H A P T E R

11

DIGITAL LOGIC CIRCUITS

Digital computers have played a prominent role in engineering and science for over half a century, performing a number of essential functions such as numerical computations and data acquisition. The elements of all digital computers are *combinational* and *sequential* logic circuits, built up from basic *logic gates*. The inputs, operations, and outputs of these circuits are described in terms of the binary number system and boolean algebra. Several practical examples are presented to demonstrate that even simple combinations of logic gates can perform useful functions in engineering practice. A number of logic modules are introduced that are described in terms of simple logic gates and yet provide more advanced functions, such as read-only memory, multiplexing, and decoding. Throughout the chapter, simple examples are given to demonstrate the usefulness of digital logic circuits in various engineering applications.

LO **Learning Objectives**

Students will learn to...

1. Understand the concepts of analog and digital signals and of quantization. *Section 11.1.*
2. Convert between decimal and binary number systems and use the hexadecimal system and BCD and Gray codes. *Section 11.2.*
3. Write truth tables, and realize logic functions from truth tables by using logic gates. *Section 11.3.*
4. Systematically design logic functions using Karnaugh maps. *Section 11.4.*
5. Study various combinational logic modules, including multiplexers, memory and decoder elements, and programmable logic arrays. *Section 11.5.*
6. Study sequential logic modules including flip-flops, latches and counters. *Section 11.6.*
7. *Understand the operation of digital counters and registers.* Counters are a very important class of digital circuits and are based on sequential logic elements. Registers are the most fundamental form of *random-access memory* (RAM). *Section 11.7.*
8. *Design simple sequential circuits using state transition diagrams.* Sequential circuits can be designed using formal design procedures employing state diagrams. *Section 11.8.*

11.1 ANALOG AND DIGITAL SIGNALS

One of the fundamental distinctions in the study of electronic circuits (and in the analysis of any signals derived from physical measurements) is that between analog and digital signals. An **analog signal** is an electric signal whose value varies in analogy with a physical quantity (e.g., temperature, force, or acceleration). For example, a voltage proportional to a measured variable pressure or to a vibration naturally varies in an analog fashion. Figure 11.1 depicts an analog function of time $f(t)$. We note immediately that for each value of time t, $f(t)$ can take any value among any of the values in the given range. For example, in the case of the output voltage of an op-amp, we expect the signal to take any value between $+V_{sat}$ and $-V_{sat}$, where V_{sat} is the supply-imposed saturation voltage.

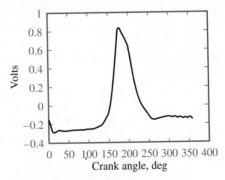

Figure 11.1 Voltage analog of internal combustion engine in-cylinder pressure

A **digital signal**, on the other hand, can take only a *finite number of values*. This is an extremely important distinction, as will be shown shortly. An example of a digital signal is a signal that allows display of a temperature measurement on a digital readout. Let us hypothesize that the digital readout is three digits long and can display numbers from 0 to 100, and let us assume that the temperature sensor is correctly calibrated to measure temperatures from 0 to 100°C. Further, the output of the sensor ranges from 0 to 5 V, where 0 V corresponds to 0°C and 5 V to 100°C. Therefore, the calibration constant of the sensor is

$$k_T = \frac{100°C - 0°C}{5\ V - 0\ V} = 20°C/V$$

Clearly, the output of the sensor is an analog signal; however, the display can show only a finite number of readouts (101, to be precise). Because the display itself can only take a value out of a discrete set of states—the integers from 0 to 100—we call it a digital display, indicating that the variable displayed is expressed in digital form.

Now, each temperature on the display corresponds to a *range of voltages:* each digit on the display represents one-hundredth of the 5-V range of the sensor, or 0.05 V = 50 mV. Thus, the display will read 0 if the sensor voltage is between 0 and 49 mV, 1 if it is between 50 and 99 mV, and so on. Figure 11.2 depicts the staircase function relationship between the analog voltage and the digital readout. This **quantization** of the sensor output voltage is in effect an approximation. If one wished to know the temperature with greater precision, a greater number of display digits could be employed.

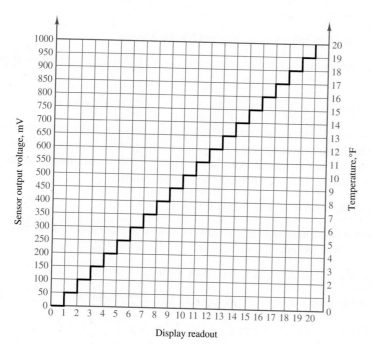

Figure 11.2 Digital representation of an analog signal

The most common digital signals are binary signals. A **binary signal** is a signal that can take only one of two discrete values and is therefore characterized by transitions between two states. Figure 11.3 displays a typical binary signal. In binary arithmetic (which we discuss in Section 11.2), the two discrete values f_1 and f_0 are

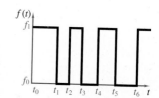

Figure 11.3 A binary signal

represented, respectively, by the numbers 1 and 0. In binary voltage waveforms, these values are represented by two voltage levels. For example, in the TTL convention (see Chapter 9), these values are (nominally) 5 and 0 V, respectively; in CMOS circuits, these values can vary substantially. Other conventions are also used, including reversing the assignment, for example, by letting a 0-V level represent a logic 1 and a 5-V level represent a logic 0. Note that in a binary waveform, knowledge of the transition between one state and another (e.g., from f_0 to f_1 at $t = t_2$) is equivalent to knowledge of the state. Thus, digital logic circuits can operate by detecting transitions between voltage levels. The transitions are often called **edges** and can be positive (f_0 to f_1) or negative (f_1 to f_0). Virtually all the signals handled by a computer are binary. From here on, whenever we speak of digital signals, you may assume that the text is referring to signals of the binary type, unless otherwise indicated.

Table 11.1 **Conversion from decimal to binary**

Decimal number n_{10}	Binary number n_2
0	0
1	1
2	10
3	11
4	100
5	101
6	110
7	111
8	1000
9	1001
10	1010
11	1011
12	1100
13	1101
14	1110
15	1111
16	10000

11.2   THE BINARY NUMBER SYSTEM

The binary number system is a natural choice for representing the behavior of circuits that operate in one of two states (on or off, 1 or 0, or the like). The diode and transistor gates and switches studied in Chapters 9 and 10 fall into this category. Table 11.1 shows the correspondence between decimal and binary number systems for integer decimal numbers up to 16.

Binary numbers are based on powers of 2, whereas the decimal system is based on powers of 10. For example, the number 372 in the decimal system can be expressed as

$$372 = (3 \times 10^2) + (7 \times 10^1) + (2 \times 10^0)$$

while the binary number 10110 corresponds to the following combination of powers of 2:

$$10110 = (1 \times 2^4) + (0 \times 2^3) + (1 \times 2^2) + (1 \times 2^1) + (0 \times 2^0)$$

It is relatively simple to see the correspondence between the two number systems if we add the terms on the right-hand side of the previous expression. Let n_2 represent the number n **base 2** (i.e., in the binary system), and let n_{10} be the same number **base 10**. Then our notation will be as follows:

$$10110_2 = 16 + 0 + 4 + 2 + 0 = 22_{10}$$

Note that a fractional number can also be similarly represented. For example, the number 3.25 in the decimal system may be represented as

$$3.25_{10} = 3 \times 10^0 + 2 \times 10^{-1} + 5 \times 10^{-2}$$

while in the binary system the number 10.011 corresponds to

$$10.011_2 = 1 \times 2^1 + 0 \times 2^0 + 0 \times 2^{-1} + 1 \times 2^{-2} + 1 \times 2^{-3}$$

$$= 2 + 0 + 0 + \frac{1}{4} + \frac{1}{8} = 2.375_{10}$$

Table 11.1 shows that it takes four binary digits, also called **bits**, to represent the decimal numbers up to 15. Usually, the rightmost bit is called the **least significant bit**, or **LSB**, and the leftmost bit is called the **most significant bit**, or **MSB**. Since binary numbers clearly require a larger number of digits than decimal numbers do, the digits are usually grouped into sets of 4, 8, or 16. Four bits are a **nibble** and eight bits are a **byte**. A **word** is the basic unit of data in a digital system. A word may be two or more bytes depending upon the particular digital architecture.

Addition and Subtraction

The operations of addition and subtraction are based on the simple rules shown in Table 11.2. Note that, just as is done in the decimal system, a carry is generated whenever the sum of two digits exceeds the largest single-digit number in the given number system, which is 1 in the binary system. The carry is treated exactly as in the decimal system. A few examples of binary addition are shown in Figure 11.4, with their decimal counterparts.

Table 11.2 Rules for addition

$0 + 0 = 0$
$0 + 1 = 1$
$1 + 0 = 1$
$1 + 1 = 0$ (with a carry of 1)

Decimal	Binary	Decimal	Binary	Decimal	Binary
5	101	15	1111	3.25	11.01
+6	+110	+20	+10100	+5.75	+101.11
11	1011	35	100011	9.00	1001.00

Figure 11.4 Examples of binary addition

The procedure for subtracting binary numbers is based on the rules of Table 11.3. A few examples of binary subtraction are given in Figure 11.5, with their decimal counterparts.

Table 11.3 Rules for subtraction

$0 - 0 = 0$
$1 - 0 = 1$
$1 - 1 = 0$
$0 - 1 = 1$ (with a borrow of 1)

Decimal	Binary	Decimal	Binary	Decimal	Binary
9	1001	16	10000	6.25	110.01
−5	−101	−3	−11	−4.50	−100.10
4	0100	13	01101	1.75	001.11

Figure 11.5 Examples of binary subtraction

Multiplication and Division

Whereas in the decimal system the multiplication table consists of $10^2 = 100$ entries, in the binary system we only have $2^2 = 4$ entries. Table 11.4 represents the complete multiplication table for the binary number system.

Division in the binary system is also based on rules analogous to those of the decimal system, with the two basic laws given in Table 11.5. Once again, we need be concerned with only two cases, and just as in the decimal system, division by zero is not contemplated.

Table 11.4 Rules for multiplication

$0 \times 0 = 0$
$0 \times 1 = 0$
$1 \times 0 = 0$
$1 \times 1 = 1$

Table 11.5 Rules for division

$0 \div 1 = 0$
$1 \div 1 = 1$

Conversion from Decimal to Binary

The conversion of a decimal number to its binary equivalent is performed by successive division of the decimal number by 2, checking for the remainder each time. Figure 11.6 illustrates this idea with an example. The result obtained in Figure 11.6 may be easily verified by performing the opposite conversion, from binary to decimal:

$$110001 = 2^5 + 2^4 + 2^0 = 32 + 16 + 1 = 49$$

Remainder
$49 \div 2 = 24 + 1$
$24 \div 2 = 12 + 0$
$12 \div 2 = 6 + 0$
$6 \div 2 = 3 + 0$
$3 \div 2 = 1 + 1$
$1 \div 2 = 0 + 1$

$$49_{10} = 110001_2$$

Figure 11.6 Example of conversion from decimal to binary

Remainder
$37 \div 2 = 18 + 1$
$18 \div 2 = 9 + 0$
$9 \div 2 = 4 + 1$
$4 \div 2 = 2 + 0$
$2 \div 2 = 1 + 0$
$1 \div 2 = 0 + 1$

$$37_{10} = 100101_2$$

$2 \times 0.53 = 1.06 \rightarrow 1$
$2 \times 0.06 = 0.12 \rightarrow 0$
$2 \times 0.12 = 0.24 \rightarrow 0$
$2 \times 0.24 = 0.48 \rightarrow 0$
$2 \times 0.48 = 0.96 \rightarrow 0$
$2 \times 0.96 = 1.92 \rightarrow 1$
$2 \times 0.92 = 1.84 \rightarrow 1$
$2 \times 0.84 = 1.68 \rightarrow 1$
$2 \times 0.68 = 1.36 \rightarrow 1$
$2 \times 0.36 = 0.72 \rightarrow 0$
$2 \times 0.72 = 1.44 \rightarrow 1$

$$0.53_{10} = 0.10000111101_2$$

Figure 11.7 Conversion from decimal to binary

The same technique can be used for converting decimal fractional numbers to their binary form, provided that the whole number is separated from the fractional part and each is converted to binary form (separately), with the results added at the end. Figure 11.7 outlines this procedure by converting the number 37.53 to binary form. The procedure is outlined in two steps. First, the integer part is converted; then, to convert the fractional part, one simple technique consists of multiplying the decimal fraction by 2 in successive stages. If the result exceeds 1, a 1 is needed to the right of the binary fraction being formed (100101 . . . , in our example). Otherwise, a 0 is added. This procedure is continued until no fractional terms are left. In this case, the decimal part is 0.53_{10}, and Figure 11.7 illustrates the succession of calculations. Stopping the procedure outlined in Figure 11.7 after 11 digits results in the following approximation:

$$37.53_{10} = 100101.10000111101$$

Greater precision could be attained by continuing to add binary digits, at the expense of added complexity. Note that an infinite number of binary digits may be required to represent a decimal number *exactly*.

Complements and Negative Numbers

To simplify the operation of subtraction in digital computers, **complements** are used almost exclusively. In practice, this corresponds to replacing the operation $X-Y$ with the operation $X + (-Y)$. This procedure results in considerable simplification since the computer hardware need include only adding circuitry. Two types of complements are used with binary numbers: the **ones complement** and the **twos complement**.

The ones complement of an n-bit binary number is obtained by subtracting the number itself from $2^n - 1$. Two examples are as follows:

$$a = 0101$$
$$\text{Ones complement of } a = (2^4 - 1) - a$$
$$= (1111) - (0101)$$
$$= 1010$$
$$b = 101101$$
$$\text{Ones complement of } b = (2^6 - 1) - b$$
$$= (111111) - (101101)$$
$$= 010010$$

The twos complement of an n-bit binary number is obtained by subtracting the number itself from 2^n. Twos complements of the same numbers a and b used in the preceding illustration are computed as follows:

$$a = 0101$$
$$\text{Twos complement of } a = 2^4 - a$$
$$= (10000) - (0101)$$
$$= 1011$$
$$b = 101101$$
$$\text{Twos complement of } b = 2^6 - b$$
$$= (1000000) - (101101)$$
$$= 010011$$

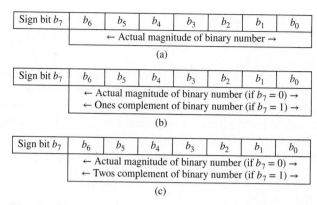

Figure 11.8 (a) An 8-bit sign-magnitude binary number; (b) an 8-bit ones complement binary number; (c) an 8-bit twos complement binary number

A simple rule that may be used to obtain the twos complement directly from a binary number is the following: Starting at the least significant (rightmost) bit, copy each bit *until the first 1 has been copied*, and then replace each successive 1 by a 0 and each 0 by a 1. You may wish to try this rule on the two previous examples to verify that it is much easier to use than subtraction from 2^n.

Different conventions exist in the binary system to represent whether a number is negative or positive. One convention, called the **sign-magnitude convention**, makes use of a *sign bit*, usually positioned at the beginning of the number, for which a value of 1 represents a minus sign and a value of 0 represents a plus sign. Thus, an 8-bit binary number would consist of 1 sign bit followed by 7 *magnitude bits*, as shown in Figure 11.8(a). In a digital system that uses 8-bit signed integer words, we could represent integer numbers (decimal) in the range

$$-(2^7 - 1) \leq N \leq + (2^7 - 1)$$

or

$$-127 \leq N \leq + 127$$

A second convention uses the ones complement notation. In this convention, a sign bit is also used to indicate whether the number is positive (sign bit $= 0$) or negative (sign bit $= 1$). However, the magnitude of the binary number is represented by the true magnitude if the number is positive and by its *ones complement* if the number is negative. Figure 11.8(b) illustrates the convention. For example, the number 91_{10} would be represented by the 7-bit binary number 1011011_2 with a leading 0 (the sign bit): $\mathbf{0}1011011_2$. On the other hand, the number -91_{10} would be represented by the 7-bit ones complement binary number 0100100_2 with a leading 1 (the sign bit): $\mathbf{1}0100100_2$.

Most digital computers use the twos complement convention in performing integer arithmetic operations. The twos complement convention represents positive numbers by a sign bit of 0, followed by the true binary magnitude; negative numbers are represented by a sign bit of 1, *followed by the twos complement of the binary number*, as shown in Figure 11.8(c). The advantage of the twos complement convention is that the algebraic sum of twos complement binary numbers is carried out very simply by adding the two numbers *including the sign bit*.

The Hexadecimal System

It should be apparent by now that representing numbers in base-2 and base-10 systems is simply a matter of convenience, given a specific application. Another frequently used base is 16, which results in the **hexadecimal system**. In the hexadecimal (or hex) code, the bits in a binary number are subdivided into groups of 4. Since there are 16 possible combinations for a 4-bit number, the natural digits in the decimal system (0 through 9) are insufficient to represent a hex digit. To solve this problem, the first six letters of the alphabet are used, as shown in Table 11.6. Thus, in hex code, an 8-bit word corresponds to just two digits; for example,

$$1010\ 0111_2 = A7_{16}$$
$$0010\ 1001_2 = 29_{16}$$

The **ASCII**[1] character code represents all alphanumeric characters, and others, commonly used in printed documents to hexadecimal values. This code is used, for example, to define the visual output associated with **char** type variables found in all computer programming languages. The 128 members of the standard ASCII character set are listed in Appendix D along with their hexadecimal equivalents.

Table 11.6 Hexadecimal code

0	0000
1	0001
2	0010
3	0111
4	0100
5	0101
6	0110
7	0111
8	1000
9	1001
A	1010
B	1011
C	1100
D	1101
E	1110
F	1111

Binary Codes

In this subsection, we describe two common binary codes that are often used for practical reasons. The first is a method of representing decimal numbers in digital logic circuits that is referred to as **binary-coded decimal**, or **BCD**, **representation**. The simplest BCD representation is just a sequence of 4-bit binary numbers that stops after the first 10 entries, as shown in Table 11.7. There are also other BCD codes, all reflecting the same principle: Each decimal digit is represented by a fixed-length binary word. One should realize that although this method is attractive because of its direct correspondence with the decimal system, it is not efficient. Consider, for example, the decimal number 68. Its binary representation by direct conversion is the 7-bit number 1000100. However, the corresponding BCD representation would require 8 bits:

$$68_{10} = 01101000_{BCD}$$

Table 11.7 BCD code

0	0000
1	0001
2	0010
3	0011
4	0100
5	0101
6	0110
7	0111
8	1000
9	1001

Table 11.8 Three-bit Gray code

Binary	Gray
000	000
001	001
010	011
011	010
100	110
101	111
110	101
111	100

Another code that finds many applications is the **Gray code**, which is simply a reshuffled binary code with the property that any two consecutive numbers differ by only 1 bit. Table 11.8 illustrates the 3-bit Gray code. The Gray code is useful in

---

[1]American Standard Code for Information Interchange.

encoding applications because a single bit reading error results in an off-by-one counting error. Thus, the impact of bit reading errors is more likely to be marginal than when using other encoding schemes.

Digital Position Encoders

Position encoders are devices that output a digital signal proportional to their (linear or angular) position. These devices are very useful in measuring instantaneous position in *motion control* applications. Motion control is a technique used when it is necessary to accurately control the motion of a moving object; examples are found in robotics, machine tools, and servomechanisms. For example, in positioning the arm of a robot to pick up an object, it is very important to know its exact position at all times. Since one is usually interested in both rotational and translational motion, two types of encoders are discussed in this example: *linear* and *angular* position encoders.

An optical position encoder consists of an *encoder pad*, which is either a strip (for translational motion) or a disk (for rotational motion) with alternating black and white areas. These areas are arranged to reproduce some binary code, as shown in Figure 11.9, where both the conventional binary and Gray codes are depicted for a 4-bit linear encoder pad. A fixed array of photodiodes (see Chapter 8) senses the reflected light from each of the cells across a row of the encoder path; depending on the amount of light reflected, each photodiode circuit will output a voltage corresponding to a binary 1 or 0. Thus, a different 4-bit word is generated for each row of the encoder.

Decimal	Binary	Decimal	Gray code
15	1111	15	1000
14	1110	14	1001
13	1101	13	1011
12	1100	12	1010
11	1011	11	1110
10	1010	10	1111
9	1001	9	1101
8	1000	8	1100
7	0111	7	0100
6	0110	6	0101
5	0101	5	0111
4	0100	4	0110
3	0011	3	0010
2	0010	2	0011
1	0001	1	0001
0	0000	0	0000

Figure 11.9 Binary and Gray code patterns for linear position encoders

Suppose the encoder pad is 100 mm in length. Then its resolution can be computed as follows. The pad will be divided into $2^4 = 16$ segments, and each segment corresponds to an increment of $100/16$ mm $= 6.25$ mm. If greater resolution were necessary, more bits could be employed: an 8-bit pad of the same length would attain a resolution of $100/256$ mm $= 0.39$ mm.

A similar construction can be employed for the 5-bit angular encoder of Figure 11.10. In this case, the angular resolution can be expressed in degrees of rotation, where $2^5 = 32$ sections correspond to 360°. Thus, the resolution is $360°/32 = 11.25°$. Once again, greater angular resolution could be obtained by employing a larger number of bits.

(Continued)

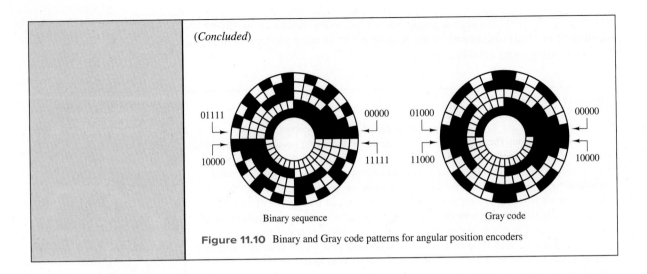

(*Concluded*)

Binary sequence   Gray code

Figure 11.10 Binary and Gray code patterns for angular position encoders

EXAMPLE 11.1   Twos Complement Operations

Problem

Perform the following subtractions, using twos complement arithmetic.

1. $X - Y = 1011100 - 1110010$
2. $X - Y = 10101111 - 01110011$

Solution

Analysis: The twos complement subtractions are performed by replacing the operation $X - Y$ with the operation $X + (-Y)$. Thus, we first find the twos complement of Y and add the result to X in each of the two cases:

$$X - Y = 1011100 - 1110010 = 1011100 + (2^7 - 1110010)$$
$$= 1011100 + 0001110 = 1101010$$

Next, we add the *sign bit* (in boldface type) in front of each number (1 in the first case since the difference $X - Y$ is a negative number):

$$X - Y = \mathbf{1}1101010$$

Repeating for the second subtraction gives

$$X - Y = 10101111 - 01110011 = 10101111 + (2^8 - 01110011)$$
$$= 10101111 + 10001101 = 00111100$$
$$= \mathbf{0}00111100$$

where the first digit is a 0 because $X - Y$ is a positive number.

EXAMPLE 11.2   Conversion From Binary to Hexadecimal

Problem

Convert the following binary numbers to hexadecimal form.

1. 100111
2. 1011101
3. 11001101
4. 101101111001
5. 100110110
6. 1101011011

Solution

Analysis: A simple method for binary to hexadecimal conversion consists of grouping each binary number into 4-bit groups and then performing the conversion for each 4-bit word following Table 11.6:

1. $100111_2 = 0010_2 0111_2 = 27_{16}$
2. $1011101_2 = 0101_2 1101_2 = 5D_{16}$
3. $11001101_2 = 1100_2 1101_2 = CD_{16}$
4. $101101111001_2 = 1011_2 0111_2 1001_2 = B79_{16}$
5. $100110110_2 = 0001_2 0011_2 0110_2 = 136_{16}$
6. $1101011011_2 = 0011_2 0101_2 1011_2 = 35B_{16}$

Comments: To convert from hexadecimal to binary, replace each hexadecimal number with the equivalent 4-bit nibble.

CHECK YOUR UNDERSTANDING

Convert the following decimal numbers to binary form.

a. 39	b. 59	c. 512	d. 0.4475
e. $\frac{25}{32}$	f. 0.796875	g. 256.75	h. 129.5625
i. 4,096.90625			

Answers: (a) 100111, (b) 111011, (c) 1000000000, (d) 0.0111001101000, (e) 0.11001, (f) 0.110011, (g) 100000000.11, (h) 10000001.1001, (i) 1000000000000.11101

CHECK YOUR UNDERSTANDING

Convert the following binary numbers to decimal.

a. 1101	b. 11011	c. 10111
d. 0.1011	e. 0.001101	f. 0.001101101
g. 111011.1011	h. 1011011.001101	i. 10110.0101011101

Answers: (a) 13, (b) 27, (c) 23, (d) 0.6875, (e) 0.203125, (f) 0.212890625, (g) 59.6875, (h) 91.203125, (i) 22.34082031125

CHECK YOUR UNDERSTANDING

Perform the following additions and subtractions. Express the answer in decimal form for (a) through (d) and in binary form for (e) through (h).

a. $1001.1_2 + 1011.01_2$
b. $100101_2 + 100101_2$
c. $0.1011_2 + 0.1101_2$
d. $1011.01_2 + 1001.11_2$
e. $64_{10} - 32_{10}$
f. $127_{10} - 63_{10}$
g. $93.5_{10} - 42.75_{10}$
h. $\left(84\frac{9}{32}\right)_{10} - \left(48\frac{5}{16}\right)_{10}$

Answer: (a) 20.75_{10}, (b) 74_{10}, (c) 1.5_{10}, (d) 21_{10}, (e) 100000_2, (f) 1000000_2, (g) 110011.11_2, (h) 100010.11_2

CHECK YOUR UNDERSTANDING

How many possible numbers can be represented in a 12-bit word?

If we use an 8-bit word with a sign bit (7 magnitude bits plus 1 sign bit) to represent voltages -5 and $+5$ V, what is the smallest increment of voltage that can be represented?

Answers: 4,096; 39 mV

CHECK YOUR UNDERSTANDING

Find the twos complement of the following binary numbers.

a. 11101001   b. 10010111
c. 1011110

Answer: (a) 00010111, (b) 01101001, (c) 0100010

CHECK YOUR UNDERSTANDING

Convert the following numbers from hexadecimal to binary or from binary to hexadecimal.

a. F83   b. 3C9
c. A6   d. 110101110_2
e. 10111001_2 f. 11011101101_2

Answers: (a) 111110000011, (b) 001111001001, (c) 10100110, (d) 1AE, (e) B9, (f) 6ED

CHECK YOUR UNDERSTANDING

Convert the following numbers from hexadecimal to binary, and find their twos complements.

a. F43   b. 2B9
c. A6

Answers: (a) 0000 1011 1101, (b) 1101 0100 0111, (c) 0101 1010

11.3   BOOLEAN ALGEBRA AND LOGIC GATES

The mathematics associated with the binary number system (and with the more general field of logic) is called *boolean*, in honor of the English mathematician George Boole, who published a treatise in 1854 entitled "An Investigation of the Laws of Thought, on Which Are Founded the Mathematical Theories of Logic and Probabilities." The development of a *logical algebra*, as Boole called it, is one of the results of his investigations. The variables in a boolean, or logic, expression can take only one of two values, usually represented by the numbers 0 and 1. These variables are sometimes referred to as true (1) and false (0). This convention is normally referred to as **positive logic**. There is also a **negative logic** convention in which the roles of logic 1 and logic 0 are reversed. In this book we employ only positive logic.

Analysis of **logic functions**, that is, functions of logical (boolean) variables, can be carried out in terms of truth tables. A truth table is a listing of all the possible values that each of the boolean variables can take and of the corresponding value of the desired function. In the following paragraphs we define the basic logic functions upon which boolean algebra is founded, and we describe each in terms of a set of rules and a truth table; in addition, we introduce **logic gates**. Logic gates are physical devices (see Chapters 9 and 10) that can be used to implement logic functions.

AND and OR Gates

The basis of **boolean algebra** lies in the operations of **logical addition**, or the **OR** operation; and **logical multiplication**, or the **AND** operation. Both of these find a correspondence in simple logic gates, as we shall presently illustrate. Logical addition, although represented by the symbol +, differs from conventional algebraic addition, as shown in the last rule listed in Table 11.9. Note that this rule also differs from the last rule of binary addition studied in Section 11.2. Logical addition can be represented by the logic gate called an **OR gate**, whose symbol and whose inputs and outputs are shown in Figure 11.11. The OR gate represents the following logical statement:

If either X or Y is true (1), then Z is true (1).   Logical OR   (11.1)

This rule is embodied in the electronic gates discussed in Chapters 9 and 10, in which a logic 1 corresponds, say, to a 5-V signal and a logic 0 to a 0-V signal.

Logical multiplication is denoted by the center dot · and is defined by the rules of Table 11.10. Figure 11.12 depicts the **AND gate**, which corresponds to this operation. The AND gate corresponds to the following logical statement:

If both X and Y are true (1), then Z is true (1).   Logical AND   (11.2)

One can easily envision logic gates (AND and OR) with an arbitrary number of inputs; three- and four-input gates are not uncommon.

The rules that define a logic function are often represented in tabular form by means of a **truth table**. Truth tables for the AND and OR gates are shown in Figures 11.11 and 11.12. A truth table is nothing more than a tabular summary of all possible outputs of a logic gate, given all possible input values. If the number of inputs is 3, the number of possible combinations grows from 4 to 8, but the basic idea is unchanged. Truth tables are very useful in defining logic functions. A typical logic design problem might specify requirements such as "the output Z shall be

Table 11.9 Rules for logical addition (OR)

$0 + 0 = 0$
$0 + 1 = 1$
$1 + 0 = 1$
$1 + 1 = 1$

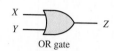
OR gate

X	Y	Z
0	0	0
0	1	1
1	0	1
1	1	1

Truth table

Figure 11.11 Logical addition and the OR gate

Table 11.10 Rules for logical multiplication (AND)

$0 \cdot 0 = 0$
$0 \cdot 1 = 0$
$1 \cdot 0 = 0$
$1 \cdot 1 = 1$

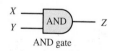
AND gate

X	Y	Z
0	0	0
0	1	0
1	0	0
1	1	1

Truth table

Figure 11.12 Logical multiplication and the AND gate

Logic gate realization of the statement "the output Z shall be logic 1 only when the condition $(X = 1 \text{ AND } Y = 1) \text{ OR } (W = 1)$ occurs, and shall be logic 0 otherwise."

X	Y	W	Z
0	0	0	0
0	0	1	1
0	1	0	0
0	1	1	1
1	0	0	0
1	0	1	1
1	1	0	1
1	1	1	1

Truth table

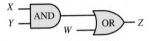

Solution using logic gates

Figure 11.13 Example of logic function implementation with logic gates

logic 1 only when the condition $(X = 1 \text{ AND } Y = 1) \text{ OR } (W = 1)$ occurs, and shall be logic 0 otherwise." The truth table for this particular logic function is shown in Figure 11.13 as an illustration. The design consists, then, of determining the combination of logic gates that exactly implements the required logic function. Truth tables can greatly simplify this procedure.

The AND and OR gates form the basis of all logic design in conjunction with the **NOT gate**. The NOT gate is essentially an inverter (which can be constructed by using bipolar or field-effect transistors, as discussed in Chapters 9 and 10, respectively), and it provides the complement of the logic variable connected to its input. The complement of a logic variable X is denoted by $\overline{X}$. The NOT gate has only one input, as shown in Figure 11.14.

To illustrate the use of the NOT gate, or inverter, we return to the design example of Figure 11.13, where we required that the output of a logic circuit be $Z = 1$ only if $X = 0 \text{ AND } Y = 1 \text{ OR }$ if $W = 1$. We recognize that except for the requirement $X = 0$, this problem would be identical if we stated it as follows: "The output Z shall be logic 1 only when the condition $(\overline{X} = 1 \text{ AND } Y = 1) \text{ OR } (W = 1)$ occurs, and shall be logic 0 otherwise." If we use an inverter to convert X to $\overline{X}$, we see that the required condition becomes $(\overline{X} = 1 \text{ AND } Y = 1) \text{ OR } (W = 1)$. The formal solution to this elementary design exercise is illustrated in Figure 11.15.

NOT gate

X	$\overline{X}$
1	0
0	1

Truth table for NOT gate

Figure 11.14 Complements and the NOT gate

X	$\overline{X}$	Y	W	Z
0	1	0	0	0
0	1	0	1	1
0	1	1	0	1
0	1	1	1	1
1	0	0	0	0
1	0	0	1	1
1	0	1	0	0
1	0	1	1	1

Truth table

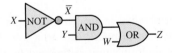

Solution using logic gates

Figure 11.15 Solution of a logic problem using logic gates

In the course of the discussion of logic gates, we make frequent use of truth tables to evaluate logic expressions. A set of basic rules will facilitate this task. Table 11.11 lists some of the rules of boolean algebra; each of these can be proved by using a truth table, as will be shown in examples and exercises. An example proof for rule 16 is given in Figure 11.16 in the form of a truth table. This technique can be employed to prove any of the laws of Table 11.11. From the simple truth table in Figure 11.16, which was obtained step by step, we can clearly see that indeed $X \cdot (X + Y) = X$. This method for proving the validity of logical equations is called **proof by perfect induction**. The 19 rules of Table 11.11 can be used to simplify logic expressions.

Table 11.11 **Rules of boolean algebra**

1.	$0 + X = X$
2.	$1 + X = 1$
3.	$X + X = X$
4.	$X + \overline{X} = 1$
5.	$0 \cdot X = 0$
6.	$1 \cdot X = X$
7.	$X \cdot X = X$
8.	$\overline{X} \cdot \overline{X} = 0$
9.	$\overline{\overline{X}} = X$
10.	$X + Y = Y + X$
11.	$X \cdot Y = Y \cdot X$
12.	$X + (Y + Z) = (X + Y) + Z$
13.	$X \cdot (Y \cdot Z) = (X \cdot Y) \cdot Z$
14.	$X \cdot (Y + Z) = X \cdot Y + X \cdot Z$
15.	$X + X \cdot Z = X$
16.	$X \cdot (X + Y) = X$
17.	$(X + Y) \cdot (X + Z) = X + Y \cdot Z$
18.	$X + \overline{X} \cdot Y = X + Y$
19.	$X \cdot Y + Y \cdot Z + \overline{X} \cdot Z = X \cdot Y + \overline{X} \cdot Z$

10, 11. }Commutative law
12, 13. }Associative law
14. Distributive law
15. Absorption law

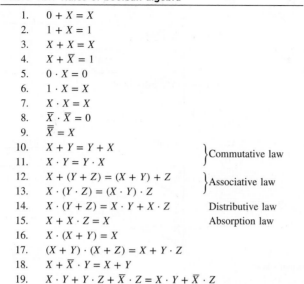

X	Y	$X + Y$	$X \cdot (X + Y)$
0	0	0	0
0	1	1	0
1	0	1	1
1	1	1	1

Figure 11.16 Proof of rule 16 by perfect induction

De Morgan's Laws

Two very important logic rules are known as **De Morgan's laws**. These laws state that AND and OR functions can be interchanged by making appropriate NOT operations. In terms of Boolean algebra these theorems are:

$$(\overline{X + Y}) = \overline{X} \cdot \overline{Y} \tag{11.3}$$

De Morgan's laws

$$(\overline{X \cdot Y}) = \overline{X} + \overline{Y} \tag{11.4}$$

Notice the **duality** that exists between AND and OR operations. One consequence of De Morgan's laws may be stated as:

Any logic function can be implemented using only OR and NOT gates or only AND and NOT gates.

De Morgan's laws can be visualized in terms of logic gates and the associated truth tables, as shown in Figure 11.17.

Another consequence of De Morgan's laws is the ability to express any logic function as a **sum of products** (SOP) and/or as a **product of sums** (POS), as shown in Figure 11.18. The two forms are logically equivalent; however, one may be simpler to implement with logic gates.

In Figure 11.18, the SOP expression can be expressed as:

$$XY + WZ = \overline{\overline{XY + WZ}} = \overline{\overline{XY} \cdot \overline{WZ}}$$

If $\overline{XY} = (A + B)$ and $\overline{WZ} = (C + D)$, then:

$$XY + WZ = \overline{(A + B) \cdot (C + D)}$$

such that the SOP form $XY + WZ$ is equivalent to the complement (or negation) of the POS form $(A + B) \cdot (C + D)$.

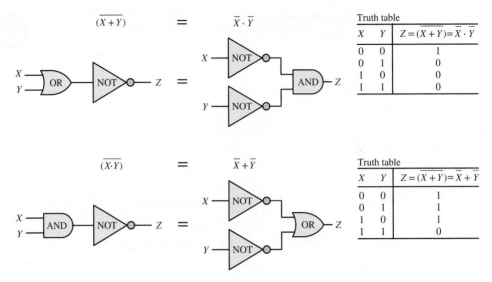

Figure 11.17 De Morgan's laws

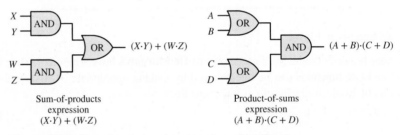

Sum-of-products
expression
$(X \cdot Y) + (W \cdot Z)$

Product-of-sums
expression
$(A + B) \cdot (C + D)$

Figure 11.18 Sum-of-products and product-of-sums logic functions

NAND and NOR Gates

In addition to the AND and OR gates we have just analyzed, the complementary forms of these gates, called NAND and NOR, are very commonly used in practice. In fact, NAND and NOR gates form the basis of most practical logic circuits. Figure 11.19 depicts these two gates and illustrates how they can be easily interpreted in terms of AND, OR, and NOT gates by virtue of De Morgan's laws. You

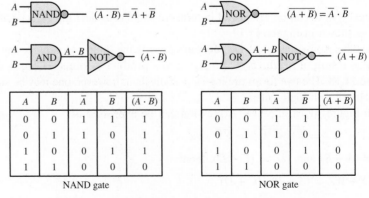

Figure 11.19 Equivalence of NAND and NOR gates with AND and OR gates

can readily verify that the logic function implemented by the NAND and NOR gates corresponds, respectively, to AND and OR gates followed by an inverter. It is very important to note that, by De Morgan's laws, the NAND gate performs a *logical addition* on the *complements* of the inputs, while the NOR gate performs a *logical multiplication* on the *complements* of the inputs. Functionally, then, any logic function could be implemented with either NOR or NAND gates only.

The next section shows how to systematically approach the design of logic functions. First, we provide a few examples to illustrate logic design with NAND and NOR gates.

The XOR (Exclusive OR) Gate

It is rather common practice for a manufacturer of integrated circuits (ICs) to provide common combinations of logic circuits in a single IC package. We review many of these common **logic modules** in Section 11.5. An example of this idea is provided by the **exclusive OR (XOR) gate**, which provides a logic function similar, but not identical, to the OR gate we have already studied. The XOR gate acts as an OR gate, except when its inputs are all logic 1s; in this case, the output is a logic 0 (thus the term *exclusive*). Figure 11.20 shows the logic circuit symbol adopted for this gate and the corresponding truth table. The logic function implemented by the XOR gate is the following: either X or Y, but not both. This description can be extended to an arbitrary number of inputs.

The symbol adopted for the exclusive OR operation is $\oplus$, and so we write

$$Z = X \oplus Y$$

to denote this logic operation. The XOR gate can be obtained by a combination of the basic gates we are already familiar with. For example, if we observe that the XOR function corresponds to $Z = X \oplus Y = (X + Y) \cdot (\overline{X \cdot Y})$, we can realize the XOR gate by means of the circuit shown in Figure 11.21.

Common IC logic gate configurations, are typically available in both of the two more common device families, TTL and CMOS.

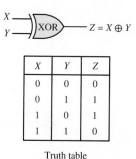

X	Y	Z
0	0	0
0	1	1
1	0	1
1	1	0

Truth table

Figure 11.20 XOR gate

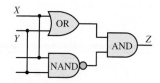

Figure 11.21 Realization of an XOR gate

Fail-Safe Autopilot Logic

This example aims to illustrate the significance of De Morgan's laws and of the duality of the sum-of-products and product-of-sums forms. Suppose that a fail-safe autopilot system in a commercial aircraft requires that, prior to initiating a takeoff or landing maneuver, the following check be passed: Two of three possible pilots must be available. The three possibilities are the pilot, the copilot, and the autopilot. Imagine further that there exist switches in the pilot and copilot seats that are turned on by the weight of the crew, and that a self-check circuit exists to verify the proper operation of the autopilot system. Let the variable X denote the pilot state (1 if the pilot is sitting at the controls), Y denote the same condition for the copilot, and Z denote the state of the autopilot, where $Z = 1$ indicates that the autopilot is functioning. Then since we wish two of these conditions to be active before the maneuver can be initiated, the logic function corresponding to "system ready" is

$$f = X \cdot Y + X \cdot Z + Y \cdot Z$$

This can also be verified by the truth table shown below.

(Continued)

FOCUS ON MEASUREMENTS

(*Concluded*)

Pilot	Copilot	Autopilot	System ready
0	0	0	0
0	0	1	0
0	1	0	0
0	1	1	1
1	0	0	0
1	0	1	1
1	1	0	1
1	1	1	1

The function f defined above is based on the notion of a *positive check*; that is, it indicates when the system is ready. Let us now apply De Morgan's laws to the function f, which is in sum-of-products form:

$$\bar{f} = g = \overline{X \cdot Y + X \cdot Z + Y \cdot Z} = (\overline{X} + \overline{Y}) \cdot (\overline{X} + \overline{Z}) \cdot (\overline{Y} + \overline{Z})$$

The function g, in product-of-sums form, conveys exactly the same information as the function f, but it performs a negative check; in other words, g verifies the *system not ready condition*. Clearly, whether one chooses to implement the function in one form or another is simply a matter of choice; the two forms give exactly the same information.

EXAMPLE 11.3 Simplification of Logical Expression

Problem

Using the rules of Table 11.11, simplify the following function.

$$f(A, B, C, D) = \overline{A} \cdot \overline{B} \cdot D + \overline{A} \cdot B \cdot D + B \cdot C \cdot D + A \cdot C \cdot D$$

Solution

Find: Simplified expression for logical function of four variables.

Analysis:

$$f = \overline{A} \cdot \overline{B} \cdot D + \overline{A} \cdot B \cdot D + B \cdot C \cdot D + A \cdot C \cdot D$$

$= \overline{A} \cdot D \cdot (\overline{B} + B) + B \cdot C \cdot D + A \cdot C \cdot D$	Rule 14
$= \overline{A} \cdot D + B \cdot C \cdot D + A \cdot C \cdot D$	Rule 4
$= (\overline{A} + A \cdot C) \cdot D + B \cdot C \cdot D$	Rule 14
$= (\overline{A} + C) \cdot D + B \cdot C \cdot D$	Rule 18
$= \overline{A} \cdot D + C \cdot D + B \cdot C \cdot D$	Rule 14
$= \overline{A} \cdot D + C \cdot D \cdot (1 + B)$	Rules 6 and 14
$= \overline{A} \cdot D + C \cdot D = (\overline{A} + C) \cdot D$	Rules 2 and 14

EXAMPLE 11.4   Realizing Logic Functions From Truth Tables

Problem

Realize the logic function described by the truth table below.

A	B	C	y
0	0	0	0
0	0	1	1
0	1	0	0
0	1	1	1
1	0	0	1
1	0	1	1
1	1	0	1
1	1	1	1

Solution

Known Quantities: Value of function $y(A, B, C)$ for each possible combination of logical variables A, B, C.

Find: Logical expression realizing the function y.

Analysis: To determine a logical expression for the function y, first we need to convert the truth table to a logical expression. We do so by expressing y as the sum of the products of the three variables for each combination that yields $y = 1$. If the value of a variable is 1, we use the uncomplemented variable. If it's 0, we use the complemented variable. For example, the second row (first instance of $y = 1$) would yield the term $\bar{A} \cdot \bar{B} \cdot C$. Thus,

$$y = \bar{A} \cdot \bar{B} \cdot C + \bar{A} \cdot B \cdot C + A \cdot \bar{B} \cdot \bar{C} + A \cdot \bar{B} \cdot C + A \cdot B \cdot \bar{C} + A \cdot B \cdot C$$
$$= \bar{A} \cdot C(\bar{B} + B) + A \cdot \bar{B} \cdot (\bar{C} + C) + A \cdot B \cdot (\bar{C} + C)$$
$$= \bar{A} \cdot C + A \cdot \bar{B} + A \cdot B = \bar{A} \cdot C + A \cdot (\bar{B} + B) = \bar{A} \cdot C + A = A + C$$

Thus, the function is a two-input OR gate, as shown in Figure 11.22.

Comments: The derivation above has made use of two rules from Table 11.11: rules 4 and 18. Could you have predicted that the variable B would not be used in the final realization?

$A + C = y$ or

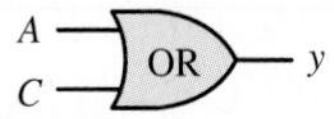

Figure 11.22 Logic gate for example 11.4.

EXAMPLE 11.5   De Morgan's Laws and Product-of-Sums Expressions

Problem

Realize the logic function $y = A + B \cdot C$ in product-of-sums form. Implement the solution, using AND, OR, and NOT gates.

Solution

Known Quantities: Logical expression for the function $y(A, B, C)$.

Find: Physical realization using AND, OR, and NOT gates.

Analysis: We use the fact that $\bar{\bar{y}} = y$ and apply De Morgan's laws as follows:

$$\bar{y} = \overline{A + (B \cdot C)} = \overline{A} \cdot \overline{(B \cdot C)} = \overline{A} \cdot (\overline{B} + \overline{C})$$

$$\bar{\bar{y}} = y = \overline{\overline{A} \cdot (\overline{B} + \overline{C})}$$

The preceding sum-of-products function is realized using complements of each variable (obtained using NOT gates) and is finally complemented as shown in Figure 11.23.

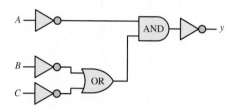

Figure 11.23 Logic gate realization of the function of Example 11.5.

Comments: It should be evident that the original sum-of-products expression, which could be implemented with just one AND and one OR gate, has a much more efficient realization.

EXAMPLE 11.6   **Realizing the AND Function With NAND Gates**

Problem

Use a truth table to show that the AND function can be realized using only NAND gates, and show the physical realization.

Solution

Known Quantities: AND and NAND truth tables.

Find: AND realization using NAND gates.

Assumptions: Consider two-input functions and gates.

Analysis: The truth table below summarizes the two functions:

A	$B (= A)$	$A \cdot B$	$\overline{(A \cdot B)}$
0	0	0	1
1	1	1	0

Figure 11.24 NAND gate as an inverter

A	B	**NAND** $\overline{A \cdot B}$	**AND** $A \cdot B$
0	0	1	0
0	1	1	0
1	0	1	0
1	1	0	1

Figure 11.25 NAND gate realization of AND logic function.

Clearly, to realize the AND function, we need to simply invert the output of a NAND gate. This is easily accomplished if we observe that a NAND gate with its inputs tied together acts as an inverter; you can verify this in the above truth table by looking at the NAND output for the input combinations 0–0 and 1–1, or by referring to Figure 11.24. The final realization is shown in Figure 11.25.

Comments: NAND (and NOR) gates are well suited to implement functions that contain complemented products. Complementary logic gates arise naturally from the inverting characteristics of transistor switches.

EXAMPLE 11.7   Realizing the AND Function With NOR Gates

Problem

Show analytically that the AND function can be realized using only NOR gates, and determine the physical realization.

Solution

Known Quantities: AND and NOR functions.

Find: AND realization using NOR gates.

Assumptions: Consider two-input functions and gates.

Analysis: We can solve this problem using De Morgan's laws. The output of an AND gate can be expressed as $f = A \cdot B$. Using De Morgan's theorem, we write

$$f = \bar{\bar{f}} = \overline{\overline{A \cdot B}} = \overline{\bar{A} + \bar{B}}$$

The above function is implemented very easily if we see that a NOR gate with its input tied together acts as a NOT gate (see Figure 11.26). Thus, the logic circuit of Figure 11.27 provides the desired answer.

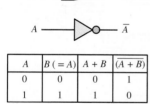

A	$B\,(=A)$	$A + B$	$\overline{(A + B)}$
0	0	0	1
1	1	1	0

Figure 11.26 NOR gate as an inverter

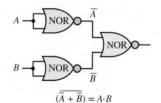

$$\overline{(\bar{A} + \bar{B})} = A \cdot B$$

Figure 11.27 NOR gate realization of AND logic function.

Comments: NOR (and NAND) gates are well suited to implement functions that contain complemented products. Complementary logic gates arise naturally from the inverting characteristics of transistor switches. As a result, such gates are commonly employed in practice.

EXAMPLE 11.8   Realizing a Function With NAND and NOR Gates

Problem

Realize the following function, using only NAND and NOR gates:

$$y = \overline{(A \cdot B)} + C$$

Solution

Known Quantities: Logical expression for *y*.

Find: Realization of *y* using only NAND and NOR gates.

Assumptions: Consider two-input functions and gates.

Analysis: Refer to Examples 11.6 and 11.7 and realize the term $Z = \overline{(A \cdot B)}$ using a two-input NAND gate, and the term $\overline{Z + C}$ using a two-input NOR gate. The solution is shown in Figure 11.28.

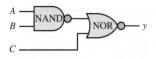

Figure 11.28 Realization of function in Example 11.8.

EXAMPLE 11.9   Half Adder

Problem

Analyze the half adder circuit of Figure 11.29.

Solution

Known Quantities: Logic circuit.

Find: Truth table, functional description.

Schematics, Diagrams, Circuits, and Given Data: Figure 11.29.

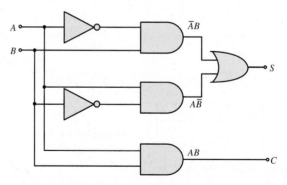

Figure 11.29 Logic circuit realization of a half adder

Analysis: The addition of two binary digits was summarized in Table 11.2. It is important to observe that when both *A* and *B* are equal to 1, the sum requires two digits: the lower digit is a 0, and there also is a carry of 1. Thus, the circuit representing this operation must give an output consisting of two digits. Figure 11.29 shows a circuit called a *half adder* that performs binary addition providing two output bits: the sum *S* and the carry *C*.

A logic statement for the rule of addition can be written as follows: *S* is 1 if *A* is 0 and *B* is 1, or if *A* is 1 and *B* is 0; *C* is 1 if *A* and *B* are 1. In terms of a logic function, we can express this statement with the following logical expressions:

$$S = \overline{A}B + A\overline{B}   \quad \text{and} \quad   C = AB$$

The circuit of Figure 11.29 implements this function using NOT, AND, and OR gates.

EXAMPLE 11.10   Full Adder

Problem

Analyze the full adder circuit of Figure 11.30.

Solution

Known Quantities: Logic circuit.

Find: Truth table, functional description.

Schematics, Diagrams, Circuits, and Given Data: Figure 11.30.

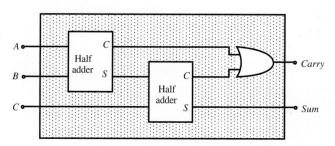

Figure 11.30 Logic circuit realization of a full adder

Analysis: To perform a complete addition we need a *full adder*, that is, a circuit capable of performing a complete 2-bit addition, including taking a carry from a preceding operation. The circuit of Figure 11.30 uses two half adders, such as the one described in Example 11.9, and an OR gate to process the addition of 2-bits, A and B, plus the possible carry from a preceding addition from another (half or full) adder circuit. The truth table below illustrates this operation.

A	0	0	0	0	1	1	1	1
B	0	0	1	1	0	0	1	1
C	0	1	0	1	0	1	0	1
Sum	0	0	0	1	0	1	1	1
Carry	0	1	1	0	1	0	0	1

Truth table for full adder

Comments: To perform the addition of two 4-bit nibbles, we would need a half adder for the first column (LSB), and a full adder for each additional column, that is, three full adders.

CHECK YOUR UNDERSTANDING

Prepare a step-by-step truth table for the following logic expressions.

a. $\overline{(X + Y + Z)} + (X \cdot Y \cdot Z) \cdot \overline{X}$ b. $\overline{X} \cdot Y \cdot Z + Y \cdot (Z + W)$

c. $(X \cdot \overline{Y} + Z \cdot \overline{W}) \cdot (W \cdot X + \overline{Z} \cdot Y)$

(*Hint:* Your truth table must have 2^n entries, where n is the number of logic variables.)

(c)

X	Y	Z	W	Result
0	0	0	0	0
0	0	0	1	0
0	0	1	0	0
0	0	1	1	0
0	1	0	0	0
0	1	0	1	0
0	1	1	0	0
0	1	1	1	0
1	0	0	0	0
1	0	0	1	1
1	0	1	0	0
1	0	1	1	1
1	1	0	0	1
1	1	0	1	0
1	1	1	0	1
1	1	1	1	0

(b)

X	Y	Z	W	Result
0	0	0	0	0
0	0	0	1	0
0	0	1	0	0
0	0	1	1	0
0	1	0	0	0
0	1	0	1	1
0	1	1	0	1
0	1	1	1	1
1	0	0	0	0
1	0	0	1	0
1	0	1	0	0
1	0	1	1	0
1	1	0	0	0
1	1	0	1	1
1	1	1	0	1
1	1	1	1	1

Answer: (a)

X	Y	Z	Result
0	0	0	1
0	0	1	0
0	1	0	0
0	1	1	0
1	0	0	0
1	0	1	0
1	1	0	0
1	1	1	0

CHECK YOUR UNDERSTANDING

Implement the three logic functions of the previous Check Your Understanding exercise using the smallest number of AND, OR, and NOT gates only.

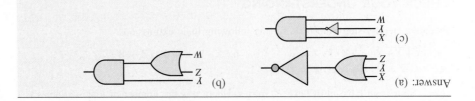

Answer: (a), **(b)**, **(c)**

CHECK YOUR UNDERSTANDING

Implement the three logic functions of the previous Check Your Understanding exercise using the least number of NAND and NOR gates only. (*Hint:* Use De Morgan's laws and the fact that $\overline{\overline{f}} = f$.)

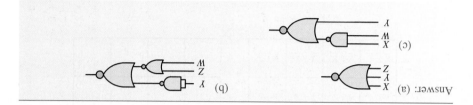

CHECK YOUR UNDERSTANDING

Show that one can obtain an OR gate by using NAND gates only. (*Hint:* Use three NAND gates.)

CHECK YOUR UNDERSTANDING

Show that the XOR function can also be expressed as $Z = X \cdot \overline{Y} + Y \cdot \overline{X}$. Realize the corresponding function using NOT, AND, and OR gates. [*Hint:* Use truth tables for the logic function Z (as defined in the exercise) and for the XOR function.]

11.4   KARNAUGH MAPS AND LOGIC DESIGN

In examining the design of logic functions by means of logic gates, we have discovered that more than one solution is usually available for the implementation of a given logic expression. It should also be clear by now that some combinations of gates can implement a given function more efficiently than others. How can we be assured of having chosen the most efficient realization? Fortunately, there is a procedure that utilizes a map describing all possible combinations of the variables present in the logic function of interest. This map is called a **Karnaugh map**, after its inventor. Figure 11.31 depicts the appearance of Karnaugh maps for two-, three-, and four-variable expressions in two different forms. As can be seen, the row and column assignments for two or more variables are arranged so that all adjacent terms change by only 1 bit. For example, in the two-variable map, the columns next to column 01 are columns 00 and 11. Also note that each map consists of 2^N **cells**, where N is the number of logic variables.

Each cell in a Karnaugh map contains a **minterm**, that is, a product of the N variables that appear in our logic expression (perhaps in complemented form). For example, for the case of three variables ($N = 3$), there are $2^3 = 8$ such combinations, or minterms: $\overline{X} \cdot \overline{Y} \cdot \overline{Z}, \overline{X} \cdot \overline{Y} \cdot Z, \overline{X} \cdot Y \cdot \overline{Z}, \overline{X} \cdot Y \cdot Z, X \cdot \overline{Y} \cdot \overline{Z}, X \cdot \overline{Y} \cdot Z, X \cdot Y \cdot \overline{Z},$ and $X \cdot Y \cdot Z$. The content of each cell—that is, the minterm—is the product of the variables appearing at the corresponding vertical and horizontal coordinates.

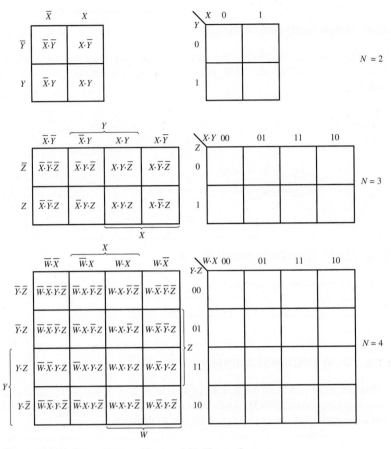

Figure 11.31 Two-, three-, and four-variable Karnaugh maps

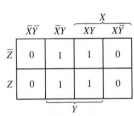

	$\overline{X}\overline{Y}$	$\overline{X}Y$	XY	$X\overline{Y}$
$\overline{Z}$	0	1	1	0
Z	0	1	1	0

Karnaugh map

X	Y	Z	Desired function
0	0	0	0
0	0	1	0
0	1	0	1
0	1	1	1
1	0	0	0
1	0	1	0
1	1	0	1
1	1	1	1

Truth table

Figure 11.32 Truth table and Karnaugh map representations of a logic function

For example, in the three-variable map, $X \cdot Y \cdot \overline{Z}$ appears at the intersection of $X \cdot Y$ and $\overline{Z}$. The map is filled by placing a value of 1 for any combination of variables for which the desired output is a 1. For example, consider the function of three variables for which we desire to have an output of 1 whenever variables X, Y, and Z have the following values:

$X = 0$	$Y = 1$	$Z = 0$
$X = 0$	$Y = 1$	$Z = 1$
$X = 1$	$Y = 1$	$Z = 0$
$X = 1$	$Y = 1$	$Z = 1$

The same truth table is shown in Figure 11.32 together with the corresponding Karnaugh map.

The Karnaugh map provides an immediate view of the values of the function in graphical form. Further, the arrangement of the cells in the Karnaugh map is such that any two adjacent cells contain minterms that vary in only one variable. This property, as will be verified shortly, is quite useful in the design of logic functions by means of logic gates, especially if we consider the map to be continuously wrapping around itself, as if the top and bottom, and right and left, edges were touching. For the three-variable map given in Figure 11.31, for example, the cell $X \cdot \overline{Y} \cdot \overline{Z}$ is adjacent to $\overline{X} \cdot \overline{Y} \cdot \overline{Z}$ if we "roll" the map so that the right edge touches the left. Note that these two cells differ only in the variable X, a property that we earlier claimed adjacent cells have.[2]

X	W	Y	Z	Desired function
0	0	0	0	1
0	0	0	1	1
0	0	1	0	0
0	0	1	1	0
0	1	0	0	0
0	1	0	1	1
0	1	1	0	1
0	1	1	1	0
1	0	0	0	0
1	0	0	1	1
1	0	1	0	1
1	0	1	1	0
1	1	0	0	0
1	1	0	1	0
1	1	1	0	0
1	1	1	1	1

Truth table for four-variable expression

	$\overline{W}\cdot\overline{X}$	$\overline{W}\cdot X$	$W\cdot X$	$W\cdot\overline{X}$
$\overline{Y}\cdot\overline{Z}$	1	0	0	0
$\overline{Y}\cdot Z$	1	1	0	1
$Y\cdot Z$	0	0	1	0
$Y\cdot\overline{Z}$	0	1	0	1

Figure 11.33 Karnaugh map for a four-variable expression

Shown in Figure 11.33 is a more complex, four-variable logic function, which will serve as an example in explaining how Karnaugh maps can be used directly to implement a logic function. First, we define a *subcube* as a set of 2^m adjacent cells with logical value 1, for $m = 1, 2, 3, \ldots, N$. Thus, a subcube can consist of 1, 2, 4, 8, 16, 32, $\ldots$ cells. All possible subcubes for the four-variable map of Figure 11.31 are shown in Figure 11.34. Note that there are no four-cell subcubes in this particular case. Note also that there is some overlap between subcubes. Examples of four-cell and eight-cell subcubes are shown in Figure 11.35 for an arbitrary expression.

In general, one tries to find the largest possible subcubes to cover all the 1 entries in the map. How do maps and subcubes help in the realization of logic functions, then? The use of maps and subcubes in minimizing logic expressions is best explained by considering the following rule of boolean algebra:

$$Y \cdot X + Y \cdot \overline{X} = Y$$

where the variable Y could represent a product of logic variables [e.g., we could similarly write $(Z \cdot W) \cdot X + (Z \cdot W) \cdot \overline{X} = Z \cdot W$ with $Y = Z \cdot W$]. This rule is easily proved by factoring Y

$$Y \cdot (X + \overline{X})$$

and observing that $X + \overline{X} = 1$ always. Then it should be clear that variable X need not appear in the expression at all.

Let us apply this rule to a more complex logic expression, to verify that it can also apply to this case. Consider the logic expression

$$\overline{W} \cdot X \cdot \overline{Y} \cdot Z + \overline{W} \cdot \overline{X} \cdot \overline{Y} \cdot Z + W \cdot \overline{X} \cdot \overline{Y} \cdot Z + W \cdot X \cdot \overline{Y} \cdot Z$$

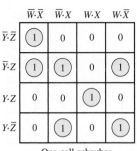

One-cell subcubes

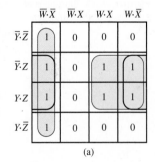

Two-cell subcubes

Figure 11.34 One- and two-cell subcubes for the Karnaugh map of Figure 11.31

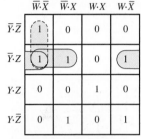

(a)

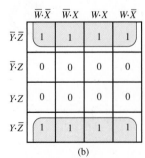

(b)

Figure 11.35 Four- and eight-cell subcubes for an arbitrary logic function

---

[2]A useful rule to remember is that in a two-variable map, there are two minterms adjacent to any given minterm; in a three-variable map, three minterms are adjacent to any given minterm; in a four-variable map, the number is four, and so on.

	$\overline{W}\cdot\overline{X}$	$\overline{W}\cdot X$	$W\cdot X$	$W\cdot\overline{X}$
$\overline{Y}\cdot\overline{Z}$	0	0	0	0
$\overline{Y}\cdot Z$	1	1	1	1
$Y\cdot Z$	0	0	0	0
$Y\cdot\overline{Z}$	0	0	0	0

Figure 11.36 Karnaugh map for the function $\overline{W}\cdot X\cdot\overline{Y}\cdot Z+\overline{W}\cdot\overline{X}\cdot\overline{Y}\cdot Z+W\cdot\overline{X}\cdot\overline{Y}\cdot Z+W\cdot X\cdot\overline{Y}\cdot Z$

	$\overline{W}\cdot\overline{X}$	$\overline{W}\cdot X$	$W\cdot X$	$W\cdot\overline{X}$
$\overline{Y}\cdot\overline{Z}$	1	1	1	1
$\overline{Y}\cdot Z$	1	1	1	1
$Y\cdot Z$	0	0	0	0
$Y\cdot\overline{Z}$	0	0	0	0

Figure 11.37 Illustration of Karnaugh map simplification.

and factor it as follows:

$$\overline{W}\cdot Z\cdot\overline{Y}\cdot(X+\overline{X})+W\cdot\overline{Y}\cdot Z\cdot(\overline{X}+X)=\overline{W}\cdot Z\cdot\overline{Y}+W\cdot\overline{Y}\cdot Z$$
$$=\overline{Y}\cdot Z\cdot(\overline{W}+W)=\overline{Y}\cdot Z$$

That is quite a simplification! If we consider, now, a map in which we place a 1 in the cells corresponding to the minterms $\overline{W}\cdot X\cdot\overline{Y}\cdot Z$, $\overline{W}\cdot\overline{X}\cdot\overline{Y}\cdot Z$, $W\cdot\overline{X}\cdot\overline{Y}\cdot Z$, and $W\cdot X\cdot\overline{Y}\cdot Z$, forming the previous expression, we obtain the Karnaugh map of Figure 11.36. It can easily be verified that the map of Figure 11.36 shows a single four-cell subcube corresponding to the term $\overline{Y}\cdot Z$.

We have not established formal rules yet, but it definitely appears that the map method for simplifying boolean expressions is a convenient tool. In effect, the map has performed the algebraic simplification automatically! We can see that in any subcube, one or more of the variables present will appear in both complemented *and* uncomplemented forms in all their combinations with the other variables. These variables can be eliminated. As an illustration, in the *eight-cell* subcube case of Figure 11.37, the full-blown expression would be

$$\overline{W}\cdot\overline{X}\cdot\overline{Y}\cdot\overline{Z}+\overline{W}\cdot X\cdot\overline{Y}\cdot\overline{Z}+W\cdot X\cdot\overline{Y}\cdot\overline{Z}+W\cdot\overline{X}\cdot\overline{Y}\cdot\overline{Z}$$
$$+\overline{W}\cdot\overline{X}\cdot\overline{Y}\cdot\overline{Z}+\overline{W}\cdot X\cdot\overline{Y}\cdot\overline{Z}+W\cdot X\cdot\overline{Y}\cdot\overline{Z}+W\cdot\overline{X}\cdot\overline{Y}\cdot\overline{Z}$$

However, if we consider the eight-cell subcube, we note that the three variables X, W, and Z appear in both complemented and uncomplemented form in all their combinations with the other variables and thus can be removed from the expression. This reduces the seemingly unwieldy expression simply to $\overline{Y}$! In logic design terms, a simple inverter with Y input is sufficient to implement the expression.

Sum-of-Products and Product-of-Sums Realizations

Logic functions can be expressed in either of two forms: sum of products (SOP) or product of sums (POS). For example, the following logic expression is in SOP form:

$$\overline{W}\cdot X\cdot\overline{Y}\cdot Z+\overline{W}\cdot\overline{X}\cdot\overline{Y}\cdot Z+W\cdot\overline{X}\cdot\overline{Y}\cdot Z+W\cdot X\cdot\overline{Y}\cdot Z$$

The following rules are useful in determining the minimal sum-of-products expression.

FOCUS ON PROBLEM SOLVING

SUM-OF-PRODUCTS REALIZATIONS

1. Begin with isolated cells. These must be used as they are, since no simplification is possible.
2. Find all cells that are adjacent to only one other cell, forming two-cell subcubes.
3. Find cells that form four-cell subcubes, eight-cell subcubes, and so forth.
4. The minimal expression is formed by the collection of the *smallest number of maximal subcubes*.

De Morgan's laws state that every SOP expression has an equivalent POS form. A simple example of a POS expression is $(W+Y)\cdot(Y+Z)$. For any particular logical expression one of the two forms may lead to a realization involving a smaller number of gates. For Karnaugh maps, the POS form is found using the following rules:

FOCUS ON PROBLEM SOLVING

PRODUCT-OF-SUMS REALIZATIONS

1. Group 0s in subcubes exactly as is done for 1s when seeking an SOP expression.

2. Produce a complemented Karnaugh map by swapping X with $\overline{X}$, Y with $\overline{Y}$, and Z with $\overline{Z}$.

3. Each subcube of 0s represents a *sum* of the complemented Karnaugh map's elements.

4. Form the product of those sums.

An alternate POS realization method is to represent each subcube of 0s as the product of the Karnaugh map elements, form the sum of these products, and complement the entire summation. After some manipulation using De Morgan's laws, the result will yield an equivalent POS form. It should be noted that the POS form yielded by this alternate method may not *appear* to be equivalent to the form found using the highlighted method above. In such cases, it is always possible to show that each form yields the same Karnaugh map. Example 11.16 is a good example of this point. Examples 11.16 and 11.17 illustrate how one form may result in a more efficient solution than the other.

Don't-Care Conditions

Another simplification technique may be employed whenever the value of a logic function is permitted to be either 1 or 0 for certain combinations of the input variables. This situation often arises in problem specifications. A good example is the binary-coded decimal system, in which the six four-bit combinations [1010], [1011], [1100], [1101], [1110], and [1111] are not permitted. The algorithm used to determine the value of the BCD nibble should be indifferent to these six combinations. On the other hand, an error checking algorithm should not be; it should detect an erroneous input nibble!

Whenever it does not matter whether a position in the map is filled by a 1 or a 0, a **don't-care** entry is used, denoted by an *x*. When forming subcubes in the Karnaugh map, each don't-care entry can be treated as either a 1 or a 0 as necessary to yield the smallest number of subcubes and therefore the greatest simplification.

Safety Circuit for Operation of a Stamping Press

FOCUS ON MEASUREMENTS

Problem:

In this example, the techniques illustrated in the preceding examples are applied to a practical situation. To operate a stamping press, an operator must press two buttons (b_1 and b_2) 1 m apart from each other and away from the press (this ensures that the operator's hands cannot be caught in the press). When the buttons are pressed, the logical variables b_1 and b_2 are equal to 1. Thus, we can define a new variable $A = b_1 \cdot b_2$; when

(Continued)

(Continued)

$A = 1$, the operator's hands are safely away from the press. In addition to the safety requirement, however, other conditions must be satisfied before the operator can activate the press. The press is designed to operate on one of two workpieces, part I and part II, but not both. Thus, acceptable logic states for the press to be operated are "part I is in the press, but not part II" and "part II is in the press, but not part I." If we denote the presence of part I in the press by the logical variable $B = 1$ and the presence of part II by the logical variable $C = 1$, we can then impose additional requirements on the operation of the press. For example, a robot used to place either part in the press could activate a pair of switches (corresponding to logical variables B and C) indicating which part, if any, is in the press. Finally, for the press to be operable, it must be "ready," meaning that it has to have completed any previous stamping operation. Let the logical variable $D = 1$ represent the ready condition. We have now represented the operation of the press in terms of four logical variables, summarized in the truth table of Table 11.12. Note that only two combinations of the logical variables will result in operation of the press: $ABCD = 1011$ and $ABCD = 1101$. You should verify that these two conditions correspond to the desired operation of the press. Using a Karnaugh map, realize the logic circuitry required to implement the truth table shown.

Table 11.12 **Conditions for operation of stamping press**

(A) $b_1 \cdot b_2$*	(B) Part I is in press	(C) Part II is in press	(D) Press is operable	Press operation 1 = pressing 0 = not pressing
0	0	0	0	0
0	0	0	1	0
0	0	1	0	0
0	0	1	1	0
0	1	0	0	0
0	1	0	1	0
0	1	1	0	0
0	1	1	1	0
1	0	0	0	0
1	0	0	1	0
1	0	1	0	0
1	0	1	1	1
1	1	0	0	0
1	1	0	1	1
1	1	1	0	0
1	1	1	1	0

*Both buttons (b_1, b_2) must be pressed for this to be a 1.

Solution:

Table 11.12 can be converted to a Karnaugh map, as shown in Figure 11.38. Since there are many more 0s than 1s in the table, the use of 0s in covering the map will lead to greater simplification. This will result in a product-of-sums expression. The four subcubes shown in Figure 11.38 yield the equation

$$A \cdot D \cdot (C + B) \cdot (\overline{C} + \overline{B})$$

By De Morgan's law, this equation is equivalent to

$$A \cdot D \cdot (C + B) \cdot (\overline{C \cdot B})$$

which can be realized by the circuit of Figure 11.39.

(Continued)

(*Concluded*)

For the purpose of comparison, the corresponding sum-of-products circuit is shown in Figure 11.40. Note that this circuit employs a greater number of gates and will therefore lead to a more expensive design.

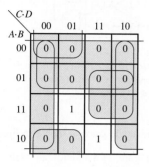

Figure 11.38 Karnaugh map for stamping press operation.

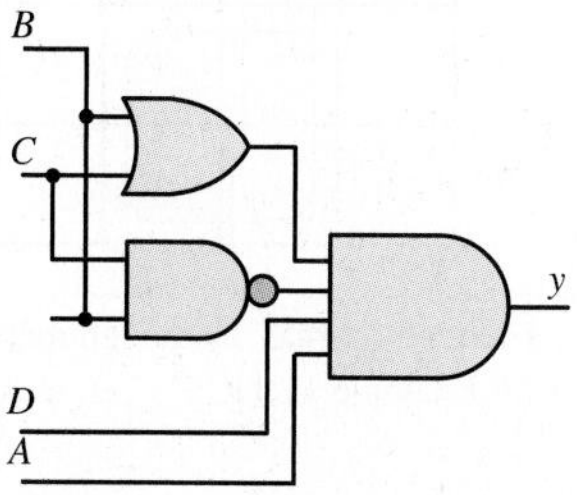

Figure 11.39 Karnaugh map realization by product-of-sums circuit.

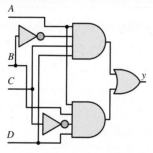

Figure 11.40 Karnaugh map realization by sum-of-products circuit.

EXAMPLE 11.11   Logic Circuit Design Using Karnaugh Maps

Problem

Design a logic circuit that implements the truth table of Figure 11.41.

Solution

Known Quantities: Truth table for $y(A, B, C, D)$.

Find: Realization of y.

Assumptions: Two-, three-, and four-input gates are available.

Analysis: We use the Karnaugh map of Figure 11.42, which is shown with values of 1 and 0 already in place. We recognize four subcubes in the map; three are four-cell subcubes, and one is a two-cell subcube. The expressions for the subcubes are $\overline{A} \cdot \overline{B} \cdot \overline{D}$ for the two-cell subcube; $\overline{B} \cdot \overline{C}$ for the subcube that wraps around the map; $\overline{C} \cdot D$ for the 4-by-1 subcube; and $A \cdot D$ for the square subcube at the bottom of the map. Thus, the expression for y is

$$y = \overline{A} \cdot \overline{B} \cdot \overline{D} + \overline{B} \cdot \overline{C} + \overline{C}D + AD$$

A	B	C	D	y
0	0	0	0	1
0	0	0	1	1
0	0	1	0	1
0	0	1	1	0
0	1	0	0	0
0	1	0	1	1
0	1	1	0	0
0	1	1	1	0
1	0	0	0	1
1	0	0	1	1
1	0	1	0	0
1	0	1	1	1
1	1	0	0	0
1	1	0	1	1
1	1	1	0	0
1	1	1	1	1

Figure 11.41 Truth table for example 11.11.

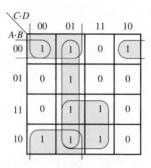

Figure 11.42 Karnaugh map for Example 11.11

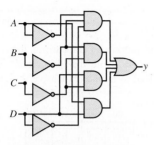

Figure 11.43 Logic circuit realization of Karnaugh map of Figure 11.42

The implementation of the above function with logic gates is shown in Figure 11.43.

Comments: The Karnaugh map highlighting of Figure 11.42 yields an SOP expression because all the 1s were highlighted.

EXAMPLE 11.12   Deriving a Sum-of-Products Expression From a Logic Circuit

Problem

Derive the truth table and minimum sum-of-products expression for the circuit of Figure 11.44.

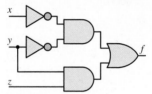

Figure 11.44 Logic circuit for Example 11.12.

Solution

Known Quantities: Logic circuit representing $f(x, y, z)$.

Find: Expression for f and corresponding truth table.

Analysis: To determine the truth table, we write the expression corresponding to the logic circuit of Figure 11.44:

$$f = \bar{x} \cdot \bar{y} + y \cdot z$$

The truth table corresponding to this expression and the corresponding Karnaugh map with sum-of-products covering are shown in Figure 11.45.

x	y	z	f
0	0	0	1
0	0	1	1
0	1	0	0
0	1	1	1
1	0	0	0
1	0	1	0
1	1	0	0
1	1	1	1

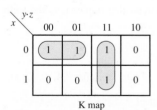

Figure 11.45 Truth table corresponding to circuit of Fig. 11.44.

Comments: If we used 0s in covering the Karnaugh map for this example, the resulting expression would be a POS. Verify that the complexity of the circuit would be unchanged. Note also that the subcube ($x = 0$, $yz = 01$, 11) is not used because it does not further minimize the solution.

EXAMPLE 11.13   **Realizing a Sum of Products Using Only NAND Gates**

Problem

Realize the following function in sum-of-products form, using only two-input NAND gates.

$$f = (\bar{x} + \bar{y}) \cdot (y + \bar{z})$$

Solution

Known quantities: $f(x, y, z)$.

Find: Logic circuit for f using only NAND gates.

Analysis: The first step is to convert the expression for f into an expression that can be easily implemented with NAND gates. We observe that direct application of De Morgan's laws yields

$$\bar{x} + \bar{y} = \overline{x \cdot y}$$
$$y + \bar{z} = \overline{z \cdot \bar{y}}$$

Thus, we can write the function as

$$f = (\overline{x \cdot y}) \cdot (\overline{z \cdot \bar{y}})$$

and implement it with five NAND gates, as shown in Figure 11.46.

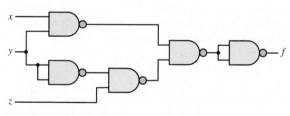

Figure 11.46

Comments: Note that we used two NAND gates as inverters—one to obtain $\bar{y}$, the other to invert the output of the fourth NAND gate, equal to $\overline{(\overline{x \cdot y}) \cdot (\overline{z \cdot \bar{y}})}$.

EXAMPLE 11.14   **Simplifying Expressions by Using Karnaugh Maps**

Problem

Simplify the following expression by using a Karnaugh map.

$$f = x \cdot y + \bar{x} \cdot z + y \cdot z$$

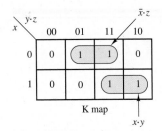

Figure 11.47 Karnaugh map for Example 11.14.

Solution

Known Quantities: $f(x, y, z)$.

Find: Minimal expression for f.

Analysis: We highlight a three-term Karnaugh map, as shown in Figure 11.47. The 1s can be covered using just two subcubes: $f = x \cdot y + \bar{x} \cdot z$. Thus, the term $y \cdot z$ is redundant.

Comments: Notice that the term $y \cdot z$ is also included implicitly by the two subcubes. Thus, in this example, the expression $x \cdot y + \bar{x} \cdot z + y \cdot z$ is equivalent to, but more complicated than, $x \cdot y + \bar{x} \cdot z$.

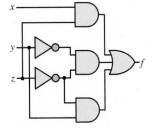

EXAMPLE 11.15   Simplifying a Logic Circuit by Using the Karnaugh Map

Problem

Derive the Karnaugh map corresponding to the circuit of Figure 11.48, and use the resulting map to simplify the expression.

Solution

Known Quantities: Logic circuit.

Find: Simplified logic circuit.

Analysis: We first determine the expression $f(x, y, z)$ from the logic circuit:

$$f = (x \cdot z) + (\bar{y} \cdot \bar{z}) + (y \cdot \bar{z})$$

This expression leads to the Karnaugh map shown in Figure 11.49. Inspection of the Karnaugh map reveals that the map could have been covered more efficiently by using four-cell subcubes. The improved map covering, corresponding to the simpler function $f = x + \bar{z}$, and the resulting logic circuit are shown in Figure 11.50.

Figure 11.48 Logic circuit for example 11.15.

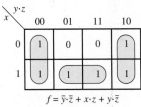

$$f = \bar{y} \cdot \bar{z} + x \cdot z + y \cdot \bar{z}$$

Figure 11.49 Karnaugh map corresponding to circuit of Fig. 11.48.

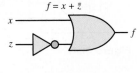

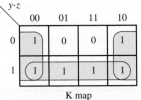

K map

Figure 11.50 Karnaugh map for Example 11.15 with improved cell coverage, and resulting logic circuit.

Comments: In general, the largest possible subcubes in a Karnaugh map correspond to the minimum possible solution.

EXAMPLE 11.16   Comparison of Sum-of-Products and
Product-of-Sums Designs

Problem

Realize the function f described by the accompanying truth table, using both 0 and 1 coverings in the Karnaugh map.

x	y	z	f
0	0	0	0
0	0	1	1
0	1	0	1
0	1	1	1
1	0	0	1
1	0	1	1
1	1	0	0
1	1	1	0

Solution

Known Quantities: Truth table for logic function.

Find: Realization in both sum-of-products and product-of-sums forms.

Analysis:

1. *Product-of-sums expression.* Product-of-sums expressions use 0s to determine the logical expression from a Karnaugh map. Figure 11.51 depicts the Karnaugh map covering with 0s, leading to the expression

$$f = (x + y + z) \cdot (\overline{x} + \overline{y})$$

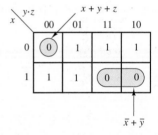

Figure 11.51 Product-of-sums
Karnaugh map.

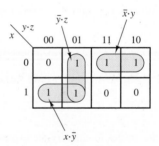

Figure 11.52 Sum-of-
products Karnaugh map.

2. *Sum-of-products expression.* Sum-of-products expressions use 1s to determine the logical expression from a Karnaugh map. Figure 11.52 depicts the Karnaugh map covering with 1s, leading to the expression

$$f = (\overline{x} \cdot y) + (x \cdot \overline{y}) + (\overline{y} \cdot z)$$

Comments: The product-of-sums solution requires the use of five gates (two OR, two NOT, and one AND), while the sum-of-products solution will use six gates (one OR, two NOT, and three AND). Thus, solution 1 leads to the simpler design.

EXAMPLE 11.17   Product-of-Sums Design

Problem

Realize the function f described by the accompanying truth table in minimal product-of-sums form. Draw the corresponding Karnaugh map.

x	y	z	f
0	0	0	1
0	0	1	0
0	1	0	1
0	1	1	0
1	0	0	1
1	0	1	0
1	1	0	0
1	1	1	0

Solution

Known Quantities: Truth table for logic function.

Find: Realization in minimal product-of-sums forms.

Analysis: We cover the Karnaugh map of Figure 11.53 using 0s, and we obtain the following function:

$$f = \bar{z} \cdot (\bar{x} + \bar{y})$$

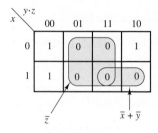

Figure 11.53 Product-of-sums Karnaugh map for Example 11.17.

Comments: What is the equivalent sum-of-products solution? Find it! Is it simpler?

EXAMPLE 11.18   Using Don't-Care Conditions to Simplify Expressions—1

Problem

Use don't-care entries to simplify the expression

$$f(A, B, C, D) = \bar{A} \cdot \bar{B} \cdot \bar{C} \cdot D + \bar{A} \cdot \bar{B} \cdot C \cdot \bar{D} + \bar{A} \cdot \bar{B} \cdot C \cdot D$$
$$+ \bar{A} \cdot B \cdot \bar{C} \cdot D + A \cdot \bar{B} \cdot C \cdot D + A \cdot B \cdot \bar{C} \cdot \bar{D}$$

Note that the x's never occur, and so they may be assigned a 1 or a 0, whichever will best simplify the expression.

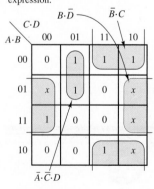

Figure 11.54

Solution

Known Quantities: Logical expression; don't-care conditions.

Find: Minimal realization.

Schematics, Diagrams, Circuits, and Given Data: Don't-care conditions: $f(A, B, C, D) = \{0100, 0110, 1010, 1110\}$.

Analysis: We cover the Karnaugh map of Figure 11.54 using 1s, and also using x entries for each don't-care condition. Treating all the x entries as 1s, we complete the covering with two four-cell subcubes and one two-cell subcube, to obtain the following simplified expression:

$$f(A, B, C, D) = B \cdot \bar{D} + \bar{B} \cdot C + \bar{A} \cdot \bar{C} \cdot D$$

Comments: Note that we could have also interpreted the don't-care entries as 0s and tried to solve in product-of-sums form. Verify that the expression obtained above is indeed the minimal one.

EXAMPLE 11.19   **Using Don't-Care Conditions to Simplify Expressions—2**

Problem

Find a minimum sum-of-products realization for the expression $f(A, B, C)$.

Solution

Known Quantities: Logical expression, don't-care conditions.

Find: Minimal realization.

Schematics, Diagrams, Circuits, and Given Data:

$$f(A, B, C) = \begin{cases} 1 & \text{for } \{A, B, C\} = \{000, 010, 011\} \\ x & \text{for } \{A, B, C\} = \{100, 101, 110\} \end{cases}$$

Analysis: We cover the Karnaugh map of Figure 11.55 using 1s, and also using x entries for each don't-care condition. By appropriately selecting two of the three don't-care entries to be equal to 1, we complete the covering with one four-cell subcube and one two-cell subcube, to obtain the following minimal expression:

$$f(A, B, C) = \overline{A} \cdot B + \overline{C}$$

Comments: Note that we have chosen to set one of the don't-care entries equal to 0, since it would not lead to any further simplification.

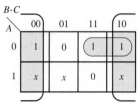

Figure 11.55

EXAMPLE 11.20   **Using Don't-Care Conditions to Simplify Expressions—3**

Problem

Find a minimum sum-of-products realization for the expression $f(A, B, C, D)$.

Solution

Known Quantities: Logical expression; don't-care conditions.

Find: Minimal realization.

Schematics, Diagrams, Circuits, and Given Data:

$$f(A, B, C) = \begin{cases} 1 & \text{for } \{A, B, C, D\} = \{0000, 0011, 0110, 1001\} \\ x & \text{for } \{A, B, C, D\} = \{1010, 1011, 1101, 1110, 1111\} \end{cases}$$

Analysis: We cover the Karnaugh map of Figure 11.56 using 1s, and using x entries for each don't-care condition. By appropriately selecting four of the five don't-care entries to be equal to 1, we complete the covering with one four-cell subcube, two two-cell subcubes, and one one-cell subcube, to obtain the following expression:

$$f(A, B, C) = \overline{A} \cdot \overline{B} \cdot \overline{C} \cdot \overline{D} + B \cdot C \cdot \overline{D} + A \cdot D + \overline{B} \cdot C \cdot D$$

Comments: Would the product-of-sums realization be simpler? Verify.

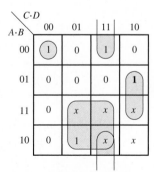

Figure 11.56

CHECK YOUR UNDERSTANDING

Simplify the following expression, using a Karnaugh map.

$$\overline{W} \cdot \overline{X} \cdot \overline{Y} \cdot \overline{Z} + \overline{W} \cdot \overline{X} \cdot Y \cdot \overline{Z} + W \cdot X \cdot \overline{Y} \cdot \overline{Z} + W \cdot \overline{X} \cdot \overline{Y} \cdot \overline{Z} + W \cdot \overline{X} \cdot Y \cdot \overline{Z}$$
$$+ W \cdot X \cdot Y \cdot \overline{Z}$$

Simplify the following expression, using a Karnaugh map.

$$\overline{W} \cdot \overline{X} \cdot \overline{Y} \cdot \overline{Z} + \overline{W} \cdot \overline{X} \cdot Y \cdot \overline{Z} + W \cdot X \cdot \overline{Y} \cdot \overline{Z} + W \cdot \overline{X} \cdot \overline{Y} \cdot \overline{Z} + W \cdot \overline{X} \cdot Y \cdot \overline{Z}$$
$$+ \overline{W} \cdot X \cdot \overline{Y} \cdot \overline{Z}$$

Answers: $W \cdot \overline{Z} + \overline{X} \cdot \overline{Y} \cdot \overline{Z}$; $\overline{X} \cdot \overline{Z} + \overline{X} \cdot \overline{Z}$

CHECK YOUR UNDERSTANDING

Verify that the product-of-sums expression for Example 11.16 can be realized with fewer gates.

CHECK YOUR UNDERSTANDING

Would a sum-of-products realization for Example 11.17 require fewer gates?

Answer: No

CHECK YOUR UNDERSTANDING

Prove that the circuit of Figure 11.53 can also be obtained from the sum of products.

CHECK YOUR UNDERSTANDING

In Example 11.18, assign a value of 0 to the don't-care terms and derive the corresponding minimal expression. Is the new function simpler than the one obtained in Example 11.18?

Answer: $f = A \cdot B \cdot \overline{C} \cdot \overline{D} + \overline{A} \cdot \overline{C} \cdot D + \overline{A} \cdot \overline{B} \cdot C + \overline{B} \cdot C \cdot D$; No

CHECK YOUR UNDERSTANDING

In Example 11.19, assign a value of 0 to the don't-care terms and derive the corresponding minimal expression. Is the new function simpler than the one obtained in Example 11.19?

In Example 11.19, assign a value of 1 to all don't-care terms and derive the corresponding minimal expression. Is the new function simpler than the one obtained in Example 11.19?

Answers: $f = \bar{A} \cdot B + A \cdot \bar{B} + \bar{A} \cdot \bar{C}$; No. $f = \bar{A} \cdot B + A \cdot \bar{B} + \bar{C}$; No.

CHECK YOUR UNDERSTANDING

In Example 11.20, assign a value of 0 to all don't-care terms and derive the corresponding minimal expression. Is the new function simpler than the one obtained in Example 11.20?

In Example 11.20, assign a value of 1 to all don't-care terms and derive the corresponding minimal expression. Is the new function simpler than the one obtained in Example 11.20?

Answers: $f = \bar{A} \cdot \bar{B} \cdot \bar{C} \cdot \bar{D} + A \cdot \bar{B} \cdot \bar{C} \cdot D + \bar{A} \cdot \bar{B} \cdot C \cdot D + \bar{A} \cdot \bar{C} \cdot D + A \cdot \bar{B} \cdot C \cdot D$; No. $f = \bar{A} \cdot \bar{B} \cdot \bar{C} + B \cdot \bar{C} \cdot \bar{D} + A \cdot D + \bar{B} \cdot \bar{C} \cdot D + A \cdot \bar{C}$; No.

11.5   COMBINATIONAL LOGIC MODULES

The basic logic gates described in the previous section are used to implement more advanced functions and are often combined to form logic modules, which, thanks to modern technology, are available in compact integrated-circuit packages. In this section we discuss a few of the more common **combinational logic modules**, illustrating how these can be used to implement advanced logic functions.

Multiplexers

Multiplexers, or **data selectors**, are combinational logic circuits that permit the selection of one of many inputs. A typical multiplexer (MUX) has 2^n **data lines**, n **address** (or **data select**) **lines**, and one output. In addition, other control inputs (e.g., enables) may exist. Standard, commercially available MUXs allow for n up to 4; however, two or more MUXs can be combined if a greater range is needed. The MUX allows for one of 2^n inputs to be selected as the data output; the selection of which input is to appear at the output is made by way of the address lines. Figure 11.57 depicts the block diagram of a four-input MUX. The input data lines are labeled D_0, D_1, D_2, and D_3; the **data select**, or address, **lines** are labeled I_0 and I_1; and the output is available in both complemented and uncomplemented form and is thus labeled F or $\bar{F}$. Finally, an **enable** input, labeled E, is also provided, as a means of enabling or disabling the MUX: if $E = 1$, the MUX is disabled; if $E = 0$, it is enabled. The negative logic (MUX off when $E = 1$ and on when $E = 0$) is represented by the small "bubble" at the enable input, which represents a complement operation (just as at the output of NAND and NOR gates). The enable input is useful whenever one is

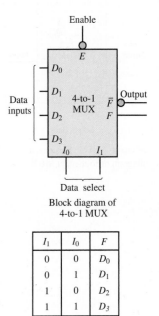

Block diagram of
4-to-1 MUX

I_1	I_0	F
0	0	D_0
0	1	D_1
1	0	D_2
1	1	D_3

Truth table of
4-to-1 MUX

Figure 11.57 4-to-1 MUX

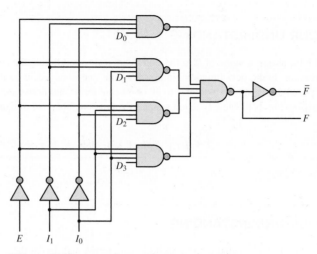

Figure 11.58 Internal structure of the 4-to-1 MUX

interested in a cascade of MUXs; this would be of interest if we needed to select a line from a large number, say, $2^8 = 256$. Then two four-input MUXs could be used to provide the data selection of 1 of 8.

The material described in previous sections is quite adequate to describe the internal workings of a multiplexer. Figure 11.58 shows the internal construction of a 4-to-1 MUX using exclusively NAND gates (inverters are also used, but the reader will recall that a NAND gate can act as an inverter if properly connected).

In the design of digital systems (e.g., microprocessors), a single line is often required to carry two or more different digital signals. However, only one signal at a time can be placed on the line. A MUX will allow us to select, at different instants, the signal we wish to place on this single line. This property is shown here for a 4-to-1 MUX. Figure 11.59 depicts the functional diagram of a 4-to-1 MUX, showing four data lines, D_0 through D_3, and two select lines, I_0 and I_1.

The data selector function of a MUX is best understood in terms of Table 11.13. In this truth table, the x's represent don't-care entries. As can be seen from the truth table, the output selects one of the data lines depending on the values of I_1 and I_0, assuming that I_0 is the least significant bit. As an example, $I_1 I_0 = 10$ selects D_2, which means that the output F will select the value of the data line D_2. Therefore $F = 1$ if $D_2 = 1$ and $F = 0$ if $D_2 = 0$.

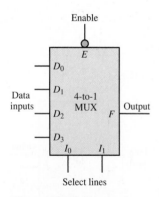

Figure 11.59 Functional diagram of four-input MUX

Table 11.13 **MUX truth table**

I_1	I_0	D_3	D_2	D_1	D_0	F
0	0	x	x	x	0	0
0	0	x	x	x	1	1
0	1	x	x	0	x	0
0	1	x	x	1	x	1
1	0	x	0	x	x	0
1	0	x	1	x	x	1
1	1	0	x	x	x	0
1	1	1	x	x	x	1

Read-Only Memory (ROM)

Another common technique for implementing logic functions uses a **read-only memory**, or **ROM**. As the name implies, a ROM is a logic circuit that holds information in storage ("memory")—in the form of binary numbers—that cannot be altered but can be "read" by a logic circuit. A ROM is an array of memory cells, each of which can store either a 1 or a 0. The array consists of $2^m \times n$ cells, where n is the number of bits in each word stored in a ROM. To access the information stored in a ROM, m address lines are required. When an address is selected, in a fashion similar to the operation of the MUX, the binary word corresponding to the address selected appears at the output, which consists of n bits, that is, the same number of bits as the stored words. In some sense, a ROM can be thought of as a MUX that has an output consisting of a word instead of a single bit.

Figure 11.60 depicts the conceptual arrangement of a ROM with $n = 4$ and $m = 2$. The ROM table has been filled with arbitrary 4-bit words, just for the purpose of illustration. In Figure 11.60, if one were to select an enable input of 0 (i.e., on) and values for the address lines of $I_0 = 0$ and $I_1 = 1$, the output word would be $W_2 = 0110$, so that $b_0 = 0$, $b_1 = 1$, $b_2 = 1$, $b_3 = 0$. Depending on the content of the ROM and the number of address and output lines, one could implement an arbitrary logic function.

Unfortunately, the data stored in read-only memories must be entered during fabrication and cannot be altered later. A much more convenient type of read-only memory is the **erasable programmable read-only memory (EPROM)**, the content of which can be easily programmed and stored and may be changed if needed. EPROMs find use in many practical applications, because of their flexibility in content and ease of programming. The following example illustrates the use of an EPROM to perform the linearization of a nonlinear function.

ROM address		ROM content (4-bit words)				
I_1	I_0	b_3	b_2	b_1	b_0	
0	0	0	1	1	0	W_0
0	1	1	0	0	1	W_1
1	0	0	1	1	0	W_2
1	1	1	1	1	1	W_3

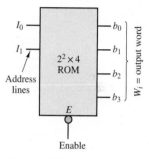

Figure 11.60 Read-only memory

Decoders and Read-Write Memory

Decoders, which are commonly used for applications such as address decoding or memory expansion, are combinational logic circuits as well. Our reason for introducing decoders is to show some of the internal organization of semiconductor memory devices.

Figure 11.61 shows the truth table for a 2-to-4 decoder. The decoder has an enable input $\overline{G}$ and select inputs B and A. It also has four outputs, Y_0 through Y_3. When the enable input is logic 1, all decoder outputs are forced to logic 1 regardless of the select inputs.

This simple description of decoders permits a brief discussion of the internal organization of an **SRAM (static random-access, or read and write, memory)**. An SRAM is internally organized to provide memory with high speed (i.e., short access time), a large bit capacity, and low cost. The memory array in this memory device has a column length equal to the number of words W and a row length equal to the number of bits per word N. To select a word, an n-to-W decoder is needed. Since the address inputs to the decoder select only one of the decoder's outputs, the decoder selects one word in the memory array. Figure 11.62 shows the internal organization of a typical SRAM.

Thus, to choose the desired word from the memory array, the proper address inputs are required. As an example, if the number of words in the memory array is 8, a 3-to-8 decoder is needed.

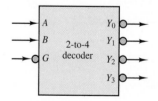

Inputs			Outputs			
Enable	Select					
$\overline{G}$	A	B	Y_0	Y_1	Y_2	Y_3
1	x	x	1	1	1	1
0	0	0	0	1	1	1
0	0	1	1	0	1	1
0	1	0	1	1	0	1
0	1	1	1	1	1	0

Figure 11.61 A 2-to-4 decoder

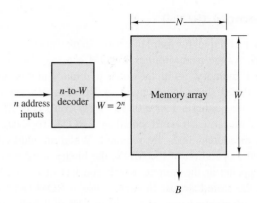

Figure 11.62 Internal organization of an SRAM

 Gate Arrays and Programmable Logic Devices

Digital logic design is performed today primarily using **programmable logic devices (PLDs)**. These are arrays of gates having interconnections that can be programmed to perform a specific logical function. PLDs are large combinational logic modules consisting of arrays of AND and OR gates that can be programmed using special programming languages called **hardware description languages (HDLs)**. Figure 11.63 shows the block diagram of one type of high-density PLD. We define three types of PLDs:

PROM (programmable read-only memory): Offers high speed and low cost for relatively small designs.

PLA (programmable logic array): Offers flexible features for more complex designs.

PAL/GAL (programmable array logic/generic array logic): Offers good flexibility and is faster and less expensive than a PLA.

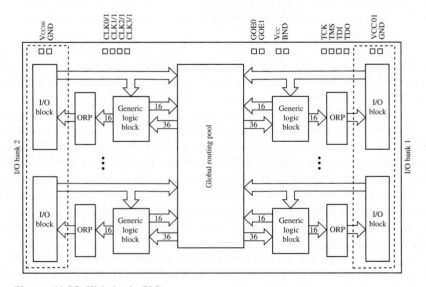

Figure 11.63 High-density PLD

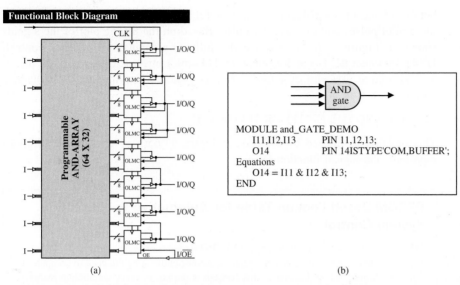

Figure 11.64 (a) The ispGAL16V8 connection diagram; (b) sample code for AND function (for ispGAL16V8)

To illustrate the concept of a logic design using a PLD, we employ a generic array logic (ispGAL16V8) to realize an output signal from three ANDed input signals. The functional block diagram of the GAL is shown in Figure 11.64(a). Notice that the device has eight input lines and eight output lines. The output lines also provide a clock input for timing purposes. The sample code is shown in Figure 11.64(b). The code first defines the inputs and outputs; and it states the equation describing the function to be implemented, O14=I11&I12&I13, defining which output and inputs are to be used, and the functional relationship. Note that the symbol & represents the logical function AND.

A second example of the use of a PLD introduces the concept of timing diagrams. Figure 11.65 depicts a timing diagram related to an automotive fuel-injection system, in which multiple injections are to be performed. Three *pilot* injections and one *primary* injection are to be performed. The *master control* line enables the entire sequence. The resulting output sequence, shown at the

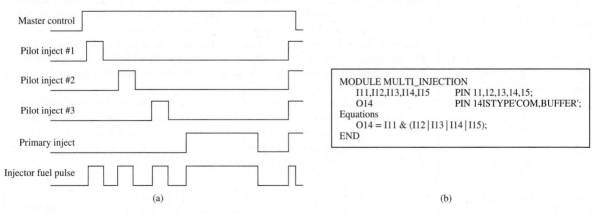

Figure 11.65 (a) Injector timing sequence; (b) sample code for multiple-injection sequence

bottom of the plot and labeled "injector fuel pulse," is the combination of the three pilot pulses and the primary pulse. Based on the timing plot of the signals shown in Figure 11.65(a), we use the following inputs: I11=master control, I12=pilot inject #1, I13=pilot inject #2, I14=pilot inject #3, I14=primary inject, and the output O14=injector fuel pulse. You should convince yourself that the required function is

I11 AND [I12 OR I13 OR I14 OR I14]

This function is realized by the code in Figure 11.65(b). Note that the symbol | represents the logical function OR.

FOCUS ON MEASUREMENTS

EPROM-Based Lookup Table for Automotive Fuel-Injection System Control

One of the most common applications of EPROMs is an *arithmetic lookup table*, which is used to store computed values of certain functions, eliminating the need to compute the function. A practical application of this concept is present in every automobile manufactured in the United States since the early 1980s, as part of the exhaust emission control system. For the catalytic converter to minimize the emissions of exhaust gases (especially hydrocarbons, oxides of nitrogen, and carbon monoxide), it is necessary to maintain the *air-to-fuel ratio A/F* as close as possible to the stoichiometric ratio of 14.7 parts of air for each part of fuel. Most modern engines are equipped with fuel-injection systems that are capable of delivering accurate amounts of fuel to each individual cylinder—thus, the task of maintaining an accurate A/F amounts to measuring the mass of air that is aspirated into each cylinder and computing the corresponding mass of fuel. Many automobiles are equipped with a *mass airflow sensor*, capable of measuring the mass of air drawn into each cylinder during each engine cycle. Let the output of the mass airflow sensor be denoted by the variable M_A, and let this variable represent the mass of air (in grams) actually entering a cylinder during a particular stroke. It is then desired to compute the mass of fuel M_F (also expressed in grams) required to achieve an A/F of 14.7. This computation is simply

$$M_F = \frac{M_A}{14.7}$$

Although this computation is a simple division, its actual calculation in a low-cost digital computer (such as would be used on an automobile) is rather complicated. It would be much simpler to tabulate a number of values of M_A, to precompute the variable M_F, and then to store the result of this computation in an EPROM. If the EPROM address were made to correspond to the tabulated values of air mass, and the content at each address to the corresponding fuel mass (according to the precomputed values of the expression $M_F = M_A/14.7$), it would not be necessary to perform the division by 14.7. For each measurement of air mass into one cylinder, an EPROM address is specified and the corresponding content is read. The content at the specific address is the mass of fuel required by that particular cylinder.

In practice, the fuel mass needs to be converted to a time interval corresponding to the duration of time during which the fuel injector is open. This final conversion factor can also be accounted for in the table. Suppose, for example, that the fuel injector is capable of injecting K_F g/s of fuel; then the time duration T_F during which the injector should be open to inject M_F g of fuel into the cylinder is given by

$$T_F = \frac{M_F}{K_F} \text{ s}$$

(Continued)

(*Concluded*)

Therefore, the complete expression to be precomputed and stored in the EPROM is

$$T_F = \frac{M_A}{14.7 \times K_F} \text{ s}$$

Figure 11.66 illustrates this process graphically.

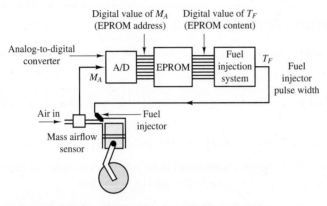

Figure 11.66 Use of EPROM lookup table in automotive fuel-injection system

To provide a numerical illustration, consider a hypothetical engine capable of aspirating air in the range $0 < M_A < 0.51$ g and equipped with fuel injectors capable of injecting at the rate of 1.36 g/s. Thus, the relationship between T_F and M_A is

$$T_F = 50 \times M_A \qquad \text{ms} = 0.05 M_A \text{ s}$$

If the digital value of M_A is expressed in decigrams (dg, or tenths of a gram), the lookup table of Figure 11.67 can be implemented, illustrating the conversion capabilities provided by the EPROM. Note that to represent the quantities of interest in an appropriate binary format compatible with the 8-bit EPROM, the units of air mass and of time have been scaled.

M_A (g) $\times 10^{-2}$	Address (digital value of M_A)	Content (digital value of T_F)	T_F (ms) $\times 10^{-1}$
0	00000000	00000000	0
1	00000001	00000101	5
2	00000010	00001010	10
3	00000011	00001111	15
4	00000100	00010100	20
5	00000101	00011001	25
⋮	⋮	⋮	⋮
51	00110011	11111111	255

Figure 11.67 Lookup table for automotive fuel-injection application

CHECK YOUR UNDERSTANDING

Which combination of the control lines will select the data line D_3 for a 4-to-1 MUX?

Show that an 8-to-1 MUX with eight data inputs (D_0 through D_7) and three control lines (I_0 through I_2) can be used as a data selector. Which combination of the control lines will select the data line D_5?

Which combination of the control lines will select the data line D_4 for an 8-to-1 MUX?

CHECK YOUR UNDERSTANDING

How many address inputs do you need if the number of words in a memory array is 16?

11.6   LATCHES AND FLIP-FLOPS

A **flip-flop** is an elementary **sequential logic** gate. Various types of flip-flops exist; however, all flip-flops share the following characteristics:

1. A flip-flop is a **bistable device**; that is, it can remain in one of two stable states (0 and 1) until appropriate conditions cause it to change state. Thus, a flip-flop can serve as a **memory element**.
2. A flip-flop has two outputs, one of which is the complement of the other.

RS Flip-Flop

It is customary to depict flip-flops by their block diagram and their output by a name, such as Q. Figure 11.68 represents the **RS flip-flop**, which has two inputs, denoted by S and R, and two outputs Q and $\bar{Q}$. The value at Q is called the binary output *state* of the flip-flop. The two inputs R and S are used to change the state of the flip-flop, according to the following rules:

1. When $R = S = 0$, Q remains unchanged from its present state.
2. When $S = 1$ and $R = 0$, the output is *set* such that $Q = 1$.
3. When $S = 0$ and $R = 1$, the output is *reset* such that $Q = 0$.
4. S and R are not permitted to be 1 simultaneously.

A **timing diagram** is a convenient means of describing the transitions that occur in the output of a flip-flop due to changes in its inputs. Figure 11.69 depicts a table of transitions for an *RS* flip-flop Q as well as the corresponding timing diagram.

It is important to note that the *RS* flip-flop is **level-sensitive**. This means that the set and reset operations are completed only after the R and S inputs have reached the appropriate levels. Thus, in Figure 11.69 the transitions in the Q output occur with a small delay relative to the transitions in the R and S inputs.

Figure 11.70 illustrates how an *RS* flip-flop could be constructed from two inverters and two NAND gates. Consider the case when $S = R = 0$ such that $\bar{S} = \bar{R} = 1$. Then the result of each NAND gate is determined entirely by $\bar{Q}$ and Q. That is, when one input to a NAND gate is set high to 1, the output of that NAND gate is the inversion of the other input (refer to the NAND gate truth table in Chapter 11). Thus, when $S = R = 0$, the outputs of the two

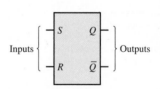

S	R	Q
0	0	Present state
0	1	Reset
1	0	Set
1	1	Disallowed

Figure 11.68 *RS* flip-flop symbol and truth table

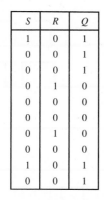

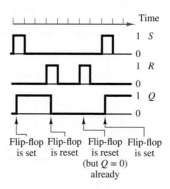

S	R	Q
1	0	1
0	0	1
0	0	1
0	1	0
0	0	0
0	0	0
0	1	0
0	0	0
1	0	1
0	0	1

Flip-flop   Flip-flop   Flip-flop   Flip-flop
is set   is reset   is reset   is set
                        (but $Q = 0$)
                        already

Figure 11.69 Timing diagram for the RS flip-flop

NAND gates in Figure 11.70 are simply $\overline{\overline{Q}} = Q$ and $\overline{Q}$. In other words, the output states of the RS flip-flop remain unchanged from their prior states whenever S and R are both set low to 0.

When S is set high to 1, the output of the upper NAND gate Q is also set high to 1. Why? Because when S is set high to 1, $\overline{S}$ is set low to 0, and when one input to a NAND gate is low, the output of the NAND gate is high *regardless* of the state of the other input. Likewise, when R is set high to 1, $\overline{Q}$ is set high to 1.

The only difficulty with the RS flip-flop occurs when both S and R are set high to 1. Clearly, it is an inherent contradiction to suppose that both Q and $\overline{Q}$ are both set high to 1 at any point in time. Why? Because $\overline{Q}$ is, by definition, the inversion of Q. Thus, $S = R = 1$ is not allowed. The RS flip-flop cannot be both set and reset at the same time. In practice, one could set $S = R = 1$, but the output will be unstable.

As is true for any logic network, it is possible to find alternate formulations of the RS flip-flop. One of De Morgan's laws states that a NAND gate is equivalent to an OR gate with inverted inputs. Make this replacement in Figure 11.70, and note that both S and R are now inverted twice prior to their OR gates. That is, two of the OR gate inputs are S and R. Of course, the other two OR gate inputs $\overline{Q}$ and Q are also inverted. These two inversions could just as well occur at the outputs of the OR gates. In other words, the two OR gates become two NOR gates with inputs $S + Q$ and $R + \overline{Q}$, as shown in Figure 11.71.

Figure 11.72 shows the same two-NOR-gate implementation of the RS flip-flop, but with an enable input E connected to two AND gates such that the R or S inputs will be effective only when $E = 1$. The enable input is often a **clock** signal used to synchronize other inputs.

Figure 11.72 also illustrates two additional features: the **preset** P and **clear** C functions. These features have no effect when set low to 0. However, when P is set high to 1, the output of the upper NOR gate $\overline{Q}$ is set low to 0 and, thus, Q is set high to 1. Notice that the preset is not controlled by the enable input. $P = 1$ always results in $Q = 1$. Likewise, when C is set high to 1, the output of the lower NOR gate Q is set low to 0. Again, notice that the clear is not controlled by the enable input. For this reason, the preset and clear are said to be **asynchronous**. It is important to realize that $P = C = 1$ is not allowed. The timing diagram of Figure 11.72 illustrates the role of the enable, preset, and clear inputs. Notice that transitions due to S and R occur only after E is set high, whereas the effects of P and C are independent of E. The

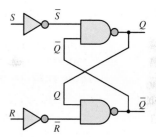

Figure 11.70 NAND gate implementation of the RS flip-flop

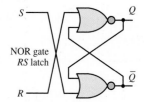

Figure 11.71 NOR gate implementation of the RS flip-flop

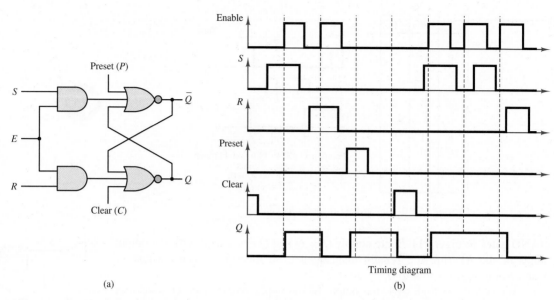

Figure 11.72 The *RS* flip-flop with enable, preset, and clear lines: (a) logic diagram, (b) example timing diagram

flip-flop can be designed so that the *P* and *C* inputs are also controlled by *E*; in fact, many commercial flip-flops are designed this way so that all inputs are synchronized with *E*.

Another extension of the *RS* flip-flop, called a **data latch**, or **delay**, is shown in Figure 11.73. In this circuit, $R = \overline{S}$ such that when $E = 1$, $Q = D$. When *E* is set low to 0, the output *Q* does not change but retains its value until *E* is set high again. In other words, *Q* is *latched* when *E* is set low and *unlatched* when *E* is set high. The timing diagram illustrates that this effect also delays the impact of *D* on *Q* until the *next* time *E* is set high.

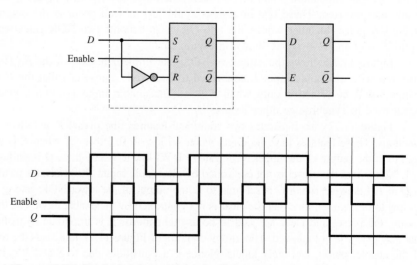

Figure 11.73 Data latch and associated timing diagram

D Flip-Flop

The **D flip-flop** is an extension of the data latch that utilizes two *RS* flip-flops, as shown in Figure 11.74(a), and a clock signal to drive their enable inputs. Note that the clock is inverted prior to E_1 such that the latch is enabled when the clock goes low. However, since Q_2 is disabled when the clock is low, the output of the *D* flip-flop will not switch to the 1 state until the clock subsequently goes high to enable the second latch and transfer the state of Q_1 to Q_2.

It is important to note the triangular "knife-edge" symbol shown at the CLK input in Figure 11.74(b). This symbol indicates that the *D* flip-flop changes state only on a positive clock *transition*; that is, a transition from low to high. Internally, Q_1 is set on a negative transition, whereas Q_2 (and therefore Q) is set on a positive transition, as shown in Figure 11.74(c). Thus, this particular *D* flip-flop is said to be positive edge–triggered, or **leading edge–triggered**, as indicated in the following truth table:

D	**CLK**	*Q*
0	↑	0
1	↑	1

where the symbol ↑ indicates a positive transition.

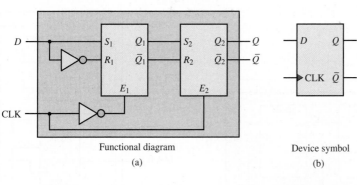

Functional diagram

(a)

Device symbol

(b)

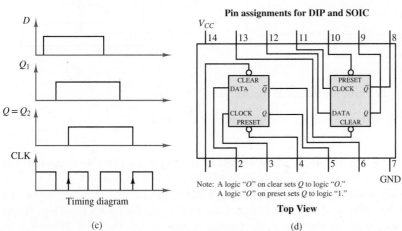

Timing diagram

(c)

Pin assignments for DIP and SOIC

Note: A logic "*O*" on clear sets Q to logic "*O*."
A logic "*O*" on preset sets Q to logic "1."

Top View

(d)

Figure 11.74 The *D* flip-flop: (a) functional diagram; (b) device symbol; (c) timing waveforms; and (d) IC schematic.

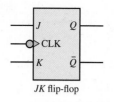

JK flip-flop

J_n	K_n	Q_{n+1}
0	0	Q_n
0	1	0 (reset)
1	0	1 (set)
1	1	$\bar{Q}_n$ (toggle)

Figure 11.75 Truth table for the *JK* flip-flop

JK Flip-Flop

The symbol and truth table of the **JK flip-flop** are shown in Figure 11.75. The *bubble* at the clock input signifies it is negative or **trailing edge–triggered**. Its operating rules are:

- When *J* and *K* are both low, no change occurs in the state of the flip-flop.
- When $J = 0$ and $K = 1$, the flip-flop is reset to 0.
- When $J = 1$ and $K = 0$, the flip-flop is set to 1.
- When both *J* and *K* are high, the flip-flop will toggle between states at every negative transition of the clock input, denoted by the symbol ↓.

The operation of the *JK* flip-flop can also be explained in terms of two *RS* flip-flops as shown in Figure 11.76. When the clock transitions from low to high, the *master* is enabled; however, the *slave* does not receive the master outputs until it is enabled during a negative clock transition. This behavior is similar to that of an *RS* flip-flop, except for the $J = 1$, $K = 1$ condition, which is allowed and results in the outputs being *toggled*.

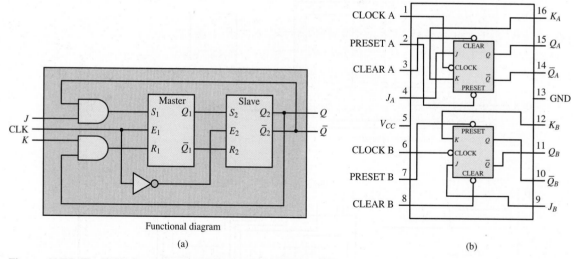

Functional diagram

(a)

(b)

Figure 11.76 The *JK* flip-flop; (a) functional diagram; and (b) IC schematic.

The *JK* flip-flop is also known as the *universal flip-flop* because it can be configured to behave as an *RS* or *D* flip-flop. When both inputs are 0, the outputs remain in their previous state during a clock transition; when one input is high and the other is low, the outputs behave as they do for an *RS* flip-flop. When the inputs are set so that $K = \bar{J}$, the outputs behave as a *D* flip-flop. When the inputs are set so that $K = J$, the outputs behave as a *T* flip-flop, which is described in Example 11.22.

EXAMPLE 11.21   *RS* Flip-Flop Timing Diagram

Problem

Determine the output of an *RS* flip-flop for the series of inputs given in the table below.

R	0	0	0	1	0	0	0
S	1	0	1	0	0	1	0

Solution

Known Quantities: *RS* flip-flop truth table (Figure 11.68).

Find: Output Q of *RS* flip-flop.

Analysis: We complete the timing diagram for the *RS* flip-flop, following the rules stated earlier to determine the output of the device; the result is summarized below.

R	0	0	0	1	0	0	0
S	1	0	1	0	0	1	0
Q	1	1	1	0	0	1	1

A sketch of the waveforms, shown below, can also be generated to visualize the transitions.

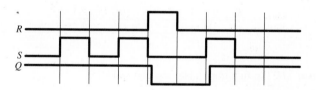

EXAMPLE 11.22   The *T* Flip-Flop

Problem

Determine the truth table and timing diagram of the **T flip-flop** of Figure 11.77. Note that the *T* flip-flop is a *JK* flip-flop with its inputs tied together.

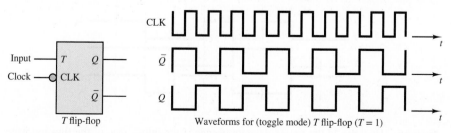

Figure 11.77 The *T* flip-flop symbol and timing waveforms

Solution

Known Quantities: *JK* flip-flop truth table (see Figure 11.75).

Find: Truth table and timing diagram for *T* flip-flop.

Analysis: The *T* flip-flop is a *JK* flip-flop with $K = J$. Thus, the flip-flop will need only a two-element truth table to describe its operation, corresponding to the top and bottom entries in the *JK* flip-flop truth table of Figure 11.75. The truth table is shown below. A timing diagram is also included in Figure 11.77.

T	CLK	Q_{k+1}
0	↓	Q_k
1	↓	$\overline{Q_k}$

Comments: The *T* flip-flop takes its name from the fact that it *toggles* between the high and low states. Note that the toggling frequency is one-half that of the clock. Thus the *T* flip-flop also acts as a *divide-by-2* counter.

EXAMPLE 11.23   The *JK* Flip-Flop Timing Diagram

Problem

Determine the output of a *JK* flip-flop for the series of inputs given in the table below. The initial state of the flip-flop is $Q_0 = 1$.

J	0	1	0	1	0	0	1
K	0	1	1	0	0	1	1

Solution

Known Quantities: *JK* flip-flop truth table (see Figure 11.75).

Find: Output of *JK* flip-flop *Q* as a function of the input transitions.

Analysis: Complete the timing diagram for the *JK* flip-flop, using the rules of Figure 11.75.

J	0	1	0	1	0	0	1
K	0	1	1	0	0	1	1
Q	1	0	0	1	1	0	1

A sketch of the waveforms, shown below, can also be generated to visualize the transitions. Each vertical line corresponds to a clock transition.

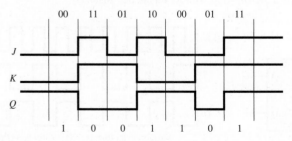

Comments: How would the timing diagram change if the initial state of the flip-flop were $Q_0 = 1$?

CHECK YOUR UNDERSTANDING

Derive the detailed truth table and draw a timing diagram for the JK flip-flop, using the model of Figure 11.76 with two flip-flops. Include each unique internal input in the table and timing diagram.

11.7   DIGITAL COUNTERS AND REGISTERS

One of the more immediate applications of flip-flops is in the design of **counters**. A counter is a sequential logic device that can take one of N possible states, stepping through these states in a sequential fashion. When the counter has reached its last state, it resets to 0 and is ready to start counting again. For example, a 3-bit **binary up counter** would have $2^3 = 8$ possible states and might appear as shown in the functional block of Figure 11.78. The clock input steps the counter through the eight states, one transition per clock pulse. This particular counter also has a reset input, which can force the counter outputs low: $b_2 b_1 b_0 = 000$.

Although binary counters are very useful in many applications, one is often interested in a **decade counter**, that is, a counter that counts from 0 to 9 and then

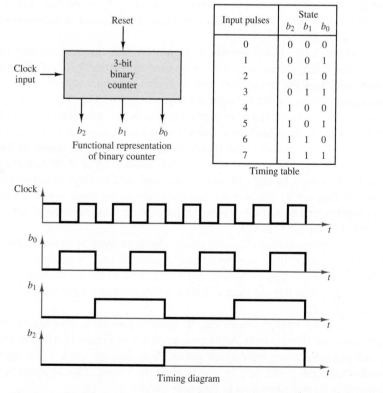

Input pulses	State		
	b_2	b_1	b_0
0	0	0	0
1	0	0	1
2	0	1	0
3	0	1	1
4	1	0	0
5	1	0	1
6	1	1	0
7	1	1	1

Timing table

Functional representation of binary counter

Timing diagram

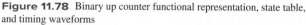

Figure 11.78 Binary up counter functional representation, state table, and timing waveforms

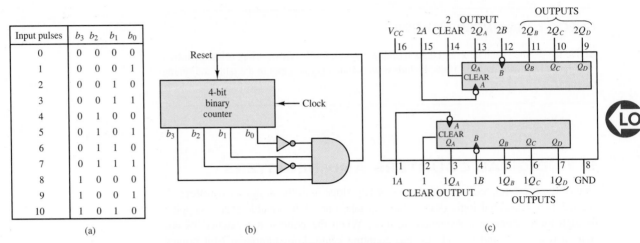

Figure 11.79 Decade counter: (a) counting sequence; (b) functional diagram; and (c) IC schematic

resets. A 4-bit binary counter can easily be configured in principle to provide this function by means of simple logic that resets the counter when it has reached the count $1001_2 = 9_{10}$. As shown in Figure 11.79, if bits b_3 and b_1 are tied to a four-input AND gate, along with $\bar{b}_2$ and $\bar{b}_0$, the output of the AND gate will reset the counter when the count reaches 10. Additional logic can provide a *carry* bit whenever a reset condition is reached, could be passed along to another decade counter, enabling counts up to 99. Decade counters can be cascaded to represent any series of decimal digits.

Although the decade counter of Figure 11.79 is attractive because of its simplicity, this configuration would never be used in practice, because of **propagation delays**, which are due to the finite response time of the internal transistors. In general, propagation delays are not the same for any two gates or flip-flops. Thus, if the reset signal—which is presumed to be applied at exactly the same time to each of the four *JK* flip-flops in the 4-bit binary counter—does not cause the flip-flops to reset at exactly the same time, then the binary word appearing at the output of the counter will change from 1001 to some other number, and the output of the four-input NAND gate will no longer be high. In such a condition, the flip-flops that have not already reset will then not be able to reset, and the counting sequence will be compromised. This problem can be addressed with the aid of **state transition diagrams**, which are discussed in the next section.

An implementation of a 3-bit binary **ripple counter** is shown in Figure 11.80. Its transition table illustrates how the Q output of each stage becomes the clock input to the next stage, while each flip-flop is held in the toggle mode. The output transitions assume that the clock (CLK) is a simple square wave (all *JK*s are negative edge–triggered).

This 3-bit ripple counter can be used to provide a divide-by-8 counter by connecting the outputs to an AND gate, as shown in Figure 11.81. The result is one output pulse for every eight clock pulses. Note that the clock input signal is also connected to the AND gate to synchronize the output. This application of ripple counters is further illustrated in Example 11.24.

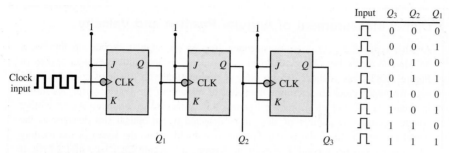

Input	Q_3	Q_2	Q_1
⊓	0	0	0
⊓	0	0	1
⊓	0	1	0
⊓	0	1	1
⊓	1	0	0
⊓	1	0	1
⊓	1	1	0
⊓	1	1	1

Figure 11.80 Ripple counter

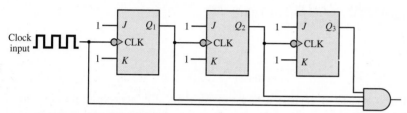

Figure 11.81 Divide-by-8 circuit

A slightly more complex version of the binary counter is the **synchronous counter**, in which the input clock drives all the flip-flops simultaneously. Figure 11.82 depicts a 3-bit synchronous counter. In this figure, we have chosen to represent each flip-flop as a T flip-flop. The clocks to all the flip-flops are incremented simultaneously. The reader should verify that Q_0 toggles to 1 first, and then Q_1 toggles to 1, and that the AND gate ensures that Q_2 will toggle only after Q_0 and Q_1 have both reached the 1 state ($Q_0 \cdot Q_1 = 1$).

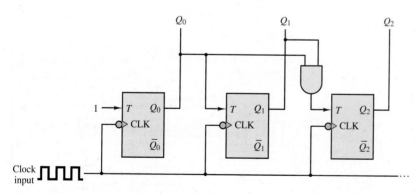

Figure 11.82 Three-bit synchronous counter

Other common counters are the **ring counter**, illustrated in Example 11.25, and the **up-down counter**, which has an additional select input that determines whether the counter counts up or down.

FOCUS ON MEASUREMENTS

Digital Measurement of Angular Position and Velocity

Another type of angular position encoder, besides the angular encoder in the Focus on Measurements box, "Digital Position Encoders," is the slotted encoder shown in Figure 11.83. This encoder can be used in conjunction with a pair of counters and a high-frequency clock to determine the speed of rotation of the slotted wheel. As shown in Figure 11.84, a clock of known frequency is connected to a counter while another counter records the number of slot pulses detected by an optical slot detector as the wheel rotates. Dividing the counter values, one could obtain the speed of the rotating wheel in radians per second. For example, assume a clocking frequency of 1.2 kHz. If both counters are started at zero and at some instant the timer counter reads 3,050 and the encoder counter reads 2,850, then the speed of the rotating encoder is found to be

$$1,200 \, \frac{\text{cycles}}{\text{s}} \cdot \frac{2,850 \text{ slots}}{3,050 \text{ cycles}} = 1,121.3 \, \frac{\text{slots}}{\text{s}}$$

and

$$1,121.3 \, \frac{\text{slots}}{\text{s}} \times 1° \text{ per slot} \times \frac{2\pi}{360} \, \frac{\text{rad}}{\text{deg}} = 19.6 \, \frac{\text{rad}}{\text{s}}$$

If this encoder is connected to a rotating shaft, it is possible to measure the angular position and velocity of the shaft. Such shaft encoders are used in measuring the speed of rotation of electric motors, machine tools, engines, and other rotating machinery.

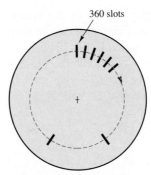

360 slots

360 slots; 1 increment = 1 degree

Figure 11.83 Slotted wheel for position encoder.

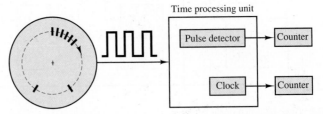

Time processing unit

Pulse detector → Counter

Clock → Counter

Figure 11.84 Calculating the speed of rotation of the slotted wheel

A typical application of the slotted encoder is to compute the ignition and injection timing in an automotive engine. In an automotive engine, information related to speed is

(Continued)

(*Concluded*)

obtained from the camshaft and the flywheel, which have known reference points. The reference points determine the timing for the ignition firing points and fuel-injection pulses and are identified by special slot patterns on the camshaft and crankshaft. Two methods are used to detect the special slots (reference points): *period measurement with additional transition detection (PMA)* and *period measurement with missing transition detection (PMM)*. In the PMA method, an additional slot (reference point) determines a known reference position on the crankshaft or camshaft. In the PMM method, the reference position is determined by the absence of a slot. Figure 11.85 illustrates a typical PMA pulse sequence, showing the presence of an additional pulse. The additional slot may be used to determine the timing for the ignition pulses relative to a known position of the crankshaft. Figure 11.86 depicts a typical PMM pulse sequence. Because the period of the pulses is known, the additional slot or the missing slot can be easily detected and used as a reference position. How would you implement these pulse sequences, using ring counters?

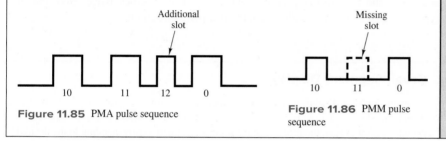

Figure 11.85 PMA pulse sequence

Figure 11.86 PMM pulse sequence

Registers

A register consists of a cascade of flip-flops that can store binary data, 1 bit in each flip-flop. The simplest type of register is the parallel input–parallel output register shown in Figure 11.87. In this register, the *load* input pulse, which acts on all clocks simultaneously, causes the parallel inputs $b_0b_1b_2b_3$ to be transferred to the respective flip-flops. The D flip-flop employed in this register allows the transfer from b_n to Q_n to occur very directly. Thus, D flip-flops are very commonly used in this type of application. The binary word $b_3b_2b_1b_0$ is now "stored," each bit being represented by the state of a flip-flop. Until the load input is applied again and a new word appears at the parallel inputs, the register will preserve the stored word.

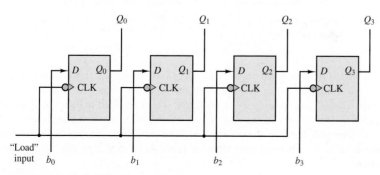

Figure 11.87 A 4-bit parallel register

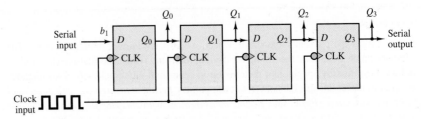

Figure 11.88 A 4-bit shift register

The construction of the parallel register presumes that the *N*-bit word to be stored is available in parallel form. However, often a binary word will arrive in serial form, that is, 1 bit at a time. A register that can accommodate this type of logic signal is called a **shift register**. Figure 11.88 illustrates how the same basic structure of the parallel register applies to the shift register, except that the input is now applied to the first flip-flop and shifted along at each clock pulse. Note that this type of register provides both a serial and a parallel output.

FOCUS ON MEASUREMENTS

Seven-Segment Display

A **seven-segment display** is a very convenient device for displaying digital data. The display is shown in Figure 11.89. Operation of a seven-segment display requires a decoder circuit to light the proper combinations of segments corresponding to the desired decimal digit.

This display, with the appropriate decoder driver, is capable of displaying values ranging from 0 to 9.

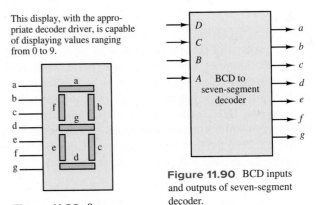

Figure 11.89 Seven-segment display

Figure 11.90 BCD inputs and outputs of seven-segment decoder.

A typical *BCD* to seven-segment decoder function block is shown in Figure 11.90, where the lowercase letters correspond to the segments shown in Figure 11.89. The decoder features four data inputs (*A, B, C, D*), which are used to light the appropriate segment(s). The outputs of the decoder are connected to the seven-segment display. The decoder will light up the appropriate segments corresponding to the incoming value. A *BCD* to seven-segment decoder function is similar to the 2-to-4 decoder function.

==

EXAMPLE 11.24   Divider Circuit

Problem

A binary ripple counter provides a means of dividing the fixed output rate of a clock by powers of 2. For example, the circuit of Figure 11.91 is a divide-by-2 or divide-by-4 counter. Draw the timing diagrams for the clock input Q_0 and Q_1 to demonstrate these functions.

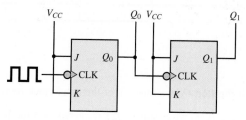

Figure 11.91

==

Solution

Known Quantities: JK flip-flop truth table (Figure 11.75).

Find: Output of each flip-flop Q as a function of the input clock transitions.

Assumptions: Assume positive edge–triggered devices. The DC supply voltage is V_{CC}.

Analysis: Following the timing diagram of Figure 11.92, we see that Q_0 switches at one-half the frequency of the clock input, and that Q_1 switches at one-half the frequency of Q_0, hence the timing diagram shown.

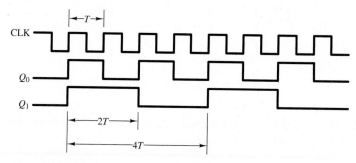

Figure 11.92 Divider circuit timing diagram

==

EXAMPLE 11.25   Ring Counter

Problem

Draw the timing diagram for the ring counter of Figure 11.93.

==

Solution

Known Quantities: JK flip-flop truth table (Figure 11.75).

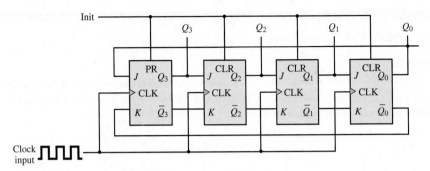

Figure 11.93 Ring counter

Find: Output of each flip-flop Q as a function of the input clock transitions.

Assumptions: The JK flip-flops are positive edge–triggered, and the Init line is set high to 1 until *after* the first positive edge transition of the clock. Then, the Init line is immediately set low to 0.

Analysis: The first positive clock transition will *set* $Q_3 = 1$ and *reset* the other three flip-flops to $Q_2 = Q_1 = Q_0 = 0$. At the second positive clock transition, $Q_3 = 1$ such that the second flip-flop is set high to $Q_2 = 1$. Both Q_1 and Q_0 remain unchanged and Q_3 is reset low to 0. The pattern continues, causing the 1 state to ripple from left to right over and over again.

CLK	Q_3	Q_2	Q_1	Q_0
↑	1	0	0	0
↑	0	1	0	0
↑	0	0	1	0
↑	0	0	0	1
↑	1	0	0	0
↑	0	1	0	0
↑	0	0	1	0

Comments: The shifting behavior of the ring counter is implemented in the shift registers discussed in the following section.

CHECK YOUR UNDERSTANDING

The speed of the rotating encoder of the Focus on Measurements box, "Digital Measurement of Angular Position and Velocity," is found to be 9,425 rad/s. The encoder timer reads 10, and the clock counter reads 300. Assuming that both the timer counter and the encoder counter started at zero, find the clock frequency.

11.8 SEQUENTIAL LOGIC DESIGN

The design of sequential circuits, just like the design of combinational circuits, can be carried out by means of a systematic procedure. A **state diagram** and its associated **state transition table** describe the logic states and their interrelationships required of the system design. Consider the 3-bit binary counter of Figure 11.94, which is made up of three T flip-flops. You can easily verify that the input equations for this counter are $T_1 = 1$, $T_2 = q_1$, and $T_3 = q_1q_2$. Knowing the inputs, the three outputs can be determined at any moment. The outputs Q_1, Q_2, and Q_3 form the **state** of the machine. It is straightforward to show that as the clock goes through a series of cycles, the counter will go through the transitions shown in Table 11.14, where we indicate the current state by lowercase q and the next state by an uppercase Q. Note that the state diagram of Figure 11.94 provides information regarding the sequence of states assumed by the counter in graphical form. In a state diagram, each state is denoted by a circle called a **node**, and the transition from one state to another is indicated by a **directed edge**, that is, a line with a directional arrow. The analysis of sequential circuits consists of determining either their transition table or their state diagram.

Table 11.14 **State transition table for 3-bit binary counter**

Current state			Input			Next state		
q_3	q_2	q_1	T_3	T_2	T_1	Q_3	Q_2	Q_1
0	0	0	0	0	1	0	0	1
0	0	1	0	1	1	0	1	0
0	1	0	0	0	1	0	1	1
0	1	1	1	1	1	1	0	0
1	0	0	0	0	1	1	0	1
1	0	1	0	1	1	1	1	0
1	1	0	0	0	1	1	1	1
1	1	1	1	1	1	0	0	0

The reverse of this analysis process is the design process. That is, how can one systematically design a sequential circuit, such as a counter, by employing state transition tables and state diagrams?

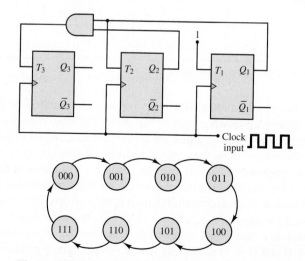

Figure 11.94 A 3-bit binary counter and state diagram

The initial design specification for a logic circuit is usually in the form of either a transition table or a state diagram. The goal of the design process is to identify one logic circuit, among many that matches those specifications. Remember that there is no single unique implementation for a given set of output specifications. Therefore, the first step is to select a flip-flop and use its truth table characteristics to define its **excitation table**. The truth and excitation tables for the *RS*, *D*, and *JK* flip-flops are given in Tables 11.15, 11.16, and 11.17, respectively.

Table 11.15 **Truth table and excitation table for *RS* flip-flop**

Truth table for RS flip-flop				Excitation table for RS flip-flop			
S	R	Q_t	Q_{t+1}	Q_t	Q_{t+1}	S	R
0	0	0	0	0	0	0	$d^\dagger$
0	0	1	1	0	1	1	0
0	1	0	0	1	0	0	1
0	1	1	0	1	1	d	0
1	0	0	1				
1	0	1	1				
1	1	x*	x				
1	1	x	x				

*An x indicates that this combination of inputs is not allowed.
†A d denotes a don't-care entry.

Table 11.16 **Truth table and excitation table for *D* flip-flop**

Truth table for D flip-flop			Excitation table for D flip-flop		
D	Q_t	Q_{t+1}	Q_t	Q_{t+1}	D
0	0	0	0	0	0
0	1	0	0	1	1
1	0	1	1	0	0
1	1	1	1	1	1

Table 11.17 **Truth table and excitation table for *JK* flip-flop**

Truth table for JK flip-flop				Excitation table for JK flip-flop			
J	K	Q_t	Q_{t+1}	Q_t	Q_{t+1}	J	K
0	0	0	0	0	0	0	$d^\dagger$
0	0	1	1	0	1	1	d
0	1	0	0	1	0	d	1
0	1	1	0	1	1	d	0
1	0	0	1				
1	0	1	1				
1	1	0	1				
1	1	1	0				

†A d denotes a don't-care entry.

The use of excitation tables will now be demonstrated in the design of a **modulo-4 binary up-down counter**. The phrase "modulo-4 binary" indicates that the counter output is limited to the integers 0 to 3 represented in binary form; that is, in bits. Of course, these four integers can be completely represented by 2 bits. The phrase "up-down" indicates that the counter will increment or decrement its output depending upon the value of a single bit input, which will be high or low (1 or 0) to increment or decrement the output, respectively. Figure 11.95 shows the state diagram for this counter, where a clockwise or counterclockwise progression is an increment or decrement, respectively. One flip-flop is required to produce the states ($Q = 0$ and $Q = 1$) of each of the two output bits. For this example design, choose two *RS* flip-flops and begin constructing the state transition table shown in

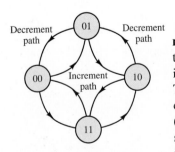

Figure 11.95 State diagram of a modulo-4 binary up-down counter

Table 11.18 **State transition table for modulo-4 binary up-down counter**

Input x	Current state q_1	Current state q_2	Next state Q_1	Next state Q_2	S_1	R_1	S_2	R_2	Output y
0	0	0	1	1	1	0	1	0	1
0	0	1	0	0	0	d	0	1	0
0	1	0	0	1	0	1	1	0	1
0	1	1	1	0	d	0	0	1	0
1	0	0	0	1	0	d	1	0	1
1	0	1	1	0	1	0	0	1	0
1	1	0	1	1	d	0	1	0	1
1	1	1	0	0	0	1	0	1	0

Table 11.18. Note immediately that for a device with a single bit input and a double bit output there are eight distinct combinations of inputs and outputs. The first five columns of Table 11.18 specify the desired *next state* $Q_1 Q_2$ for each possible input x and *current state* $q_1 q_2$. Notice that the information in these first five columns matches the information presented in Figure 11.95.

Next, match the values of each output pair (Q_t, Q_{t+1}) found in the *RS* flip-flop excitation table to each of the two pairs of counter outputs (q_1, Q_1) and (q_2, Q_2) to determine the *RS* input pairs (S_1, R_1) and (S_2, R_2). For example, the first row of the counter's state transition table is developed by matching $(q_1 = 0, Q_1 = 1)$ to the second row of the *RS* excitation table where $(Q_t = 0, Q_{t+1} = 1)$. Thus, the *RS* input pair $(S_1 = 1, R_1 = 0)$ will produce the desired relationship between the *current state* variable q_1 and the *next state* variable Q_1. For the same first row of the state transition table, since $q_2 = q_1 = 0$ and $Q_2 = Q_1 = 1$, the other *RS* input pair must also be $(S_2 = 1, R_2 = 0)$. The other rows of the state transition table are filled out in exactly the same manner. A d in the table represents a don't-care condition. Remember that for this counter $x = 0$ indicates a decrement and $x = 1$ indicates an increment.

At this point, the required logic circuit can be determined using combinational logic tools, such as the Karnaugh maps of Figure 11.96. Verify that the following expressions can be obtained from those maps.

$$S_1 = \overline{x}\,\overline{q}_1\overline{q}_2 + x\overline{q}_1 q_2 = (\overline{x}\,\overline{q}_2 + x q_2)\overline{q}_1$$
$$R_1 = \overline{x} q_1\overline{q}_2 + x q_1 q_2 = (\overline{x}\,\overline{q}_2 + x q_2)q_1$$
$$S_2 = \overline{q}_2$$
$$R_2 = q_2$$

The complete design is shown in Figure 11.97.

Programmable Logic Controllers

Sequential logic designs and state machines are found in programmable logic controllers, or PLCs, which are finite-state machines that are used in a variety of industrial applications to implement logic functions. For example, machining, packaging, material handling, and automated assembly are some of the example applications in which these systems are encountered. PLCs are specialized computers that are very effective at executing a series of complex logical decisions.

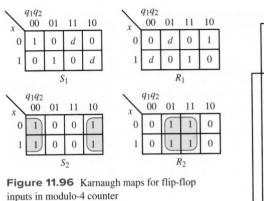

Figure 11.96 Karnaugh maps for flip-flop inputs in modulo-4 counter

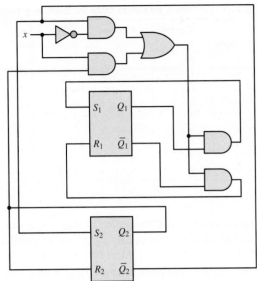

Figure 11.97 Implementation of modulo-4 counter

Conclusion

This chapter contains an overview of digital logic circuits. These circuits form the basis of all digital computers, and of most electronic devices used in industrial and consumer applications. Upon completing this chapter, you should have mastered the following learning objectives:

1. *Understand the concepts of analog and digital signals and of quantization.*
2. *Convert between decimal and binary number systems and use the hexadecimal system and BCD and Gray codes.* The binary and hexadecimal systems form the basis of numerical computing.
3. *Write truth tables, and realize logic functions from truth tables using logic gates.* Boolean algebra permits the analysis of digital circuits through a relatively simple set of rules. Digital logic gates are the means through which one can implement logic functions; truth tables permit the easy visualization of logic functions and can aid in the realization of these functions by using logic gates.
4. *Systematically design logic functions using Karnaugh maps.* The design of logic circuits can be systematically approached by using an extension of truth tables called the Karnaugh map. Karnaugh maps facilitate the simplification of logic expressions and their realization through logic gates in either sum-of-products or product-of-sums form.
5. *Study various combinational logic modules, including multiplexers, memory and decoder elements, and programmable logic arrays.* Practical digital logic circuits rarely consist of individual logic gates; gates are usually integrated into combinational logic modules that include memory elements and gate arrays.

6. *Analyze the operation of flip-flops and latches*, which are the building blocks of a sequential logic circuits. *Feedback* from outputs to inputs creates outputs whose future values depend upon their present values. In other words, these circuits possess *memory*. The operation flip-flops and latches are described by state transition tables and state diagrams.

7. *Understand the operation of digital counters and registers*. Counters are a very important class of digital circuits and are based on sequential logic elements. Registers are the most fundamental form of *random-access memory* (RAM).

8. *Design simple sequential circuits using state transition diagrams*. Sequential circuits can be designed using formal design procedures employing state diagrams.

HOMEWORK PROBLEMS

Section 11.2: The Binary Number System

11.1 Convert the following base-10 numbers to hexadecimal and binary:

   a. 303   b. 275   c. 18   d. 43   e. 87

11.2 Convert the following hexadecimal numbers to base-10 and binary:

   a. C b. 44   c. 28   d. 59   e. 14

11.3 Convert the following base-10 numbers to binary:

   a. 231.45   b. 58.78   c. 21.22   d. 93.375

11.4 Convert the following binary numbers to hexadecimal and base 10:

   a. 1101   b. 1000100   c. 1111100   d. 1110000
   e. 10000   f. 101010

11.5 Perform the following additions, all in the binary system:

   a. 10101111 + 10100
   b. 111100001 + 111000
   c. 111001011 + 111001

11.6 Perform the following subtractions, all in the binary system:

   a. 11010001 − 11100
   b. 11111100 − 101010
   c. 100110110 − 1001100

11.7 Assuming that the most significant bit is the sign bit, find the decimal value of the following sign-magnitude form 8-bit binary numbers:

   a. 10100111   b. 01010110   c. 11111100

11.8 Find the sign-magnitude form binary representation of the following decimal numbers:

   a. 122   b. −110   c. −87   d. 40

11.9 Find the twos complement of the following binary numbers:

   a. 1110   b. 1100101   c. 1110000   d. 11100

11.10 Assuming you have 10 fingers, including thumbs:

   a. How high can you count on your fingers in a binary (base 2) number system?

   b. How high can you count on your fingers in base 6, using one hand to count units and the other hand for the carries?

Section 11.3: Boolean Algebra and Logic Gates

11.11 Use the truth table to prove that
$$\overline{A} + AB = \overline{A} + B.$$

11.12 Realize the logic function:
$$Y = (A + \overline{B}) \cdot [(\overline{C} \cdot D) + A]$$
using the logic gates and compute the truth table.

11.13 Using the method of proof by perfect induction, show that
$$(X + Y) \cdot (\overline{X} + X \cdot Y) = Y$$

11.14 Simplify the expression
$$Y = \overline{A} \cdot \overline{B} \cdot C + \overline{A} \cdot B \cdot C + \overline{A} \cdot \overline{C}$$
using boolean algebra, and then draw the logic circuit using logic gates.

11.15   Simplify the expression

$$Y = A \cdot \overline{B} \cdot \overline{C} + A \cdot \overline{B} \cdot C + \overline{A} \cdot B \cdot C + A \cdot B \cdot C$$

using the boolean algebra.

11.16   Simplify the expression

$$Y = \overline{A} \cdot \overline{B} \cdot \overline{C} + \overline{A} \cdot \overline{B} \cdot C + \overline{A} \cdot B \cdot \overline{C}$$

using the boolean algebra.

11.17   Find the logic function defined by the truth table given in Figure P11.17.

A	B	C	F
0	0	0	0
0	0	1	1
0	1	0	0
0	1	1	1
1	0	0	1
1	0	1	1
1	1	0	1
1	1	1	1

Figure P11.17

11.18   Determine the boolean function describing the operation of the circuit shown in Figure P11.18 and simplify it using boolean algebra.

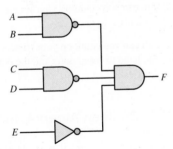

Figure P11.18

11.19   Use a truth table to show when the output of the circuit of Figure P11.19 is 1.

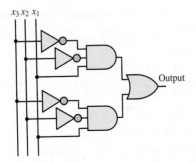

Figure P11.19

11.20   Baseball is a complicated game, and often the manager has a difficult time keeping track of all the rules of thumb that guide decisions. To assist your favorite baseball team, you have been asked to design a logic circuit that will flash a light when the manager should give the steal sign. The rules have been laid out for you by a baseball fan with limited knowledge of the game as follows: Give the steal sign if there is a runner on first base and

a. There are no other runners, the pitcher is right-handed, and the runner is fast; or

b. There is one other runner on third base, and one of the runners is fast; or

c. There is one other runner on second base, the pitcher is left-handed, and both runners are fast.

Under no circumstances should the steal sign be given if all three bases have runners. Design a logic circuit that implements these rules to indicate when the steal sign should be given.

11.21   A small county board is composed of three commissioners. Each commissioner votes on measures presented to the board by pressing a button indicating whether the commissioner votes for or against a measure. If two or more commissioners vote for a measure, it passes. Design a logic circuit that takes the three votes as inputs and lights either a green or a red light to indicate whether a measure passed.

11.22   A water purification plant uses one tank for chemical sterilization and a second, larger tank for settling and aeration. Each tank is equipped with two sensors that measure the height of water in each tank and the flow rate of water into each tank. When the height of water or the flow rate is too high, the sensors produce a logic high output. Design a logic circuit that sounds an alarm whenever the height of water in both tanks is too high and either of the flow rates is too high, or whenever both flow rates are too high and the height of water in either tank is also too high.

11.23   Many automobiles incorporate logic circuits to alert the driver to problems or potential problems. In one particular car, a buzzer is sounded whenever the ignition key is turned and either a door is open or a seat belt is not fastened. The buzzer also sounds when the key is not turned but the lights are on. In addition, the car will not start unless the key is in the ignition, the car is in park, and all doors are closed and seat belts fastened. Design a logic circuit that takes all the inputs listed and sounds the buzzer and starts the car when appropriate.

11.24   An on/off start-up signal governs the compressor motor of a large commercial air conditioning unit. In general, the start-up signal should be on whenever the output of a temperature sensor S exceeds a reference temperature. However, you are asked to limit the compressor start-ups to certain hours of the day and also enable service technicians to start up or shut down the compressor through a manual override. A time-of-day indicator D is available with on/off outputs, as is a manual override switch M. A separate timer T prohibits a compressor start-up within 10 min of a previous shutdown. Design a logic diagram that incorporates the state of all four devices (S, D, M, and T) and produces the correct on/off condition for the motor start-up.

11.25   NAND gates require one less transistor than AND gates. They are often used exclusively to construct logic circuits. One such logic circuit that uses three-input NAND gates is shown in Figure P11.25.

    a. Determine the truth table for this circuit.

    b. Give the logic equation that represents the circuit (you do not need to reduce it).

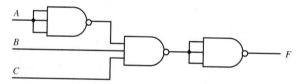

Figure P11.25

11.26   Draw a logic circuit that will accomplish the equation:

$$F = (A + \overline{B}) \cdot \overline{(C + \overline{A})} \cdot B$$

11.27   The circuit shown in Figure P11.27 is called a half adder for two single bit inputs, giving a two-bit sum as outputs. Build a truth table and verify that it indeed acts as a summer.

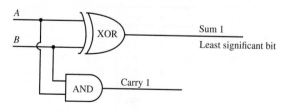

Figure P11.27

11.28   Draw a logic circuit that will accomplish the equation

$$F = \left[(A + C \cdot \overline{B}) + A \cdot \overline{B} \cdot \overline{C}\right] \cdot \overline{(B + C)}$$

11.29   Determine the truth table (F given A, B, C, and D) and the logical expression for the circuit of Figure P11.29.

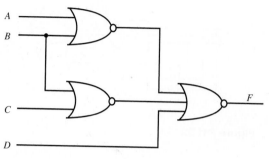

Figure P11.29

11.30   Determine the truth table (F given A, B, and C) and the logical expression for the circuit of Figure P11.30.

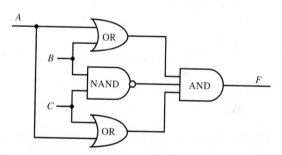

Figure P11.30

11.31   A "vote taker" logic circuit forces its output to agree with a majority of its inputs. Such a circuit is shown in Figure P11.31 for the three voters, A, B, and C. Write the logic expression for the output of this circuit in terms of its inputs. Also create a truth table for the output in terms of the inputs.

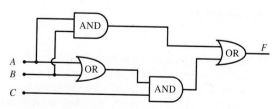

Figure P11.31

11.32   A "consensus indicator" logic circuit is shown in Figure P11.32. Write the logical expression for the output of this circuit in terms of its input. Also create a truth table for the output in terms of the inputs.

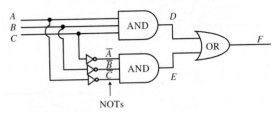

NOTs

Figure P11.32

11.33   A half-adder circuit is shown in Figure P11.33. Write the logical expression for the outputs of this circuit in terms of its inputs. Also create a truth table for the outputs in terms of the inputs.

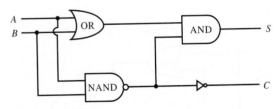

Figure P11.33

11.34   For the logic circuit shown in Figure P11.34, write the logical expression for the outputs of this circuit in terms of its inputs, and create a truth table for the outputs in terms of the inputs, including any required intermediate variables.

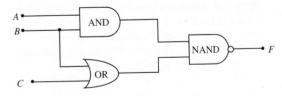

Figure P11.34

11.35   For the logic circuit in Figure P11.35, write the logical expression for the outputs of this circuit in terms of its inputs, and create a truth table for the outputs in terms of the inputs, including any required intermediate variables.

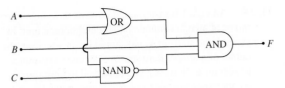

Figure P11.35

11.36   Determine the minimum expression for the following logic function, simplifying the expression:

$$f(A, B, C) = (A + B) \cdot A \cdot B + \bar{A} \cdot C + A \cdot \bar{B} \cdot C + \bar{B} \cdot \bar{C}$$

11.37   Complete the truth table for the circuit of Figure P11.37.

a.   What mathematical function does this circuit perform, and what do the outputs signify?

b.   How many standard 14-pin ICs would it take to construct this circuit?

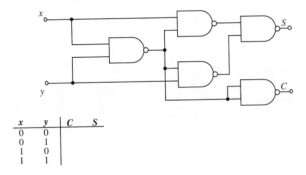

x	y	C	S
0	0		
0	1		
1	0		
1	1		

Figure P11.37

Section 11.4: Karnaugh Maps and Logic Design

11.38   Find the logic function corresponding to the truth table of Figure P11.38 in the simplest sum-of-products form.

A	B	C	F
0	0	0	1
0	0	1	0
0	1	0	0
0	1	1	0
1	0	0	1
1	0	1	0
1	1	0	1
1	1	1	1

Figure P11.38

11.39   Find the minimum expression for the output of the logic circuit shown in Figure P11.39.

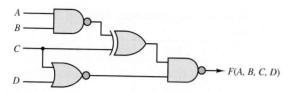

Figure P11.39

11.40   Build the Karnaugh map of the function
$Y = \overline{A \cdot B \cdot C}$ and verify it using boolean algebra.

11.41   Use a Karnaugh map to minimize the function
$Y = \overline{C} \cdot \overline{B} \cdot A + \overline{C} \cdot B \cdot \overline{A} + \overline{C} \cdot \overline{B} \cdot \overline{A}$

11.42   Fill in the Karnaugh map for the function
$Y = f(A, B, C)$ defined by the truth table of
Figure P11.42, and find the minimum expression
for the function.

A	B	C	Y
0	0	0	0
0	0	1	0
0	1	0	1
0	1	1	1
1	0	0	1
1	0	1	1
1	1	0	0
1	1	1	0

Figure P11.42

11.43   A function F is defined such that it equals 1
when a 4-bit input code is equivalent to any of the
decimal numbers 3, 6, 9, 12, or 15. Function F is 0 for
input codes 0, 2, 8, and 10. Other input values cannot
occur. Use a Karnaugh map to determine a minimal
expression for this function. Design and sketch a
circuit to implement this function, using only AND
and NOT gates.

11.44   Design the circuit of the function
$Y = f(A, B, C)$ described in Figure P11.44.

A	B	C	Y
0	0	0	0
0	0	1	0
0	1	0	0
0	1	1	1
1	0	0	0
1	0	1	0
1	1	0	1
1	1	1	x

Figure P11.44

11.45   Design a logic circuit that will produce the ones
complement of an 8-bit signed binary number.

11.46   Construct the Karnaugh map for the logic
function defined by the truth table of Figure P11.46,
and find the minimum expression for the function.

A	B	C	D	F
0	0	0	0	1
0	0	0	1	0
0	0	1	0	1
0	0	1	1	0
0	1	0	0	0
0	1	0	1	0
0	1	1	0	0
0	1	1	1	1
1	0	0	0	1
1	0	0	1	0
1	0	1	0	1
1	0	1	1	0
1	1	0	0	1
1	1	0	1	1
1	1	1	0	1
1	1	1	1	0

Figure P11.46

11.47   Use a Karnaugh map to minimize the function
$Y = (A + \overline{B}) \cdot [(\overline{C} \cdot D) + \overline{A}]$

11.48   Find the minimum output expression for the
circuit of Figure P11.48.

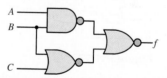

Figure P11.48

11.49   Design a combinational logic circuit that will add
two 4-bit binary numbers.

11.50   Minimize the expression described in the truth
table of Figure P11.50, and draw the circuit.

A	B	C	F
0	0	0	1
0	0	1	1
0	1	0	0
0	1	1	1
1	0	0	1
1	0	1	1
1	1	0	1
1	1	1	0

Figure P11.50

11.51   Find the minimum expression for the output of the
logic circuit of Figure P11.51.

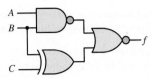

Figure P11.51

11.52   The objective of this problem is to design a combinational logic circuit that will aid in determination of the acceptability of emergency blood transfusions. It is known that human blood can be categorized into four types: A, B, AB, and O. Persons with type A blood can donate to both A and AB types and can receive blood from both A and O types. Persons with type B blood can donate to both B and AB types and can receive from both B and O types. Persons with type AB blood can donate only to type AB but can receive from any type. Persons with type O blood can donate to any type but can receive only from type O. Make appropriate variable assignments, and design a circuit that will approve or disapprove any particular transfusion based on these conditions.

11.53   Find the minimum expression for the logic function at the output of the logic circuit of Figure P11.53.

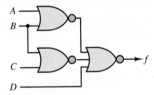

Figure P11.53

11.54   Determine the minimum boolean logic expression associated with the Karnaugh map in Figure P11.54 and create (realize) the logic circuit.

$C \cdot D$ \ $A \cdot B$	00	01	11	10
00	1	0	0	1
01	1	1	1	1
11	0	0	1	0
10	0	1	0	0

Figure P11.54

11.55

a. Construct a Karnaugh map associated with the truth table of Figure P11.55.

b. What is the minimum expression for the function?

c. Draw the logic circuit, using AND, OR, and NOT gates.

A	B	C	$f(A, B, C)$
0	0	0	1
0	0	1	1
0	1	0	0
0	1	1	1
1	0	0	1
1	0	1	1
1	1	0	1
1	1	1	0

Figure P11.55

11.56   Fill in the Karnaugh map for the logic function defined by the truth table of Figure P11.56. What is the minimum expression for the function?

A	B	C	D	F
0	0	0	0	1
0	0	0	1	0
0	0	1	0	1
0	0	1	1	0
0	1	0	0	0
0	1	0	1	0
0	1	1	0	0
0	1	1	1	1
1	0	0	0	1
1	0	0	1	0
1	0	1	0	1
1	0	1	1	0
1	1	0	0	1
1	1	0	1	1
1	1	1	0	1
1	1	1	1	0

Figure P11.56

11.57   Fill in the Karnaugh map for the logic function defined by the truth table of Figure P11.57. What is the minimum expression for the function? Realize the function using only NAND gates.

A	B	C	D	F
0	0	0	0	1
0	0	0	1	1
0	0	1	0	1
0	0	1	1	1
0	1	0	0	0
0	1	0	1	1
0	1	1	0	0
0	1	1	1	1
1	0	0	0	1
1	0	0	1	1
1	0	1	0	0
1	0	1	1	0
1	1	0	0	0
1	1	0	1	1
1	1	1	0	1
1	1	1	1	0

Figure P11.57

11.58 Design a circuit with a 4-bit input representing the binary number $A_3 A_2 A_1 A_0$. The output should be 1 if the input value is divisible by 3. Assume that the circuit is to be used only for the digits 0 through 9 (thus, values for 10 to 15 can be don't-care conditions).

a. Draw the Karnaugh map and truth table for the function.

b. Determine the minimum expression for the function.

c. Draw the circuit, using only AND, OR, and NOT gates.

11.59 Find the simplified sum-of-products representation of the function from the Karnaugh map shown in Figure P11.59. Note that x is the don't care term.

$C \cdot D$ \\ $A \cdot B$	00	01	11	10
00	0	1	0	0
01	1	1	0	0
11	0	x	1	0
10	0	0	1	0

Figure P11.59

11.60 Can the circuit for Problem 11.54 be simplified if it is known that the input represents a BCD (binary-coded decimal) number, that is, if it can never be greater than 9_{10}? If not, explain why not. Otherwise, design the simplified circuit.

11.61 Find the simplified sum-of-products representation of the function from the Karnaugh map shown in Figure P11.61.

$C \cdot D$ \\ $A \cdot B$	00	01	11	10
00	0	1	x	0
01	0	1	x	0
11	0	1	0	1
10	x	x	1	0

Figure P11.61

11.62 One method of ensuring reliability in data transmission systems is to transmit a parity bit along with every nibble, byte, or word of binary data transmitted. The parity bit confirms whether an even or odd number of 1s were transmitted in the data. In even-parity systems, the parity bit is set to 1 when the number of 1s in the transmitted data is odd. Odd-parity systems set the parity bit to 1 when the number of 1s in the transmitted data is even. Assume that a parity bit is transmitted for every nibble of data. Design a logic circuit that checks the nibble of data and transmits the proper parity bit for both even- and odd-parity systems.

11.63 Assume that a parity bit is transmitted for every nibble of data. Design two logic circuits that check a nibble of data and its parity bit to determine if there may have been a data transmission error. Assume first an even-parity system, then an odd-parity system.

11.64 Design a logic circuit that takes a 4-bit Gray code input from an optical encoder and translates it into two 4-bit nibbles of BCD.

11.65 Design a logic circuit that takes a 4-bit Gray code input from an optical encoder and determines if the input value is a multiple of 3.

11.66 The 4221 code is a base 10–oriented code that assigns the weights 4221 to each of 4 bits in a nibble of data. Design a logic circuit that takes a BCD nibble as input and converts it to its 4221 equivalent. The logic circuit should also report an error in the BCD input if its value exceeds 1001.

11.67 The 4-bit digital output of each of two sensors along an assembly line conveyor belt is proportional to the number of parts that pass by on the conveyor belt in a 30-s period. Design a logic circuit that reports an error if the outputs of the two sensors differ by more than one part per 30-s period.

Section 11.5: Combinational Logic Modules

11.68 A function F is defined such that it equals 1 when a 4-bit input code is equivalent to any of the decimal numbers 3, 6, 9, 12, or 15. F is 0 for input codes 0, 2, 8, and 10. Other input values cannot occur. Use a Karnaugh map to determine a minimal expression for this function. Design and sketch a circuit to implement this function using only AND and NOT gates.

11.69 Fill in the Karnaugh map for the logic function defined by the truth table of Figure P11.69. What is the minimum expression for the function? Realize the function using a 1-of-8 multiplexer.

A	B	C	D	$f(A, B, C, D)$
0	0	0	0	1
0	0	0	1	0
0	0	1	0	1
0	0	1	1	1
0	1	0	0	0
0	1	0	1	1
0	1	1	0	0
0	1	1	1	0
1	0	0	0	0
1	0	0	1	1
1	0	1	0	0
1	0	1	1	0
1	1	0	0	1
1	1	0	1	0
1	1	1	0	1
1	1	1	1	1

Figure P11.69

11.70   Fill in the truth table for the multiplexer circuit shown in Figure P11.70. What binary function is performed by these multiplexers?

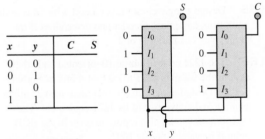

x	y	C	S
0	0		
0	1		
1	0		
1	1		

Figure P11.70

11.71   The circuit of Figure P11.71 can operate as a 4-to-16 decoder. Terminal EN denotes the enable input. Describe the operation of the 4-to-16 decoder. What is the role of logic variable A?

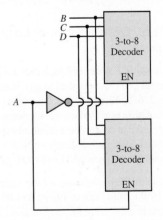

Figure P11.71

11.72   Show that the circuit given in Figure P11.72 converts 4-bit binary numbers to 4-bit Gray code.

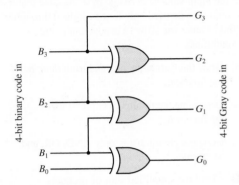

Figure P11.72

11.73   Suppose one of your classmates claims that the following boolean expressions represent the conversion from 4-bit Gray code to 4-bit binary numbers:

$$B_3 = G_3$$
$$B_2 = G_3 \oplus G_2$$
$$B_1 = G_3 \oplus G_2 \oplus G_1$$
$$B_0 = G_3 \oplus G_2 \oplus G_1 \oplus G_0$$

a. Show that your classmate's claim is correct.

b. Draw the circuit that implements the conversion.

11.74   Select the proper inputs for a four-input multiplexer to implement the function $f(A, B, C) = \overline{A}B\overline{C} + A\overline{B}C + AC$. Assume inputs $I_0, I_1, I_2,$ and I_3 correspond to $\overline{A}\overline{B}, \overline{A}B, A\overline{B},$ and AB, respectively, and that each input may be 0, 1, $\overline{C}$, or C.

11.75   Select the proper inputs for an 8-bit multiplexer to implement the function $f(A, B, C, D) = \sum(2, 5, 6, 8, 9, 10, 11, 13, 14)_{10}$. Assume the inputs I_0 through I_7 correspond to $\overline{A}\overline{B}\overline{C}, \overline{A}\overline{B}C, \overline{A}B\overline{C}, \overline{A}BC, A\overline{B}\overline{C}, A\overline{B}C, A B\overline{C},$ and ABC, respectively, and that each input may be 0, 1, $\overline{D}$, or D.

11.76   Use a 3-to-8 decoder and an N-input OR gate to implement the logic function $f(x, y, z) = xy + x\overline{y} + \overline{xyz}$. Draw the logic diagram and create the associated truth table.

Section 11.6: Latches and Flip-Flops

11.77   The input to the circuit of Figure P11.77 is a square wave having a period of 2 s, maximum value of 5 V, and minimum value of 0 V. Assume all flip-flops are initially in the reset state.

a. Explain what the circuit does.

b. Sketch the timing diagram, including the input and all four outputs.

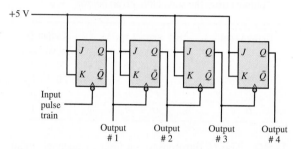

Figure P11.77

11.78 Suppose a circuit is constructed from three D-type flip-flops, one input I, with

$$D_1 = Q_2, \qquad D_0 = Q_1$$
$$D_2 = I \cdot (Q_1 \cdot \overline{Q_0} + \overline{Q_1} \cdot Q_0) + \overline{I} \cdot (Q_1 \cdot Q_0$$
$$+ \overline{Q_1} \cdot \overline{Q_0}) = \alpha \oplus I \qquad \text{with } \alpha = Q_1 \oplus Q_0$$

a. Draw the circuit diagram.

b. Assume the circuit starts with all flip-flops set. Sketch a table that shows the outputs of all three flip-flops.

11.79 Suppose that you want to use a JK flip-flop for a laboratory experiment. However, you have only D flip-flops. Assuming that you have all the logic gates available, make a JK flip-flop using a D flip-flop and some logic gate(s).

11.80 Draw a timing diagram (four complete clock cycles) for A_0, A_1, and A_2 for the circuit of Figure P11.80. Assume that all initial values are 0. Note that all flip-flops are negative edge–triggered.

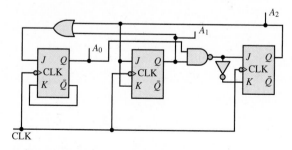

Figure P11.80

11.81 Assume that the slotted encoder shown in Figure P11.81 has a length of 1 m and a total of 1,000 slots (i.e., there is one slot per millimeter). If a

counter is incremented by 1 each time a slot goes past a sensor, design a digital counting system that determines the speed of the moving encoder (in meters per second).

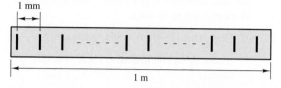

Figure P11.81

11.82 Given the sequential circuit of Figure P11.82, determine the output Y when input A is [1 0 1 1].

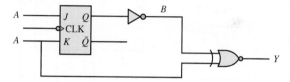

Figure P11.82

11.83 Write the truth table for an RS flip-flop with enable (E), preset (P), and clear (C) lines.

11.84 A JK flip-flop is wired as shown in Figure P11.84 with a given input signal. Assuming that Q is at logic 0 initially and the trailing-edge triggering is effective, sketch the output Q.

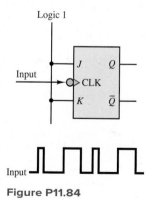

Figure P11.84

11.85 With reference to the JK flip-flop of Problem 11.84, assume that the output at the Q terminal is made to serve as the input to a second JK flip-flop wired exactly as the first. Sketch the Q output of the second flip-flop.

Section 11.7: Digital Counters and Registers

11.86 A binary pulse counter can be constructed by interconnecting T-type flip-flops in an appropriate manner. Assume it is desired to construct a counter that can count up to 100_{10}.

a. How many flip-flops would be required?

b. Sketch the circuit needed to implement this counter.

11.87 Explain what the circuit of Figure P11.87 does and how it works. (*Hint*: This circuit is called a 2-bit synchronous binary up-down counter.)

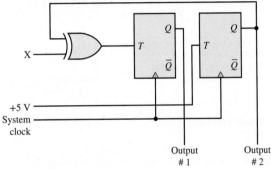

Figure P11.87

11.88 Describe how the ripple counter works. Why is it so named? What disadvantages can you think of for this counter?

Section 11.8: Sequential Logic Design

11.89 Using necessary logic gates and D-type flip-flops, create a sequential circuit (one input–one output) from the state table given below.

Current state Q'_n	Next state $D = Q'_{n+1}$		Output Q	
	$I = 0$	$I = 1$	$I = 0$	$I = 1$
A	A	B	0	0
B	B	A	0	1
C	C	B	0	0
D	D	A	0	1

11.90 Use JK flip-flops to construct a sequential circuit with the state diagram shown in Figure P11.90.

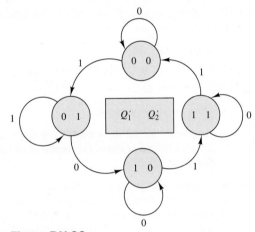

Figure P11.90

PART III

ELECTROMECHANICS

yuyanga/Getty Images

wi6995/Shutterstock

C H A P T E R

12

PRINCIPLES OF
ELECTROMECHANICS

he objective of this chapter is to introduce the fundamental notions of electromechanical energy conversion, leading to an understanding of the operation of various electromechanical transducers. The chapter also serves as an introduction to the material on electric machines to be presented in Chapter 13. The foundations for the material introduced in this chapter may be found in the circuit analysis chapters (1 through 6).

The subject of electromechanical energy conversion is one that should be of particular interest to the *non–electrical* engineer, because it forms one of the important points of contact between electrical engineering and other engineering disciplines. Electromechanical transducers are commonly used in the design of industrial and aerospace control systems and in biomedical applications, and they form the basis of many common appliances. In the course of our exploration of electromechanics, we illustrate the operation of practical devices, such as loudspeakers, relays, solenoids, and sensors for the measurement of position and velocity.

> **(LO)** **Learning Objectives**
>
> *Students will learn to...*
> 1. Review the basic principles of electricity and magnetism. *Section 12.1.*
> 2. Use the concepts of reluctance and magnetic circuit equivalents to compute magnetic flux and currents in simple magnetic structures. *Section 12.2.*
> 3. Understand the properties of magnetic materials and their effects on magnetic circuit models. *Section 12.3.*
> 4. Use magnetic circuit models to analyze transformers. *Section 12.4.*
> 5. Model and analyze force generation in electromagnetomechanical systems. Analyze moving-iron transducers (electromagnets, solenoids, relays) and moving-coil transducers (electrodynamic shakers, loudspeakers, and seismic transducers). *Section 12.5.*

12.1   ELECTRICITY AND MAGNETISM

The notion that the phenomena of electricity and magnetism are interconnected was first proposed in the early 1800s by H. C. Oersted, a Danish physicist. Oersted showed that an electric current produces magnetic effects (more specifically, a magnetic field). Soon after, the French scientist André Marie Ampère expressed this relationship by means of a precise formulation known as *Ampère's law*. A few years later, the English scientist Faraday illustrated how the converse of Ampère's law also holds true, that is, that a magnetic field can generate an electric field; in short, *Faraday's law* states that a changing magnetic field gives rise to a voltage.

As is explained in the next few sections, the magnetic field forms a necessary connection between electrical and mechanical energy. Ampère's and Faraday's laws formally illustrate the relationship between electric and magnetic fields, but it should already be evident from your own individual experience that the magnetic field can also convert magnetic energy to mechanical energy (e.g., by lifting a piece of iron with a magnet). In effect, the devices we commonly refer to as *electromechanical* should more properly be referred to as electro*magneto*mechanical, since they almost invariably operate through a conversion from electrical to mechanical energy (or vice versa) by means of a magnetic field. Chapters 12 and 13 are concerned with the use of electricity and magnetic materials for the purpose of converting electrical to mechanical energy, and back.

The Magnetic Field and Faraday's Law

The quantities used to quantify the strength of a magnetic field are the **magnetic flux ϕ**, in units of **webers** (Wb); and the **magnetic flux density B**, in units of webers per square meter (Wb/m^2), or **teslas** (T). The latter quantity and the associated **magnetic field intensity H** (in units of amperes per meter, or A/m) are vectors.[1] Thus, the density of the magnetic flux and its intensity are in general described in

---

[1]We will use the boldface symbols **B** and **H** to denote the vector forms of B and H; the standard typeface will represent the scalar flux density or field intensity in a given direction.

vector form, in terms of the components present in each spatial direction (e.g., on the *x*, *y*, and *z* axes). In discussing magnetic flux density and field intensity in this chapter and Chapter 13, we almost always assume that the field is a *scalar field*, that is, that it lies in a single spatial direction. This will simplify many explanations.

It is customary to represent the magnetic field by means of the familiar *lines of force* (a concept also due to Faraday); we visualize the strength of a magnetic field by observing the density of these lines in space. You probably know from a previous course in physics that such lines are closed in a magnetic field; that is, they form continuous loops exiting at a magnetic north pole (by definition) and entering at a magnetic south pole. The relative strengths of the magnetic fields generated by two magnets could be depicted as shown in Figure 12.1.

Magnetic fields are generated by electric charge in motion, and their effect is measured by the force they exert on a moving charge. As you may recall from previous physics courses, the vector force **f** exerted on a charge of *q* moving at velocity **u** in the presence of a magnetic field with flux density **B** is given by

$$\mathbf{f} = q\mathbf{u} \times \mathbf{B} \tag{12.1}$$

where the symbol × denotes the (vector) cross product. If the charge is moving at a velocity **u** in a direction that makes an angle θ with the magnetic field, then the magnitude of the force is given by

$$f = quB\sin\theta \tag{12.2}$$

and the direction of this force is at right angles with the plane formed by the vectors **B** and **u**. This relationship is depicted in Figure 12.2.

The magnetic flux ϕ is then defined as the integral of the flux density over some surface area. For the simplified (but often useful) case of magnetic flux lines perpendicular to a cross-sectional area *A*, the flux is given by:

$$\phi = \int_A B \, dA \tag{12.3}$$

in webers, where the subscript *A* indicates that the integral is evaluated over surface *A*. Furthermore, if the flux were to be uniform over the cross-sectional area *A* (a useful simplification), the preceding integral is approximated by:

$$\boxed{\phi = B \cdot A} \tag{12.4}$$

Figure 12.3 illustrates this idea, by showing hypothetical magnetic flux lines traversing a surface, delimited in the figure by a thin conducting wire.

Faraday's law states that if the imaginary surface *A* were bounded by a conductor—for example, the thin wire of Figure 12.3—then a *changing* magnetic field would induce a voltage, and therefore a current, in the conductor. More precisely, Faraday's law states that a time-varying flux causes an induced **electromotive force**, or **emf**, *e* as follows:

$$e = -\frac{d\phi}{dt} \tag{12.5}$$

A little discussion is necessary at this point to explain the meaning of the minus sign in equation 12.5. Consider the one-turn coil of Figure 12.4, which forms

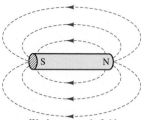

Weaker magnetic field

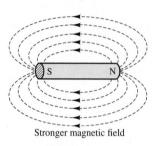

Stronger magnetic field

Figure 12.1 Lines of force in a magnetic field

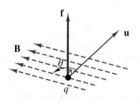

Figure 12.2 Charge moving in a constant magnetic field

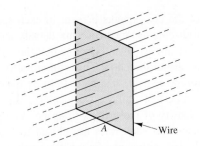

Figure 12.3 Magnetic flux lines crossing a surface bounded by a thin conducting wire

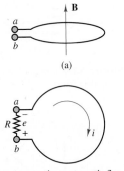

Current generating a magnetic flux opposing the increase in flux due to **B**

(b)

Figure 12.4 Flux direction

a circular cross-sectional area, in the presence of a magnetic field with flux density **B** oriented in a direction perpendicular to the plane of the coil. If the magnetic field, and therefore the flux within the coil, is constant, no voltage will exist across terminals a and b; if, however, the flux were increasing and terminals a and b were connected—for example, by means of a resistor, as indicated in Figure 12.4(b)—current would be generated in the coil such that *the magnetic flux generated by the current would oppose the increasing flux*. Thus, the flux induced by such a current would be in the direction opposite to that of the original flux density vector **B**. This principle is known as **Lenz's law**. The reaction flux would then point downward in Figure 12.4(a), or into the page in Figure 12.4(b). Now, by virtue of the **right-hand rule**, this reaction flux would induce a current clockwise in Figure 12.4(b), that is, a current out of terminal b and into terminal a. The resulting voltage across the hypothetical resistor R would then be negative. If, on the other hand, the original flux were decreasing, current would be induced in the coil so as to reestablish the initial flux; but this would mean that the current would have to generate a flux in the upward direction in Figure 12.4(a) [or out of the page in Figure 12.4(b)]. Thus, the resulting voltage would change sign.

The polarity of the induced voltage can usually be determined from physical considerations; therefore the minus sign in equation 12.5 can be left out. We use this convention throughout the chapter.

In practical applications, the size of the voltages induced by the changing magnetic field can be significantly increased if the conducting wire is coiled so as to multiply the area crossed by the magnetic flux lines many times over. For an N-turn coil with cross-sectional area A, for example, we have the emf

$$e = N\frac{d\phi}{dt} \qquad \text{(minus sign is understood)} \qquad (12.6)$$

CHECK YOUR UNDERSTANDING

A coil having 100 turns is immersed in a magnetic field that is varying uniformly from 80 to 30 mWb in 2 s. Find the induced voltage in the coil.

Answer: $e = -2.5$ V

Figure 12.5 shows an *N*-turn coil *linking* a certain amount of magnetic flux; you can see that if *N* is very large and the coil is tightly wound (as is usually the case in the construction of practical devices), it is not unreasonable to presume that each turn of the coil links the same flux. It is convenient, in practice, to define the **flux linkage** λ as

Right-hand rule

$$\lambda = N\phi \qquad (12.7)$$

so that

$$\boxed{e = \frac{d\lambda}{dt}} \qquad (12.8)$$

Note that equation 12.8, relating the derivative of the flux linkage to the induced emf, is analogous to the equation describing current as the derivative of charge:

$$i = \frac{dq}{dt} \qquad (12.9)$$

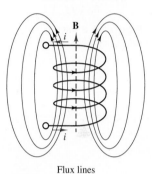

Flux lines

Figure 12.5 Concept of flux linkage

In other words, flux linkage can be viewed as the dual of charge in circuit analysis provided that we are aware of the simplifying assumptions just stated in the preceding paragraphs, namely, a uniform magnetic field perpendicular to the area delimited by a tightly wound coil. These assumptions are not at all unreasonable when applied to the inductor coils commonly employed in electric circuits.

What, then, are the physical mechanisms that can cause magnetic flux to change, and therefore to induce an electromotive force? Two such mechanisms are possible. The first consists of physically moving a permanent magnet in the vicinity of a coil, for example, so as to create a time-varying flux. The second requires a time varying current to produce a time-varying magnetic field. The latter method is more practical in many circumstances, since it does not require the use of permanent magnets and allows variation of field strength by varying the applied current; however, the former method is conceptually simpler to visualize. The voltages induced by a moving magnetic field are called **motion voltages**; those generated by a time-varying magnetic field are termed **transformer voltages**. We are interested in both in this chapter, for different applications.

In the analysis of linear circuits, as in Chapter 3, it is assumed that the relationship between flux linkage and current is linear:

$$\lambda = Li \qquad (12.10)$$

so that the effect of a time-varying current is to induce a transformer voltage across an inductor coil, according to the expression

$$v = L\frac{di}{dt} \qquad (12.11)$$

This is, in fact, the defining equation for the ideal **self-inductance** *L*. In addition to self-inductance, however, it is important to consider the **magnetic coupling** that can occur between neighboring circuits. Self-inductance measures the voltage induced in a circuit by the magnetic field generated by a current flowing in the same circuit. It is also possible that a second circuit in the vicinity of the first may experience an induced voltage as a consequence of the magnetic field generated in the first circuit. As explained in Section 12.4, this principle underlies the operation of all transformers.

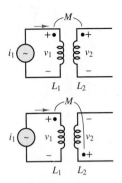

Figure 12.6 Mutual
inductance

Self- and Mutual Inductance

Figure 12.6 depicts a pair of coils, one of which, L_1, is excited by a current i_1 and therefore develops a magnetic field and a resulting induced voltage v_1. The second coil, L_2, is not energized by a current, but links some of the flux generated by current i_1 around L_1 because of its close proximity to the first coil. The magnetic coupling between the coils established by virtue of their proximity is described by a quantity called **mutual inductance** and defined by the symbol M. The mutual inductance is defined by the equation

$$v_2 = M \frac{di_1}{dt} \tag{12.12}$$

The dots shown in the two drawings indicate the polarity of the coupling between the coils. If the dots are at the same end of the coils, the voltage induced in coil 2 by a current in coil 1 has the same polarity as the voltage induced by the same current in coil 1; otherwise, the voltages are in opposition, as shown in the lower part of Figure 12.6. Thus, the presence of such dots indicates that magnetic coupling is present between two coils. It should also be pointed out that if a current (and therefore a magnetic field) were present in the second coil, an additional voltage would be induced across coil 1. The voltage induced across a coil is, in general, equal to the sum of the voltages induced by self-inductance and mutual inductance.

In practical electromagnetic circuits, the self-inductance of a circuit is not necessarily constant; in particular, the inductance parameter L is not constant, in general, but depends on the strength of the magnetic field intensity, so that it will not be possible to use such a simple relationship as $v = L \, di/dt$, with L constant. If we revisit the definition of the transformer voltage

$$e = N \frac{d\phi}{dt} \tag{12.13}$$

we see that in an inductor coil, the inductance is given by

$$L = \frac{N\phi}{i} = \frac{\lambda}{i} \tag{12.14}$$

This expression implies that the relationship between current and flux in a magnetic structure is linear if the inductance L is constant (the inductance being the slope of the line). In fact, the properties of ferromagnetic materials are such that the flux–current relationship is nonlinear, so that the simple linear inductance parameter used in electric circuit analysis is not adequate to represent the behavior of the magnetic circuits of this chapter. In any practical situation, the relationship between the flux linkage λ and the current is nonlinear, and might be described by a curve similar to that shown in Figure 12.7. Whenever the i-λ curve is not a straight line, it is more convenient to analyze the magnetic system in terms of energy calculations, since the corresponding circuit equation would be nonlinear.

In a magnetic system, the energy stored in the magnetic field is equal to the integral of the instantaneous power, which is the product of voltage and current, just as in a conventional electric circuit:

$$W_m = \int ei \, dt' \tag{12.15}$$

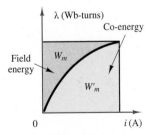

Figure 12.7 Relationship between flux linkage, current, energy, and co-energy

However, in this case, the voltage corresponds to the induced emf, according to Faraday's law,

$$e = \frac{d\lambda}{dt} = N\frac{d\phi}{dt} \tag{12.16}$$

and is therefore related to the rate of change of the magnetic flux. The energy stored in the magnetic field could therefore be expressed in terms of the current by:

$$W_m = \int ei \, dt' = \int \frac{d\lambda}{dt} i \, dt' = \int i \, d\lambda' \tag{12.17}$$

It should be straightforward to recognize that this energy is equal to the area above the λ-i curve of Figure 12.7. From the same figure, it is also possible to define a fictitious (but useful) quantity called **co-energy**, equal to the area under the curve and identified by the symbol W'_m. From the figure, it is also possible to see that the co-energy can be expressed in terms of the stored energy by:

$$W'_m = i\lambda - W_m \tag{12.18}$$

Example 12.1 illustrates the calculation of energy, co-energy, and induced voltage, using the concepts developed in these paragraphs.

The calculation of the energy stored in the magnetic field around a magnetic structure will be particularly useful later in the chapter when the discussion turns to practical electromechanical transducers and it will be necessary to actually compute the forces generated in magnetic structures.

EXAMPLE 12.1   Energy and Co-Energy Calculation for an Inductor

Problem

Compute the energy, co-energy, and incremental linear inductance for an iron-core inductor with a given λ-i relationship. Also compute the voltage across the terminals, given the current through the coil.

Solution

Known Quantities: λ-i relationship; nominal value of λ; coil resistance; coil current.

Find: W_m; W'_m; L_Δ; v.

Schematics, Diagrams, Circuits, and Given Data: $i = (\lambda + 0.5\lambda^2)$ A; $\lambda_0 = 0.5$ V-s; $R = 1$ Ω; $i(t) = 0.625 + 0.01 \sin(400t)$.

Assumptions: Assume that the magnetic equation can be linearized, and use the linear model in all circuit calculations.

Analysis:

1. *Calculation of energy and co-energy.* From equation 12.17, we calculate the energy as follows.

$$W_m = \int_0^\lambda i(\lambda') \, d\lambda' = \frac{\lambda^2}{2} + \frac{\lambda^3}{6}$$

The above expression is valid in general; in our case, the inductor is operating at a nominal flux linkage $\lambda_0 = 0.5$ V-s, and we can therefore evaluate the energy to be

$$W_m(\lambda = \lambda_0) = \left(\frac{\lambda^2}{2} + \frac{\lambda^3}{6}\right)\bigg|_{\lambda=0.5} = 0.1458 \text{ J}$$

Thus, after equation 12.18, the co-energy is given by

$$W'_m = i\lambda - W_m$$

where

$$i = \lambda + 0.5\lambda^2 = 0.625 \text{ A}$$

and

$$W'_m = i\lambda - W_m = (0.625)(0.5) - (0.1458) = 0.1667 \text{ J}$$

2. *Calculation of incremental inductance.* If we know the nominal value of flux linkage (i.e., the operating point), we can calculate a linear inductance L_Δ, valid around values of λ close to the operating point λ_0. This incremental inductance is defined by the expression

$$L_\Delta = \left(\frac{di}{d\lambda}\right)^{-1}\bigg|_{\lambda=\lambda_0}$$

and can be computed to be

$$L_\Delta = \left(\frac{di}{d\lambda}\right)^{-1}\bigg|_{\lambda=\lambda_0} = (1 + \lambda)^{-1}\big|_{\lambda=\lambda_0} = \frac{1}{1 + \lambda}\bigg|_{\lambda=0.5} = 0.667 \text{ H}$$

The above expressions can be used to analyze the circuit behavior of the inductor when the flux linkage is around 0.5 V-s, or, equivalently, when the current through the inductor is around 0.625 A.

3. *Circuit analysis using linearized model of inductor.* We can use the incremental linear inductance calculated above to compute the voltage across the inductor in the presence of a current $i(t) = 0.625 + 0.01\sin(400t)$. Using the basic circuit definition of an inductor with series resistance R, the voltage across the inductor is given by

$$v = iR + L_\Delta \frac{di}{dt} = [0.625 + 0.01\sin(400t)] \times 1 + 0.667 \times 4\cos(400t)$$

$$= 0.625 + 0.01\sin(400t) + 2.668\cos(400t)$$

$$= 0.625 + 2.668\sin(400t + 89.8°) \quad \text{V}$$

Comments: The linear approximation in this case is not a bad one: the small sinusoidal current is oscillating around a much larger average current. In this type of situation, it is reasonable to assume that the inductor behaves linearly. This example explains why the linear inductor model introduced in Chapter 3 is an acceptable approximation in most circuit analysis problems.

CHECK YOUR UNDERSTANDING

The relation between the flux linkages and the current for a magnetic material is given by $\lambda = 6i/(2i + 1)$ Wb-turns. Determine the energy stored in the magnetic field for $\lambda = 2$ Wb-turns.

Linear Variable Differential Transformer (LVDT)

The linear variable differential transformer (LVDT) is a displacement transducer based on the mutual inductance concept just discussed. Figure 12.8 represents an LVDT as a primary coil subject to AC excitation (v_{ex}) and of a pair of identical secondary coils, which are connected so that:

$$v_{out} = v_1 - v_2$$

The ferromagnetic core between the primary and secondary coils can be displaced in proportion to some external motion x and determines the magnetic coupling between primary and secondary coils. Intuitively, as the core is displaced upward, greater coupling will occur between the primary coil and the top secondary coil, thus inducing a greater voltage in the top secondary coil. Hence, $v_{out} > 0$ for positive displacements. The converse is true for negative displacements. More formally, if the primary coil has resistance R_p and self-inductance L_p, we can write

$$iR_p + L_p \frac{di}{dt} = v_{ex}$$

and the voltages induced in the secondary coils are given by

$$v_1 = M_1 \frac{di}{dt}$$

$$v_2 = M_2 \frac{di}{dt}$$

so that

$$v_{out} = (M_1 - M_2) \frac{di}{dt}$$

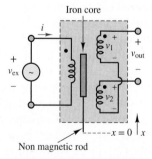

Figure 12.8 Linear variable differential transformer

where M_1 and M_2 are the mutual inductances between the primary and the respective secondary coils. It should be apparent that each of the mutual inductances is dependent on the position of the iron core. For example, with the core at the *null position*, $M_1 = M_2$ and $v_{out} = 0$. The LVDT is typically designed so that $M_1 - M_2$ is linearly related to the displacement of the core x.

(Continued)

(Concluded)

Because the excitation is by necessity an AC signal (why?), the output voltage is actually given by the difference of two sinusoidal voltages at the same frequency and is therefore itself a sinusoid, whose amplitude and phase depend on the displacement x. Thus, v_{out} is an *amplitude-modulated* (AM) signal, similar to the one discussed in the Focus on Measurements box, "Capacitive Displacement Transducer and Microphone," in Chapter 3. To recover a signal proportional to the actual displacement, it is therefore necessary to use a demodulator circuit, such as the one discussed in the Focus on Measurements box, "Peak Detector Circuit for Capacitive Displacement Transducer," in Chapter 8.

Ampère's Law

As explained in the previous section, Faraday's law is one of two fundamental laws relating electricity to magnetism. The second relationship, which forms a counterpart to Faraday's law, is **Ampère's law**. Qualitatively, Ampère's law states that the magnetic field intensity **H** in the vicinity of a conductor is related to the current carried by the conductor; thus Ampère's law establishes a dual relationship with Faraday's law.

In the previous section, the magnetic field is described by its flux density **B** and flux ϕ. To explain Ampère's law and the behavior of magnetic materials, we define the magnetic field intensity **H** as:

$$\mathbf{B} = \mu\mathbf{H} = \mu_r\mu_0\mathbf{H} \qquad \text{Wb/m}^2 \text{ or T} \tag{12.19}$$

where μ is a scalar constant for a particular physical medium (at least, for the applications we consider here) and is called the **permeability** of the medium. The permeability of a material can be factored as the product of the permeability of free space $\mu_0 = 4\pi \times 10^{-7}$ H/m, and the relative permeability μ_r, which varies greatly according to the medium. For example, for air and for most electrical conductors and insulators, μ_r is equal to 1. For ferromagnetic materials, μ_r can take values ranging from 10^3 to 10^6. The size of μ_r represents a measure of the magnetic properties of the material. A consequence of Ampère's law is that the larger the value of μ, the smaller the current required to produce a large flux density in an electromagnetic structure. Consequently, many electromechanical devices make use of ferromagnetic materials, called iron cores, to enhance their magnetic properties. Table 12.1 gives approximate values of μ_r for some common materials.

The reason for introducing the magnetic field intensity is that it is independent of the properties of the materials employed in the construction of magnetic circuits. Thus, a given magnetic field intensity **H** will give rise to different flux densities in different materials. It will therefore be useful to define *sources* of magnetic energy in terms of the magnetic field intensity, so that different magnetic structures and materials can then be evaluated or compared for a given source. As stated earlier, both the magnetic flux density and the field intensity are vector quantities; however, for ease of analysis, scalar fields will be chosen by appropriately selecting the orientation of the fields, wherever possible.

Table 12.1 **Relative permeabilities for common materials**

Material	μ_r
Air	1
Water	1
Copper	1
Nickel	100–600
Ferrite (manganese-zinc)	640
Steel	100
Electrical steel	4,000
Iron	5,000
Permalloy	8,000
Mu-metal	20,000
Nanoperm	80,000
Metglas	1,000,000

Ampère's law states that the integral of the vector magnetic field intensity **H** around a closed path is equal to the total current linked by the closed path i:

$$\oint \mathbf{H} \cdot d\mathbf{l} = \sum i \tag{12.20}$$

where $d\mathbf{l}$ is an increment in the direction of the closed path. If at every point along the path the magnetic field is parallel to the path, we can use scalar quantities to write:

$$\int H \, dl = \sum i \tag{12.21}$$

Figure 12.9 illustrates the case of a wire carrying a current i and of a circular path of radius r surrounding the wire. In this simple case, you can see that the magnetic field intensity **H** is determined by the familiar right-hand rule. This rule states that if the direction of current i points in the direction of the thumb of one's right hand, the resulting magnetic field encircles the conductor in the direction in which the other four fingers would encircle it. Thus, in the case of Figure 12.9, the closed-path integral becomes equal to $H \cdot 2\pi r$, since the path and the magnetic field are in the same direction, and therefore the magnitude of the magnetic field intensity is given by

$$H = \frac{i}{2\pi r} \tag{12.22}$$

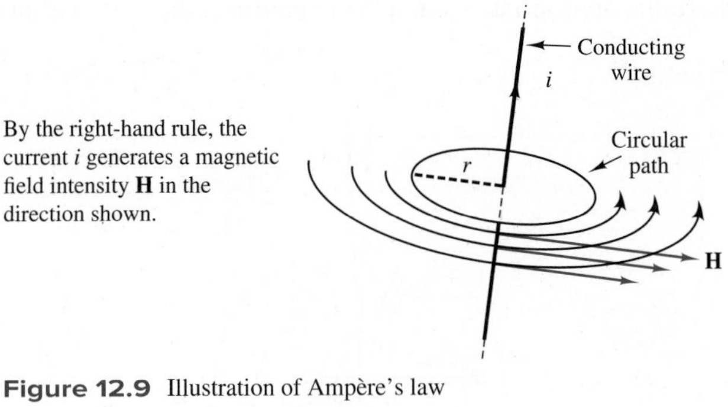

By the right-hand rule, the current i generates a magnetic field intensity **H** in the direction shown.

Figure 12.9 Illustration of Ampère's law

CHECK YOUR UNDERSTANDING

The magnitude of **H** at a radius of 0.5 m from a long linear conductor is 1 A-m^{-1}. Find the current in the wire.

Answer: $I = \pi$ A

Now, the magnetic field intensity $\boldsymbol{H}$ is unaffected by the material surrounding the conductor, but the flux density $\boldsymbol{B}$ depends on the material properties. The density of flux lines around the conductor would be far greater in the presence of a magnetic material than if the conductor were surrounded by air. The field generated by a single conducting wire is relatively weak; however, if the wire is a tightly wound coil with many turns, the strength of the magnetic field is increased greatly. For a coil with N turns, one can verify visually that the lines of force associated with the magnetic field link all the turns of the conducting coil, so that we have effectively increased the current linked by the flux lines N-fold. The product $N \cdot i$ is a useful quantity in electromagnetic circuits and is called the **magnetomotive force**,[2] $\mathcal{F}$ (or **mmf**), in analogy with the electromotive force.

$$\mathcal{F} = Ni \quad \text{A-turns} \qquad \text{Magnetomotive force} \qquad\qquad (12.23)$$

Figure 12.10 illustrates the magnetic flux lines in the vicinity of a coil. The magnetic field generated by the coil can be made to generate a much greater flux density if the coil encloses a magnetic material. The most common ferromagnetic materials are steel and iron; in addition to these, many alloys and oxides of iron—as well as nickel—and some artificial ceramic materials called **ferrites** exhibit magnetic properties. In recent years, rare earth magnets have found increasing use, especially in the design of high-performance electric motors. The two most common rare earth materials are neodymium and samarium (lanthanides), which are used in compounds that include transition metals, such as iron, nickel, and cobalt. Such magnets can produce magnetic fields of strength two to three times greater than ferrites. Winding a coil around a ferromagnetic material accomplishes two useful tasks at once: It forces the magnetic flux to be concentrated within the coil and—if the shape of the magnetic material is appropriate—completely confines the flux within the magnetic

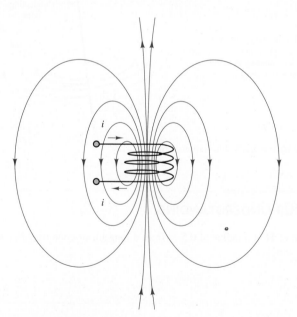

Figure 12.10 Magnetic flux lines in the vicinity of a current-carrying coil

---

[2]Note that although they are dimensionally equal to amperes, the units of magnetomotive force are ampere-turns.

material, thus forcing the closed path for the flux lines to be almost entirely enclosed within the ferromagnetic material. Typical arrangements are the iron-core inductor and the toroid of Figure 12.11. The flux densities for these inductors are given by

$$B = \frac{\mu N i}{l} \qquad \text{Flux density for tightly wound circular coil} \qquad (12.24)$$

$$B = \frac{\mu N i}{2\pi r_2} \qquad \text{Flux density for toroidal coil} \qquad (12.25)$$

In equation 12.24, l represents the length of the coil wire; Figure 12.11 defines the parameter r_2 in equation 12.25.

Intuitively, the presence of a high-permeability material near a source of magnetic flux causes the flux to preferentially concentrate in the high-μ material, rather than in air, much as a conducting path concentrates the current produced by an electric field in an electric circuit. Figure 12.12 depicts an example of a simple electromagnetic structure which forms the basis of the practical transformer.

Table 12.2 summarizes the variables introduced thus far in the discussion of electricity and magnetism.

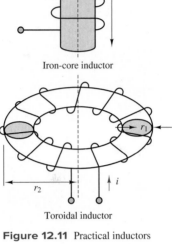

Iron-core inductor

Toroidal inductor

Figure 12.11 Practical inductors

Table 12.2 **Magnetic variables and units**

Variable	Symbol	Units
Current	I	A
Magnetic flux density	B	Wb/m^2 = T
Magnetic flux	ϕ	Wb
Magnetic field intensity	H	A/m
Electromotive force	e	V
Magnetomotive force	$\mathcal{F}$	A-turns
Flux linkage	λ	Wb-turns

12.2   MAGNETIC CIRCUITS

It is possible to analyze the operation of electromagnetic devices such as the one depicted in Figure 12.12 by means of magnetic equivalent circuits, similar in many respects to the equivalent electric circuits of earlier chapters. Before we can present this technique, however, we need to make a few simplifying approximations. The first of these approximations assumes that there exists a **mean path** for the magnetic flux and that the corresponding mean flux density is approximately constant over the cross-sectional area of the magnetic structure. Using equation 12.4, we see that a coil wound around a core with cross-sectional area A will have flux density

$$B = \frac{\phi}{A} \qquad (12.26)$$

where A is assumed to be perpendicular to the direction of the flux lines. Figure 12.12 illustrates such a mean path and the cross-sectional area A. Knowing the flux density, we obtain the field intensity:

$$H = \frac{B}{\mu} = \frac{\phi}{A\mu} \qquad (12.27)$$

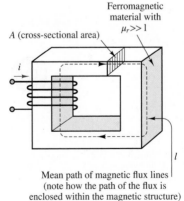

Mean path of magnetic flux lines
(note how the path of the flux is
enclosed within the magnetic structure)

Figure 12.12 A simple electromagnetic structure

But then, knowing the field intensity, we can relate the mmf of the coil $\mathcal{F}$ to the product of the magnetic field intensity H and the length of the magnetic (mean) path l; we can use equations 12.24 and 12.19 to derive

$$\mathcal{F} = N \cdot i = H \cdot l \tag{12.28}$$

In summary, the mmf is equal to the magnetic flux times the length of the magnetic path, divided by the permeability of the material times the cross-sectional area:

$$\mathcal{F} = \phi \frac{l}{\mu A} \tag{12.29}$$

A review of this formula reveals that the magnetomotive force $\mathcal{F}$ may be viewed as being analogous to the voltage source in a series electric circuit, and that the flux ϕ is then equivalent to the electric current in a series circuit and the term $l/\mu A$ to the *"magnetic resistance"* of one leg of the magnetic circuit. You will note that the term $l/\mu A$ is very similar to the term describing the resistance of a cylindrical conductor of length l and cross-sectional area A, where the permeability μ is analogous to the conductivity σ. The term $l/\mu A$ occurs frequently enough to be assigned the name of **reluctance** and the symbol $\mathcal{R}$. It is also important to recognize the *relationship between the reluctance of a magnetic structure and its inductance*. This can be derived easily starting from equation 12.14:

$$L = \frac{\lambda}{i} = \frac{N\phi}{i} = \frac{N}{i} \frac{Ni}{\mathcal{R}} = \frac{N^2}{\mathcal{R}} \quad \text{H} \tag{12.30}$$

In summary, when an N-turn coil carrying a current i is wound around a magnetic core such as the one indicated in Figure 12.12, the mmf $\mathcal{F}$ generated by the coil produces a flux ϕ that is *mostly* concentrated within the core and is assumed to be uniform across the cross section. Within this simplified picture, then, the analysis of a magnetic circuit is analogous to that of resistive electric circuits. This analogy is illustrated in Table 12.3 and in the examples in this section.

Table 12.3 **Analogy between electric and magnetic circuits**

Electrical quantity	Magnetic quantity
Electrical field intensity E, V/m	Magnetic field intensity H, A-turns/m
Voltage v, V	Magnetomotive force $\mathcal{F}$, A-turns
Current i, A	Magnetic flux ϕ, Wb
Current density J, A/m^2	Magnetic flux density B, Wb/m^2
Resistance R, Ω	Reluctance $\mathcal{R} = l/\mu A$, A-turns/Wb
Conductivity σ, 1/Ω-m	Permeability μ, Wb/A-m

The usefulness of the magnetic circuit analogy can be emphasized by analyzing a magnetic core similar to that of Figure 12.12, but with a slightly modified geometry. Figure 12.13 depicts the magnetic structure and its equivalent-circuit analogy. In the figure, we see that the mmf $\mathcal{F} = Ni$ excites the magnetic circuit, which is composed of four legs: two of mean path length l_1 and cross-sectional area $A_1 = d_1 w$, and the

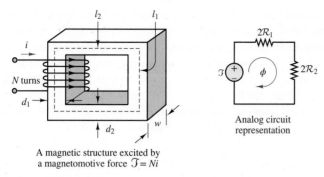

A magnetic structure excited by
a magnetomotive force $\mathcal{F} = Ni$

Figure 12.13 Analogy between magnetic and electric circuits

other two of mean length l_2 and cross-sectional area $A_2 = d_2 w$. Thus, the reluctance encountered by the flux in its path around the magnetic core is given by the quantity $\mathcal{R}_{\text{series}}$, with

$$\mathcal{R}_{\text{series}} = 2\mathcal{R}_1 + 2\mathcal{R}_2$$

and

$$\mathcal{R}_1 = \frac{l_1}{\mu A_1} \qquad \mathcal{R}_2 = \frac{l_2}{\mu A_2}$$

It is important at this stage to review the assumptions and simplifications made in analyzing the magnetic structure of Figure 12.13:

1. All the magnetic flux is linked by all the turns of the coil.
2. The flux is confined exclusively within the magnetic core.
3. The density of the flux is uniform across the cross-sectional area of the core.

You can probably see intuitively that the first of these assumptions might not hold true near the ends of the coil, but that it is more reasonable if the coil is tightly wound. The second assumption is equivalent to stating that the relative permeability of the core is infinitely higher than that of air (presuming that this is the medium surrounding the core); if this were the case, the flux would indeed be confined within the core. It is worthwhile to note that we make a similar assumption when we treat wires in electric circuits as perfect conductors: The conductivity of copper is substantially greater than that of free space, by a factor of approximately 10^{15}. In the case of magnetic materials, however, even for the best alloys, we have a relative permeability only on the order of 10^3 to 10^5. Thus, an approximation that is quite appropriate for electric circuits is not nearly as good in the case of magnetic circuits. The flux in a structure, such as those of Figures 12.12 and 12.13, not confined within the core is usually referred to as **leakage flux**. Finally, the assumption that the flux is uniform across the core cannot hold for a finite-permeability medium, but it is very helpful in giving an approximate *mean* behavior of the magnetic circuit.

The magnetic circuit analogy is therefore far from exact. However, short of employing the tools of electromagnetic field theory and of vector calculus, or advanced numerical simulation software, it is the most convenient tool at the engineer's disposal for the analysis of magnetic structures. In the remainder of this chapter, the approximate analysis based on the electric circuit analogy is used to obtain approximate solutions to problems involving a variety of useful magnetic circuits. Among these are the loudspeaker, solenoids, automotive fuel injectors, and sensors for the measurement of linear and angular velocity and position.

EXAMPLE 12.2 Analysis of Magnetic Structure and Equivalent Magnetic Circuit

Problem

Calculate the flux, flux density, and field intensity on the magnetic structure of Figure 12.14.

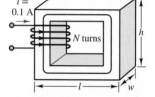

$l = 0.1$ m, $h = 0.1$ m, $w = 0.01$ m

Figure 12.14 Magnetic structure for Example 12.2.

Solution

Known Quantities: Relative permeability; number of coil turns; coil current; structure geometry.

Find: ϕ; B; H.

Schematics, Diagrams, Circuits, and Given Data: $\mu_r = 1,000$; $N = 500$ turns; $i = 0.1$ A. The cross-sectional area is $A = w^2 = (0.01)^2 = 0.0001$ m^2. The magnetic circuit geometry is defined in Figures 12.14 and 12.15.

Assumptions: All magnetic flux is linked by the coil; the flux is confined to the magnetic core; the flux density is uniform.

Analysis:

1. *Calculation of magnetomotive force.* From equation 12.28, we calculate the magnetomotive force:

$$\mathcal{F} = \text{mmf} = Ni = (500 \text{ turns})(0.1 \text{ A}) = 50 \text{ A-turns}$$

2. *Calculation of mean path.* Next, we estimate the mean path of the magnetic flux. On the basis of the assumptions, we can calculate a mean path that runs through the geometric center of the magnetic structure, as shown in Figure 12.15. The path length is

$$l_c = 4 \times 0.09 \text{ m} = 0.36 \text{ m}$$

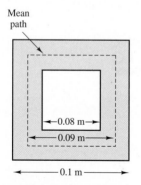

Figure 12.15 Cross section of magnetic structure for Example 12.2.

Mean path

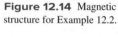

0.08 m

0.09 m

0.1 m

3. *Calculation of reluctance.* Knowing the magnetic path length and cross-sectional area, we can calculate the reluctance of the circuit:

$$\mathcal{R} = \frac{l_c}{\mu A} = \frac{l_c}{\mu_r \mu_0 A} = \frac{0.36}{1,000 \times 4\pi \times 10^{-7} \times 0.0001}$$

$$= 2.865 \times 10^6 \text{ A-turns/Wb}$$

The corresponding equivalent magnetic circuit is shown in Figure 12.16.

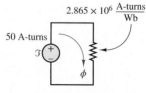

$2.865 \times 10^6 \dfrac{\text{A-turns}}{\text{Wb}}$

50 A-turns

$\mathcal{F}$

ϕ

Figure 12.16 Equivalent magnetic circuit for Example 12.2.

4. *Calculation of magnetic flux, flux density, and field intensity.* On the basis of the assumptions, we can now calculate the magnetic flux

$$\phi = \frac{\mathcal{F}}{\mathcal{R}} = \frac{50 \text{ A-turns}}{2.865 \times 10^6 \text{ A-turns/Wb}} = 1.75 \times 10^{-5} \text{ Wb}$$

the flux density

$$B = \frac{\phi}{A} = \frac{\phi}{w^2} = \frac{1.75 \times 10^{-5} \text{ Wb}}{0.0001 \text{ m}^2} = 0.175 \text{ Wb/m}^2$$

and the magnetic field intensity

$$H = \frac{B}{\mu} = \frac{B}{\mu_r \mu_0} = \frac{0.175 \text{ Wb/m}^2}{1{,}000 \times 4\pi \times 10^{-7} \text{ H/m}} = 139 \text{ A-turns/m}$$

Comments: This example illustrates all the basic calculations that pertain to magnetic structures. Remember that the assumptions stated in this example (and earlier in the chapter) simplify the problem and make its approximate numerical solution possible in a few simple steps. In reality, flux leakage, fringing, and uneven distribution of flux across the structure would require the solution of three-dimensional equations using finite-element methods. These methods are not discussed in this book, but are necessary for practical engineering designs.

The usefulness of these approximate methods is that you can, for example, quickly calculate the approximate magnitude of the current required to generate a given magnetic flux or flux density. These calculations can be used to determine electromagnetic energy and magnetic forces in practical structures.

The methodology described in this example is summarized in the following Focus on Problem Solving box.

CHECK YOUR UNDERSTANDING

Determine the equivalent reluctance of the structure of Figure 12.17 as seen by the "source" if μ_r for the structure is 1,000, $l = 5$ cm, and all the legs are 1 cm on a side.

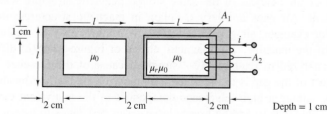

Figure 12.17 Magnetic structure with two loops.

FOCUS ON PROBLEM SOLVING

MAGNETIC STRUCTURES AND EQUIVALENT MAGNETIC CIRCUITS

Direct Problem

Given—The structure geometry and the coil parameters.

Calculate—The magnetic flux in the structure.

1. Compute the mmf.
2. Determine the length and cross section of the magnetic path for each continuous *leg* or section of the path.
3. Calculate the equivalent reluctance of the *leg*.
4. Generate the equivalent magnetic circuit diagram, and calculate the total equivalent reluctance.
5. Calculate the flux, flux density, and magnetic field intensity, as needed.

Inverse Problem

Given—The desired flux or flux density and structure geometry.

Calculate—The necessary coil current and number of turns.

1. Calculate the total equivalent reluctance of the structure from the desired flux.
2. Generate the equivalent magnetic circuit diagram.
3. Determine the mmf required to establish the required flux.
4. Choose the coil current and number of turns required to establish the desired mmf.

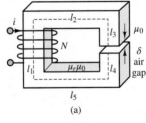

(a)

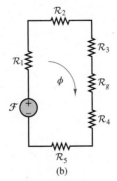

(b)

Figure 12.18 (a) Magnetic circuit with air gap; (b) its equivalent magnetic circuit

Consider the analysis of the same simple magnetic structure when an **air gap** is present. Air gaps are very common in magnetic structures; in rotating machines, for example, air gaps are necessary to allow for free rotation of the inner core of the machine. The magnetic circuit of Figure 12.18(a) differs from the circuit analyzed in Example 12.2 simply because of the presence of an air gap; the effect of the gap is to break the continuity of the high-permeability path for the flux, adding a high-reluctance component to the equivalent circuit. The situation is analogous to adding a very large series resistance to a series electric circuit. It should be evident from Figure 12.18(a) that the basic concept of reluctance still applies, although now two different permeabilities must be taken into account.

The equivalent circuit for the structure of Figure 12.18(a) may be drawn as shown in Figure 12.18(b), where $\mathcal{R}_n$ is the reluctance of path l_n, for $n = 1, 2, \ldots, 5$, and $\mathcal{R}_g$ is the reluctance of the air gap. The reluctances can be expressed as follows, if we assume that the magnetic structure has a uniform cross-sectional area A:

$$\mathcal{R}_1 = \frac{l_1}{\mu_r \mu_0 A} \qquad \mathcal{R}_2 = \frac{l_2}{\mu_r \mu_0 A} \qquad \mathcal{R}_3 = \frac{l_3}{\mu_r \mu_0 A}$$

$$\mathcal{R}_4 = \frac{l_4}{\mu_r \mu_0 A} \qquad \mathcal{R}_5 = \frac{l_5}{\mu_r \mu_0 A} \qquad \mathcal{R}_g = \frac{\delta}{\mu_0 A_g} \qquad (12.31)$$

Note that in computing $\mathcal{R}_g$, the length of the gap is given by δ and the permeability is given by μ_0, as expected, but A_g is different from the cross-sectional area A of the structure. This is so because the flux lines exhibit a phenomenon known as **fringing** as they cross an air gap. The flux lines actually *bow out* of the gap defined by the cross section A, not being contained by the high-permeability material any longer. Thus, it is customary to define an area A_g that is greater than A, to account for this phenomenon. Example 12.3 describes in greater detail the procedure for finding A_g and also discusses the phenomenon of fringing.

EXAMPLE 12.3   Magnetic Structure With Air Gaps

Problem

Compute the equivalent reluctance of the magnetic circuit of Figure 12.19 and the flux density established in the bottom bar of the structure.

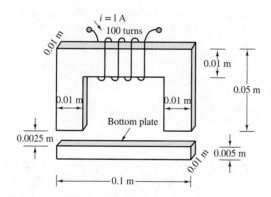

Figure 12.19 Electromagnetic structure with air gaps

Solution

Known Quantities: Relative permeability; number of coil turns; coil current; structure geometry.

Find: $\mathcal{R}_{eq}$; B_{bar}.

Schematics, Diagrams, Circuits, and Given Data: $\mu_r = 10{,}000$; $N = 100$ turns; $i = 1$ A.

Assumptions: All magnetic flux is linked by the coil; the flux is confined to the magnetic core; the flux density is uniform.

Analysis:

1. *Calculation of magnetomotive force.* From equation 12.28, we calculate the magnetomotive force:

$$\mathcal{F} = \text{mmf} = Ni = (100 \text{ turns})(1 \text{ A}) = 100 \text{ A-turns}$$

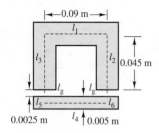

2. *Calculation of mean path.* Figure 12.20 depicts the geometry. The path length is

$$l_c = l_1 + l_2 + l_3 + l_4 + l_5 + l_6 + l_g + l_g$$

Figure 12.20

However, the path must be broken into three legs: the upside-down U-shaped element, the air gaps, and the bar. We cannot treat these three parts as one because the relative permeability of the magnetic material is very different from that of the air gap. Thus, we define the following three paths, neglecting the very small (half bar thickness) lengths l_5 and l_6:

$$l_U = l_1 + l_2 + l_3 \qquad l_{\text{bar}} = l_4 + l_5 + l_6 \approx l_4 \qquad l_{\text{gap}} = l_g + l_g$$

where

$$l_U = 0.18 \text{ m} \qquad l_{\text{bar}} = 0.09 \text{ m} \qquad l_{\text{gap}} = 0.005 \text{ m}$$

Next, we compute the cross-sectional area. For the magnetic structure, we calculate the U-shaped element cross section to be $A_U = w^2 = (0.01)^2 = 0.0001 \text{ m}^2$ and the cross section of the bar to be $A_{\text{bar}} = (0.01 \times 0.005) = 0.0005 \text{ m}^2$. For the air gap, we will make an empirical adjustment to account for the phenomenon of *fringing*, that is, to account for the tendency of the magnetic flux lines to bow out of the magnetic path, as illustrated in Figure 12.21. A rule of thumb used to account for fringing is to add the length of the gap to each dimension of the actual cross-sectional area. Thus

$$A_{\text{gap}} = (0.01 \text{ m} + l_g)^2 = (0.0125)^2 = 0.15625 \times 10^{-3} \text{ m}^2$$

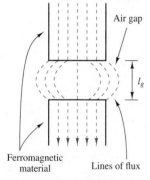

Figure 12.21 Fringing effects in air gap

3. *Calculation of reluctance.* Knowing the magnetic path length and cross-sectional area, we can calculate the reluctance of each leg of the circuit:

$$\mathcal{R}_U = \frac{l_U}{\mu_U A_U} = \frac{l_U}{\mu_r \mu_0 A_U} = \frac{0.18}{10{,}000 \times 4\pi \times 10^{-7} \times 0.0001}$$
$$= 1.43 \times 10^5 \text{ A-turns/Wb}$$

$$\mathcal{R}_{\text{bar}} = \frac{l_{\text{bar}}}{\mu_{\text{bar}} A_{\text{bar}}} = \frac{l_{\text{bar}}}{\mu_r \mu_0 A_{\text{bar}}} = \frac{0.09}{10{,}000 \times 4\pi \times 10^{-7} \times 0.0005}$$
$$= 143.2 \times 10^3 \text{ A-turns/Wb}$$

$$\mathcal{R}_{\text{gap}} = \frac{l_{\text{gap}}}{\mu_{\text{gap}} A_{\text{gap}}} = \frac{l_{\text{gap}}}{\mu_0 A_{\text{gap}}} = \frac{0.005}{4\pi \times 10^{-7} \times 0.156 \times 10^{-3}} = 25.5 \times 10^6 \text{ A-turns/Wb}$$

Note that the reluctance of the air gap is dominant with respect to that of the magnetic structure, in spite of the small dimension of the gap. This is so because the relative permeability of the air gap is much smaller than that of the magnetic material.

The equivalent reluctance of the structure is

$$\mathcal{R}_{\text{eq}} = \mathcal{R}_U + \mathcal{R}_{\text{bar}} + \mathcal{R}_{\text{gap}} = 1.43 \times 10^5 + 143.2 \times 10^3 + 2.55 \times 10^7$$
$$= 25.8 \times 10^6 \text{ A-turns/Wb}$$

Thus,

$$\mathcal{R}_{\text{eq}} \approx \mathcal{R}_{\text{gap}}$$

Since the gap reluctance is two orders of magnitude greater than the reluctance of the magnetic structure, it is reasonable to neglect the magnetic structure reluctance and work only with the gap reluctance in calculating the magnetic flux.

4. *Calculation of magnetic flux and flux density in the bar.* From the result of the preceding subsection, we calculate the flux

$$\phi = \frac{\mathcal{F}}{\mathcal{R}_{eq}} \approx \frac{\mathcal{F}}{\mathcal{R}_{gap}} = \frac{100 \text{ A-turns}}{2.55 \times 10^7 \text{ A-turns/Wb}} = 3.9 \times 10^{-6} \text{ Wb}$$

and the flux density in the bar

$$B_{bar} = \frac{\phi}{A} = \frac{3.92 \times 10^{-6} \text{ Wb}}{0.00005 \text{ m}^2} = 78.5 \times 10^{-3} \text{ Wb/m}^2$$

Comments: It is very common to neglect the reluctance of the magnetic material sections in these approximate calculations. We shall make this assumption very frequently in the remainder of the chapter.

CHECK YOUR UNDERSTANDING

Find the equivalent reluctance of the magnetic circuit shown in Figure 12.22 if μ_r of the structure is infinite, $\delta = 2$ mm, and the physical cross section of the core is 1 cm². Do not neglect fringing.

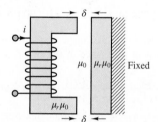

Figure 12.22 Magnetic structure.

Answer: $\mathcal{R}_{eq} = 22 \times 10^6$ A-turns/Wb

EXAMPLE 12.4   Magnetic Structure of Electric Motor

Problem

Figure 12.23 depicts the configuration of an electric motor. The electric motor consists of a *stator* and a *rotor*. Compute the air gap flux and flux density. Neglect fringing.

Solution

Known Quantities: Relative permeability; number of coil turns; coil current; structure geometry.

Find: ϕ_{gap}; B_{gap}.

Schematics, Diagrams, Circuits, and Given Data: $\mu_r \to \infty$; $N = 1{,}000$ turns; $i = 10$ A; $l_{gap} = 0.01$ m; $A_{gap} = 0.1$ m². The magnetic circuit geometry is defined in Figure 12.23.

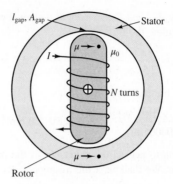

Figure 12.23 Cross-sectional view of synchronous motor

Assumptions: All magnetic flux is linked by the coil; the flux is confined to the magnetic core; the flux density is uniform. The reluctance of the magnetic structure is negligible.

Analysis:

1. *Calculation of magnetomotive force.* From equation 12.28, we calculate the magneto-motive force:

$$\mathcal{F} = \text{mmf} = Ni = (1{,}000 \text{ turns})(10 \text{ A}) = 10{,}000 \text{ A-turns}$$

2. *Calculation of reluctance.* Knowing the magnetic path length and cross-sectional area, we can calculate the equivalent reluctance of the two gaps:

$$\mathcal{R}_{\text{gap}} = \frac{l_{\text{gap}}}{\mu_{\text{gap}} A_{\text{gap}}} = \frac{l_{\text{gap}}}{\mu_0 A_{\text{gap}}} = \frac{0.01}{4\pi \times 10^{-7} \times 0.1} = 7.96 \times 10^4 \text{ A-turns/Wb}$$

$$\mathcal{R}_{\text{eq}} = 2\mathcal{R}_{\text{gap}} = 1.59 \times 10^5 \text{ A-turns/Wb}$$

3. *Calculation of magnetic flux and flux density.* From the results of steps 1 and 2, we calculate the flux

$$\phi = \frac{\mathcal{F}}{\mathcal{R}_{\text{eq}}} = \frac{10{,}000 \text{ A-turns}}{1.59 \times 10^5 \text{ A-turns/Wb}} = 0.0628 \text{ Wb}$$

and the flux density

$$B_{\text{bar}} = \frac{\phi}{A} = \frac{0.0628 \text{ Wb}}{0.1 \text{ m}^2} = 0.628 \text{ Wb/m}^2$$

Comments: Note that the flux and flux density in this structure are significantly larger than those in Example 12.3 because of the larger mmf and larger gap area of this magnetic structure. The subject of electric motors is formally approached in Chapter 13.

EXAMPLE 12.5   Equivalent Circuit of Magnetic Structure With Multiple Air Gaps

Problem

Figure 12.24 depicts the configuration of a magnetic structure with two air gaps. Determine the equivalent circuit of the structure.

Solution

Known Quantities: Structure geometry.

Find: Equivalent-circuit diagram.

Assumptions: All magnetic flux is linked by the coil; the flux is confined to the magnetic core; the flux density is uniform. The reluctance of the magnetic structure is negligible.

Analysis:

1. *Calculation of magnetomotive force.*

$$\mathcal{F} = \text{mmf} = Ni$$

2. *Calculation of reluctance.* Knowing the magnetic path length and cross-sectional area, we can calculate the equivalent reluctance of the two gaps:

$$\mathcal{R}_{\text{gap}-1} = \frac{l_{\text{gap}-1}}{\mu_{\text{gap}-1}A_{\text{gap}-1}} = \frac{l_{\text{gap}-1}}{\mu_0 A_{\text{gap}-1}}$$

$$\mathcal{R}_{\text{gap}-2} = \frac{l_{\text{gap}-2}}{\mu_{\text{gap}-2}A_{\text{gap}-2}} = \frac{l_{\text{gap}-2}}{\mu_0 A_{\text{gap}-2}}$$

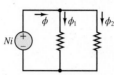

Figure 12.24 Magnetic structure with two air gaps

3. *Calculation of magnetic flux and flux density.* Note that the flux must now divide between the two legs, and that a different air-gap flux will exist in each leg. Thus

$$\phi_1 = \frac{Ni}{\mathcal{R}_{\text{gap}-1}} = \frac{Ni\mu_0 A_{\text{gap}-1}}{l_{\text{gap}-1}}$$

$$\phi_2 = \frac{Ni}{\mathcal{R}_{\text{gap}-2}} = \frac{Ni\mu_0 A_{\text{gap}-2}}{l_{\text{gap}-2}}$$

and the total flux generated by the coil is $\phi = \phi_1 + \phi_2$.

The equivalent circuit is shown in the bottom half of Figure 12.24.

Comments: Note that the two legs of the structure act as resistors in a parallel circuit.

CHECK YOUR UNDERSTANDING

Find the equivalent magnetic circuit of the structure of Figure 12.25 if μ_r is infinite. Give expressions for each of the circuit values if the physical cross-sectional area of each of the legs is given by

$$A = l \times w$$

Do not neglect fringing.

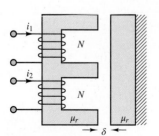

Figure 12.25 Magnetic structure.

Answer: $\mathcal{R}_g = \mathcal{R}_1 = \mathcal{R}_2 = \mathcal{R}_3 = \delta/\mu_0(l + \delta)(w + \delta);\ \mathcal{F}_1 = Ni_1;\ \mathcal{F}_2 = Ni_2$

EXAMPLE 12.6   Inductance, Stored Energy, and Induced Voltage

Problem

1. Find the inductance and the magnetic energy stored in the structure of Figure 12.18(a). The structure is identical to that of Example 12.2 except for the air gap. Ignore fringing.

2. Assume that the flux density in the air gap varies sinusoidally as $B(t) = B_0 \sin(\omega t)$. Determine the induced voltage across the coil e.

Solution

Known Quantities: Relative permeability; number of coil turns; coil current; structure geometry; flux density in air gap.

Find: L; W_m; e.

Schematics, Diagrams, Circuits, and Given Data: $\mu_r \to \infty$; $N = 500$ turns; $i = 0.1$ A. The magnetic circuit geometry is defined in Figures 12.14 and 12.15. The air gap has $l_g = 0.002$ m. $B_0 = 0.6$ Wb/m^2.

Assumptions: All magnetic flux is linked by the coil; the flux is confined to the magnetic core; the flux density is uniform. The reluctance of the magnetic structure is negligible.

Analysis:

1. Use equation 12.30 to calculate the inductance of the magnetic structure.

$$L = \frac{N^2}{\mathcal{R}}$$

To calculate the reluctance, assume that the reluctance of the structure is negligible.

$$\mathcal{R}_{\text{gap}} = \frac{l_{\text{gap}}}{\mu_{\text{gap}} A_{\text{gap}}} = \frac{l_{\text{gap}}}{\mu_0 A_{\text{gap}}} = \frac{0.002}{4\pi \times 10^{-7} \times 0.0001} = 1.59 \times 10^7 \text{ A-turns/Wb}$$

and

$$L = \frac{N^2}{\mathcal{R}} = \frac{500^2}{1.59 \times 10^7} = 0.157 \text{ H}$$

Finally, calculate the stored magnetic energy as follows:

$$W_m = \frac{1}{2} L i^2 = \frac{1}{2} \times (0.157 \text{ H}) \times (0.1 \text{ A})^2 = 0.785 \times 10^{-3} \text{ J}$$

2. To calculate the induced voltage due to a time-varying magnetic flux at the frequency of 60 Hz (377 rad/s), we use equation 12.16:

$$e = \frac{d\lambda}{dt} = N \frac{d\phi}{dt} = NA \frac{dB}{dt} = NAB_0 \omega \cos(\omega t)$$

$$= 500 \times 0.0001 \times 0.6 \times 377 \cos(377t) = 11.31 \cos(377t) \quad \text{V}$$

Comments: The voltage induced across a coil in an electromagnetic transducer is a very important quantity called the *back electromotive force*, or back emf.

Magnetic Reluctance Position Sensor

A simple magnetic structure, very similar to those examined in the previous examples, finds very common application in the variable-reluctance position sensor, which, in turn, finds widespread application in a variety of configurations for the measurement of linear and angular velocity. Figure 12.26 depicts one particular configuration that is used in many applications. In this structure, a permanent magnet with a coil of wire wound around it forms the sensor; a steel disk (typically connected to a rotating shaft) has a number of tabs that pass between the pole pieces of the sensor. The area of the tab is assumed equal to the area of the cross section of the pole pieces and is equal to a^2. The reason for the name *variable-reluctance sensor* is that the reluctance of the magnetic structure is variable, depending on whether a ferromagnetic tab lies between the pole pieces of the magnet.

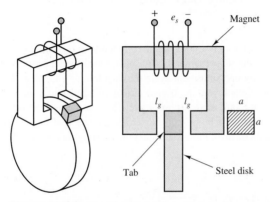

Figure 12.26 Variable-reluctance position sensor

The principle of operation of the sensor is that an electromotive force e_S is induced across the coil by the change in magnetic flux caused by the passage of the tab between the pole pieces when the disk is in motion. As the tab enters the volume between the pole pieces, the flux will increase, because of the lower reluctance of the configuration, until it reaches a maximum when the tab is centered between the poles of the magnet. Figure 12.27 depicts the approximate shape of the resulting voltage, which, according to Faraday's law, is given by

$$e_S = -\frac{d\phi}{dt}$$

The rate of change of flux is dictated by the geometry of the tab and of the pole pieces and by the speed of rotation of the disk. It is important to note that, since the flux is changing only if the disk is rotating, this sensor cannot detect the static position of the disk.

One common application of this concept is in the measurement of the speed of rotation of rotating machines, including electric motors and internal combustion engines. In these applications, use is made of a *60-tooth wheel*, which permits the conversion of the speed rotation directly to units of revolutions per minute. The output of a variable-reluctance position sensor magnetically coupled to a rotating disk equipped with 60 tabs

(Continued)

(Concluded)

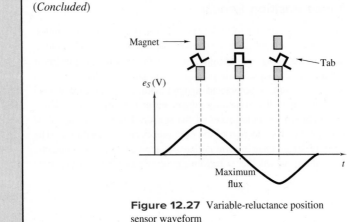

Figure 12.27 Variable-reluctance position sensor waveform

(teeth) is processed through a comparator or Schmitt trigger circuit. The voltage waveform generated by the sensor is nearly sinusoidal when the teeth are closely spaced, and it is characterized by one sinusoidal cycle for each tooth on the disk. If a negative zero-crossing detector is employed, the trigger circuit will generate a pulse corresponding to the passage of each tooth, as shown in Figure 12.28. If the time between any two pulses is measured by means of a high-frequency clock, the speed of the engine can be directly determined in units of revolutions per minute (r/min) by means of a digital counter (see Chapter 11).

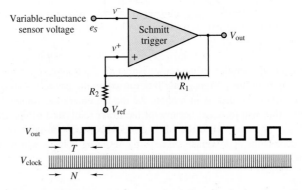

Figure 12.28 Signal processing for a 60-tooth wheel rpm sensor

---

Voltage Calculation in Magnetic Reluctance Position Sensor

Problem:

This example illustrates the calculation of the voltage induced in a magnetic reluctance sensor by a rotating toothed wheel. In particular, we will find an approximate expression for the reluctance and the induced voltage for the position sensor shown in Figure 12.29,

(Continued)

(*Continued*)

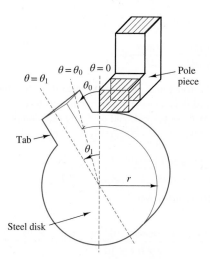

Figure 12.29 Reluctance sensor for
measurement of angular position

and we will show that the induced voltage is speed dependent. It will be assumed that
the reluctance of the core and fringing at the air gaps are both negligible.

Solution:

From the geometry shown in the preceding Focus on Measurements box, the equivalent
reluctance of the magnetic structure is twice that of one gap, since the permeability of
the tab and the magnetic structure are assumed infinite (i.e., they have negligible
reluctance). When the tab and the poles are aligned, the angle θ is zero, as shown in
Figure 12.29, and the area of the air gap is maximum. For angles greater than $2\theta_0$, the
magnetic length of the air gaps is so large that the magnetic field may reasonably be
taken as zero.

To model the reluctance of the gaps, we assume the following simplified expression, where the area of overlap of the tab with the magnetic poles is assumed proportional to the angular displacement:

$$\mathcal{R} = \frac{2l_g}{\mu_0 A} = \frac{2l_g}{\mu_0 ar(\theta_1 - \theta)} \qquad \text{for } 0 < \theta < \theta_1$$

Naturally, this is an approximation; however, the approximation captures the essential
idea of this transducer, namely, that the reluctance will decrease with increasing
overlap area until it reaches a minimum, and then the reluctance will increase as the
overlap area decreases. For $\theta = \theta_1$, that is, with the tab outside the magnetic pole
pieces, we have $\mathcal{R}_{\max} \to \infty$. For $\theta = 0$, that is, with the tab perfectly aligned with
the pole pieces, we have $\mathcal{R}_{\min} = 2l_g/\mu_0 ar\theta_1$. The flux ϕ may therefore be computed
as follows:

$$\phi = \frac{Ni}{\mathcal{R}} = \frac{Ni\mu_0 ar(\theta_1 - \theta)}{2l_g}$$

(*Continued*)

(*Concluded*)

The induced voltage e_S is found by

$$e_S = -\frac{d\phi}{dt} = -\frac{d\phi}{d\theta}\frac{d\theta}{dt} = -\frac{Ni\mu_0 ar}{2l_g}\omega$$

where $\omega = d\theta/dt$ is the rotational speed of the steel disk. It should be evident that the induced voltage is speed dependent. For $a = 1$ cm, $r = 10$ cm, $l_g = 0.1$ cm, $N = 1,000$ turns, $i = 10$ mA, $\theta_1 = 6° \approx 0.1$ rad, and $\omega = 400$ rad/s (approximately 3,800 r/min), we have

$$\mathcal{R}_{max} = \frac{2 \times 0.1 \times 10^{-2}}{4\pi \times 10^{-7} \times 1 \times 10^{-2} \times 10 \times 10^{-2} \times 0.1}$$

$$= 1.59 \times 10^7 \text{ A-turns/Wb}$$

$$e_{S\,peak} = \frac{1,000 \times 10 \times 10^{-3} \times 4\pi \times 10^{-7} \times 1 \times 10^{-2} \times 10^{-1}}{2 \times 0.1 \times 10^{-2}} \times 400$$

$$= 2.5 \text{ mV}$$

That is, the peak amplitude of e_S will be 2.5 mV.

12.3   MAGNETIC MATERIALS AND *B-H* CURVES

In the analysis of magnetic circuits presented in the previous sections, the relative permeability μ_r was treated as a constant. In fact, the relationship between the magnetic flux density **B** and the associated field intensity **H** is:

$$\mathbf{B} = \mu\mathbf{H} \tag{12.32}$$

and is characterized by the fact that the relative permeability of magnetic materials is not a constant but is a function of the magnetic field intensity. In effect, all magnetic materials exhibit a phenomenon called **saturation**, whereby the flux density increases in proportion to the field intensity until it cannot do so any longer. Figure 12.30 illustrates the general behavior of all magnetic materials. You will note that since the *B-H* curve shown in the figure is nonlinear, the value of μ (which is the slope of the curve) depends on the intensity of the magnetic field.

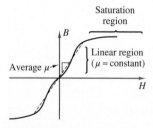

Figure 12.30 Permeability and magnetic saturation effects

To understand the reasons for the saturation of a magnetic material, we need to briefly review the mechanism of magnetization. The basic idea behind magnetic materials is that the spin of electrons constitutes motion of charge, and therefore leads to magnetic effects, as explained in the introductory section of this chapter. In most materials, the electron spins cancel out, on the whole, and no net effect remains. In ferromagnetic materials, on the other hand, atoms can align so that the electron spins cause a net magnetic effect. In such materials, there exist small regions with strong magnetic properties, called **magnetic domains**, the effects of which are neutralized in unmagnetized material by other, similar regions that are oriented differently, in a random pattern. When the material is magnetized, the magnetic domains tend to align with one another, to a degree that is determined by the intensity of the applied magnetic field.

In effect, a large number of miniature magnets within the material are aligned (*polarized*) by the applied magnetic field. As the field increases, more and more domains become aligned. When all the domains have become aligned, any further increase in magnetic field intensity does not yield an increase in flux density beyond the increase that would be caused in a nonmagnetic material. Thus, the relative permeability μ_r approaches 1 in the saturation region. It should be apparent that an exact value of μ_r cannot be determined; the value of μ_r used in the earlier examples is to be interpreted as an average permeability, for intermediate values of flux density. For example, commercial magnetic steels saturate at flux densities of a few teslas.

There are two more features that cause magnetic materials to further deviate from the ideal model of the linear *B-H* relationship: **eddy currents** and **hysteresis**. The first phenomenon consists of currents that are caused by any time-varying flux in the core material. As you know, a time-varying flux will induce a voltage, and therefore a current. When this happens inside the magnetic core, the induced voltage will cause *eddy* currents (the terminology should be self-explanatory) in the core, which depend on the resistivity of the core. Figure 12.31 illustrates the phenomenon of eddy currents. The effect of these currents is to dissipate energy in the form of heat. Eddy currents are reduced by selecting high-resistivity core materials, or by *laminating* the core, introducing tiny, discontinuous air gaps between core layers (see Figure 12.31). Lamination of the core reduces eddy currents greatly without affecting the magnetic properties of the core.

Hysteresis is another loss mechanism in magnetic materials; it displays a rather complex behavior, related to the magnetization properties of a material. The curve of Figure 12.32 reveals that the *B-H* curve for a magnetic material during magnetization (as *H* is increased) is displaced with respect to the curve that is measured when the material is demagnetized. To understand the hysteresis process, consider a core that has been energized for some time, with a field intensity of H_1 A-turns/m. As the current required to sustain the mmf corresponding to H_1 is decreased, we follow the hysteresis curve from the point α to the point β. When the mmf is exactly zero, the material displays the **remanent** (or **residual**) **magnetization** B_r. To bring the flux density to zero, we must further decrease the mmf (i.e., produce a negative current) until the field intensity reaches the value $-H_0$ (point γ on the curve). As the mmf is made more negative, the curve eventually reaches the point α'. If the excitation current to the coil is now increased, the magnetization curve will follow the path $\alpha' = \beta' = \gamma' = \alpha$, eventually returning to the original point in the *B-H* plane, but via a different path.

The result of this process, by which an *excess mmf* is required to magnetize or demagnetize the material, is a net energy loss. It is difficult to evaluate this loss; however, it can be shown that it is related to the area between the curves of Figure 12.32. Experimental techniques exist that measure these losses.

Figure 12.33 depicts magnetization curves for three very common ferromagnetic materials: cast iron, cast steel, and sheet steel.

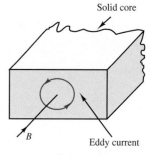

Solid core

B

Eddy current

Laminated core
(the laminations are separated
by a thin layer of insulation)

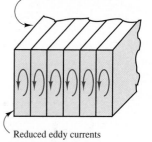

Reduced eddy currents

Figure 12.31 Eddy currents in magnetic structures

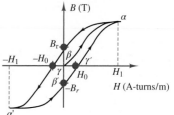

Figure 12.32 Hysteresis in magnetization curves

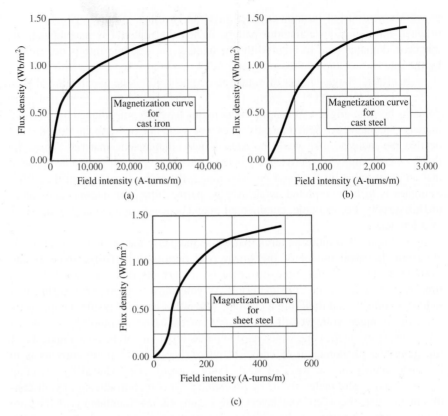

Figure 12.33 Magnetization curves for (a) cast iron, (b) cast steel, and (c) sheet steel

12.4   TRANSFORMERS

One of the more common magnetic structures in everyday applications is the transformer. The ideal transformer was introduced in Chapter 6 as a device that can step an AC voltage up or down by a fixed ratio, with a corresponding decrease or increase in current. The structure of a simple magnetic transformer is shown in Figure 12.34, which illustrates that a transformer is very similar to the magnetic circuits described earlier in this chapter. Coil L_1 represents the input side of the transformer, while coil L_2 is the output coil; both coils are wound around the same magnetic structure, which we show here to be similar to the "square doughnut" of the earlier examples.

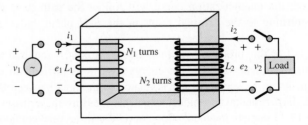

Figure 12.34 Structure of a transformer

The ideal transformer is defined by the same set of assumptions made earlier. The flux is confined to the core, the flux links all turns of both coils, and the permeability of the core is infinite. The last assumption is equivalent to stating that an arbitrarily small mmf is sufficient to establish a flux in the core. In addition, we assume that the ideal transformer coils offer negligible resistance to current.

A time-varying voltage applied to the primary side of the transformer results in a corresponding time-varying current in L_1. This current acts as an mmf and causes a time-varying flux in the structure. This flux will induce an emf across the secondary coil! Without the need for a direct electrical connection, the transformer can couple a source voltage across to the primary winding to the secondary winding, which is connected to a load; the coupling occurs by means of the magnetic field acting on both coils. Thus, a transformer operates by converting electric energy to magnetic, and then back to electric. The following derivation illustrates this viewpoint in the ideal case (no loss of energy) and compares the result with the definition of the ideal transformer in Chapter 6.

If a time-varying voltage source is connected to the input side, then by virtue of Faraday's law, a corresponding time-varying flux $d\phi/dt$ is established in coil L_1:

$$e_1 = N_1 \frac{d\phi}{dt} = v_1 \tag{12.33}$$

But since the flux thus produced also links coil L_2, an emf is induced across the output coil as well:

$$e_2 = N_2 \frac{d\phi}{dt} = v_2 \tag{12.34}$$

This induced emf can be measured as the voltage v_2 at the output terminals, and one can readily see that the ratio of the open-circuit output voltage to input-terminal voltage is

$$\frac{v_2}{v_1} = \frac{N_2}{N_1} = N \tag{12.35}$$

A load current i_2, and its corresponding mmf $\mathcal{F}_2 = N_2 i_2$, is produced when a load is connected to the output terminals in Figure 12.34. The mmf would cause the flux in the core to change; however, this is not possible since a change in ϕ would cause a corresponding change in the voltage induced across the input coil. But this voltage is determined (fixed) by the source v_1 so that the input coil is forced to generate a **counter-mmf** to oppose the mmf of the output coil drawing a current i_1 from the source v_1 such that:

$$i_1 N_1 = i_2 N_2 \tag{12.36}$$

or

$$\frac{i_2}{i_1} = \frac{N_1}{N_2} = \alpha = \frac{1}{N} \tag{12.37}$$

where α is the ratio of primary to secondary turns (the transformer ratio) and N_1 and N_2 are the primary and secondary turns, respectively. If there were any net difference between the input and output mmf, the flux balance required by the input voltage source would not be satisfied. Thus, the two magnetomotive forces must be equal.

As you can easily verify, these results are the same as in Chapter 6; in particular, the ideal transformer does not dissipate any power, since

$$v_1 i_1 = v_2 i_2 \tag{12.38}$$

Note the distinction we have made between the induced voltages (emf's) e and the terminal voltages v. In general, these are not the same.

The results obtained for the ideal case do not completely represent the physical nature of transformers. A number of loss mechanisms need to be included in a practical transformer model, to account for the effects of leakage flux, for various magnetic core losses (e.g., hysteresis), and for the unavoidable resistance of the wires that form the coils.

Commercial transformer ratings are usually given on the **nameplate**, which indicates the following normal operating conditions:

- Primary-to-secondary voltage ratio
- Design frequency of operation
- (Apparent) rated output power

For example, a typical nameplate might read 480:240 V, 60 Hz, 2 kVA. The voltage ratio can be used to determine the turns ratio, while the rated output power represents the continuous power level that can be sustained without overheating. It is important that this power be rated as the apparent power in kilovoltamperes, rather than real power in kilowatts, since a load with low power factor would still draw current and therefore operate near rated power. Another important performance characteristic of a practical transformer is its **power efficiency**, defined by:

$$\text{Power efficiency } \eta = \frac{\text{Output power}}{\text{Input power}} \tag{12.39}$$

EXAMPLE 12.7   Transformer Nameplate

Problem

Determine the turns ratio and the rated currents of a transformer from nameplate data.

Solution

Known Quantities: Nameplate data.

Find: $\alpha = N_1/N_2$; I_1; I_2.

Schematics, Diagrams, Circuits, and Given Data: Nameplate data: 120 V/480 V; 48 kVA; 60 Hz.

Assumptions: Assume an ideal transformer.

Analysis: The first element in the nameplate data is a pair of voltages, indicating the primary and secondary voltages for which the transformer is rated. The ratio α is found as follows:

$$\alpha = \frac{N_1}{N_2} = \frac{480}{120} = 4$$

To find the primary and secondary currents, we use the kilovoltampere rating (apparent power) of the transformer:

$$I_1 = \frac{|S|}{V_1} = \frac{48 \text{ kVA}}{480 \text{ V}} = 100 \text{ A} \qquad I_2 = \frac{|S|}{V_2} = \frac{48 \text{ kVA}}{120 \text{ V}} = 400 \text{ A}$$

Comments: In computing the rated currents, we have assumed that no losses take place in the transformer; in fact, there will be losses due to coil resistance and magnetic core effects. These losses result in heating of the transformer and limit its rated performance.

CHECK YOUR UNDERSTANDING

The high-voltage side of a transformer has 500 turns, and the low-voltage side has 100 turns. When the transformer is connected as a step-down transformer, the load current is 12 A. Calculate: (a) the turns ratio α; and (b) the primary current. Then, (c) Calculate the turns ratio if the transformer is used as a step-up transformer.

The output of a transformer under certain conditions is 12 kW. The copper losses are 189 W, and the core losses are 52 W. Calculate the efficiency of this transformer.

EXAMPLE 12.8   Impedance Transformer

Problem

Find the equivalent load impedance seen by the voltage source (i.e., reflected from secondary to primary) for the transformer of Figure 12.35.

Solution

Known Quantities: Transformer turns ratio α.

Find: Reflected impedance Z_2'.

Assumptions: Assume an ideal transformer.

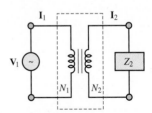

Figure 12.35 Ideal transformer

Analysis: By definition, the load impedance is equal to the ratio of secondary phasor voltage and current:

$$Z_2 = \frac{V_2}{I_2}$$

To find the reflected impedance, we can express the above ratio in terms of the primary voltage and current:

$$Z_2 = \frac{V_2}{I_2} = \frac{V_1/\alpha}{\alpha I_1} = \frac{1}{\alpha^2}\frac{V_1}{I_1}$$

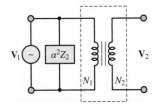

Figure 12.36 Equivalent reflected circuit for impedance transformer.

where the ratio $\mathbf{V}_1/\mathbf{I}_1$ is the impedance seen by the source at the primary coil, that is, the *reflected load impedance* seen by the primary (source) side of the circuit. Thus, we can write the load impedance Z_2 in terms of the primary circuit voltage and current; we call this the *reflected impedance* Z'_2:

$$Z_2 = \frac{1}{\alpha^2}\frac{\mathbf{V}_1}{\mathbf{I}_1} = \frac{1}{\alpha^2}Z_1 = \frac{1}{\alpha^2}Z'_2$$

Thus, $Z'_2 = \alpha^2 Z_2$. Figure 12.36 depicts the equivalent circuit with the load impedance reflected back to the primary.

Comments: The equivalent reflected circuit calculations are convenient because all circuit elements can be referred to a single set of variables (i.e., only primary or secondary voltages and currents).

CHECK YOUR UNDERSTANDING

The output impedance of a servo amplifier is 250 Ω. The servomotor that the amplifier must drive has an impedance of 2.5 Ω. Calculate the turns ratio of the transformer required to match these impedances.

Answer: $\alpha = 10$

12.5 ELECTROMECHANICAL ENERGY CONVERSION

From the material developed thus far, it should be apparent that electromagneto-mechanical devices are capable of converting mechanical forces and displacements to electromagnetic energy, and that the converse is also possible. The objective of this section is to formalize the basic principles of energy conversion in electromagnetomechanical systems, and to illustrate its usefulness and potential for application by presenting several examples of **energy transducers**. A transducer is a device that can convert electric to mechanical energy (in this case, it is often called an **actuator**), or vice versa (in which case it is called a **sensor**).

Several physical mechanisms permit conversion of electric to mechanical energy and back, including the **piezoelectric effect**,[3] consisting of the generation of a change in electric field in the presence of strain in certain crystals (e.g., quartz), and **electrostriction** and **magnetostriction**, in which changes in the dimension of certain materials lead to a change in their electrical (or magnetic) properties. This chapter is concerned only with transducers in which electric energy is converted to mechanical energy through the coupling of a magnetic field. It is important to note that all rotating machines (motors and generators) fit the basic definition of electromechanical transducers we have just given.

---

[3]See the Focus on Measurements box, "Charge Amplifiers," in Chapter 7.

Forces in Magnetic Structures

Mechanical forces can be converted to electric signals, and vice versa, by means of the coupling provided by energy stored in the magnetic field. In this subsection, we discuss the computation of mechanical forces and of the corresponding electromagnetic quantities of interest; these calculations are of great practical importance in the design and application of electromechanical actuators. For example, a problem of interest is the computation of the current required to generate a given force in an electromechanical structure. This is the kind of application that is likely to be encountered by the engineer in the selection of an electromechanical device for a given task.

As already seen in this chapter, an electromechanical system includes an electrical system, interacting through a magnetic field. Figure 12.37 illustrates the coupling between the electrical and mechanical systems. In the mechanical system, energy loss can occur because of the heat developed as a consequence of *friction*, while in the electrical system, analogous losses are incurred because of *resistance*. Loss mechanisms are also present in the magnetic coupling medium, since *eddy current losses* and *hysteresis losses* are unavoidable in ferromagnetic materials. Either system can supply energy, and either system can store energy. Thus, the figure depicts the flow of energy from the electrical to the mechanical system, accounting for these various losses. The same flow could be reversed if mechanical energy were converted to electrical form.

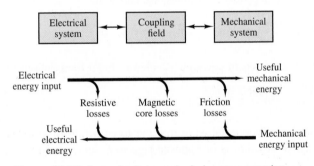

Figure 12.37 Losses in electromechanical energy conversion.

Moving-Iron Transducers

One important class of electromagnetomechanical transducers is that of **moving-iron transducers**, which include common devices such as electromagnets, solenoids, and relays. The simplest example of a moving-iron transducer is the of Figure 12.38, in which the U-shaped element is fixed and the bar is movable. In the following paragraphs, we shall derive a relationship between the current applied to the coil, the displacement of the movable bar, and the magnetic force acting in the air gap.

The principle that will be applied throughout the section is that for a mass to be displaced, some work needs to be done; this work corresponds to a change in the energy stored in the electromagnetic field, which causes the mass to be displaced. With reference to Figure 12.38, let f_e represent the magnetic force acting on the bar and x the displacement of the bar, in the direction shown. Then the net work W_m into the electromagnetic field is equal to the sum of the work done by

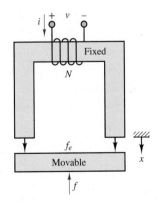

Figure 12.38 Basic electromagnet.

the electric circuit plus the work done by the mechanical system. Therefore, for an incremental amount of work, we can write

$$dW_m = ei\,dt - f_e\,dx \tag{12.40}$$

where e is the electromotive force across the coil and the minus sign is due to the sign convention indicated in Figure 12.38. Recalling that the emf e is equal to the derivative of the flux linkage (equation 12.16), we can further expand equation 12.40 to obtain

$$dW_m = ei\,dt - f_e\,dx = i\frac{d\lambda}{dt}dt - f_e\,dx = i\,d\lambda - f_e\,dx \tag{12.41}$$

or

$$f_e\,dx = i\,d\lambda - dW_m \tag{12.42}$$

Now, observe that the flux in the magnetic structure of Figure 12.38 depends on two variables, which are in effect independent: the current through the coil and the displacement of the bar. Each of these variables can cause the magnetic flux to change. Similarly, the energy stored in the electromagnetic field is also dependent on both current and displacement. Thus we can rewrite equation 12.42 as follows:

$$f_e\,dx = i\left(\frac{\partial\lambda}{\partial i}di + \frac{\partial\lambda}{\partial x}dx\right) - \left(\frac{\partial W_m}{\partial i}di + \frac{\partial W_m}{\partial x}dx\right) \tag{12.43}$$

Since i and x are independent variables, we can write

$$f_e = i\frac{\partial\lambda}{\partial x} - \frac{\partial W_m}{\partial x} \quad \text{and} \quad 0 = i\frac{\partial\lambda}{\partial i} - \frac{\partial W_m}{\partial i} \tag{12.44}$$

From the first expression in equation 12.44 we obtain the relationship

$$f_e = \frac{\partial}{\partial x}(i\lambda - W_m) = \frac{\partial}{\partial x}(W'_m) \tag{12.45}$$

where W'_m is the co-energy. Observe that the force acting to *push* the bar toward the electromagnet structure is of opposite sign to f_e, and assuming that $W_m = W'_m$, we can write

$$f = -f_e = -\frac{\partial}{\partial x}(W'_m) = -\frac{\partial W_m}{\partial x} \tag{12.46}$$

Equation 12.46 includes a very important assumption: the energy is equal to the co-energy. If you refer to Figure 12.7, you will realize that in general this is not true. Energy and co-energy are equal only if the λ-i relationship is linear. Thus, the useful result of equation 12.46, stating that the magnetic force acting on the moving iron is proportional to the rate of change of stored energy with displacement, applies only for *linear magnetic structures*.

Thus, to determine the forces present in a magnetic structure, it is necessary to compute the energy stored in the magnetic field. To simplify the analysis, we assume hereafter that the structures analyzed are magnetically linear. This is, of course, only an approximation, in that it neglects a number of practical aspects of electromechanical systems (e.g., the nonlinear λ-i curves described earlier, and the core losses typical of magnetic materials), but it permits relatively simple analysis of many useful magnetic structures. Thus, although the analysis method presented in this section is only approximate, it will serve the purpose of providing a feeling for the direction and the magnitude of the forces and currents present in electromechanical devices. On the basis of a linear approximation, it can be shown that the stored energy in a magnetic structure is given by

$$W_m = \frac{\phi \mathcal{F}}{2} \tag{12.47}$$

and since the flux and the mmf are related by the expression

$$\phi = \frac{Ni}{\mathcal{R}} = \frac{\mathcal{F}}{\mathcal{R}} \tag{12.48}$$

the stored energy can be related to the reluctance of the structure according to

$$W_m = \frac{\phi^2 \mathcal{R}(x)}{2} \tag{12.49}$$

where the reluctance has been explicitly shown to be a function of displacement, as is the case in a moving-iron transducer. Finally, then, we use the following approximate expression to compute the magnetic force acting on the moving iron:

$$f = -\frac{dW_m}{dx} = -\frac{\phi^2}{2}\frac{d\mathcal{R}(x)}{dx} \qquad \text{Magnetic force} \tag{12.50}$$

Examples 12.9, 12.10, and 12.12 illustrate the application of this approximate technique for the computation of forces and currents (the two problems of practical engineering interest to the user of such electromechanical systems) in some common devices. The Focus on Problem Solving box outlines the solution techniques for these classes of problems.

FOCUS ON PROBLEM SOLVING

ANALYSIS OF MOVING-IRON ELECTROMECHANICAL TRANSDUCERS

Calculation of current required to generate a given force

1. Derive an expression for the reluctance of the structure as a function of air gap displacement: $\mathcal{R}(x)$.

2. Express the magnetic flux in the structure as a function of the mmf (i.e., of the current I) and of the reluctance $\mathcal{R}(x)$:

$$\phi = \frac{\mathcal{F}(i)}{\mathcal{R}(x)}$$

3. Compute an expression for the force, using the known expressions for the flux and for the reluctance:

$$|f| = \frac{\phi^2}{2}\frac{d\mathcal{R}(x)}{dx}$$

(*Continued*)

FOCUS ON PROBLEM SOLVING

(*Concluded*)

4. Solve the expression in step 3 for the unknown current *i*.

Calculation of force generated due to transducer geometry and mmf

Repeat steps 1 through 3 above, substituting the known current to solve for the force *f*.

EXAMPLE 12.9   An Electromagnet

Problem

An electromagnet is used to collect and support a solid piece of steel, as shown in Figure 12.38. Calculate the *starting current* required to lift the load and the *holding current* required to keep the load in place once it has been lifted and is attached to the magnet. Assume that the cross-sectional areas of the electromagnet, load (bar), and air gap are equal.

Solution

Known Quantities: Geometry, magnetic permeability, number of coil turns, mass, acceleration of gravity, initial position of steel bar.

Find: Current required to lift the bar; current required to hold the bar in place.

Schematics, Diagrams, Circuits, and Given Data:

$$N = 500$$
$$\mu_0 = 4\pi \times 10^{-7}$$
$$\mu_r = 10^4 \text{ (equal for electromagnet and load)}$$

Initial distance (air gap) = 0.5 m

Magnetic path length of electromagnet = l_1 = 0.60 m

Magnetic path length of movable load = l_2 = 0.30 m

Gap cross-sectional area = 3×10^{-4} m^2

$$m = \text{mass of load} = 5 \text{ kg}$$
$$g = 9.8 \text{ m/s}^2$$

Assumptions: None.

Analysis: To compute the current we need to derive an expression for the force in the air gap. We use the equation

$$f_{\text{mech}} = \frac{\phi^2}{2} \frac{\partial \mathcal{R}(x)}{\partial x}$$

and calculate the reluctance, flux and force as follows:

$$\mathcal{R}(x) = \mathcal{R}_{Fe} + \mathcal{R}_{\text{gap}}$$
$$= \frac{2x}{\mu_0 A} + \frac{l_1 + l_2}{\mu_0 \mu_r A}$$

$$\phi = \frac{\mathcal{F}}{\mathcal{R}(x)} = \frac{Ni}{\left(\frac{2x}{\mu_0 A} + \frac{l_1+l_2}{\mu_0 \mu_r A}\right)}$$

$$\frac{\partial \mathcal{R}(x)}{\partial x} = \frac{2}{\mu_0 A} \Rightarrow f_{\text{mag}} = \frac{\phi^2}{2}\frac{\partial \mathcal{R}(x)}{\partial x} = \frac{(Ni)^2}{\left(\frac{2x}{\mu_0 A} + \frac{l_1+l_2}{\mu_0 \mu_r A}\right)^2}\frac{1}{\mu_0 A}$$

With this expression we can now calculate the current required to overcome the gravitational force when the load is 0.5 m away. The force we must overcome is $mg = 49$ N.

$$f_{\text{mag}} = \frac{(Ni)^2}{\left(\frac{2x}{\mu_0 A} + \frac{l_1+l_2}{\mu_0 \mu_r A}\right)^2}\frac{1}{\mu_0 A} = f_{\text{gravity}}$$

$$i^2 = f_{\text{gravity}}\frac{\mu_0 A\left(\frac{2x}{\mu_0 A} + \frac{l_1+l_2}{\mu_0 \mu_r A}\right)^2}{N^2} = 520 \times 10^3\,\text{A}^2 \qquad i = 721\ \text{A}$$

Finally, we calculate the holding current by letting $x = 0$:

$$f_{\text{mag}} = \frac{(Ni)^2}{\left(\frac{l_1+l_2}{\mu_0 \mu_r A}\right)^2}\frac{1}{\mu_0 A} = f_{\text{gravity}}$$

$$i^2 = f_{\text{gravity}}\frac{\mu_0 A\left(\frac{l_1+l_2}{\mu_0 \mu_r A}\right)^2}{N^2} = 4.21 \times 10^{-3}\,\text{A}^2$$

$$i = 64.9\ \text{mA}$$

Comments: Note how much smaller the holding current is than the lifting current.

One of the more common practical applications of the concepts discussed in this section is the solenoid. Solenoids find application in a variety of electrically controlled valves. The action of a solenoid valve is such that when it is energized, the plunger moves in such a direction as to permit the flow of a fluid through a conduit, as shown schematically in Figure 12.39.

Examples 12.10 and 12.11 illustrate the calculations involved in the determination of forces and currents in a solenoid.

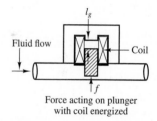

Force acting on plunger
with coil energized

Figure 12.39 Application of the solenoid as a valve

EXAMPLE 12.10   A Solenoid

Problem

Figure 12.40 depicts a simplified representation of a solenoid. The restoring force for the plunger is provided by a spring.

1. Derive a general expression for the force exerted on the plunger as a function of the plunger position x.

2. Determine the mmf required to pull the plunger to its end position ($x = a$).

Solution

Known Quantities: Geometry of magnetic structure; spring constant.

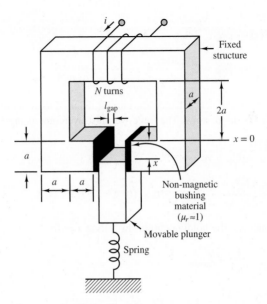

Figure 12.40 A solenoid

Find: f; mmf.

Schematics, Diagrams, Circuits, and Given Data: $a = 0.01$ m; $l_{gap} = 0.001$ m; $k = 10$ N/m.

Assumptions: Assume that the reluctance of the iron is negligible; neglect fringing. At $x = 0$ the plunger is in the gap by an infinitesimal displacement ε.

Analysis:

1. *Force on the plunger.* To compute a general expression for the magnetic force exerted on the plunger, we need to derive an expression for the force in the air gap. Using equation 12.50, we see that we need to compute the reluctance of the structure and the magnetic flux to derive an expression for the force.

   Since we are neglecting the iron reluctance, we can write the expression for the reluctance as follows. Note that the area of the gap is variable, depending on the position of the plunger, as shown in Figure 12.41.

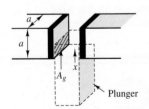

Figure 12.41 Detail of solenoid structure.

$$\mathcal{R}_{gap}(x) = 2 \times \frac{l_{gap}}{\mu_0 A_{gap}} = \frac{2 l_{gap}}{\mu_0 a x}$$

The derivative of the reluctance with respect to the displacement of the plunger can then be computed to be

$$\frac{d\mathcal{R}_{gap}(x)}{dx} = \frac{-2 l_{gap}}{\mu_0 a x^2}$$

Knowing the reluctance, we can calculate the magnetic flux in the structure as a function of the coil current:

$$\phi = \frac{Ni}{\mathcal{R}(x)} = \frac{Ni \mu_0 a x}{2 l_{gap}}$$

The force in the air gap is given by

$$f_{gap} = \frac{\phi^2}{2} \frac{d\mathcal{R}(x)}{dx} = \frac{(Ni \mu_0 a x)^2}{8 l_{gap}^2} \frac{-2 l_{gap}}{\mu_0 a x^2} = -\frac{\mu_0 a (Ni)^2}{4 l_{gap}} = kX$$

Thus, the force in the gap is proportional to the square of the current and does not vary with plunger displacement.

2. *Calculation of magnetomotive force.* To determine the required magnetomotive force, we observe that the magnetic force must overcome the mechanical (restoring) force generated by the spring. Thus, $f_{gap} = kx = ka$. For the stated values, $f_{gap} = (10 \text{ N/m}) \times (0.01 \text{ m}) = 0.1$ N, and

$$Ni = \sqrt{\frac{4l_{gap}f_{gap}}{\mu_0 a}} = \sqrt{\frac{4 \times 0.001 \times 0.1}{4\pi \times 10^{-7} \times 0.01}} = 178 \text{ A-turns}$$

The required mmf can be most effectively realized by keeping the current value relatively low and using a large number of turns.

Comments: The same mmf can be realized with an infinite number of combinations of current and number of turns; however, there are tradeoffs involved. If the current is very large (and the number of turns small), the required wire diameter will be very large. Conversely, a small current will require a small wire diameter and a large number of turns. A homework problem explores this tradeoff.

CHECK YOUR UNDERSTANDING

A solenoid is used to exert force on a spring. Estimate the position of the plunger if the number of turns in the solenoid winding is 1,000 and the current going into the winding is 40 mA. Use the same values as in Example 12.10 for all other variables.

ɯɯ ϛˑ0 = x :ɹǝʍsu∀

EXAMPLE 12.11   Transient Response of a Solenoid

Problem

Analyze the current response of the solenoid of Example 12.10 to a step change in excitation voltage. Plot the force and current as a function of time.

Solution

Known Quantities: Coil inductance and resistance; applied current.

Find: Current and force response as a function of time.

Schematics, Diagrams, Circuits, and Given Data: See Example 12.10. $N = 1,000$ turns. $V = 12$ V. $R_{coil} = 5$ Ω.

Assumptions: The inductance of the solenoid is approximately constant and is equal to the midrange value (plunger displacement equal to $a/2$).

Analysis: From Example 12.10, we have an expression for the reluctance of the solenoid:

$$\mathcal{R}_{gap}(x) = \frac{2l_{gap}}{\mu_0 a x}$$

Using equation 12.30 and assuming $x = a/2$, we calculate the inductance of the structure:

$$L \approx \frac{N^2}{\mathcal{R}_{\text{gap}}|_{x=a/2}} = \frac{N^2 \mu_0 a^2}{4 l_{\text{gap}}} = \frac{10^6 \times 4\pi \times 10^{-7} \times 10^{-4}}{4 \times 10^{-3}} = 31.4 \text{ mH}$$

The equivalent solenoid circuit is shown in Figure 12.42. When the switch is closed, the solenoid current rises exponentially with time constant $\tau = L/R_{\text{coil}} = 6.3$ ms. As shown in Chapter 4, the response is of the form

$$i(t) = \frac{V}{R_{\text{coil}}}(1 - e^{-t/\tau}) = \frac{V}{R_{\text{coil}}}(1 - e^{-R_{\text{coil}}/L}) = \frac{12}{5}(1 - e^{-t/6.3\times10^{-3}}) \quad \text{A}$$

To determine how the magnetic force responds during the turn-on transient, we return to the expression for the force derived in Example 12.10:

$$f_{\text{gap}}(t) = \frac{\mu_0 a (Ni)^2}{4 l_{\text{gap}}} = \frac{4\pi \times 10^{-7} \times 0^{-2} \times 10^6}{4 \times 10^{-3}} i^2(t) = \pi i^2(t)$$

$$= \pi \left[\frac{12}{5}(1 - e^{-t/6.3\times10^{-3}}) \right]^2$$

The two curves are plotted in Figure 12.42(b).

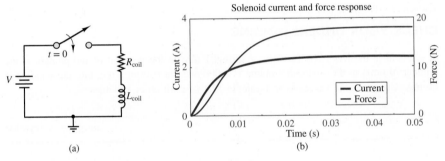

Figure 12.42 Solenoid equivalent electric circuit and step response

Comments: The assumption that the inductance is approximately constant is not quite accurate. The reluctance (and therefore the inductance) of the structure will change as the plunger moves into position. However, allowing for the inductance to be a function of plunger displacement causes the problem to become nonlinear and requires numerical solution of the differential equation (i.e., the transient response results of Chapter 4 no longer apply).

Practical Facts About Solenoids

Solenoids can be used to produce linear or rotary motion, in either the *push* or the *pull* mode. The most common solenoid types are listed here:

1. *Single-action linear* (push or pull). Linear stroke motion, with a restoring force (e.g., from a spring), to return the solenoid to the neutral position.

2. *Double-acting linear.* Two solenoids back to back can act in either direction. The restoring force is provided by another mechanism (e.g., a spring).

3. *Mechanical latching solenoid* (bistable). An internal latching mechanism holds the solenoid in place against the load.

(Continued)

(*Concluded*)

4. *Keep solenoid.* Fitted with a permanent magnet so that no power is needed to hold the load in the pulled-in position. Plunger is released by applying a current pulse of opposite polarity to that required to pull in the plunger.

5. *Rotary solenoid.* Constructed to permit rotary travel. Typical range is 25 to 95°. Return action via mechanical means (e.g., a spring).

6. *Reversing rotary solenoid.* Rotary motion is from one end to the other; when the solenoid is energized again, it reverses direction.

Solenoid power ratings are dependent primarily on the current required by the coil, and on the coil resistance. The I^2R is the primary power sink, and solenoids are therefore limited by the heat they can dissipate. Solenoids can operate in continuous or pulsed mode. The power rating depends on the mode of operation, and can be increased by adding *hold-in resistors* to the circuit to reduce the *holding current* required for continuous operation. The hold resistor is switched into the circuit once the *pull-in* current required to pull the plunger has been applied and the plunger has

moved into place. The holding current can be significantly smaller than the pull-in current.

A common method to reduce the solenoid holding current employs a normally closed (NC) switch in parallel with a hold-in resistor. In Figure 12.43, when the pushbutton (PB) closes the circuit, full voltage is applied to the solenoid coil, bypassing the resistor through the NC switch. When the solenoid closes, the NC switch opens, connecting the resistor in series with the coil. The resistor will now limit the current to the value required to hold the solenoid in position. Note the diode "snubber" circuit to shunt the reverse current when the solenoid is deenergized.

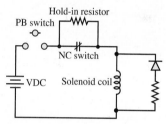

Figure 12.43 Circuit used to reduce solenoid holding current.

Another electromechanical device that finds common application in industrial practice is the relay. The relay is an electromechanical switch that permits the opening and closing of electrical contacts by means of an electromagnetic structure similar to those discussed earlier in this section.

A relay such as would be used to start a high-voltage single-phase motor is shown in Figure 12.44. The magnetic structure has dimensions equal to 1 cm on all sides, and the transverse dimension is 8 cm. The relay works as follows. When the pushbutton is pressed, an electric current flows through the coil and generates a field in the magnetic structure. The resulting force draws the movable part toward the fixed part, causing an electrical contact to be made. The advantage of the relay is that a relatively low-level current can be used to control the opening and closing of a circuit that can carry large currents. In this particular example, the relay is energized by a 120-VAC contact, establishing a connection in a 240-VAC circuit. Such relay circuits are commonly employed to remotely switch large industrial loads.

Circuit symbols for relays are shown in Figure 12.45. An example of the calculations that would typically be required in determining the mechanical and electrical characteristics of a simple relay are given in Example 12.12.

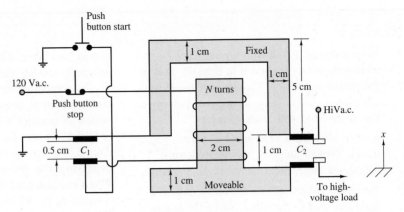

Figure 12.44 A relay

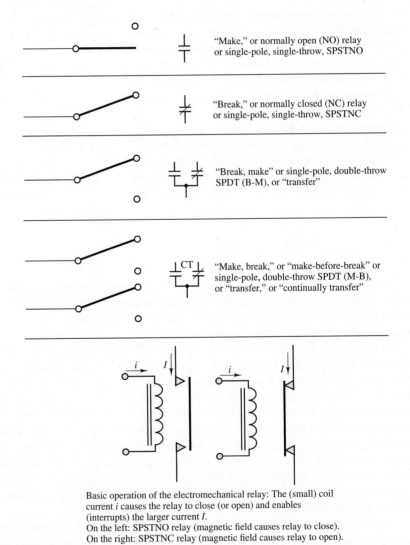

Basic operation of the electromechanical relay: The (small) coil
current i causes the relay to close (or open) and enables
(interrupts) the larger current I.
On the left: SPSTNO relay (magnetic field causes relay to close).
On the right: SPSTNC relay (magnetic field causes relay to open).

Figure 12.45 Circuit symbols and basic operation of relays

EXAMPLE 12.12 A Relay

Problem

Figure 12.46 depicts a simplified representation of a relay. Determine the current required for the relay to make contact (i.e., pull in the ferromagnetic plate) from a distance x.

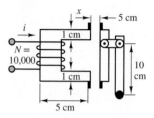

Figure 12.46 Relay circuit for Example 12.12.

Solution

Known Quantities: Relay geometry; restoring force to be overcome; distance between bar and relay contacts; number of coil turns.

Find: i.

Schematics, Diagrams, Circuits, and Given Data: $A_{\text{gap}} = (0.01 \text{ m})^2$; $x = 0.05$ m; $f_{\text{restore}} = 5$ N; $N = 10{,}000$.

Assumptions: Assume that the reluctance of the iron is negligible; neglect fringing.

Analysis:

$$\mathcal{R}_{\text{gap}}(x) = \frac{2x}{\mu_0 A_{\text{gap}}}$$

The derivative of the reluctance with respect to the displacement of the plunger can then be computed as

$$\frac{d\mathcal{R}_{\text{gap}}(x)}{dx} = \frac{2}{\mu_0 A_{\text{gap}}}$$

Knowing the reluctance, we can calculate the magnetic flux in the structure as a function of the coil current:

$$\phi = \frac{Ni}{\mathcal{R}(x)} = \frac{Ni\mu_0 A_{\text{gap}}}{2x}$$

and the force in the air gap is given by

$$f_{\text{gap}} = \frac{\phi^2}{2} \frac{d\mathcal{R}(x)}{dx} = \frac{(Ni\mu_0 A_{\text{gap}})^2}{8x^2} \frac{2}{\mu_0 A_{\text{gap}}} = \frac{\mu_0 A_{\text{gap}} (Ni)^2}{4x^2}$$

The magnetic force must overcome a mechanical holding force of 5 N; thus,

$$f_{\text{gap}} = \frac{\mu_0 A_{\text{gap}} (Ni)^2}{4x^2} = f_{\text{restore}} = 5 \text{ N}$$

or

$$i = \frac{1}{N} \sqrt{\frac{4x^2 f_{\text{restore}}}{\mu_0 A_{\text{gap}}}} = \frac{1}{10{,}000} \sqrt{\frac{4(0.05)^2 5}{4\pi \times 10^{-7} \times 0.0001}} = \pm 2 \text{ A}$$

Comments: The current required to close the relay is much larger than that required to hold the relay closed, because the reluctance of the structure is much smaller once the gap is reduced to zero.

Moving-Coil Transducers

Another important class of electromagnetomechanical transducers is that of **moving-coil transducers**. This class of transducers includes a number of common devices, such as microphones, loudspeakers, and all electric motors and generators. The aim of this section is to explain the relationship between a fixed magnetic field, the emf across the moving coil, and the forces and motions of the moving element of the transducer.

The basic principle of operation of electromechanical transducers is that a magnetic field exerts a force on a charge moving through it. The equation describing this effect is

$$\mathbf{f} = q\mathbf{u} \times \mathbf{B} \qquad (12.51)$$

which is a vector equation, as explained earlier. To correctly interpret equation 12.51, we must recall the right-hand rule and apply it to the transducer, illustrated in Figure 12.47, depicting a structure consisting of a sliding bar which makes contact with a fixed conducting frame. Although this structure does not represent a practical actuator, it will be a useful aid in explaining the operation of moving-coil transducers such as motors and generators. In Figure 12.47, and in all similar figures in this section, a small cross represents the "tail" of an arrow pointing into the page, while a dot represents an arrow pointing out of the page; this convention will be useful in visualizing three-dimensional pictures.

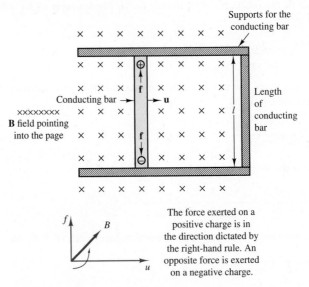

Figure 12.47 A simple electromechanical motion transducer

CHECK YOUR UNDERSTANDING

In the circuit in Figure 12.47, the conducting bar is moving with a velocity of 6 m/s. The flux density is 0.5 Wb/m², and $l = 1.0$ m. Find the magnitude of the resulting induced voltage.

Motor Action

A moving-coil transducer can act as a motor when an externally supplied current flowing through the electrically conducting part of the transducer is converted to a force that can cause the moving part of the transducer to be displaced. Such a current would flow, for example, if the support of Figure 12.47 were made of conducting material, so that the conductor and the right-hand side of the support "rail" were to form a loop (in effect, a one-turn coil). To understand the effects of this current flow in the conductor, one must consider the fact that a charge moving at a velocity u' (along the conductor and perpendicular to the velocity of the conducting bar, as shown in Figure 12.48) corresponds to a current $i = dq/dt$ along the length l of the conductor. This fact can be explained by considering the current i along a differential element dl and writing

$$i\,dl = \frac{dq}{dt} \cdot u'\,dt \tag{12.52}$$

since the differential element dl would be traversed by the current in time dt at a velocity u'. Thus we can write

$$i\,dl = dq\,u' \tag{12.53}$$

or

$$il = qu' \tag{12.54}$$

for the geometry of Figure 12.48. From Section 12.1, the force developed by a charge moving in a magnetic field is, in general, given by

$$\mathbf{f} = q\mathbf{u} \times \mathbf{B} \tag{12.55}$$

For the term $q\mathbf{u}'$ we can substitute $i\mathbf{l}$, to obtain

$$\mathbf{f}' = i\mathbf{l} \times \mathbf{B} \tag{12.56}$$

Using the right-hand rule, we determine that the force $\mathbf{f}'$ generated by the current i is in the direction that would push the conducting bar to the left. The magnitude of this force is $f' = Bli$ if the magnetic field and the direction of the current are perpendicular. If they are not, then we must consider the angle γ formed by $\mathbf{B}$ and $\mathbf{l}$; in the more general case,

$$\boxed{f' = Bli \sin\gamma = Bli \text{ if } \gamma = 90° \qquad Bli \text{ law}} \tag{12.57}$$

The phenomenon we have just described is sometimes referred to as the **Bli law**.

Generator Action

The other mode of operation of a moving-coil transducer occurs when an external force causes the coil (i.e., the moving bar, in Figure 12.47) to be displaced. This external force is converted to an emf across the coil, as will be explained in the following paragraphs.

Since positive and negative charges are forced in opposite directions in the transducer of Figure 12.47, a potential difference will appear across the conducting bar; this potential difference is the electromotive force, or emf. The emf must be equal to the force exerted by the magnetic field. In short, the electric force per unit

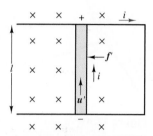

Figure 12.48 Simplified structure of moving-coil transducer.

charge (or electric field) e/l must equal the magnetic force per unit charge $f/q = Bu$. Thus, the relationship

$$e = Blu \qquad Blu \text{ law}$$ (12.58)

holds whenever **B**, **l**, and **u** are mutually perpendicular, as in Figure 12.49. If equation 12.58 is analyzed in greater depth, it can be seen that the product lu (length times velocity) is the area crossed per unit time by the conductor. If one visualizes the conductor as "cutting" the flux lines into the base in Figure 12.48, it can be concluded that the electromotive force is equal to the *rate at which the conductor "cuts" the magnetic lines of flux*. It will be useful for you to carefully absorb this notion of conductors cutting lines of flux, since this greatly simplifies the understanding of the material in this section and in Chapter 13.

In general, **B**, **l**, and **u** are not necessarily perpendicular. In this case one needs to consider the angles formed by the magnetic field with the normal to the plane containing **l** and **u**, and the angle between **l** and **u**. The former is angle α of Figure 12.49; the latter is angle β in the same figure. It should be apparent that the optimum values of α and β are 0° and 90°, respectively. Thus, most practical devices are constructed with these values of α and β. Unless otherwise noted, it will be tacitly assumed that this is the case. The **Blu** law just illustrated explains how a moving conductor in a magnetic field can generate an electromotive force.

To summarize the electromechanical energy conversion that takes place in the simple device of Figure 12.47, we must note now that the presence of a current in the loop formed by the conductor and the rail requires that the conductor move to the right at a velocity u (Blu law), thus cutting the lines of flux and generating the emf that gives rise to current i. On the other hand, the same current causes a force f' to be exerted on the conductor (Bli law) in the direction opposite to the movement of the conductor. Thus, it is necessary that an *externally applied force* f_{ext} exists to cause the conductor to move to the right with a velocity u. The external force must overcome the force f'. This is the basis of electromechanical energy conversion.

An additional observation we must make at this point is that the current i flowing around a closed loop generates a magnetic field, as explained in Section 12.1. Since this additional field is generated by a one-turn coil in our illustration, it is reasonable to assume that it is negligible with respect to the field already present (perhaps established by a permanent magnet). Finally, we must consider that this coil links a certain amount of flux, which changes as the conductor moves from left to right. The area crossed by the moving conductor in time dt is

$$dA = lu \, dt$$ (12.59)

so that if the flux density B is uniform, the rate of change of the flux linked by the one-turn coil is

$$\frac{d\phi}{dt} = B \frac{dA}{dt} = Blu$$ (12.60)

In other words, the *rate of change* of the flux linked by the conducting loop is equal to the emf generated in the conductor. You should realize that this statement simply confirms Faraday's law.

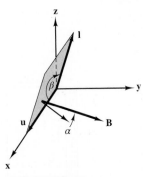

Figure 12.49 When magnetic flux, current and velocity vectors are mutually perpendicular, $e = Blu$.

It was briefly mentioned that the *Blu* and *Bli* laws indicate that, thanks to the coupling action of the magnetic field, a conversion of mechanical to electric energy—or the converse—is possible. The simple structures of Figures 12.47 and 12.48 can, again, serve as an illustration of this energy conversion process, although we have not yet indicated how these idealized structures can be converted to a practical device. In this section we begin to introduce some physical considerations. Before we proceed any further, we should try to compute the power—electric and mechanical—that is generated (or is required) by our ideal transducer. The electric power is given by

$$P_E = ei = Blui \quad \text{W} \tag{12.61}$$

while the mechanical power required, say, to move the conductor from left to right is given by the product of force and velocity:

$$P_M = f_{\text{ext}} u = Bliu \quad \text{W} \tag{12.62}$$

The principle of conservation of energy states that in this ideal (lossless) transducer we can convert a given amount of electric energy to mechanical energy, or vice versa. We can utilize the structure of Figure 12.47 to illustrate this reversible action. If the closed path containing the moving conductor is now formed from a closed circuit containing a resistance R and a battery V_B, as shown in Figure 12.50, the externally applied force f_{ext} generates a positive current i into the battery provided that the emf is greater than V_B. When $e = Blu > V_B$, the ideal transducer acts as a *generator*. For any given set of values of B, l, R, and V_B, there will exist a velocity u for which the current i is positive. If the velocity is lower than this value—that is, if $e = Blu < V_B$—then the current i is negative, and the conductor is forced to move to the right. In this case the battery acts as a source of energy and the transducer acts as a *motor* (i.e., electric energy drives the mechanical motion).

In practical transducers, we must be concerned with the inertia, friction, and elastic forces that are invariably present on the mechanical side of the transducer. Similarly, on the electrical side we must account for the inductance of the circuit, its resistance, and possibly some capacitance. Consider the structure of Figure 12.51. In the figure, the conducting bar has been placed on a surface with a coefficient of sliding friction b; it has a mass m and is attached to a fixed structure by means of a spring with spring constant k. The equivalent circuit representing the coil inductance and resistance is also shown.

If we recognize that $u = dx/dt$ in the figure, we can write the equation of motion for the conductor as

$$m\frac{du}{dt} + bu + \frac{1}{k}\int u\,dt = f = Bli \tag{12.63}$$

where the *Bli* term represents the driving input that causes the mass to move. The driving input in this case is provided by the electric energy source v_S; thus the transducer acts as a motor, and f is the electromechanical force acting on the mass of the conductor. On the electrical side, the circuit equation is

$$v_S - L\frac{di}{dt} - Ri = e = Blu \tag{12.64}$$

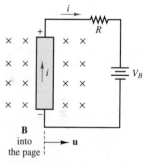

Figure 12.50 Motor and generator action in an ideal transducer

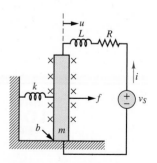

Figure 12.51 A more realistic representation of the transducer of Figure 12.50

Equations 12.63 and 12.64 could then be solved by knowing the excitation voltage v_S and the physical parameters of the mechanical and electric circuits. For example, if the excitation voltage were sinusoidal, with

$$v_S(t) = V_S \cos \omega t$$

and the field density were constant

$$B = B_0$$

then we could postulate sinusoidal solutions for the transducer velocity u and current i:

$$u = U \cos(\omega t + \theta_u) \qquad i = I \cos(\omega t + \theta_i) \tag{12.65}$$

and use phasor notation to solve for the unknowns (U, I, θ_u, θ_i).

The results obtained in the present section apply directly to transducers that are based on translational (linear) motion. These basic principles of electromechanical energy conversion and the analysis methods developed in the section are next applied to practical transducers in a few examples. A Focus on Problem Solving box outlines the analysis procedure for moving-coil transducers.

FOCUS ON PROBLEM SOLVING

ANALYSIS OF MOVING-COIL ELECTROMECHANICAL TRANSDUCERS

1. Apply KVL to write the differential equation for the electrical subsystem, including the back emf ($e = Blu$) term.

2. Apply Newton's second law to write the differential equation for the mechanical subsystem, including the magnetic force $f = Bli$ term.

3. Use a Laplace transform on the two coupled differential equations to formulate a system of linear algebraic equations, and solve for the desired mechanical and electrical variables.

EXAMPLE 12.13   A Loudspeaker

Problem

A loudspeaker, shown in Figure 12.52, uses a permanent magnet and a moving coil to produce the vibrational motion that generates the pressure waves we perceive as sound. Vibration of the loudspeaker is caused by changes in the input current to a coil; the coil is, in turn, coupled to a magnetic structure that can produce time-varying forces on the speaker diaphragm. A simplified model for the mechanics of the speaker is also shown in Figure 12.52. The force exerted on the coil is also exerted on the mass of the speaker diaphragm, as shown in Figure 12.53, which depicts a free-body diagram of the forces acting on the loudspeaker diaphragm.

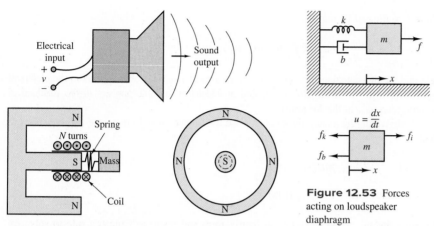

Figure 12.52 Loudspeaker

Figure 12.53 Forces acting on loudspeaker diaphragm

The force exerted on the mass f_i is the magnetic force due to current flow in the coil. The electric circuit that describes the coil is shown in Figure 12.54, where L represents the inductance of the coil, R represents the resistance of the windings, and e is the emf induced by the coil moving through the magnetic field.

Determine the frequency response $U(j\omega)/V(j\omega)$ of the speaker.

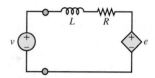

Figure 12.54 Model of transducer electrical side

Solution

Known Quantities: Circuit and mechanical parameters; magnetic flux density; number of coil turns; coil radius.

Find: Frequency response of loudspeaker $U(j\omega)/V(j\omega)$.

Schematics, Diagrams, Circuits, and Given Data: Coil radius = 0.05 m; $L = 10$ mH; $R = 8$ Ω; $m = 0.01$ kg; $b = 22.75$ N-s^2/m; $k = 5 \times 10^4$ N/m; $N = 47$; $B = 1$ T.

Analysis: To determine the frequency response of the loudspeaker, we write the differential equations that describe the electrical and mechanical subsystems. We apply KVL to the electric circuit, using the circuit model of Figure 12.54, in which we have represented the *Blu* term (motional voltage) in the form of a *back electromotive force e*:

$$v - L\frac{di}{dt} - Ri - e = 0$$

or

$$L\frac{di}{dt} + Ri + Blu = v$$

Next, we apply Newton's second law to the mechanical system, consisting of a lumped mass representing the mass of the moving diaphragm m; an elastic (spring) term, which represents the elasticity of the diaphragm k; and a damping coefficient b, representing the frictional losses and aerodynamic damping affecting the moving diaphragm.

$$m\frac{du}{dt} = f_i - f_d - f_k = f_i - bu - kx$$

where $f_i = Bli$ and therefore

$$-Bli + m\frac{du}{dt} + bu + k\int_{-\infty}^{t} u(t')\,dt' = 0$$

Note that the two equations are *coupled*; that is, a mechanical variable appears in the electrical equation (velocity u in the Blu term), and an electrical variable appears in the mechanical equation (current i in the Bli term).

To derive the frequency response, we use the Laplace transform on the two equations to obtain

$$(sL + R)I(s) + BlU(s) = V(s)$$

$$-BlI(s) + \left(sm + b + \frac{k}{s}\right)U(s) = 0$$

We can write the above equations in matrix form and resort to Cramer's rule to solve for $U(s)$ as a function of $V(s)$:

$$\begin{bmatrix} sL + R & Bl \\ -Bl & sm + b + \dfrac{k}{s} \end{bmatrix}\begin{bmatrix} I(s) \\ U(s) \end{bmatrix} = \begin{bmatrix} V(s) \\ 0 \end{bmatrix}$$

with solution

$$U(s) = \frac{\det\begin{bmatrix} sL + R & V(s) \\ -Bl & 0 \end{bmatrix}}{\det\begin{bmatrix} sL + R & Bl \\ -Bl & sm + b + \dfrac{k}{s} \end{bmatrix}}$$

or

$$\frac{U(s)}{V(s)} = \frac{Bl}{(sL + R)(sm + b + k/s) + (Bl)^2}$$

$$= \frac{Bls}{(Lm)s^3 + (Rm + Lb)s^2 + \left[Rb + kL + (Bl)^2\right]s + kR}$$

To determine the frequency response of the loudspeaker, we let $s \to j\omega$ in the above expression:

$$\frac{U(j\omega)}{V(j\omega)} = \frac{jBl\omega}{kR - (Rm + Lb)\omega^2 + j\{[Rb + kL + (Bl)^2]\omega - (Lm)\omega^3\}}$$

where $l = 2\pi Nr$, and substitute the appropriate numerical parameters:

$$\frac{U(j\omega)}{V(j\omega)} = \frac{j14.8\omega}{4 \times 10^5 - (0.08 + 0.2275)\omega^2 + j\left[(182 + 500 + 218)\omega - (10^{-4})\omega^3\right]}$$

$$= \frac{j14.8\omega}{4 \times 10^5 - 0.3075\omega^2 + j\left[(900)\omega - (10^{-4})\omega^3\right]}$$

The resulting frequency response is plotted in Figure 12.55.

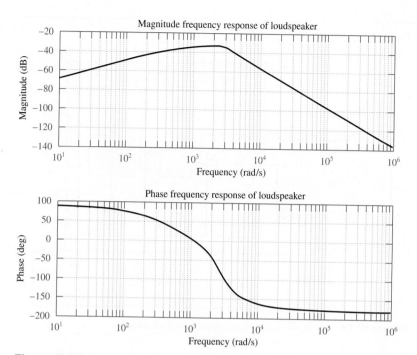

Figure 12.55 Frequency response of loudspeaker

CHECK YOUR UNDERSTANDING

In Example 12.13, we examined the frequency response of a loudspeaker. However, over time, permanent magnets may become demagnetized. Find the frequency response of the same loudspeaker if the permanent magnet has lost its strength to a point where $B = 0.95$ T.

$$\text{Answer: } U(j\omega)/V(j\omega) = \frac{j(14.03\omega)}{(4 \times 10^5 - 0.3075\omega^2) + j(889 \times 10^{-4}\omega^3)}$$

Seismic Transducer

Problem:

The device shown in Figure 12.56 is called a *seismic transducer* and can be used to measure the displacement, velocity, or acceleration of a body. The permanent magnet of mass m is supported on the case by a spring k, and there is some viscous damping b between the magnet and the case; the coil is fixed to the case. You may assume that the coil has length l and resistance and inductance R_{coil} and L_{coil}, respectively; the magnet exerts a magnetic field B. Find the transfer function between the output voltage v_{out} and the velocity of the body dx_c/dt. Note that $x(t)$ is not equal to zero when the system is at rest. We ignore this offset displacement.

FOCUS ON MEASUREMENTS

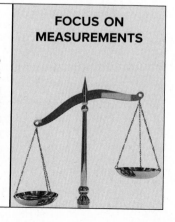

(Continued)

(Concluded)

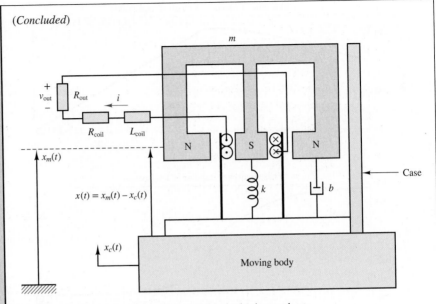

Figure 12.56 An electro-magneto-mechanical seismic transducer

Solution:

First we apply KVL around the electric circuit to write the differential equation describing the electrical systems:

$$L\frac{di}{dt} + (R_{\text{coil}} + R_{\text{out}})i + Bl\frac{dx}{dt} = 0$$

Also note that $v_{\text{out}} = -R_{\text{out}}i$. Next, we observe that the displacement of the magnet, x_m, is equal to the sum of the case displacement, x_c, and the relative displacement between the magnet and the case, $x(t)$: $x_m = x + x_c$. Apply Newton's second law to the mass of the magnet, m, we obtain

$$m\frac{d^2x_m}{dt^2} = -k(x_m - x_c) - b\left(\frac{dx_m}{dt} - \frac{dx_c}{dt}\right) - Bli$$

Substituting the relation $x_m = x + x_c$, we obtain

$$m\left(\frac{d^2x}{dt^2} + \frac{d^2x_c}{dt^2}\right) + kx + b\frac{dx}{dt} = -Bli$$

From this expression we can now derive the transfer function between the displacement of the case, $X_c(s)$, and the output voltage, $V_{\text{out}}(s)$. Let $R = R_{\text{coil}} + R_{\text{out}}$. Then

$$(Ls + R)I(s) + BlsX(s) = 0$$

$$-BlI(s) + (ms^2 + bs + k)X(s) = -ms^2X_c(s)$$

$$I(s) = \frac{Blms^3X_c(s)}{mLs^3 + (bL + mR)s^2 + (kL + Rb + B^2l^2)s + kR}$$

Now, let the velocity of the case be $U_c(s) = sX_c(s)$; since $V_{\text{out}}(s) = -R_{\text{out}}I(s)$, the transfer function from case velocity to output voltage becomes

$$\frac{V_{\text{out}}(s)}{U_c(s)} = -\frac{BlmR_{\text{out}}s^2}{mLs^3 + (bL + mR)s^2 + (kL + Rb + B^2l^2)s + kR}$$

Conclusion

This chapter introduces electromechanical systems. Electromechanical devices include a variety of sensors and transducers that find common engineering application in many fields. All electromechanical devices use the coupling between mechanical and electrical systems provided by a magnetic field. This magnetic coupling makes it possible to convert energy from electric to mechanical form, and back. Devices that convert electric to mechanical energy include all forms of electromagnetomechanical actuators, such as electromagnets, solenoids, relays, electrodynamic shakers, linear motors, and loudspeakers. Conversion from mechanical to electric energy results in generators, and various sensors that can detect mechanical displacement, velocity, or acceleration. Upon completing this chapter, you should have mastered the following learning objectives:

1. *Review the basic principles of electricity and magnetism.* The basic laws that govern electromagnetomechanical energy conversion are Faraday's law, stating that a changing magnetic field can induce a voltage, and Ampère's law, stating that a current flowing through a conductor generates a magnetic field.

2. *Use the concepts of reluctance and magnetic circuit equivalents to compute magnetic flux and currents in simple magnetic structures.* The two fundamental variables in the analysis of magnetic structures are the magnetomotive force and the magnetic flux; if some simplifying approximations are made, these quantities are linearly related through the reluctance parameter, in much the same way as voltage and current are related through resistance according to Ohm's law. This simplified analysis permits approximate calculation of forces and currents in electromagnetomechanical structures.

3. *Understand the properties of magnetic materials and their effects on magnetic circuit models.* Magnetic materials are characterized by a number of nonideal properties, which must be considered in a detailed analysis of any electromechanical transducer. The most important phenomena are saturation, eddy currents, and hysteresis.

4. *Use magnetic circuit models to analyze transformers.* One of the most common magnetic structures in use in electric power systems is the transformer. The methods developed in the earlier sections provide all the tools needed to perform an analysis of these important devices.

5. *Model and analyze force generation in electromagnetomechanical systems.* Analyze moving-iron transducers (electromagnets, solenoids, relays) and moving-coil transducers (electrodynamic shakers, loudspeakers, and seismic transducers). Electromagnetomechanical transducers can be broadly divided into two categories: moving-iron transducers, which include all electromagnets, solenoids, and relays; and moving-coil transducers, which include loudspeakers, electrodynamic shakers, and all electric motors. Section 12.5 develops analysis and design methods for these devices.

HOMEWORK PROBLEMS

Section 12.1: Electricity and Magnetism

12.1   For the electromagnet of Figure P12.1:

a. Find the flux density in the core.

b. Sketch the magnetic flux lines and indicate their direction.

c. Indicate the north and south poles of the magnet.

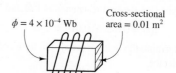

Figure P12.1

12.2 A single loop of wire carrying current I_2 is placed near the end of a solenoid having N turns and carrying current I_1, as shown in Figure P12.2. The solenoid is fastened to a horizontal surface, but the single coil is free to move. With the currents directed as shown, is there a resultant force on the single coil? If so, in what direction? Why?

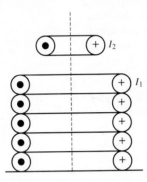

Figure P12.2

12.3 A practical LVDT is typically connected to a resistive load. Derive the LVDT equations in the presence of a resistive load R_L connected across the output terminals, using the results of the Focus on Measurements box, "Linear Variable Differential Transformer." Let R_S, L_S be the secondary coil parameters.

12.4 On the basis of the equations of the Focus on Measurements box "Linear Variable Differential Transformer," and of the results of Problem 12.3, derive the frequency response of the LVDT, and determine the range of frequencies for which the device will have maximum sensitivity for a given excitation. (*Hint:* Compute dv_{out}/dv_{ex}, and set the derivative equal to zero to determine the maximum sensitivity.)

12.5 An iron-core inductor has the following characteristic:

$$i = \frac{\lambda}{0.5 + \lambda}$$

a. Determine the energy, co-energy, and incremental inductance for $\lambda = 1$ V-s.

b. Given that the coil resistance is 1 Ω and that

$$i(t) = 0.625 + 0.01 \sin 400t \quad A$$

determine the voltage across the terminals on the inductor.

12.6 Repeat Problem 12.5 if

$$i = \frac{\lambda^2}{0.5 + \lambda^2}$$

12.7 An iron-core inductor has the characteristic shown in Figure P12.7:

a. Determine the energy and the incremental inductance for $i = 1.0$ A.

b. Given that the coil resistance is 2 Ω and that $i(t) = 0.5 \sin 2\pi t$, determine the voltage across the terminals of the inductor.

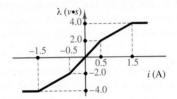

Figure P12.7

12.8 Determine the reluctance of the structure of Figure 12.12 in the text if the cross-sectional area is $A = 0.1$ m^2 and $\mu_r = 2{,}000$. Assume that each leg of the mean magnetic path is 0.1 m in length and that it runs through the exact center of the structure.

Section 12.2: Magnetic Circuits

12.9

a. Find the reluctance of a magnetic circuit if a magnetic flux $\phi = 4.2 \times 10^{-4}$ Wb is established by an impressed mmf of 400 A-turns.

b. Find the magnetizing force H in SI units if the magnetic circuit is 6 in long.

12.10 For the circuit shown in Figure P12.10:

a. Determine the reluctance values and show the magnetic circuit, assuming that $\mu = 3{,}000\mu_0$.

b. Determine the inductance of the device.

c. The inductance of the device can be modified by cutting an air gap in the magnetic structure. If a gap of 0.1 mm is cut in the arm of length l_3, what is the new value of inductance?

d. As the gap is increased in size (length), what is the limiting value of inductance? Neglect leakage flux and fringing effects.

12.11 The magnetic circuit shown in Figure P12.11 has two parallel paths. Find the flux and flux density in each leg of the magnetic circuit. Neglect fringing at the air gaps and any leakage fields. $N = 1{,}000$ turns, $i = 0.2$ A, $l_{g1} = 0.02$ cm, and $l_{g2} = 0.04$ cm. Assume the reluctance of the magnetic core to be negligible.

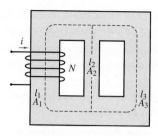

$N = 100$ turns $A_2 = 25$ cm²

$l_1 = 30$ cm $l_3 = 30$ cm

$A_1 = 100$ cm² $A_3 = 100$ cm²

$l_2 = 10$ cm

Figure P12.10

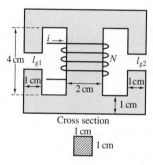

Cross section

Figure P12.11

12.12 Find the current necessary to establish a flux of $\phi = 3 \times 10^{-4}$ Wb in the series magnetic circuit of Figure P12.12. Here $l_{iron} = l_{steel} = 0.3$ m, area (throughout) $= 5 \times 10^{-4}$ m², and $N = 100$ turns. Assume $\mu_r = 5{,}195$ for cast iron and $\mu_r = 1{,}000$ for cast steel.

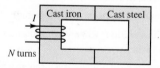

Figure P12.12

12.13 Find the magnetic flux ϕ established in the series magnetic circuit of Figure P12.13.

12.14

a. Find the current I required to establish a flux $\phi = 2.4 \times 10^{-4}$ Wb in the magnetic circuit of Figure P12.14. Here area(throughout) $= 2 \times 10^{-4}$ m², $l_{ab} = l_{ef} = 0.05$ m, $l_{af} = l_{be} = 0.02$ m, $l_{bc} = l_{dc}$, and the material is sheet steel.

b. Compare the mmf drop across the air gap to that across the rest of the magnetic circuit. Discuss your results, using the value of μ for each material.

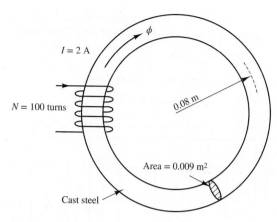

$I = 2$ A

$N = 100$ turns

0.08 m

Area = 0.009 m²

Cast steel

Figure P12.13

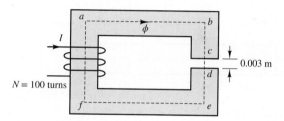

0.003 m

$N = 100$ turns

Figure P12.14

12.15 For the series-parallel magnetic circuit of Figure P12.15, find the value of I required to establish a flux in the gap of $\phi = 2 \times 10^{-4}$ Wb. Here, $l_{ab} = l_{bg} = l_{gh} = l_{ha} = 0.2$ m, $l_{bc} = l_{fg} = 0.1$ m, $l_{cd} = l_{ef} = 0.099$ m, and the material is sheet steel.

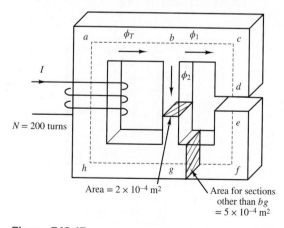

Area = 2×10^{-4} m²

Area for sections other than bg = 5×10^{-4} m²

Figure P12.15

12.16 Refer to the actuator of Figure P12.16. The entire device is made of sheet steel. The coil has 2,000 turns. The armature is stationary so that the length of the air gaps, $g = 10$ mm, is fixed. A direct current passing through the coil produces a flux density of 1.2 T in the gaps. Assume $\mu_r = 4,000$ for sheet steel. Determine:

a. The coil current.

b. The energy stored in the air gaps.

c. The energy stored in the steel.

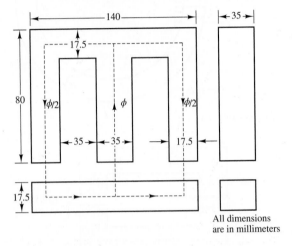

All dimensions are in millimeters

Figure P12.16

12.17 A core is shown in Figure P12.17, with $\mu_r = 2,000$ and $N = 100$. Find:

a. The current needed to produce a flux density of 0.4 Wb/m² in the center leg.

b. The current needed to produce a flux density of 0.8 Wb/m² in the center leg.

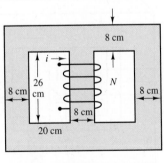

Cross section
8 cm

8 cm

Figure P12.17

Section 12.4: Transformers

12.18 For the transformer shown in Figure P12.18, $N = 1,000$ turns, $l_1 = 16$ cm, $A_1 = 4$ cm², $l_2 = 22$ cm,

$A_2 = 4$ cm², $l_3 = 5$ cm, and $A_3 = 2$ cm². The relative permeability of the material is $\mu_r = 1,500$.

a. Construct the equivalent magnetic circuit, and find the reluctance associated with each part of the circuit.

b. Determine the self-inductance and mutual inductance for the pair of coils (that is, L_{11}, L_{22}, and $M = L_{12} = L_{21}$).

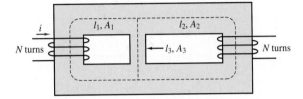

Figure P12.18

12.19 A transformer is delivering power to a 300-Ω resistive load. To achieve the desired power transfer, the turns ratio is chosen so that the resistive load referred to the primary is 7,500 Ω. The parameter values, *referred to the secondary winding*, are:

$$r_1 = 20\ \Omega \qquad L_1 = 1.0\ \text{mH} \qquad L_m = 25\ \text{mH}$$
$$r_2 = 20\ \Omega \qquad L_2 = 1.0\ \text{mH}$$

Core losses are negligible.

a. Determine the turns ratio.

b. Determine the input voltage, current, and power and the efficiency when this transformer is delivering 12 W to the 300-Ω load at a frequency $f = 10,000/2\pi$ Hz.

12.20 A 220/20-V transformer has 50 turns on its low-voltage side. Calculate:

a. The number of turns on its high side.

b. The turns ratio α when it is used as a step-down transformer.

c. The turns ratio α when it is used as a step-up transformer.

12.21 The high-voltage side of a transformer has 750 turns, and the low-voltage side has 50 turns. When the high side is connected to a rated voltage of 120 V, 60 Hz, a rated load of 40 A is connected to the low side. Calculate:

a. The turns ratio.

b. The secondary voltage (assuming no internal transformer impedance voltage drops).

c. The resistance of the load.

12.22 A transformer is to be used to match an 8-Ω loudspeaker to a 500-Ω audio line. What is the turns ratio of the transformer, and what are the voltages at the primary and secondary terminals when 10 W of audio power is delivered to the speaker? Assume that the speaker is a resistive load and that the transformer is ideal.

12.23   The high-voltage side of a step-down transformer has 800 turns, and the low-voltage side has 100 turns. A voltage of 240 VAC is applied to the high side, and the load impedance is 3 Ω (low side). Find:

a. The secondary voltage and current.

b. The primary current.

c. The primary input impedance from the ratio of primary voltage to current.

d. The primary input impedance.

12.24   Calculate the transformer ratio of the transformer in Problem 12.23 when it is used as a step-up transformer.

12.25   A 2,300/240-V, 60-Hz, 4.6-kVA transformer is designed to have an induced emf of 2.5 V/turn. Assuming an ideal transformer, find:

a. The numbers of high-side turns N_h and low-side turns N_l.

b. The rated current of the high-voltage side I_h.

c. The transformer ratio when the device is used as a step-up transformer.

Section 12.5: Electromechanical Energy Conversion

12.26   Calculate the current required to lift the load for the electromagnet of Example 12.9. Calculate the holding current required to keep the load in place once it has been lifted and is attached to the magnet. Assume: $N = 700$; $\mu_0 = 4\pi \times 10^{-7}$; $\mu_r = 10^4$ (equal for electromagnet and load); initial distance (air gap) = 0.5 m; magnetic path length of electromagnet = $l_1 = 0.80$ m; magnetic path length of movable load = $l_2 = 0.40$ m; gap cross-sectional area = 5×10^{-4} m^2; m = mass of load = 10 kg; $g = 9.8$ m/s^2.

12.27   For the electromagnet of Example 12.9:

a. Calculate the current required to keep the bar in place. (*Hint:* The air gap becomes zero, and the iron reluctance cannot be neglected.) Assume $\mu_r = 1,000$, $L = 1$ m.

b. If the bar is initially 0.1 m away from the electromagnet, what initial current would be required to lift the magnet?

12.28   The electromagnet of Figure P12.28 has reluctance given by $\mathcal{R}(x) = 7 \times 10^8(0.002 + x)$ H^{-1}, where x is the length of the variable gap in meters. The coil has 980 turns and 30-Ω resistance. For an applied voltage of 120 VDC, find:

a. The energy stored in the magnetic field for $x = 0.005$ m.

b. The magnetic force for $x = 0.005$ m.

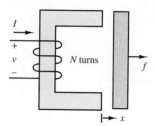

Figure P12.28

12.29   With reference to Example 12.10, determine the best combination of current magnitude and wire diameter to reduce the volume of the solenoid coil to a minimum. Will this minimum volume result in the lowest possible resistance? How does the power dissipation of the coil change with the wire gauge and current value? To solve this problem, you will need to find a table of wire gauge diameter, resistance, and current ratings. Table 1.2 in this book contains some information. The solution can only be found numerically.

12.30   Derive the same result obtained in Example 12.10, using equation 12.46 and the definition of inductance given in equation 12.30. You will first compute the inductance of the magnetic circuit as a function of the reluctance, then compute the stored magnetic energy, and finally write the expression for the magnetic force given in equation 12.46.

12.31   Derive the same result obtained in Example 12.11, using equation 12.46 and the definition of inductance given in equation 12.30. You will first compute the inductance of the magnetic circuit as a function of the reluctance, then compute the stored magnetic energy, and finally write the expression for the magnetic force given in equation 12.46.

12.32   With reference to Example 12.11, generate a simulation program (e.g., using Simulink^{TM}) that accounts for the fact that the solenoid inductance is not constant but is a function of plunger position. Compare graphically the current and force step responses of the constant-L simplified solenoid model to the step responses obtained in Example 12.11. Assume $\mu_r = 1,000$.

12.33   With reference to Example 12.12, calculate the required holding current to keep the relay closed. The mass of the moving element is $m = 0.05$ kg. Neglect damping. The initial position is $x = \epsilon = 0.001$ m.

12.34   The relay circuit shown in Figure P12.34 has the following parameters: $A_{gap} = 0.001$ m²; $N = 500$ turns; $L = 0.02$ m; $\mu = \mu_0 = 4\pi \times 10^{-7}$ (neglect the iron reluctance); $k = 1,000$ N/m; $R = 18$ Ω. What is the minimum DC supply voltage v for which the relay will make contact when the electrical switch is closed?

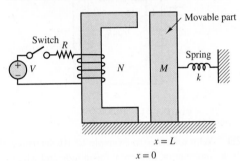

Figure P12.34

12.35   The magnetic circuit shown in Figure P12.35 is a very simplified representation of devices used as *surface roughness sensors*. The stylus is in contact with the surface and causes the plunger to move along with the surface. Assume that the flux ϕ in the gap is given by the expression $\phi = \beta/\mathcal{R}(x)$, where β is a known constant and $\mathcal{R}(x)$ is the reluctance of the gap. The emf e is measured to determine the surface profile. Derive an expression for the displacement x as a function of the various parameters of the magnetic circuit and of the measured emf. (Assume a frictionless contact between the moving plunger and the magnetic structure and that the plunger is restrained to vertical motion only. The cross-sectional area of the plunger is A.)

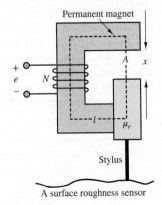

Figure P12.35   A surface roughness sensor

12.36   A cylindrical solenoid is shown in Figure P12.36. The plunger may move freely along its axis. The air gap between the shell and the plunger is uniform and equal to 1 mm, and the diameter d is 25 mm. If the exciting coil carries a current of 7.5 A, find the force acting on the plunger when $x = 2$ mm. Assume $N = 200$ turns, and neglect the reluctance of the steel shell. Assume l_g is negligible.

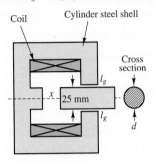

Figure P12.36

12.37   The double-excited electromechanical system shown in Figure P12.37 moves horizontally. Assume that resistance, magnetic leakage, and fringing are negligible; the permeability of the core is very large; and the cross section of the structure is $w \times w$. Find:

a. The reluctance of the magnetic circuit.

b. The magnetic energy stored in the air gap.

c. The force on the movable part as a function of its position.

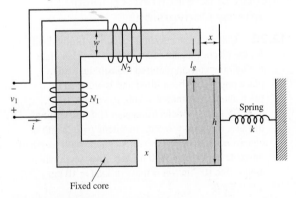

Figure P12.37

12.38   Determine the force F between the faces of the poles (stationary coil and plunger) of the solenoid pictured in Figure P12.38 when it is energized. When energized, the plunger is drawn into the coil and comes to rest with only a negligible air gap separating the two. The flux density in the cast steel pathway is 1.1 T. The diameter of the plunger is 10 mm. Assume that the reluctance of the steel is negligible.

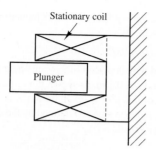

Figure P12.38

12.39 An electromagnet is used to support a solid piece of steel, as shown in Example 12.9. A force of 10,000 N is required to support the weight. The cross-sectional area of the magnetic core (the fixed part) is 0.01 m². The coil has 1,000 turns. Determine the minimum current that can keep the weight from falling for $x = 1.0$ mm. Assume negligible reluctance in steel and negligible fringing in the air gaps.

12.40 The armature, frame, and core of a 12-VDC control relay are made of sheet steel. The average length of the magnetic circuit is 12 cm when the relay is energized, and the average cross section of the magnetic circuit is 0.60 cm². The coil is wound with 250 turns and carries 50 mA. Determine:

a. The flux density $\mathcal{B}$ in the magnetic circuit of the relay when the coil is energized.

b. The force $\mathcal{F}$ exerted on the armature to close it when the coil is energized.

12.41 A relay is shown in Figure P12.41. Find the differential equations describing the system.

12.42 A solenoid having a cross section of 10 cm² is shown in Figure P12.42.

a. Calculate the force exerted on the plunger when the distance x is 2 cm and the current in the coil (where $N = 100$ turns) is 5 A. Assume that the fringing and leakage effects are negligible. The relative permeabilities of the magnetic material and the non-magnetic sleeve are 2,000 and 1.

b. Develop a set of differential equations governing the behavior of the solenoid.

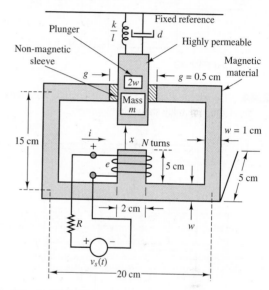

Figure P12.42

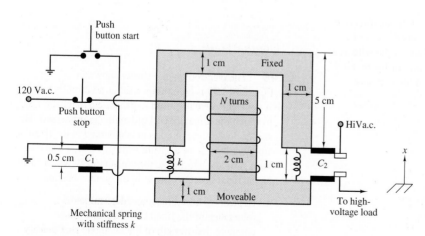

Figure P12.41

12.43 Derive the differential equations (electrical and mechanical) for the relay shown in Figure P12.43. Do not assume that the inductance is fixed; it is a function of x. You may assume that the iron reluctance is negligible.

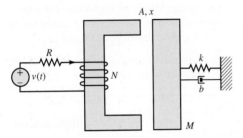

Figure P12.43

12.44 Derive the complete set of differential equations describing the relay shown in Figure P12.44.

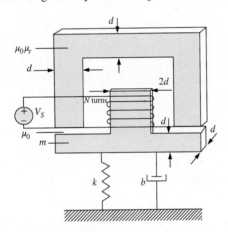

Figure P12.44

12.45 A wire of length 20 cm vibrates in one direction in a constant magnetic field with a flux density of 0.1 T; see Figure P12.45. The position of the wire as a function of time is given by $x(t) = 0.1 \sin 10t$ m. Find the induced emf across the length of the wire as a function of time.

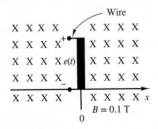

Figure P12.45

12.46 The wire of Problem 12.45 induces a time-varying emf of

$$e_1(t) = 0.02 \cos 10t$$

A second wire is placed in the same magnetic field but has a length of 0.1 m, as shown in Figure P12.46. The position of this wire is given by $x(t) = 1 - 0.1 \sin 10t$. Find the induced emf $e(t)$ defined by the difference between $e_1(t)$ and $e_2(t)$.

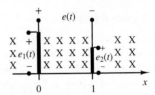

Figure P12.46

12.47 A conducting bar shown in Figure 12.48 in the text is carrying 4 A of current in the presence of a magnetic field $B = 0.3$ Wb/m². Find the magnitude and direction of the force induced on the bar.

12.48 A wire, shown in Figure P12.48, is moving in the presence of a magnetic field $B = 0.4$ Wb/m². Find the magnitude and direction of the induced voltage in the wire.

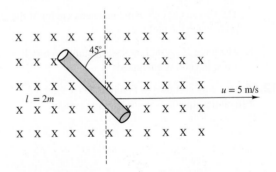

Figure P12.48

12.49 The electrodynamic shaker shown in Figure P12.49 is commonly used as a vibration tester. A constant current is used to generate a magnetic field in which the armature coil of length l is immersed. The shaker platform with mass m is mounted in the fixed structure by way of a spring with stiffness k. The platform is rigidly attached to the armature coil, which slides on the fixed structure thanks to frictionless bearings.

a. Neglecting iron reluctance, determine the reluctance of the fixed structure, and hence compute the strength of the magnetic flux density B in which the armature coil is immersed.

b. Knowing B, determine the dynamic equations of motion of the shaker, assuming that the moving coil has resistance R and inductance L.

c. Derive the transfer function and frequency response function of the shaker mass *velocity* in response to the input voltage V_S.

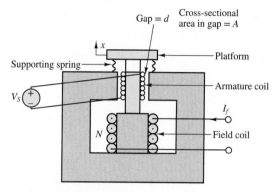

Figure P12.49 Electrodynamic shaker

12.50   The electrodynamic shaker of Figure P12.49 is used to perform vibration testing of an electrical connector. The connector is placed on the test platform (with mass m), and it may be assumed to have negligible mass when compared to the platform. The test consists of shaking the connector at the frequency $\omega = 2\pi \times 100$ rad/s.

Given the parameter values $B = 1{,}000$ Wb/m^2, $l = 5$ m, $k = 1{,}000$ N/m, $m = 1$ kg, $b = 5$ N-s/m, $L = 0.8$ H, and $R = 0.5\ \Omega$, determine the peak amplitude of the sinusoidal voltage V_S required to generate an acceleration of $5g$ (49 m/s^2).

12.51   Derive and sketch the frequency response of the loudspeaker of Example 12.13 for (1) $k = 50{,}000$ N/m and (2) $k = 5 \times 10^6$ N/m. Describe qualitatively how the loudspeaker frequency response changes as the spring stiffness k increases and decreases. What will the frequency response be in the limit as k approaches zero? What kind of speaker would this condition correspond to?

12.52   The loudspeaker of Example 12.13 has a midrange frequency response. Modify the mechanical parameters of the loudspeaker (mass, damping, and spring rate) so as to obtain a loudspeaker with a bass response centered on 400 Hz. Demonstrate that your design accomplishes the intended task, using frequency response plots. *Note: This is an open-ended design problem.*

12.53   The electrodynamic shaker shown in Figure P12.53 is used to perform vibration testing of an electronic circuit. The circuit is placed on a test table with mass m, and is assumed to have negligible mass when compared to the table. The test consists of shaking the circuit at the frequency $\omega = 2\pi(100)$ rad/s.

a. Write the dynamic equations for the shaker. Clearly indicate system input(s) and output(s).

b. Find the frequency response function of the table acceleration in response to the applied voltage.

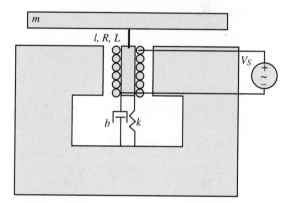

Figure P12.53

c. Given the following parameter values:

$$B = 200 \text{ Wb/m}^2 \qquad l = 5 \text{ m} \qquad k = 100 \text{ N/m}$$
$$m = 0.2 \text{ kg} \qquad b = 5 \text{ N-s/m}$$
$$L = 8 \text{ mH} \qquad R = 0.5\ \Omega$$

Determine the peak amplitude of the sinusoidal voltage V_S required to generate an acceleration of $5g$ (49 m/s^2) under the stated test conditions.

Design Credits: Mini DVI cable adapter isolated on white background: Robert Lehmann/Alamy Stock Photo; Balance scale: Alex Slobodkin/E+/Getty Images; Icon for "Focus on measurements" weighing scales: Media Bakery.

CHAPTER

13

INTRODUCTION TO ELECTRIC MACHINES

The objective of this chapter is to introduce the basic operation of rotating electric machines. The operation of the three major classes of electric machines—DC, synchronous, and induction—is described as intuitively as possible, building on the material presented in Chapter 12.

The emphasis of this chapter is on explaining the properties of each type of machine, with its advantages and disadvantages with regard to other types; and on classifying these machines in terms of their performance characteristics and preferred field of application.

13.1   ROTATING ELECTRIC MACHINES

This introductory section is aimed at explaining the common properties of all rotating electric machines. We begin our discussion with reference to Figure 13.1, in which a hypothetical rotating machine is depicted in a cross-sectional view. In the figure, a box with a cross inscribed in it indicates current flowing into the page, while a dot represents current out of the plane of the page.

In Figure 13.1, we identify a **stator**, of cylindrical shape, and a **rotor**, which, as the name indicates, rotates inside the stator, separated from the latter by means of an air gap. The rotor and stator each consist of a magnetic core, some electrical insulation, and the windings necessary to establish a magnetic flux (unless this is created by a permanent magnet). The rotor is mounted on a bearing-supported shaft, which can be connected to *mechanical loads* (if the machine is a motor) or to a *prime mover* (if the machine is a generator) by means of belts, pulleys, chains, or other mechanical couplings. The windings carry the electric currents that generate the magnetic fields and flow to the electrical loads, and also provide the closed loops in which voltages will be induced (by virtue of Faraday's law, as discussed in Chapter 12).

Basic Classification of Electric Machines

An immediate distinction can be made between different types of windings characterized by the nature of the current they carry. If the current serves the sole purpose of providing a magnetic field and is independent of the load, it is called a *magnetizing*, or excitation, current, and the winding is termed a **field winding**. Field currents are nearly always direct current (DC) and are of relatively low power, since their only purpose is to magnetize the core (recall the important role of high-permeability cores in generating large magnetic fluxes from relatively small currents). On the other hand, if the winding carries only the load current, it is called an **armature**. In DC and alternating-current (AC) synchronous machines, separate windings exist to carry field and armature currents. In the induction motor, the magnetizing and load currents flow in the same winding, called the *input winding*, or *primary*; the output winding is then called the *secondary*. As we shall see, this terminology, which is reminiscent of transformers, is particularly appropriate for induction motors, which bear a significant analogy to the operation of the transformers studied in Chapters 6 and 12. Table 13.1 characterizes the principal machines in terms of their field and armature configuration.

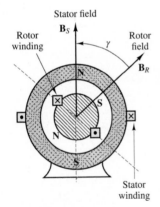

Figure 13.1 A rotating electric machine

Table 13.1 **Configurations of the three types of electric machines**

Machine type	Winding	Winding type	Location	Current
DC	Input and output	Armature	Rotor	AC (winding) DC (at brushes)
	Magnetizing	Field	Stator	DC
Synchronous	Input and output	Armature	Stator	AC
	Magnetizing	Field	Rotor	DC
Induction	Input	Primary	Stator	AC
	Output	Secondary	Rotor	AC

It is also useful to classify electric machines in terms of their energy conversion characteristics. A machine acts as a **generator** if it converts mechanical energy from a prime mover, say, an internal combustion engine, to electric energy. Examples of generators are the large machines used in power generating plants, or the common automotive alternator. A machine is classified as a **motor** if it converts electric energy to mechanical form. The latter class of machines is probably of more direct interest to you, because of its widespread application in engineering design. Electric motors are used to provide forces and torques to generate motion in countless industrial applications. Machine tools, robots, punches, presses, mills, and propulsion systems for electric vehicles are but a few examples of the application of electric machines in engineering.

Note that in Figure 13.1 we have explicitly shown the direction of two magnetic fields: that of the rotor $\mathbf{B}_R$ and that of the stator $\mathbf{B}_S$. Although these fields are generated by different means in different machines (e.g., permanent magnets, alternating currents, direct currents), the presence of these fields is what causes a rotating machine to turn and enables the generation of electric power. In particular, we see that in Figure 13.1 the north pole of the rotor field will seek to align itself with the south pole of the stator field. It is this magnetic attraction force that permits the generation of torque in an electric motor; conversely, a generator exploits the laws of electromagnetic induction to convert a changing magnetic field to an electric current.

To simplify the discussion in later sections, we now introduce some basic concepts that apply to all rotating electric machines. Referring to Figure 13.2, which depicts a permanent-magnet DC machine, note that the force on a wire is given by the expression:

$$\mathbf{f} = i_w \mathbf{l} \times \mathbf{B} \qquad (13.1)$$

where i_w is the current in the wire, $\mathbf{l}$ is a vector along the direction of the wire, and × denotes the cross product of two vectors. Then the torque for a multiturn coil is:

$$T = KBi_w \sin \alpha \qquad (13.2)$$

where:

B = magnetic flux density caused by stator field

K = constant depending on coil geometry

α = angle between $\mathbf{B}$ and normal to plane of coil

In the machine of Figure 13.2, there are two magnetic fields: one generated within the stator, the other within the rotor windings. Either (but not both) of these fields

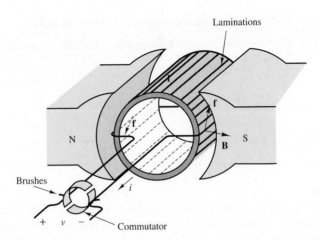

Figure 13.2 Stator and rotor fields and the force acting on a
rotating permanent-magnet DC machine

could be generated by a current or by a permanent magnet. Thus, we could replace
the permanent-magnet stator of Figure 13.2 with a suitably arranged winding to
generate a stator field in the same direction. If the stator were made of a toroidal coil
of radius R (see Chapter 12), then the magnetic field of the stator would generate a
flux density B, where:

$$B = \mu H = \mu \frac{Ni}{2\pi R} \tag{13.3}$$

and where N is the number of turns and i is the coil current. The direction of the
torque is always the direction determined by the rotor and stator fields as they seek
to align to each other (i.e., counterclockwise in the diagram of Figure 13.1).

It is important to note that Figure 13.2 is only one example of the major features
and characteristics of rotating machines. A variety of configurations exist, depending
on whether each of the fields is generated by a current in a coil or by a permanent
magnet and whether the load and magnetizing currents are direct or alternating. The
type of excitation (AC or DC) provided to the windings permits a first classification
of electric machines (see Table 13.1). According to this classification, one can define
the following types of machines:

- *DC machines:* Direct current in both stator and rotor (the
  stator could also be realized by a permanent magnet, as in
  Figure 13.2)
- *Synchronous machines:* Alternating current in one stator,
  direct current in the rotor (the rotor could alternatively
  consist of a permanent magnet)
- *Induction machines:* Alternating current in both

In most industrial applications, the induction machine is the preferred choice,
because of the simplicity of its construction. However, the analysis of the performance of an induction machine is rather complex. On the other hand, DC machines
are quite complex in their construction but can be analyzed relatively simply with

the analytical tools we have already acquired. Therefore, the progression of this chapter is as follows. We start with a section that discusses the physical construction of DC machines, both motors and generators. Then we continue with a discussion of synchronous machines, in which one of the currents is now alternating, since these can easily be understood as an extension of DC machines. Finally, we consider the case where both rotor and stator currents are alternating, and we analyze the induction machine.

Performance Characteristics of Electric Machines

As already stated earlier in this chapter, electric machines are **energy conversion devices**, and we are therefore interested in their energy conversion **efficiency**. Typical applications of electric machines as motors or generators must take into consideration the energy losses associated with these devices. Figure 13.3(a) and (b) represents the various loss mechanisms you must consider in analyzing the efficiency of an electric machine for the case of DC machines. It is important for you to keep in mind this conceptual flow of energy when analyzing electric machines. The sources of loss in a rotating machine can be separated into three fundamental groups: electrical (I^2R) losses, core losses, and mechanical losses.

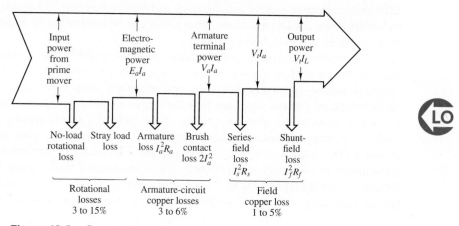

Figure 13.3a Generator losses, direct current

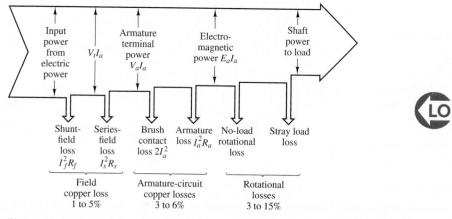

Figure 13.3b Motor losses, direct current

Usually I^2R losses are computed on the basis of the DC resistance of the windings at 75°C; in practice, these losses vary with operating conditions. The difference between the nominal and actual I^2R loss is usually lumped under the category of *stray-load loss*. In DC machines, it is also necessary to account for the *brush contact loss* associated with slip rings and commutators.

Mechanical losses are due to *friction* (mostly in the bearings) and *windage*, that is, the air drag force that opposes the motion of the rotor. In addition, if external devices (e.g., blowers) are required to circulate air through the machine for cooling purposes, the energy expended by these devices is included in the mechanical losses.

Open-circuit core losses consist of *hysteresis* and *eddy current* losses, with only the excitation winding energized (see Chapter 12 for a discussion of hysteresis and eddy currents). Often these losses are summed with friction and windage losses to give rise to the *no-load rotational loss*. The latter quantity is useful if one simply wishes to compute efficiency. Since open-circuit core losses do not account for the changes in flux density caused by the presence of load currents, an additional magnetic loss is incurred that is not accounted for in this term. *Stray-load losses* are used to lump the effects of nonideal current distribution in the windings and of the additional core losses just mentioned. Stray-load losses are difficult to determine exactly and are often assumed to be equal to 1.0 percent of the output power for DC machines; these losses can be determined by experiment in synchronous and induction machines.

The performance of an electric machine can be quantified in a number of ways. In the case of an electric motor, it is usually portrayed in the form of a graphical **torque–speed characteristic** and **efficiency map**. The torque–speed characteristic of a motor describes how the torque supplied by the machine varies as a function of the speed of rotation of the motor for steady speeds. As we shall see in later sections, the torque–speed curves vary in shape with the type of motor (DC, induction, synchronous) and are very useful in determining the performance of the motor when connected to a mechanical load. Figure 13.4(a) depicts the torque–speed curve of induction motor. Figure 13.4(b) depicts a typical efficiency map for a permanent-magnet synchronous motor. In most

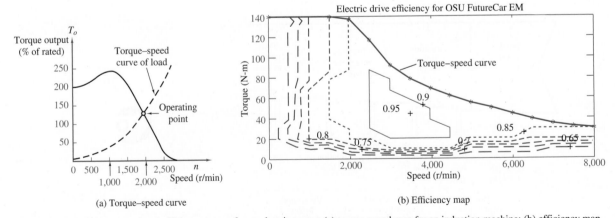

Figure 13.4 Torque–speed and efficiency curves for an electric motor: (a) torque-speed map for an induction machine; (b) efficiency map for an electric drive system for a hybrid-electric vehicle

engineering applications, it is quite likely that the engineer is required to make a decision regarding the performance characteristics of the motor best suited to a specified task. In this context, the torque–speed curve of a machine is a very useful piece of information.

The first feature we note of the torque–speed characteristic is that it bears a strong resemblance to the i-v characteristics used in earlier chapters to represent the behavior of electrical sources. It should be clear that, according to this torque–speed curve, the motor is not an ideal source of torque (if it were, the curve would appear as a horizontal line across the speed range). One can readily see, for example, that the induction motor represented by the curves of Figure 13.4(a) would produce maximum torque in the range of speeds between approximately 800 and 1,400 r/min. What determines the actual speed of the motor (and therefore its output torque and power) is the torque–speed characteristic of the load connected to it, much as a resistive load determines the current drawn from a voltage source. In the figure, we display the torque–speed curve of a load, represented by the dashed line; the operating point of the motor-load pair is determined by the intersection of the two curves.

Another important observation pertains to the fact that the motor of Figure 13.4(a) produces a nonzero torque at zero speed. This fact implies that as soon as electric power is connected to the motor, the latter is capable of supplying a certain amount of torque; this zero-speed torque is called the **starting torque**. If the load requires less than the starting torque the motor can provide, then the motor can accelerate the load until the motor speed and torque settle to a stable value, at the operating point. As we discuss each type of machine in greater detail, we shall devote some time to the discussion of its torque–speed curve.

The efficiency of an electric machine is also an important design and performance characteristic. The 2005 Department of Energy's Energy Policy Act, also known as EPACT, has required electric motor manufacturers to guarantee a minimum efficiency. The efficiency of an electric motor is usually described using a contour plot of the efficiency value (a number between 0 and 1) in the torque–speed plane. This representation permits a determination of the motor efficiency as a function of its performance and operating conditions. Figure 13.4(b) depicts the efficiency map of an electric drive used in a hybrid-electric vehicle—a 20-kW permanent-magnet AC synchronous machine. Note that the peak efficiency can be as high as 0.95 (95 percent), but that the efficiency decreases significantly away from the optimum point (around 3,500 r/min and 45 N-m), to values as low as 0.65.

The most common means of conveying information regarding electric machines is the *nameplate*. Typical information conveyed by the nameplate includes

1. Type of device (e.g., DC motor, alternator)
2. Manufacturer
3. Rated voltage and frequency
4. Rated current and voltamperes
5. Rated speed and horsepower

The **rated voltage** is the terminal voltage for which the machine was designed, and which will provide the desired magnetic flux. Operation at higher voltages will increase magnetic core losses, because of excessive core saturation. The **rated current** and **rated voltamperes** are an indication of the typical current and power levels at the terminal that will not cause undue overheating due to copper losses (I^2R losses) in the windings. These ratings are not absolutely precise, but they give an indication of the range of excitations for which the motor will perform without overheating. Other name plate characteristics are introduced in Example 13.2.

Peak power operation in a motor may exceed rated torque, power, or currents by a substantial factor (up to as much as 6 or 7 times the rated value); however, continuous operation of the motor above the rated performance will cause the machine to overheat and eventually to sustain damage. Thus, it is important to consider both peak and continuous power requirements when selecting a motor for a specific application. An analogous discussion is valid for the speed rating: While an electric machine may operate above rated speed for limited periods of time, the large centrifugal forces generated at high rotational speeds will eventually cause undesirable mechanical stresses, especially in the rotor windings.

Another important feature of electric machines is the **regulation** of the machine speed or voltage, depending on whether it is used as a motor or as a generator, respectively. Regulation is the ability to maintain speed or voltage constant in the face of load variations. The ability to closely regulate speed in a motor or voltage in a generator is an important feature of electric machines; regulation is often improved by means of feedback control mechanisms, some of which are briefly introduced in this chapter. We take the following definitions as being adequate for the intended purpose of this chapter:

$$\text{Speed regulation} = \text{SR} = \frac{\text{Speed at no load} - \text{Speed at rated load}}{\text{Speed at rated load}} \qquad (13.4)$$

$$\text{Voltage regulation} = \text{VR} = \frac{\text{Voltage at no load} - \text{Voltage at rated load}}{\text{Voltage at rated load}} \qquad (13.5)$$

Please note that the rated value is usually taken to be the nameplate value, and that the meaning of *load* changes depending on whether the machine is a motor, in which case the load is mechanical, or a generator, in which case the load is electrical.

EXAMPLE 13.1 Regulation

Problem

Find the percentage of speed regulation of a shunt DC motor.

Solution

Known Quantities: No-load speed; speed at rated load.

Find: Percentage speed regulation, denoted by SR%.

Schematics, Diagrams, Circuits, and Given Data:

n_{nl} = no-load speed = 1,800 r/min

n_{rl} = rated load speed = 1,760 r/min

Analysis:

$$SR\% = \frac{n_{nl} - n_{rl}}{n_{rl}} \times 100 = \frac{1,800 - 1,760}{1,760} \times 100 = 2.27\%$$

Comments: Speed regulation is an intrinsic property of a motor; however, external speed controls can be used to regulate the speed of a motor to any (physically achievable) desired value. Some motor control concepts are discussed later in this chapter.

CHECK YOUR UNDERSTANDING

The percentage of speed regulation of a motor is 10 percent. If the full-load speed is 50π rad/s, find (a) the no-load speed in radians per second, and (b) the no-load speed in revolutions per minute. Finally, (c) if the percentage of voltage regulation for a 250-V generator is 10 percent, find the no-load voltage of the generator.

Answer: (a) $\omega = 55\pi$ rad/s; (b) $n = 1,650$ r/min; (c) $V_{no-load} = 275$ V

Table 13.2 summarizes important unit conversions that relate SI to English units, as the latter are still used in nameplate data in the United States.

Table 13.2 **Unit conversions for electric machines**

Quantity	SI unit	English unit
Length	1 m	3.281 ft
Mass	1 kg	2.205 lb (mass)
Force	1 N	0.224 lb (force)
Torque	1 N-m	0.738 lb-ft
		8.85 lb-in
Power	1 kW	1.341 hp
Moment of inertia	1 kg-m^2	23.73 lb-ft^2

EXAMPLE 13.2   Nameplate Data

Problem

Discuss the nameplate data, shown below, of a typical induction motor.

Solution

Known Quantities: Nameplate data.

Find: Motor characteristics.

Schematics, Diagrams, Circuits, and Given Data: The nameplate appears below.

MODEL	19308 J-X		
TYPE	CJ4B	FRAME	324TS
VOLTS	230/460	°C AMB.	40
		INS. CL.	B
FRT. BRG	210SF	EXT. BRG	312SF
SERV FACT	1.0	OPER INSTR	C-517
PHASE | 3	Hz | 60	CODE | G	WDGS | 1
H.P.	40		
R.P.M.	3,565		
AMPS	106/53		
NEMA NOM.	EFF		
NOM. P.F.			
DUTY	CONT.	NEMA DESIGN	B

Analysis: The nameplate of a typical induction motor is shown in the preceding table. The model number (sometimes abbreviated as MOD) uniquely identifies the motor to the manufacturer. It may be a style number, a model number, an identification number, or an instruction sheet reference number.

The term *frame* (sometimes abbreviated as FR) refers principally to the physical size of the machine, as well as to certain construction features.

Ambient temperature (abbreviated as AMB, or MAX. AMB) refers to the maximum ambient temperature in which the motor is capable of operating. Operation of the motor in a higher ambient temperature may result in shortened motor life and reduced torque.

Insulation class (abbreviated as INS. CL.) refers to the type of insulation used in the motor. The classes most often used are class A (105°C) and class B (130°C).

The duty (DUTY), or time rating, denotes the length of time the motor is expected to be able to carry the rated load under usual service conditions. "CONT." means that the machine can be operated continuously.

The "CODE" letter sets the limits of starting kilovoltamperes per horsepower for the machine. There are 19 levels, denoted by the letters A through V, excluding I, O, and Q.

Service factor (abbreviated as SERV FACT) is a term defined by NEMA (the National Electrical Manufacturers Association) as follows: "The service factor of a general-purpose alternating-current motor is a multiplier which, when applied to the rated horsepower, indicates a permissible horsepower loading which may be carried under the conditions specified for the service factor."

The voltage figure given on the nameplate refers to the voltage of the supply circuit to which the motor should be connected. Sometimes two voltages are given, for example, 230/460. In this case, the machine is intended for use on either a 230-V or a 460-V circuit. Special instructions will be provided for connecting the motor for each of the voltages.

The term "BRG" indicates the nature of the bearings supporting the motor shaft.

CHECK YOUR UNDERSTANDING

The nameplate of a three-phase induction motor indicates the following values:

H.P. = 10	Volt = 220 V
R.P.M. = 1,750	Service factor = 1.15
Temperature rise = 60°C	Amp = 30 A

Find the rated torque, rated voltamperes, and maximum continuous output power.

Answer: $I_{rated} = 40.7$ N-m; rated VA = 11,431 VA; $P_{max} = 11.5$ hp.

EXAMPLE 13.3   Torque–Speed Curves

Problem

Discuss the significance of the torque–speed curve of an electric motor.

Solution

An induction motor has a torque output that varies directly with speed; hence, the power output varies directly with the speed. Motors with this characteristic are commonly used with fans, blowers, and centrifugal pumps. Figure 13.5 shows typical torque–speed curves for this type of motor. Superimposed on the motor torque–speed curve is the torque–speed curve for a typical fan where the input power to the fan varies as the cube of the fan speed. Point A is the actual operating point, which could be determined graphically by plotting the load line and the motor torque–speed curve on the same graph, as illustrated in Figure 13.5. The fan will operate at the speed corresponding to the intersection of the two curves.

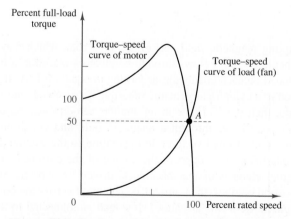

Figure 13.5 Torque–speed curves of electric motor and load

CHECK YOUR UNDERSTANDING

A motor having the characteristics shown in Figure 13.4(a) is to drive a load; the load has a linear torque–speed curve and requires 150 percent of rated torque at 1,500 r/min. Find the operating point for this motor-load pair.

Answer: 170 percent of rated torque; 1,700 r/min.

Basic Operation of All Rotating Machines

We have already seen in Chapter 12 how the magnetic field in electromechanical devices provides a form of coupling between electrical and mechanical systems. Intuitively, one can identify two aspects of this coupling, both of which play a role in the operation of electric machines:

1. Magnetic attraction and repulsion forces generate mechanical torque.

2. The magnetic field can induce a voltage in the machine windings (coils) by virtue of Faraday's law.

Thus, an electric machine can serve either as a motor or a generator, depending on whether the input power is electric and mechanical power is produced (motor action), or the input power is mechanical and the output power is electric (generator action). Figure 13.6 illustrates the two cases graphically.

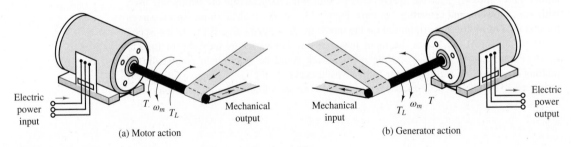

Electric power input
T ω_m T_L
Mechanical output
(a) Motor action

Mechanical input
T_L ω_m T
Electric power output
(b) Generator action

Figure 13.6 Generator (a) and motor (b) action in an electric machine

The coupling magnetic field performs a dual role, which may be explained as follows. When a current i flows through conductors placed in a magnetic field, a force is produced on each conductor, according to equation 13.1. If these conductors are attached to a cylindrical structure, a torque is generated; and if the structure is free to rotate, then it will rotate at an angular velocity ω_m. As the conductors rotate, however, they move through a magnetic field and cut through flux lines, thus generating an electromotive force in opposition to the excitation. This emf is also called *counter-emf*, as it opposes the source of the current i. If, on the other hand, the rotating element of the machine is driven by a prime mover (e.g., an internal combustion engine), then an emf is generated across the coil that is rotating in the magnetic field (the armature). If a load is connected to the armature, a current i will flow to the load, and this current flow will in turn cause a reaction torque on the armature that opposes the torque imposed by the prime mover.

You see, then, that for energy conversion to take place, two elements are srequired:

1. A coupling field **B**; generated in the field winding or by a permanent magnet.
2. An armature winding that supports the load current i and the emf e.

Magnetic Poles in Electric Machines

Before discussing the actual construction of a rotating machine, we should spend a few paragraphs to illustrate the significance of **magnetic poles** in an electric machine. In an electric machine, torque is developed as a consequence of magnetic forces of attraction and repulsion between magnetic poles on the stator and on the rotor; these poles produce a torque that accelerates the rotor and a reaction torque on the stator. It is also important to observe that the number of poles must be even, since there have to be equal numbers of north and south poles.

The motion and associated electromagnetic torque of an electric machine are the result of two magnetic fields that are trying to align with each other so that the south pole of one field attracts the north pole of the other. Figure 13.7 illustrates this action by analogy with two permanent magnets, one of which is allowed to rotate about its center of mass.

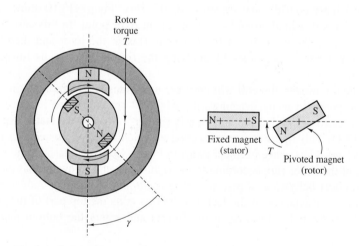

Figure 13.7 Alignment action of poles

Figure 13.8 depicts a two-pole machine in which the stator poles are constructed in such a way as to project closer to the rotor than to the stator structure. This type of construction is rather common, and poles constructed in this fashion are called **salient poles**. Note that the rotor could also be constructed to have salient poles.

To understand magnetic polarity, we need to consider the direction of the magnetic field in a coil carrying current. Figure 13.9 shows how the *right-hand rule* can be employed to determine the direction of the magnetic flux. If one were to grasp the coil with the right hand, with the fingers curling in the direction of current flow, then the thumb would be pointing in the direction of the magnetic flux. Magnetic flux by convention is viewed as entering the south pole and exiting from the north pole. Thus, to determine whether a magnetic pole is north or

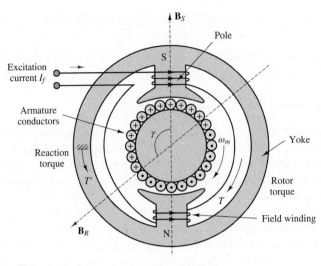

Cross section of DC machine

Figure 13.8 A two-pole machine with salient stator poles

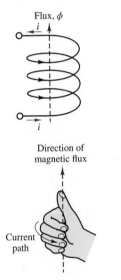

Figure 13.9 Right-hand rule

south, we must consider the direction of the flux. Figure 13.10 shows a cross section of a coil wound around a pair of salient rotor poles. In this case, one can readily identify the direction of the magnetic flux in the rotor and therefore the magnetic polarity of the poles by applying the right-hand rule, as illustrated in the figure.

Often, however, the coil windings are not arranged as simply as in the case of salient poles. In many machines, the windings are embedded in slots cut into the stator or rotor, so that the situation is similar to that of the stator depicted in Figure 13.11. This figure is a cross section in which the wire connections between "crosses" and "dots" have been cut away. In Figure 13.11, the dashed line indicates the axis of the stator flux according to the right-hand rule, showing that the slotted stator in effect behaves as a pole pair. The north and south poles indicated in the figure are a consequence of the fact that the flux exits the top part of the structure (thus, the north pole indicated in the figure) and enters the bottom half of the

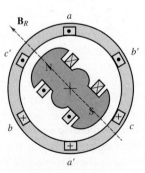

Figure 13.10 Magnetic field in a salient rotor winding

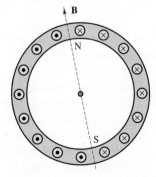

Figure 13.11 Magnetic field of stator

structure (thus, the south pole). In particular, if you consider that the windings are arranged so that the current entering the right-hand side of the stator (to the right of the dashed line) flows through the back end of the stator and then flows outward from the left-hand side of the stator slots (left of the dashed line), you can visualize the windings in the slots as behaving in a manner similar to the coils of Figure 13.10, where the flux axis of Figure 13.11 corresponds to the flux axis of each of the coils of Figure 13.10. The actual circuit that permits current flow is completed by the front and back ends of the stator, where the wires are connected according to the pattern a-a', b-b', c-c', as depicted in the figure.

Another important consideration that facilitates understanding of the operation of electric machines pertains to the use of alternating currents. It should be apparent by now that if the current flowing into the slotted stator is alternating, the direction of the flux will also alternate, so that in effect the two poles will reverse polarity every time the current reverses direction, that is, every half-cycle of the sinusoidal current. Further—since the magnetic flux is approximately proportional to the current in the coil—as the amplitude of the current oscillates in a sinusoidal fashion, so will the flux density in the structure. Thus, *the magnetic field developed in the stator changes both spatially and in time.*

This property is typical of AC machines, where a *rotating magnetic field* is established by energizing the coil with an alternating current. As explained in Section 13.2, the principles underlying the operation of DC and AC machines are quite different: In a direct-current machine, there is no rotating field, but a mechanical switching arrangement (the *commutator*) makes it possible for the rotor and stator magnetic fields to always align at right angles to each other.

The book website includes two-dimensional "animations" of the most common types of electric machines. You might wish to explore these animations to better understand the basic concepts described in this section.

13.2   DIRECT-CURRENT MACHINES

As explained in the introductory section, DC machines are easier to analyze than their AC counterparts although their actual construction is made rather complex by the need to have a commutator, which switches the load winding connection to the source so as to always maintain an angle close to 90° between the stator and the rotor magnetic fields. The objective of this section is to describe the major construction features and the operation of DC machines, as well as to develop simple circuit models that are useful in analyzing the performance of this class of machines.

Physical Structure of DC Machines

A representative DC machine was depicted in Figure 13.8, with the magnetic poles clearly identified, for both the stator and the rotor. Figure 13.12 is a photograph of the same type of machine. Note the salient pole construction of the stator and the slotted rotor. As previously stated, the torque developed by the machine is a consequence of the magnetic forces between stator and rotor poles. This torque is maximum when the angle γ between the rotor and stator poles is 90°. Also, as you can see from the figure, in a DC machine the armature circuit is on the rotor, and the field winding is on the stator.

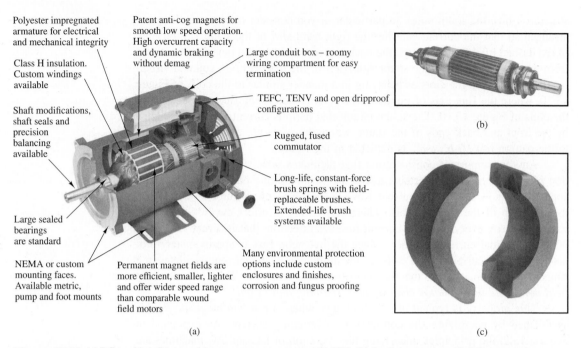

To keep this torque angle close to 90° as the rotor spins on its shaft, ame-chanical switch, called a **commutator**, is configured so the rotor poles are consistently close to 90° with respect to the fixed stator poles. In a DC machine, the magnetizing current is DC so that there is no spatial alternation of the stator poles due to time-varying currents. To understand the operation of the commutator, consider the simplified diagram of Figure 13.13. In the figure, the brushes are fixed, and the rotor revolves at an angular velocity ω_m; the instantaneous position of the rotor is given by the expression $\theta = \omega_m t - \gamma$.

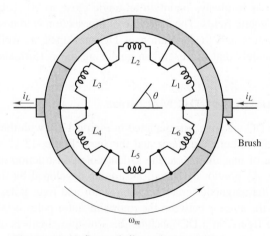

Figure 13.13 Rotor winding and commutator

The commutator is fixed to the rotor and is made up in this example of six segments that are made of electrically conducting material but are insulated from one another. Further, the rotor windings are configured so that they form six coils, connected to the commutator segments as shown in Figure 13.13.

As the commutator rotates counterclockwise, the rotor magnetic field rotates with it up to $\theta = 30°$. At that point, the direction of the current changes in coils L_3 and L_6 as the brushes make contact with the next segment. Now the direction of the magnetic field is $-30°$. As the commutator continues to rotate, the direction of the rotor field will again change from $-30°$ to $+30°$, and it will switch again when the brushes switch to the next pair of segments. In this machine, then, the torque angle γ is not always $90°$, but can vary by as much as $\pm30°$; the actual torque produced by the machine would fluctuate by as much as ±14 percent, since the torque is proportional to $\sin \gamma$. As the number of segments increases, the torque fluctuation produced by the commutation is greatly reduced. In a practical machine, for example, one might have as many as 60 segments, and the variation of γ from $90°$ would be only $\pm3°$, with a torque fluctuation of less than 1 percent. Thus, the DC machine can produce a nearly constant torque (as a motor) or voltage (as a generator).

Configuration of DC Machines

The DC machine of Figure 13.12 employs a permanent magnet to generate a constant magnetic field in the stator. However, in DC machines, the field excitation that provides the magnetizing current may be provided by an external source, in which case the machine is said to be **separately excited** [Figure 13.14(a)]. More often, the field excitation is derived from the armature voltage, and the machine is said to be **self-excited**. The latter configuration does not require the use of a separate source for the field excitation and is therefore frequently preferred. If a machine is in the separately excited configuration, an additional source V_f is required. In the self-excited case, one method used to provide the field excitation is to connect the field in parallel with the armature; since the field winding typically has significantly higher resistance than the armature circuit (remember that it is the armature that carries the load current), this will not draw excessive current from the armature. Further, a series resistor can be added to the field circuit to provide the means for adjusting the field current independent of the armature voltage. This configuration is called a **shunt-connected** machine and is depicted in Figure 13.14(b). Another method for self-exciting a DC machine consists of connecting the field in series with the armature, leading to the **series-connected** machine, depicted in Figure 13.14(c); in this case, the field winding will support the entire armature current, and thus the field coil must have low resistance (and therefore relatively few turns). This configuration is rarely used for generators, since the generated voltage and the load voltage must always differ by the voltage drop across the field coil, which varies with the load current. Thus, a series generator would have poor (large) regulation. However, series-connected motors are commonly used in traction applications.

DC Machine Models

As stated earlier, it is relatively easy to develop a simple model of a DC machine, which is well suited to performance analysis, without the need to resort to the

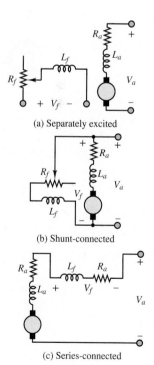

(a) Separately excited

(b) Shunt-connected

(c) Series-connected

Figure 13.14

details of the construction of the machine itself. This section illustrates the development of such models in two steps. First, algebraic equations relating field and armature currents and voltages to speed and torque are introduced; second, the differential equations describing the dynamic behavior of DC machines are derived.

When a field excitation is established, a magnetic flux ϕ is generated by the field current I_f. From equation 13.2, we know that the torque acting on the rotor is proportional to the product of the magnetic field and the current in the load-carrying wire; the latter current is the armature current I_a (i_w in equation 12.2). Assuming that, by virtue of the commutator, the torque angle γ is kept very close to 90°, and therefore $\sin \gamma = 1$, we obtain the following expression for the torque (in units of newton-meters) in a DC machine:

$$T = k_T \phi I_a \qquad \text{for } \gamma = 90° \qquad \text{DC machine torque} \qquad (13.6)$$

You may recall that this is simply a consequence of the *Bli* law of Chapter 12. The mechanical power generated (or absorbed) is equal to the product of the machine torque and the mechanical speed of rotation ω_m rad/s, and is therefore given by

$$P_m = \omega_m T = \omega_m k_T \phi I_a \qquad (13.7)$$

Recall now that the rotation of the armature conductors in the field generated by the field excitation causes a **back emf** E_b in a direction that opposes the rotation of the armature. According to the *Blu* law (see Chapter 12), then, this back emf is given by

$$E_b = k_a \phi \omega_m \qquad \text{DC machine back emf} \qquad (13.8)$$

where k_a is called the **armature constant** and is related to the geometry and magnetic properties of the structure. The voltage E_b represents a countervoltage (opposing the DC excitation) in the case of a motor and the generated voltage in the case of a generator. Thus, the electric power dissipated (or generated) by the machine is given by the product of the back emf and the armature current:

$$P_e = E_b I_a \qquad (13.9)$$

The constants k_T and k_a in equations 13.6 and 13.8 are related to geometry factors, such as the dimension of the rotor and the number of turns in the armature winding; and to properties of materials, such as the permeability of the magnetic materials. Note that in the ideal energy conversion case $P_m = P_e$, and therefore $k_a = k_T$. We shall in general assume such ideal conversion of electric to mechanical energy (or vice versa) and will therefore treat the two constants as being identical: $k_a = k_T$. The constant k_a is given by

$$k_a = \frac{pN}{2\pi M} \qquad (13.10)$$

where:

p = number of magnetic poles

N = number of conductors per coil

M = number of parallel paths in armature winding

An important observation concerning the units of angular speed must be made at this point. The equality (under the no-loss assumption) between the constants k_a and k_T in equations 13.6 and 13.8 results from the choice of consistent units, namely, volts and amperes for the electrical quantities and newton-meters and radians per second for the mechanical quantities. You should be aware that it is fairly common practice to refer to the speed of rotation of an electric machine in units of revolutions per minute (r/min).[1] In this book, we uniformly use the symbol n to denote angular speed in revolutions per minute; the following relationship should be committed to memory:

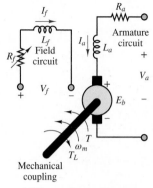

(a) Motor reference direction

$$n\,(\text{r/min}) = \frac{60}{2\pi}\omega_m \qquad \text{rad/s} \tag{13.11}$$

If the speed is expressed in revolutions per minute, the armature constant changes as follows:

$$E_b = k'_a \phi n \tag{13.12}$$

where

$$k'_a = \frac{pN}{60M} \tag{13.13}$$

Having introduced the basic equations relating torque, speed, voltages, and currents in electric machines, we may now consider the interaction of these quantities in a DC machine at steady state, that is, operating at constant speed and field excitation. Figure 13.15 depicts the electric circuit model of a separately excited DC machine, illustrating both motor and generator action. It is very important to note the reference direction of armature current flow, and of the developed torque, in order to make a distinction between the two modes of operation. The field excitation is shown as a voltage V_f generating the field current I_f that flows through a variable resistor R_f and through the field coil L_f. The variable resistor permits adjustment of the field excitation. The armature circuit, on the other hand, consists of a voltage source representing the back emf E_b, the armature resistance R_a, and the armature voltage V_a. This model is appropriate both for motor and for generator action. When $V_a < E_b$, the machine acts as a generator (I_a flows out of the machine). When $V_a > E_b$, the machine acts as a motor (I_a flows into the machine). Thus, according to the circuit model of Figure 13.15, the operation of a DC machine at steady state (i.e., with the inductors in the circuit replaced by short-circuits) is described by the following equations:

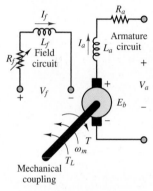

(b) Generator reference direction

Figure 13.15 Electric circuit model of a separately excited DC machine

$$-I_f + \frac{V_f}{R_f} = 0 \quad \text{and} \quad V_a - R_a I_a - E_b = 0 \quad \text{(motor action)}$$
$$\tag{13.14}$$
$$-I_f + \frac{V_f}{R_f} = 0 \quad \text{and} \quad V_a + R_a I_a - E_b = 0 \quad \text{(generator action)}$$

Equation 13.14 together with equations 13.6 and 13.8 may be used to determine the steady-state operating condition of a DC machine.

---

[1]Note that the abbreviation *rpm* although certainly familiar to the reader, is not a standard unit, and its use should be discouraged.

The circuit model of Figure 13.15 permits the derivation of a simple set of differential equations that describe the *dynamic* analysis of a DC machine. The dynamic equations describing the behavior of a separately excited DC machine are as follows:

$$V_a(t) - I_a(t)R_a - L_a\frac{dI_a(t)}{dt} - E_b(t) = 0 \qquad \text{(armature circuit)} \qquad (13.15a)$$

$$V_f(t) - I_f(t)R_f - L_f\frac{dI_f(t)}{dt} = 0 \qquad \text{(field circuit)} \qquad (13.15b)$$

These equations can be related to the operation of the machine in the presence of a load. If we assume that the motor is rigidly connected to an inertial load with moment of inertia J and that the friction losses in the load are represented by a viscous friction coefficient b, then the torque developed by the machine (in the motor mode of operation) can be written as

$$T(t) = T_L + b\omega_m(t) + J\frac{d\omega_m(t)}{dt} \qquad (13.16)$$

where T_L is the load torque. Typically T_L is either constant or some function of speed ω_m in a motor. In the case of a generator, the load torque is replaced by the torque supplied by a prime mover, and the machine torque $T(t)$ opposes the motion of the prime mover, as shown in Figure 13.15. Since the machine torque is related to the armature and field currents by equation 13.6, equations 13.16 and 13.17 are coupled to each other; this coupling may be expressed as follows:

$$T(t) = k_a\phi I_a(t) \qquad (13.17)$$

or

$$k_a\phi I_a(t) = T_L + b\omega_m(t) + J\frac{d\omega_m(t)}{dt} \qquad (13.18)$$

The dynamic equations described in this section apply to any DC machine. In the case of a *separately excited* machine, a further simplification is possible, since the flux is established by virtue of a separate field excitation, and therefore

$$\phi = \frac{N_f}{\mathcal{R}}I_f = k_f I_f \qquad (13.19)$$

where N_f is the number of turns in the field coil, $\mathcal{R}$ is the reluctance of the structure, and I_f is the field current.

DC Machine Steady-State Equations

The equations that describe the steady-state behavior of DC motors and generators are summarized below. The key to interpreting these equations is in correctly evaluating the expression for the flux ϕ for each of the four cases of interest in this chapter: field generated by a separate excitation, field generated by a shunt connection, field generated by a series connection, field generated by a permanent magnet (constant field). See Figure 13.14 for a reference to the first three configurations.

DC Motor Steady-State Equations

$$E_b = k_a \phi \omega_m \quad \text{V}$$

$$T = k_a \phi I_a \quad \text{N-m}$$

In a *separately excited* machine [Figure 13.14(a)]:

$$V_s = E_b - I_a R_a \quad \text{V}$$

$$\phi = k_f I_f = k_f \frac{V_f}{R_f}$$

where V_s is the external source voltage.

In a *shunt-connected* machine [Figure 13.14(b)]:

$$V_s = E_b - I_a R_a \quad \text{V}$$

$$\phi = \phi_{\text{shunt}} = k_f I_f = k_f \frac{V_a}{R_f}$$

In a *series-connected* machine [Figure 13.14(c)]:

$$V_s = E_b - I_a R_a - I_a R_f \quad \text{V}$$

$$\phi = \phi_{\text{series}} = k_f I_f = k_f I_a$$

Finally, in a *permanent-magnet* machine, where the field excitation is provided by a permanent magnet:

$$V_s = E_b - I_a R_a \quad \text{V}$$

$$\phi = \phi_{\text{PM}} = \text{constant}$$

DC Generator Steady-State Equations

$$E_b = k_a \phi \omega_m \quad \text{V}$$

$$T = \frac{P}{\omega_m} = \frac{E_b I_a}{\omega_m} = k_a \phi I_a \quad \text{N-m}$$

where V_g is the generator open-circuit output voltage, with no load connected.
In a *separately excited* machine [Figure 13.14(a)]:

$$V_g = E_b - I_a R_a \quad \text{V}$$

$$\phi = k_f I_f = k_f \frac{V_f}{R_f} \quad \text{Wb}$$

In a *shunt-connected* machine [Figure 13.14(b)]:

$$V_g = E_b - I_a R_a \quad \text{V}$$

$$\phi = \phi_{\text{shunt}} = k_f I_f = k_f \frac{V_g}{R_f} \quad \text{Wb}$$

In a *series-connected* machine [Figure 13.14(b)]:

$$V_g = E_b - I_a R_a - I_a R_f \quad \text{V}$$

$$\phi = \phi_{\text{series}} = k_f I_f = k_f I_a \quad \text{Wb}$$

Finally, in a *permanent-magnet* machine:

$$V_g = E_b - I_a R_a \quad \text{V}$$

$$\phi = \phi_{\text{PM}} = \text{constant} \quad \text{Wb}$$

13.3   DIRECT-CURRENT MOTORS

DC motors are widely used in applications requiring accurate speed or torque control, for example in servo systems. In the preceding section, we had introduced the analysis of a separately excited DC machine; in this section we extend that analysis to include a review of the other three commonly used configurations (*shunt, series* and *permanent-magnet motors*), to study their torque–speed characteristics and dynamic behavior.

The Shunt Motor

In a shunt motor [see Figure 13.14(b)], the armature current is found by dividing the net voltage across the armature circuit (source voltage minus back emf) by the armature resistance:

$$I_a = \frac{V_s - k_a \phi \omega_m}{R_a} \tag{13.20}$$

An expression for the armature current may also be obtained from equation 13.17, as follows:

$$I_a = \frac{T}{k_a \phi} \tag{13.21}$$

It is then possible to relate the torque requirements to the speed of the motor by substituting equation 13.20 in equation 13.21:

$$\frac{T}{k_a \phi} = \frac{V_s - k_a \phi \omega_m}{R_a} \tag{13.22}$$

Equation 13.22 describes the steady-state torque–speed characteristic of the shunt motor. To understand this performance equation, we observe that if V_s, k_a, ϕ, and R_a are fixed in equation 13.22 (the flux is essentially constant in the shunt motor for a fixed V_s), then the speed of the motor is directly related to the armature current. Now consider the case where the load applied to the motor is suddenly increased, causing the speed of the motor to drop. As the speed decreases, the armature current increases, according to equation 13.20. The excess armature current causes the motor to develop additional torque, according to equation 13.21 until a new equilibrium is reached between the higher armature current and developed torque and the lower speed of rotation. The equilibrium point is dictated by the balance of mechanical and electric power, in accordance with the relation:

$$E_b I_a = T \omega_m \tag{13.23}$$

Thus, the shunt DC motor will adjust to variations in load by changing its speed to preserve this power balance. The torque–speed curves for the shunt motor may be obtained by rewriting the equation relating the speed to the armature current:

$$\omega_m = \frac{V_s - I_a R_a}{k_a \phi} = \frac{V_s}{k_a \phi} - \frac{R_a T}{(k_a \phi)^2} \qquad \begin{array}{l} T\text{-}\omega \text{ curve for} \\ \text{shunt motor} \end{array} \tag{13.24}$$

To interpret equation 13.24, one can start by considering the motor operating at rated speed and torque. As the load torque is reduced, the armature current will also decrease, causing the speed to increase in accordance with equation 13.24. The increase in speed depends on the extent of the voltage drop across the armature resistance $I_a R_a$. The change in speed will be on the same order of magnitude as this drop; it typically takes values around 10 percent. This corresponds to a relatively good speed regulation, which is an attractive feature of the shunt DC motor (recall the discussion of regulation in Section 13.1). The dynamic behavior of the shunt motor is described by equations 13.15 through 13.18, with the additional relation:

$$I_a(t) = I_s(t) - I_f(t) \tag{13.25}$$

Series Motors

The series motor [see Figure 13.14(c)] behaves somewhat differently from the shunt and separately excited motors because the flux is established solely by virtue of the series current flowing through the armature. It is relatively simple to derive an expression for the emf and torque equations for the series motor if we approximate the relationship between flux and armature current by assuming that the motor operates in the linear region of its magnetization curve. Then we can write

$$\phi = k_S I_a \tag{13.26}$$

and the emf and torque equations become, respectively,

$$E_b = k_a \omega_m \phi = k_a \omega_m k_S I_a \tag{13.27}$$

$$T = k_a \phi I_a = k_a k_S I_a^2 \tag{13.28}$$

The circuit equation for the series motor becomes

$$V_s = E_b + I_a(R_a + R_S) = (k_a \omega_m k_S + R_T)I_a \tag{13.29}$$

where R_a is the armature resistance, R_S is the series field winding resistance, and R_T is the total series resistance. From equation 13.29, we can solve for I_a and substitute in the torque expression (equation 13.28) to obtain the following torque–speed relationship:

$$\boxed{T = k_a k_S \frac{V_s^2}{(k_a \omega_m k_S + R_T)^2} \qquad \text{T-ω curve for series DC motor}} \tag{13.30}$$

which indicates the inverse squared relationship between torque and speed in the series motor. This expression describes a behavior that can, under certain conditions, become unstable. Since the speed increases when the load torque is reduced, one can readily see that if one were to disconnect the load altogether, the speed would tend to increase to dangerous values. To prevent excessive speeds, series motors are always mechanically coupled to the load. This feature is not necessarily a drawback, though, because series motors can develop very high torque at low speeds and therefore can serve very well for traction-type loads (e.g., conveyor belts or vehicle propulsion systems).

The differential equation for the armature circuit of the motor can be given as

$$
\begin{aligned}
V_s &= I_a(t)(R_a + R_S) + L_a\frac{dI_a(t)}{dt} + L_S\frac{dI_a(t)}{dt} + E_b \\
&= I_a(t)(R_a + R_S) + L_a\frac{dI_a(t)}{dt} + L_S\frac{dI_a(t)}{dt} + k_a k_S I_a \omega_m
\end{aligned}
\tag{13.31}
$$

Permanent-Magnet DC Motors

Permanent-magnet (PM) DC motors have become increasingly common in applications requiring relatively low torques and efficient use of space. The construction of PM DC motors differs from that of the motors considered thus far in that the magnetic field of the stator is produced by suitably located poles made of magnetic materials. Thus, the basic principle of operation, including the idea of commutation, is unchanged with respect to the wound-stator DC motor. What changes is that there is no need to provide a field excitation, whether separately or by means of the self-excitation techniques discussed in the preceding sections. Therefore, the PM motor is intrinsically simpler than its wound-stator counterpart.

The equations that describe the operation of the PM motor follow. The torque produced is related to the armature current by a torque constant k_{PM}, which is determined by the geometry of the motor:

$$
T = k_{T,PM} I_a
\tag{13.32}
$$

As in the conventional DC motor, the rotation of the rotor produces the usual count or back emf E_b, which is linearly related to speed by a voltage constant $k_{a,PM}$:

$$
E_b = k_{a,PM} \omega_m
\tag{13.33}
$$

The equivalent circuit of the PM motor is particularly simple, since we need not model the effects of a field winding. Figure 13.16 shows the circuit model and the torque–speed curve of a PM motor.

We can use the circuit model of Figure 13.16 to derive the torque–speed curve shown in the same figure as follows. From the circuit model, for a constant speed (and therefore constant current), we may consider the inductor a short-circuit and write the equation:

$$
\begin{aligned}
V_s &= I_a R_a + E_b = I_a R_a + k_{a,PM} \omega_m \\
&= \frac{T}{k_{T,PM}} R_a + k_{a,PM} \omega_m
\end{aligned}
\tag{13.34}
$$

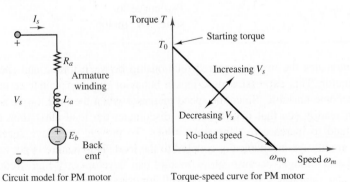

Circuit model for PM motor   Torque-speed curve for PM motor

Figure 13.16 Circuit model and torque–speed curve of PM motor

thus obtaining the equations relating speed and torque:

$$\boxed{\omega_m = \frac{V_s}{k_{a,\text{PM}}} - \frac{T R_a}{k_{a,\text{PM}}} k_{T,\text{PM}}} \qquad \begin{array}{l} \textit{T-}\omega \text{ curve for} \\ \text{PM DC motor} \end{array} \qquad (13.35)$$

and

$$T = \frac{V_s}{R_a} k_{T,\text{PM}} - \frac{\omega_m}{R_a} k_{a,\text{PM}} k_{T,\text{PM}} \qquad (13.36)$$

From these equations, one can extract the stall torque T_0, that is, the zero-speed torque:

$$T_0 = \frac{V_s}{R_a} k_{T,\text{PM}} \qquad (13.37)$$

and the no-load speed ω_{m0}:

$$\omega_{m0} = \frac{V_s}{k_{a,\text{PM}}} \qquad (13.38)$$

Under dynamic conditions, assuming an inertia plus viscous friction load, the torque produced by the motor can be expressed as

$$T = k_{T,\text{PM}} I_a(t) = T_{\text{load}}(t) + b\omega_m(t) + J\frac{d\omega_m(t)}{dt} \qquad (13.39)$$

The differential equation for the armature circuit of the motor is therefore given by

$$V_s = I_a(t) R_a + L_a\frac{dI_a(t)}{dt} + E_b$$

$$= I_a(t) R_a + L_a\frac{dI_a(t)}{dt} + k_{a,\text{PM}}\omega_m(t) \qquad (13.40)$$

The fact that the airgap flux is constant in a PM DC motor makes its characteristics somewhat different from those of the wound DC motor. A direct comparison of PM and wound-field DC motors reveals the following advantages and disadvantages of each configuration.

Comparison of Wound-Field and PM DC Motors

1. PM motors are smaller and lighter than wound motors for a given power rating. Further, their efficiency is greater because there are no field winding losses.
2. An additional advantage of PM motors is their essentially linear speed–torque characteristic, which makes analysis (and control) much easier. Reversal of rotation is also accomplished easily, by reversing the polarity of the source.
3. A major disadvantage of PM motors is that they can become demagnetized by exposure to excessive magnetic fields, application of excessive voltage, or operation at excessively high or low temperatures.
4. A less obvious drawback of PM motors is that their performance is subject to greater variability from motor to motor than is the case for wound motors, because of variations in the magnetic materials.

EXAMPLE 13.4   DC Shunt Motor Analysis

Problem

Find the speed and torque generated by a four-pole DC shunt motor.

Solution

Known Quantities: Motor ratings; circuit and magnetic parameters.

Find: ω_m, T.

Schematics, Diagrams, Circuits, and Given Data:
Motor ratings: 3 hp, 240 V, 120 r/min.
Circuit and magnetic parameters: $I_S = 30$ A; $I_f = 1.4$ A; $R_a = 0.6$ Ω; $\phi = 20$ mWb;
$N = 1{,}000$; $M = 4$ (see equation 13.10).

Analysis: We convert the power to SI units:

$$P_{\text{RATED}} = 3 \text{ hp} \times 746 \frac{\text{W}}{\text{hp}} = 2{,}238 \text{ W}$$

Next we compute the armature current as the difference between source and field current
(equation 13.25):

$$I_a = I_s - I_f = 30 - 1.4 = 28.6 \text{ A}$$

The no-load armature voltage E_b is given by:

$$E_b = V_s - I_a R_a = 240 - 28.6 \times 0.6 = 222.84 \text{ V}$$

and equation 13.10 can be used to determine the armature constant:

$$k_a = \frac{pN}{2\pi M} = \frac{4 \times 1{,}000}{2\pi \times 4} = 159.15 \frac{\text{V-s}}{\text{Wb-rad}}$$

Knowing the motor constant, we can calculate the speed, after equation 13.25:

$$\omega_m = \frac{E_a}{k_a \phi} = \frac{222.84 \text{ V}}{(159.15 \text{ V-s/Wb-rad})(0.02 \text{ Wb})} = 70 \frac{\text{rad}}{\text{s}}$$

Finally, the torque developed by the motor can be found as the ratio of the power to the
angular velocity:

$$T = \frac{P}{\omega_m} = \frac{2{,}238 \text{ W}}{70 \text{ rad/s}} = 32 \text{ N-m}$$

CHECK YOUR UNDERSTANDING

A 200-V DC shunt motor draws 10 A at 1,800 r/min. The armature circuit resistance is
0.15 Ω, and the field winding resistance is 350 Ω. What is the torque developed by the
motor?

Answer: $T = \dfrac{P}{\omega_m} = 9.93$ N-m

EXAMPLE 13.5 DC Shunt Motor Analysis

Problem

Determine the following quantities for the DC shunt motor, connected as shown in the circuit of Figure 13.17:

1. Field current required for full-load operation.
2. No-load speed.
3. Plot of the speed–torque curve of the machine in the range from no-load torque to rated torque.
4. Power output at rated torque.

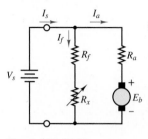

Figure 13.17 Shunt motor configuration

Solution

Known Quantities: Magnetization curve, rated current, rated speed, circuit parameters.

Find: I_f; $n_{\text{no-load}}$; T-n curve, P_{rated}.

Schematics, Diagrams, Circuits, and Given Data:
Figure 13.18 (magnetization curve)
Motor ratings: 8 A, 120 r/min
Circuit parameters: $R_a = 0.2\ \Omega$; $V_s = 7.2$ V; N = number of coil turns in winding = 200

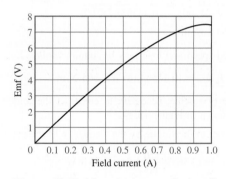

Figure 13.18 Magnetization curve for a small DC motor

Analysis:

1. To find the field current, we must find the generated emf since R_f is not known. Writing KVL around the armature circuit, we obtain

$$V_s = E_b + I_a R_a$$
$$E_b = V_s - I_a R_a = 7.2 - 8(0.2) = 5.6 \text{ V}$$

Having found the back emf, we can find the field current from the magnetization curve. At $E_b = 5.6$ V, we find that the field current and field resistance are

$$I_f = 0.6 \text{ A} \quad \text{and} \quad R_f = \frac{7.2}{0.6} = 12\ \Omega$$

2. To obtain the no-load speed, we use the equations:

$$E_b = k_a \phi \frac{2\pi n}{60} \qquad T = k_a \phi I_a$$

leading to

$$V_s = I_a R_a + E_b = I_a R_a + k_a \phi \frac{2\pi}{60} n$$

or

$$n = \frac{V_s - I_a R_a}{k_a \phi (2\pi/60)}$$

At no load, and assuming no mechanical losses, the torque is zero, and we see that the current I_a must also be zero in the torque equation ($T = k_a \phi I_a$). Thus, the motor speed at no load is given by

$$n_{\text{no-load}} = \frac{V_s}{k_a \phi (2\pi/60)}$$

We can obtain an expression for $k_a \phi$, knowing that, at full load:

$$E_b = 5.6 \text{ V} = k_a \phi \frac{2\pi n}{60}$$

so that, for constant field excitation:

$$k_a \phi = E_b \left(\frac{60}{2\pi n} \right) = 5.6 \left[\frac{60}{2\pi(120)} \right] = 0.44563 \frac{\text{V-s}}{\text{rad}}$$

Finally, we may solve for the no-load speed.

$$n_{\text{no-load}} = \frac{V_s}{k_a \phi (2\pi/60)} = \frac{7.2}{(0.44563)(2\pi/60)}$$
$$= 154.3 \text{ r/min}$$

3. The torque at rated speed and load may be found as follows:

$$T_{\text{rated load}} = k_a \phi I_a = (0.44563)(8) = 3.565 \text{ N-m}$$

Now we have the two points necessary to construct the torque–speed curve for this motor, which is shown in Figure 13.19.

4. The power is related to the torque by the frequency of the shaft:

$$P_{\text{rated}} = T \omega_m = (3.565) \left(\frac{120}{60} \right) (2\pi) = 44.8 \text{ W}$$

or, equivalently:

$$P = 44.8 \text{ W} \times \frac{1}{746} \frac{\text{hp}}{\text{W}} = 0.06 \text{ hp}$$

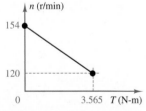

Figure 13.19 Torque-speed curve for motor of Example 13.5.

![LO] **EXAMPLE 13.6   DC Series Motor Analysis**

Problem

Determine the torque developed by a DC series motor when the current supplied to the motor is 60 A.

Solution

Known Quantities: Motor ratings; operating conditions.

Find: T_{60}, torque delivered at 60-A series current.

Schematics, Diagrams, Circuits, and Given Data:
Motor ratings: 10 hp, 115 V, full-load speed = 1,800 r/min
Operating conditions: motor draws 40 A

Assumptions: The motor operates in the linear region of the magnetization curve.

Analysis: Within the linear region of operation, the flux per pole is directly proportional to the current in the field winding. That is,

$$\phi = k_S I_a$$

The full-load speed is

$$n = 1,800 \text{ r/min}$$

or

$$\omega_m = \frac{2\pi n}{60} = 60\pi \qquad \text{rad/s}$$

Rated output power is

$$P_{\text{rated}} = 10 \text{ hp} \times 746 \text{ W/hp} = 7,460 \text{ W}$$

and full-load torque is

$$T_{40\,A} = \frac{P_{\text{rated}}}{\omega_m} = \frac{7,460}{60\pi} = 39.58 \text{ N-m}$$

Thus, the machine constant may be computed from the torque equation for the series motor:

$$T = k_a k_S I_a^2 = K I_a^2$$

At full load:

$$K = k_a k_S = \frac{39.58 \text{ N-m}}{40^2 \text{ A}^2} = 0.0247 \frac{\text{N-m}}{\text{A}^2}$$

and we can compute the torque developed for a 60-A supply current to be

$$T_{60\,A} = K I_a^2 = 0.0247 \times 60^2 = 88.92 \text{ N-m}$$

CHECK YOUR UNDERSTANDING

A series motor draws a current of 25 A and develops a torque of 100 N-m. Find (a) the torque when the current rises to 30 A if the field is unsaturated and (b) the torque when the current rises to 30 A and the increase in current produces a 10 percent increase in flux.

Answer: (a) 144 N-m; (b) 132 N-m

EXAMPLE 13.7   Dynamic Response of PM DC Motor

Problem

Develop a set of differential equations and a transfer function describing the dynamic response of the motor angular velocity of a PM DC motor connected to a mechanical load.

Solution

Known Quantities: PM DC motor circuit model; mechanical load model.

Find: Differential equations and transfer functions of electromechanical system.

Analysis: The dynamic response of the electromechanical system can be determined by applying KVL to the electric circuit (Figure 13.16) and Newton's second law to the mechanical system. These equations will be coupled to one another, as you shall see, because of the nature of the motor back emf and torque equations.

Applying KVL and equation 13.33 to the electric circuit, we obtain

$$V_L(t) - R_a I_a(t) - L_a \frac{dI_a(t)}{dt} - E_b(t) = 0$$

or

$$L_a \frac{dI_a(t)}{dt} + R_a I_a(t) + K_{a,\text{PM}} \omega_m(t) = V_L(t)$$

Applying Newton's second law and equation 13.32 to the load inertia, we obtain

$$J \frac{d\omega(t)}{dt} = T(t) - T_{\text{load}}(t) - b\omega$$

or

$$-K_{T,\text{PM}} I_a(t) + J \frac{d\omega(t)}{dt} + b\omega(t) = -T_{\text{load}}(t)$$

These two differential equations are coupled because the first depends on ω_m and the second on I_a. Thus, they need to be solved simultaneously.

To derive the transfer function, we use the Laplace transform on the two equations to obtain

$$(sL_a + R_a)I_a(s) + K_{a,\text{PM}}\Omega(s) = V_L(s)$$
$$-K_{T,\text{PM}} I_a(s) + (sJ + b)\Omega(s) = -T_{\text{load}}(s)$$

We can write the above equations in matrix form and resort to Cramer's rule to solve for $\Omega_m(s)$ as a function of $V_L(s)$ and $T_{\text{load}}(s)$.

$$\begin{bmatrix} sL_a + R_a & K_{a,\text{PM}} \\ -K_{T,\text{PM}} & sJ + b \end{bmatrix} \begin{bmatrix} I_a(s) \\ \Omega_m(s) \end{bmatrix} = \begin{bmatrix} V_L(s) \\ -T_{\text{load}}(s) \end{bmatrix}$$

with solution:

$$\Omega_m(s) = \frac{\det \begin{bmatrix} sL_a + R_a & V_L(s) \\ K_{T,\text{PM}} & -T_{\text{load}}(s) \end{bmatrix}}{\det \begin{bmatrix} sL_a + R_a & K_{a,\text{PM}} \\ -K_{T,\text{PM}} & sJ + b \end{bmatrix}}$$

or

$$\Omega_m(s) = -\frac{sL_a + R_a}{(sL_a + R_a)(sJ + b) + K_{a,\text{PM}}K_{T,\text{PM}}} T_{\text{load}}(s)$$

$$+ \frac{K_{T,\text{PM}}}{(sL_a + R_a)(sJ + b) + K_{a,\text{PM}}K_{T,\text{PM}}} V_L(s)$$

Comments: Note that the dynamic response of the motor angular velocity depends on both the input voltage and the load torque. This problem is explored further in the homework problems.

DC Drives and DC Motor Speed Control

The advances made in power semiconductors have made it possible to realize low-cost speed control systems for DC motors. In this section we describe some of the considerations that are behind the choice of a specific drive type, and some of the loads that are likely to be encountered.

Constant-torque loads are quite common and are characterized by a need for constant torque over the entire speed range. This need is usually due to friction; the load will demand increasing horsepower at higher speeds, since power is the product of speed and torque. Thus, the power required will increase linearly with speed. This type of loading is characteristic of conveyors, extruders, and surface winders.

Another type of load is one that requires *constant horsepower* over the speed range of the motor. Since torque is inversely proportional to speed with constant horsepower, this type of load will require higher torque at low speeds. Examples of constant-horsepower loads are machine tool spindles (e.g., lathes). This type of application requires very high starting torques.

Variable-torque loads are also common. In this case, the load torque is related to the speed in some fashion, either linearly or geometrically. For some loads, for example, torque is proportional to the speed (and thus horsepower is proportional to speed squared); examples of loads of this type are positive displacement pumps. More common than the linear relationship is the squared-speed dependence of inertial loads such as centrifugal pumps, some fans, and all loads in which a flywheel is used for energy storage.

To select the appropriate motor and adjustable-speed drive for a given application, we need to examine how each method for speed adjustment operates on a DC motor. Armature voltage control serves to smoothly adjust speed from 0 to 100 percent of the nameplate rated value (i.e., base speed), provided that the field excitation is also equal to the rated value. Within this range, it is possible to fully control motor speed for a constant-torque load, thus providing a linear increase in horsepower, as shown in Figure 13.20. Field weakening allows for increases in speed of up to several times the base speed; however, field control changes the characteristics of the DC motor from constant torque to constant horsepower, and therefore the torque output drops with speed, as shown in Figure 13.20. Operation above base speed requires special provision for field control, in addition to the circuitry required for armature voltage control, and is therefore more complex and costly.

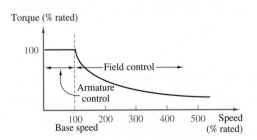
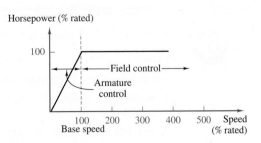

Figure 13.20 Speed control in DC motors

CHECK YOUR UNDERSTANDING

Describe the cause-and-effect behavior of the speed control method of changing armature voltage for a shunt DC motor.

reached between motor and the motor runs at constant speed.
armature current to drop and the motor torque to decrease until a balance condition is
speed to increase as well. The corresponding increase in back emf, however, causes the
Consequently, the motor torque increases until it exceeds the load torque, causing the
Answer: Increasing the armature voltage leads to an increase in armature current.

13.4   DIRECT-CURRENT GENERATORS

The same analysis and equations used in the preceding section can be applied to DC generators, with the understanding that in a motor, the external voltage V_s is a DC supply that enables the motor to generate a torque, while in a generator the torque provided by a prime mover results in the motor rotating at a speed Ω, which in turn generates an open-circuit voltage V_g. When the generator is connected to a load, armature current flows, and a load voltage V_L is generated. Figure 13.21 depicts the configuration of a separately excited DC generator, and Figure 13.22 depicts a magnetization curve for a generator that can be used to calculate the back emf (generator open-circuit voltage) as a function of field current. Two examples follow, to illustrate methods of analysis for DC generators.

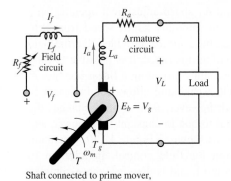

Shaft connected to prime mover,
providing torque T resulting in motor
angular velocity ω_m

Figure 13.21 Separately excited DC generator

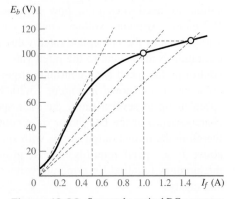

Figure 13.22 Separately excited DC generator magnetization curve

EXAMPLE 13.8   Separately Excited DC Generator

Problem

A separately excited DC generator is characterized by the magnetization of Figure 13.22.

1. If the prime mover is driving the generator at 800 r/min, what is the no-load terminal voltage V_a?

2. If a 1-Ω load is connected to the generator, what is the generator voltage?

3. Assume steady-state operation.

Solution

Known Quantities: Generator magnetization curve and ratings.

Find: Terminal voltage with no load and 1-Ω load.

Schematics, Diagrams Circuits and Green Data: Generator ratings: 100 V, 100 A, 1,000 r/min. Circuit parameters: $R_a = 0.14\ \Omega$; $V_f = 100$ V; $R_f = 100\ \Omega$.

Analysis:

1. The field current in the machine at steady state is

$$I_f = \frac{V_f}{R_f} = \frac{100\ \text{V}}{100\ \Omega} = 1\ \text{A}$$

From the magnetization curve, it can be seen that this field current will produce 100 V at a speed of 1,000 r/min. Since this generator is actually running at 800 r/min, the induced emf may be found by assuming a linear relationship between speed and emf. This approximation is reasonable, provided that the departure from the nominal operating condition is small. Let n_0 and E_{b0} be the nominal speed and emf, respectively (that is, 1,000 r/min and 100 V). Then:

$$\frac{E_b}{E_{b0}} = \frac{n}{n_0}$$

and therefore:

$$E_b = \frac{n}{n_0} E_{b0} = \frac{800\ \text{r/min}}{1,000\ \text{r/min}} \times 100\ \text{V} = 80\ \text{V}$$

The open-circuit (output) terminal voltage of the generator is equal to the emf from the circuit model of Figure 13.15: therefore:

$$V_a = E_b = 80\ \text{V}$$

2. When a load resistance is connected to the circuit (the practical situation), the terminal (or load) voltage is no longer equal to E_b, since there will be a voltage drop across the armature winding resistance. The armature (or load) current may be determined from

$$I_a = I_L = \frac{E_b}{R_a + R_L} = \frac{80\ \text{V}}{(0.14 + 1)\ \Omega} = 70.2\ \text{A}$$

where $R_L = 1\ \Omega$ is the load resistance. The terminal (load) voltage is therefore given by

$$V_L = I_L R_L = 70.2 \times 1 = 70.2\ \text{V}$$

CHECK YOUR UNDERSTANDING

A 24-coil, two-pole DC generator has 16 turns per coil in its armature winding. The field excitation is 0.05 Wb per pole, and the armature angular velocity is 180 rad/s. Find the machine constant and the total induced voltage.

Answer: $k_a = 5.1$; $E_b = 45.9$ V

EXAMPLE 13.9   **Separately Excited DC Generator**

Problem

Determine the following quantities for a separately excited DC:

1. Induced voltage
2. Machine constant
3. Torque developed at rated conditions
4. Assume steady-state operation

Solution

Known Quantities: Generator ratings and machine parameters.

Find: E_b, k_a, T.

Schematics, Diagrams, Circuits, and Given Data: Generator ratings: 1,000 kW, 2,000 V, 3,600 r/min. Circuit parameters: $R_0 = 0.1\ \Omega$; flux per pole $\phi = 0.5$ Wb.

Analysis:

1. The armature current may be found by observing that the rated power is equal to the product of the terminal (load) voltage and the current. Then:

$$I_a = \frac{P_{\text{rated}}}{V_L} = \frac{1,000 \times 10^3}{2,000} = 500\ \text{A}$$

The generated voltage is equal to the sum of the terminal voltage and the voltage drop across the armature resistance (see Figure 13.14):

$$E_b = V_a + I_a R_a = 2,000 + 500 \times 0.1 = 2,050\ \text{V}$$

2. The speed of rotation of the machine in units of radians per second is

$$\omega_a = \frac{2\pi n}{60} = \frac{2\pi \times 3,600\ \text{r/min}}{60\ \text{r/min}} = 377\ \text{rad/s}$$

Thus, the machine constant is found to be

$$L_a = \frac{E_b}{\phi \omega_m} = \frac{2.050\ \text{V}}{0.5\ \text{Wb} \times 377\ \text{rad/s}} = 10.876 \frac{\text{V-s}}{\text{Wb-rad}}$$

3. The torque developed is found from equation 13.6:

$$T = k_a \phi I_a = 10.876\ \text{V-s/Wb-rad} \times 0.5\ \text{Wb} \times 500\ \text{A} = 2.718.9\ \text{N-m}$$

Comments: In many practical cases, it is not actually necessary to know the armature constant and the flux separately, but it is sufficient to know the value of the product $k_a\phi$. For example, suppose that the armature resistance of a DC machine is known and that, given a known field excitation, the armature current, load voltage, and speed of the machine can be measured. Then the product $k_a\phi$ may be determined from equation 13.8, as follows:

$$k_a\phi = \frac{E_b}{\omega_m} = \frac{V_L + (R_a + R_s)}{\omega_m}$$

where V_L, I_a and ω_m are measured quantities for given operating conditions.

CHECK YOUR UNDERSTANDING

A 1,000-kW, 1,000-V, 2,400 r/min separately excited DC generator has an armature circuit resistance of 0.04 Ω. The flux per pole is 0.4 Wb. Find (a) the induced voltage, (b) the machine constant, and (c) the torque developed at the rated conditions.

Answer: (a) $E_b = 1,040$ V; (b) $k_a = 10.34 \frac{\text{V-s}}{\text{Wb-rad}}$; (c) $T = 4,138$ N-m

CHECK YOUR UNDERSTANDING

A 100-kW, 250-V shunt generator has a field circuit resistance of 50 Ω and an armature circuit resistance of 0.05 Ω. Find (a) the full-load line current flowing to the load, (b) the field current, (c) the armature current, and (d) the full-load generator voltage.

Answer: (a) 400 A; (b) 5 A; (c) 405 A; (d) 270.25 V

13.5 ALTERNATING-CURRENT MACHINES

AC machines represent the vast majority of industrial applications. The objective of this section is to explain the basic operation of both synchronous and induction machines and to outline their performance characteristics. In doing so, we also point out the relative advantages and disadvantages of these machines in comparison with DC machines.

Rotating Magnetic Fields

As mentioned in Section 13.1, the fundamental principle of operation of AC machines is the generation of a rotating magnetic field, which causes the rotor to turn at a speed that depends on the speed of rotation of the magnetic field. We now explain how a rotating magnetic field can be generated in the stator and air gap of an AC machine by means of alternating currents.

Consider the stator shown in Figure 13.23, which supports windings $a\text{-}a'$, $b\text{-}b'$, and $c\text{-}c'$. The coils are geometrically spaced 120° apart, and a three-phase voltage is applied to the coils. As you may recall from the discussion of AC power in Chapter 6, the currents generated by a three-phase source are also spaced by 120°, as illustrated in Figure 13.24. The phase voltages referenced to the neutral terminal would then be given by the expressions

$$v_a = A \cos(\omega_e t)$$
$$v_b = A \cos\left(\omega_e t - \frac{2\pi}{3}\right)$$
$$v_c = A \cos\left(\omega_e t + \frac{2\pi}{3}\right)$$

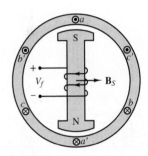

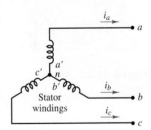

Figure 13.23 Two-pole three-phase stator

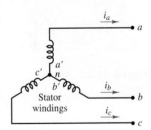

Figure 13.24 Three-phase stator winding currents

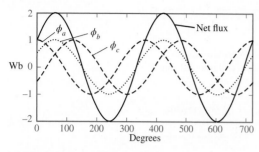

Figure 13.25 Flux distribution in a three-phase stator winding as a function of angle of rotation

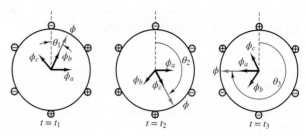

Figure 13.26 Rotating flux in a three-phase machine

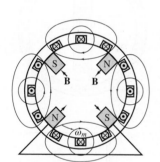

Figure 13.27 Four-pole stator

where ω_e is the frequency of the AC supply, or line frequency. The coils in each winding are arranged in such a way that the flux distribution generated by any one winding is approximately sinusoidal. Such a flux distribution may be obtained by appropriately arranging groups of coils for each winding over the stator surface. Since the coils are spaced 120° apart, the flux distribution resulting from the sum of the contributions of the three windings is the sum of the fluxes due to the separate windings, as shown in Figure 13.25. Thus, the flux in a three-phase machine rotates in space according to the vector diagram of Figure 13.26, and the flux is constant in amplitude. A stationary observer on the machine's stator would see a sinusoidally varying flux distribution, as shown in Figure 13.25.

Since the resultant flux of Figure 13.25 is generated by the currents of Figure 13.24, the speed of rotation of the flux must be related to the frequency of the sinusoidal phase currents. In the case of the stator of Figure 13.23, the number of magnetic poles resulting from the winding configuration is 2; however, it is also possible to configure the windings so that they have more poles. For example, Figure 13.27 depicts a simplified view of a four-pole stator.

In general, the speed of the rotating magnetic field is determined by the frequency of the excitation current f and by the number of poles present in the stator p according to

$$n_s = \frac{120f}{p} \text{ r/min} \qquad \text{Synchronous speed}$$

or

$$\omega_s = \frac{2\pi n_S}{60} = \frac{2\pi \times 2f}{p} \qquad \text{Synchronous speed}$$

(13.41)

where n_s (or ω_s) is usually called the **synchronous speed**.

Now, the structure of the windings in the preceding discussion is the same whether the AC machine is a motor or a generator; the distinction between the two depends on the direction of power flow. In a generator, the electromagnetic torque is a reaction torque that opposes rotation of the machine; this is the torque against which the prime mover does work. In a motor, on the other hand, the rotational (motional) voltage generated in the armature opposes the applied voltage; this voltage is the counter- (or back) emf. Thus, the description of the rotating magnetic field given thus far applies to both motor and generator action in AC machines.

As described a few paragraphs earlier, the stator magnetic field rotates in an AC machine, and therefore the rotor cannot "catch up" with the stator field and is in constant pursuit of it. The speed of rotation of the rotor will therefore depend on the number of magnetic poles present in the stator and in the rotor. The magnitude of the torque produced in the machine is a function of the angle γ between the stator and rotor magnetic fields; precise expressions for this torque depend on how the magnetic fields are generated and will be given separately for the two cases of synchronous and induction machines. What is common to all rotating machines is that the number of stator and rotor poles must be identical if any torque is to be generated. Further, the number of poles must be even, since for each north pole there must be a corresponding south pole.

One important desired feature in an electric machine is an ability to generate a constant electromagnetic torque. With a constant-torque machine, one can avoid torque pulsations that could lead to undesired mechanical vibration in the motor itself and in other mechanical components attached to the motor (e.g., mechanical loads, such as spindles or belt drives). A constant torque may not always be achieved although it will be shown that it is possible to accomplish this goal when the excitation currents are multiphase. A general rule of thumb, in this respect, is that it is desirable, insofar as possible, to produce a constant flux per pole.

13.6   THE ALTERNATOR (SYNCHRONOUS GENERATOR)

One of the most common AC machines is the **synchronous generator**, or **alternator**. In this machine, the field winding is on the rotor, and the connection is made by means of brushes, in an arrangement similar to that of the DC machines studied earlier. The rotor field is obtained by means of a direct current provided to the rotor winding, or by permanent magnets. The rotor is then connected to a mechanical source of power and rotates at a speed that we will consider constant to simplify the analysis.

Figure 13.28 depicts a two-pole three-phase synchronous machine. Figure 13.29 depicts a four-pole three-phase alternator, in which the rotor poles are generated

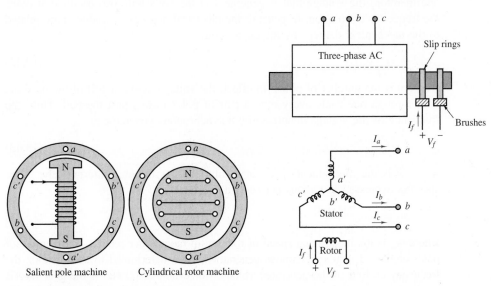

Figure 13.28 Two-pole synchronous machine

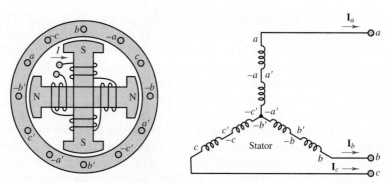

Figure 13.29 Four-pole three-phase alternator

by means of a wound salient pole configuration and the stator poles are the result of windings embedded in the stator according to the simplified arrangement shown in the figure, where each of the pairs a/a', b/b', and so on contributes to the generation of the magnetic poles, as follows. The group a/a', b/b', c/c' produces a sinusoidally distributed flux corresponding to one of the pole pairs, while the group $-a/-a'$, $-b/-b'$, $-c/-c'$ contributes the other pole pair. The connections of the coils making up the windings are also shown in Figure 13.29. Note that the coils form a wye connection (see Chapter 6). The resulting flux distribution is such that the flux completes two sinusoidal cycles around the circumference of the air gap. Note also that each arm of the three-phase wye connection has been divided into two coils, wound in different locations, according to the schematic stator diagram of Figure 13.29. One could then envision analogous configurations with greater numbers of poles, obtained in the same fashion, that is, by dividing each arm of a wye connection into more windings.

The arrangement shown in Figure 13.29 requires that a further distinction be made between mechanical degrees θ_m and electrical degrees θ_e. In the four-pole alternator, the flux will see two complete cycles during one rotation of the rotor, and therefore the voltage that is generated in the coils will also oscillate at twice the frequency of rotation. In general, the electrical degrees (or radians) are related to the mechanical degrees by the expression:

$$\theta_e = \frac{p}{2}\theta_m \tag{13.42}$$

where p is the number of poles. In effect, the voltage across a coil of the machine goes through one cycle every time a pair of poles moves past the coil. Thus, the frequency of the voltage generated by a synchronous generator is

$$f = \frac{p}{2}\frac{n}{60} \quad \text{Hz} \tag{13.43}$$

where n is the mechanical speed in revolutions per minute. Alternatively, if the speed is expressed in radians per second, we have

$$\omega_e = \frac{p}{2}\omega_m \tag{13.44}$$

where ω_m is the mechanical speed of rotation in radians per second. The number of poles employed in a synchronous generator is then determined by two factors: the frequency desired of the generated voltage (for example, 60 Hz, if the generator is used to produce AC power) and the speed of rotation of the prime mover. In the latter respect, there is a significant difference, for example, between the speed of rotation of a steam turbine generator and that of a hydroelectric generator, the former being much greater.

A common application of the alternator is seen in automotive battery-charging systems, in which, however, the generated AC voltage is rectified to provide the DC required for charging the battery. Figure 13.30 depicts an automotive alternator.

Figure 13.30 Automotive alternator (*BorgWarner*)

CHECK YOUR UNDERSTANDING

A synchronous generator has a multipolar construction that permits changing its synchronous speed. If only two poles are energized, at 50 Hz, the speed is 3,000 r/min. If the number of poles is progressively increased to 4, 6, 8, 10, and 12, find the synchronous speed for each configuration. Draw the complete equivalent circuit of a synchronous generator and its phasor diagram.

13.7   THE SYNCHRONOUS MOTOR

Synchronous motors are virtually identical to synchronous generators with regard to their construction, except for an additional winding for helping start the motor and minimizing motor speed over- and undershoots. The principle of operation is, of course, the opposite: An AC excitation provided to the armature generates a magnetic field in the air gap between stator and rotor, resulting in a mechanical torque. To generate the rotor magnetic field, some direct current must be provided to the field windings; this is often accomplished by means of an **exciter**, which consists of a small DC generator propelled by the motor itself, and therefore mechanically connected to it. It was mentioned earlier that to obtain a constant torque in an electric motor, it is necessary to keep the rotor and stator magnetic fields constant relative to each other. This means that the electromagnetically rotating field in the stator and the mechanically rotating rotor field should be aligned at all times. The only condition for which this is possible occurs if both fields are rotating at the synchronous speed $n_s = 120f/p$. Thus, synchronous motors are by their very nature constant-speed motors, if the excitation frequency is constant.

For a non–salient pole (cylindrical rotor) synchronous machine, the torque can be written in terms of the stator alternating current $i_S(t)$ and the rotor direct current, I_f:

$$T = ki_S(t)I_f \sin(\gamma) \qquad \text{Synchronous motor torque} \qquad (13.45)$$

where γ is the angle between the stator and rotor fields (see Figure 13.7). Let the angular speed of rotation be

$$\omega_m = \frac{d\theta_m}{dt} \qquad \text{rad/s} \qquad (13.46)$$

where $\omega_m = 2\pi n/60$, and let ω_e be the electrical frequency of $i_S(t)$, where $i_S(t) = \sqrt{2}\, I_S \sin(\omega_e t)$. Then the torque may be expressed as

$$T = k\sqrt{2}\, I_S \sin(\omega_e t) I_f \sin(\gamma) \qquad (13.47)$$

where k is a machine constant, I_S is the rms value of the stator current, and I_f is the rotor direct current. Now, the rotor angle γ can be expressed as a function of time by

$$\gamma = \gamma_0 + \omega_m t \qquad (13.48)$$

where γ_0 is the angular position of the rotor at $t = 0$; the torque expression then becomes

$$T = k\sqrt{2}\, I_S I_f \sin(\omega_e t) \sin(\omega_m t + \gamma_0)$$
$$= k\frac{\sqrt{2}}{2} I_S I_f \cos[(\omega_m - \omega_e)t - \gamma_0] - \cos[(\omega_m + \omega_e)t + \gamma_0] \qquad (13.49)$$

It is a straightforward matter to show that the average value of this torque, denoted by $\langle T \rangle$, is different from zero only if $\omega_m = \pm\omega_e$, that is, only if the motor is turning at the synchronous speed. The resulting average torque is then given by

$$\langle T \rangle = k\sqrt{2}\, I_S I_f \cos(\gamma_0) \qquad (13.50)$$

Note that equation 13.49 corresponds to the sum of an average torque plus a fluctuating component at twice the original electrical (or mechanical) frequency. The fluctuating component results because, in the foregoing derivation, a single-phase current was assumed. The use of multiphase currents reduces the torque fluctuation to zero and permits the generation of a constant torque.

A per-phase circuit model describing the synchronous motor is shown in Figure 13.31, where the rotor circuit is represented by a field winding equivalent resistance and inductance, R_f and L_f, respectively, and the stator circuit is represented by equivalent stator winding inductance and resistance, L_S and R_S, respectively, and by the induced emf E_b. From the exact equivalent circuit as given in Figure 13.31, we have

$$V_S = E_b + I_S(R_S + jX_S) \qquad (13.51)$$

where X_S is known as the *synchronous reactance* and includes magnetizing reactance.

The motor power is

$$P_{\text{out}} = \omega_S T = |V_S||I_S|\cos(\theta) \qquad (13.52)$$

for each phase, where T is the developed torque and θ is the angle between the stator voltage and current, V_S and I_S.

When the phase winding resistance R_S is neglected, the circuit model of a synchronous machine can be redrawn as shown in Figure 13.32. The input power (per phase) is equal to the output power in this circuit, since no power is dissipated in the circuit:

$$P_\phi = P_{\text{in}} = P_{\text{out}} = |\mathbf{V}_S||\mathbf{I}_S|\cos(\theta) \qquad (13.53)$$

Also by inspection of Figure 13.32, we have

$$d = |\mathbf{E}_b|\sin(\delta) = |\mathbf{I}_S|X_S\cos(\theta) \qquad (13.54)$$

Then:

$$|\mathbf{E}_b||\mathbf{V}_S|\sin(\delta) = |\mathbf{V}_S||\mathbf{I}_S|X_S\cos(\theta) = X_S P_\phi \qquad (13.55)$$

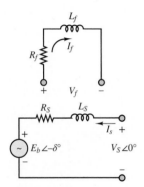

Figure 13.31 Per-phase circuit model

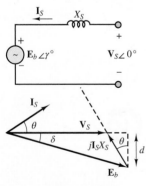

Figure 13.32 Simplified per-phase circuit model and associated vector diagram

The total power of a three-phase synchronous machine is then given by

$$P = 3\frac{|\mathbf{V}_S||\mathbf{E}_b|}{X_S}\sin(\delta) \tag{13.56}$$

Because of the dependence of the power upon the angle δ, this angle has come to be called the **power angle**. If δ is zero, the synchronous machine cannot develop useful power. The developed power has its maximum value at δ equal to 90°. If we assume that $|\mathbf{E}_b|$ and $|\mathbf{V}_S|$ are constant, we can draw the curve shown in Figure 13.33, relating the power and power angle in a synchronous machine.

A synchronous generator is usually operated at a power angle varying from 15° to 25°. For synchronous motors and small loads, δ is close to 0°, and the motor torque is just sufficient to overcome its own windage and friction losses; as the load increases, the rotor field falls further out of phase with the stator field (although the two are still rotating at the same speed) until δ reaches a maximum at 90°. If the load torque exceeds the maximum torque, which is produced for $\delta = 90°$, the motor is forced to slow down below synchronous speed. This condition is undesirable, and provisions are usually made to shut down the motor automatically whenever synchronism is lost. The maximum torque is called the **pull-out torque** and is an important measure of the performance of the synchronous motor.

Accounting for each of the phases, the total torque is given by

$$T = \frac{m}{\omega_s}|\mathbf{V}_S||\mathbf{I}_S|\cos(\theta) \tag{13.57}$$

where m is the number of phases. From Figure 13.32, we have $E_b\sin(\delta) = X_S I_S\cos(\theta)$. Therefore, for a three-phase machine, the developed torque is

$$T = \frac{P}{\omega_s} = \frac{3}{\omega_s}\frac{|\mathbf{V}_S||\mathbf{E}_b|}{X_S}\sin(\delta) \quad \text{N-m} \tag{13.58}$$

Typically, analysis of multiphase motors is performed on a per-phase basis, as illustrated in Examples 13.10 and 13.11.

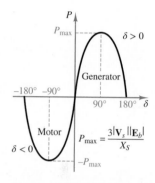

Figure 13.33 Power versus power angle for a synchronous machine

---

EXAMPLE 13.10 Synchronous Motor Analysis

Problem

Find the kilovoltampere rating, the induced voltage, and the power angle of the rotor for a fully loaded synchronous motor.

Solution

Known Quantities: Motor ratings; motor synchronous impedance.

Find: S; $\mathbf{E}_b$; δ.

Schematics, Diagrams, Circuits, and Given Data: Motor ratings: 460 V; three-phase; power factor = 0.707 lagging; full-load stator current: 12.5 A. $Z_S = 1 + j12$ Ω.

Assumptions: Use per-phase analysis.

Analysis: The circuit model for the motor is shown in Figure 13.34. The per-phase current in the wye-connected stator winding is

$$I_S = |\mathbf{I}_S| = 12.5 \text{ A}$$

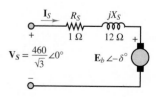

Figure 13.34

The per-phase voltage is

$$V_S = |\mathbf{V}_S| = \frac{460 \text{ V}}{\sqrt{3}} = 265.58 \text{ V}$$

The kilovoltampere rating of the motor is expressed in terms of the apparent power S (see Chapter 6):

$$S = 3V_S I_S = 3 \times 265.58 \text{ V} \times 12.5 \text{ A} = 9,959 \text{ W}$$

From the equivalent circuit, we have

$$\begin{aligned}\mathbf{E}_b &= \mathbf{V}_S - \mathbf{I}_S(R_S + jX_S)\\ &= 265.58 - (12.5\angle{-45°} \text{ A}) \times (1 + j12 \text{ } \Omega) = 179.31\angle{-32.83°} \text{ V}\end{aligned}$$

The induced line voltage is defined to be

$$V_{\text{line}} = \sqrt{3}\,E_b = \sqrt{3} \times 179.31 \text{ V} = 310.57 \text{ V}$$

From the expression for $\mathbf{E}_b$, we can find the power angle:

$$\delta = -32.83°$$

Comments: The minus sign indicates that the machine is in the motor mode.

EXAMPLE 13.11   Synchronous Motor Analysis

Problem

Find the stator current, the line current, and the induced voltage for a synchronous motor, with reference to Figure 13.34, where $Z_S = R_S + jX_S$

Solution

Known Quantities: Motor ratings; motor synchronous impedance.

Find: $\mathbf{I}_S$; $\mathbf{I}_{\text{line}}$; $\mathbf{E}_b$.

Schematics, Diagrams, Circuits, and Given Data: Motor ratings: 208 V; three-phase; 45 kVA; 60 Hz; power factor = 0.8 leading; $Z_S = 0 + j2.5 \text{ } \Omega$. Friction and windage losses: 1.5 kW; core losses: 1.0 kW; load power: 15 hp.

Assumptions: Use per-phase analysis.

Analysis: The output power of the motor is 15 hp; that is,

$$P_{\text{out}} = 15 \text{ hp} \times 0.746 \text{ kW/hp} = 11.19 \text{ kW}$$

The electric power supplied to the machine is

$$\begin{aligned}P_{\text{in}} &= P_{\text{out}} + P_{\text{mech}} + P_{\text{core loss}} + P_{\text{elec loss}}\\ &= 11.19 \text{ kW} + 1.5 \text{ kW} + 1.0 \text{ kW} + 0 \text{ kW} = 13.69 \text{ kW}\end{aligned}$$

As discussed in Chapter 6, the resulting line current is

$$I_{\text{line}} = \frac{P_{\text{in}}}{\sqrt{3}V\cos\theta} = \frac{13,690 \text{ W}}{\sqrt{3} \times 208 \text{ V} \times 0.8} = 47.5 \text{ A}$$

Because of the delta connection, the armature current is

$$\mathbf{I}_S = \frac{1}{\sqrt{3}}\mathbf{I}_{\text{line}} = 27.4\angle 36.87°\,\text{A}$$

The emf may be found from the equivalent circuit and KVL:

$$\mathbf{E}_b = \mathbf{V}_S - jX_S\mathbf{I}_S$$

$$= 208\angle 0° - (j2.5\,\Omega)(27.4\angle 36.87°\,\text{A}) = 255\angle -12.4°\,\text{V}$$

The power angle is

$$\delta = -12.4°$$

CHECK YOUR UNDERSTANDING

Find an expression for the maximum pull-out torque of the synchronous motor.

Answer: $T_{\text{max}} = \dfrac{3V_S E_b}{\omega_m X_S}$

Synchronous motors are not very commonly used in practice, for various reasons, among which are that they are essentially required to operate at constant speed (unless a variable-frequency AC supply is available) and that they are not self-starting. Further, separate AC and DC supplies are required. It will be seen shortly that the induction motor overcomes most of these drawbacks.

13.8   THE INDUCTION MOTOR

The induction motor is the most widely used electric machine, because of its relative simplicity of construction. The stator winding of an induction machine is similar to that of a synchronous machine; thus, the description of the three-phase winding of Figure 13.23 also applies to induction machines. The primary advantage of the induction machine, which is almost exclusively used as a motor (its performance as a generator is not very good), is that no separate excitation is required for the rotor. The rotor typically consists of one of two arrangements: a **squirrel cage** or a **wound rotor**. The former contains conducting bars short-circuited at the end and embedded within it; the latter consists of a multiphase winding similar to that used for the stator, but electrically short-circuited.

In either case, the induction motor operates by virtue of currents induced from the stator field in the rotor. In this respect, its operation is similar to that of a transformer, in that currents in the stator (which acts as a primary coil) induce currents in the rotor (acting as a secondary coil). In most induction motors, no external electrical connection is required for the rotor, thus permitting a simple, rugged construction without the need for slip rings or brushes. Unlike the synchronous motor, the induction motor operates not at synchronous speed, but at a somewhat lower speed, which is dependent on the load. Figure 13.35 illustrates the appearance of a squirrel cage induction motor. The following discussion focuses mainly on this very common configuration.

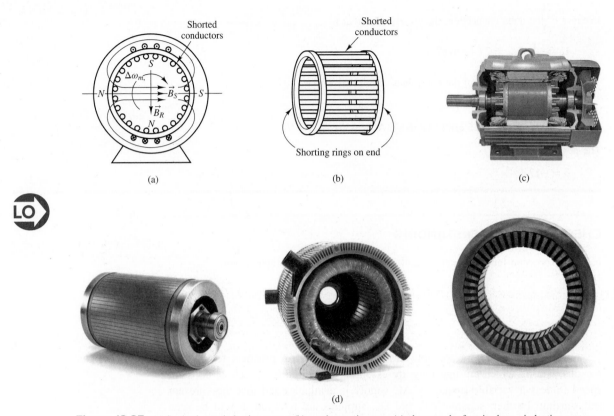

Figure 13.35 (a) Squirrel cage induction motor; (b) conductors in rotor; (c) photograph of squirrel cage induction motor; (d) views of Smokin' Buckeye motor: rotor, stator, and cross section of stator ((c) *Normal Life/Shutterstock;* (d) *Courtesy: David H. Koether Photography*)

By now you are acquainted with the notion of a rotating stator magnetic field. Imagine now that a squirrel cage rotor is inserted in a stator in which such a rotating magnetic field is present. The stator field will induce voltages in the cage conductors, and if the stator field is generated by a three-phase source, the resulting rotor currents—which circulate in the bars of the squirrel cage, with the conducting path completed by the shorting rings at the end of the cage—are also three-phase and are determined by the magnitude of the induced voltages and by the impedance of the rotor. Since the rotor currents are induced by the stator field, the number of poles and the speed of rotation of the induced magnetic field are the same as those of the stator field, *if the rotor is at rest*. Thus, when a stator field is initially applied, the rotor field is synchronous with it, and the fields are stationary with respect to one another. Thus, according to the earlier discussion, a *starting torque* is generated.

If the starting torque is sufficient to cause the rotor to start spinning, the rotor will accelerate up to its operating speed. However, an induction motor can never reach synchronous speed; if it did, the rotor would appear to be stationary with respect to the rotating stator field, since it would be rotating at the same speed. But in the absence of relative motion between the stator and rotor fields, no voltage would be induced in the rotor. Thus, an induction motor is limited to speeds somewhere below the synchronous speed n_s. Let the speed of rotation of the rotor be n; then the rotor is losing ground with respect to the rotation of the stator field at a speed $n_s - n$. In effect, this is equivalent to backward motion of the rotor at the **slip speed**, defined by $n_s - n$. The **slip** s is usually defined as a fraction of n_s:

$$s = \frac{n_s - n}{n_s} \qquad \text{Slip in induction machine} \tag{13.59}$$

which leads to the following expression for the rotor speed:

$$n = n_s(1 - s) \tag{13.60}$$

The slip s is a function of the load, and the amount of slip in a given motor is dependent on its construction and rotor type (squirrel cage or wound rotor). Since there is a relative motion between the stator and rotor fields, voltages will be induced in the rotor at a frequency called the **slip frequency**, $f_R = sf$, where f is the frequency of the sinusoidal excitation related to the relative speed of the two fields. This gives rise to an interesting phenomenon: The rotor field travels relative to the rotor at the slip speed sn_s, but the rotor is mechanically traveling at the speed $(1 - s)n_s$, so that the net effect is that the rotor field travels at the speed:

$$sn_s + (1 - s)n_s = n_s \tag{13.61}$$

that is, at synchronous speed. The fact that the rotor field rotates at synchronous speed—although the rotor itself does not—is extremely important because it means that the stator and rotor fields will continue to be stationary with respect to each other, and therefore a net torque can be produced.

As in the case of DC and synchronous motors, important characteristics of induction motors are the starting torque, the maximum torque, and the torque–speed curve. These will be discussed shortly, after some analysis of the induction motor is performed.

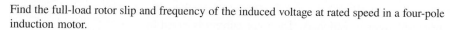

EXAMPLE 13.12   Induction Motor Analysis

Problem

Find the full-load rotor slip and frequency of the induced voltage at rated speed in a four-pole induction motor.

Solution

Known Quantities: Motor ratings.

Find: s; f_R.

Schematics, Diagrams, Circuits, and Given Data: Motor ratings: 230 V; 60 Hz; full-load speed: 1,725 r/min.

Analysis: The synchronous speed of the motor is

$$n_s = \frac{120f}{p} = \frac{60f}{p/2} = \frac{60\,\text{s/min} \times 60\,\text{r/s}}{4/2} = 1{,}800 \text{ r/min}$$

The slip is

$$s = \frac{n_s - n}{n_s} = \frac{1{,}800 \text{ r/min} - 1{,}725 \text{ r/min}}{1{,}800 \text{ r/min}} = 0.0417$$

The rotor frequency f_R is

$$f_R = sf = 0.0417 \times 60 \text{ Hz} = 2.5 \text{ Hz}$$

CHECK YOUR UNDERSTANDING

A three-phase induction motor has six poles. (a) If the line frequency is 60 Hz, calculate the speed of the magnetic field in revolutions per minute. (b) Repeat the calculation if the frequency is changed to 50 Hz.

Answer: (a) $n = 1,200$ r/min; (b) $n = 1,000$ r/min.

The induction motor can be described by means of an equivalent circuit, which is essentially that of a rotating transformer. (See Chapter 6 for a circuit model of the transformer.) Figure 13.36 depicts such a circuit model, where:

R_S = stator resistance per phase $\qquad$ R_R = rotor resistance per phase

X_S = stator reactance per phase $\qquad$ X_R = rotor reactance per phase

X_m = magnetizing (mutual) reactance

R_C = equivalent core-loss resistance

E_S = per-phase induced voltage in stator windings

E_R = per-phase induced voltage in rotor windings

The primary internal stator voltage $\mathbf{E}_S$ is coupled to the secondary rotor voltage $\mathbf{E}_R$ by an ideal transformer with an effective turns ratio of α. For the rotor circuit, the induced voltage at any slip will be

$$\mathbf{E}_R = s\mathbf{E}_{R0} \tag{13.62}$$

where $\mathbf{E}_{R0}$ is the induced rotor voltage at the condition in which the rotor is stationary. Also, $X_R = \omega_R L_R = 2\pi f_R L_R = 2\pi s f L_R = sX_{R0}$, where $X_{R0} = 2\pi f L_R$ is the reactance when the rotor is stationary. The rotor current is given by

$$\mathbf{I}_R = \frac{\mathbf{E}_R}{R_R + jX_R} = \frac{s\mathbf{E}_{R0}}{R_R + jsX_{R0}} = \frac{\mathbf{E}_{R0}}{R_R/s + jX_{R0}} \tag{13.63}$$

The resulting rotor equivalent circuit is shown in Figure 13.37.

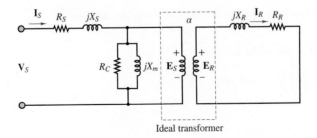

Ideal transformer

Figure 13.36 Circuit model for induction machine

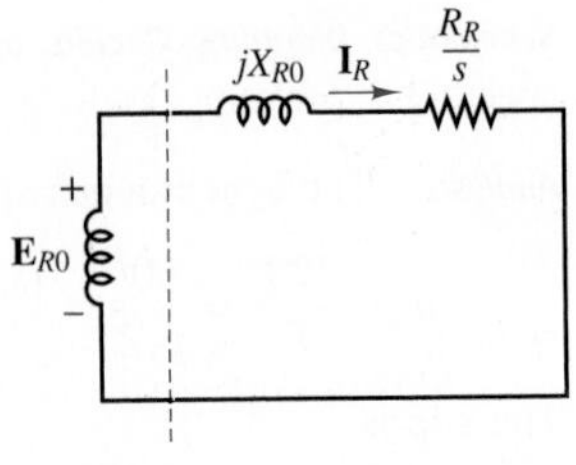

Figure 13.37 Rotor circuit

The voltages, currents, and impedances on the secondary (rotor) side can be reflected to the primary (stator) by means of the effective turns ratio. When this transformation is effected, the transformed rotor voltage is given by

$$\mathbf{E}_2 = \mathbf{E}_R' = \alpha \mathbf{E}_{R0} \tag{13.64}$$

The transformed (reflected) rotor current is

$$\mathbf{I}_2 = \frac{\mathbf{I}_R}{\alpha} \tag{13.65}$$

The transformed rotor resistance can be defined as

$$R_2 = \alpha^2 R_R \tag{13.66}$$

and the transformed rotor reactance can be defined by

$$X_2 = \alpha^2 X_{R0} \tag{13.67}$$

The final per-phase equivalent circuit of the induction motor is shown in Figure 13.38.

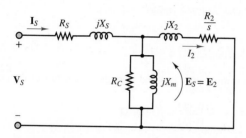

Figure 13.38 Equivalent circuit of an induction machine

Examples 13.13 and 13.14 illustrate the use of the circuit model in determining the performance of the induction motor.

EXAMPLE 13.13   Induction Motor Analysis

Problem

Determine the following quantities for an induction motor, using the circuit model of Figures 13.36 to 13.38.

1. Speed
2. Stator current
3. Power factor
4. Output torque

Solution

Known Quantities: Motor ratings; circuit parameters.

Find: n; ω_m; $\mathbf{I}_S$; power factor (pf); T.

Schematics, Diagrams, Circuits, and Given Data: Motor ratings: 460 V; 60 Hz; four poles; $s = 0.022$; $P_{out} = 14$ hp; $R_S = 0.641\ \Omega$; $R_2 = 0.332\ \Omega$; $X_S = 1.106\ \Omega$; $X_2 = 0.464\ \Omega$; $X_m = 26.3\ \Omega$

Assumptions: Use per-phase analysis. Neglect core losses ($R_C = 0$).

Analysis:

1. The per-phase equivalent circuit is shown in Figure 13.38. The synchronous speed is found to be

$$n_s = \frac{120f}{p} = \frac{60\ \text{s/min} \times 60\ \text{r/s}}{4/2} = 1{,}800\ \text{r/min}$$

or

$$\omega_s = 1{,}800\frac{\text{r}}{\text{min}} \times \frac{2\pi\ \text{rad}}{60\ \text{s/min}} = 188.5\ \text{rad/s}$$

The rotor mechanical speed is

$$n = (1 - s)n_s = 1{,}760\ \text{r/min}$$

or

$$\omega_m = (1 - s)\omega_s = 184.4\ \text{rad/s}$$

2. The reflected rotor impedance is found from the parameters of the per-phase circuit to be

$$Z_2 = \frac{R_2}{s} + jX_2 = \frac{0.332}{0.022} + j0.464\ \Omega$$

$$= 15.09 + j0.464\ \Omega$$

The combined magnetization plus rotor impedance is therefore equal to

$$Z = \frac{1}{1/jX_m + 1/Z_2} = \frac{1}{-j0.038 + 0.0662\angle -1.76°} = 12.93\angle 31.2°\ \Omega$$

and the total impedance is

$$Z_{total} = Z_S + Z = 0.641 + j1.106 + 11.06 + j6.69$$
$$= 11.70 + j7.8 = 14.06\angle 33.7°\ \Omega$$

Finally, the stator current is given by

$$\mathbf{I}_S = \frac{\mathbf{V}_S}{Z_{total}} = \frac{460/\sqrt{3}\angle 0°\ \text{V}}{14.07\angle 33.6°\ \Omega} = 18.88\angle -33.7°\ \text{A}$$

3. The power factor is

$$\text{pf} = \cos 33.6° = 0.832\ \text{lagging}$$

4. The output power P_{out} is

$$P_{out} = 14\ \text{hp} \times 746\ \text{W/hp} = 10.444\ \text{kW}$$

and the output torque is

$$T = \frac{P_{out}}{\omega_m} = \frac{10{,}444\ \text{W}}{184.4\ \text{rad/s}} = 56.64\ \text{N-m}$$

CHECK YOUR UNDERSTANDING

A four-pole induction motor operating at a frequency of 60 Hz has a full-load slip of 4 percent. Find the frequency of the voltage induced in the rotor (a) at the instant of starting and (b) at full load.

EXAMPLE 13.14   Induction Motor Analysis

Problem

Determine the following quantities for a three-phase induction motor, using the circuit model of Figure 13.38.

1. Stator current
2. Power factor
3. Full-load electromagnetic torque

Solution

Known Quantities: Motor ratings; circuit parameters.

Find: $\mathbf{I}_S$; pf; T.

Schematics, Diagrams, Circuits, and Given Data: Motor ratings: 500 V; three-phase; 50 Hz; $p = 8$; $s = 0.05$; $P = 14$ hp.

Circuit parameters: $R_S = 0.13\ \Omega$; $R'_R = 0.32\ \Omega$; $X_S = 0.6\ \Omega$; $X'_R = 1.48\ \Omega$; $Y_m = G_C + jB_m = $ magnetic branch admittance describing core loss and mutual inductance $= 0.004 - j0.05\ \Omega^{-1}$; stator/rotor turns ratio $= 1:\alpha = 1:1.57$.

Assumptions: Use per-phase analysis. Neglect mechanical losses.

Analysis: The approximate equivalent circuit of the three-phase induction motor on a per-phase basis is shown in Figure 13.39. The parameters of the model are calculated as follows:

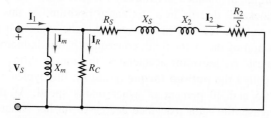

Figure 13.39 Per-phase equivalent circuit of induction motor

$$R_2 = R'_R \times \left(\frac{1}{\alpha}\right)^2 = 0.32 \times \left(\frac{1}{1.57}\right)^2 = 0.13 \ \Omega$$

$$X_2 = X'_R \times \left(\frac{1}{\alpha}\right)^2 = 1.48 \times \left(\frac{1}{1.57}\right)^2 = 0.6 \ \Omega$$

$$Z = R_S + \frac{R_2}{s} + j(X_S + X_2)$$

$$= 0.13 + \frac{0.13}{0.05} + j(0.6 + 0.6) = 2.73 + j1.2 \ \Omega$$

Using the approximate circuit, we have

$$\mathbf{I}_2 = \frac{\mathbf{V}_S}{Z} = \frac{(500/\sqrt{3})\angle 0° \ \text{V}}{2.73 + j1.2 \ \Omega} = 88.6 - 38.9 \ \text{A}$$

$$\mathbf{I}_R = \mathbf{V}_S G_C = 288.7 \ \text{V} \times 0.004 \ \Omega^{-1} = 1.15 \ \text{A}$$

$$\mathbf{I}_m = -j\mathbf{V}_S B_m = 288.7 \ \text{V} \times (-j0.05)\Omega = -j14.4 \ \text{A}$$

$$\mathbf{I}_1 = \mathbf{I}_2 + \mathbf{I}_R + \mathbf{I}_m = 89.75 - j53.3 \ \text{A}$$

$$\text{Input power factor} = \frac{\text{Re}[\mathbf{I}_1]}{|\mathbf{I}_1|} = \frac{89.95}{104.6} = 0.86 \ \text{lagging}$$

$$\text{Torque} = \frac{3P}{\omega_S} = \frac{3 I_2^2 R_2 / s}{4\pi f / p} = 931 \ \text{N-m}$$

CHECK YOUR UNDERSTANDING

A four-pole, 1,746 r/min, 220-V, three-phase, 60-Hz, 10-hp, Y-connected induction machine has the following parameters: $R_S = 0.4 \ \Omega$, $R_2 = 0.14 \ \Omega$, $X_m = 16 \ \Omega$, $X_S = 0.35 \ \Omega$, $X_2 = 0.35 \ \Omega$, $R_C = \infty$. Using Figure 13.38 find (a) the stator current, (b) the rotor current, (c) the motor power factor, and (d) the total stator power input.

Answer: (a) $25.92\angle{-222.43°}$ A; (b) $24.35\angle{-6.51°}$ A; (c) 0.9243; (d) 9,129 W

Performance of Induction Motors

The performance of induction motors can be described by torque–speed curves similar to those already used for DC motors. Figure 13.40 depicts an induction motor torque–speed curve, with five torque ratings marked *a* through *e*. Point *a* is the *starting torque*, also called **breakaway torque**, and is the torque available with the rotor "locked," that is, in a stationary position. At this condition, the frequency of the voltage induced in the rotor is highest, since it is equal to the frequency of rotation of the stator field; consequently, the inductive reactance of the rotor is greatest. As the rotor accelerates, the torque drops off, reaching a minimum value called the **pull-up torque** (point *b*); this typically occurs somewhere between 25 and 40 percent of synchronous speed. As the rotor speed continues to increase, the rotor reactance decreases further (since the frequency of the induced voltage is determined by the relative speed of rotation of the rotor with respect to the stator field). The torque becomes a maximum when the rotor

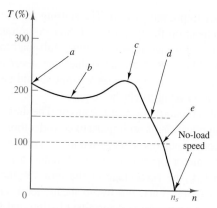

Figure 13.40 Performance curve for induction motor

Figure 13.41 Induction motor classification

inductive reactance is equal to the rotor resistance; maximum torque is also called **breakdown torque** (point *c*). Beyond this point, the torque drops off until it is zero at synchronous speed, as discussed earlier. Also marked on the curve are the *150 percent torque* (point *d*) and the *rated torque* (point *e*).

A general formula for the computation of the induction motor steady-state torque–speed characteristic is

$$T = \frac{1}{\omega_e} \frac{m V_S^2 R_R / s}{(R_S + R_R/s)^2 + (X_S + X_R)^2} \qquad \begin{array}{l} T\text{-}\omega \text{ equation for} \\ \text{induction machine} \end{array}$$

(13.68)

where *m* is the number of phases.

Different construction arrangements permit the design of induction motors with different torque–speed curves, thus permitting the user to select the motor that best suits a given application. Figure 13.41 depicts the four basic classifications—classes A, B, C, and D—as defined by NEMA. The determining features in the classification are the locked-rotor torque and current, the breakdown torque, the pull-up torque, and the percentage of slip. Class A motors have a higher breakdown torque than class B motors, and a slip of 5 percent or less. Motors in this class are often designed for a specific application. Class B motors are general-purpose motors; this is the most commonly used type of induction motor, with typical values of slip of 3 to 5 percent. Class C motors have a high starting torque for a given starting current, and a low slip. These motors are typically used in applications demanding high starting torque but having relatively normal running loads, once the running speed has been reached. Class D motors are characterized by high starting torque, high slip, low starting current, and low full-load speed. A typical value of slip is around 13 percent.

Factors that should be considered in the selection of an AC motor for a given application are the *speed range*, both minimum and maximum, and the speed variation. For example, it is important to determine whether constant speed is required; what variation might be allowed, either in speed or in torque; or whether variable-speed operation is required, in which case a variable-speed drive will be needed.

The torque requirements are obviously important as well. The starting and running torque should be considered; they depend on the type of load. Starting torque can vary from a small percentage of full-load torque to several times full-load torque. Furthermore, the excess torque available at start-up determines the *acceleration characteristics* of the motor. Similarly, *deceleration characteristics* should be considered, to determine whether external braking might be required.

Another factor to be considered is the *duty cycle* of the motor. The duty cycle, which depends on the nature of the application, is an important consideration when the motor is used in repetitive, noncontinuous operation, such as is encountered in some types of machine tools. If the motor operates at zero or reduced load for periods of time, the duty cycle—that is, the percentage of the time the motor is loaded—is an important selection criterion. Last, but by no means least, are the *thermal properties* of a motor. Motor temperature is determined by internal losses and by ventilation; motors operating at a reduced speed may not generate sufficient cooling, and forced ventilation may be required.

Thus far, we have not considered the dynamic characteristics of induction motors. Among the integral-horsepower induction motors (i.e., motors with horsepower rating greater than 1), the most common dynamic problems are associated with starting and stopping and with the ability of the motor to continue operation during supply system transient disturbances. Dynamic analysis methods for induction motors depend to a considerable extent on the nature and complexity of the problem and the associated precision requirements. When the electric transients in the motor are to be included as well as the motion transients, and especially when the motor is an important element in a large network, the simple transient equivalent circuit of Figure 13.42 provides a good starting approximation. There, X'_S is called the *transient reactance*. The voltage E'_S is called the *voltage behind the transient reactance* and is assumed to be equal to the initial value of the induced voltage, at the start of the transient. The stator resistance is R_S. The dynamic analysis problem consists of selecting a sufficiently simple but reasonably realistic representation that will not unduly complicate the dynamic analysis, particularly through the introduction of non-linearities.

It should be remarked that the basic equations of the induction machine, as derived from first principles, are quite nonlinear. Thus, an accurate dynamic analysis of the induction motor, without any linearizing approximations, requires the use of computer simulation.

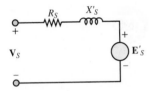

Figure 13.42 Simplified induction motor dynamic model

AC Motor Speed and Torque Control

As explained in an earlier section, AC machines are constrained to fixed-speed or near fixed-speed operation when supplied by a constant-frequency source. Several simple methods exist to provide limited speed control in AC induction machines; more complex methods, involving the use of advanced power electronics circuits, can be used if the intended application requires wide-bandwidth control of motor speed or torque. In this subsection we provide a general overview of available solutions.

Pole Number Control

The (conceptually) easiest method to implement speed control in an induction machine is by *varying the number of poles*. Equation 13.41 explains the dependence of synchronous speed in an AC machine on the supply frequency and on the number of poles. For machines operated at 60 Hz, the following speeds can be achieved by varying the number of magnetic poles in the stator winding:

Number of poles	2	4	6	8	12
n (r/min)	3,600	1,800	1,200	800	600

While for machines operating at 50 Hz, the speeds are

Number of poles	2	4	6	8	12
n (r/min)	3,000	1,500	1,000	667	500

Motor stators can be wound so that the number of pole pairs in the stators can be varied by switching between possible winding connections. Such switching requires that care be taken in timing it to avoid damage to the machine.

Slip Control

Since the rotor speed is inherently dependent on the slip, *slip control* is a valid means of achieving some speed variation in an induction machine. Since motor torque falls with the square of the voltage (see equation 13.68), it is possible to change the slip by changing the motor torque through a change in motor voltage. This procedure allows for speed control over the range of speeds that allow for stable motor operation. With reference to Figure 13.40, this is possible only above point *c*, that is, above the *breakdown torque*.

Rotor Control

For motors with wound rotors, it is possible to connect the rotor slip rings to resistors; adding resistance to the rotor increases the losses in the rotor and therefore causes the rotor speed to decrease. This method is also limited to operation above the *breakdown torque* although it should be noted that the shape of the motor torque–speed characteristic changes when the rotor resistance is changed.

Frequency Regulation

The last two methods cause additional losses to be introduced in the machine. If a variable-frequency supply is used, motor speed can be controlled without any additional losses. As seen in equation 13.41, the motor speed is directly dependent on the supply frequency, as the supply frequency determines the speed of the rotating magnetic field. However, to maintain the same motor torque characteristics over a range of speeds, the motor voltage must change with frequency, to maintain a constant torque. Thus, generally, the volts/hertz ratio should be held constant. This condition is difficult to achieve at start-up and at very low frequencies, in which cases the voltage must be raised above the constant volts/hertz ratio that will be appropriate at higher frequency.

Variable-Frequency Drives

AC machines are capable of variable-speed operation if equipped with a **variable-frequency drive**. Variable-frequency drives are capable of producing a variable-frequency sinusoidal output starting from a fixed-frequency AC input, in which case they are called AC-AC converters, or from a DC input, in which case they are called DC-AC converters, or **inverters**. Figure 13.43 depicts the general configuration of a variable-speed drive for an AC machine. The top half of the figure illustrates the case of a three-phase AC source, wherein a three-phase rectifier provides a DC output, which is conditioned by a DC link circuit (often, a capacitor) to smooth any ripple that may be present in the rectified AC output; the output of the DC link is a conditioned DC input that is appropriate for DC-AC conversion, through an **inverter**, capable of generating a three-phase sinusoidal output of variable output frequency and voltage; the controller provides the necessary control functions, for example, speed or torque control, for the electric machine that is connected to the output of the inverter. In the bottom half of the figure a similar diagram is shown, but in this case the input is a DC source, for example a solar array or a battery. We focus the remainder of this section on DC-AC conversion, that is, on the inverter.

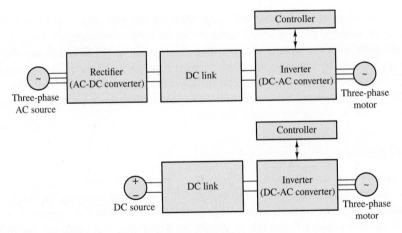

Figure 13.43 Variable-speed drive for AC machine

The basic topology of the inverter is shown in Figure 13.44. The inverter consists of an array of power switches (IGBTs shown in the figure), that are switched on and off so as to produce a pulse-width-modulated (PWM) waveform that, when filtered by the inductance of the electric machine, has nearly sinusoidal behavior. This technique is called **sinusoidal PWM**. The switching frequency for the power switches (the IGBTs in Figure 13.44) determines the frequency of the three-phase sinusoidal output. A microcontroller receives information from the electric machine (speed of rotation, current, voltage) and computes the appropriate switching pattern to obtain the desired motor speed and torque. The process by which a train of pulses is converted into a sinusoidal waveform (sinusoidal PWM) is conceptually depicted in Figure 13.45: if one were able to perfectly filter an appropriately generated pulse train, the resulting waveform would be a sinusoidal waveform. In fact, the electric machine, through its inductance, provides the needed filtering. In practice, an actual sinusoidal PWM waveform is more complicated than what is shown in the figure.

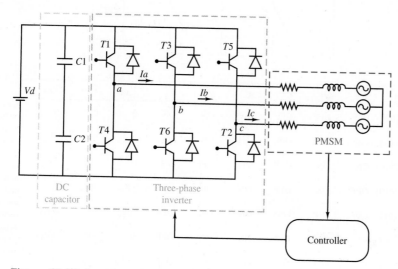

Figure 13.44 Inverter connected to an electric machine

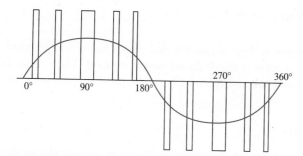

Figure 13.45 Conceptual sketch of sinusoidal PWM waveform generation

Another modulation technique that is gaining in popularity is the so-called **space vector PWM**, summarized in Figure 13.46, where six power switches are shown in the form of ideal switches, and the three-phase voltages are shown. The table that accompanies the figure illustrates the switching law that would result in a three-phase quasi-sinusoidal output.

In many applications, for example, electric traction drives in electric and hybrid vehicles, one is often interested in controlling the torque output of the machine. The most common method employed to achieve torque control is based on **vector control**. One type of vector control is called **field-oriented control** and is applicable to both synchronous and induction machines, and it uses PWM to

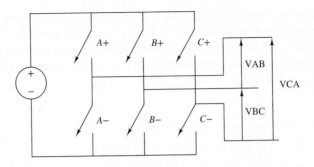

Vector	A^+	B^+	C^+	A^-	B^-	C^-	V_{AB}	V_{BC}	V_{CA}	
$V_0 = \{000\}$	OFF	OFF	OFF	ON	ON	ON	0	0	0	Zero vector
$V_1 = \{100\}$	ON	OFF	OFF	OFF	ON	ON	$+V_{dc}$	0	$-V_{dc}$	Active vector
$V_2 = \{110\}$	ON	ON	OFF	OFF	OFF	ON	0	$+V_{dc}$	$-V_{dc}$	Active vector
$V_3 = \{010\}$	OFF	ON	OFF	ON	OFF	ON	$-V_{dc}$	$+V_{dc}$	0	Active vector
$V_4 = \{011\}$	OFF	ON	ON	ON	OFF	OFF	$-V_{dc}$	0	$+V_{dc}$	Active vector
$V_5 = \{001\}$	OFF	OFF	ON	ON	ON	OFF	0	$-V_{dc}$	$+V_{dc}$	Active vector
$V_6 = \{101\}$	ON	OFF	ON	OFF	ON	OFF	$+V_{dc}$	$-V_{dc}$	0	Active vector
$V_7 = \{111\}$	ON	ON	ON	OFF	OFF	OFF	0	0	0	Zero vector

Figure 13.46 Space vector PWM

control the motor voltage magnitude and phase angle, and frequency. A second vector control method is called **direct torque control**. These methods are beyond the scope of this book and are usually the subject of a graduate course in electric drive control.

Conclusion

This chapter introduces the most common classes of rotating electric machines. These machines, which can range in power from the milliwatt to the megawatt range, find common application in virtually every field of engineering, from consumer products to heavy-duty industrial applications. The principles introduced in this chapter can give you a solid basis from which to build upon.

Upon completing this chapter, you should have mastered the following learning objectives:

1. *Understand the basic principles of operation of rotating electric machines, their classification, and basic efficiency and performance characteristics.* Electric machines are defined in terms of their mechanical characteristics (torque–speed curves, inertia, friction and windage losses) and their electrical characteristics (current and voltage requirements). Losses and efficiency are an important part of the operation of electric machines, and it should be recognized that machines will suffer from electrical, mechanical, and magnetic core losses. All machines are based on the principle of establishing a magnetic field in the stationary part of the machine (stator) and a magnetic field in the moving part of the machine (rotor); electric machines can then be classified according to how the stator and rotor fields are established.

2. *Understand the operation and basic configurations of separately excited, permanent-magnet, shunt and series DC machines.* Direct-current machines, operated from a DC supply, are among the most common electric machines. The rotor (armature) circuit is connected to an external DC supply via a commutator. The stator electric field can be established by an external circuit (separately excited machines), by a permanent magnet (PM machines), or by the same supply used for the armature (self-excited machines).

3. *Analyze DC motors under steady-state and dynamic operation.* DC motors are commonly used in a variety of variable-speed applications (e.g., electric vehicles, servos) which require speed control; thus, their dynamics are also of interest.

4. *Analyze DC generators at steady state.* DC generators can be used to supply a variable direct current and voltage when propelled by a prime mover (engine, or other thermal or hydraulic machine).

5. *Understand the operation and basic configuration of AC machines, including the synchronous motor and generator and the induction machine.* AC machines require an alternating-current supply. The two principal classes of AC machines are the synchronous and induction types. Synchronous machines rotate at a predetermined speed, which is equal to the speed of a rotating magnetic field present in the stator, called the *synchronous speed.* Induction machines also operate based on a rotating magnetic field in the stator; however, the speed of the rotor is dependent on the operating conditions of the machine and is always less than the synchronous speed. Variable-speed AC machines require more sophisticated electric power supplies that can provide variable voltage/current and variable frequency. As the cost of power electronics is steadily decreasing, variable-speed AC drives are becoming increasingly common.

HOMEWORK PROBLEMS

Section 13.1: Rotating Electric Machines

13.1   The power rating of a motor can be modified to account for different ambient temperature, according to the following table:

Ambient temperature	30°C	35°C	40°C
Variation of rated power	+8%	+5%	0
Ambient temperature	45°C	50°C	55°C
Variation of rated power	−5%	−12.5%	−25%

A motor with $P_e = 10$ kW is rated up to 85°C. Find the actual power for each of the following conditions:

a. Ambient temperature is 50°C.

b. Ambient temperature is 30°C.

13.2   The speed–torque characteristic of an induction motor has been empirically determined as follows:

Speed (r/min)	1,470	1,440	1,410	1,300	1,100
Torque (N-m)	3	6	9	13	15

Speed (r/min)	900	750	350	0
Torque (N-m)	13	11	7	5

The motor will drive a load requiring a starting torque of 4 N-m and increase linearly with speed to 8 N-m at 1,500 r/min.

a. Find the steady-state operating point of the motor.

b. Equation 13.68 predicts that the motor speed can be regulated in the face of changes in load torque by adjusting the stator voltage. Find the change in voltage required to maintain the speed at the operating point of part a if the load torque increases to 10 N-m.

Section 13.2: Direct-Current Machines

13.3   Calculate the force exerted by each conductor, 6 in long, on the armature of a DC motor when it carries a current of 90 A and lies in a field the density of which is 5.2×10^{-4} Wb/in^2.

13.4   In a DC machine, the air gap flux density is 4 Wb/m^2. The area of the pole face is 2 cm $\times$ 4 cm. Find the flux per pole in the machine.

Section 13.3: Direct-Current Motors

13.5   A 220-V shunt motor has an armature resistance of 0.32 Ω and a field resistance of 110 Ω. At no load the armature current is 6 A and the speed is 1,800 r/min. Assume that the flux does not vary with load, and calculate

a. The speed of the motor when the line current is 62 A (assume a 2-V brush drop).

b. The speed regulation of the motor.

13.6   A 50-hp, 550-V shunt motor has an armature resistance, including brushes, of 0.36 Ω. When operating at rated load and speed, the armature takes 75 A. What resistance should be inserted in the armature circuit to obtain a 20 percent speed reduction when the motor is developing 70 percent of rated torque? Assume that there is no flux change.

13.7   A shunt DC motor has a shunt field resistance of 400 Ω and an armature resistance of 0.2 Ω. The motor nameplate rating values are 440 V, 1,200 r/min, 100 hp, and full-load efficiency of 90 percent. Find

a. The motor line current.

b. The field and armature currents.

c. The counter-emf at rated speed.

d. The output torque.

13.8   A 240-V series motor has an armature resistance of 0.42 Ω and a series-field resistance of 0.18 Ω. If the speed is 500 r/min when the current is 36 A, what will be the motor speed when the load reduces the line current to 21 A? (Assume a 3-V brush drop and that the flux is proportional to the current.)

13.9   A 220-V DC shunt motor [see Figure 13.14(b)] has an armature resistance of 0.2 Ω and a rated armature current of 50 A. Find

a. The voltage generated in the armature.

b. The power developed.

13.10   A 550-V series motor takes 112 A and operates at 820 r/min when the load is 75 hp. If the effective armature-circuit resistance is 0.15 Ω, calculate the horsepower output of the motor when the current drops to 84 A, assuming that the flux is reduced by 15 percent.

13.11   A 200-V DC shunt motor has the following parameters:

$$R_a = 0.1\ \Omega \qquad R_f = 100\ \Omega$$

When running at 1,100 r/min with no load connected to the shaft, the motor draws 4 A from the line. Find E and the rotational losses at 1,100 r/min (assuming that the stray-load losses can be neglected).

13.12   A 230-V DC shunt motor has the following parameters:

$$R_a = 0.5\ \Omega \qquad R_f = 75\ \Omega$$
$$P_{\text{rot}} = 500\ \text{W} \qquad \text{at } 1,120\ \text{r/min}$$

When loaded, the motor draws 46 A from the line. Find

a. The speed, P_{dev}, and T_{sh}.

b. If $L_f = 25$ H, $L_a = 0.008$ H, and the terminal voltage has a 115-V change, find $i_a(t)$ and $\omega_m(t)$.

13.13   A 200-VDC shunt motor with an armature resistance of 0.1 Ω and a field resistance of 100 Ω draws a line current of 5 A when running with no load at 955 r/min. Determine the motor speed, the motor efficiency, the total losses (i.e., rotational and I_2R losses), and the load torque T_{sh} that will result when the motor draws 40 A from the line. Assume rotational power losses are proportional to the square of shaft speed.

13.14   A 50-hp, 230-V shunt motor has a field resistance of 17.7 Ω and operates at full load when the line current is 181 A at 1,350 r/min. To increase the speed of the motor to 1,600 r/min, a resistance of 5.3 Ω is "cut in" via the field rheostat; the line current then increases to 190 A. Calculate

a. The power loss in the field and its percentage of the total power input for the 1,350 r/min speed.

b. The power losses in the field and the field rheostat for the 1,600 r/min speed.

c. The percent losses in the field and in the field rheostat at 1,600 r/min.

13.15   A 10-hp, 230-V shunt-wound motor has a rated speed of 1,000 r/min and full-load efficiency of 86 percent. Armature circuit resistance is 0.26 Ω; field-circuit resistance is 225 Ω. If this motor is operating under rated load and the field flux is very quickly reduced to 50 percent of its normal value, what will be the effect upon counter-emf, armature current, and torque? What effect will this change have upon the operation of the motor, and what will be its speed when stable operating conditions have been regained?

13.16   The machine of Example 13.5 is to be used in a
series connection. That is, the field coil is connected in
series with the armature, as shown in Figure P13.16.
The machine is to be operated under the same
conditions as in Example 13.5, that is, $n = 120$ r/min
and $I_a = 8$ A. In the operating region, $\phi = kI_f$ and
$k = 200$. The armature resistance is 0.2 Ω, and the
resistance of the field winding is negligible.

a. Find the number of field winding turns necessary
    for full-load operation.

b. Find the torque output for the following speeds:

    1. $n' = 2n$ 3. $n' = n/2$

    2. $n' = 3n$ 4. $n' = n/4$

c. Plot the speed–torque characteristic for the
    conditions of part b.

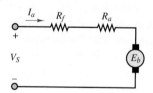

Figure P13.16

13.17   With reference to Example 13.7, assume that the
load torque applied to the PM DC motor is zero.
Determine the speed response of the motor speed to
a step change in input voltage. Derive expressions
for the natural frequency and damping ratio of the
second-order system. What determines whether the
system is over- or underdamped?

13.18   A motor with polar moment of inertia J develops
torque according to the relationship $T = a\omega + b$. The
motor drives a load defined by the torque–speed
relationship $T_L = c\omega^2 + d$. If the four coefficients are
all positive constants, determine the equilibrium
speeds of the motor-load pair, and whether these
speeds are stable.

13.19   Assume that a motor has known friction and
windage losses described by the equation $T_{FW} = b\omega$.
Sketch the T-ω characteristic of the motor if the load
torque T_L is constant, and the T_L-ω characteristic if the
motor torque is constant. Assume that T_{FW} at full
speed is equal to 30 percent of the load torque.

13.20   A PM DC motor is rated at 6 V, 3,350 r/min and
has the following parameters: $r_a = 7$ Ω, $L_a = 120$ mH,
$k_T = 7 \times 10^{-3}$ N-m/A, $J = 1 \times 10^{-6}$ kg-m^2. The no-load
armature current is 0.15 A.

a. In the steady-state no-load condition, the magnetic
    torque must be balanced by an internal damping
    torque; find the damping coefficient b. Now sketch
    a model of the motor, write the dynamic equations,
    and determine the transfer function from armature
    voltage to motor speed. What is the approximate
    3-dB bandwidth of the motor?

b. Now let the motor be connected to a pump with
    inertia $J_L = 1 \times 10^{-4}$ kg-m^2, damping coefficient
    $b_L = 5 \times 10^{-3}$ N-m-s, and load torque
    $T_L = 3.5 \times 10^{-3}$ N-m. Sketch the model describing
    the motor-load configuration, and write the
    dynamic equations for this system; determine the
    new transfer function from armature voltage to
    motor speed. What is the approximate 3-dB
    bandwidth of the motor/pump system?

13.21   A PM DC motor with torque constant k_{PM} is used
to power a hydraulic pump; the pump is a positive
displacement type and generates a flow proportional to
the pump velocity: $q_p = k_p\omega$. The fluid travels through
a conduit of negligible resistance; an accumulator is
included to smooth out the pulsations of the pump. A
hydraulic load (modeled by a fluid resistance R) is
connected between the pipe and a reservoir (assumed
at zero pressure). Sketch the motor-pump circuit.
Derive the dynamic equations for the system, and
determine the transfer function between motor voltage
and the pressure across the load.

13.22   The shunt motor in Figure P13.22 is
characterized by a field coefficient $k_f = 0.12$ V-s/A-rad,
such that the back emf is given by the expression
$E_b = k_f I_f \omega$ and the motor torque by the expression
$T = k_f I_f I_a$. The motor drives an inertia/viscous
friction load with parameters $J = 0.8$ kg-m^2 and
$b = 0.6$ N-m-s/rad. The field equation may be
approximated by $V_S = R_f I_f$. The armature resistance
is $R_a = 0.75$ Ω, and the field resistance is $R_f = 60$ Ω.
The system is perturbed around the nominal operating
point $V_{S0} = 150$ V, $\omega_0 = 200$ rad/s, and $I_{a0} = 186.67$ A.

a. Derive the dynamic system equations in *symbolic
    form*.

b. Linearize the equations you obtained in part a.

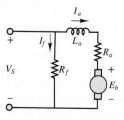

Figure P13.22

13.23 A PM DC motor is rigidly coupled to a fan; the fan load torque is described by the expression $T_L = 5 + 0.05\omega + 0.001\ \omega^2$, where torque is in newton-meters and speed in radians per second. The motor has $k_a\phi = k_T\phi = 2.42$; $R_a = 0.2\ \Omega$, and the inductance is negligible. If the motor voltage is 50 V, what is the speed of rotation of the motor and fan?

13.24 A separately excited DC motor has the following parameters:

$$R_a = 0.1\ \Omega \qquad R_f = 100\ \Omega \qquad L_a = 0.2\ \text{H}$$
$$L_f = 0.02\ \text{H} \qquad K_a = 0.8 \qquad K_f = 0.9$$

An inertial load has $J = 0.5$ kg-m^2 and $b = 2$ N-m-s/rad. No external load torque is applied.

a. Sketch a diagram of the system and derive the (three) differential equations.

b. Sketch a simulation block diagram of the system (you should have three integrators).

c. Code the diagram, using Simulink.

d. Run the following simulations:
    Armature control. Assume a constant field with $V_f = 100$ V; now simulate the response of the system when the armature voltage changes in step fashion from 50 to 75 V. Save and plot the current and angular speed responses.
    Field control. Assume a constant armature voltage with $V_a = 100$ V; now simulate the response of the system when the field voltage changes in step fashion from 75 to 50 V. This procedure is called *field weakening*. Save and plot the current and angular speed responses.

13.25 Determine the transfer functions from *input voltage* to *angular velocity* and from *load torque* to *angular velocity* for a PM DC motor rigidly connected to an inertial load. Assume resistance and inductance parameters R_a, L_a let the armature constant be k_a. Assume ideal energy conversion, so that $k_a = k_T$. The motor has inertia J_m and damping coefficient b_m, and it is rigidly connected to an inertial load with inertia J and damping coefficient b. The load torque T_L acts on the load to oppose the magnetic torque.

13.26 Assume that the coupling between the motor and the inertial load of Problem 13.25 is flexible (e.g., a long shaft). This can be modeled by adding a torsional spring between the motor inertia and the load inertia. Now we can no longer lump together the two inertias and damping coefficients as if they were one; we need to write separate equations for the two inertias. In total, there will be three equations in this system: the

motor electrical equation, the motor mechanical equation (J_m and B_m), and the load mechanical equation (J and B).

a. Sketch a diagram of the system.

b. Use free-body diagrams to write each of the two mechanical equations. Set up the equations in matrix form.

c. Compute the transfer function from input voltage to load speed, using the method of determinants.

13.27 A wound DC motor is connected in both a shunt and a series configuration. Assume generic resistance and inductance parameters R_a, R_f, L_a, L_f; let the field magnetization constant be k_f and the armature constant be k_a. Assume ideal energy conversion, so that $k_a = k_T$. The motor has inertia J_m and damping coefficient b_m, and it is rigidly connected to an inertial load with inertia J and damping coefficient b.

a. Sketch a system-level diagram of the two configurations that illustrates both the mechanical and electrical systems.

b. Write an expression for the torque–speed curve of the motor in each configuration.

c. Write the differential equations of the motor-load system in each configuration.

d. Determine whether the differential equations of each system are linear; if one (or both) is (are) nonlinear, could they be made linear with some simple assumption? Explain clearly under what conditions this would be the case.

13.28 Derive the differential equations describing the electrical and mechanical dynamics of a shunt-connected DC motor, shown in Figure P13.28, and draw a simulation block diagram of the system. The motor constants are k_a, k_T = armature and torque reluctance and k_f = field flux.

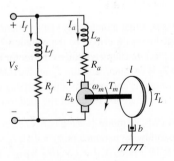

Figure P13.28

13.29 Derive the differential equations describing the electrical and mechanical dynamics of a series-connected DC motor, shown in Figure P13.29, and draw a simulation block diagram of the system. The motor constants are k_a, k_T = armature and torque reluctance and k_f = field flux.

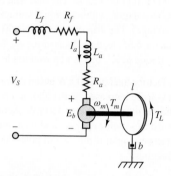

Figure P13.29

13.30 Develop a Simulink simulator for the shunt-connected DC motor of Problem 13.28. Assume the following parameter values: $L_a = 0.15$ H; $L_f = 0.05$ H; $R_a = 1.8\ \Omega$; $R_f = 0.2\ \Omega$; $k_a = 0.8$ V-s/rad; $k_T = 20$ N-m/A; $k_f = 0.20$ Wb/A; $b = 0.1$ N-m-s/rad; $J = 1$ kg-m^2.

13.31 Develop a Simulink simulator for the series-connected DC motor of Problem 13.29. Assume the following parameter values: $L = L_a + L_f = 0.2$ H; $R = R_a + R_f = 2\ \Omega$; $k_a = 0.8$ V-s/rad; $k_T = 20$ N-m/A; $k_f = 0.20$ Wb/A; $b = 0.1$ N-m-s/rad; $J = 1$ kg-m^2.

Section 13.4: Direct-Current Generators

13.32 A 120-V. 10-A shunt generator has an armature resistance of 0.6 Ω. The shunt field current is 2 A. Determine the voltage regulation of the generator.

13.33 A 20-kW. 230-V separately excited generator has an armature resistance of 0.2 Ω and a load current of 100 A. Find

a. The generated voltage when the terminal voltage is 230 V.

b. The output power.

13.34 A 10-kW, 120-VDC series generator has an armature resistance of 0.1 Ω and a series field resistance of 0.05 Ω. Assuming that it is delivering rated current at rated power, find (a) the armature current and (b) the generated voltage.

13.35 The armature resistance of a 30-kW, 440-V shunt generator is 0.1 Ω. Its shunt field resistance is 200 Ω. Find

a. The power developed at rated load.

b. The load, field, and armature currents.

c. The electric power loss.

13.36 A four-pole, 450-kW, 4.6-kV shunt generator has armature and field resistances of 2 and 333 Ω. The generator is operating at the rated speed of 3.600 r/min. Find the no-load voltage of the generator and terminal voltage at half load.

13.37 A 30-kW, 240-V generator is running at half load at 1,800 r/min with an efficiency of 85 percent. Find the total losses and input power.

13.38 A self-excited DC shunt generator is delivering 20 A to a 100-V line when it is driven at 200 rad/s. The magnetization characteristic is shown in Figure P13.38. It is known that $R_a = 1.0\ \Omega$ and $R_f = 100\ \Omega$. When the generator is disconnected from the line, the drive motor speeds up to 220 rad/s. What is the terminal voltage?

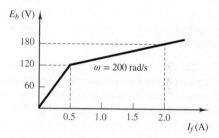

Figure P13.38

13.39 A high-pressure supply and a hydraulic motor are used as a prime mover to generate electricity through a DC generator. The system diagram is sketched in Figure P13.39. Assume that an ideal pressure source P_S is available and that a hydraulic motor is connected to it through a linear "fluid resistor," used to regulate the average flow rate. An accumulator is inserted just upstream of the hydraulic motor to smooth pressure pulsations. The combined inertia of the hydraulic motor and of the DC generator is represented by the parameter J. The DC generator is of the permanent-magnet type and has armature constants $k_a = k_T$. The permanent magnet flux is ϕ. Assume a resistive load for the generator R_L.

a. Derive the system differential equations.

b. Compute the transfer function of the system from supply pressure P_S to load voltage V_L.

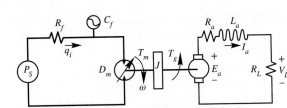

Figure P13.39

Section 13.6: The Alternator (Synchronous Generator)

13.40   An automotive alternator is rated 500 VA and 20 V. It delivers its rated voltamperes at a power factor of 0.85. The resistance per phase is 0.05 Ω, and the field takes 2 A at 12 V. If the friction and windage loss is 25 W and the core loss is 30 W, calculate the percent efficiency under rated conditions.

13.41   It has been determined by test that the synchronous reactance X_s and armature resistance r_a of a 2,300-V, 500-VA, three-phase synchronous generator are 8.0 and 0.1 Ω, respectively. If the machine is operating at rated load and voltage at a power factor of 0.867 lagging, find the generated voltage per phase and the torque angle.

13.42   The circuit of Figure P13.42 represents a voltage regulator for a car alternator. Unlike other alternators, a car alternator is *not* driven at constant speed. Briefly, explain the function of Q, D, Z, and SCR.

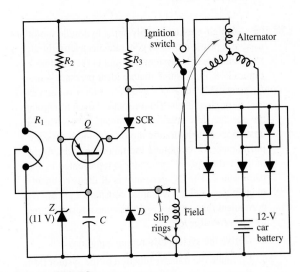

Figure P13.42

Section 13.7: The Synchronous Motor

13.43   A non–salient pole, Y-connected, three-phase, two-pole synchronous machine has a synchronous reactance of 7 Ω and negligible resistance and rotational losses. One point on the open-circuit characteristic is given by $V_o = 400$ V (phase voltage) for a field current of 3.32 A. The machine is to be operated as a motor, with a terminal voltage of 400 V (phase voltage). The armature current is 50 A, with power factor 0.85, leading. Determine E_b, field current, torque developed, and power angle δ.

13.44   A factory load of 900 kW at 0.6 power factor lagging is to be increased by the addition of a synchronous motor that takes 450 kW. At what power factor must this motor operate, and what must be its kilovoltampere input if the overall power factor is to be 0.9 lagging?

13.45   A non–salient pole, Y-connected, three-phase, two-pole synchronous generator is connected to a 400-V (line to line), 60-Hz, three-phase line. The stator impedance is $0.5 + j1.6$ Ω (per phase). The generator is delivering rated current (36 A) at unity power factor to the line. Determine the power angle for this load and the value of E_b for this condition. Sketch the phasor diagram, showing $\mathbf{E}_b$, $\mathbf{I}_S$, and $\mathbf{V}_S$.

13.46   A non–salient pole, three-phase, two-pole synchronous motor is connected in parallel with a three-phase, Y-connected load so that the per-phase equivalent circuit is as shown in Figure P13.46. The parallel combination is connected to a 220-V (line to line), 60-Hz, three-phase line. The load current $\mathbf{I}_L$ is 25 A at a power factor of 0.866 inductive. The motor has $X_S = 2$ Ω and is operating with $I_f = 1$ A and $T = 50$ N-m at a power angle of −30°. (Neglect all losses for the motor.) Find $\mathbf{I}_S$, P_{in} (to the motor), the overall power factor (i.e., angle between $\mathbf{I}_1$ and $\mathbf{V}_S$), and the total power drawn from the line.

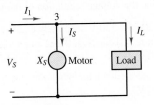

Figure P13.46

13.47   A four-pole, three-phase, Y-connected, non–salient pole synchronous motor has a synchronous reactance of 10 Ω. This motor is connected to a $230\sqrt{3}$ V (line to line), 60-Hz, three-phase line and is driving a load such that $T_{shaft} = 30$ N-m. The line current is 15 A, leading the phase voltage. Assuming that all losses can be neglected, determine the power angle δ and E for this condition. If the load is removed, what is the line current, and is it leading or lagging the voltage?

13.48   A 10-hp, 230-V, 60 Hz, three-phase, Y-connected synchronous motor delivers full load at a power factor of 0.8 leading. The synchronous reactance is 6 Ω, the rotational loss is 230 W, and the field loss is 50 W. Find

a. The armature current.

b. The motor efficiency.

c. The power angle.

Neglect the stator winding resistance.

13.49   A 2,000-hp, unity power factor, three-phase, Y-connected, 2,300-V, 30-pole, 60-Hz synchronous motor has a synchronous reactance of 1.95 Ω per phase. Neglect all losses. Find the maximum power and torque.

13.50   A 1,200-V, three-phase, Y-connected synchronous motor takes 110 kW (exclusive of field winding loss) when operated under a certain load at 1,200 r/min. The back emf of the motor is 2,000 V. The synchronous reactance is 10 Ω per phase, with negligible winding resistance. Find the line current and the torque developed by the motor.

13.51   The per-phase impedance of a 600-V, three-phase, Y-connected synchronous motor is $5 + j50$ Ω. The motor takes 24 kW at a leading power factor of 0.707. Determine the induced voltage and the power angle of the motor.

Section 13.8: The Induction Motor

13.52   A 74.6-kW, three-phase, 440-V (line to line), four-pole, 60-Hz induction motor has the following (per-phase) parameters referred to the stator circuit (see Figure 13.36):

$$R_S = 0.06\ \Omega \qquad X_S = 0.3\ \Omega \qquad X_m = 5\ \Omega$$
$$R_R = 0.08\ \Omega \qquad X_R = 0.3\ \Omega$$

The no-load power input is 3,240 W at a current of 45 A. Determine the line current, input power, developed torque, shaft torque, and efficiency at $s = 0.02$.

13.53   A 60-Hz, four-pole, Y-connected induction motor is connected to a 400-V (line to line), three-phase, 60-Hz line. The equivalent circuit parameters are

$$R_S = 0.2\ \Omega \qquad R_R = 0.1\ \Omega$$
$$X_S = 0.5\ \Omega \qquad X_R = 0.2\ \Omega$$
$$X_m = 20\ \Omega$$

When the machine is running at 1,755 r/min, the total rotational and stray-load losses are 800 W. Determine the slip, input current, total input power, mechanical power developed, shaft torque, and efficiency.

13.54   A three-phase, 60-Hz induction motor has eight poles and operates with a slip of 0.05 for a certain load. Determine

a. The speed of the rotor with respect to the stator.

b. The speed of the rotor with respect to the stator magnetic field.

c. The speed of the rotor magnetic field with respect to the rotor.

d. The speed of the rotor magnetic field with respect to the stator magnetic field.

13.55   A three-phase, two-pole, 400-V (per phase), 60-Hz induction motor develops 37 kW (total) of mechanical power P_m at a certain speed. The rotational loss at this speed is 800 W (total). (Stray-load loss is negligible.)

a. If the total power transferred to the rotor is 40 kW, determine the slip and the output torque.

b. If the total power into the motor P_{in} is 45 kW and R_S is 0.5 Ω, find I_S and the power factor.

13.56   The nameplate speed of a 25-Hz induction motor is 720 r/min. If the speed at no load is 745 r/min, find

a. The slip.

b. The percent regulation.

13.57   The nameplate of a squirrel cage four-pole induction motor has the following information: 25 hp, 220 V, three-phase, 60 Hz, 830 r/min, 64-A line current. If the motor draws 20,800 W when operating at full load, calculate

a. Slip.

b. Percent regulation if the no-load speed is 895 r/min.

c. Power factor.

d. Torque.

e. Efficiency.

13.58   A 60-Hz, four-pole, Y-connected induction motor is connected to a 200-V (line to line), three-phase, 60-Hz line. The equivalent circuit parameters are

$$R_S = 0.48\ \Omega \qquad \text{Rotational loss torque} = 3.5\text{ N-m}$$
$$X_S = 0.8\ \Omega \qquad R_R = 0.42\ \Omega \text{ (referred to stator)}$$
$$X_m = 30\ \Omega \qquad X_R = 0.8\ \Omega \text{ (referred to stator}$$

The motor is operating at slip $s = 0.04$. Determine the input current, input power, mechanical power, and shaft torque (assuming that stray-load losses are negligible).

13.59

a. A three-phase, 220-V, 60-Hz induction motor runs at 1,140 r/min. Determine the number of poles (for minimum slip), the slip, and the frequency of the rotor currents.

b. To reduce the starting current, a three-phase squirrel cage induction motor is started by reducing the line voltage to $V_s/2$. By what factor are the starting torque and the starting current reduced?

13.60   A six-pole induction motor for vehicle traction has a 50-kW input electric power rating and is 85 percent efficient. If the supply is 220 V at 60 Hz, compute the motor speed and torque at a slip of 0.04.

13.61   An AC induction machine has six poles and is designed for 60-Hz, 240-V (rms) operation. When the machine operates with 10 percent slip, it produces 60 N-m of torque.

a. The machine is now used in conjunction with a friction load that opposes a torque of 50 N-m. Determine the speed and slip of the machine when used with the above-mentioned load.

b. If the machine has an efficiency of 92 percent, what minimum rms current is required for operation with the load of part a?

(*Hint:* You may assume that the speed–torque curve is approximately linear in the region of interest.)

13.62   A blocked-rotor test was performed on a 5-hp, 220-V, four-pole, 60-Hz, three-phase induction motor. The following data were obtained: $V = 48$ V, $I = 18$ A, $P = 610$ W. Calculate

a. The equivalent stator resistance per phase R_S.

b. The equivalent rotor resistance per phase R_R.

c. The equivalent blocked-rotor reactance per phase X_R.

13.63   Calculate the starting torque of the motor of Problem 13.62 when it is started at

a. 220 V

b. 110 V

The starting torque equation is

$$T = \frac{m}{\omega_e} \cdot V_S^2 \cdot \frac{R_R}{(R_R + R_S)^2 + (X_R + X_S)^2}$$

13.64   A four-pole, three-phase induction motor drives a turbine load. At a certain operating point the machine

has 4 percent slip and 87 percent efficiency. The motor drives a turbine with torque–speed characteristic given by $T_L = 20 + 0.006\omega^2$. Determine the torque at the motor-turbine shaft and the total power delivered to the turbine. What is the total power consumed by the motor?

13.65   A four-pole, three-phase induction motor rotates at 1,700 r/min when the load is 100 N-m. The motor is 88 percent efficient.

a. Determine the slip at this operating condition.

b. For a constant-power, 10-kW load, determine the operating speed of the machine.

c. Sketch the motor and load torque–speed curves on the same graph. Show numerical values.

d. What is the total power consumed by the motor?

13.66   Find the speed of the rotating field of a six-pole, three-phase motor connected to (a) a 60-Hz line and (b) a 50-Hz line, in revolutions per minute and radians per second.

13.67   A six-pole, three-phase, 440-V, 60-Hz induction motor has the following model impedances:

$$R_S = 0.8 \ \Omega \qquad X_S = 0.7 \ \Omega$$
$$R_R = 0.3 \ \Omega \qquad X_R = 0.7 \ \Omega$$
$$X_m = 35 \ \Omega$$

Calculate the input current and power factor of the motor for a speed of 1,200 r/min.

13.68   An eight-pole, three-phase, 220-V, 60-Hz induction motor has the following model impedances:

$$R_S = 0.78 \ \Omega \qquad X_S = 0.56 \ \Omega \qquad X_m = 32 \ \Omega$$
$$R_R = 0.28 \ \Omega \qquad X_R = 0.84 \ \Omega$$

Find the input current and power factor of this motor for $s = 0.02$.

13.69   A nameplate is given in Example 13.2. Find the rated torque, rated voltamperes, and maximum continuous output power for this motor.

13.70   A three-phase induction motor, at rated voltage and frequency, has a starting torque of 140 percent and a maximum torque of 210 percent of full-load torque. Neglect stator resistance and rotational losses and assume constant rotor resistance. Determine

a. The slip at full load.

b. The slip at maximum torque.

c. The rotor current at starting as a percentage of full-load rotor current.

13.71 A 60-Hz, four-pole, three-phase induction motor delivers 35 kW of mechanical (output) power. At a certain operating point the machine has 4 percent slip and 87 percent efficiency. Determine the torque delivered to the load and the total electric (input) power consumed by the motor.

13.72 A four-pole, three-phase induction motor rotates at 16,800 rev/min when the load is 140 N-m. The motor is 85 percent efficient.

a. Determine the slip at this operating condition.

b. For a constant-power, 20-kW load, determine the operating speed of the machine.

c. Sketch the motor and load torque–speed curves for the load of part b. on the same graph. Show numerical values.

13.73 An AC induction machine has six poles and is designed for 60-Hz, 240-V (rms) operation. When the machine operates with 10 percent slip, it produces 60 N-m of torque.

a. The machine is now used in conjunction with an 800-W constant power load. Determine the speed and slip of the machine when used with the above-mentioned load.

b. If the machine has an efficiency of 89 percent, what minimum rms current is required for operation with the load of part a?

(*Hint:* You may assume that the speed–torque curve is approximately linear in the region of interest.)

LINEAR ALGEBRA AND COMPLEX NUMBERS

A.1 SOLVING SIMULTANEOUS LINEAR EQUATIONS, CRAMER'S RULE, AND MATRIX EQUATION

The solution of simultaneous equations, such as those that are often seen in circuit theory, may be obtained relatively easily by using Cramer's rule. This method applies to 2×2 or larger systems of equations. Cramer's rule requires the use of the concept of determinant. Linear, or matrix, algebra is valuable because it is systematic, general, and useful in solving complicated problems. A determinant is a scalar defined on a square array of numbers, or matrix, such as

$$\det(A) = |A| = \begin{vmatrix} a_{11} & a_{12} \\ a_{21} & a_{22} \end{vmatrix} \tag{A.1}$$

In this case the matrix is a 2×2 array with two rows and two columns, and its determinant is defined as

$$\det = a_{11}a_{22} - a_{12}a_{21} \tag{A.2}$$

A third-order, or 3×3, determinant such as

$$\det(A) = \begin{vmatrix} a_{11} & a_{12} & a_{13} \\ a_{21} & a_{22} & a_{23} \\ a_{31} & a_{32} & a_{33} \end{vmatrix} \tag{A.3}$$

is given by

$$\begin{aligned} \det = {} & a_{11}(a_{22}a_{33} - a_{23}a_{32}) - a_{12}(a_{21}a_{33} - a_{23}a_{31}) \\ & + a_{13}(a_{21}a_{32} - a_{22}a_{31}) \end{aligned} \tag{A.4}$$

For higher-order determinants, you may refer to a linear algebra book. To illustrate Cramer's method, a set of two equations in general form will be solved here. A set of two linear simultaneous algebraic equations in two unknowns can be written in the form:

$$\begin{aligned} a_{11}x_1 + a_{12}x_2 &= b_1 \\ a_{21}x_1 + a_{22}x_2 &= b_2 \end{aligned} \tag{A.5}$$

where x_1 and x_2 are the two unknowns. The coefficients a_{11}, a_{12}, a_{21}, and a_{22} are known quantities. The two quantities on the right-hand sides, b_1 and b_2, are also known (these are typically the source currents and voltages in a circuit problem). The set of equations can be arranged in matrix form, as shown in equation A.6.

$$\begin{bmatrix} a_{11} & a_{12} \\ a_{21} & a_{22} \end{bmatrix} \begin{bmatrix} x_1 \\ x_2 \end{bmatrix} = \begin{bmatrix} b_1 \\ b_2 \end{bmatrix} \tag{A.6}$$

In equation A.6, a coefficient matrix multiplied by a vector of unknown variables is equated to a right-hand-side vector. Cramer's rule can then be applied to find x_1 and x_2, using the following formulas:

$$x_1 = \frac{\begin{vmatrix} b_1 & a_{12} \\ b_2 & a_{22} \end{vmatrix}}{\begin{vmatrix} a_{11} & a_{12} \\ a_{21} & a_{22} \end{vmatrix}} \qquad x_2 = \frac{\begin{vmatrix} a_{11} & b_1 \\ a_{21} & b_2 \end{vmatrix}}{\begin{vmatrix} a_{11} & a_{12} \\ a_{21} & a_{22} \end{vmatrix}} \tag{A.7}$$

Thus, the solution is given by the ratio of two determinants: the denominator is the determinant of the coefficient matrix, while the numerator is the determinant of the same matrix with the right-hand-side vector ($[b_1 \ b_2]^T$ in this case) substituted in place of the column of the coefficient matrix corresponding to the desired variable (i.e., first column for x_1, second column for x_2, etc.). In a circuit analysis problem, the coefficient matrix is formed by the resistance (or conductance) values, the vector of unknowns is composed of the mesh currents (or node voltages), and the right-hand-side vector contains the source currents or voltages.

In practice, many calculations involve solving higher-order systems of linear equations. Therefore, a variety of computer software packages are often used to solve higher-order systems of linear equations.

CHECK YOUR UNDERSTANDING

A.1   Use Cramer's rule to solve the system

$$5v_1 + 4v_2 = 6$$
$$3v_1 + 2v_2 = 4$$

A.2   Use Cramer's rule to solve the system

$$i_1 + 2i_2 + i_3 = 6$$
$$i_1 + i_2 - 2i_3 = 1$$
$$i_1 - i_2 + i_3 = 0$$

A.3   Convert the following system of linear equations into a matrix equation as shown in equation A.6, and find matrices A and b.

$$2i_1 - 2i_2 + 3i_3 = -10$$
$$-3i_1 + 3i_2 - 2i_3 + i_4 = -2$$
$$5i_1 - i_2 + 4i_3 - 4i_4 = 4$$
$$i_1 - 4i_2 + i_3 + 2i_4 = 0$$

Answers: A1: $v_1 = 2$, $v_2 = -1$; A2: $i_1 = 1$, $i_2 = 2$, $i_3 = 1$; A3:
$$A = \begin{bmatrix} 2 & -2 & 3 & 0 \\ -3 & 3 & -2 & 1 \\ 5 & -1 & 4 & -4 \\ 1 & -4 & 1 & 2 \end{bmatrix}, \quad b = \begin{bmatrix} -10 \\ -2 \\ 4 \\ 0 \end{bmatrix}$$

A.2   INTRODUCTION TO COMPLEX ALGEBRA

From your earliest training in arithmetic, you have dealt with real numbers such as 4, -2, $\frac{5}{9}$, π, e, etc., which may be used to measure distances in one direction or another from a fixed point. However, a number that satisfies the equation:

$$x^2 + 9 = 0 \tag{A.8}$$

is not a real number. Imaginary numbers were introduced to solve equations such as equation A.8. Imaginary numbers add a new dimension to our number system. To deal with imaginary numbers, a new element, j, is added to the number system having the property:

$$j^2 = -1 \tag{A.9}$$

or

$$j = \sqrt{-1}$$

Thus, we have $j^3 = -j$, $j^4 = 1$, $j^5 = j$, etc. Using equation A.9, you can see that the solutions to equation A.8 are $\pm j3$. In mathematics, the symbol i is used for the imaginary unit, but this might be confused with current in electrical engineering. Therefore, the symbol j is used in this book.

A complex number (indicated in boldface notation) is an expression of the form:

$$\mathbf{A} = a + jb \tag{A.10}$$

where a and b are real numbers. The complex number $\mathbf{A}$ has a real part a and an imaginary part b, which can be expressed as

$$a = \text{Re } \mathbf{A}$$
$$b = \text{Im } \mathbf{A} \tag{A.11}$$

It is important to note that a and b are both real numbers. The complex number $a + jb$ can be represented on a rectangular coordinate plane, called the *complex plane*, by interpreting it as a point (a, b). That is, the horizontal coordinate is a on the real axis, and the vertical coordinate is b on the imaginary axis, as shown in Figure A.1. The complex number $\mathbf{A} = a + jb$ can also

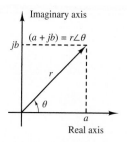

Figure A.1 Polar form representation of complex numbers

be uniquely located in the complex plane by specifying the distance r along a straight line from the origin and the angle θ, which this line makes with the real axis, as shown in Figure A.1. From the right triangle of Figure A.1, we can see that:

$$
\begin{aligned}
r &= \sqrt{a^2 + b^2} \\
\theta &= \tan^{-1}\left(\frac{b}{a}\right) \\
a &= r\cos\theta \\
b &= r\sin\theta
\end{aligned}
\tag{A.12}
$$

Then we can represent a complex number by the expression:

$$
\mathbf{A} = re^{j\theta} = r\angle\theta
\tag{A.13}
$$

which is called the polar form of the complex number. The number r is called the magnitude (or amplitude), and the number θ is called the angle (or argument). The two numbers are usually denoted by $r = |\mathbf{A}|$ and $\theta = \arg\mathbf{A} = \angle\mathbf{A}$.

Given a complex number $\mathbf{A} = a + jb$, the *complex conjugate* of $\mathbf{A}$, denoted by the symbol $\mathbf{A}^*$, is defined by the following equalities:

$$
\begin{aligned}
\operatorname{Re}\mathbf{A}^* &= \operatorname{Re}\mathbf{A} \\
\operatorname{Im}\mathbf{A}^* &= -\operatorname{Im}\mathbf{A}
\end{aligned}
\tag{A.14}
$$

That is, the sign of the imaginary part is reversed in the complex conjugate.

Finally, two complex numbers are equal *if and only if* the real parts are equal and the imaginary parts are equal, which is equivalent to stating that two complex numbers are equal only if their magnitudes are equal and their arguments are equal.

The following examples and exercises should help clarify these explanations.

EXAMPLE A.1

Convert the complex number $\mathbf{A} = 3 + j4$ to its polar form.

Solution:

$$
r = \sqrt{3^2 + 4^2} = 5 \qquad \theta = \tan^{-1}\left(\frac{4}{3}\right) = 53.13°
$$

$$
\mathbf{A} = 5\angle 53.13°
$$

EXAMPLE A.2

Convert the number $\mathbf{A} = 4\angle(-60°)$ to its complex form.

Solution:

$$
\begin{aligned}
a &= 4\cos(-60°) = 4\cos(60°) = 2 \\
b &= 4\sin(-60°) = -4\sin(60°) = -2\sqrt{3}
\end{aligned}
$$

Thus, $\mathbf{A} = 2 - j2\sqrt{3}$.

Addition and *subtraction* of complex numbers are governed by the following rules:

$$(a_1 + jb_1) + (a_2 + jb_2) = (a_1 + a_2) + j(b_1 + b_2)$$
$$(a_1 + jb_1) - (a_2 + jb_2) = (a_1 - a_2) + j(b_1 - b_2)$$

(A.15)

Multiplication of complex numbers in polar form follows the law of exponents. That is, the magnitude of the product is the product of the individual magnitudes, and the angle of the product is the sum of the individual angles, as shown below.

$$\mathbf{AB} = (Ae^{j\theta})(Be^{j\phi}) = ABe^{j(\theta+\phi)} = AB\angle(\theta + \phi)$$

(A.16)

If the numbers are given in rectangular form and the product is desired in rectangular form, it may be more convenient to perform the multiplication directly, using the rule that $j^2 = -1$, as illustrated in equation A.17.

$$(a_1 + jb_1)(a_2 + jb_2) = a_1 a_2 + ja_1 b_2 + ja_2 b_1 + j^2 b_1 b_2$$
$$= (a_1 a_2 + j^2 b_1 b_2) + j(a_1 b_2 + a_2 b_1)$$
$$= (a_1 a_2 - b_1 b_2) + j(a_1 b_2 + a_2 b_1)$$

(A.17)

Division of complex numbers in polar form follows the law of exponents. That is, the magnitude of the quotient is the quotient of the magnitudes, and the angle of the quotient is the difference of the angles, as shown in equation A.18.

$$\frac{\mathbf{A}}{\mathbf{B}} = \frac{Ae^{j\theta}}{Be^{j\phi}} = \frac{A\angle\theta}{B\angle\phi} = \frac{A}{B} \angle(\theta - \phi)$$

(A.18)

Division in the rectangular form can be accomplished by multiplying the numerator and denominator by the complex conjugate of the denominator. Multiplying the denominator by its complex conjugate converts the denominator to a real number and simplifies division. This is shown in Example A.4. Powers and roots of a complex number in polar form follow the laws of exponents, as shown in equations A.19 and A.20.

$$\mathbf{A}^n = (Ae^{j\theta})^n = A^n e^{jn\theta} = A^n \angle n\theta$$

(A.19)

$$\mathbf{A}^{1/n} = (Ae^{j\theta})^{1/n} = A^{1/n} e^{j1/n\theta}$$
$$= \sqrt[n]{A}\angle\left(\frac{\theta + k2\pi}{n}\right) \qquad k = 0, \pm 1, \pm 2, \ldots$$

(A.20)

EXAMPLE A.3

Perform the following operations, given that $\mathbf{A} = 2 + j3$ and $\mathbf{B} = 5 - j4$.

(a) $\mathbf{A} + \mathbf{B}$ (b) $\mathbf{A} - \mathbf{B}$ (c) $2\mathbf{A} + 3\mathbf{B}$

Solution:

$$\mathbf{A} + \mathbf{B} = (2 + 5) + j[3 + (-4)] = 7 - j$$
$$\mathbf{A} - \mathbf{B} = (2 - 5) + j[3 - (-4)] = -3 + j7$$

For part c, $2\mathbf{A} = 4 + j6$ and $3\mathbf{B} = 15 - j12$. Thus, $2\mathbf{A} + 3\mathbf{B} = (4 + 15) + j[6 + (-12)] = 19 - j6$

EXAMPLE A.4

Perform the following operations in both rectangular and polar form, given that $\mathbf{A} = 3 + j3$ and $\mathbf{B} = 1 + j\sqrt{3}$.

(a) $\mathbf{AB}$ (b) $\mathbf{A} \div \mathbf{B}$

Solution:

(a) In rectangular form:

$$\mathbf{AB} = (3 + j3)(1 + j\sqrt{3}) = 3 + j3\sqrt{3} + j3 + j^2 3\sqrt{3}$$
$$= \left(3 + j^2 3\sqrt{3}\right) + j(3 + 3\sqrt{3})$$
$$= (3 - 3\sqrt{3}) + j(3 + 3\sqrt{3})$$

To obtain the answer in polar form, we need to convert $\mathbf{A}$ and $\mathbf{B}$ to their polar forms:

$$\mathbf{A} = 3\sqrt{2}e^{j45°} = 3\sqrt{2}\angle 45°$$
$$\mathbf{B} = \sqrt{4}e^{j60°} = 2\angle 60°$$

Then

$$\mathbf{AB} = \left(3\sqrt{2}e^{j45°}\right)\sqrt{4}e^{j60°} = 6\sqrt{2}\angle 105°$$

(b) To find $\mathbf{A} \div \mathbf{B}$ in rectangular form, we can multiply $\mathbf{A}$ and $\mathbf{B}$ by $\mathbf{B}^*$.

$$\frac{\mathbf{A}}{\mathbf{B}} = \frac{3 + j3}{1 + j\sqrt{3}}\frac{1 - j\sqrt{3}}{1 - j\sqrt{3}}$$

Then

$$\frac{\mathbf{A}}{\mathbf{B}} = \frac{(3 + 3\sqrt{3}) + j(3 - 3\sqrt{3})}{4}$$

In polar form, the same operation may be performed as follows:

$$\frac{\mathbf{A}}{\mathbf{B}} = \frac{3\sqrt{2}\angle 45°}{2\angle 60°} = \frac{3\sqrt{2}}{2}\angle(45° - 60°) = \frac{3\sqrt{2}}{2}\angle(-15°)$$

Euler's Identity

Euler's formula extends the usual definition of the exponential function to allow for complex numbers as arguments.

$$e^{j\theta} = \cos\theta + j\sin\theta \tag{A.21}$$

All the standard trigonometry formulas in the complex plane are direct consequences of Euler's formula. The two important formulas are

$$\cos\theta = \frac{e^{j\theta} + e^{-j\theta}}{2} \qquad \sin\theta = \frac{e^{j\theta} - e^{-j\theta}}{2j} \tag{A.22}$$

EXAMPLE A.5

Using Euler's formula, show that

$$\cos\theta = \frac{e^{j\theta} + e^{-j\theta}}{2}$$

Solution:

Using Euler's formula gives

$$e^{j\theta} = \cos\theta + j\sin\theta$$

Extending the above formula, we can obtain

$$e^{-j\theta} = \cos(-\theta) + j\sin(-\theta) = \cos\theta - j\sin\theta$$

Thus,

$$\cos\theta = \frac{e^{j\theta} + e^{-j\theta}}{2}$$

CHECK YOUR UNDERSTANDING

A.4   In a certain AC circuit, $V = IZ$, where $Z = 7.75\angle 90°$ and $I = 2\angle -45°$. Find V.

A.5   In a certain AC circuit, $V = IZ$, where $Z = 5\angle 82°$ and $V = 30\angle 45°$. Find I.

A.6   Show that the polar form of AB in Example A.4 is equivalent to its rectangular form.

A.7   Show that the polar form of $A \div B$ in Example A.4 is equivalent to its rectangular form.

A.8   Using Euler's formula, show that $\sin\theta = (e^{j\theta} - e^{-j\theta})/2j$.

Answers: A4: $V = 15.5\angle 45°$; A5: $I = 6\angle(-37°)$

A P P E N D I X

B

THE LAPLACE TRANSFORM

T he transient analysis methods illustrated in Chapter 4 for first- and second-order circuits can become rather cumbersome when applied to higher-order circuits. More-over, solving the differential equations directly does not reveal the strong connection that exists between the transient response and the frequency response of a circuit. The aim of this appendix is to introduce an alternate solution method based on the concepts of complex frequency and of the **Laplace transform**. The concepts presented will demon-strate that the frequency response of linear circuits is but a special case of the general transient response of the circuit, when analyzed by means of Laplace methods. In addition, the use of the Laplace transform method reveals *systems* concepts, such as poles, zeros, and transfer functions.

B.1   COMPLEX FREQUENCY

In Chapter 3, we considered circuits with sinusoidal excitations such as

$$v(t) = A \cos(\omega t + \phi) \tag{B.1}$$

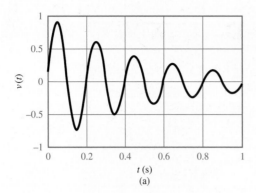
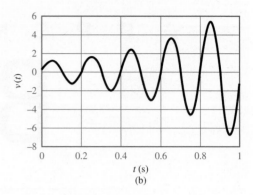

Figure B.1 Damped sinusoid: (*a*) exponential decay, negative σ; (*b*) exponential growth, positive σ

which we also wrote in the equivalent phasor form:

$$\mathbf{V}(j\omega) = A e^{j\phi} = A\angle\phi \tag{B.2}$$

The two expressions just given are related by

$$v(t) = \text{Re}(\mathbf{V}\,e^{j\omega t}) \tag{B.3}$$

As was shown in Chapter 3, phasor notation is extremely useful in solving AC steady-state circuits, in which the voltages and currents are *steady-state sinusoids*. We now consider a different class of waveforms, useful in the transient analysis of circuits, namely, *damped sinusoids*. The most general form of a damped sinusoid is

$$v(t) = A e^{\sigma t}\cos(\omega t + \phi) \tag{B.4}$$

As one can see, a damped sinusoid is a sinusoid multiplied by a real exponential $e^{\sigma t}$. The constant σ is real and is usually zero or negative in most practical circuits. Figure B.1(a) and (b) depict the case of a damped sinusoid with negative σ and with positive σ, respectively. Note that the case of $\sigma = 0$ corresponds exactly to a sinusoidal waveform. The definition of phasor voltages and currents given in Chapter 3 can easily be extended to account for the case of damped sinusoidal waveforms by defining a new variable s, called the *complex frequency*:

$$s = \sigma + j\omega \tag{B.5}$$

Note that the special case of $\sigma = 0$ corresponds to $s = j\omega$, that is, the familiar steady-state sinusoidal (phasor) case. We shall now refer to the complex variable $\mathbf{V}(s)$ as the **complex frequency domain** representation of $v(t)$. It should be observed that from the viewpoint of circuit analysis, the use of the Laplace transform is analogous to phasor analysis; that is, substituting the variable s wherever $j\omega$ was used is the only step required to describe a circuit using the new notation.

CHECK YOUR UNDERSTANDING

B.1   Find the complex frequencies that are associated with

a. $5e^{-4t}$ b. $\cos 2\omega t$ c. $\sin(\omega t + 2\theta)$ d. $4e^{-2t}\sin(3t-50°)$ e. $e^{-3t}(2 + \cos 4t)$

B.2   Find s and $\mathbf{V}(s)$ if $v(t)$ is given by

a. $5e^{-2t}$ b. $5e^{-2t}\cos(4t + 10°)$ c. $4\cos(2t - 20°)$

B.3   Find $v(t)$ if

a. $s = -2$, $\mathbf{V} = 2\angle 0°$ b. $s = j2$, $\mathbf{V} = 12\angle -30°$ c. $s = -4 + j3$, $\mathbf{V} = 6\angle 10°$

Answers: **B.1:** a. −4; b. ±j2ω; c. ±jω; d. −2 ± j3; e. −3 and −3 ± j4. **B.2:** a. −2, 5∠0°; b. −2 + j4, 5∠10°; c. j2. **B.3:** a. $2e^{-2t}$; b. $12\cos(2t - 30°)$; c. $6e^{-4t}\cos(3t + 10°)$.

All the concepts and rules used in AC network analysis (see Chapter 3), such as impedance, admittance, KVL, KCL, and Thévenin's and Norton's theorems, carry over to the damped sinusoid case exactly. In the complex frequency domain, the current $\mathbf{I}(s)$ and voltage $\mathbf{V}(s)$ are related by the expression:

$$\mathbf{V}(s) = \mathbf{Z}(s)\mathbf{I}(s) \tag{B.6}$$

where $\mathbf{Z}(s)$ is the familiar impedance, with s replacing $j\omega$. We may obtain $\mathbf{Z}(s)$ from $\mathbf{Z}(j\omega)$ by simply replacing $j\omega$ by s. For a resistance R, the impedance is

$$\mathbf{Z}_R(s) = R \tag{B.7}$$

For an inductance L, the impedance is

$$\mathbf{Z}_L(s) = sL \tag{B.8}$$

For a capacitance C, it is

$$\mathbf{Z}_C(s) = \frac{1}{sC} \tag{B.9}$$

Impedances in series or parallel are combined in exactly the same way as in the AC steady-state case, since we only replace $j\omega$ by s.

EXAMPLE B.1   Complex Frequency Notation

Problem:

Use complex impedance ideas to determine the response of a series RL circuit to a damped exponential voltage.

Solution:

Known Quantities: Source voltage, resistor, inductor values.

Find: The time-domain expression for the series current $i_L(t)$.

Schematics, Diagrams, Circuits, and Given Data: $v_s(t) = 10e^{-2t}\cos(5t)$ V; $R = 4$ Ω; $L = 2$ H.

Assumptions: None.

Analysis: The input voltage phasor can be represented by the expression

$$\mathbf{V}(s) = 10\angle 0 \text{ V}$$

The impedance seen by the voltage source is

$$\mathbf{Z}(s) = R + sL = 4 + 2s$$

Thus, the series current is

$$\mathbf{I}(s) = \frac{\mathbf{V}(s)}{\mathbf{Z}(s)} = \frac{10}{4 + 2s} = \frac{10}{4 + 2(-2 + j5)} = \frac{10}{j10} = j1 = 1\angle\left(-\frac{\pi}{2}\right)$$

Finally, the time-domain expression for the current is

$$i_L(t) = e^{-2t}\cos(5t - \pi/2) \qquad A$$

Comments: The phasor analysis method illustrated here is completely analogous to the method introduced in Chapter 3, with the complex frequency $j\omega$ (steady-state sinusoidal frequency) replaced by s (damped sinusoidal frequency).

Transfer functions $H(s)$ can be defined as a ratio of a voltage to a current, a ratio of a voltage to a voltage, a ratio of a current to a current, or a ratio of a current to a voltage. The transfer function $H(s)$ is a function of network elements and their interconnections. Using the transfer function and knowing the input (voltage or current) to a circuit, we can find an expression for the output either in the complex frequency domain or in the time domain. As an example, suppose $\mathbf{V}_i(s)$ and $\mathbf{V}_o(s)$ are the input and output voltages to a circuit, respectively, in complex frequency notation. Then

$$H(s) = \frac{\mathbf{V}_o(s)}{\mathbf{V}_i(s)} \tag{B.10}$$

from which we can obtain the output in the complex frequency domain by computing

$$\mathbf{V}_o(s) = H(s)\mathbf{V}_i(s) \tag{B.11}$$

If $\mathbf{V}_i(s)$ is a known damped sinusoid, we can then proceed to determine $v_o(t)$ by means of the method illustrated earlier in this section.

CHECK YOUR UNDERSTANDING

B.4   Given the transfer function $H(s) = 3(s + 2)/(s^2 + 2s + 3)$ and the input $\mathbf{V}_i(s) = 4\angle 0°$, find the forced response $v_o(t)$ if

a. $s = -1$ b. $s = -1 + j1$ c. $s = -2 + j1$

B.5   Given the transfer function $H(s) = 2(s + 4)/(s^2 + 4s + 5)$ and the input $\mathbf{V}_i(s) = 6\angle 30°$, find the forced response $v_o(t)$ if

a. $s = -4 + j1$ b. $s = -2 + j2$

B.2   THE LAPLACE TRANSFORM

The Laplace transform, named after the French mathematician and astronomer Pierre Simon de Laplace, is defined by

$$\mathcal{L}[f(t)] = F(s) = \int_0^\infty f(t)e^{-st}\,dt \tag{B.12}$$

The function $F(s)$ is the Laplace transform of $f(t)$ and is a function of the complex frequency $s = \sigma + j\omega$, considered earlier in this section. Note that the function $f(t)$ is defined only for $t \geq 0$. This definition of the Laplace transform applies to what is known as the **one-sided** or **unilateral Laplace transform**, since $f(t)$ is evaluated only for positive t. To conveniently express arbitrary functions only for positive time, we introduce a special function called the **unit-step function** $u(t)$, defined by the expression:

$$u(t) = \begin{cases} 0 & t < 0 \\ 1 & t > 0 \end{cases} \tag{B.13}$$

EXAMPLE B.2   Computing a Laplace Transform

Problem:

Find the Laplace transform of $f(t) = e^{-at} u(t)$.

Solution:

Known Quantities: Function to be Laplace-transformed.

Find: $F(s) = \mathcal{L}[f(t)]$.

Schematics, Diagrams, Circuits, and Given Data: $f(t) = e^{-at} u(t)$.

Assumptions: None.

Analysis: From equation B.12,

$$F(s) = \int_0^\infty e^{-at} e^{-st} dt = \int_0^\infty e^{-(s+a)t} dt = \frac{1}{s+a} e^{-(s+a)t} \Big|_0^\infty = \frac{1}{s+a}$$

Comments: Table B.1 contains a list of common Laplace transform pairs.

EXAMPLE B.3   Computing a Laplace Transform

Problem:

Find the Laplace transform of $f(t) = \cos(\omega t) u(t)$.

Solution:

Known Quantities: Function to be Laplace-transformed.

Find: $F(s) = \mathcal{L}[f(t)]$.

Schematics, Diagrams, Circuits, and Given Data: $f(t) = \cos(\omega t) u(t)$.

Assumptions: None.

Analysis: Using equation B.12 and applying Euler's identity to $\cos(\omega t)$ give:

$$F(s) = \int_0^\infty \frac{1}{2}\left(e^{j\omega t} + e^{-j\omega t}\right)e^{-st}\,dt = \frac{1}{2}\int_0^\infty \left(e^{(-s+j\omega)t} + e^{(-s-j\omega)t}\right)dt$$

$$= \frac{1}{-s+j\omega}e^{-(s+j\omega)t}\Big|_0^\infty + \frac{1}{-s-j\omega}e^{-(s-j\omega)t}\Big|_0^\infty$$

$$= \frac{1}{-s+j\omega} + \frac{1}{-s-j\omega} = \frac{s}{s^2+\omega^2}$$

Comments: Table B.1 contains a list of common Laplace transform pairs.

Table B.1 **Laplace transform pairs**

$f(t)$	$F(s)$
$\delta(t)$ (unit impulse)	1
$u(t)$ (unit step)	$\dfrac{1}{s}$
$e^{-at}u(t)$	$\dfrac{1}{s+a}$
$\sin \omega t\, u(t)$	$\dfrac{\omega}{s^2+\omega^2}$
$\cos \omega t\, u(t)$	$\dfrac{s}{s^2+\omega^2}$
$e^{-at}\sin \omega t\, u(t)$	$\dfrac{\omega}{(s+a)^2+\omega^2}$
$e^{-at}\cos \omega t\, u(t)$	$\dfrac{s+a}{(s+a)^2+\omega^2}$
$tu(t)$	$\dfrac{1}{s^2}$

CHECK YOUR UNDERSTANDING

B.6   Find the Laplace transform of the following functions:

a. $u(t)$ b. $\sin(\omega t)\, u(t)$ c. $tu(t)$

B.7   Find the Laplace transform of the following functions:

a. $e^{-at}\sin \omega t\, u(t)$ b. $e^{-at}\cos \omega t\, u(t)$

Answers: **B.6:** a. $\dfrac{1}{s}$; b. $\dfrac{\omega}{s^2+\omega^2}$; c. $\dfrac{1}{s^2}$. **B.7:** a. $\dfrac{\omega}{(s+a)^2+\omega^2}$; b. $\dfrac{s+a}{(s+a)^2+\omega^2}$

From what has been said so far about the Laplace transform, it is obvious that we may compile a lengthy table of functions and their Laplace transforms by repeated application of equation B.12 for various functions of time $f(t)$. Then we could obtain a wide variety of inverse transforms by matching entries in the table. Table B.1 lists some of the more common **Laplace transform pairs**. The computation of the **inverse Laplace transform** is in general rather complex if one wishes to consider arbitrary functions of s. In many practical cases, however, it is possible to use combinations of known transform pairs to obtain the desired result.

EXAMPLE B.4   Computing an Inverse Laplace Transform

Problem:

Find the inverse Laplace transform of

$$F(s) = \frac{2}{s+3} + \frac{4}{s^2+4} + \frac{4}{s}$$

Solution:

Known Quantities: Function to be inverse Laplace-transformed.

Find: $f(t) = \mathcal{L}^{-1}[F(s)]$.

Schematics, Diagrams, Circuits, and Given Data:

$$F(s) = \frac{2}{s+3} + \frac{4}{s^2+4} + \frac{4}{s} = F_1(s) + F_2(s) + F_3(s)$$

Assumptions: None.

Analysis: Using Table B.1, we can individually inverse-transform each of the elements of $F(s)$:

$$f_1(t) = 2\mathcal{L}^{-1}\left(\frac{1}{s+3}\right) = 2e^{-3t}u(t)$$

$$f_2(t) = 2\mathcal{L}^{-1}\left(\frac{2}{s^2+2^2}\right) = 2\sin(2t)\,u(t)$$

$$f_3(t) = 4\mathcal{L}^{-1}\left(\frac{1}{s}\right) = 4u(t)$$

Thus

$$f(t) = f_1(t) + f_2(t) + f_3(t) = (2e^{-3t} + 2\sin 2t + 4)\,u(t)$$

EXAMPLE B.5   Computing an Inverse Laplace Transform

Problem:

Find the inverse Laplace transform of

$$F(s) = \frac{2s+5}{s^2+5s+6}$$

Solution:

Known Quantities: Function to be inverse Laplace–transformed.

Find: $f(t) = \mathcal{L}^{-1}[F(s)]$.

Assumptions: None.

Analysis: A direct entry for the function cannot be found in Table B.1. In such cases, one must compute a *partial fraction expansion* of the function $F(s)$ and then individually transform each term in the expansion. A partial fraction expansion is the inverse operation of obtaining a common denominator and is illustrated below.

$$F(s) = \frac{2s+5}{s^2+5s+6} = \frac{A}{s+2} + \frac{B}{s+3}$$

To obtain the constants A and B, we multiply the above expression by each of the denominator terms:

$$(s+2)F(s) = A + \frac{(s+2)B}{s+3}$$

$$(s+3)F(s) = \frac{(s+3)A}{s+2} + B$$

From the above two expressions, we can compute A and B as follows:

$$A = (s+2)F(s)|_{s=-2} = \left.\frac{2s+5}{s+3}\right|_{s=-2} = 1$$

$$B = (s+3)F(s)|_{s=-3} = \left.\frac{2s+5}{s+2}\right|_{s=-3} = 1$$

Finally,

$$F(s) = \frac{2s+5}{s^2+5s+6} = \frac{1}{s+2} + \frac{1}{s+3}$$

and using Table B.1, we compute

$$f(t) = (e^{-2t} + e^{-3t})u(t)$$

CHECK YOUR UNDERSTANDING

B.8 Find the inverse Laplace transform of each of the following functions:

a. $F(s) = \dfrac{1}{s^2+5s+6}$ b. $F(s) = \dfrac{s-1}{s(s+2)}$

c. $F(s) = \dfrac{3s}{(s^2+1)(s^2+4)}$ d. $F(s) = \dfrac{1}{(s+2)(s+1)^2}$

Answer: a. $f(t) = (e^{-2t} - e^{-3t})u(t)$; b. $f(t) = (\frac{3}{2}e^{-2t} - \frac{1}{2})u(t)$;
c. $f(t) = (\cos t - \cos 2t)u(t)$; d. $f(t) = (e^{-2t} + te^{-t} - e^{-t})u(t)$

B.3   TRANSFER FUNCTIONS, POLES, AND ZEROS

It should be clear that the Laplace transform is a convenient tool for analyzing the transient response of a circuit. The Laplace variable s is an extension of the steady-state frequency response variable $j\omega$ already encountered in this appendix. Thus, it is possible to describe the input-output behavior of a circuit by using Laplace transform ideas in the same way in which we used frequency response ideas earlier. Now we can define voltages and currents in the complex frequency domain as $\mathbf{V}(s)$ and $\mathbf{I}(s)$, and we denote impedances by the notation $\mathbf{Z}(s)$, where s replaces the familiar $j\omega$. We define an extension of the frequency response of a circuit, called the *transfer function,* as the ratio of any output variable to any input variable, i.e.,

$$H_1(s) = \frac{\mathbf{V}_o(s)}{\mathbf{V}_i(s)} \quad \text{or} \quad H_2(s) = \frac{\mathbf{I}_o(s)}{\mathbf{V}_i(s)} \quad \text{etc.} \tag{B.14}$$

As an example, consider the circuit of Figure B.2. We can analyze it by using a method analogous to phasor analysis by defining impedances:

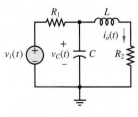
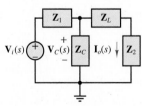

Figure B.2 A circuit and its Laplace transform domain equivalent

$$Z_1 = R_1 \qquad Z_C = \frac{1}{sC} \qquad Z_L = sL \qquad Z_2 = R_2 \tag{B.15}$$

Then we can use mesh analysis to determine

$$I_o(s) = V_i(s) \frac{Z_C}{(Z_L + Z_2)Z_C + (Z_L + Z_2)Z_1 + Z_1 Z_C} \tag{B.16}$$

or, upon simplifying and substituting the relationships of equation B.15,

$$H_2(s) = \frac{I_o(s)}{V_i(s)} = \frac{1}{R_1 L C s^2 + (R_1 R_2 C + L)s + R_1 + R_2} \tag{B.17}$$

If we were interested in the relationship between the input voltages and, say, the capacitor voltage, we could similarly calculate

$$H_1(s) = \frac{V_C(s)}{V_i(s)} = \frac{sL + R_2}{R_1 L C s^2 + (R_1 R_2 C + L)s + R_1 + R_2} \tag{B.18}$$

Note that a transfer function consists of a *ratio of polynomials;* this ratio can also be expressed in factored form, leading to the discovery of additional important properties of the circuit. Let us, for the sake of simplicity, choose numerical values for the components of the circuit of Figure B.2. For example, let $R_1 = 0.5\ \Omega$, $C = \frac{1}{4}$ F, $L = 0.5$ H, and $R_2 = 2\ \Omega$. Then we can substitute these values into equation B.18 to obtain

$$H_1(s) = \frac{0.5s + 2}{0.0625\,s^2 + 0.375s + 2.5} = 8\left(\frac{s + 4}{s^2 + 6s + 40}\right) \tag{B.19}$$

Equation B.19 can be factored into products of first-order terms as follows:

$$H_1(s) = 8\left[\frac{s + 4}{(s - 3.0000 + j5.5678)(s - 3.0000 - j5.5678)}\right] \tag{B.20}$$

where it is apparent that the response of the circuit has very special characteristics for three values of s: $s = -4$; $s = +3.0000 - j5.5678$; and $s = +3.0000 + j5.5678$. In the first case, at the complex frequency $s = -4$, the numerator of the transfer function becomes zero, and the response of the circuit is zero, regardless of how large the input voltage is. We call this particular value of s a **zero** of the transfer function. In the latter two cases, for $s = +3.0000 \pm j5.5678$, the response of the circuit becomes infinite, and we refer to these values of s as **poles** of the transfer function.

It is customary to represent the response of electric circuits in terms of poles and zeros, since knowledge of the location of these poles and zeros is equivalent to knowing the transfer function and provides complete information regarding the response of the circuit. Further, if the poles and zeros of the transfer function of a circuit are plotted in the complex plane, it is possible to visualize the response of the circuit very effectively. Figure B.3 depicts the pole-zero plot of the circuit of Figure B.2; in plots of this type it is customary to denote zeros by a small circle and poles by an "×."

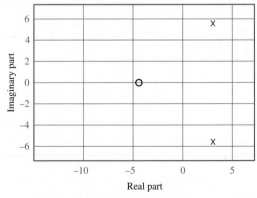

Figure B.3 Zero–pole plot for the circuit of Figure B.2

The poles of a transfer function have a special significance, in that they are equal to the roots of the natural response of the system. They are also called the **natural frequencies** of the circuit. Example B.6 illustrates this point.

EXAMPLE B.6   Poles of a Second-Order Circuit

Problem:

Determine the poles of a parallel *RLC* circuit. Express the homogeneous equation using i_L as the independent variable.

Solution:

Known Quantities: Values of resistor, inductor, and capacitor.

Find: Poles of the circuit.

Assumptions: None.

Analysis: The differential equation describing the natural response of the parallel *RLC* circuit is

$$\frac{d^2 i}{dt^2} + \frac{R}{L}\frac{di}{dt} + \frac{1}{LC}i = 0$$

with the characteristic equation given by

$$s^2 + \frac{R}{L}s + \frac{1}{LC} = 0$$

Now, let us determine the transfer function of the circuit, say, $\mathbf{V}_L(s)/\mathbf{V}_S(s)$. Applying the voltage divider rule, we can write

$$\frac{\mathbf{V}_L(s)}{\mathbf{V}_S(s)} = \frac{sL}{1/sC + R + sL}$$

$$= \frac{s^2}{s^2 + (R/L)s + 1/LC}$$

The denominator of this function, which determines the poles of the circuit, is identical to the characteristic equation of the circuit: The poles of the transfer function are identical to the roots of the characteristic equation!

$$s_{1,2} = -\frac{R}{2L} \pm \frac{1}{2}\sqrt{\left(\frac{R}{L}\right)^2 - \frac{4}{LC}}$$

Comments: Describing a circuit by means of its transfer function is completely equivalent to representing it by means of its differential equation. However, it is often much easier to derive a transfer function by basic circuit analysis than it is to obtain the differential equation of a circuit.

A P P E N D I X

C

FUNDAMENTALS OF ENGINEERING (FE) EXAMINATION

C.1 INTRODUCTION

The *Fundamentals of Engineering* (FE) examination[1] is one of four steps to be completed toward registering as a Professional Engineer (PE). Each of the 50 states in the United States has laws that regulate the practice of engineering; these laws are designed to ensure that registered professional engineers have demonstrated sufficient competence and experience. Each state's Board of Registration administers the exam and supplies information and registration forms.

The FE exam is offered throughout the year, except during the months of March, June, September, and December.

An examinee handbook is freely available through the NCEES website. The handbook contains information about eligibility, registration, fees, accommodations, what to bring to the exam, the calculator policy, and answers to other questions and issues that are likely to occur. **Additional information is available on the NCEES website**.

Four steps are required to become a Professional Engineer:

1. *Education.* Usually this requirement is satisfied by completing a B.S. degree in engineering from an accredited college or university.

---

[1]This exam used to be called *Engineer in Training (EIT)*.

2. *Fundamentals of Engineering examination.* One must pass a discipline-specific examination described in Section C.2.

3. *Experience.* Following successful completion of the Fundamentals of Engineering examination, several years of engineering experience are required.

4. *Principles and practices of engineering examination.* One must pass a second examination, also known as the Professional Engineer (PE) examination, which requires in-depth knowledge of one particular branch of engineering.

This appendix provides a review of the background material in electrical engineering required in three of the discipline-specific FE exams. Those exams are prepared by the National Council of Examiners for Engineering and Surveying[2] (NCEES).

C.2   EXAM FORMAT AND CONTENT

The FE exam is offered in six specific engineering disciplines:

1. Chemical

2. Civil

3. Electrical and Computer

4. Environmental

5. Industrial and Systems

6. Mechanical

A seventh Other Disciplines exam is also offered. The 6-h, 110-question exam is offered year-round at NCEES-approved test centers. Detailed specifications of each exam can be found online at http://ncees.org/engineering/fe/.

The passing score on the FE exam is not published by NCEES because it varies slightly across the discipline-specific exams and over time. However, data on passing rates is published and available on the NCEES website.

Of the seven exams, only three cover material presented in this book. Naturally, the Electrical and Computer Engineering exam covers nearly all the material. The Mechanical Engineering exam covers five areas of Electricity and Magnetism, namely:

- Charge, current, voltage, power, and energy

- Ohm's law and Kirchhoff's current and voltage laws

- Equivalent circuits (series and parallel)

- AC circuits

- Motors and generators

The Other Disciplines exam covers a similar set of topics as well as additional material on measuring devices, sensors, data acquisition, and data processing.

C.3   PRACTICE QUESTIONS ON ELECTRICITY AND MAGNETISM

What follows is a series of typical and relevant practice questions on FE exam material related to Electricity and Magnetism, including extensions of the theory to circuits, electronics, logic, instrumentation, communications, and electromechanics. The questions are ordered as they would be encountered in a typical engineering curriculum. Answers to the questions are provided at the end of this appendix.

---

[2]P.O. Box 1686 (1826 Seneca Road), Clemson, SC 29633-1686.

Students preparing for the FE exam should keep in mind that the actual exam time is 5 hours and 20 minutes, or 320 minutes. With 110 questions on the exam the average time per question is slightly less than 3 minutes. Therefore, it is important to develop techniques for quickly arriving at correct answers or likely correct answers. In other words, it is not advisable to approach the FE exam as one would a typical undergraduate engineering exam. Rather it is worthwhile to develop skill at eliminating answers that are unreasonable or unlikely to be correct. For example, one can often eliminate answers due to the unreasonable scale of the answer or due to a units mismatch. It is also worthwhile to develop skill at approximating solutions. Remember, the average time per question is less than 3 minutes. The exam questions are designed with this limitation in mind. What does that tell you about the nature of many of the exam questions? When solving the practice questions below, look for ways in which you could have found the correct answer more quickly, more approximately, and/or with greater probability of correctness. And limit yourself to 3 minutes each!

Finally, the exam score is based solely on the number of correct answers. There are no deductions for wrong answers, so when in doubt, guess!

CHECK YOUR UNDERSTANDING

C.1 Determine the total charge entering a circuit element between $t = 1$ s and $t = 2$ s if the current passing through the element is $i = 5t$.

C.2 A lightbulb sees a 3-A current for 15 s. The lightbulb generates 3 kJ of energy in the form of light and heat. What is the voltage drop across the lightbulb?

C.3 How much energy does a 75-W electric bulb consume in 6 hours?

C.4 Find the voltage drop v_{ab} required to move a charge q from point a to point b if $q = -6$ C and it takes 30 J of energy to move the charge.

C.5 Two 2-C charges are separated by a dielectric with a thickness of 4 mm and with a dielectric constant $\varepsilon = 10^{-12}$ F/m. What is the force exerted by each charge on the other?

C.6 The magnitude of the force on a particle of charge q placed in the empty space between two infinite parallel plates with a spacing d and a potential difference V is proportional to:

  a. qV/d^2 b. qV/d c. qV^2/d d. q^2V/d e. q^2V^2/d

C.7 Assuming the connecting wires and the battery have negligible resistance, the voltage across the 25-Ω resistance in Figure C.7 is

  a. 25 V   b. 60 V   c. 50 V   d. 15 V   e. 12.5 V

C.8 Assuming the connecting wires and the battery have negligible resistance, the voltage across the 6-Ω resistor in Figure C.8 is

  a. 6 V   b. 3.5 V   c. 12 V   d. 8 V   e. 3 V

C.9 A 125-V battery charger is used to charge a 75-V battery with internal resistance of 1.5 Ω, as shown in Figure C.9. If the charging current is not to exceed 5 A, the minimum resistance in series with the charger must be

  a. 10 Ω b. 5 Ω c. 38.5 Ω d. 41.5 Ω e. 8.5 Ω

C.10 A coil with an inductance of 1 H and negligible resistance carries the current shown in Figure C.10. The maximum energy stored in the inductor is

  a. 2 J   b. 0.5 J   c. 0.25 J   d. 1 J   e. 0.2 J

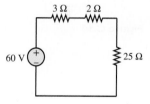

Figure C.7

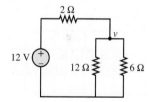

Figure C.8

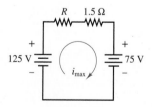

Figure C.9

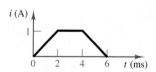

Figure C.10

C.11   The maximum voltage that will appear across the coil is

   a. 5 V   b. 100 V   c. 250 V   d. 500 V   e. 5,000 V

C.12   A voltage sine wave of peak value 100 V is in phase with a current sine wave of peak value 4 A. When the phase angle is 60° later than a time at which the voltage and the current are both zero, the instantaneous power is most nearly

   a. 300 W   b. 200 W   c. 400 W   d. 150 W   e. 100 W

C.13   A sinusoidal voltage whose amplitude is $20\sqrt{2}$ V is applied to a 5-Ω resistor. The root-mean-square value of the current is

   a. 5.66 A   b. 4 A   c. 7.07 A   d. 8 A   e. 10 A

C.14   The magnitude of the steady-state root-mean-square voltage across the capacitor in the circuit of Figure C.14 is

   a. 30 V   b. 15 V   c. 10 V   d. 45 V   e. 60 V

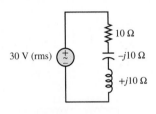

Figure C.14

The next set of questions (Exercises C.15 to C.19) pertain to single-phase AC power calculations and refer to the single-phase electrical network shown in Figure C.15. In this figure, $\mathbf{E}_S = 480\angle 0°$ V; $\mathbf{I}_S = 100\angle -15°$ A; $\omega = 120\pi$ rad/s. Further, load A is a bank of single-phase induction machines. The bank has an efficiency η of 80 percent, a power factor of 0.70 lagging, and a load of 20 hp. Load B is a bank of overexcited single-phase synchronous machines. The machines draw 15 kVA, and the load current leads the line voltage by 30°. Load C is a lighting (resistive) load and absorbs 10 kW. Load D is a proposed single-phase capacitor that will correct the source power factor to unity. This material is covered in Sections 13.1 and 13.2.

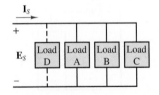

Figure C.15

C.15   The root-mean-square magnitude of load A current, denoted by I_A, is most nearly

   a. 44.4 A   b. 31.08 A   c. 60 A   d. 38.85 A   e. 55.5 A

C.16   The phase angle of $\mathbf{I}_A$ with respect to the line voltage $\mathbf{E}_S$ is most nearly

   a. 36.87°   b. 60°   c. 45.6°   d. 30°   e. 48°

C.17   The power absorbed by synchronous machines is most nearly

   a. 20,000 W   b. 7,500 W   c. 13,000 W   d. 12,990 W   e. 15,000 W

C.18   The power factor of the system before load D is installed is most nearly

   a. 0.70 lagging   b. 0.866 leading   c. 0.866 lagging
   d. 0.966 leading   e. 0.966 lagging

C.19   The capacitance of the capacitor that will give a unity power factor of the system is most nearly

   a. 219 μF   b. 187 μF   c. 132.7 μF   d. 240 μF   e. 132.7 pF

Answers: C.1: $q = \int_{t=2}^{t=1} i\,dt = \int_{t=2}^{t=1} 5t\,dt = \left(\frac{5t^2}{2}\right)\Big|_{t=2}^{t=1} = 7.5$ C

C.2: The total charge is $\Delta q = i\Delta t = 3 \times 15 = 45$ C. The voltage drop is

$$v = \frac{\Delta w}{\Delta q} = \frac{45}{3 \times 10^3} = 66.67 \text{ V}$$

C.3: The energy used is

$w = pt = 75$ [W] $\times$ 6 [h] $= 75$ [W] $\times$ 6 $\times$ 3,600 [s] $(450) = 1.62$ MJ

C.4: The voltage drop is $v_{ab} = \dfrac{m}{q} = \dfrac{30}{-6} = -5$ V

C.5: $F = \dfrac{q_1 q_2}{4\pi\varepsilon r^2} = \dfrac{(2\times10^{-3})\times(2\times10^{-3})}{4\pi \times 10^{-12}\times(4\times10^{-3})^2} = 2\times10^{10}$ N

C.6: Answer is a, since this is the only term that has a distance squared term in the denominator.

C.7: This problem calls for application of the voltage divider rule, discussed in Section 2.6. Applying the voltage divider rule to the circuit of Figure C.7, we have

$$v_{25\,\Omega} = 60\left(\frac{25}{3+2+25}\right) = 50 \text{ V}$$

Thus, the answer is c.

C.8: This problem can be solved most readily by applying nodal analysis (Section 3.1), since one of the node voltages is already known. Applying KCL at the node v, we obtain

$$\frac{12-v}{2} = \frac{v}{6} + \frac{v}{12}$$

This equation can be solved to show that $v = 8$ V. Note that it is also possible to solve this problem by mesh analysis (Section 3.2). You are encouraged to try this method as well.

C.9: The circuit of Figure C.9 describes the charging arrangement. Applying KVL to the circuit of Figure C.9, we obtain

$$i_{max}R + 1.5i_{max} - 125 + 75 = 0$$

and using $i = i_{max} = 5$ A, we can find R from the following equation:

$$5R + 7.5 - 125 + 75 = 0$$

$$R = 8.5 \ \Omega$$

Thus, e is the correct answer.

C.10: The energy stored in an inductor is $W = \frac{1}{2}Li^2$ (see Section 4.1). Since the maximum current is 1 A, the maximum energy will be $W_{max} = \frac{1}{2}Li_{max}^2 = \frac{1}{2}$ J. Thus, b is the correct answer.

C.11: Since the voltage across an inductor is given by $v = L(di/dt)$, we need to find the maximum (positive) value of di/dt. This will occur anywhere between $t = 0$ and $t = 2$ ms. The corresponding slope is

$$\left.\frac{di}{dt}\right|_{max} = \frac{1}{2 \times 10^{-3}} = 500$$

Therefore $v_{max} = 1 \times 500 = 500$ V, and the correct answer is d.

C.12: As discussed in Section 7.1, the instantaneous AC power $p(t)$ is

$$p(t) = \frac{VI}{2}\cos\theta + \frac{VI}{2}\cos(2\omega t + \theta_V + \theta_I)$$

In this problem, when the phase angle is 60° later than a "zero crossing," we have $\theta_V = \theta_I = 0, \theta = \theta_V - \theta_I = 0, 2\omega t = 120°$. Thus, we can compute the power at this instant as

$$p = \frac{100\times4}{2} + \frac{100\times4}{2}\cos(120°) = 300 \text{ W}$$

The correct answer is a.

Answers: C.13: From Section 4.2, we know that

$$V_{rms} = \frac{V}{\sqrt{2}} = \frac{20\sqrt{2}}{\sqrt{2}} = 20 \text{ V}$$

Thus, $I_{rms} = 20/5 = 4$ A. Therefore, b is the correct answer.

C.14: This problem requires the use of impedances (Section 4.4). Using the voltage divider rule for impedances, we write the voltage across the capacitor as

$$\mathbf{V} = 30\angle 0° \times \frac{-j10}{10 - j10 + j10}$$

$$= 30\angle 0° \times \frac{10 - j10}{10} = 30\angle 0° \times 1\angle -90° = 30\angle 90°$$

Thus, the rms amplitude of the voltage across the capacitor is 30 V, and a is the correct answer. Note the importance of the phase angle in this kind of problem.

C.15: The output power P_o of the single-phase induction motor is $P_o = 20 \times 746 = 14,920$ W. The input electric power P_{in} is

$$P_{in} = \frac{P_o}{u} = \frac{14,920}{0.80} = 18,650 \text{ W}$$

P_{in} can be expressed as

$$P_{in} = E_S I_A \cos\theta_A$$

Therefore, the rms magnitude of the current $\mathbf{I}_A$ is found as

$$I_A = \frac{P_{in}}{E_S \cos\theta_A} = \frac{18,650}{480 \times 0.70} = 55.5015 \approx 55.5 \text{ A}$$

Thus, the correct answer is e.

C.16: The phase angle between $\mathbf{I}_A$ and $\mathbf{E}_S$ is

$$\theta = \cos^{-1} 0.70 = 45.57° \approx 45.6°$$

The correct answer is c.

C.17: The apparent power S is known to be 15 kVA, and θ is 30°. From the power triangle, we have

$$P = S \cos\theta$$

Therefore, the power drawn by the bank of synchronous motors is

$$P = 15,000 \times \cos 30° = 12,990.38 \approx 12.99 \text{ kW}$$

The answer is d.

C.18: From the expression for the current $\mathbf{I}_S$, we have

$$pf = \cos\theta = \cos[0° - (-15°)] = \cos 15° = 0.966 \text{ lagging}$$

The correct answer is e.

C.19: The reactive power Q_A in load A is

$$\theta_A = \cos^{-1} 0.70 = 45.57°$$

$$Q_A = P_A \times \tan\theta_A$$

Therefore,

$$Q_A = 18,650 \times \tan 45.57° = 19,025 \text{ VAR}$$

The total reactive power Q_B in load B is

$$Q_B = S \times \sin\theta_B = 15,000 \times \sin(-30°) = -7,500 \text{ VAR}$$

The total reactive power Q is

$$Q = Q_A + Q_B = 19,025 - 7,500 = 11,525 \text{ VAR}$$

To cancel this reactive power, we set

$$Q_c = -Q = -11,525 \text{ VAR}$$

and

$$Q_c = -\frac{E_S^2}{X_c} \quad \text{and} \quad X_c = -\frac{1}{\omega C}$$

Therefore, the capacitance required to obtain a power factor of unity is

$$C = -\frac{Q_c}{\omega E_S^2} = \frac{11,525}{120\pi \times 480^2} = 132.7 \ \mu F$$

The correct answer is c.

A P P E N D I X

D

ASCII CHARACTER CODE

I n addition to the codes described elsewhere in the book (binary, octal, hexadecimal, binary-coded decimal), a character encoding convention adopted by all computer manufacturers is **ASCII**,[1] which maps a unique numeric value to each of 128 graphic or control characters commonly used in the display of text. The complete code is shown in Table D.1. Notice that the numeric values are shown in hexadecimal. An additional 128 nonstandard characters are often defined for any particular font implemented using the ASCII code, for a total of 256 characters in a typical font. It is no accident that 256 characters are often defined since that is the number of items that can be uniquely mapped by 8 bits or 1 byte of memory.

---

[1]American Standard Code for Information Interchange.

Table D.1 **ASCII**

Graphic or control	ASCII (hex)	Graphic or control	ASCII (hex)	Graphic or control	ASCII (hex)
NUL	00	+	2B	V	56
SOH	01	,	2C	W	57
STX	02	−	2D	X	58
ETX	03	.	2E	Y	59
EOT	04	/	2F	Z	5A
ENQ	05	0	30	[	5B
ACK	06	1	31	\	5C
BEL	07	2	32	]	5D
BS	08	3	33	↑	5E
HT	09	4	34	←	5F
LF	0A	5	35	`	60
VT	0B	6	36	a	61
FF	0C	7	37	b	62
CR	0D	8	38	c	63
SO	0E	9	39	d	64
SI	0F	:	3A	e	65
DLE	10	;	3B	f	66
DC1	11	<	3C	g	67
DC2	12	=	3D	h	68
DC3	13	>	3E	i	69
DC4	14	?	3F	j	6A
NAK	15	@	40	k	6B
SYN	16	A	41	l	6C
ETB	17	B	42	m	6D
CAN	18	C	43	n	6E
EM	19	D	44	o	6F
SUB	1A	E	45	p	70
ESC	1B	F	46	q	71
FS	1C	G	47	r	72
GS	1D	H	48	s	73
RS	1E	I	49	t	74
US	1F	J	4A	u	75
SP	20	K	4B	v	76
!	21	L	4C	w	77
"	22	M	4D	x	78
#	23	N	4E	y	79
$	24	O	4F	z	7A
%	25	P	50	{	7B
&	26	Q	51			7C
'	27	R	52	}	7D
(	28	S	53	~	7E
)	29	T	54	DEL	7F
*	2A	U	55		

Index